A Specialist Periodical Report

Amino-acids, Peptides, and Proteins

Volume 5

A Review of the Literature Published during 1972

Senior Reporter

R. C. Sheppard, *MRC Laboratory of Molecular Biology, Cambridge*

ISBN: 0 85186 044 3

Library of Congress Catalog Card No. 72-92548

The Chemical Society
Burlington House, London, W1V 0BN

Reporters

P. M. Bayley, *National Institute of Medical Research, Mill Hill, London*
T. L. Blundell, *University of Sussex*
J. Bridgen, *MRC Laboratory of Molecular Biology, Cambridge*
B. W. Bycroft, *University of Nottingham*
R. D. Gillard, *University of Kent*
P. M. Hardy, *University of Exeter*
R. W. Hay, *University of Stirling*
C. E. Johnson, *University of Liverpool*
J. H. Jones, *University of Oxford*
G. L. Kellett, *University of York*
P. Knowles, *University of Leeds*
G. L. E. Koch, *MRC Laboratory of Molecular Biology, Cambridge*
S. H. Laurie, *Leicester Polytechnic*
H. D. Law, *Glasgow College of Technology*
R. H. Pain, *University of Newcastle upon Tyne*
G. R. Penzer, *University of York*
H. W. E. Rattle, *Portsmouth Polytechnic*
B. Ridge, *University of Exeter*
B. Robson, *University of Newcastle upon Tyne*
R. D. Ryder, *University of Newcastle upon Tyne*
R. M. Stephens, *Portsmouth Polytechnic*

Organic formulae composed by Wright's Symbolset method

Printed in Great Britain by John Wright and Sons Ltd. at The Stonebridge Press, Bristol BS4 5NU

Preface

This fifth Report reviews papers relevant to the chemistry of amino-acids, peptides, and proteins published in the main journals during 1972. The subject matter of Chapter 5 (Chemical Structure and Biological Activity) has not been covered in this series since Volume 1, and therefore this article also includes some references from the 1971 and earlier literature. Chapter 6 (Metal Derivatives of Amino-acids, Peptides, and Proteins) is a biennial survey and covers papers appearing during 1971 and 1972. In all, some 2700 references are cited.

As in some previous volumes, we have taken the opportunity of reprinting (in Chapter 7) new and revised recommendations of the I.U.P.A.C.–I.U.B. Commission on Biochemical Nomenclature. This year we reproduce revised versions of two previously published recommendations, *viz.* 'Definitive Rules for Naming Synthetic Modifications of Natural Peptides' and 'Abbreviated Nomenclature of Synthetic Polypeptides (Polymerised Amino-acids)'. Also included are definitive rules for 'A One-letter Notation for Amino-acid Sequences' not previously published in this series. Chapter 2 includes two examples of the use of this one-letter notation for the presentation of large amounts of data in a relatively compact and, to the experienced eye, a reasonably easily assimilated form. In general, however, the conventional three-letter symbols are preferred throughout these Reports.

All who have in the past found these Reports a useful guide to the current literature will wish to thank my predecessor, Dr. G. T. Young, for establishing this series in its present format. My personal thanks are also due to the many contributors to the present volume for their time and effort willingly devoted to its preparation.

R. C. SHEPPARD

Contents

Abbreviations

Abbreviations for amino-acids and their use in the formulation of derivatives follow with some exceptions the revised (1971) Recommendations of the I.U.P.A.C.–I.U.B. Commission on Biochemical Nomenclature, which are reprinted in Chapter 5 of Volume 4.

Other abbreviations which have been used without definition are:

Adoc	adamantyloxycarbonyl
Aoc	t-amyloxycarbonyl
Asu	α-aminosuberic acid
Asx	aspartic acid or asparagine (not yet determined)
ATP	adenosine 5′-triphosphate
Bpoc	2-(4-biphenylyl)-isopropoxycarbonyl
BSA	bovine serum albumin
c.d.	circular dichroism
Cha	cyclohexylamine
Cm	carboxymethyl
Cmc	*S*-carboxymethylcysteine
Dce	2,2-diethoxycarbonyl
Dcha	dicyclohexylamine
DMF	*NN*-dimethylformamide
DMSO	dimethyl sulphoxide
DNA	deoxyribonucleic acid
Dnp	2,4-dinitrophenyl
Dns	1-dimethylaminonaphthalene-5-sulphonyl (dansyl)
Dopa	3,4-dihydroxyphenylalanine
DP	degree of polymerization
Ec	ethylcarbamoyl
edta	ethylenediamine tetra-acetate
e.p.r.	electron paramagnetic resonance
e.s.r.	electron spin resonance
Gal	galactose
g.l.c.	gas–liquid chromatography
Glc	glucose
Glp	pyrrolid-2-one-5-carboxylic acid
Glx	glutamic acid or glutamine (not yet determined)
GTP	guanosine 5′-triphosphate
i.r.	infrared

Man	mannose
NAD	nicotinamide-adenine dinucleotide (NAD^+, oxidized; NADH, reduced)
n.m.r.	nuclear magnetic resonance
ONSu	succinimido-oxy
OPcp	pentachlorophenoxy
OPic	4-picolyloxy
o.r.d.	optical rotatory dispersion
OTcp	2,4,5-trichlorophenoxy
Pipoc	piperidino-oxycarbonyl
Pth-Gly	the phenylthiohydantoin derived from glycine, *etc.*
RNA	ribonucleic acid
SDS	sodium dodecyl sulphate
Ser(P)	*O*-phosphorylserine
t.l.c.	thin-layer chromatography
u.v.	ultraviolet
Ztf	1-benzyloxycarbonylamino-2,2,2-trifluoro-ethyl

1

Amino-acids

BY P. M. HARDY

The general format and topics covered in this chapter remain as in previous volumes. The purely biological effects of amino-acids or their simple derivatives have only been reported where there is some definite chemical interest involved, and biosynthetic studies as such are not included. Synthetic and degradative work is only discussed where novel methods or intermediates are involved. Amino-acid derivatives useful for peptide synthesis are dealt with in Chapter 3. Crystal structures are listed but not described. Reactions of protein- or peptide-bound amino-acid residues are only dealt with if they are also of interest in connection with free amino-acids or their simple derivatives.

1 Naturally Occurring Amino-acids

Occurrence of Known Amino-acids.—As is customary in this section, only papers dealing with amino-acids which are rarely encountered or which are of particular current interest are mentioned.

The idea has been put forward that some compounds, such as the non-protein amino-acids, may be synthesized by many more plants, or even by all of them, than has hitherto been recognized. However, they may be formed in amounts that fall below the threshold concentration that can be recognized in plant extracts by routine analytical procedures.[1] An attempt has been made to test this concept by studying the nitrogenous fraction produced by the large-scale processing of sugar beet. Many non-protein amino-acids and derivatives are in fact present in this mixture in very small quantities, too small to have been recognized after a small-scale extraction. These include L-azetidine-2-carboxylic acid, ε-*N*-acetyl-lysine, γ-L-glutamyl-γ-aminobutyric acid, and γ-*N*-acetyl-L-α,γ-diaminobutyric acid; two new natural amino-acids from this source are reported below (p. 3). Fowden concludes that the particular genetic complexes necessary for the formation of individual secondary products may be more widely distributed in plants than often thought. Different patterns of product accumulation may reflect differences in the degree to which genes are 'switched on'.[1] ε-*N*-Acetyl-lysine is a new plant product, but further studies on its appearance in histones are reported.[2, 3]

[1] L. Fowden, *Phytochemistry*, 1972, **11**, 2271.
[2] W. F. Marzluff, D. M. Miller, and K. S. McCarty, *Arch. Biochem. Biophys.*, 1972, **152**, 472.
[3] W. F. Marzluff and K. S. McCarty, *Biochemistry*, 1972, **11**, 2677.

$N^GN'^G$-Dimethylarginine and N^GN^G-dimethylarginine have been identified in the basic A1 protein from bovine myelin. These residues produce methylamine and dimethylamine, respectively, on alkaline hydrolysis of the protein; these amines were identified by a g.l.c.–m.s. method as dimethylamine is not detected on amino-acid analysis.[4] Dityrosine, the fluorescent *oo'*-biphenol analogue of tyrosine, has now been isolated from bovine ligamentum nuchae;[5] it also occurs in uterine protein.[6] It can be simply synthesized by the action of horseradish peroxidase and hydrogen peroxide on tyrosine.[7] A survey of the protein of medulla cells from hair or quill of a number of mammalian species has been made, and the occurrence of the ε-(γ-glutamyl)lysine cross-link seems to be a general phenomenon. This cross-link contributes to the insolubility of these proteins.[8]

The antibiotic edeine D has been found to contain a residue of β-phenyl-β-alanine. This replaces a residue of β-tyrosine in the close analogue edeine A.[9] Participation of an ε-amino-group in a peptide bond in racemomycin-C has been confirmed by the isolation of ε-*N*-(L-β-lysyl)-L-β-lysine from a partial acid hydrolysate. The possibility of transpeptidation occurring during hydrolysis was ruled out by N-derivatization experiments.[10] The fungus *Boletus satanas* has been shown to contain γ-hydroxynorvaline. G.l.c. analysis of the *N*-[(*S*)-α-methoxypropionyl]-lactones derived from two isomers separated by ion-exchange chromatography showed one lactone to be the pure (2*S*, 4*R*)-isomer and the other to be partly racemized, (2*S*, 4*S*) : (2*R*, 4*R*) = 3 : 2.[11]

New Natural Free Amino-acids.—The antimetabolite rhizobotoxine, first isolated in 1965 from the root nodules produced by *Rhizobium japonicum* in the soybean *Glycine max* (L.) Merr.,[12] has now been identified as the enol-ether 2-amino-4-(2-amino-3-hydroxypropoxy)-*trans*-but-3-enoic acid, $CH_2OH{\cdot}CH(NH_2)CH_2{\cdot}O{\cdot}CH{:}CH{\cdot}CH(NH_2)CO_2H$.[13] Its dihydro-derivative *O*-(2-amino-3-hydroxypropyl)homoserine is also produced by the same bacterium.[14] This is the first instance of an alkyl-ether derivative of homoserine being found in the absence of added alcohol.

4 S. W. Brostoff, A. Rosegay, and W. J. A. Vandenheuvel, *Arch. Biochem. Biophys.*, 1972, **148**, 156.
5 F. W. Keeley and F. S. LaBella, *Biochim. Biophys. Acta*, 1972, **263**, 52.
6 J. W. Downie, F. S. LaBella, and M. West, *Biochim. Biophys. Acta*, 1972, **263**, 604.
7 A. J. Gross and I. W. Sizer, *J. Biol. Chem.*, 1959, **234**, 1611.
8 M. W. J. Harding and G. E. Rogers, *Biochim. Biophys. Acta*, 1972, **257**, 37.
9 M. Wojciechowska, J. Ciarkowski, H. Chmara, and E. Borowski, *Experientia*, 1972, **28**, 1423.
10 H. Tamiyama, Y. Sawada, K. Miyazeki, S. Tanaka, F. Miyoshi, and K. Hiraoka, *Chem. and Pharm. Bull.* (*Japan*), 1972, **20**, 1432.
11 P. Matsinger, Ch. Catalformo, and C. H. Eugster, *Helv. Chim. Acta*, 1972, **55**, 1478.
12 L. D. Owens and D. A. Wright, *Plant Physiol.*, 1965, **40**, 927.
13 L. D. Owens, J. F. Thompson, R. G. Pitcher, and T. Williams, *J.C.S. Chem. Comm.*, 1972, 714.
14 L. D. Owens, J. F. Thompson, and P. V. Fennessey, *J.C.S. Chem. Comm.*, 1972, 715.

Four new amino-acids have been characterized from members of the Leguminosae; their structures, however, differ widely. A (1*S*, 3*S*) isoquinoline derivative (1) occurs in a variety of the velvet bean. It can readily

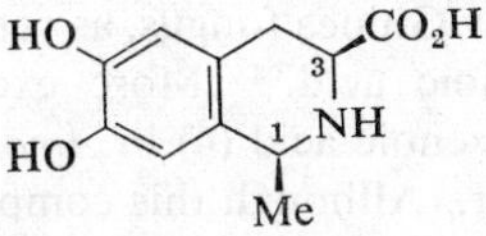

(1)

be synthesized from L-dopa and acetaldehyde in a largely stereoselective condensation.[15] 4,5-Dihydroxy-L-pipecolic acid has been isolated from *Calliandra haematocephala* Hassk. and synthesized by *cis*-hydroxylation of natural 4,5-dehydropipecolic acid with osmium tetroxide, but its stereochemistry is not yet fully known.[16] An amino-acid previously synthesized as an analogue of phenylalanine, *p*-aminophenylalanine, has now turned up in the form of its L-isomer in *Vigna vexillata*,[17] and *S*-(2-hydroxy-2-carboxyethanethiomethyl)-L-cysteine, $HO_2C\cdot CH(OH)\cdot CH_2S\cdot CH_2\cdot SCH_2CH(NH_2)$-$CO_2H$, has been reported to occur in *Acacia georginae* seed.[18] The latter can be prepared from L-L-djenkolic acid by treatment with a limited quantity of sodium nitrite in acetic acid ($NH_2 \rightarrow OH$), but racemization during this substitution is almost complete.[18]

As far as other plants are concerned, Fowden and his colleagues have added a new cyclopropane amino-acid (2) and a new unsaturated amino-acid (3) to their series of such compounds isolated from the Sapidaceae

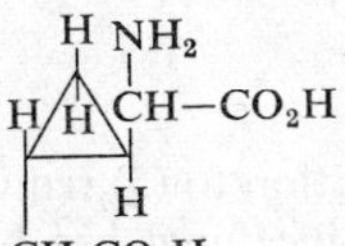

(2)

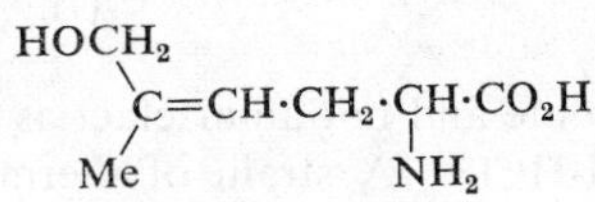

(3)

and Hippocastanaceae,[19] and the seeds of the Kentucky coffee tree have yielded two stereoisomers of β-hydroxy-γ-methylglutamic acid and other non-protein amino-acids as yet unidentified.[20] The sugar beet, *Beta vulgaris*, has been found to contain two new amino-acid derivatives, ε-*N*-acetyl-*allo*-δ-hydroxy-L-lysine and γ-*N*-lactyl-L-$\alpha\gamma$-diaminobutyric

[15] M. E. Daxenbichler, R. Kleiman, D. Weisleder, C. M. Van Etten, and K. D. Carlson, *Tetrahedron Letters*, 1972, 1801.
[16] M. Marlier, G. A. Dardenne, and J. Casimir, *Phytochemistry*, 1972, **11**, 2597.
[17] G. A. Dardenne, M. Marlier, and J. Casimir, *Phytochemistry*, 1972, **11**, 2567.
[18] K. Ito and L. Fowden, *Phytochemistry*, 1972, **11**, 2541.
[19] L. Fowden, C. M. Macgibbon, F. A. Mellon, and R. C. Sheppard, *Phytochemistry*, 1972, **11**, 1105.
[20] G. A. Dardenne, J. Casimir, E. A. Bell, and J. R. Nulu, *Phytochemistry*, 1972, **11**, 787.

acid.[1] Details of the isolation of 4-methylene-DL-proline from *Eriobotrya japonica* have now been published.[21]

Four unsaturated amino-acids from fungi have been reported. The fruit bodies of *Tricholomopsis rutilans* contain L-2-amino-4-methyl-5-hexenoic acid,[22] and a New Guinea fungus, as yet only tentatively identified, yields L-2-aminohex-4-ynoic acid.[23] More exotic is the occurrence of *trans*-2-amino-5-chlorohexenoic acid (4) in *Amanita solitaria*, a mushroom with a chlorine-like odour. Although this compound is readily obtainable from the allenic amino-acid (5) known to occur in this fungus, it is probably

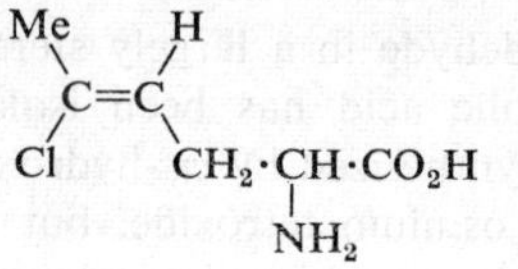

(4)

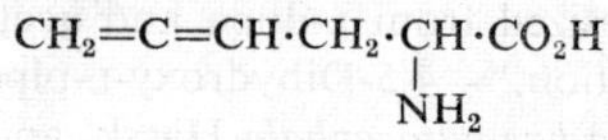

(5)

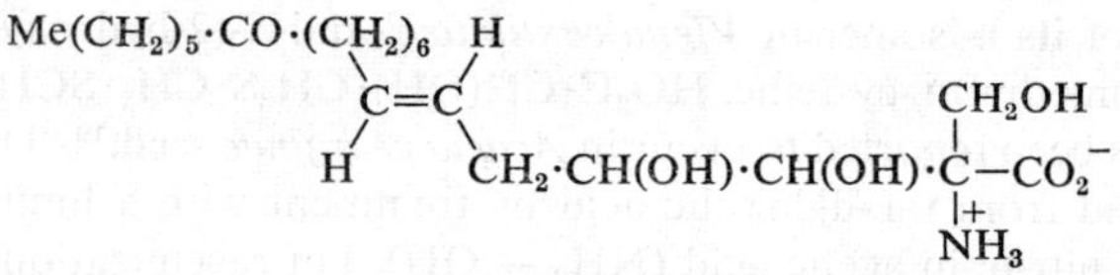

(6)

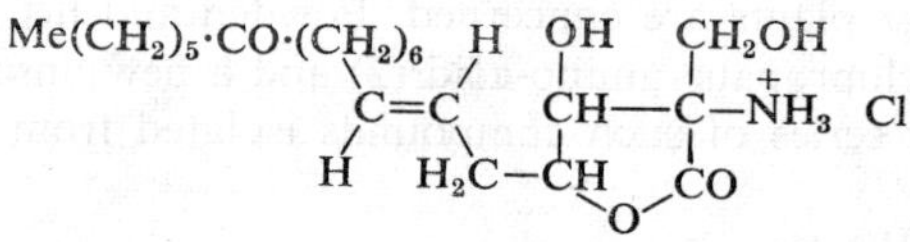

(7)

a true metabolite and not an artefact as addition to (5) requires treatment with hot 1M-HCl.[24] A strain of thermophilic fungi has been shown to produce the rather complex amino-acid thermozymocidin (6), an anti-fungal material. Methanolic hydrogen chloride converts thermozymocidin into the α-amino-γ-lactone hydrochloride (7). Although the stereochemistry has not been fully elucidated, it seems that the adjacent hydroxymethyl and secondary hydroxy-groups in the lactone have the *trans* configuration as ketal or acetal derivatives cannot be prepared.[25]

A new guanidino-compound, L-thalassemine, has been isolated from the body-wall muscle of the echuiroid worm *Thalassema neptuni*. This amino-acid, guanidoethylphospho-*O*-(α-*NN*-dimethyl)serine, occurs

[21] D. P. Gray and L. Fowden, *Phytochemistry*, 1972, **11**, 745.
[22] R. Rudzats, E. Gellart, and B. Halpern, *Biochem. Biophys. Res. Comm.*, 1972, **47**, 290.
[23] S.-I. Hatanaka, Y. Niimura, and K. Taniguchi, *Phytochemistry*, 1972, **11**, 3327.
[24] W. S. Chilton and G. Tsou, *Phytochemistry*, 1972, **11**, 2853.
[25] F. Aragozzini, P. L. Manachini, R. Cravieri, B. Rindone, and C. Scholastico, *Tetrahedron*, 1972, **28**, 5493.

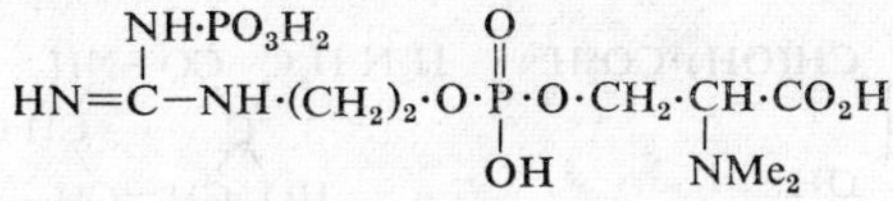

(8)

together with its N^1-phosphoryl-derivative (8); the occurrence of *NN*-dimethylserine either free or combined has not been previously noted.[26]

New Amino-acids from Hydrolysates.—Acid fission of the antibacterial compound ristocetin A yields a number of unusual amino-acids. One of these contains a diphenyl ether grouping with hydroxyl and $-CH(NH_2)$-CO_2H units on one ring and methyl and $-CH(NH_2)CO_2H$ units on the other.[27, 28] The n.m.r. coupling pattern and nuclear Overhauser experiments favour the substitution patterns (9) and (10) for the two ring systems.[27]

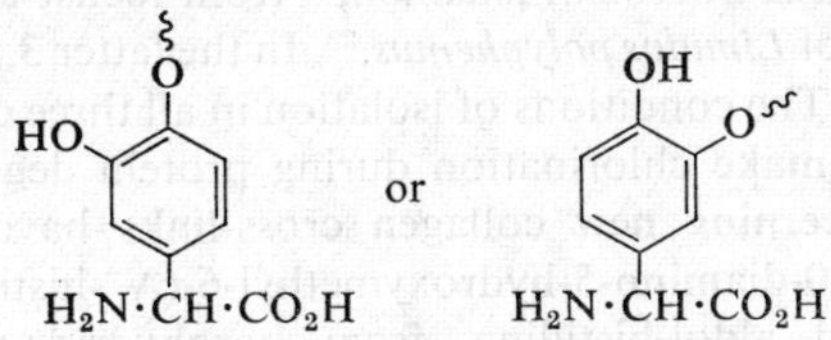

(9)

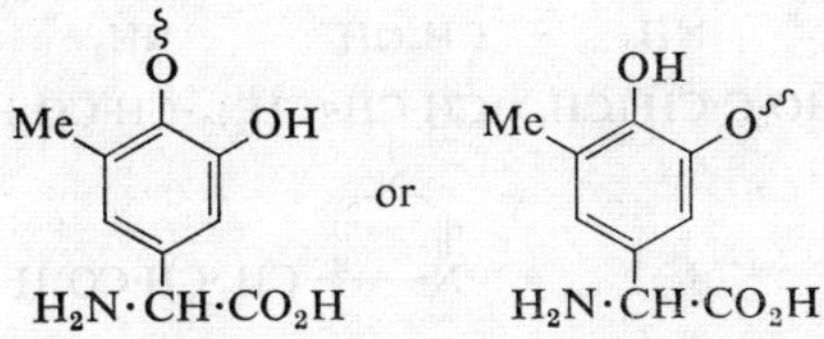

(10)

One of the co-occurring amino-acids seems to be identical except that it lacks the methyl group.[27, 28] The structure of the depsipeptide detoxin-D has been elucidated. It contains the novel pyrrolidine amino-acid (11); the o.r.d. spectrum of the *N*-valyl lactone derivative of (11) formed upon alkaline hydrolysis of detoxin-D shows the ring substituents to be *cis*.[29] A hitherto unidentified amino-acid produced by acid hydrolysis of chlorosis-inducing toxins from the plant pathogen *Pseudomonas* has now been characterized as 3-aminomethyl-6-carboxy-3-hydroxy-2-piperidone (12).[30]

[26] N. van Thoai, Y. Robin, and Y. Guillon, *Biochemistry*, 1972, **21**, 3890.

[27] J. R. Fehlner, R. E. J. Hutchinson, D. S. Tarbell, and J. R. Schenk, *Proc. Nat. Acad. Sci. U.S.A.*, 1972, **69**, 2420.

[28] N. N. Lomakina, M. S. Yurina, V. N. Sheinker, and K. F. Turchin, *Antibiotki Moscow*, 1972, **17**, 488.

[29] K. Kakinuma, N. Otake, and H. Yonemara, *Tetrahedron Letters*, 1972, **69**, 2420.

[30] P. A. Taylor, H. K. Schnoes, and R. D. Durbin, *Biochim. Biophys. Acta*, 1972, **286**, 107.

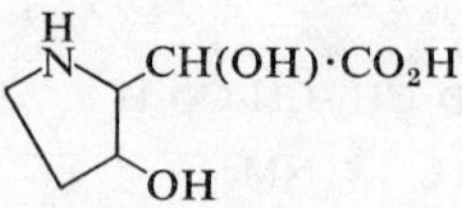

(11)

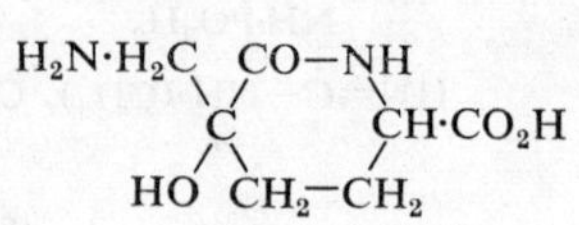

(12)

Hydrolysis of the antitubercular peptides tuberactinomycins A and N gives *threo*-γ-hydroxy-β-lysine if hydrochloric acid is used, but the *erythro*-isomer if sulphuric acid is used. It is therefore not clear which stereoisomer exists in the original peptide.[30a]

Although chlorination of the aromatic ring of tyrosine is known to occur under some conditions during acid hydrolysis with hydrochloric acid, 3-chlorotyrosine has not previously been reported to occur naturally. Three reports have appeared in 1972 of its isolation, from the scleroprotein of the whelk *Buccinum undatum*,[31] from locust cuticular protein,[32] and in the cuticle of *Limulus polyphemus*.[33] In the latter 3,5-dichlorotyrosine was also present. The conditions of isolation in all three cases were thought to be such as to make chlorination during protein degradation unlikely. Two papers concerning new collagen cross-links have been published. Isolation of 2,10-diamino-5-hydroxymethyl-6-(N^{τ}-histidyl)-undecandioic acid (13), termed aldol-histidine, from borohydride-reduced cow-skin

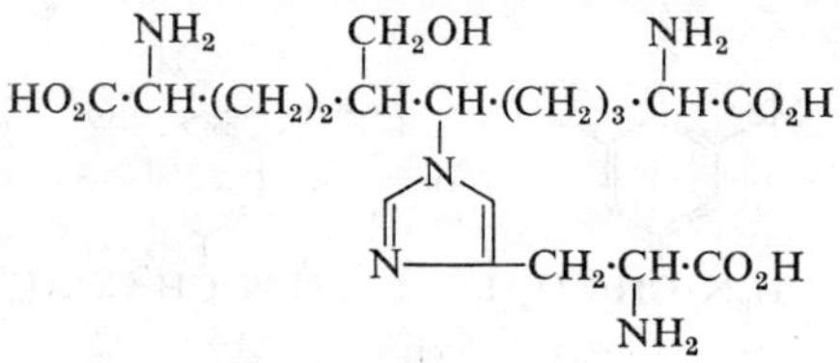

(13)

insoluble collagen provides the first evidence for a histidine-containing cross-link. Its biosynthesis probably involves a Michael addition of the histidine imidazole to the known collagen component 2,10-diamino-5-formyl-5-undecendioic acid; the product on reduction would yield (13).[34] After reduction of calf insoluble collagen with NaB^3H_4, a labelled degradation product has been identified as *N*-(δ-hydroxynorleucino)-2,3,4,5,6-pentahydroxyhexylamine. This no doubt arises by the reduction of the Schiff base formed by interaction of the carbonyl group of a hexose sugar and the ε-amino-group of δ-hydroxylysine. The structure of the amino-acid was established by mass spectral comparison with N^{ε}-galactosyl-

[30a] T. Wakamiya, T. Shiba, and T. Kaneka, *Bull. Chem. Soc. Japan*, 1972, **45**, 3668.
[31] S. Hunt, *F.E.B.S. Letters*, 1972, **24**, 109.
[32] S. O. Anderson, *Acta Chem. Scand.*, 1972, **26**, 3097.
[33] S. Welinder, *Biochim. Biophys. Acta*, 1972, **279**, 491.
[34] R. B. Fairweather, M. L. Tanzer, and P. M. Gallop, *Biochem. Biophys. Res. Comm.*, 1972, **48**, 1311.

hydroxylysine, but the specific hexose configuration remains to be determined.[35]

2 The Chemical Synthesis and Resolution of Amino-acids

General Methods of Synthesis.—The preparation of α-amino-acids from α-keto-acids continues to be explored. A detailed study of the temperature dependence of the hydrogenolytic asymmetric transamination of ethyl pyruvate by several optically active amines has now appeared. Amines of the *R* configuration produce (*R*)-alanine (D-alanine) at lower temperatures, but as the reaction temperature increases (*S*)-alanine is formed. Using (*R*)-α-methylbenzylamine, for example, configurational inversion occurs between 10 and 20 °C. The highest optical purity recorded for the alanine was 68%.[36] Reduction of α-keto-acids by (*R*)-(+)-α-phenethylamineborane or its enantiomer gives better yields of α-amino-acids than does the use of sodium cyanoborohydride, but the optical purity of the products is low. This is ascribed to the relatively large distance between the asymmetric carbon of the reducing agent and the developing tetrahedral carbon of the product in the transition state.[37] The imine obtained from *R*- or *S*-α-methylbenzylamine and ethyl glyoxylate reacts with enneacarbonyldi-iron to form two diastereoisomeric complexes of the structure $Fe(CO)_4$-($PhCHMeN{:}CH{\cdot}CO_2Et$) which can be readily separated. Treatment of these with halogeno-compounds such as benzyl bromide or ethyl bromoacetate followed by catalytic hydrogenation leads to phenylalanine and aspartic acid, respectively. Optical purities of 77 and 78% were obtained. Treatment of the iron complex with acetyl chloride gives some alanine as well as threonine; the intermediate acyl complex apparently undergoes partial decarbonylation.[38]

Di-isopinocamphenylborane has been used as the chiral reagent in the asymmetric synthesis of α-amino-acids from nitriles. The imino-borane intermediate [i.r. shows it to be a mixture of the monomer and dimer (14)] is treated with acetone cyanohydrin as a source of HCN, and subsequent cleavage from the boron with methanol liberates the α-aminonitrile ready for hydrolysis. Convenience is claimed as an advantage of this method. Valine prepared from 2-methylpropionitrile in this way had an optical purity of 12.4%.[39]

Two papers concerning the use of dehydroamino-acid derivatives as α-amino-acid precursors have appeared. Reduction of α-acylamino-acrylic acids using catalysts prepared from Rh(hexa-1,5-diene)Cl_2 and *o*-anisylmethylcyclohexylphosphine (*i.e.* a phosphine in which the chirality

[35] M. L. Tanzer, R. B. Fairweather, and P. M. Gallop, *Arch. Biochem. Biophys.*, 1972, **151**, 137.
[36] K. Harada and T. Yoshida, *J. Org. Chem.*, 1972, **37**, 4366.
[37] R. F. Borch and S. R. Levitan, *J. Org. Chem.*, 1972, **37**, 2347.
[38] J. Y. Chenard, D. Commereuc, and Y. Chauvin, *J.C.S. Chem. Comm.*, 1972, 750.
[39] U. E. Diner, M. Worsley, J. R. Lown, and J. Forsythe, *Tetrahedron Letters*, 1972, 3145.

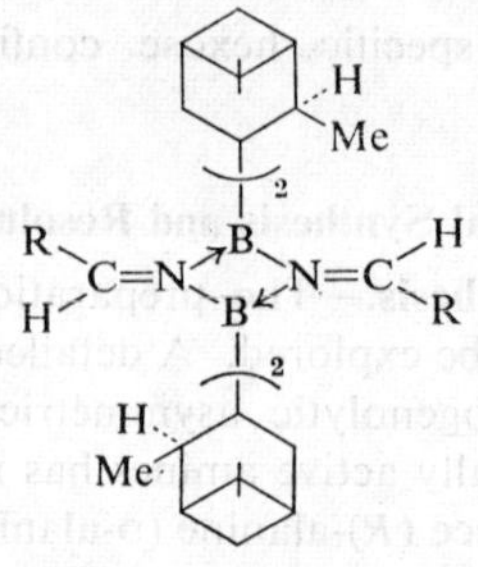

(14)

is on the phosphorus itself) gave products of up to 90% optical purity.[40] Reductive hydrolysis of unsaturated azlactones with Raney nickel in alcoholic ammonia under 2—3 atmospheres of hydrogen gives acylamino-acid amides and hence α-amino-acids in yields often better than those obtained from such azlactones by other methods.[41]

An admirably brief communication reports a new general method of α-amino-acid synthesis through amination of α-lithiated carboxylic acid salts. Phenylacetic acid, for example, on successive treatment with lithium isopropylamine and *O*-methylhydroxylamine gave, after work-up, a 55.5% yield of α-phenylglycine. This is the first report of the conversion of carboxylic acids into α-amino-acids by a one-stage procedure.[42] A way in which the optical purity of intermediates in the asymmetric synthesis of α-amino-acids can be established has been put forward,[43] and the preparation of α-amino-acids by the C-alkylation of glycine as its bis-(*N*-salicylidene glycinato)cobaltate(III) complex by acrylonitrile, methyl acrylate, and acetaldehyde has been described.[44]

Synthesis under Simulated Prebiotic Conditions.—It has been pointed out that reliance on *Rf* values or column chromatographic elution positions alone for the identification of amino-acids produced under simulated prebiotic conditions is insufficient for unambiguous identification. Examination of products from the action of an electric discharge on a mixture of methane, nitrogen, water, and traces of ammonia by g.l.c.–m.s. has shown that a peak at the isoleucine position on the amino-acid analyser is in fact α-hydroxy-γ-aminobutyric acid. The spectrum of products also indicates that no selective synthesis of the branched-chain amino-acids present in proteins occurs.[45] All the non-protein amino-acids found in the Murchison

[40] W. S. Knowles, M. J. Sabacky, and B. D. Vineyard, *J.C.S. Chem. Comm.*, 1972, 10.
[41] A. Badshak, N. H. Khan, and A. R. Kidwai, *J. Org. Chem.*, 1972, **37**, 2916.
[42] S. Yamada, T. Oguri, and T. Shioiri, *J.C.S. Chem. Comm.*, 1972, 623.
[43] J. C. Fiaud and A. Horeau, *Tetrahedron Letters*, 1972, 2565.
[44] Yu. N. Belokon, N. I. Kuznetsova, R. M. Murtazin, and M. M. Dolgaya, *Izvest. Akad. Nauk S.S.S.R., Ser. khim.*, 1972, 2772.
[45] J. E. Van-Trump and S. L. Miller, *Science*, 1972, **178**, 859.

meteorite are produced in this electric discharge reaction, and the pattern of relative abundances is on the whole quite similar.[46]

The hitherto neglected prebiotic synthesis of the sulphur amino-acids has now received attention. Addition of hydrogen sulphide to the gas mixture detailed in the preceding paragraph leads to the formation of methionine; a synthetic scheme has been proposed (Scheme 1). The presence

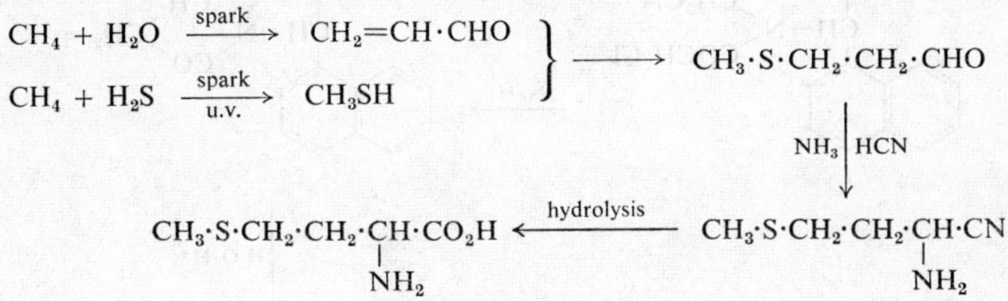

Scheme 1

of acrolein in the products was established, and this is thought to be a key intermediate in the formation of glutamic acid, homocysteine, α,γ-diaminobutyric acid, and α-hydroxy-γ-aminobutyric acid as well as methionine.[47]

The prebiotic synthesis of amino-acids by the attack of cyanide ion on nitriles (Scheme 2) does not appear to be feasible under the dilute conditions

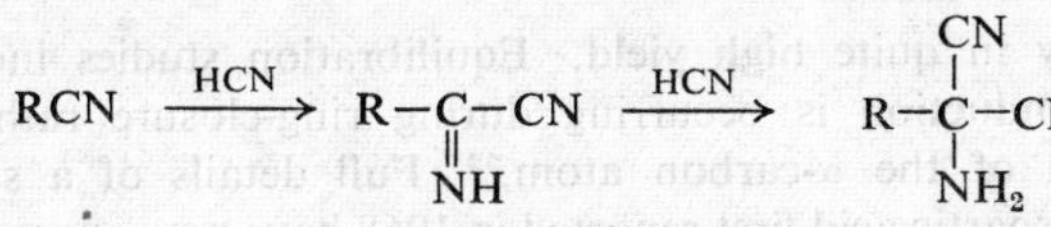

Scheme 2

considered reasonable except in the cases of aspartic and glutamic acids. Only in these instances does the inductive effect of the side-chain group R result in a rapid enough addition.[48] Morphological and compositional resemblances between material occurring in precambrian rock from South Africa and the products of simulated prebiotic syntheses have been noted.[49]

Two papers concerning the study of prehistoric amino-acids have appeared. The racemization of amino-acids near neutral pH has been considered and related to organic geochemistry,[50] and the dating of fossil bones using the extent of racemization of isoleucine has been proposed.[51]

[46] D. Ring, Y. Wolman, N. Friedmann, and S. L. Miller, *Proc. Nat. Acad. Sci. U.S.A.*, 1972, **69**, 765.
[47] Y. Wolman, W. J. Haverland, and S. L. Miller, *Proc. Nat. Acad. Sci. U.S.A.*, 1972, **69**, 809.
[48] Y. Wolman and S. L. Miller, *Tetrahedron Letters*, 1972, 1199.
[49] C. Simonescu, F. Denes, and E. Schreiber, *Compt. rend.*, 1972, **275**, *D*, 703.
[50] J. L. Bada, *J. Amer. Chem. Soc.*, 1972, **94**, 1371.
[51] J. L. Bada, *Earth Planet Sci. Letters*, 1972, **15**, 273.

Protein and other Naturally Occurring Amino-acids.—The preparation of aspartic acid through β-lactams derived from *N*-alkyl-*N*-chloracetamido-acetonitriles has been studied in detail. Derivatives of (*R*)(+)-α-(1-naphthyl)ethylamine (Scheme 3) gave aspartic acid of the highest (54—67%)

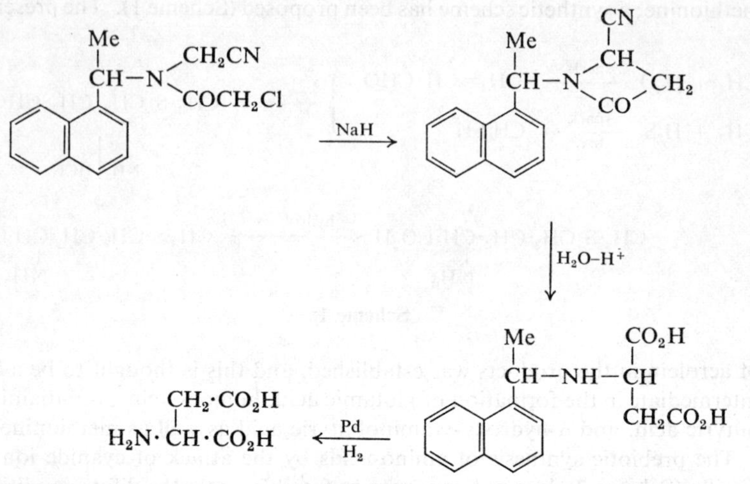

Scheme 3

optical purity in quite high yield. Equilibration studies indicated that asymmetric induction is occurring during ring-closure rather than by epimerization of the α-carbon atom.[52] Full details of a stereospecific synthesis of aspartic acid first reported in 1968 have now appeared (Scheme 4). In this method L-*erythro*-1,2-diphenylethanolamine is used to form an

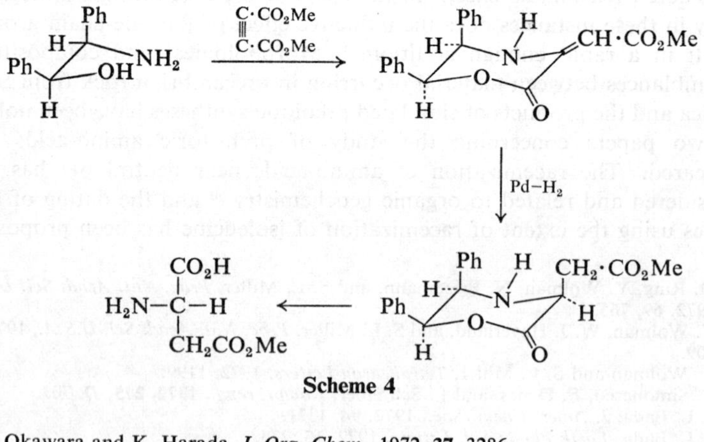

Scheme 4

[52] T. Okawara and K. Harada, *J. Org. Chem.*, 1972, **37**, 3286.

oxazinone which is hydrogenated and hydrolysed to give a 98% yield of L-aspartic acid.[53] DL-Glutamic acid has been prepared in 20% yield by a reaction sequence involving the addition of copper salicylideneglycinate to methyl acrylate in the presence of base. The corresponding Schiff base with acetaldehyde gave 4% of glutamic acid in a similar reaction.[54]

A convenient synthetic route to the reduction products of several collagen cross-links and cross-link precursors has been developed. The key intermediate 5-(but-3-enyl)-3-phenylhydantoin (15) can be used as a source of

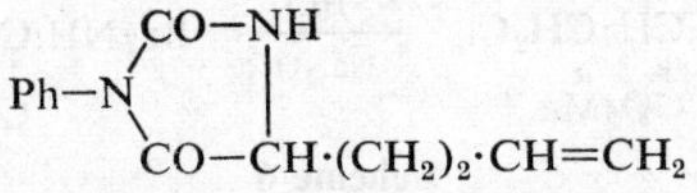

(15)

hydroxynorleucine, 5,6-dihydroxynorleucine, hydroxylysinonorleucine, and hydroxylysinohydroxynorleucine.[55] A new synthesis of 1-aminocyclopropane-1-carboxylic acid from methyl α-acetamidoacrylate involves the addition of diazomethane to give a pyrazoline which loses nitrogen on pyrolysis (Scheme 5).[56] A novel method of preparing L-(−)-α-methyl-3,4-

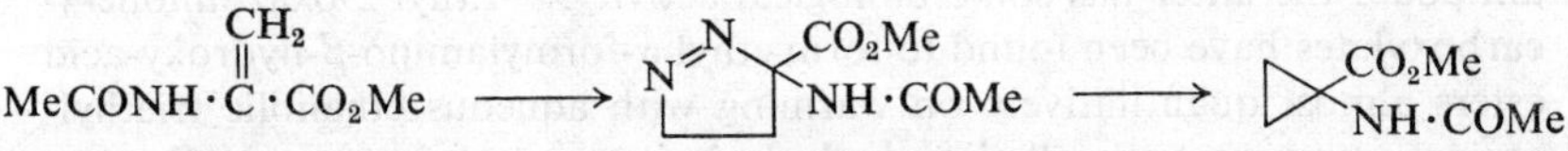

Scheme 5

dimethoxyphenylalanine by an Ugi reaction involving L-(−)-α-methylbenzylamine has been developed. The resulting diastereoisomeric L(or D)-α-methylbenzyl-DL-amino-acids can be easily separated.[57]

***C*-Alkyl and Substituted *C*-Alkyl Amino-acids.**—The preparation of 27 *C*-alkyl amino-acids, 14 of which are novel, has been reported; four of these are methionine analogues. Five of these materials are toxic to *Escherischia coli*.[58] L-β-Chloroalanine has been prepared from L-aspartic acid through its β-hydrazide (Scheme 6).[59]

Amino-acids with Aliphatic Hydroxy-groups in the Side-chain.—The synthesis of both enantiomers of δ-hydroxy-β-lysine from D-galacturonic acid and 3-amino-3-deoxy-D-glucose *via* the lactone of 3,6-diacetamido-5-hydroxyhexanoic acid (16) has led to the synthesis of negamycin (17) and its optical

[53] J. P. Vigneron, H. Kagan, and A. Horeau, *Bull. Soc. chim. France*, 1972, 3836.
[54] Yu. N. Belokon, V. M. Belokon, N. I. Kuznetsova, and M. M. Dolgaya, *Bull. Acad. Sci. U.S.S.R.*, 1972, 1288.
[55] N. R. Davis and A. J. Bailey, *Biochem. J.*, 1972, **129**, 91.
[56] I. Bregovec and T. Jakovcic, *Monatsh.*, 1972, **103**, 288.
[57] K. Freter, M. Gotz, and K. Grozinger, *J. Medicin. Chem.*, 1972, **15**, 1072.
[58] C. J. Abshire and G. Planet, *J. Medicin. Chem.*, 1972, **15**, 226.
[59] K. Okumura, T. Iwasaki, T. Okawara, and K. Matsumoto, *Bull. Inst. Chem. Res. Kyoto Univ.*, 1972, **50**, 209.

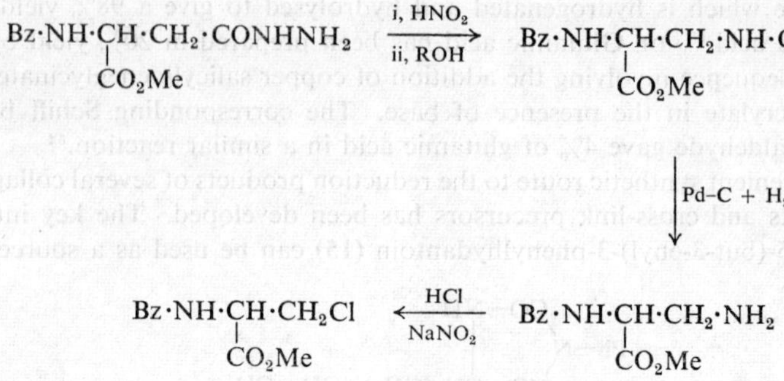

Scheme 6

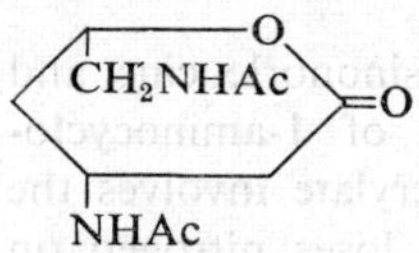

(16)

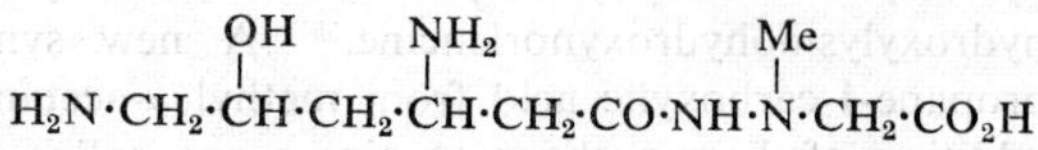

(17)

antipode; the latter has some biological activity.[60] Ethyl 2-oxazolinone-4-carboxylates have been found to form ethyl α-formylamino-β-hydroxy-acid esters almost quantitatively on warming with aqueous ethanolic triethylamine, although such alkaline hydrolysis is unusual for 2-oxazolinones. Since ethyl 2-oxazolinone-4-carboxylates are readily available from ethyl isocyanoalkanoates and carbonyl compounds in basic media, this affords a new synthetic route to β-hydroxy-α-amino-acids (Scheme 7).[61, 62]

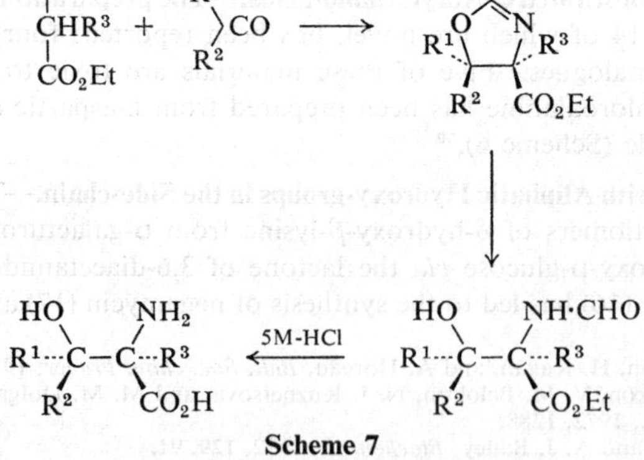

Scheme 7

[60] S. Shibahara, S. Kando, K. Maeda, H. Umezawa, and M. Ohno, *J. Amer. Chem. Soc.* 1972, **94**, 4353.
[61] D. Hoppe and U. Schollkopf, *Annalen*, 1972, **763**, 1.
[62] D. Hoppe and U. Schollkopf, *Angew. Chem. Internat. Edn.*, 1972, **11**, 432.

Aromatic and Heterocyclic Amino-acids.—L-Tryptophan has been enzymically prepared from indole, ammonia, and sodium pyruvate in the presence of tryptophanase in good yield; 5-hydroxytryptophan can be made if 5-hydroxyindole is used in place of indole.[63] An electrolytic synthesis of DL-tryptophan from (18) has been reported,[64] and an improved synthesis of indole-3-acetylaspartic acid *via* the *p*-nitrophenyl ester of indole-3-acetic acid has been described.[65] A thieno-[3,2-*b*]pyrrole analogue of tryptophan

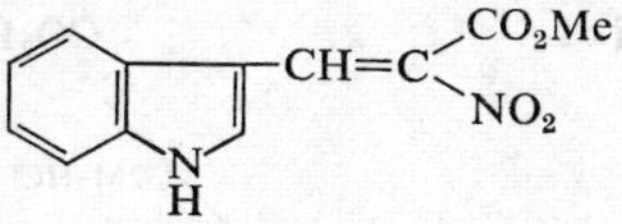

(18)

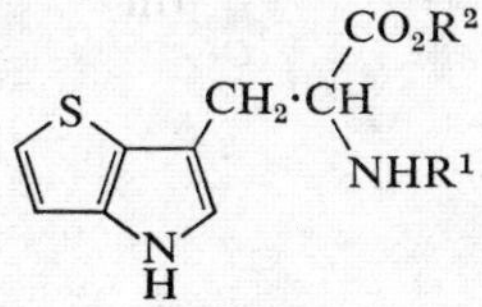

(19)

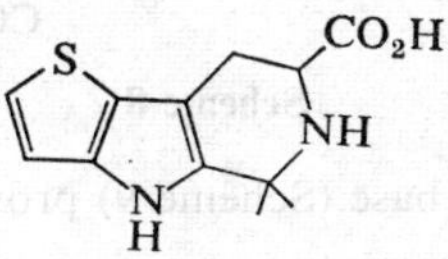

(20)

(19; $R^1 = R^2 = H$) has been prepared and found to be less stable than the indole compound. Hydrolytic deacylation of a precursor (19; R^1 = COMe; R^2 = Me) proved difficult; attempts to cleave with hydrazine led to the formation of (20) as well as the desired amino-acid. The *gem*-dimethyl group arose from the use of acetone to precipitate the product and consume excess hydrazine.[66]

Alkylation of the methyl ester of $N^{\alpha}N^{\pi}$-dibenzoyl-L-histidine with trimethyloxonium fluoroborate provides a convenient route to N^{τ}-methyl-L-histidine. Hydrolysis of the resulting ester affords 75% of the desired compound.[67] Pyrazole will add on to α-acetamidoacrylic acid in boiling acetic acid to give an excellent yield of the acyl derivative of β-(pyrazolyl-*N*)-DL-alanine; this is a convenient new synthesis for this amino-acid.[68] A new preparation of 3-(3,4-dihydroxyphenyl)-L-alanine involving oxidation of an *N*-acyl-L-tyrosine methyl ester with benzoyl peroxide in chloroform has been published.[69] 1,4-Cyclohexadiene-L-alanine hydrate, a naturally occurring inhibitor found in certain bacteria, has been found to undergo

[63] H. Nakazawa, H. Enei, S. Okumura, H. Yoshida, and H. Yamada, *F.E.B.S. Letters*, 1972, **25**, 43.
[64] E. V. Zaparazhets, I. A. Aurutskaya, and M. Y. Fioshin, *Electrokhimiya*, 1972, **8**, 1809.
[65] R. C. Mollan, D. M. X. Donnelly, and M. A. Harmey, *Phytochemistry*, 1972, **11**, 1485.
[66] A. J. Humphries, R. L. Keener, K. Yano, F. S. Skelton, and E. Freiter, *J. Org. Chem.*, 1972, **37**, 3626.
[67] H. C. Beyerman, L. Moat, and A. Van-Zou, *Rec. Trav. chim.*, 1972, **91**, 246.
[68] I. Murakoshi, S. Ohmiya, and J. Haginawa, *Chem. and Pharm. Bull.* (*Japan*), 1972, **20**, 409.
[69] H. Varbrüggen and K. Krolikiewicz, *Chem. Ber.*, 1972, **105**, 1168.

dehydrogenation in the solid state in the presence of oxygen to form L-phenylalanine. All attempts to desiccate the hydrate have led to this dehydrogenation; this and other evidence suggest that the water of crystallization may have a role in this reaction.[70] The carboxylation of α-lithiobenzyl isocyanide enables α-phenylglycine to be rapidly synthesized (Scheme 8),[71] and the formation of oxazoles from acid chlorides simply

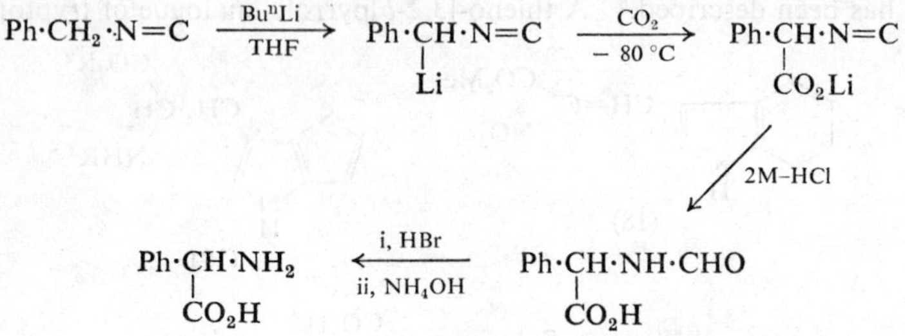

Scheme 8

using triethylamine as the base (Scheme 9) provides a convenient route to *C*-aroyl amino-acids.[72]

Azetidine-3-carboxylic acid has been prepared for the first time. Its p*K* values are similar to those of proline, and an *X*-ray study shows the ring to be puckered by less than 1°. In the last step in its synthesis (hydrogenolysis of an *N*-benzhydryl group) use of Pd–C as catalyst effected no cleavage, but palladium hydroxide gave a quantitative yield of the desired material.[73] In the preparation of aziridine-2-carboxamides by the reduction of 2*H*-azirine-2-carboxamides, sodium borohydride is an effective agent but lithium aluminium hydride is not.[74] A new synthesis and resolution of

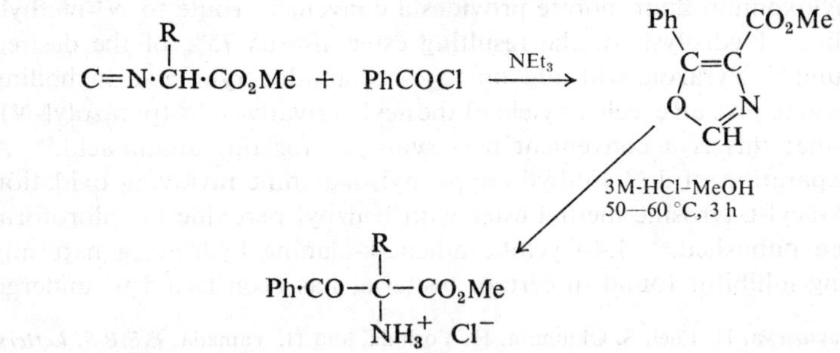

Scheme 9

[70] C. Ressler, *J. Org. Chem.*, 1972, **37**, 2933.
[71] W. Vallburg, J. Strating, M. G. Waldring, and H. Wynberg, *Synthetic Comm.*, 1972, **2**, 423.
[72] M. Suzuki, T. Iwasaki, K. Matsumoto, and K. Okumura, *Synthetic Comm.*, 1972, **2**, 237.
[73] A. G. Anderson and R. Lok, *J. Org. Chem.*, 1972, **37**, 3953.
[74] T. Nishiwaki and F. Fujiyama, *Synthetic Comm.*, 1972, **2**, 569.

cis- and *trans*-5-methylproline has been reported,[75] and homologues of proline containing 12—15-membered rings have been prepared. These latter syntheses involve the base-catalysed rearrangement of α-halogeno-ω-aminolactams prepared from lactams derived from oximes by the Beckmann rearrangement.[76]

N-Substituted Amino-acids.—The first synthesis of N^{δ}-hydroxyornithine has been reported. The preparation involves the tosylation of *O*-benzylhydroxylamine, alkylation with 1,3-dibromopropane to give γ-(*N*-tosyl-*N*-benzyloxy)aminopropyl bromide, and then treatment with diethyl sodioacetamidomalonate as in a conventional amino-acid synthesis. N^{δ}-Tosyl-N^{δ}-benzyloxy-L-ornithine was prepared by a papain-catalysed resolution.[77] A reaction of general utility for the synthesis of cyclic *N*-hydroxyamino-acids and *N*-hydroxy-*N*-alkylamino-acids has been developed. The reaction sequence involves cyanoethylation, N-oxidation, and a Cope elimination of acrylonitrile (Scheme 10). 1-Hydroxyproline prepared by this method

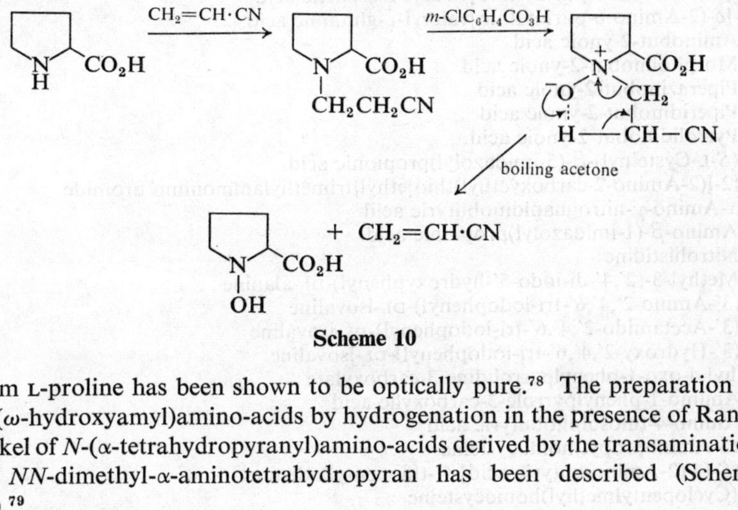

Scheme 10

from L-proline has been shown to be optically pure.[78] The preparation of *N*-(ω-hydroxyamyl)amino-acids by hydrogenation in the presence of Raney nickel of *N*-(α-tetrahydropyranyl)amino-acids derived by the transamination of *NN*-dimethyl-α-aminotetrahydropyran has been described (Scheme 11).[79]

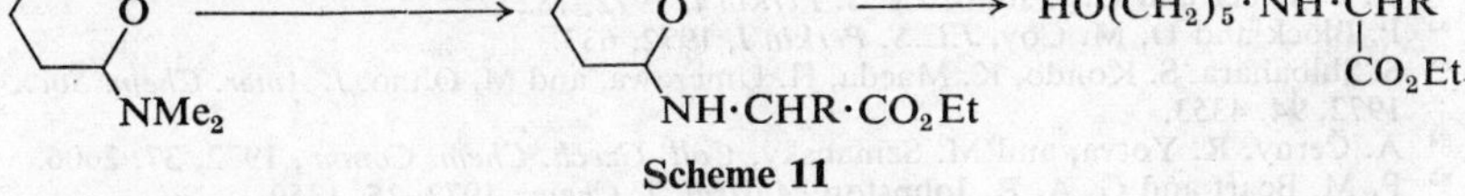

Scheme 11

[75] C. G. Overberger, K. H. David, and J. A. Moore, *Macromolecules*, 1972, **5**, 368.
[76] J. A. Elberling and H. T. Nagasawa, *J. Heterocyclic Chem.*, 1972, **9**, 411.
[77] Y. Isowa, T. Takashima, M. Ohmori, H. Kurita, M. Sato, and K. Mori, *Bull. Chem. Soc. Japan*, 1972, **45**, 1461.
[78] H. T. Nagasawa, J. G. Kohlhoff, P. S. Fraser, and A. A. Mikhail, *J. Medicin. Chem.*, 1972, **15**, 483.
[79] G. Glacet and J. Tesse, *Compt. rend.*, 1972, **275**, *C*, 147.

α-Amino-acids containing Sulphur.—The copper-catalysed alkaline autoxidation of homocysteine is a useful source of homocysteinesulphinic acid.[80]

A List of α-Amino-acids which have been Synthesized for the First Time.—

Compound	*Ref.*
2,6-Diamino-4-oxohexanoic acid	81
4,5-Dihydroxy-L-pipecolic acid	16
S-(2-Hydroxy-2-carboxyethanethiomethyl)-L-cysteine	18
2-Amino-3-(6-thieno[3,2-*b*]pyrrolyl)propionic acid	66
Azetidine-3-carboxylic acid	73
3,3′,5,5′-Tetramethyl-L-thyronine	82
δ-Hydroxy-β-lysine	83
L-2-Aminohex-4-ynoic acid	23
N^{δ}-Hydroxyornithine	77
N-[δ-(2-Amino-6-purinylthio]valerylglycine	84
N-[δ-(2-Amino-6-purinylthio]valeryl-DL-valine	84
N-[δ-(2-Amino-6-purinylthio]valeryl-L-leucine	84
N-[δ-(2-Amino-6-purinylthio]valeryl-L-aspartic acid	84
N-[δ-(2-Amino-6-purinylthio]valeryl-L-glutamic acid	84
4-Aminobut-2-ynoic acid	85
4-Morpholinobut-2-ynoic acid	85
4-Piperazinobut-2-ynoic acid	85
4-Piperidinobut-2-ynoic acid	85
4-Pyrrolidinobut-2-ynoic acid	85
α-(*S*-L-Cysteinyl)-β-(5-imidazolyl)propionic acid	86
L-{2-[(2-Amino-2-carboxyethyl)thio]ethyl}trimethylammonium bromide	87
L-α-Amino-γ-nitroguanidinobutyric acid	88
α-Amino-β-(1-imidazolyl)propionic acid	89
4-Nitrohistidine	89
2-Methyl-3-(2′,4′-di-iodo-5′-hydroxyphenyl)-DL-alanine	89
4-(3′-Amino-2′,4′,6′-tri-iodophenyl)-DL-isovaline	90
4-(3′-Acetamido-2′,4′,6′-tri-iodophenyl)-DL-isovaline	90
4-(3′-Hydroxy-2′,4′,6′-tri-iodophenyl)-DL-isovaline	90
Ethyl 4-oxo-1-phenylpyrrolidine-3-carboxylate	91
4-Anilino-1-phenylpyrrole-3-carboxylic acid	91
2-Amino-4-thiosulphobutyric acid	92
S-(3-Aminopropyl)homocysteine	93
9-{*S*-[4-(2-Amino)butyric acid]-5′-thiopentyl}adenine	93
S-(Cyclopentylmethyl)homocysteine	93

80 P. Lucki and C. de Marco, *Analyt. Biochem.*, 1972, **45**, 236.
81 R. C. Hider and D. I. John, *J.C.S. Perkin I*, 1972, 1825.
82 P. Block and D. M. Coy, *J.C.S. Perkin I*, 1972, 633.
83 S. Shibahara, S. Kondo, K. Maeda, H. Umezawa, and M. Ohno, *J. Amer. Chem. Soc.*, 1972, **94**, 4353.
84 A. Černy, R. Yotva, and M. Semanský, *Coll. Czech. Chem. Comm.*, 1972, **37**, 2606.
85 P. M. Beart and G. A. R. Johnston, *Austral. J. Chem.*, 1972, **25**, 1359.
86 Y. A. Yankeelov and G. J. Jolley, *Biochemistry*, 1972, **11**, 159.
87 H. A. Itano and E. A. Robinson, *J. Biol. Chem.*, 1972, **247**, 4819.
88 J. W. Van Nispen and G. I. Tesser, *Synthetic Comm.*, 1972, **2**, 207.
89 G. E. Trout, *J. Medicin. Chem.*, 1972, **15**, 1259.
90 G. Shtacher and S. Dayagi, *J. Medicin. Chem.*, 1972, **15**, 1174.
91 V. J. Bauer and S. R. Safir, *J. Medicin. Chem.*, 1972, **15**, 440.
92 C. De Marco and P. Lucki, *Analyt. Biochem.*, 1972, **48**, 346.
93 J. K. Coward and W. D. Sweet, *J. Medicin. Chem.*, 1972, **15**, 381.

Compound	*Ref.*
5-Amino-4-chloro-6-[3-(homocysteinyl)-n-propyl]aminopyrimidine	93
L-3-{[*p*-(Hydroxymethyl)-αα-diphenylbenzyl]thio}alanine	94
L-3-{[αα-Bis-(α-hydroxy-*p*-tolyl)-benzyl]thio}alanine	94
L-3-{[(4-Biphenylyl)diphenylmethyl]thio}alanine	94
L-3-[(αα-Diphenylpiperonyl)thio]alanine	94
N-[Tris(hydroxymethyl)methyl]alanine	95
2-Ethylamino-DL-alanine	96
2-Propylamino-DL-alanine	96
2-Butylamino-DL-alanine	96
2-Amylamino-DL-alanine	96
2,2-Diethylamino-DL-alanine	96
2-(2′-Hydroxyethyl)amino-DL-alanine	96
5′-Guanosylhomocysteine	97
12—15-membered-ring homologues of proline	76
14 new *C*-alkyl amino-acids with aliphatic side-chains	58

Labelled Amino-acids.—A simple and general procedure for labelling α-amino-acids at their α-C atoms by deuterium (or tritium) by a catalytic exchange in $MeCO_2D$–D_2O containing DCl at 120 °C has been developed. That such a reaction is of utility demonstrates that there is a limitation to the conditions under which such labelled compounds can be used for tracer studies.[98] The preparations of deuterium- and tritium-labelled (3*S*)- and (3*R*)-phenylalanines [99] and (2*R*, 3*S*)-[3-$^3H_1$]- and (2*S*, 3*R*)-[3-$^3H_1$]-tyrosines [100] have been described. These materials have been used to investigate the stereochemical courses of elimations to αβ-unsaturated acids catalysed by L-phenylalanine ammonia-lyase [99] and enzymes from maize and potatoes.[100] The formation of L-αγ-diamino[4-^{3}H]butyric acid, L-[5-^{3}H]ornithine, and DL-[6-^{3}H]lysine by the catalytic tritiation of β-, γ-, and δ-cyano-compounds has been detailed,[101] and a systematic study of the catalytic microtritiation of αβ-unsaturated amino-acid derivatives has been made.[102] A re-evaluation of the loss of tritium labels from positions 3 and 5 of the aromatic ring of tyrosine on acid hydrolysis indicates that such exchange has been hitherto underestimated. Similar losses occur from the aromatic nucleus of tryptophan.[103]

DL-3-[2′,3′,4′,5′-$^{14}C_4$]Phenylalanine has been prepared by the co-cyclotrimerization of [$^{14}C_2$]acetylene and ethyl propargylformamidomalonate

[94] K. Y. Zee-Cheng and C. C. Cheng, *J. Medicin. Chem.*, 1972, **15**, 13.
[95] J. Galsonias, C. Frezou, and P. Vieles, *Compt. rend.*, 1972, **274**, *C*, 1392.
[96] R. S. Asquith and P. Carthew, *Tetrahedron*, 1972, **28**, 4769.
[97] J. A. Elberling and H. T. Nagasawa, *J. Heterocyclic Chem.*, 1972, **9**, 411.
[98] J. L. Garrett, B. Halpern, and R. S. Kenyon, *J.C.S. Chem. Comm.*, 1972, 135.
[99] R. M. Wightman, J. Staunton, A. R. Battersby, and K. R. Hanson, *J.C.S. Perkin I*, 1972, 2355.
[100] P. G. Strange, J. Staunton, M. R. Wiltshire, A. R. Battersby, K. R. Hanson, and E. A. Mavir, *J.C.S. Perkin I*, 1972, 2364.
[101] S. Thyagarajan, I. Mezö, I. Teplán, and J. Márton, *Acta Chim. Acad. Sci. Hung.*, 1972, **73**, 23.
[102] J. Márton and A. Kovács, *Acta Chim. Acad. Sci. Hung.*, 1972, **73**, 11.
[103] L. A. Holt and B. Milligan, *Photochem. and Photobiol.*, 1972, **16**, 511.

(Scheme 12; * indicates labelled atoms). The best catalyst for this reaction is $(Ph_3P)_2Ni(CO)_2$.[104] [3-^{11}C]-β-Alanine has been synthesized from chloracetic acid and $^{11}CN^-$ with subsequent catalytic reduction,[105] and [4-^{13}C]-1-methyl-2-amino-2-imidazolin-4-one from [1-^{13}C]glycine.[106]

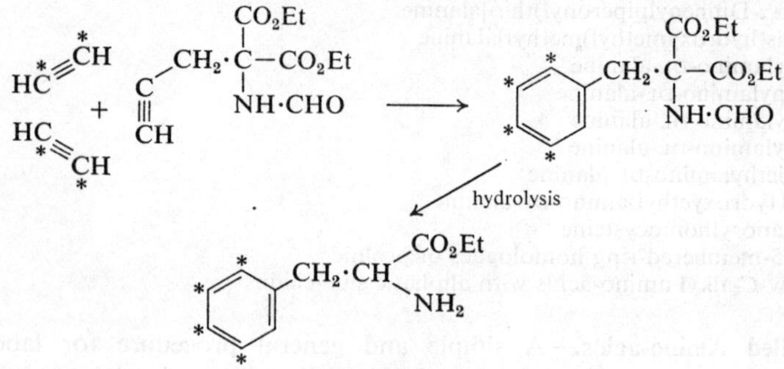

Scheme 12

An economical preparation of L-[^{35}S]methionine of high specific activity by extraction from yeast grown in the presence of $Na_2{}^{35}SO_4$ has been described,[107] and also a biosynthetic method.[108] *p*-[^{18}F]Fluorophenylalanine has been made by an exchange reaction with 4-(2′,2′-diethoxycarbonyl-2′-acetamidomethyl)phenyldiazonium tetrafluoroborate before carrying out a Schiemann reaction and subsequent hydrolysis.[109]

Resolution of α-Amino-acids.—Cyclohexyl *N*-trifluoroacetyl-L-α-amino-butyryl-L-α-aminobutyrate has been found useful as a stationary phase for the g.l.c. separation of the enantiomers of *N*-trifluoroacetyl-α-amino-acid n-propyl esters. Only glycine and L-threonine are not completely separated.[110] Separation of the enantiomers of t-leucine by g.l.c. of derivatives has not hitherto been achieved. Following up an idea that optically active stationary phases derived from α-amino-acids with less bulky side-chains might prove better for this purpose, it has been found that a resolution was possible using *N*-trifluoroacetyl-L-norvalyl-L-norvaline cyclohexyl ester.[111]

The labile metal complex (*N*-carboxymethyl-L-valine)copper(II) has been chemically bound to a polystyrene–0.8% divinylbenzene polymer to give a resin (21) which has been used to partially resolve several amino-acids. In all cases the D-isomers tend to elute first, and the degree of resolution increased as the bulkiness of the side-chain attached to the α-carbon increased.[112] D-Methionine-DL-*S*-oxide linked to a 5% cross-linked poly-

104 L. Pichat, P. N. Liem, and J. P. Guermont, *Bull. Soc. chim. France*, 1972, 4224.
105 H. Elias, H. F. Lotterbos, and K. Weiner, *Chem. Ber.*, 1972 **105**, 3754.
106 G. L. Rowley and G. L. Kenyon, *J. Heterocyclic Chem.*, 1972, **9**, 203.
107 R. Graham and W. M. Stanley, *Analyt. Biochem.*, 1972, **47**, 505.
108 M. S. Bretscher and A. E. Smith, *Analyt. Biochem.*, 1972, **47**, 310.
109 R. W. Goulding and A. J. Palmer, *Internat. J. Appl. Radiat. Isotopes*, 1972, **23**, 133.
110 W. Parr and P. Y. Howard, *Angew. Chem. Internat. Edn.*, 1972, **11**, 529.
111 W. Parr and P. Y. Howard, *J. Chromatog.*, 1972, **66**, 141.
112 R. V. Snyder, R. J. Angelici, and R. B. Meck, *J. Amer. Chem. Soc.*, 1972, **94**, 260.

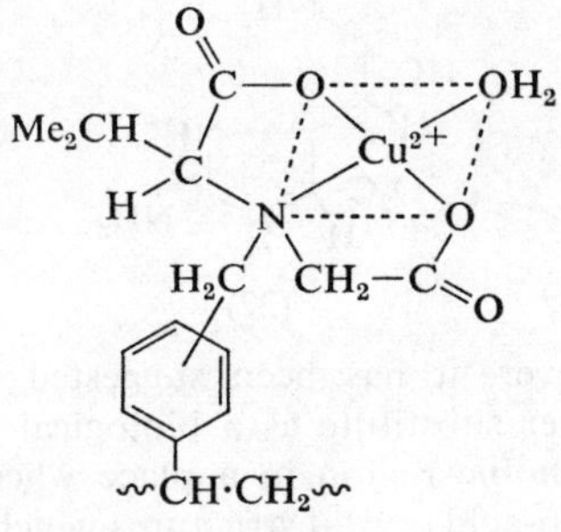

(21)

styrene and complexed with Cu^{2+} or Ni^{2+} has proved to be even more selective in the sorption of optical isomers. Conditions have been found which allow the L-isomers of isoleucine, proline, and *allo*threonine to be retained while the D-isomers are eluted.[113]

The preparation of all four stereoisomers of α,β-diaminobutyric acid from threonine has been reported.[114]

3 Physical and Stereochemical Studies of Amino-acids

Determination of Absolute Configuration.—A c.d. study of the *allo*threonine derived from the telomycin-related antibiotic LL-A-0341 has established that it has the D-configuration. This supports the 'rule of alpha epimerization' previously put forward by Bodanszky, and indicates that it originates from L-threonine. The 3-hydroxyproline and β-hydroxyleucine residues of this antibiotic were also investigated and found to have the L-configuration.[115] The peptide alkaloid lasiodine B contains a β-hydroxyleucine residue bound up as an aryl ether. The molecule can be degraded to (*p*-tolyloxy)leucine, and stereospecific syntheses of *threo*- and *erythro*-β-(*p*-tolyloxy)leucines have enabled the configuration of this residue in lasiodine to be determined.[16] The strongly basic amino-acid streptolidine, isolated on acid hydrolysis of the antibiotics streptolin, streptothricin, geomycin, roseothrycin, and racemomycin, has had its chirality established by a complete *X*-ray analysis of its dihydrochloride (22).[117]

Crystal Structures of Amino-acids.—A stereoscopic atlas of amino-acid structures has been published.[118] The *X*-ray crystal structure of β-(pyrazolyl-3)-L-alanine shows it to be, for most chemical purposes, isosteric

113 S. V. Rogozkin, V. A. Davankov, I. A. Yamskov, and V. P. Kabanov, *J. Gen. Chem. U.S.S.R.*, 1972, **42**, 1605.
114 E. Atherton and J. Meienhofer, *J. Antibiotics* (*Tokyo*), 1972, **25**, 539.
115 A. Bodanszky, M. Bodanszky, K. L. Perlman, and D. Perlman, *J. Antibiotics* (*Tokyo*), 1972, **25**, 325.
116 J. Marchand, F. Rocchiccioli, M. Pais, and F. X. Jarreau, *Bull. Soc. chim. France*, 1972, 4699.
117 B. W. Bycroft and T. J. King, *J.C.S. Chem. Comm.*, 1972, 652.
118 W. C. Hamilton, M. N. Frey, L. Golic, T. F. Koetzle, and M. S. Lehmann, *Mater. Res. Bull.*, 1972, **7**, 1225.

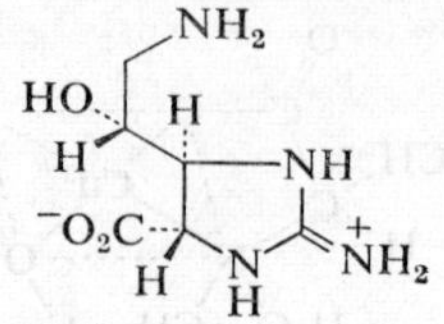

(22)

with histidine. However, it has been suggested that β-(pyrazolyl-4)-L-alanine may be a better substitute as a biological probe as it avoids the insertion of a hydrophobic region in a place where it is not present in Nature.[119] Other amino-acid crystal structures which have been established this year include homotaurine,[119a] γ-guanidobutyric acid hydrochloride,[120] L-histidine hydrochloride monohydrate,[121] the orthorhombic[122] and monoclinic[123] forms of L-histidine, and L-arginine phosphate monohydrate.[124] Amino-acid derivatives such as hippuric acid,[125] L-pyroglutamic acid *N*-methylamide,[126] *O*-(β-D-xylopyranosyl)-L-serine,[127] *N*-acetyl-L-histidine monohydrate,[128] and isostructural *N*-(halogenoacetyl)-L-phenylalanine ethyl esters[129] have also received attention.

The first papers in two projected series of neutron diffraction studies of α-amino-acids have appeared. These studies are aimed at producing more definitive stereochemical information on the hydrogen atom positions and better framework geometry than is available from *X*-ray studies. So far work concerning L-asparagine monohydrate,[130, 131] L-lysine hydrochloride dihydrate,[132, 133] L-glutamic acid hydrochloride,[134] L-histidine,[135] L-alanine,[136] and L-proline monohydrate[137] has been published.

119 N. C. Seeman, E. L. McGandy, and R. D. Rosenstein, *J. Amer. Chem. Soc.*, 1972, **94**, 1717.
119a S. Veoka, T. Fujiwara, and K. Tomita, *Bull. Chem. Soc. Japan*, 1972, **45**, 3634.
120 T. Maeda, T. Fujiwara, and K. Tomita, *Bull. Chem. Soc. Japan*, 1972, **45**, 3628.
121 K. Oda and H. Koyama, *Acta Cryst.*, 1972, **B28**, 639.
122 J. J. Madden, E. L. McGandy, and N. C. Seeman, *Acta Cryst.*, 1972, **B28**, 2377.
123 J. J. Madden, E. L. McGandy, N. C. Seeman, and M. M. Harding, *Acta Cryst.*, 1972, **B28**, 2382.
124 W. Saenger and K. C. Wagner, *Acta Cryst.*, 1972, **B28**, 2237.
125 W. Harrison, S. Rittig, and J. Trotter, *J.C.S. Perkin II*, 1972, 1036.
126 A. Aubry, M. Marraud, J. Protas, and J. Neel, *Compt. rend.*, 1972, **274**, *C*, 1378.
127 L. T. J. Delbaere, M. Higham, B. Kamenar, P. W. Kent, and C. K. Prout, *Biochim. Biophys. Acta*, 1972, **286**, 441.
128 T. J. Kistenmacher, D. J. Hunt, and P. E. Marsh, *Acta Cryst.*, 1972, **B28**, 3352.
129 C. H. Wei, D. G. Doherty, and J. R. Einstein, *Acta Cryst.*, 1972, **B28**, 907.
130 J. J. Verbist, M. S. Lehmann, T. F. Koetzle, and W. C. Hamilton, *Acta Cryst.*, 1972, **B28**, 3006.
131 M. Ramandham, S. K. Sikha, and R. Chidambaram, *Acta Cryst.*, 1972, **B28**, 3000.
132 R. R. Bugagong, A. Sequiera, and R. Chidambaram, *Acta Cryst.*, 1972, **B28**, 3214.
133 T. F. Koetzle, M. S. Lehmann, J. J. Verbist, and W. C. Hamilton, *Acta Cryst.*, 1972, **B28**, 3207.
134 A. Sequiera, H. Rajagopal, and R. Chidambaram, *Acta Cryst.*, 1972, **B28**, 2514.
135 M. S. Lehmann, T. F. Koetzle, and W. C. Hamilton, *Internat. J. Protein Res.*, 1972, **4**, 229.
136 M. S. Lehmann, T. F. Koetzle, and W. C. Hamilton, *J. Amer. Chem. Soc.*, 1972, **94**, 2657.
137 J. J. Verbist, M. S. Lehmann, T. F. Koetzle, and W. C. Hamilton, *Nature*, 1972, **235**, 328.

Optical Rotatory Dispersion and Circular Dichroism.—The use of some chiroptical probes for the configurational analysis of α-amino-acids has been reviewed.[138] Several reports this year concern the N-derivatives of α-amino-acids. A u.v.–visible and c.d. appraisal of *N*-(3-nitro-2-pyridyl)-amino-acids has been made, and as a result of this work a method of correlating the sign of the 420 nm c.d. band and the absolute configuration of the *N*-(3-nitro-2-pyridyl)-terminal amino-acid liberated by cleavage of the *N*-terminal peptide bond under mild acidic conditions has been proposed. The 3-nitro-2-pyridyl chromophore exhibits its long-wavelength optically active transition in a spectral range which is transparent for all natural amino-acids.[139] Since the reaction between 2-fluoro-3-nitropyridine and cysteinyl residue side-chains is selective and quantitative if suitable conditions are chosen for the reaction, a c.d. method is available for the analytical determination of thiol groups in peptides.[140]

N-Dimedonyl-α-amino-acids, derivatives which possess a vinylogous amide chromophore, have been spectroscopically re-investigated. Some similarity to the o.r.d. curves of some *N*-(2-pyridyl *N*-oxide)amino-acids was noted.[141] Another o.r.d. study of *N*-2,4-dinitrophenyl (DNP) α-amino-acids has been reported. The magnitude of the Cotton effects of di-DNP amino-acids is sensitive to solvent as well as to the difference in the chain length which separates the two DNP-groups. The spatial arrangement of the two DNP-groups is also a very important factor.[142] The c.d. spectra of intramolecularly hydrogen-bonded *N*-acetyl amino-acids are reported to have characteristics reminiscent of α-helical polypeptides.[143] The c.d. spectra of sultam derivatives of amino-acids [144] and pyridinium analogues of α-amino-acids [145] have also received attention.

Temperature-dependence measurements of the c.d. of solutions of L-cystine suggest that a fixed cystine residue in a protein molecule can have a bigger c.d. intensity than that expected from the spectrum of L-cystine measured at about 20 °C.[146] C.d. and n.m.r. studies of the N-methylation, N-acylation, and amidation of L-cystine have been interpreted in terms of rotamer preferences.[147] A study of the effects of temperature and hydrogen-bonding on the u.v. and c.d. spectra of *N*-acetyl-L-tyrosine n-hexyl ester, its *O*-methyl derivative, and *N*-acetyl-*O*-methyltyrosine has appeared. Variable-temperature c.d. of this sort may enable the motility of the tyrosyl side-chains in proteins to be investigated.[148] Work on the effects of

138 C. Toniolo and A. Signor, *Experientia*, 1972, **28**, 753.
139 C. Toniolo, D. Nisato, L. Biondi, and A. Signor, *J.C.S. Perkin I*, 1972, 1179.
140 C. Toniolo, D. Nisato, L. Biondi, and A. Signor, *J.C.S. Perkin I*, 1972, 1182.
141 V. Tortorella, G. Bettani, B. Halpern, and P. Crabbé, *Tetrahedron*, 1972, **28**, 2991.
142 U. Nagai, M. Kurumi, and T. Umemura, *Tetrahedron*, 1972, **28**, 4959.
143 J. R. Cann, *Biochemistry*, 1972, **11**, 2654.
144 G. Snatzke and S. H. Doss, *Tetrahedron*, 1972, **28**, 2539.
145 T. Gronneberg and K. Undheim, *Acta Chem. Scand.*, 1972, **26**, 2267.
146 T. Takagi and N. Iro, *Biochem. Biophys. Acta*, 1972, **257**, 1.
147 J. P. Casey and R. B. Martin, *J. Amer. Chem. Soc.*, 1972, **94**, 6141.
148 E. Strickland, M. Wilchek, J. Horwitz, and C. Billups, *J. Biol. Chem.*, 1972, **247**, 572.

alkaline pH on the difference spectra of *N*-acetyltyrosine and its ethyl ester now permits the differential spectrophotometric assay for chymotrypsin-like enzymes to be extended to this pH range.[149]

Nuclear Magnetic Resonance Spectra.—The 220 MHz ^1H n.m.r. spectra of the peptide alkaloids frangulanine and discorines A and B have been analysed to show that the β-hydroxyleucine residue is of the *erythro*-configuration. This assignment has been supported by chemical degradation studies.[150] The n.m.r. spectra of six sulphur-containing α-amino-acids have been interpreted in terms of the apparent fractional populations of individual rotamers;[151] similar studies have been made on aspartic acid and asparagine.[152] Use of an n.m.r. shift reagent has made it possible to establish that the configuration about the double bond in naturally occurring L-γ-ethylideneglutamic acid is *cis*,[153] and examination of the n.m.r. spectra of substituted β-lactams has shown that, in contrast to previous assumptions, they exist in four isomeric forms.[154]

Mononuclear INDOR (internuclear double resonance) spectroscopy has now been applied to free amino-acids and their derivatives.[155] This technique simplifies the usual n.m.r. spectra and gives more chemical information. It is also a powerful method for examining total conformation in polypeptides. Deuterium nuclear quadrupole relaxation times have been obtained from proton lineshape analysis of the CHD resonances of [$^2H_1$]glycine and used to calculate correlation times for the molecular motions of the compound.[156] N.m.r. studies of polycrystalline aliphatic α-amino-acids have revealed two distinct second-moment transitions between about −180 and +110 °C. These transitions originate from hindering reorientation of CH_3 and NH_3^+ groups.[157]

Mass Spectrometry.—The mass spectral fragmentation mechanisms of twenty sulphur-containing amino-acids, including *S*-alkyl-L-cysteines, their sulphoxides, *S*-alkyl-2-methyl-DL-cysteines, and cyclic amino-acids, have been discussed. Using a direct inlet system, most compounds showed large molecular ion abundances and gave reproducible fragmentations.[158] Field-desorption mass spectrometry is a very sensitive method for the analysis of compounds of low volatility and thermal instability; it is therefore very suitable for the study of amino-acids. Recent work shows that, with the exception of methionine or cysteine, all amino-acids so far

149 C. W. Ward, *Biochim. Biophys. Acta*, 1972, **271**, 87.
150 M. G. Sierra, O. A. Moscaretti, F. J. Diaz, E. A. Ruveda, C. J. Chang, E. W. Hagaman, and E. Wenkert, *J.C.S. Chem. Comm.*, 1972, 915.
151 K. D. Bartle, D. W. Jones, and R. L'Amie, *J.C.S. Perkin II*, 1972, 646.
152 K. D. Bartle, D. W. Jones, and R. L'Amie, *J.C.S. Perkin II*, 1972, 650.
153 J. R. Nulu and E. A. Bell, *Photochemistry*, 1972, **11**, 2573.
154 H. Sterk, G. Uray, and E. Ziegler, *Monatsh.*, 1972, **103**, 544.
155 W. A. Gibbons, H. Alms, J. Sogu, and H. R. Wyssbrod, *Proc. Nat. Acad. Sci. U.S.A.*, 1972, **69**, 1261.
156 J. P. Behr and J. M. Lehn, *J.C.S. Perkin II*, 1972, 1488.
157 S. Ratkovic, *Chem. Phys. Letters*, 1972, **17**, 623.
158 H. Nishimura, S. Takara, O. Okuyama, and T. Mitzutani, *Tetrahedron*, 1972, **28**, 4503.

studied yield molecular or quasi-molecular ions. Arginine and cystine behave normally, in contrast to results obtained by electron-impact or chemical ionization studies. Less fragmention occurs in the field-desorption method, and rearrangement peaks are not found.[159]

The variation of fragmentation with ring size in a series of cyclic α-amino-acids (23; n = 3—6) has been investigated. In the compounds with the

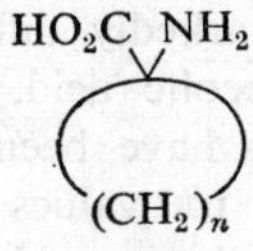

(23)

larger ring sizes about half of the total ion current is carried by $M - CO_2H^+$ ions; this is analogous to the fragmentation of the open-chain naturally occurring amino-acids. However, in the cyclobutane and cyclopropane derivatives cleavage characteristic of the ring is observed (Scheme 13).[160]

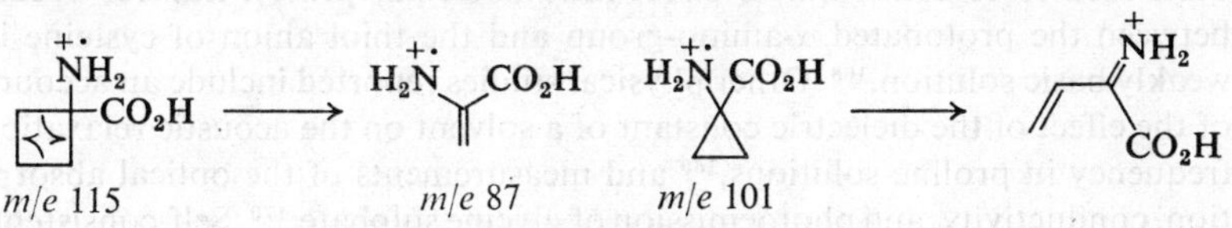

Scheme 13

Two further reports on the mass spectra of methyl- and phenyl-thiohydantoin derivatives have been published. One of these deals with the major metastable ions and points to their utility in identifying glycine, whose molecular ions are not unique, in the presence of other MTH's or PTH's. They also serve to clear up ambiguities occurring, for example, in mixtures of leucine and/or isoleucine.[161] The other paper discusses the use of low ionizing voltages; the spectra observed are less complex than the ones obtained by more conventional methods, and sensitivity is at a maximum at about 20 eV.[162]

An examination of derivatives of α-lysine, β-lysine, and various peptides containing these residues indicates that it is possible to distinguish between these isomeric amino-acids by m.s. The most intense peak for α-lysine occurs at m/e 84 and is assigned to (24); in contrast, β-lysine shows its most prominent peak at m/e 70 (25). β-Lysine peptides do not yield the aldimine fragments containing a *C*-terminal lysine that occur in the corresponding α-lysine compounds.[163]

[159] H. Winkler and H. D. Beckey, *Org. Mass Spectrometry*, 1972, **6**, 655.
[160] A. W. Coulter and C. C. Fenselau, *Org. Mass Spectrometry*, 1972, **6**, 105.
[161] T. Sun and R. E. Lovins, *Org. Mass. Spectrometry*, 1972, **6**, 39.
[162] T. Sun and R. E. Lovins, *Analyt. Biochem.*, 1972, **45**, 176.
[163] L. I. Rostovseva and A. A. Kiryushin, *Org. Mass. Spectrometry*, 1972, **6**, 1.

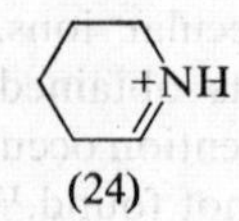
(24)

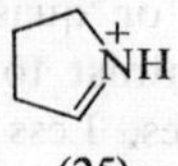
(25)

Other Physical and Stereochemical Studies.—Polarographic studies of cysteine in an ammoniacal buffer containing cobalt(III) have shown a catalytic double wave with markedly different peak potentials.[164] The dissociation constants of pipecolic acid, its 4-hydroxy-derivative, and dihydroxyphenylalanine (dopa) have been determined,[165] and a kinetic determination of α-amino-acid p*K* values based on their reactions with 1-fluoro-2,4-dinitrobenzene has been made. In this latter study the pH-dependent second-order rate constants were used to calculate the p*K*'s; the amides of glycine and valine were found to have relatively high nucleophilicities.[166] A polarized i.r. study of the vibrational spectrum of a single crystal of (−)-alanine has been made; the spectrum was correlated with a C_2 molecular point group.[167] Ultrasound absorption measurements have been used to establish that a direct intramolecular proton transfer occurs between the protonated α-amino-group and the thiol anion of cysteine in weakly basic solution.[168] Other physical studies reported include an account of the effect of the dielectric constant of a solvent on the acoustic relaxation frequency in proline solutions,[169] and measurements of the optical absorption, conductivity, and photoemission of glycine sulphate.[170] Self-consistent-field calculations of glycine have also been made.[171]

A solvent able to separate the *cis*- and *trans*-isomers of 4-hydroxymethylproline has been used to establish that loquat seeds contain not only the 'normal' *cis*-L-isomer but also the *trans*-D- and *trans*-L-forms.[172] A simple and efficient new resolution of racemic *N*-acetyl-β-(3,4-dimethoxyphenyl)-alanine with quinine has been reported,[173] and a quantitative estimation of the D- and L-enantiomers of leucine using a g.l.c. separation of *N*-trifluoroacetyl-L-prolyl-leucine methyl esters developed. Contrary to previous assertions, both the formation and use in the presence of triethylamine of *N*-trifluoroacetyl-L-prolyl chloride may be attended by extensive racemization. During coupling this can be prevented by adding the triethylamine slowly in very dilute solution in dichloromethane at the temperature of dry ice.[174]

[164] P. Anzerbacher and U. Kalous, *Coll. Czech. Chem. Comm.*, 1972, **37**, 3209.
[165] A. Brun and R. Rossel, *Compt. rend.*, 1972, **274**, *C*, 1810.
[166] J. G. Ghazarian, *Arch. Biochem. Biophys.*, 1972, **150**, 72.
[167] R. Adamowicz and E. Fishman, *Spectrochim. Acta*, 1972, **28A**, 889.
[168] G. Maass and F. Peters, *Angew. Chem. Internat. Edn.*, 1972, **11**, 428.
[169] M. K. Ul'masova, *Izvest. Akad. Nauk S.S.S.R., Ser. fiz.-mat. Nauk*, 1972, **16**, 71.
[170] G. Royal, B. Marlon, and G. Godfrey, *Compt. rend.*, 1972, **275**, *B*, 353.
[171] J. A. Ryan and J. L. Whittle, *J. Amer. Chem. Soc.*, 1972, **94**, 2396.
[172] D. O. Gray, *Phytochemistry*, 1972, **11**, 751.
[173] J. P. M. Houbiers, *Synthetic Comm.*, 1972, **2**, 143.
[174] W. A. Bonner, *J. Chromatog. Sci.*, 1972, **10**, 159.

The physicochemical properties of several surface-active *N*-acyl glutamates and their sodium salts have been examined.[175]

4 Chemical Studies of Amino-acids

Oxidation and Reduction.—In the presence of mild bases such as triethylamine, 4-alkylidene-2-phenyl-2-oxazolin-5-ones, *e.g.* (26), rapidly absorb atmospheric oxygen and eliminate carbon dioxide, forming a mixture of imides (27) and (28). The formation of these and two other acidic by-products (29) and (30) can be accounted for by postulating the decomposition of a hydroperoxide intermediate.[176] The decomposition of tryptophan

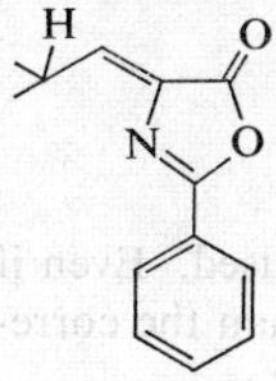

(26)

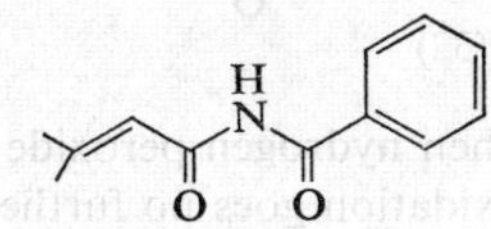

(27)

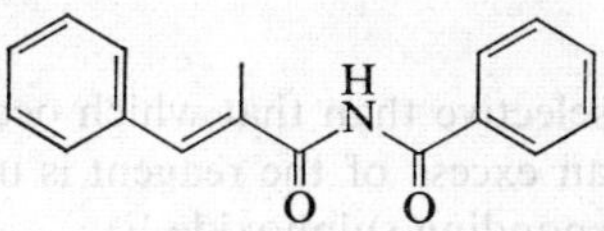

(28)

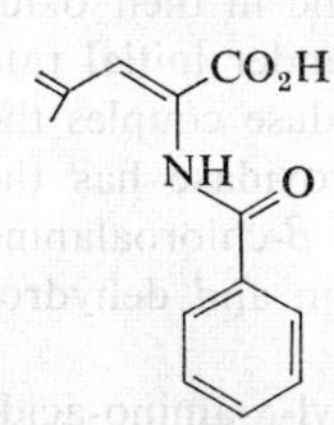

(29)

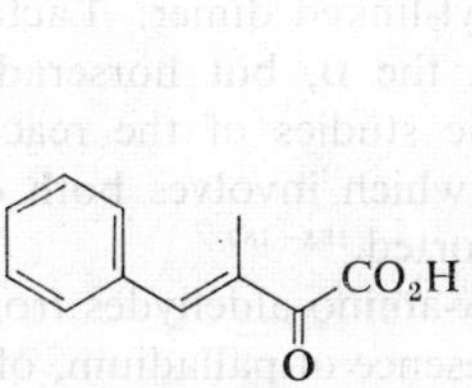

(30)

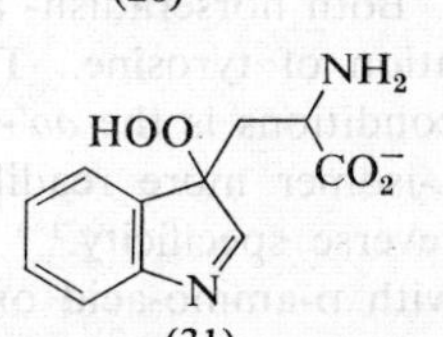

(31)

under both alkaline and acidic conditions has been re-evaluated. In alkali in the presence of oxygen a free-radical autoxidation involving the hydroperoxide (31) is proposed; this is supported by the relative stability of 1-methyltryptophan. Initiation of the reaction appears to be due to impurities in the sodium hydroxide and not to the base itself. This accords with the greater apparent stability of tryptophan in solutions of barium hydroxide.[177] The more extensive destruction that occurs in hydrochloric acid is probably due to autoxidation of the 1-protonated form initiated by impurities in the glass of the containing vessels.[178]

Attempts to prepare 4-oxolysine from 4-hydroxylysine, which is readily available from lysine by photochlorination and subsequent hydrolysis, have proved largely unsuccessful. Alkaline permanganate gave some of the desired product, but it seemed very susceptible to further oxidation;

[175] M. Takehara, H. Mariyuku, I. Yoshimura, and R. Yoshida, *J. Amer. Oil Chemists' Soc.*, 1972, **49**, 143.
[176] R. Bisson, R. B. Yeats, and E. W. Warnhoff, *Canad. J. Chem.*, 1972, **50**, 2851.
[177] M. Stewart and C. H. Nicholls, *Austral. J. Chem.*, 1972, **25**, 1595.
[178] M. Stewart and C. H. Nicholls, *Austral. J. Chem.*, 1972, **25**, 2139.

some glycine and aspartic acid are formed. *NN'*-Diphthaloyl-*threo*-4-hydroxy-L-lysineamide proved resistant to oxidation under acidic conditions, probably owing to its existence as the cyclic hemiacetal tautomer (32).[81] The oxidation of *N*-benzoylmethionine by 3-hydroperoxyindolenines, reagents thought likely to exist in biological systems, is more

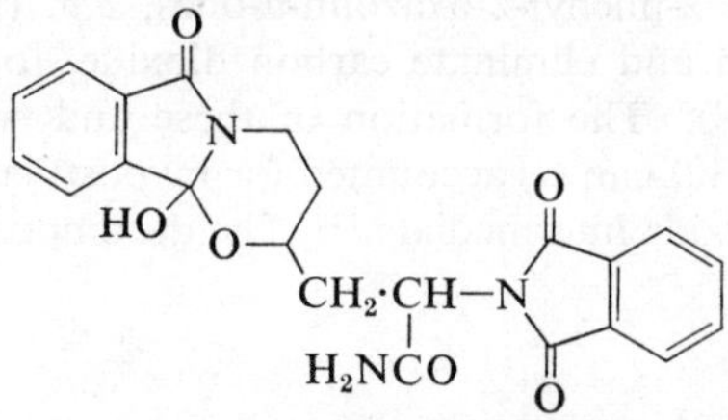

(32)

selective than that which occurs when hydrogen peroxide is used. Even if an excess of the reagent is used, oxidation goes no further than the corresponding sulphoxide.[179]

Both horseradish- and lacto-peroxidase are stereospecific in their oxidation of tyrosine. The product formed in both cases under initial rate conditions is the *oo'*-diphenyl-linked dimer. Lactoperoxidase couples the L-isomer more readily than the D, but horseradish peroxidase has the reverse specificity.[180] Kinetic studies of the reaction of β-chloroalanine with D-amino-acid oxidase, which involves both oxidation and dehydrohalogenation, have been reported.[181–183]

The formation of *N*-acyl-α-amino-aldehydes from *N*-acyl-α-amino-acids by hydrogenolysis, in the presence of palladium, of their mixed anhydrides with ethyl chloroformate has been investigated. There is considerable racemization, and the presence of a little acetic acid in the reaction mixture favours the highest yield of aldehyde. If triethylamine is present, however, much more of the corresponding alcohol is formed.[184] In an alkaline medium, amino-acids condense with pyridoxal to give Schiff bases which can be reduced to pyridoxyl amino-acids by sodium borohydride (Scheme 14). These derivatives can be used to determine the original amino-acid concentration. Spectrophotometry will detect 2×10^{-8} mol of the derivative, fluorimetry 5×10^{-10} mol, and, if $NaBT_4$ is used, quantitative determination on a picomole scale is possible.[185]

179 M. Nakagawa, T. Suzuki, T. Kawashima, and T. Hino, *Chem. and Pharm. Bull.* (*Japan*), 1972, **20**, 2413.
180 G. S. Bayse, A. W. Michaels, and M. Morrison, *Biochim. Biophys. Acta*, 1972, **284**, 30, 34.
181 J. G. Voet, D. J. T. Porter, and H. J. Bright, *Z. Naturforsch.*, 1972, **27b**, 1054.
182 Y. Miyaki, T. Abe, and T. Yamamo, *Z. Naturforsch.*, 1972, **27b**, 1376.
183 D. J. T. Porter, J. G. Voet, and H. J. Bright, *Biochem. Biophys. Res. Comm.*, 1972, **49**, 257.
184 A. Seki, K. Koga, and S. Yamada, *Chem. and Pharm. Bull.* (*Japan*), 1972, **20**, 361.
185 N. Lustenberger, H.-W. Lange, and K. Hempel, *Angew. Chem. Internat. Edn.*, 1972, **11**, 227.

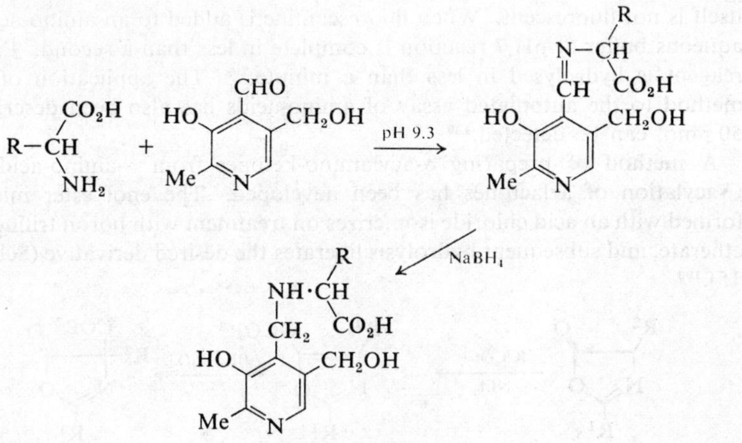

Scheme 14

General Reactions.—Details of the investigation of the reaction of phenylalanine with ninhydrin to generate fluorescent compounds from amino-acids or amines have been published. The key intermediate involved is phenylacetaldehyde; this reacts with any amino-acid present to give the major fluophor (33), a minor fluophor (34), and a non-fluorescent minor component (35).[186, 187] Further work has established that the fluophor (33) can be generated directly by reaction of an amino-acid with 4-phenylspiro[furan-2(3*H*),1′-phthalan]-3,3′-dione (36; 'fluorescamine'); the reagent

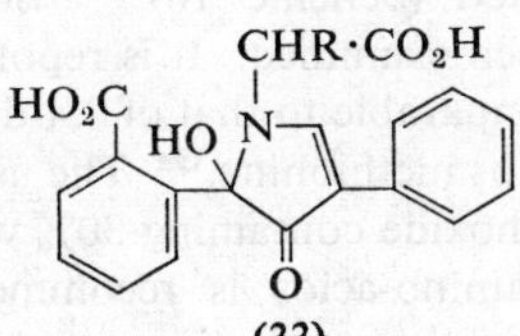

(33)

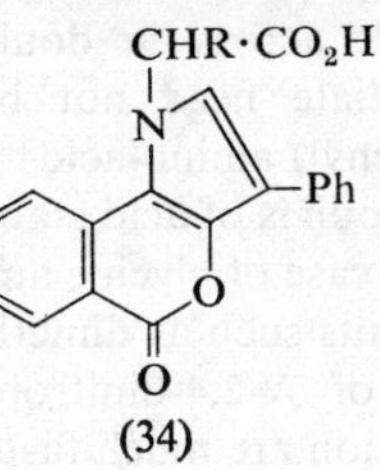

(34)

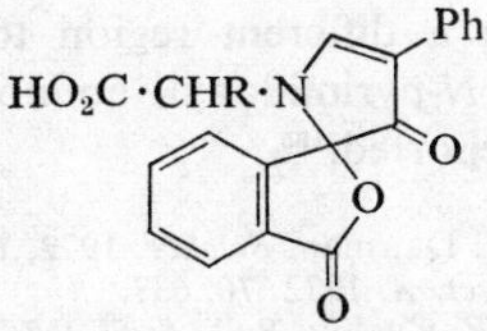

(35)

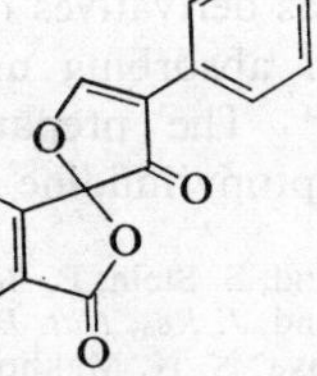

(36)

[186] M. Weigele, J. F. Blount, J. P. Tengi, R. C. Czarjowski, and W. Leimgruber, *J. Amer. Chem. Soc.*, 1972, **94**, 4052.

[187] M. Weigele, S. C. De Bernardo, J. P. Tengi, and W. Leimgruber, *J. Amer. Chem. Soc.*, 1972, **94**, 5927.

itself is not fluorescent. When fluorescamine is added to an amino-acid in aqueous buffer at pH 7 reaction is complete in less than a second. Excess reagent is hydrolysed in less than a minute.[188] The application of this method to the automated assay of amino-acids has also been described; 50 pmol can be detected.[189]

A method of preparing α-acylamino-ketones from α-amino-acids by C-acylation of azlactones has been developed. The enol ester initially formed with an acid chloride isomerizes on treatment with boron trifluoride etherate, and subsequent hydrolysis liberates the desired derivative (Scheme 15).[190]

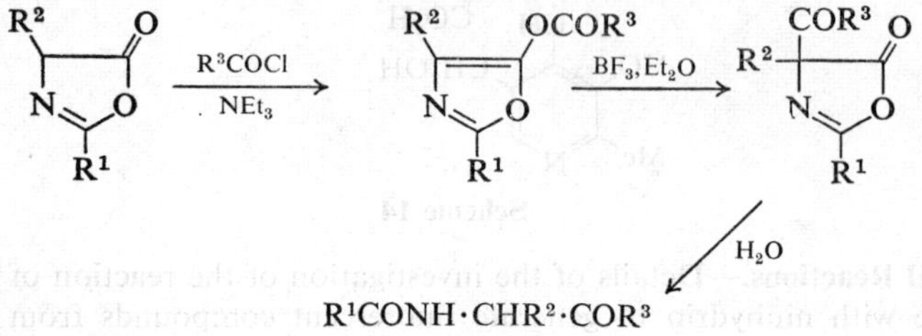

Scheme 15

A number of papers concerning further exploration of N-derivatives of general utility are most conveniently considered here. A novel route to carbohydrate derivatives involves generating α-nitro-olefinic sugars from β-nitroacetoxy precursors by treating with triethylamine. Amino-acid esters will add across the double bond to give the N-substituted product; the intermediate need not be isolated (Scheme 16).[191] Some *N*-(3,5-dinitro-2-thienyl) amino-acids have been examined. It is reported that the protecting group is of acid stability comparable to that of 2,4-dinitrophenyl except in the case of glycine and perhaps methionine.[192] The use of dipolar aprotic solvents such as dimethyl sulphoxide containing 30% water for the preparation of *N*-2,4-dinitrophenyl amino-acids is recommended. The rates of reaction are much faster than in the conventional aqueous ethanol and the yields are higher.[193] *p*-Aminobenzoyl amino-acids have been investigated as derivatives for isoelectric focusing; this acyl group has the advantage of absorbing u.v. light in a different region to the carrier ampholyte.[194] The preparation of *N*-pyrimidonyl amino-acids using methylmercaptopyrimidine has been reported.[195]

188 S. Udenfriend, S. Stein, P. Bohlen, and W. Dairman, *Science*, 1972, **178**, 871.
189 S. Udenfriend, *J. Res. Nat. Bur. Stand., Sect. A*, 1972, **76**, 637.
190 N. I. Aronova, N. N. Makhova, and S. I. Zav'yalov, *Bull. Acad. Sci. U.S.S.R.*, 1972, **21**, 349.
191 F. J. M. Rajabalee, *Synthesis*, 1972, 318.
192 L. H. Hellberg, M. J. Prodanovich, and F. Stults, *J. Heterocyclic Chem.*, 1972, **9**, 401.
193 J. A. Vinson and L. A. Pepper, *Analyt. Chim. Acta*, 1972, **56**, 245.
194 N. Catsimpoolas and B. E. Campbell, *Analyt. Biochem.*, 1972, **46**, 674.
195 S. Hoffmann, H. Schubert, and K. Nitsche, *Z. Chem.*, 1972, **12**, 21.

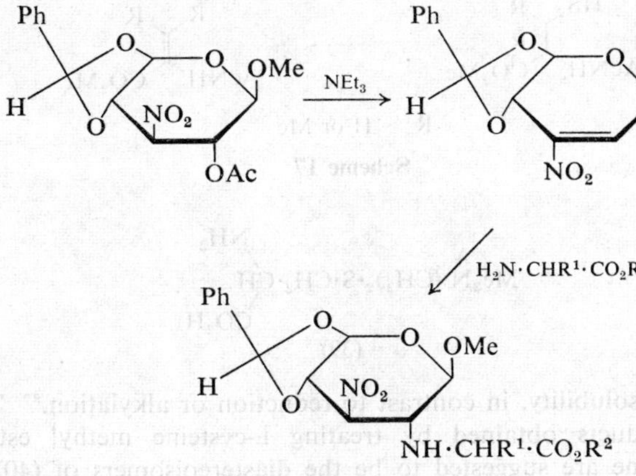

Scheme 16

A study of the solubility of amino-acids in aqueous guanidinium thiocyanate indicates that this solution is more effective than urea, guanidinium chloride, or organic solvents in decreasing the free energy of transfer of hydrophobic amino-acid side-chains from water to these solvents.[196] The standard enthalpies and entropies for protonating the anions of some biologically important amino-acids have been measured in 3M sodium perchlorate at 25 °C,[197] and a further study of the silylation of amino-acids has appeared.[198]

Specific Reactions.—The observation that the perhydro-1,4-thiazepine derivative (37) undergoes ring-opening on refluxing with silver oxide in

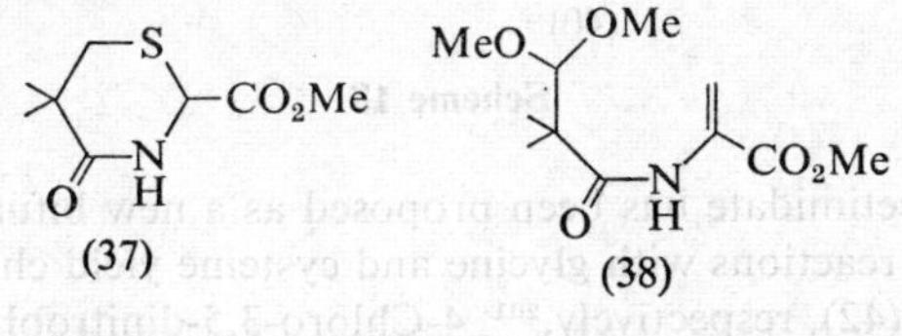

(37) (38)

methanol to form the α-acylamino-acrylate (38) has been extended to allow the preparation of α-acetamidoacrylates from various cysteine and penicillamine derivatives (Scheme 17). This reaction therefore appears to be a general one.[199] Trimethylaminoethylation of cysteine with (2-bromoethyl)trimethylammonium bromide leads to the new amino-acid thialaminine (39). This material was prepared in connection with protein modification studies; such derivatives of cysteine residues in proteins

[196] K. H. Dooley and F. J. Castellino, *Biochemistry*, 1972, **11**, 1870.
[197] R. D. Graham, D. R. Williams, and P. A. Yeo, *J.C.S. Perkin II*, 1972, 1876.
[198] H. A. Kricheldorf, *Annalen*, 1972, **763**, 17.
[199] D. Gravel, R. Gauthier, and C. Berse, *J.C.S. Chem. Comm.*, 1972, 1322.

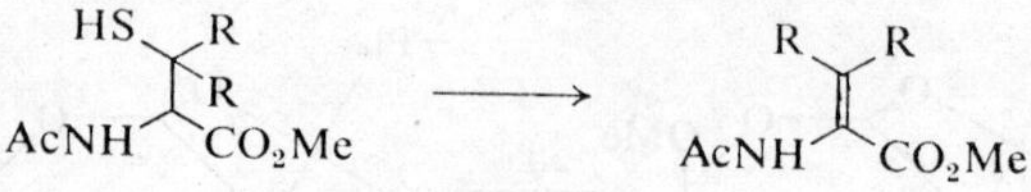

R = H or Me

Scheme 17

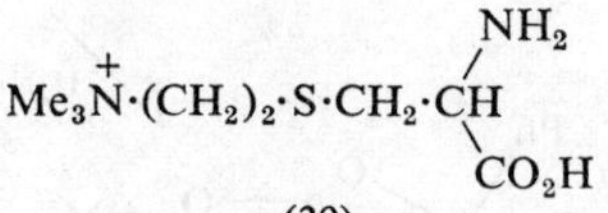

(39)

confer water solubility, in contrast to reduction or alkylation.[87] The two isomeric products obtained by treating L-cysteine methyl ester with *p*-benzoquinone are suggested to be the diastereoisomers of (40), which arise by oxidative coupling of (41) (Scheme 18).[200]

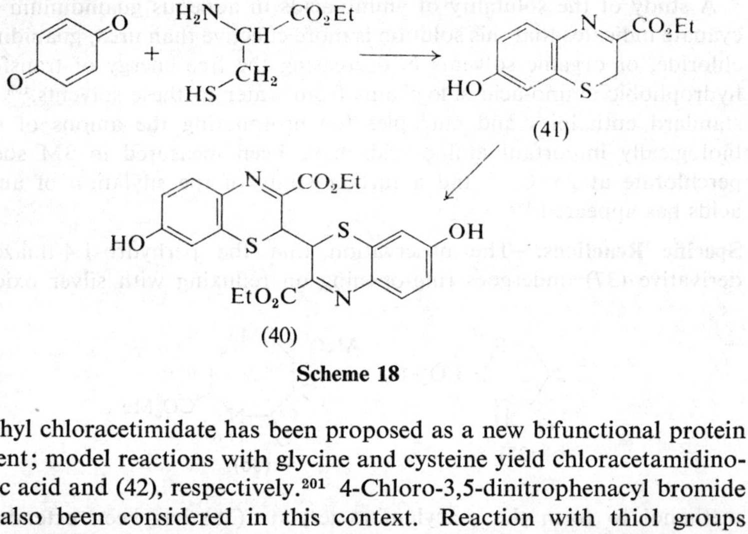

(40)

Scheme 18

Ethyl chloracetimidate has been proposed as a new bifunctional protein reagent; model reactions with glycine and cysteine yield chloracetamidinoacetic acid and (42), respectively.[201] 4-Chloro-3,5-dinitrophenacyl bromide has also been considered in this context. Reaction with thiol groups occurs readily, but the amino-groups of all amino-acids, except phenylalanine, tyrosine, and tryptophan, are not attacked. These exceptions are

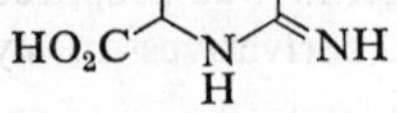

(42)

[200] G. Prota and E. Ponsiglione, *Tetrahedron Letters*, 1972, 1327.
[201] M. Olomucki and J. Diopak, *Biochim. Biophys. Acta*, 1972, **263**, 213.

ascribed to the formation of charge-transfer complexes between the aromatic rings and the reagent which facilitate reactions.[202] The modification of methionine with BBr_3 or BI_3 has been studied. After 167 h at 22 °C 94% of the methionine had reacted, largely to produce homocysteic acid, homoserine, and homoserine lactone. In experiments with proteins, other amino-acids were unaffected.[203] In the presence of ferrous or manganous ions and sulphite, methionine decomposes to free methanethiol and dimethyl sulphide. This degradation is thought to proceed through the initial formation of 3-(methylthio)propional.[204] A study of the reaction of methionylsulphonium salts with thiols has established that the methionyl residue can be regenerated in good yields.[205]

The influence of structure on the β-elimination reaction of protected *O*-tosyl β-hydroxy-α-amino-acids to give aziridinecarboxylic acid derivatives has been investigated. Ring formation of this sort is in general easier with threonine peptides.[206] It has been found possible to acylate selectively the hydroxy-groups of serine, threonine, and tyrosine without affecting the α-amino-groups by treating with carboxylic acid chlorides in anhydrous trifluoroacetic acid.[207] The conversion of DL-threonine into DL-*trans*-3-benzoyl-5-methyl-2-oxo-oxazolidine-4-carboxylic acid has been further detailed,[208] and a study has been made of the reaction products of thionyl chloride and a series of *erythro*- and *threo*-isomers of some *N*-benzoyl-β-arylserine methyl esters.[209]

A detailed investigation of the nitrous acid deamination of (*R*)-α-methylphenylalanine methyl ester in acetic acid has revealed products of elimination (68%), substitution (21%), hydrogen migration (8%), and phenyl migration (3%). The stereochemical courses of these reactions have also been examined.[210] Under the same conditions, L-valine benzyl ester gives a similar spectrum of products, but free L-valine yields only the substitution product.[211] The normal conditions used for the preparation of phenylthiohydantoins fail when applied to azetidine-2-carboxylic acid. This derivative can, however, be obtained in good yield by heating the *p*-nitrophenyl ester of the corresponding 2-phenyliminothiazolidin-5-one (Scheme 19).[212] Contrary to the views of earlier workers, the interaction

[202] J. Diopak and M. Olomucki, *Biochim. Biophys. Acta*, 1972, **263**, 220.
[203] M. Z. Atassi and M. T. Perlstein, *Tetrahedron Letters*, 1972, 1861.
[204] T. Wainwright, J. F. McMahon, and T. McDowell, *J. Sci. Food Agric.*, 1972, **23**, 911.
[205] F. Naider and Z. Bohak, *Biochemistry*, 1972, **11**, 3208.
[206] Y. Nakagawa, T. Tsuno, K. Nakajima, M. Iwai, H. Kawai, and K. Okawa, *Bull. Chem. Soc. Japan*, 1972, **45**, 1162.
[207] A. Previero, L.-G. Barry, and M.-A. Coletti-Previero, *Biochim. Biophys. Acta*, 1972, **263**, 7.
[208] T. Inui, *Bull. Chem. Soc. Japan*, 1972, **45**, 1254.
[209] S. H. Pines and M. A. Kozlowski, *J. Org. Chem.*, 1972, **37**, 292.
[210] M. Kobayashi, K. Koga, and S. Yamada, *Chem. and Pharm. Bull. (Japan)*, 1972, **20**, 1898.
[211] M. Taniguchi, K. Koga, and S. Yamada, *Chem. and Pharm. Bull. (Japan)*, 1972, **20**, 1438.
[212] H. T. Nagasawa, P. S. Fraser, and J. A. Elberling, *J. Org. Chem.*, 1972, **37**, 516.

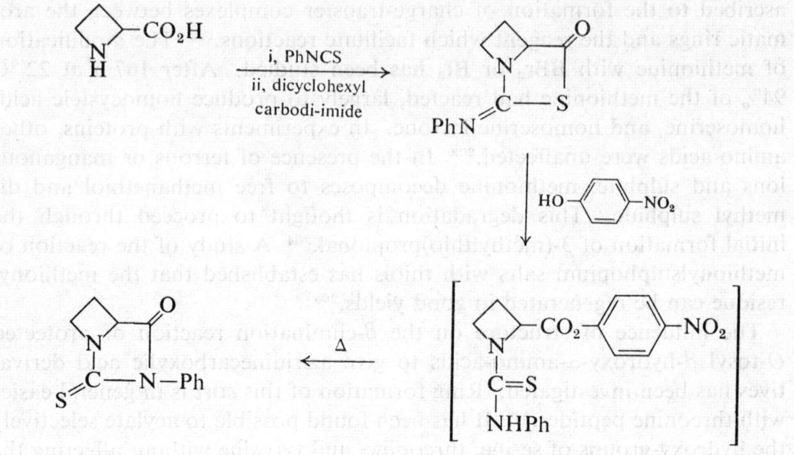

Scheme 19

between proline and other imino-acids and benzoquinone has been found to be due to charge-transfer complexation rather than chemical addition.[213] The photochemical formation of 2-isobutylpyrido[2,3-*d*]imidazole 1-oxide from 3-nitro-2-pyridyl-DL-leucine has been investigated spectrophotometrically.[214] Nicotinylglycine can be conveniently prepared by acylating glycine methyl ester using nicotinic anhydride and then saponifying. The nicotinic acid produced in the first stage is insoluble in non-polar solvents and readily separated.[215] L-Proline has been used for the spectrophotometric determination of Cu^{II}.[216]

A new cyclic imino-acid derivative of homopyridoxal can be formed from *N*-α-acetyl-L-lysine and 5-(2-chloroethyl)-3-hydroxy-2-methylpyridine-4-carboxaldehyde (Scheme 20). This product serves as a useful model of pyridoxal-P enzymes in which the cofactor is usually bound in an imine link with the ε-amino-group of a lysine residue.[216a]

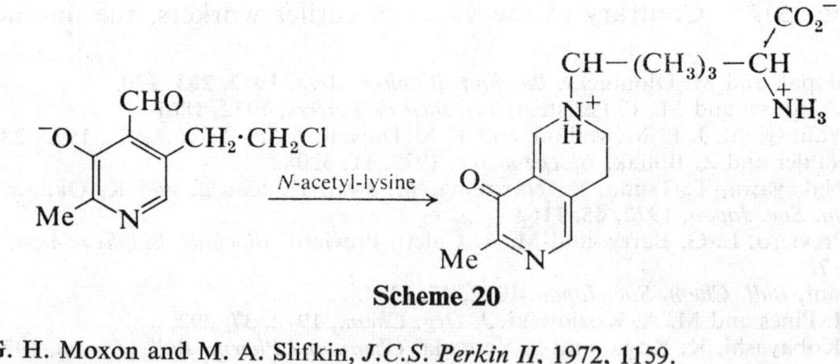

Scheme 20

[213] G. H. Moxon and M. A. Slifkin, *J.C.S. Perkin II*, 1972, 1159.
[214] G. G. Aloisi, E. Bordignan, and A. Signor, *J.C.S. Perkin II*, 1972, 2218.
[215] M. T. Wu and R. E. Lyle, *J. Pharm. Sci.*, 1972, **61**, 141.
[216] E. Campi and P. Mirti, *Analyt. Chim. Acta*, 1972, **58**, 239.
[216a] E. W. Miles, H. M. Fales, and J. B. Gin, *Biochemistry*, 1972, **11**, 4945.

An improved method for preparing L-aspartic anhydride directly from L-aspartic acid using phosphorus halides has been evolved,[217] and the facile cyclization of glutamic acid α-t-butyl derivatives re-emphasized.[218] Two studies on the hydrolysis of amino-acid esters have been reported this year; one concerns the cobalt(II)-catalysed hydrolysis of diethyl aspartate and some dipeptide ethyl esters,[219] and the other deals with the chymotrypsin-catalysed decompositions of L-tryptophan and L-tyrosine methyl esters. The mechanism of the latter reactions appears to be the same as that for acylamino-esters, but the ionizability of the free α-amino-group produces a markedly different pH dependency.[220] Although the imino-groups of peptides react with t-butyl hypochlorite, the α-amino-groups of amino-acids are largely unaffected by this reagent.[221] The thermal decomposition of a solid complex of aminomalonic acid and β-alanine has been studied in detail.[222]

Effects of Electromagnetic Radiation on Amino-acids.—The radiation-induced reactions of amino-acids have been reviewed.[223] ENDOR studies of the effect of ionizing radiation on single crystals of acetylglycine show that the peptide link is involved in the reduction process thus initiated.[224] A similar study of DL-serine at 77 K indicated that the initial species produced are anion and cation radicals. At room temperature these decay to form two stable radicals.[225] Three investigations of the effect of u.v. radiation on specific amino-acids have been made. Irradiation of single crystals of cystine dihydrochloride with monochromatic u.v. is thought to lead to two processes (Scheme 21),[226] and photolysis of aqueous solutions

$$\mathrm{RSSR} + h\nu \longrightarrow \mathrm{RS}\overset{+}{\dot{\mathrm{S}}}\mathrm{R} + e^-$$
$$e^- + \mathrm{RSSR} \longrightarrow \mathrm{RS}\cdot + \mathrm{RS}^-$$

Scheme 21

of arginine leads to fourteen ninhydrin-positive products, including ornithine, aspartic acid, and glycine.[227] The primary products of the flash photolysis of tryptophan have also been investigated.[228]

217 Y. Arioyoshi, T. Yamatomi, N. Uchiyama, and N. Sato, *Bull. Chem. Soc. Japan*, 1972, **45**, 2208.
218 M. Hollósi, M. Kajtár, Z. Ráthanyi, and J. Tomasz, *Acta Chim. Acad. Sci. Hung.*, 1972, **71**, 101.
219 A. Y. Girgis and J. I. Legg, *J. Amer. Chem. Soc.*, 1972, **94**, 8420.
220 F. J. Kézdy, S. P. Jindal, and M. L. Bender, *J. Biol. Chem.*, 1972, **247**, 5746.
221 A. Matsushima, S. Yamazaki, K. Shibata, and Y. Imada, *Biochim. Biophys. Acta*, 1972, **271**, 243.
222 Y. Nishijo, I. Imanishi, and G. Hashizume, *Bull. Chem. Soc. Japan*, 1972, **45**, 2070.
223 G. M. Warren, *Radiat. Res. Rev.*, 1972, **3**, 305.
224 H. C. Box, E. E. Budzinski, and K. T. Lilga, *J. Chem. Phys.*, 1972, **57**, 4295.
225 B. W. Castleman and G. C. Moulton, *J. Chem. Phys.*, 1972, **57**, 1095.
227 H. D. Pathak, P. C. Joshi, and U. N. Pande, *Indian J. Biochem. Biophys.*, 1972, **9**, 221.
228 R. Santus and L. I. Grossweiner, *Photochem. and Photobiol.*, 1972, **15**, 101.

5 Analytical Methods

Gas–Liquid Chromatography.—Analysis of biological substances and geochemical samples requires such sensitivity that contamination is an important problem. A g.l.c. study of necessary chemical reagents and other sources of contamination has been made, and rigorous techniques are suggested so that analysis of samples containing 1—10 mg of amino-acids per gram is readily achievable. Analysis of 3—4 billion-year-old Onverwacht chert has been used to demonstrate the application of the techniques to a geochemically important sample.[229] An improved procedure for extraction after lyophilization has allowed as little as 10^{-11} mol of protein amino-acids to be determined in sea-water samples. The extracted samples were converted into their trimethylsilyl (TMS) derivatives for g.l.c. analysis.[230] A separation of twenty amino-acids by g.l.c. of their *N*-TMS-*O*-n-butyl esters in less than thirty-five minutes has been described. Numerous extraneous peaks detected along the base line are due to incomplete esterification of the amino-acids. This occurs because of some water produced under the conditions of esterification affecting the equilibrium position.[231] Using these same protecting groups, the elution positions of twenty-six non-protein amino-acid derivatives have been established. Only in two cases does co-elution with a protein amino-acid occur; *N*-amidinoalanine overlaps cystine, and 6-aminohexanoic acid is not resolved from aspartic acid.[232]

A g.l.c.–m.s. study of the isopropylation of amino-acids shows that, under the conditions used, the *N*,*O*-di-isopropyl derivatives are not always obtained in high yield. Some amino-acids do not react, and others show multiple peaks. Arginine forms a derivative thought to be the lactam (43).[233] The molar responses of *N*-trifluoroacetyl amino-acid methyl esters with a flame ionization detector have been determined.[234] Various oxazolin-5-one derivatives of leucine have been examined for their potential in g.l.c. They all proved thermally stable and can be prepared from the free amino-acids in a single step.[235] The successful separation by g.l.c. of a mixture of

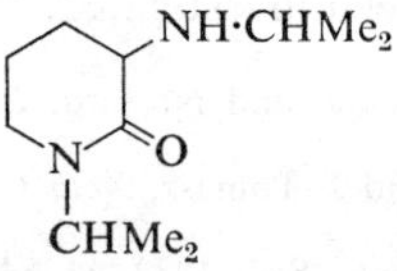

(43)

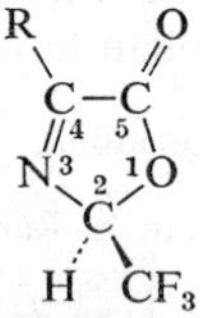

(44)

[229] J. J. Rosh, C. W. Gehrke, R. W. Zumwalt, K. C. Kuo, K. A. Kvenvolden, and D. L. Stalling, *J. Chromatog. Sci.*, 1972, **10**, 444.
[230] R. Pocklington, *Analyt. Biochem.*, 1972, **45**, 403.
[231] J. P. Hardy and S. L. Kerrin, *Analyt. Chem.*, 1972, **44**, 1497.
[232] F. Raulin, P. Shapshak, and B. N. Khare, *J. Chromatog.*, 1972, **73**, 35.
[233] B. Blessington and N. I. Y. Fiagbe, *J. Chromatog.*, 1972, **68**, 259.
[234] A. Islam and A. Darbre, *J. Chromatog.*, 1972, **71**, 223.
[235] O. Grahl-Nielson and E. Solheim, *J. Chromatog.*, 1972, **69**, 366.

ten amino-acid 2-trifluoromethyloxazolin-5-ones has subsequently been reported; the chirality of the α-C atom is lost in these derivatives but C-2 of the ring becomes asymmetric (44).[236] An improved g.l.c. analysis of methyl- and phenyl-thiohydantoins of amino-acids has been described,[237] and it has proved possible to assign configurations to amino-acid enantiomers by analysis of the g.l.c. elution positions of their *N*-trifluoroacetyl esters when the optically active stationary phase carbonylbis-(*N*-L-valine isopropyl ester) is used.[238] Lysine in acid hydrolysates of wheat and rice seeds can be conveniently determined by g.l.c. as its bisneopentylidene ethyl ester derivative (45). This is prepared by treating lysine ethyl ester with pivaldehyde.[239]

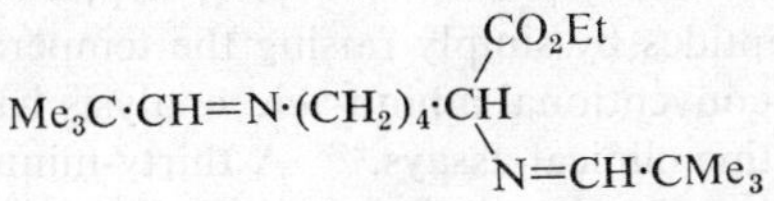

(45)

Ion-exchange Chromatography.—A study of the influence of acidity on the reaction of ninhydrin with amino-acids indicates that the optimum pH for colour development depends only on the pK_2 of the amino-acid. According to an equation proposed, histidine and tryptophan are anomalous in aqueous methyl cellosolve but not in aqueous dimethyl sulphoxide. In the latter solvent hydrogen-bonding is thought to prevent interaction of the NH of the rings with the α-amino-groups. The authors conclude that the pH of 5 as normally used represents the best compromise.[240] It has been reported that the specific radioactivity of individual ^{14}C-labelled amino-acids isolated by ion-exchange chromatography is different in consecutive fractions containing the same amino-acid. This is attributed to the resolution of [^{14}C]- and [^{12}C]-amino-acids during the ion-exchange. The specific radioactivity increases on progressive elution from cation-exchange resins, but decreases from anion-exchange resins. [^{14}C]-γ-Aminobutyric acid is not quantitatively eluted from ion-exchange resins by citrate buffers; this is attributed to adsorption on to the glass walls of containers. Recovery varies with the pH of the buffer.[241]

The hydrolysis of proteins and peptides by a mixture of sepharose-bound peptidases gives results comparable with acid hydrolysis for the acid-stable amino-acids, and has the advantage that good values are obtained also for tryptophan, asparagine, and glutamine. The peptidases involved are prolidase, aminopeptidase M, trypsin, and chymotrypsin.[242]

236 O. Grahl-Nielson and E. Solheim, *J.C.S. Chem. Comm.*, 1972, 1092.
237 J. J. Pisano, T. J. Bronzert, and H. B. Brewer, *Analyt. Biochem.*, 1972, **45**, 43.
238 B. Feibush, E. Gil-Av, and T. Tamari, *J.C.S. Perkin II*, 1972, 1197.
239 F. P. Zscheile and B. L. Brannaman, *Analyt. Biochem.*, 1972, **49**, 442.
240 P. J. Lamotte and P. G. McCormick, *Analyt. Chem.*, 1972, **44**, 821.
241 M. K. Gaitande and R. W. K. Nixey, *Analyt. Biochem.*, 1972, **50**, 416.
242 H. P. J. Bennett, D. F. Elliott, B. E. Evans, P. J. Lowry, and C. McMartin, *Biochem. J.*, 1972, **129**, 695.

Further information on the elution positions of non-protein amino-acids relative to protein amino-acids on ion-exchange chromatography has been published.[243] An automated chromatographic system for the combined analysis of amino-acids and urinary peptides has been recommended,[244] and an automatic computer-compatible digital data-acquisition system for amino-acid analysis has been developed.[245]

An improved resolution of sulphur-containing amino-acids in physiological fluids on a new 10% cross-linked resin has been described. The longer elution times occurring as a result of this high degree of cross-linking can be offset by using a higher temperature and six buffers on a two-column system.[246] A urinary hydroxyproline assay involving adsorption on a strong cation-exchange resin, a washing step, and then hydrolysis of the resin-bound peptides by simply raising the temperature before eluting and estimating by conventional amino-acid analysis has been described as well suited for routine clinical assays.[247] A thirty-minute automated assay of hydroxyproline has also been advocated.[248] Two single-column systems for the separation of desmosine from other amino-acids have been reported;[249, 250] a one-column system for the determination of ^{14}C-labelled lysine and hydroxylysine is also now available.[251]

About 8% of the *trans*-4-hydroxyproline in collagen is epimerized on acid hydrolysis. Hitherto this has not been detected, as *cis*-4-hydroxyproline elutes on amino-acid analysis at the same time as threonine. However, if the primary amino-acids in the hydrolysate are deaminated with nitrous acid before carrying out ion-exchange chromatography the *cis*-isomer can be determined.[252] Addition of 3-(3-indolyl)propionic acid to proteins or peptides before acid hydrolysis has been found to improve the recovery of tryptophan. The use of reducing agents such as thioglycollic acid in a similar way, however, affects the recovery of some other amino-acids.[253] An ion-exchange system which separates the oxidation products of tryptophan in a two-hour run has been described.[254]

Thin-layer Chromatography.—A t.l.c. method for determining the optical purity of labelled amino-acids by converting them into dipeptides with the *N*-carboxanhydride of L-leucine has proved generally useful for at least

[243] P. Shapshak and M. Okaji, *J. Chromatog.*, 1972, **64**, 178.
[244] J. A. Klosse, D. Y. Huistra, P. K. de Bree, S. K. Wadman, and J. F. G. Vliegenthart, *Clin. Chim. Acta*, 1972, **42**, 409.
[245] M. L. Johnson, E. A. Khairallah, and D. A. Yphantis, *Analyt. Biochem.*, 1972, **50**, 364.
[246] J. O. Jeppsson and I. M. Karlsson, *J. Chromatog.*, 1972, **72**, 93.
[247] B. C. Goverde and F. J. N. Veenkamp, *Clin. Chim. Acta*, 1972, **41**, 29.
[248] P. X. Callahan, J. A. Shepard, and S. Ellis, *Analyt. Biochem.*, 1972, **49**, 155.
[249] G. E. Gerbb and G. D. Kemp, *J. Chromatog.*, 1972, **71**, 361.
[250] D. P. Thornhill, *Analyt. Biochem.*, 1972, **46**, 119.
[251] R. S. Askenasi and N. A. Kefalides, *Analyt. Biochem.*, 1972, **47**, 67.
[252] D. D. Dzietviatowski, V. C. Haxall, and R. L. Riolo, *Analyt. Biochem.*, 1972, **49**, 550.
[253] L. C. Gruen and P. W. Nicholls, *Analyt. Biochem.*, 1972, **47**, 348.
[254] P. W. Nicholls and D. E. Rivett, *J. Chromatog.*, 1972, **65**, 565.

nineteen labelled amino-acids.[255] Methods for the determination of methionine and cysteine in legume seeds have been developed. Methionine is converted into its sulphone and, after t.l.c. separation, developed with ninhydrin and quantitated by reflectance densitometry. Cysteine is subjected to hydrazinolysis and the hydrogen sulphide liberated estimated colorimetrically as bismuth sulphide.[256] The separation of free amino-acids on cellulose sulphate-impregnated cellulose [257] and silica layers bound with agar-agar [258] have also been detailed.

The quantitative determination down to 10^{-14} moles of dansyl amino-acids prepared from ^{14}C-labelled dansyl chloride has been claimed. The derivatives are run on polyamide layers and the spots visualized by micro-autoradiography followed by scanning microscope photometry.[259] The separation of dansyl amino-acids on a layer composed of three sorbents using a single solvent system has been described,[260] and the t.l.c. of diphenylindenonesulphonyl amino-acids investigated.[261]

Other Methods.—Two systems for separating the cross-link amino-acids from acid hydrolysates of proteins such as elastin have been reported. Column chromatography on a polyacrylamide gel is recommended for large-scale separations,[262] and high-voltage paper electrophoresis has proved reliable and reproducible but does not separate desmosine and isodesmosine.[263] The separation of dansyl amino-acids on a polyamide column [264] and the copper complexes of amino-acids on a Sephadex anion-exchange column [265] have also proved successful. The latter system was developed for the qualitative analysis of cheese.

Determination of Specific Amino-acids.—Cysteine can be estimated by reacting it with ninhydrin at pH 10. The colour is due to the formation of hydrindantin, with a maximum at 470 nm. Other amino-acids do not interfere, but reducing substances such as glucose do if present at high concentration.[266] α-Substituted cysteines and cystines can be determined titrimetrically with *N*-bromosuccinimide using Bordeaux red as the indicator.[267] Aspartic acid and glutamic acid can be determined using guanidine carbonate,[268] and a radiometric technique for measuring L-asparagine in

[255] A. V. Barooshian, M. J. Lautenschlager, and J. M. Greenwood, *Analyt. Biochem.*, 1972, **49**, 602; A. V. Barooshian, M. J. Lautenschlager, and W. G. Harris, *ibid.*, p. 569.
[256] H. E. Herrick, J. M. Lawrence, and D. R. Coalvan, *Analyt. Biochem.*, 1972, **48**, 353.
[257] K. Nagasawa, A. Ogamo, and M. Sekiguchi, *Chem. and Pharm. Bull.* (*Japan*), 1972, **20**, 1006.
[258] M. Jellinek, *J. Chromatog.*, 1972, **69**, 402.
[259] M. Weise and G. M. Eisenbach, *Experientia*, 1972, **28**, 245.
[260] Z. Deyl and J. Rosmus, *J. Chromatog.*, 1972, **67**, 368.
[261] Ch. P. Ivanov and Y. Vladovska-Yukhanovska, *J. Chromatog.*, 1972, **71**, 111.
[262] D. P. Thornhill, *Biochim. Biophys. Acta*, 1972, **279**, 1.
[263] R. B. Moczar and L. Robert, *Analyt. Biochem.*, 1972, **45**, 422.
[264] Z. Deyl and J. Rosmus, *J. Chromatog.*, 1972, **69**, 129.
[265] K. P. Polzhofer and K. H. Ney, *Tetrahedron*, 1972, **28**, 1721.
[266] M. B. Patil, S. O. Jeyakumar, S. Roberts, and G. D. Kalyankov, *Indian J. Biochem. Biophys.*, 1972, **9**, 217.
[267] P. O. Schneider, R. J. Thibert, and R. J. Walton, *Mikrochim. Acta*, 1972, 925.
[268] A. K. Saxena, *Microchem. J.*, 1972, **17**, 91.

picomole quantities has been described. This is designed for samples of biological origin. Enzymic decarboxylation is first carried out to remove L-glutamate and L-aspartate, then incubation with L-asparaginase generates L-aspartic acid from the asparaginine. Subsequent enzymic transamination with 2-oxo[1-^{14}C]glutarate gives L-[1-^{14}C]glutamate, which is determined by scintillation counting after removing excess reagent.[269]

Histidine can now be determined by titration with lead nitrate using xylenol orange or pyrogallol red as indicator,[270] and an automated procedure for determining the fluophor produced by the interaction of histidine and *o*-phthalaldehyde in alkaline solution has been developed for blood analysis.[271] A sensitive fluorimetric assay of lysine has been based on its reaction with *o*-diacetylbenzene in the presence of 2-mercaptoethanol at pH 10; ornithine and glycine also undergo this reaction.[272] Whereas lysine and ornithine give a similar colour yield with ninhydrin at pH 5, at an acid pH ornithine develops five or six times as much colour.[273] The fluorescent product of the reaction of phenanthrene quinone with arginine has now been identified as 2-amino-1*H*-phenanthro[9,10-*d*]imidazole (46).[274]

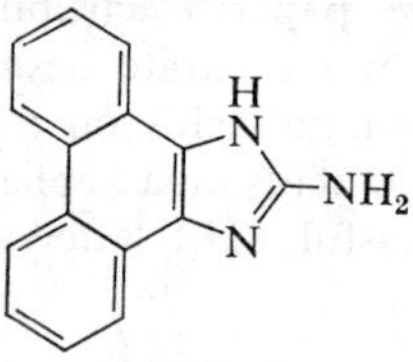

(46)

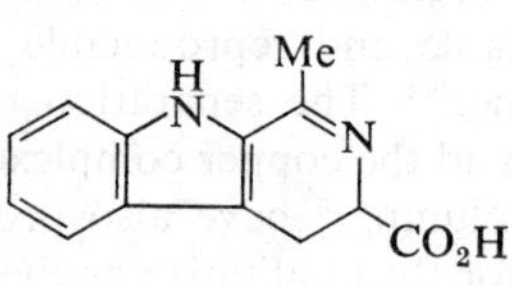

(47)

Carboxylic acid chlorides and anhydrides have been found to C-acylate the indole ring of tryptophan efficiently in anhydrous trifluoroacetic acid. Acetyl chloride, for example, yields 1-methyl-3,4-dihydro-β-carboline-3-carboxylic acid (47). This reaction has been used as the basis for a new spectrophotometric technique for the analysis of tryptophan.[275] Phosphoric acid can be used for detecting tryptophan; after four minutes at 110 °C a purple colour develops, but the mechanism is not known. Indole itself gives an orange colour.[276] Tryptophan is also the only naturally occurring amino-acid which shows a positive m.c.d. band. Furthermore, this absorption is almost completely free from overlapping contributions by other bands. This method of assay is likely to find its principal use in determining tryptophan in intact proteins. Tyrosine gives an intense m.c.d.

[269] D. A. Cooney and H. A. Milman, *Biochem. J.*, 1972, **129**, 953.
[270] O. C. Saxena, *Microchem. J.*, 1972, **17**, 210.
[271] H. D. Hill, G. K. Summer, and D. A. Newton, *Clin. Chim. Acta*, 1972, **36**, 105.
[272] M. Roth and L. Jeanneret, *Z. physiol. Chem.*, 1972, **353**, 1607.
[273] R. L. Davies, *J. Chromatog.*, 1972, **71**, 564.
[274] H. A. Itano and S. Yamada, *Analyt. Biochem.*, 1972, **48**, 483.
[275] A. Previero, G. Prota, and M.-A. Colleti-Previero, *Biochim. Biophys. Acta*, 1972, **285**, 269.
[276] F. N. Boctor, *J. Chromatog.*, 1972, **67**, 371.

band, but the error this can cause through overlap is small, *e.g.* 1% when the tryptophan : tyrosine ratio is 1 : 1.[277] A fluorimetric assay now enables 5-hydroxytryptophan to be determined in the presence of 5-hydroxyindole-acetic acid.[278] Other methods of estimation published include ones for pipecolic acid in serum and urine,[279] α-aminolaevulic acid in urine,[280] and *N*-[9-(β-D-ribofuranosyl)purin-6-ylcarbamoyl]threonine at the picomole level in transfer RNA.[281]

277 G. Barth, W. Voelter, E. Bunnenberg, and C. Djerassi, *J. Amer. Chem. Soc.*, 1972, **94**, 1293.
278 C. A. Fischer and H. H. Aprison, *Analyt. Biochem.*, 1972, **46**, 67.
279 R. F. Grimble, *Clin. Chim. Acta*, 1972, **38**, 113.
280 Gy. Berkó and I. Durkó, *Clin. Chim. Acta*, 1972, **37**, 443.
281 J. P. Miller and M. P. Schweizer, *Analyt. Biochem.*, 1972, **50**, 3275.

2
Structural Investigations of Peptides and Proteins

BY J. BRIDGEN, G. L. E. KOCH, G. ALLEN, T. L. BLUNDELL, AND R. H. PAIN

PART I: Primary Structure and Chemical Modification *by J. Bridgen and G. L. E. Koch*

1 Introduction

The contents of this Report are a review of the literature of 1972 pertinent to the primary structure and chemical modification of proteins. Not unexpectedly, it has been necessary to exercise a considerable amount of selectivity in the choice of subjects but the overall scope is similar to that of preceding Reports in this series. The availability and continuing production of reviews covering proteins such as the haemoglobins and cytochromes *c*, which are subjects in themselves, has led us to exclude these topics. On the other hand, it is clear that other proteins, such as those from membranes and the nervous system as well as those proteins which bind to nucleic acids, have started to attract an increasing amount of attention. These subjects have therefore been treated in a more comprehensive manner.

2 Methods

Two new volumes of 'Methods in Enzymology', devoted to the determination of enzyme structures, have been published.[1] These excellent reference books contain detailed information not generally found in research papers. Volume 25, 'Enzyme Structure, Part B', deals with primary structure determination methods and chemical modification. Volume 26, 'Enzyme Structure, Part C', is concerned more with physical methods and contains a particularly useful chapter on molecular-weight determination.

Improved procedures for comparing homologous protein or nucleic acid sequences have been presented.[2] Comparison tables are used which correct for multiple hits at the same base site, back mutation, multiply hit codons, genetic code degeneracy, and chance identity of two homologous sites. The concept of 'minimum mutational distance' is criticized on the grounds that the evolutionary mutational pathway is the one requiring the fewest mutagenic events and this may grossly underestimate the actual number of these events. Application of these methods [3] to haemoglobins, myoglobins,

[1] 'Methods in Enzymology', Academic Press, New York, 1972, Vols. 25, 26.
[2] R. Holmquist, C. Cantor, and T. H. Jukes, *J. Mol. Biol.*, 1972, **64**, 145.
[3] T. H. Jukes and R. Holmquist, *J. Mol. Biol.*, 1972, **64**, 163.

cytochromes *c*, and constant regions of immunoglobulins indicated that there were more replacements requiring two nucleotide substitutions than would be predicted if these replacements occurred at random over the entire genome. To account for the disparity, a model is proposed classifying amino-acids as invariable or variable. Variable amino-acids are defined as those which have changed as a result of random nucleotide replacements in the corresponding DNA. The total number of these replacements will increase with time but as this appears to be characteristic of the pathway leading to a specific organism this parameter is not always suitable as an 'evolutionary clock', since each protein will have its own characteristic rate of evolutionary change.

The theory that proteins have evolved by duplication of a small number of basic sequences has been criticized.[4] In a survey of over 50 different proteins, many of them of known structure, no good evidence of internal duplication of the peptide chain could be found. It is concluded that most proteins have probably evolved by addition of chains to the outer surface surrounding an invariable core near the active site.

Polyacrylamide Gel Electrophoresis.—Tables have been prepared which describe the making of preparative gels of concentrations in the range 2—40%,[5] and the parameters which affect protein recovery from such gels have also been investigated.[6] Not surprisingly, sample-load, pH, and gel concentration all have an effect, but also polymerization of the gel in the presence of extraneous protein can improve recoveries by up to 90%.

A new method of staining for glycoproteins with Alcian Blue has been described.[7] The method works on paper as well as on gels but the protein must be in the oxidized form.

The use of Bromophenol Blue as a tracker dye to monitor the progress of the electrophoretic run has been criticized. Two reports have appeared [8, 9] that this dye may interact with proteins to give multiple activity bands or affect the resolution of proteins on the gel. One way around this problem may be to prestain the proteins with Remazol dyes.[10] Although these are less sensitive than the popular Coomassie Blue (25 μg of protein are needed) it saves time needed for staining and destaining the gels and avoids the use of marker dyes. Alternatively, if greater sensitivity is required, the proteins may be dansylated prior to electrophoresis (see last year's Report) and the bands visualized under u.v. light. Greater sensitivity can also be obtained by reducing the volume of the gel. A method has been described [11] for

[4] A. D. McLachlan, *J. Mol. Biol.*, 1972, **64**, 417.
[5] D. P. Blatter, F. Garner, K. Van Slyke, and A. Bradley, *J. Chromatog.*, 1972, **64**, 147.
[6] G. Kapadia and A. Chrambach, *Analyt. Biochem.*, 1972, **48**, 90.
[7] A. H. Wardi and W. S. Allen, *Analyt. Biochem.*, 1972, **48**, 621.
[8] M. Hiebert, J. Gauldie, and B. L. Hillcoat, *Analyt. Biochem.*, 1972, **46**, 433.
[9] G. D. Burford and B. T. Pickering, *Biochem. J.*, 1972, **128**, 941.
[10] I. P. Griffith, *Analyt. Biochem.*, 1972, **46**, 402.
[11] L. I. Korochkin, N. P. Mertvetsov, N. M. Matveeva, and O. L. Serov, *F.E.B.S. Letters*, 1972, **22**, 213.

disc electrophoresis using capillary tubes of 25 × 50 μm whereby haemoglobin constituents of a single erythrocyte, and LDH isozymes from a single neuron, can be detected. A similar technique, but using isoelectric focusing, has been claimed [12] to be able to detect proteins in the nanogram range. Finally, micro-electrophoresis in 5λ capillary tubes has allowed the quantitative assay of less than one picogram of glucose-6-phosphate dehydrogenase.[13]

Electrophoresis in gels containing the denaturing detergent sodium dodecyl sulphate (SDS) is now in general use as a method of molecular-weight determination. However, anomalies still occur and the relative importance of the factors affecting the electrophoretic mobility of proteins in these gels has been discussed.[14] A proposed method to validate molecular-weight estimates [15] utilizes Ferguson's equation:

$$\log(M) = \log(M_0) - K_R T$$

where M = mobility in gel of acrylamide concentration T, M_0 describes the migration in the absence of a sieving medium, *i.e.* is a function of size and charge, and K_R is the retardation coefficient, depending solely on size. Proteins whose molecular weights are found from SDS gels should obey this equation over a range of acrylamide concentrations. Put more simply, the protein mobility should be proportional to the acrylamide concentration in the gel. An alternative method [16] uses a 'Gradipore' gel system, thus avoiding inaccuracies due to differences in detergent binding. Unfortunately, this technique is only applicable to globular proteins since molecules with an extended structure, *e.g.* fibrinogen, behave anomalously.

The factors involved in isoelectric focusing on polyacrylamide gels have been valuably discussed,[17] with the conclusion that slab-gels are preferable to tube-gels. A new rapid method for isoelectric focusing on slab-gels has been described [18] where equilibrium can be reached in only 30 minutes. An obviously important factor determining the successful application of this technique is the stability of the pH-gradient formed, and in a general discussion [19] of this problem it was concluded that instability increased with viscosity and ampholyte concentration. Finally, a new stain fixative based on Coomassie Brilliant Blue has been described [20] which may be used directly in the presence of ampholytes.

Protein Determination.—The year has produced the usual collection of modifications to the celebrated Lowry method. Firstly,[21] a technique has

[12] U. Grossbach, *Biochem. Biophys. Res. Comm.*, 1972, **49**, 667.
[13] T. Cremer, W. Dames, and V. Neuhoff, *Z. physiol. Chem.*, 1972, **353**, 1317.
[14] J.-S. Tung and C. A. Knight, *Analyt. Biochem.*, 1972, **48**, 153.
[15] G. A. Banker and C. W. Cotman, *J. Biol. Chem.*, 1972, **247**, 5856.
[16] L.-O. Anderson, H. Borg, and M. Mikaelsson, *F.E.B.S. Letters*, 1972, **20**, 199.
[17] O. Vesterberg, *Biochim. Biophys. Acta*, 1972, **257**, 11.
[18] J. Söderholm, P. Allestam, and T. Wadström, *F.E.B.S. Letters*, 1972, **24**, 89.
[19] L. E. Miles, J. E. Simmons, and A. Chrambach, *Analyt. Biochem.*, 1972, **49**, 109.
[20] N. Malik and A. Berrie, *Analyt. Biochem.*, 1972, **49**, 173.
[21] E. F. Hartree, *Analyt. Biochem.*, 1972, **48**, 422.

been described which gives a linear photometric response over the range 15—100 μg protein. Secondly,[22] it has been reported that addition of hydrogen peroxide to the alkaline Cu^{2+} solution containing the protein, followed by heating for 10 minutes at 50 °C before adding phenol, allows accurate protein determination in the presence of thiol groups. Thirdly,[23] for those of us involved in blood-protein determination, a note has appeared which claims that hematin interferes with the standard Lowry method.

Kirschenbaum has extended his useful compilation of molar absorptivity and $A^{1\%}_{1\,cm}$ values.[24a]

Amino-acid Analysis. (See also Chapter 1 of this Report.)—The amino-acid analyses of 148 proteins have been collected [24b] and supplement an earlier compilation (see last year's Report).

Complete hydrolysis of proteins has been achieved [25] using a column of Sepharose-bound trypsin, chymotrypsin, prolidase, and aminopeptidase M. The method has the advantage of allowing the direct determination of tryptophan, asparagine, and glutamine by subsequent amino-acid analysis. New chromatographic procedures have been described for the estimation of tryptophan oxidation products [26] and hydroxyproline,[27] as well as a method [28] for reducing haemin accumulation in glycoprotein hydrolysates. Mercaptosuccinic acid (20 μg) is recommended for the latter problem although presumably any mercaptan will do.

Magnetic circular dichroism has been proposed [29] as a method for measuring tyrosine : tryptophan ratios in proteins although the sophistication of the equipment will preclude general use. Equally sophisticated is a method of amino-acid analysis claimed to be capable of quantitation in the 10 femtomole range.[30] ^{14}C-dansyl amino-acids are separated by thin-layer chromatography, the thin-layer plate is autoradiographed, and the density of the spot is measured by scanning microscope photometry.

New computer programs for correcting unknown amino-acid hydrolysis losses [31] and for the integration of amino-acid analysis chromatograms [32–34] have also been described.

Separation Techniques.—A very comprehensive review containing preparative procedures, examples, and specific uses of affinity chromatography

[22] P. J. Geiger and S. P. Bessman, *Analyt. Biochem.*, 1972, **49**, 467.
[23] F. N. Boctor, *Analyt. Biochem.*, 1972, **50**, 500.
[24a] D. M. Kirschenbaum, *Internat. J. Protein Res.*, 1971, **3**, 109, 157, 237, 329.
[24b] D. M. Kirschenbaum, *Analyt. Biochem.*, 1972, **49**, 248.
[25] H. P. J. Bennett, D. F. Elliot, B. E. Evans, P. J. Lowry, and C. McMartin, *Biochem. J.*, 1972, **129**, 695.
[26] P. W. Nicholls and D. E. Rivett, *J. Chromatog.*, 1972, **65**, 565.
[27] P. X. Callahan, J. A. Shepard, and S. Ellis, *Analyt. Biochem.*, 1972, **49**, 155.
[28] L. B. James, *J. Chromatog.*, 1972, **68**, 123.
[29] G. Barth, E. Bunnenberg, and C. Djerassi, *Analyt. Biochem.*, 1972, **48**, 471.
[30] M. Weise and G. M. Eisenbach, *Experientia*, 1972, **28**, 245.
[31] E. J. Robel and A. B. Crane, *Analyt. Biochem.*, 1972, **48**, 233.
[32] H. L. Back, P. J. Buttery, and K. Gregson, *J. Chromatog.*, 1972, **68**, 103.
[33] H. D. Spitz, G. Henyon, and J. N. Sivertson, *J. Chromatog.*, 1972, **68**, 111.
[34] M. L. Johnson, E. A. Khairallah, and D. A. Yphantis, *Analyt. Biochem.*, 1972, **50**, 364.

has appeared[35] and this technique will not be discussed in detail here. Model studies have now demonstrated the potential of the method for mechanistic investigations as well as purification of multi-substrate enzymes.[36] Although these latter authors doubt the validity of general-ligand elution other workers refute this, and as evidence have demonstrated[37, 38] the separation of glyceraldehyde-3-phosphate dehydrogenase from lactic dehydrogenase and of lactic dehydrogenase from alcohol dehydrogenase by application of NADH and NAD^+ gradients, respectively. *N*-(3-Carboxypropionyl)aminodecylagarose, which contains ionic and hydrophobic groups, has been used for the successful chromatography of lipophilic proteins.[39] Affinity chromatography has also been employed for the isolation of specifically modified peptides on columns of Sepharose containing bound DNP-antibody. In one case[40] nitrotyrosyl-lysozyme was reduced to the amino-derivative and dinitrophenylated at pH 4.8, and after tryptic digestion a peptide containing the modified tyrosine (tyrosine-23) was specifically adsorbed on to the column. In the second case[41] 2,4-DNP-sulphenyl chloride is proposed as a tryptophan reagent suitable for the isolation of tryptophan peptides by this method.

Two new methods of protein purification have been reported. The first of these[42] utilizes the reverse of the familiar ammonium sulphate precipitation, namely solubilization of precipitated proteins with a decreasing gradient of ammonium sulphate concentration. The recommended procedure is ammonium sulphate precipitation of the protein in the presence of Celite 545 inert support, package of the insoluble material into a column, and application of a gradient of about 0.1% saturation decrease in ammonium sulphate per ml. The method has been successfully used in separating the two major allergens from ragweed pollen, antigens ε and κ.

The second procedure[43] uses hydroxyapatite columns, but run in buffers containing SDS. α- and β-chains of haemoglobin could be separated by this method and it may also be applicable to large peptides. The limitation, of course, is whether the protein can be renatured from the SDS solution, although methods have been presented for this (see last year's Report). A timely investigation of the nature of the protein-binding sites on hydroxyapatite has appeared[44] and rules are presented for predicting the elution behaviour of acidic and basic proteins.

The major problem in sequence analysis is still the preparation and purification of large peptide fragments and, unfortunately, there have been

35 P. Cuatrecasas, *Adv. Enzymology*, 1972, **36**, 39.
36 P. O'Carra and S. Barry, *F.E.B.S. Letters*, 1972, **21**, 281.
37 R. Ohlsson, P. Brodelius, and K. Mosbach, *F.E.B.S. Letters*, 1972, **25**, 234.
38 K. Mosbach, H. Guilford, R. Ohlsson, and M. Scott, *Biochem. J.*, 1972, **127**, 625.
39 R. J. Yon, *Biochem. J.*, 1972, **126**, 765.
40 M. Bustin and D. Givol, *Biochim. Biophys. Acta*, 1972, **263**, 459.
41 M. Wilchek and T. Miron, *Biochim. Biophys. Acta*, 1972, **278**, 1.
42 T. P. King, *Biochemistry*, 1972, **11**, 367.
43 B. Moss and E. N. Rosenblum, *J. Biol. Chem.*, 1972, **247**, 5194.
44 G. Bernardi, M.-G. Giro, and C. Gaillard, *Biochem. Biophys. Acta*, 1972, **278**, 409.

few recent advances in this field. One report[45] suggests the use of salt buffers with phosphocellulose and triethylaminoethylcellulose ion-exchange columns. Although monitoring of the column at 215 nm is possible, each peak must be desalted, a disadvantage not encountered with the volatile ammonium bicarbonate and ammonium acetate buffers in popular use.

For the smaller peptides two fairly conventional automatic peptide analysers have been described,[46, 47] and there has been a suggestion[48] that trinitrobenzenesulphonate could replace ninhydrin as a detection method in such a machine. No particular advantages seem to follow from such a change. Finally, a peptide analyser has been described[49] which is claimed to be of very high sensitivity. The apparatus uses a conventional ninhydrin detection system and the high sensitivity (less than one nanomole is detected) is attained by using very narrow-bore tubing and slow pumps.

Cleavage of Peptide Bonds.—A study has been made[50] of the specificity of trypsin and chymotrypsin on substrates containing methylated lysines. Contrary to popular belief no difference in digestion rate was found between oxidized ribonuclease and the enzyme methylated with formaldehyde–sodium borohydride. Similar results were obtained with polylysine and calf thymus arginine-rich histone. It is difficult to correlate these results with primary-structure work on histones (see Chromosomal Proteins section), where methylated lysines have commonly been found to resist tryptic cleavage.

The specificity of an exopeptidase from yeast, Yeast Proteinase C, has been investigated.[51] The enzyme appears to resemble carboxypeptidase C in that it liberates acidic, basic, and neutral *C*-terminal amino-acids from the peptide chain. Prolyl bonds are also cleaved.

A peptidase from ovine erythrocytes has been described[52] and although it cleaves hydrophobic regions of small peptides it will also, perhaps more usefully, remove *N*-formylmethionine from the *N*-terminus of peptides providing the next residue is not proline or a polar amino-acid.

Two extracellular proteases which, because of their limited specificity, may find great application in primary structure determination have been purified and their properties investigated. One of these, Myxobacter AL-1 protease II,[53] cleaves exclusively on the NH_2-side of lysine residues and thus becomes a valuable tool for overlapping lysine-blocked tryptic

[45] C. C. Q. Chin and F. Wold, *Analyt. Biochem.*, 1972, **46**, 585.
[46] R. W. Hartley, *Analyt. Biochem.*, 1972, **46**, 676.
[47] M. A. Atassi, M. T. Perlstein, M. C. Rosemblatt, and P. Rocek, *Analyt. Biochem.*, 1972, **49**, 164.
[48] R. Delaney, *Analyt. Biochem.*, 1972, **46**, 413.
[49] J. J. T. Gerding, H. J. M. Kempen, B. J. M. Lamers, and M. H. Gerding, *J. Chromatog.*, 1972, **66**, 145.
[50] W. K. Paik and S. Kim, *Biochemistry*, 1972, **11**, 2589.
[51] R. Hayashi and T. Hata, *Biochim. Biophys. Acta*, 1972, **263**, 673.
[52] J. Witheiler and D. B. Wilson, *J. Biol. Chem.*, 1972, **247**, 2217.
[53] M. Wingard, G. Matsueda, and R. S. Wolfe, *J. Bacteriol.*, 1972, **112**, 940.

peptides. Specificity has been checked on insulin B-chain, cytochrome c, lysozyme, vasopressin, and glutamine synthetase, and appears to be absolute although *C*-terminal lysine, dilysine, and trilysine bonds are not cleaved. The second protease was isolated from *Staphylococcus aureus*[54] and specifically cleaves on the *C*-terminal side of aspartyl and glutamyl bonds in phosphate buffer at pH 7.8. If the buffer is changed to ammonium bicarbonate, pH 7.8, or ammonium acetate, pH 4.0, only glutamyl bonds are split, although in this case there is also slow cleavage of Asn-Gly bonds.[55] This enzyme should prove particularly useful for preparing peptides for solid-phase Edman degradation (see next section) where, prior to degradation, fragments are specifically attached to a resin *via* their carboxy-groups, and also for the assignment of peptide amide groups.

Cyanogen bromide cleavage at methionine residues remains the only generally useful chemical fragmentation method. Unfortunately, oxidized forms such as methionine sulphone and sulphoxide are not cleaved, and so an investigation[56] of the reversal of this reaction using mercaptoethanol is particularly welcome. Apparently, reduction occurs rapidly at 25 °C, in the pH range 7—9, with thiol concentrations of 5×10^{-3} to 10^{-1} mol l^{-1}, conditions under which most proteins are stable.

Sequence Determination.—Pentafluorophenyl isothiocyanate has been proposed[57] as an alternative to the normal phenyl isothiocyanate as the coupling reagent in the Edman procedure. Its main advantage appears to be in conjunction with an electron-capture detector for g.l.c. analysis of phenylthiohydantoins. More useful perhaps is fluorescamine (1), which

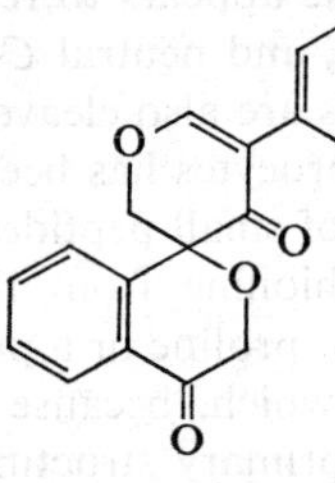

(1)

reacts with primary amines giving highly fluorescent products detectable in the picomole range.[58] The reagent has the added bonus that neither it nor its degradation products are fluorescent and that the reaction product with ammonia fluoresces only slightly. Possible applications include automated amino-acid and peptide analysis, protein assay, and the detection of amino-

[54] J. Houmard and G. R. Drapeau, *Proc. Nat. Acad. Sci. U.S.A.*, 1972, **69**, 3506.
[55] G. R. Drapeau, Y. Boily, and J. Houmard, *J. Biol. Chem.*, 1972, **247**, 6720.
[56] F. Naider and Z. Bohak, *Biochemistry*, 1972, **11**, 3208.
[57] R. M. Lequin and H. D. Niall, *Biochim. Biophys. Acta*, 1972, **257**, 76.
[58] S. Udenfriend, S. Stein, P. Böhlen, D. Airman, W. Leimgruber, and N. Weigele, *Science*, 1972, **178**, 871.

acids and peptides on paper or t.l.c. chromatograms, at levels of less than one nmol. Unfortunately proline does not react.

Now that the automatic sequenator is becoming more common, less emphasis is being placed on determining *N*-terminal sequences of proteins by hand. However, to show what can be done, a method has appeared [59] for the manual determination of up to 20 residues, although in this case 600 nmol of protein were used. If less material is available, Edman degradation in the presence of SDS has been claimed [60] to produce protein sequences of up to ten residues using only a few nmol of protein. In other words, proteins eluted from standard SDS-acrylamide gels can now be characterized by their amino-terminal sequences.

The solid-phase Edman degradation takes a further step forward with the use of *p*-phenylene di-isothiocyanate to attach tryptic peptides to an aminopolystyrene resin (Figure 1).[61] Lysine peptides are coupled *via* their ε-NH_2 groups but arginine must first be deguanidated with hydrazine to form an ornithine residue. Up to 100 nmol of peptide can be attached to 35 mg of resin and coupling yields of 80—100% are reported. The method is obviously of value and already a report has appeared of a ribosomal protein (see Section 10) whose sequence has been determined largely by this method. Significantly, this protein contained only one arginine residue, and a drawback to the method might be side-reactions, such as internal bond cleavages, occurring during the hydrazine treatment. New methods such as these are particularly valuable now that commercial versions of this machine are becoming available.

Developments also continue to be made with the liquid-phase sequenator. One paper,[62] recommended reading for all users, describes the replacement of Edman's quadrol buffer by the volatile dimethylbenzylamine. This eliminates the need for an ethyl acetate wash of the protein or peptide film, often the cause of high extractive losses. Also included are methods for preparation of the sample, addition of thiols to stabilize the serine and threonine PTH derivatives, and details of programme refinements. A description is included of the degradation of thermolysin and its cyanogen bromide fragments.

The complete sequence of β-lactoglobulin has been determined by sequencer analysis of the intact protein and its tryptic peptides.[63] For the peptides, sulphonated phenyl isothiocyanates were used as the coupling reagents (see last year's Report), thus minimizing extractive losses during the organic solvent washes. Using this procedure peptides as small as five residues have been successfully degraded.[64] However, the sequencer is

[59] M. E. Percy and B. A. Buchwald, *Analyt. Biochem.*, 1972, **45**, 60.
[60] A. M. Weiner, T. Platt, and K. Weber, *J. Biol. Chem.*, 1972, **247**, 3242.
[61] R. A. Laursen, M. J. Horn, and A. G. Bonner, *F.E.B.S. Letters*, 1972, **21**, 67.
[62] M. A. Hermodson, L. H. Ericsson, K. Titani, H. Neurath, and K. A. Walsh, *Biochemistry*, 1972, **11**, 4493.
[63] G. Braunitzer, R. Chen, B. Schrank, and A. Stangl, *Z. physiol. Chem.*, 1972, **353**, 832.
[64] J. K. Inman, J. E. Hannon, and E. Appella, *Biochem. Biophys. Res. Comm.*, 1972, **46**, 2075.

still best suited to proteins and large fragments. For those who cannot afford to buy a sequencer another do-it-yourself model has been described.[65] Perhaps it is not generally realized that the annual running costs for the commercial model can be almost as high as the initial outlay for the machine.

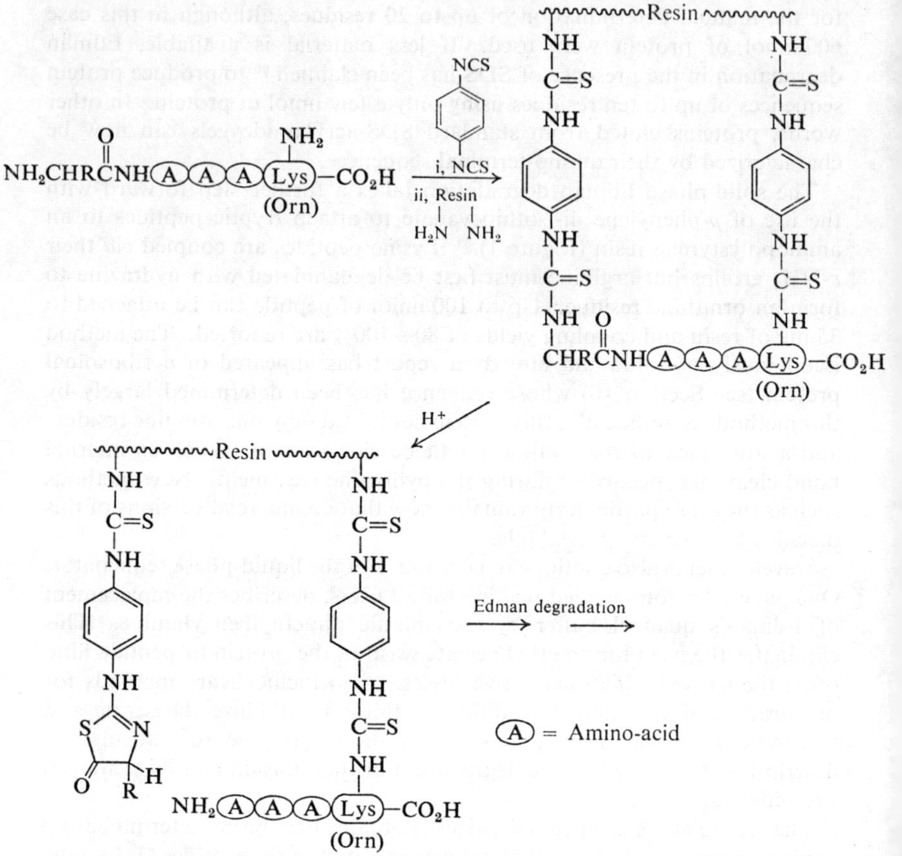

Figure 1 (Reproduced by permission from *F.E.B.S. Letters*, 1972, **21**, 67)

Whether one uses liquid-phase or solid-phase degradation, the problem of identifying the released PTH amino-acids remains. Hence, a paper[66] describing advances in the uses of g.l.c. for analysis of both phenyl- and methyl-thiohydantoins is particularly welcome. A new blend of liquid phases is recommended, which although giving better resulution also requires a long 40-minute programme. Unless one uses an automatic sample-injector this is too slow to keep up with the 22 residues a day

[65] J. D. Lynn and J. C. Bennett, *Analyt. Biochem.*, 1972, **45**, 498.
[66] J. J. Pisano, T. J. Bronzert, and H. B. Brewer, *Analyt. Biochem.*, 1972, **45**, 43.

output of the Beckman Sequencer. More useful is a method of 'on-column' silylation where the silylating agent, *NO*-bis(trimethylsilyl)-acetamide, is withdrawn into the injection syringe along with the sample. Silylation literally takes place on the column and enables the polar PTH -Asn, -Gln, -Asp, and -Glu derivatives to be chromatographed.

Of course these techniques are of no value when proteins or peptides without free *N*-terminal amino-groups are encountered. For these, isolation of the blocked peptide followed by either carboxypeptidase digestion and electrophoresis with suitable standards,[67] or mass-spectrometric sequence determination (see below), may be used. For molecules having *N*-terminal pyrrolidonecarboxylic acid, methanol–HCl treatment to form the γ-methyl ester of glutamic acid has been recommended.[68] Expected side-reactions are loss of side-chain amides, esterification of carboxy-groups, and internal chain cleavages.

The use of dipeptidyl aminopeptidase 1 (Cathepsin C) in sequence analysis continues to attract attention. The method uses a dipeptidase to degrade completely the peptide or protein chain from the *N*-terminus, after which the released dipeptides are fractionated and identified. Removal of one amino-acid from the *N*-terminus by Edman degradation then allows a second set of dipeptides to be generated and overlapped with the first. A report on the use and specificity of the enzyme[69] indicates that comparison of the two sets of generated dipeptides gives two possible sequences[70] for the 21-residue insulin A-chain. Obviously, the number of possibilities will increase rapidly for longer peptides but the main obstacle to practical application will be fractionating, and preferably quantitating, the dipeptides liberated. One possibility is combined use of g.l.c.–mass spectrometry.[71] As there are only about 400 possible dipeptides a library of their mass spectra could be accumulated, allowing automatic computer-deciphering of the results. Another rather ingenious solution to this problem might be membrane diffusion.[72] Using this procedure substrate and enzyme are kept on one side of an Amicon UM-2 Diaflow membrane while liberated dipeptides flow through and are identified by t.l.c. monitoring of the effluent. Unfortunately, the enzyme does not cleave prolyl bonds or bonds on the *N*-terminal side of arginine or lysine residues. The latter problem can of course be minimized by the use of tryptic peptides.

The original promise of mass spectrometry as a rapid, sensitive technique of sequence determination still remains unfulfilled. Nevertheless each year brings new advances, and mixture analysis in particular (see last year's Report) appears very promising. Indeed it may well be the only field where mass spectrometry offers real advantages over conventional

[67] A. Yoshida, *Analyt. Biochem.*, 1972, **49**, 320.
[68] I. Kawasaki and H. A. Itano, *Analyt. Biochem.*, 1972, **48**, 546.
[69] H. Lindley, *Biochem. J.*, 1972, **126**, 683.
[70] R. J. Rowlands and H. Lindley, *Biochem. J.*, 1972, **126**, 685.
[71] Yu. A. Ovchinnikov and A. A. Kiryushkin, *F.E.B.S. Letters*, 1972, **21**, 300.
[72] P. X. Callahan, J. K. McDonald, and S. Ellis, *Fed. Proc.*, 1972, **31**, 1105.

methods. One problem in this work is that ambiguities may occasionally result from possible sequences of identical mass or, particularly in mixture analysis, from the incorrect interpretation of non-sequence ions. Accordingly, the use of deuteriated derivatives in peptide mixture analysis has been suggested [73] as a means of minimizing the number of alternative sequences. For example, the peptide mixture Leu-Val-Gln-Ala and Gly-Pro-Ala-Thr gave a mass-spectral peak corresponding to the sequence Gly-Pro-Ala-Met, but this possibility could be eliminated after comparison with the spectrum of the deuteriated derivative. Methods of derivatization of peptide mixtures have been investigated [74] and the recommended procedure, acetylation in aqueous acetic anhydride followed by permethylation in a DMSO solution of dimsylsodium, was used to obtain the sequence of a mixture of three peptides from cheese.

One interesting advance is field-desorption mass spectrometry. This is suggested [75] as an alternative to current techniques for use on peptides of low volatility, for instance those containing arginine or histidine. The method has the advantage of requiring no prior treatment of the peptide to increase its volatility and this enables smaller initial amounts to be used. A simple procedure for the amino-acid sequencing of small peptides (2—8 residues) has also been described.[76] This is based on the quantitative pyrolytic conversion of a peptide trimethylanilinium salt into its corresponding methyl ester in the mass spectrometer probe.

Techniques are now available for dealing with almost all of the 'difficult' amino-acids (principally cysteine, histidine, arginine, and methionine). Unfortunately, many of these methods suffer from lack of practical applicability and have been tested only on milligram quantities of synthetic di- and tri-peptides. Hence, a paper [77] describing a rapid practical method for permethylation of histidine-containing peptides in 100 nmol quantities is particularly welcome. Reaction time with the methylating reagent is reduced to one minute and this prevents formation of a non-volatile quaternary derivative. As this author points out, other histidine derivatization methods [78] (and see last year's Report) which involve quantitation of the permethylating reagents are very difficult to apply in practice. Fragmentation patterns of β-lysyl [79] and Tyr-Trp [80] sequences, as well as various derivatives of sulphur-containing amino-acids,[81] have also been investigated.

[73] P. Roepstorff and K. Brunfeldt, *F.E.B.S. Letters*, 1972, **21**, 320.
[74] R. Hodges, S. B. G. Kent, and B. C. Richardson, *Biochim. Biophys. Acta*, 1972, **257**, 54.
[75] H. D. Winkler and H. D. Beckey, *Biochem. Biophys. Res. Comm.*, 1972, **46**, 391.
[76] G. M. Schier, B. Halpern, and J. Karth, *Tetrahedron Letters*, 1972, 4621.
[77] H. R. Morris, *F.E.B.S. Letters*, 1972, **22**, 257.
[78] M. L. Polan, W. J. McMurray, S. R. Lipsky, and S. Lande, *J. Amer. Chem. Soc.*, 1972, **94**, 2847.
[79] L. I. Rostovtsera and A. A. Kiryushkin, *Org. Mass Spectrometry*, 1972, **6**, 1.
[80] B. C. Das and R. D. Schmid, *F.E.B.S. Letters*, 1972, **25**, 253.
[81] H. Nishimura, S. Takara, H. Okuyama, and J. Mitzutani, *Tetrahedron*, 1972, **28**, 4503.

Finally, work continues [82, 83] on the mass spectrometric identification of the thiohydantoin products from the standard Edman degradation. The advantage over g.l.c. is that derivatives of the basic amino-acids can be identified but, as with the parent amino-acids, it is impossible to distinguish leucyl- from isoleucyl-thiohydantoin.

3 Enzymes

Glycolytic Enzymes.—The subunit composition of rabbit muscle phosphoglucomutase has been reinvestigated.[84] The enzyme, previously thought to be a monomer of molecular weight 62 000, gave both lysine and valine end-groups after dansylation. SDS-gel electrophoresis showed a single band corresponding to a subunit molecular weight of 32 000, indicating that the enzyme was in fact dimeric. Interestingly, this result could only be obtained when the gels were pre-run to remove persulphate.

Two new methods have been reported for the isolation of phosphofructokinase (PFK) from human erythrocytes. The first of these [85] produces an enzyme which is stable for only two weeks at −20 °C and gives two active bands on acrylamide gel; the authors consider that this may be due to dissociation of the enzyme at low protein concentrations. The second procedure [86] has additional heat-treatment and gel-filtration steps but produces an enzyme of higher specific activity which is stable for three weeks at 0 °C. Elution from Sepharose 4B again gave multiple activity peaks but SDS gel-electrophoresis showed that all of these could be ascribed to a subunit of molecular weight 104 000. The dimer, tetramer, hexamer, octamer, and decamer were all found.[87] Multiple forms of PFK, this time considered to be isozymes, have also been found in various rat tissues and tumours.[88] Electrophoresis on cellulose acetate membranes has revealed at least four forms of the enzyme:

Muscle	Type 1 only
Brain	Type 2 only
Kidney, Spleen, Erythrocytes	Types 2 and 3
Liver	Type 4
Yoshida ascites hepatoma and Yoshida sarcoma tumours	Type 2 and sometimes Type 3
Morris hepatoma tumours	Type 4

This situation is similar to that found for the human and rabbit enzyme. It has recently proved possible [89] to hybridize rabbit muscle and liver PFK

[82] H. Tschesche, M. Schneider, and E. Wachter, *F.E.B.S. Letters*, 1972, **23**, 367.
[83] S. Ellis, T. Fairwell, and R. E. Lovins, *Biochem. Biophys. Res. Comm.*, 1972 **49**, 1407.
[84] H. W. Duckworth and B. O. Sanwal, *Biochemistry*, 1972, **11**, 3182.
[85] L. M. Y. Lee, *Arch. Biochem. Biophys.*, 1972, **148**, 607.
[86] K.-W. Wenzel, J. Gauer, G. Zimmermann, and E. Hofmann, *F.E.B.S. Letters*, 1972, **19**, 281.
[87] K.-W. Wenzel, G. Zimmermann, J. Gauer, W. Diezel, St. Liebe, and E. Hofmann, *F.E.B.S. Letters*, 1972, **19**, 285.
[88] N. Kurata, T. Matsushima, and T. Sugimura, *Biochem. Biophys. Res. Comm.*, 1972, **48**, 473.
[89] M. Y. Tsai and R. G. Kemp, *Arch. Biochem. Biophys.*, 1972, **150**, 407.

to give a five-membered hybrid set, which suggests that the isozymes are tetrameric and of similar tertiary structure.

By binding rabbit muscle aldolase to Sepharose and removing non-attached subunits with 8M-urea, it has been claimed [90] that active subunits of the enzyme may be prepared. Comparison of this aldolase with correspondingly bound tetrameric aldolase indicated that, whereas the subunit form was more susceptible to thermal, alkali, and pH denaturation, many of its kinetic properties remained unchanged. Hence subunit interactions appear to stabilize this enzyme against denaturation but have little effect on the catalytic process. Although this method is ingenious, it is very difficult to demonstrate that only subunits remain bound to the Sepharose and that there is no interaction between them. Hybridization of calf-brain aldolase with class 1 aldolase from *Drosophila* blocked at the active site has been shown [91] to produce a set of active hybrid enzymes, demonstrating that each subunit makes an independent contribution to the enzymic activity. This result is similar to that reported previously for native and succinylated rabbit muscle aldolase.[92]

In an investigation of the nature of the aldehyde-binding site of rabbit muscle aldolase, evidence has been presented [93] for hemithioacetal formation between the substrate aldehyde group and a specific cysteine residue. Whether this is the same cysteine specifically alkylated by chloroacetal phosphate [94] is not yet clear, but these results are at variance with work [95] on the enzyme from sturgeon muscle, where titration with iodoacetate and DTNB indicated that thiol groups played little or no part in catalysis.

Isoelectric focusing has been used to demonstrate heterogeneity in various mammalian and avian aldolases.[96] Five major activity peaks were found from rat, pigeon, rabbit, and human enzymes and three peaks from that of hen. A possible explanation stems from the observation [97] that limited chymotryptic digestion of rabbit muscle aldolase releases two peptides,

α- Ile-Ser-Asn-His-Ala-Tyr
β- Ile-Ser-Asp-His-Ala-Tyr

which, as reported in the 1970 Report, are *C*-terminal in the protein. Pulse-labelling experiments [98] have indicated a half-life for the amidated form of about eight days, which is precisely that of the enzyme *in vivo*. Deamidation of the α_4 form thus produces $\alpha_3\beta$, $\alpha_2\beta_2$, $\alpha\beta_3$, and β_4 forms, which may explain the five peaks produced in the isoelectric focusing experiments. The amino-acid sequence around the active-site lysine

90 W. W.-C. Chan and H. M. Mawer, *Arch. Biochem. Biophys.*, 1972, **149**, 136.
91 O. Brenner-Holzach and F. Leuthardt, *European J. Biochem.*, 1972, **31**, 423.
92 E. A. Meighen and H. K. Schachman, *Biochemistry*, 1970, **9**, 1163.
93 J. Wagner, C. Y. Lai, and B. L. Horecker, *Arch. Biochem. Biophys.*, 1972, **152**, 398.
94 M. C. Paterson, I. L. Norton, and F. C. Hartman, *Biochemistry*, 1972, **11**, 4435.
95 P. J. Anderson, *Canad. J. Biochem.*, 1972, **50**, 111.
96 Y. Ikehara, S. Yanagi, and T. Kamiya, *J. Biochem.*, 1972, **72**, 203.
97 C. F. Midelfort and A. H. Mehler, *J. Biol. Chem.*, 1972, **247**, 3618.
98 C. F. Midelfort and A. H. Mehler, *Proc. Nat. Acad. Sci. U.S.A.*, 1972, **69**, 1816.

Rabbit muscle	Ala-Leu-Ser -Asn-His-His-Ile -Tyr-Leu-Gln-Gly-Thr-Leu-Leu-Lys*-Pro-Asn-Met -Val-Thr-Pro-Gly-His -Ala-Cys-Thr-Glu-Lys
Rabbit liver } Beef liver }	Ala-Leu-Asn-Asp-His-His-Val-Tyr-Leu-Glu-Gly-Thr-Leu-Leu-Lys*-Pro-Asn-Met -Val-Thr-Ala-Gly-His-Ala -Cys-Thr-Lys-
Rabbit brain	Ala-Leu-Ser (Asx,His,His,Ile ,Tyr,Val ,Glx,Ala,Thr,Leu,Leu,Lys*,Pro,Glx)Met -Val-Thr-Pro-Gly-Asx-Ala-Cys-Thr-Glx-Lys
Codfish	Ala-Leu-Ser -Asp-His-His-Val-Tyr-Leu-Gln-Gly-Thr-Leu-Leu-Lys*-Pro-Asn-Met -Val-Thr-Ala-Gly-His -Ser -Cys-Ser -His -Lys
Frog	Ala(Leu,Ser ,Asx,His,His,Val,Tyr,Leu,Glx,Gly,Thr,Leu,Leu,Lys*,Pro,Asx,Met)(Val,Thr,Ala,Gly,Asx,Ala,Cys,Thr,Glx)Lys
Sturgeon[99]	Ala-Leu-Ser -Asp-His-His-Val-Tyr-Leu-Glu-Gly-Thr-Leu-Leu-Lys*-Pro-Asn-Met -Val-Thr-Ala-Gly-Gln-Ala-Lys-Thr-Lys-Lys
Coelacanth[101]	Ala-Leu-Ser -Asx-His(His,Ile ,Tyr,Leu,Glx,Gly,Thr,Leu,Leu,Lys*,Pro,Asx,Met, Val,Thr,Pro,Gly,His ,Ser ,Cys,Thr,Pro)Lys
Lobster	Ala-Leu-Asx-Asx-His-His-Val-Phe-Leu-Glx-Gly-Thr-Leu-Leu-Lys*-Pro-Asn-Met -Val-Thr-Pro-Gly-Asx-Ala-Cys- Ser-Gly-Lys
Pigeon[100]	Val-Thr-Pro-Gly-His -Ala-Cys-Pro-Lys-Lys
Spinach	Leu-Leu-Lys*-Pro-Ser

Figure 2 *Amino-acid sequences around the substrate binding lysine residue* (Lys*) *in various aldolases*

residue of four different species of sturgeon muscle aldolase has been determined.[99] Surprisingly, several differences were found, although the corresponding sequences from pig, ox, and rabbit muscle were all identical. The sequences determined for this region from various species are shown in Figure 2.

A further report [102] on the trimeric 2-keto-3-deoxy-6-phosphogluconate aldolase from *Pseudomonas putida* has shown that two thiol groups per

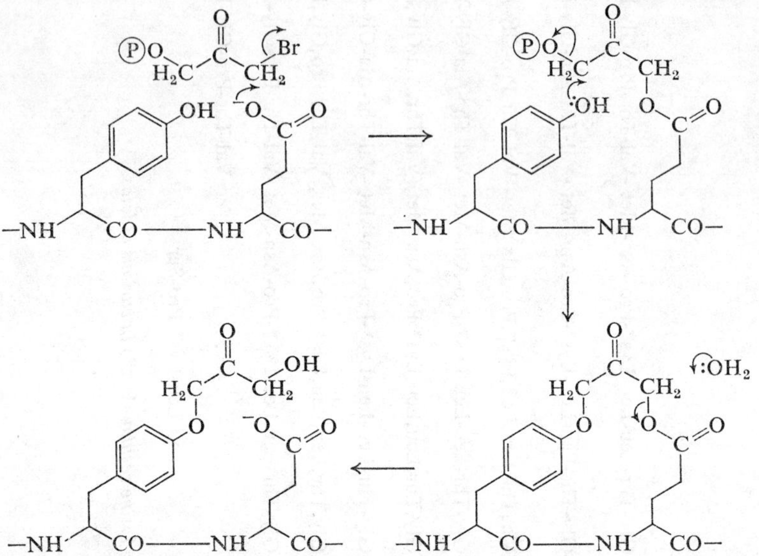

Figure 3 *A possible mechanism for the migration of active-site label from glutamic acid to tyrosine in rabbit muscle triose phosphate isomerase*
(Reproduced by permission from *Biochem. J.*, 1972, **129**, 311)

subunit can be titrated with DTNB with no loss of activity, but that the remaining two only react in the presence of 8M-urea. Up to 90% of the activity can be regained from 8M-urea solution.

Finally, a comprehensive review of the molecular and catalytic properties of class 1 and 2 aldolases, as well as a discussion of the role of aldolase as a model for structure–function relationships, has appeared.[103]

A new method for the preparation of triose phosphate isomerase (TIM) from chicken muscle has been described.[104] Values of K_{cat} and K_m for the enzyme-catalysed reaction in both directions were obtained and in all cases

[99] I. Gibbons, R. N. Perham, and P. J. Anderson, *Nature New Biol.*, 1972, **238**, 173.
[100] Y. Edwards and I. Gibbons, *F.E.B.S. Letters*, 1972, **28**, 221.
[101] R. S. Jack, Ph.D. Thesis, Univ. of Cambridge, 1972.
[102] H. Möhler, K. Decker, and W. A. Wood, *Arch. Biochem. Biophys.*, 1972, **151**, 251.
[103] C. Y. Lai and B. L. Horecker, *Essays in Biochem.*, 1972, **8**, 149.
[104] S. J. Putnam, A. F. W. Coulson, I. R. T. Farley, B. Riddleston, and J. R. Knowles, *Biochem. J.*, 1972, **129**, 301.

simple Michaelis–Menten kinetics were obeyed. The pH-dependence of the reaction has also been studied.[105] Reaction of the enzyme with the active-site-directed inhibitor bromohydroxyacetone phosphate caused total loss of catalytic acitvity,[106] the inhibitor attaching itself to the glutamic acid residue in the sequence Ala-Tyr-Glu-Pro-Val-Trp. However, unless the inhibitor–enzyme bond was immediately reduced with borohydride the phosphate group was lost and the label migrated to the adjacent tyrosine residue (Figure 3). This is similar to the migration of the acetyl group in mono-acetylated glyceraldehyde-3-phosphate dehydrogenase from the active-site cysteine to a lysine residue.[107] A peptide of identical sequence has also been isolated [108] from yeast TIM after inactivation with 3-chloro-acetal phosphate.

The almost complete amino-acid sequence of glyceraldehyde-3-phosphate dehydrogenase (GPD) from baker's yeast has been presented.[109] 277 of 330 residues can be placed directly and the remainder may be positioned by comparison with sequences from the pig and lobster muscle enzymes (Figure 4). Overall homology between the three species is clearly very high, no less than 60% of the residues being identical. The active-centre peptide of the same enzyme isolated from the thermophilic bacterium *Bacillus stearothermophilus* has been purified [110] and its amino-acid sequence determined as

Ala-His-His-Ile-Val-Ser-Asn-Ala-Ser-*Cys*-Thr-Thr-Asn-Cys-Leu-Ala-Pro-Phe-Ala-Lys

The primary structure around the reactive cysteine (in italics) is completely conserved (see Figure 4 and Volume 1). Preliminary crystallographic data have now appeared for this enzyme and it is to be hoped that the sequences shown in Figure 4 will soon be seen in three dimensions.

In contrast, the complete structure of lactate dehydrogenase (LDH) awaits the determination of the primary structure and there have been no new developments in this area. A previous report [111] that LDH from beef heart is an octamer with a subunit molecular weight of 18 000 has been shown [112, 113] to be incorrect, possibly because of the use of impure guanidinium hydrochloride in the ultracentrifuge experiments. The enzyme was found by a variety of methods to be tetrameric with subunit molecular weight 34 500. Chemical modification with butanedione [114] has

[105] B. Plaut and J. R. Knowles, *Biochem. J.*, 1972, **129**, 311.
[106] S. de la Mare, A. F. W. Coulson, J. R. Knowles, J. D. Priddle, and R. E. Offord, *Biochem. J.*, 1972, **129**, 321.
[107] E. Mathew, B. P. Meriwether, and J. H. Park, *J. Biol. Chem.*, 1967, **242**, 5024.
[108] I. L. Norton and F. C. Hartman, *Biochemistry*, 1972, **11**, 4435.
[109] G. M. Thellwall-Jones and J. I. Harris, *F.E.B.S. Letters*, 1972, **22**, 185.
[110] J. Bridgen, J. I. Harris, P. W. McDonald, R. E. Amelunxen, and J. R. Kimmel, *J. Bacteriol.*, 1972, **111**, 797.
[111] D. B. Millar, V. Frattali, and G. E. Willick, *Biochemistry*, 1969, **8**, 2416.
[112] J. S. Huston, W. W. Fish, K. G. Mann, and C. Tanford, *Biochemistry*, 1972, **11**, 1609.
[113] G. J. Fosmine and S. N. Timasheff, *Biochemistry*, 1972, **11**, 2455.
[114] P. C. Yang and G. W. Schwert, *Biochemistry*, 1972, **11**, 2218.

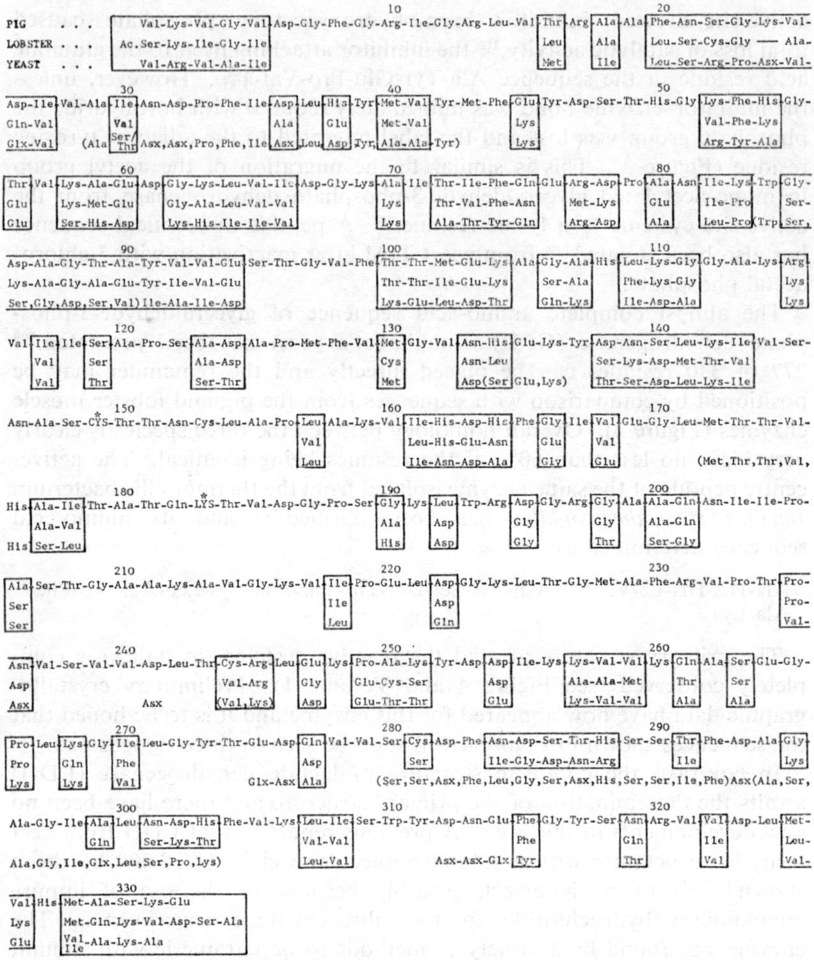

Figure 4 *Comparison of the amino-acid sequence of GPDH from pig muscle, lobster muscle, and baker's yeast. Established sequence differences are shown in the boxed regions. Sequences not experimentally determined for the yeast chain are given within parentheses and in a* provisional *order that maximizes the sequence homology between the yeast and muscle enzymes.* Asx *and* Glx *denote residues in the yeast chain for which amide assignments were not unambiguously established.* CYS* (149) *and* LYS* (183) *are residues implicated in the active site*
(Reproduced by permission from *F.E.B.S. Letters*, 1972, **22**, 185)

led to the conclusion that an arginyl, as well as the lysyl residue found previously, is involved in substrate binding.

In an investigation of the role of thiol groups in yeast alcohol dehydrogenase (ADH) it has been reported[115] that after blocking of 4.5 groups per mole protein, ADH activity is lost but 85% of the NADH-binding capacity is retained. Kinetic interpretation of the results showed that the enzyme's ability to form ternary complexes was destroyed. A similar study on horse-liver ADH using spin-labelled iodoacetamide indicated[116] that at least two thiol groups were important for activity and that these were close to the zinc-binding sites. Structural studies have now been commenced on human[117] and rat[118] liver ADH. Both enzymes are dimeric with subunit molecular weights of about 40 000, and both show extensive sequence homology with the horse-liver enzyme. The preliminary sequences available, some 20—25% of each molecule, indicate that ADH is evolving faster than GPD and also that the isoenzymes found in both the human and equine proteins evolved after separation of the ancestral species. Interestingly, no isoenzymes have been found for the rat ADH and it appears to have a unique sequence.[119] Genetic evidence that human liver ADH is formed by three genetically distinct loci has been proposed[120] and since the enzyme is dimeric this would account for six isoenzymes. However, another report[121] indicates that there are at least seven isoenzymes characterizable by electrophoresis, kinetics, and fluorimetry.

Proteolytic Enzymes.—*Serine Proteases.* A study of crystalline methylchymotrypsin has been carried out in the hope that the low activity of this form of the enzyme together with its intact substrate-binding site might permit the investigation of enzyme–substrate complexes.[122] The modification was effected with methyl *p*-nitrobenzene sulphonate, a reagent which permits the specific methylation of His-57 in chymotrypsin. However, the structure of methylchymotrypsin does not seem to be consistent with the properties of the enzyme in solution and raises the question whether the enzymatically active molecule was represented in the crystal. In this connection it is worth noting a report that methylchymotrypsin is considerably less stable than unmodified chymotrypsin.[123] Rather more success has been obtained with iodinated chymotrypsin and chymotrypsinogen.[124] Iodination was effected in the crystal with solutions of

[115] F. M. Dickinson, *Biochem. J.*, 1972, **126**, 133.
[116] J. E. Spallholz and L. H. Piette, *Arch. Biochem. Biophys.*, 1972, **148**, 596.
[117] H. Jörnvall and R. Pietruszko, *European J. Biochem.*, 1972, **25**, 283.
[118] H. Jörnvall and O. Markovic, *European J. Biochem.*, 1972, **29**, 167.
[119] H. Jörnvall, *F.E.B.S. Letters*, 1972, **28**, 32.
[120] M. Smith, D. A. Hopkinson, and H. Harris, *Ann. Human Genet. London*, 1972, **35**, 243.
[121] R. Pietruszko, H. Theorell, and C. de Zalenski, *Arch. Biochem. Biophys.*, 1972, **153**, 279.
[122] C. S. Wright, G. P. Hess, and D. M. Blow, *J. Mol. Biol.*, 1972, **63**, 295.
[123] V. K. Antonov and T. J. Vorotyntseva, *F.E.B.S. Letters*, 1972, **23**, 361.
[124] H. T. Wright and B. H. Weber, *European J. Biochem.*, 1972, **24**, 583.

I_3^- and three major sites of iodination were identified, two on tyrosine-146 and one on tyrosine-171. All other iodine was non-covalently bound. The results have led to the proposal that the ionization of a tyrosine residue, probably tyrosine-146, influences the equilibrium between the α- and γ-conformations of chymotrypsin.

It has also been reported that the inactivation of α-chymotrypsin by L-1-tosylamido-2-phenylethyl chloromethyl ketone (TPCK) is not a direct consequence of the modification of the active site of the enzyme.[125] This conclusion was based on the observation that titration of the inactivated protein with diazonium-1-*H*-tetrazole yields too much unsubstituted imidazole, although amino-acid analyses of the TPCK-treated enzyme do not show such a large discrepancy unless *S*-amino-ethylated prior to hydrolysis. It is claimed that the inactivation is not solely the result of modification of His-57 but that other sites must be involved.

An interesting affinity label for proteins in general and chymotrypsin in particular has been devised.[126] This exploits the rod-like properties of acetylenic bonds in the reagent 6-bromo-1-phenylhex-4-yn-3-one. Hopes that the selectivity of the alkylating function would be increased by its spatial orientation in the reagent were realized when it was found that Met-192 was 100 times more reactive than His-57. The specificity of the reaction was established by peptide mapping although the relevant peptide was not actually isolated.

A detailed study of the reaction of *p*-nitrophenyl cyanate with chymotrypsinogen A and chymotrypsin A has been carried out to determine the potential of this reagent in the field of enzyme modification.[127] Under mild conditions the reagent appears to be specific, reacting with Ser-195 and Cys-1 in chymotrypsin Aπ and Aδ, but only with Ser-195 in chymotrypsin Aα. The carbamylchymotrypsin Aα was crystallized in a form isomorphous with native and tosylchymotrypsin Aα, and crystallographic studies were carried out at 2.5 Å resolution.

A novel use of chemical modification of proteins called competitive labelling has been devised to determine the reactivities and pK_a's of specific functional groups in proteins. The theory and methodology of the process have been described previously and it has now been applied to the three amino-termini of α-chymotrypsin.[128] The method has shown that the half cystine-1 amino-terminus of the A-chain has a pK_a of 7.9 and is fully exposed on the surface of the protein; the isoleucine-16 amino-terminus of the B-chain has a pK_a of 8.9 and is buried whereas the alanine-149 amino-terminus of the B-chain has a pK_a of 8.5 and is slightly buried. The method has also been used to determine the ionization constants and reactivities of individual histidines, particularly in chymotrypsin.[129] In this

[125] T. T. Blair, M. A. Marini, and C. J. Martin, *F.E.B.S. Letters*, 1972, **20**, 41.
[126] J. B. Jones and D. W. Hysert, *Biochemistry*, 1972, **11**, 2726.
[127] G. T. Robillard, J. C. Powers, and P. E. Wilcox, *Biochemistry*, 1972, **11**, 1773.
[128] H. Kaplan, *J. Mol. Biol.*, 1972, **72**, 14.
[129] W. H. Cruikshank and H. Kaplan, *Biochem. Biophys. Res. Comm.*, 1972, **46**, 2134.

case tritiated FDNB was used and it was shown that His-57 in α-chymotrypsin has a pK_a (apparent) of 6.8 and a reactivity about ten times greater than α-*N*-acetyl-L-histidine.

Not unexpectedly, few studies have been reported on the primary structure of chymotrypsin. However, the isolation and sequencing of the A-chain from performic acid-oxidized human chymotrypsin II have been reported.[130] The sequence was found to be

$$\text{Cys-Gly-Val-Pro-Ala-Ile-}\begin{pmatrix}\text{Asn}\\ \text{Glu}\end{pmatrix}\text{-Pro-Val-Leu-Ser-Gly-Leu-Ser}$$

which is one residue longer than the A-chain of bovine δ-chymotrypsin and one less than the A-chain of bovine π-chymotrypsin, but identical with the first fourteen residues of bovine chymotrypsinogen A. The absence of the *C*-terminal arginine is attributed to the action of carboxypeptidase B.

The activation of chymotrypsinogen has also come under close scrutiny recently. A new species, κ-chymotrypsin, has been studied[131] and the *N*-terminal amino-acids assigned to the molecule are cystine, isoleucine, and threonine. This species is produced from δ-chymotrypsin at pH 3.1 and it can be converted into both γ-chymotrypsin and α-chymotrypsin by varying the crystallizing conditions. The main conclusion drawn from the study was that it supported the revised activation scheme for chymotrypsinogen in which it is proposed that α-chymotrypsin is produced from neochymotrypsinogen rather than δ-chymotrypsin (Figure 5). In the revised scheme κ-chymotrypsin lies between δ-chymotrypsin and γ-chymotrypsin. The proposal that Arg-145 is important in the activation process has been contradicted by the isolation of a three-chain neochymotrypsinogen in which one chain contains Ser-14 to Tyr-146.[132] Removal of Tyr-146 and Arg-145 with carboxypeptidases A and B does not impair activation of the molecule, showing that Arg-145 is not necessary in the process.

The effect of di-isopropyl phosphorofluoridate on trypsinogen and chymotrypsinogen has been investigated.[133] The reagent was found to react with both zymogens and to inhibit their potential activity, although the rate of reaction was about four orders of magnitude slower than that for the active enzymes. It was shown that in the case of trypsinogen most of the label was associated with the active-site peptides and the conclusion was drawn that the active-centre serine residues are also in an activated state in the zymogen.

Trypsin. The attention paid to trypsin in the literature contrasts sharply with the avalanche of publications on chymotrypsin. Even the study on anhydro-trypsin[134] which is prepared by base elimination of the phenyl

[130] J. Travis, *Biochem. Biophys. Res. Comm.*, 1972, **48**, 1111.
[131] D. D. Miller, T. A. Horbett, and D. C. Teller, *Biochemistry*, 1971, **10**, 4641.
[132] M. Rovery and J. Bianchetta, *Biochem. Biophys. Acta*, 1972, **268**, 212.
[133] P. H. Morgan, N. C. Robinson, K. A. Walsh, and H. Neurath, *Proc. Nat. Acad. Sci. U.S.A.*, 1972, **69**, 3312.
[134] H. Ako, R. J. Foster, and C. A. Ryan, *Biochem. Biophys. Res. Comm.*, 1972, **42**, 1402.

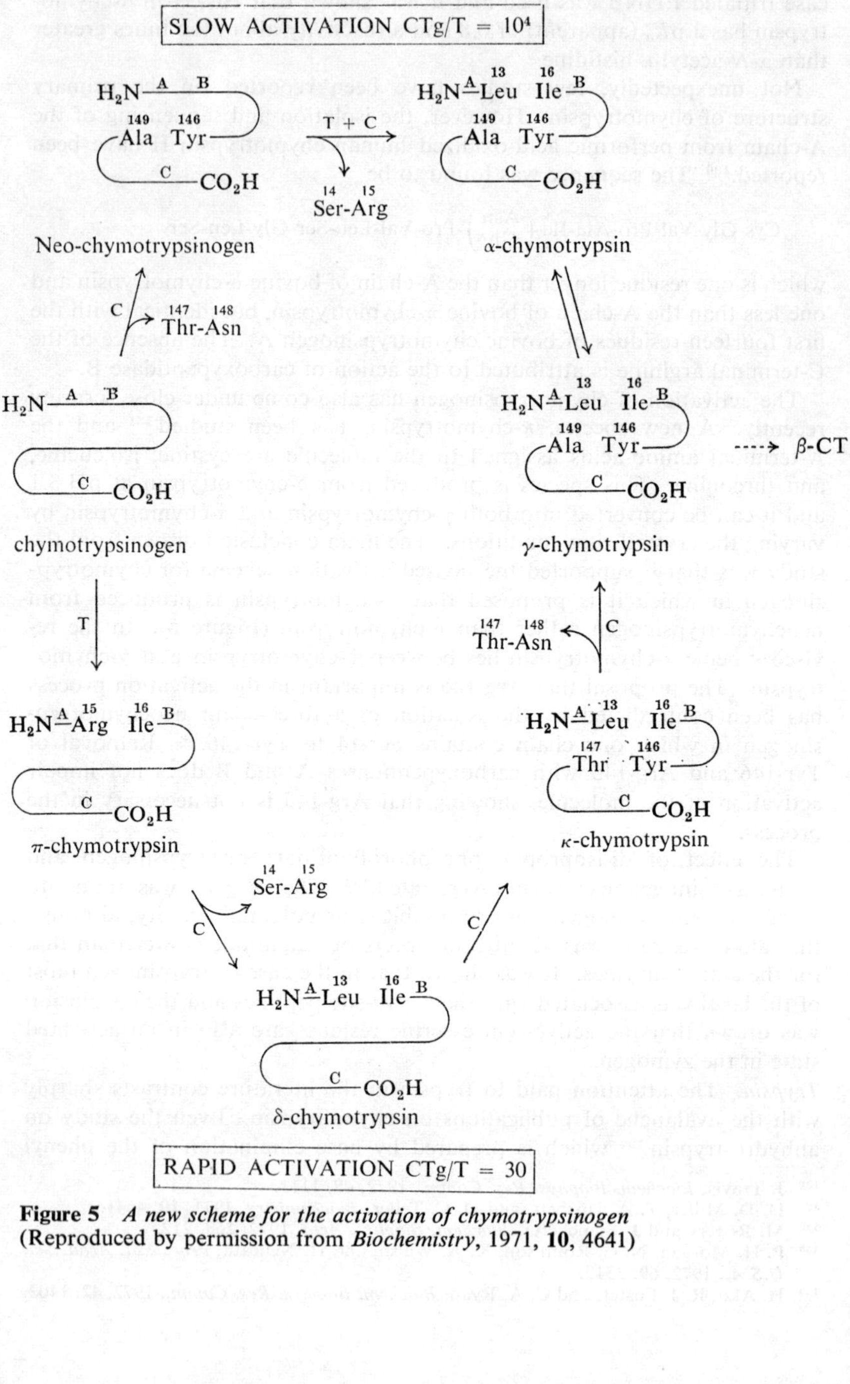

Figure 5 *A new scheme for the activation of chymotrypsinogen*
(Reproduced by permission from *Biochemistry*, 1971, **10**, 4641)

methyl sulphenyl (PMS) group from PMS-trypsin is an extension of the work described previously with anhydro-chymotrypsin. Several naturally occurring trypsin inhibitors were found to bind strongly to the anhydro-trypsin, confirming that in trypsin too, the covalent acyl attachment of the naturally occurring inhibitors to the enzyme through the reactive serine is not necessary for strong binding. The modification of trypsin by diazo-benzenesulphonic acid has been studied and it was found that tyrosine and lysine are rapidly modified whereas histidine reacts slowly.[135] The modified enzyme appeared to be less active towards the hydrolysis of *N*-α-benzyl-L-arginine *p*-nitroanilide than towards *N*-α-benzylarginine methyl ester.

An interesting report on the activation peptide from lungfish trypsinogen has been published.[136] This organism lies between the vertebrates and invertebrates in the evolutionary scale and this is reflected at the molecular level too since its trypsin is stable at neutral pH in the absence of calcium, like invertebrate trypsins, as well as being stable at pH 3, like the mammalian trypsins. However, the unusual feature lies in the activation peptide, which has the sequence

Phe-Pro-Ile-Glu-Glu-Asp-Lys

and therefore differs from all previous trypsinogens in lacking the tetra-aspartyl sequence. This sequence matches the composition of the *N*-terminal heptapeptide and since activation of lungfish trypsinogen is also promoted by Ca^{2+} it confirms the view that only three acidic residues are needed for the calcium response.

Subtilisin. The complete amino-acid sequence of subtilisin from *Bacillus amylosacchariticus* has been reported during the year.[137] Ordering of the peptides was aided by comparison with the sequence of subtilisin BPN, which appears more closely related evolutionarily to subtilisin amylosacchariticus than does subtilisin Carlsberg. If the number of amino-acid differences is taken as the sole criterion of evolutionary distance it seems that subtilisin amylosacchariticus and subtilisin BPN are about the same distance from subtilisin Carlsberg since there are about 85 substitutions in both cases. The high degree of conservation of glycine between subtilisin amylosacchariticus and subtilisin BPN lends some support to an earlier suggestion that this amino-acid is particularly important in the conformation of the protein. The most conservative regions are from residues 218—240 and 64—75, which contain the reactive residues serine-221 and histidine-64, respectively. Some support for the view that aspartic acid-32 is a functionally important residue comes from the fact that it is contained in a region (residues 9—42) which shows no differences between subtilisins amylosacchariticus and BPN. The existence of repetitions of sequence within subtilisin BPN and Carlsberg led to the original hypothesis that subtilisins may have evolved by serial duplication and fusion.

135 C. A. Bauer and G. Ehzensvard, *Acta Chem. Scand.*, 1972, **26**, 1209.
136 G. R. Reeck and H. Neurath, *Biochemistry*, 1972, **11**, 503.
137 F. S. Markland, D. M. Brown, and E. L. Smith, *J. Biol. Chem.*, 1972, **247**, 5596.

Similar repetitions also appear in the sequence of subtilisin amylosaccharicus. However, it has also been shown that the repetitions observed could have arisen by chance alone,[4] illustrating the danger of employing mere inspection of sequences for this purpose.

The low activity of subtilisin amylosacchariticus towards aromatic ester substrates previously suggested that the mechanism of the enzyme might be different to that of other subtilisins. It has now been found that the enzyme is also less sensitive to the bromomethyl ketone of benzyloxycarboxy-L-phenylalanine (ZPBK), which reacts specifically with a single histidine residue in other subtilisins.[138] However, there is a corresponding loss of esterase activity, which leads to the view that a catalytically active histidine is present in the molecule, but that its orientation is such as to make it less reactive towards aromatic amino-acid esters and ZPBK.

It has been reported that PMS-subtilisin produced as a result of the reaction of phenylmethylsulphonyl fluoride with subtilisin can be reactivated by either formohydroxamic acid or hydroxylamine.[139] Complete reactivation was effected by incubating PMS-subtilisin with 0.5M formohydroxamic acid but, in contrast to diethylphosphoryl-α-chymotrypsin, di-isopropylphosphorylsubtilisin was not reactivated by treatment with this acid.

Carboxypeptidases. Carboxypeptidase A. The structure and function of carboxypeptidase A are the subjects of two excellent reviews which have appeared recently.[140, 141] There have also been several reports of exceptional quality on the properties of the chemically modified bovine carboxypeptidase A. A study on the coupling of crystals of the enzyme with diazotized-*p*-arsanilic acid is exemplary in its combined use of some of the most sophisticated techniques in protein and peptide chemistry.[142] The diazonium salt was used because it introduces an environmentally sensitive chromophore which can be used in studies on structure–function relationships. The object of the study was to determine which residues were labelled by the reagent and to establish their yields. In order to circumvent the losses normally experienced in peptide purification, use was made of affinity chromatography with specific antiarsanilazotyrosyl antibodies. The protein was labelled in the crystal, and spectrophotometric analysis showed one residue of arsanilazotyrosine per molecule of enzyme. The amino-acid analyses showed no changes in the content of histidine and lysine, testifying to the specificity of the method. The cyanogen bromide fragments were prepared and separated and the peptide containing arsanilazotyrosine was solubilized by succinylation and redigested with chymotrypsin. The chymotryptic digest was then adsorbed on to the antibody–

138 F. S. Markland, M. Kurihara, and E. L. Smith, *J. Biol. Chem.*, 1972, **247**, 5619.
139 N. Uemitsu, M. Sugiyama, and H. Matsumuja, *Biochim. Biophys. Acta*, 1972, **258**, 562.
140 W. N. Lipscomb, *Chem. Soc. Rev.*, 1972, **1**, 319.
141 F. A. Quiocho and W. N. Lipscomb, *Adv. Protein. Chem*, 1971, **25**, 1.
142 J. I. Johansen, D. M. Livingston, and B. L. Vallee ,*Biochemistry*, 1972, **11**, 2584.

Sepharose column and the arsanilazotyrosine peptide eluted in an almost pure state from the column. Further purification was carried out by ion-exchange chromatography. The success of this approach is evident from the fact that the arsanilazotyrosine peptide was obtained with a yield of over 90%. The analysis of the peptide showed that it corresponds to the sequence 246—257:

Thr-Ile-Tyr-Gln-Ala-Ser-Gly-Gly-Ser-Ile-Asp-Trp

which implicates Tyr-248 as the only site labelled during the modification. However, it should be borne in mind that this exceptional specificity is due to the fact that the modification is carried out on the crystalline protein. An independent study on carboxypeptidase A in solution shows that tyrosines-19, -277, and -248 are all modified, albeit partially. In addition to carboxypeptidase A, which was used in the above study, carboxypeptidase Aα has been similarly investigated.[143] The significance of these data is not entirely clear at the moment as the results from the two forms of the enzyme are not identical.

Carboxypeptidase B. Although the entire sequence of the enzyme has not yet been determined, partial sequences have suggested that it has a common evolutionary origin with carboxypeptidase A. This is also reflected in the similarity between the two enzymes in their behaviour towards modifying reagents. When bovine carboxypeptidase B reacts with *N*-bromoacetyl-*N*-methyl-L-phenylalanine (L-BAMP) the enzyme is irreversibly inactivated and there is a parallel loss of both esterase and peptidase activity.[144] The competitive inhibitor ε-aminocaproate protects against this inactivation, suggesting that the active site is involved in the modification. A thermolysin digest of the enzyme modified with radioactive L-BAMP gave a single radioactive dipeptide which was purified and found to contain one residue of glutamic acid and one of phenylalanine. A single Edman degradation removed most of the phenylalanine, indicating that the sequence was Phe-Glu and that the inactivation results from the specific alkylation of the γ-carboxylate of the glutamic acid. The same result was obtained from similar studies with carboxypeptidase A, where it was possible to identify positively the modified residue as Glu-270 since there is only one Phe-Glu sequence in the enzyme. The absence of the entire sequence for carboxypeptidase B precludes a positive assignment but there is no doubt that the similarity between the enzymes is striking. Modification of porcine carboxypeptidase B with tetranitromethane has also been carried out and the loss of activity correlated with the nitration of a single tyrosyl residue.[145] The nitrotyrosyl-containing chymotryptic peptide was isolated and found to have the sequence

Thr-Ile-Tyr-Pro-Ala

143 F. A. Quiocho, C. H. McMurray, and W. N. Lipscomb, *Proc. Nat. Acad. Sci. U.S.A.*, 1972, **69**, 2850.

144 G. M. Haas, M. A. Govier, D. T. Grahn, and H. Neurath, *Biochemistry*, 1972, **11**, 3787.

145 M. Sokolovsky, *European J. Biochem.*, 1972, **25**, 267.

which is homologous with the peptide containing Tyr-248 in the active-centre sequence of bovine carboxypeptidase A. Porcine carboxypeptidase B has been treated with phenylglyoxal and the modified enzyme found to have a decreased peptidase and esterase activity.[146] It appears that the loss of activity correlates with the modification of about one arginyl residue in the protein. However, the modified residue was not identified so it is not known whether the modification is specific.

Studies on the primary structures of proteolytic enzymes from the African Lungfish, *Protopterus acthiapicus*, have been extended to carboxy-peptidase B.[147] The molecular weight and amino-acid composition closely resemble those from mammalian carboxypeptidases B, and the *N*-terminal sequence displays considerable similarity to those of bovine carboxy-peptidases B and A.

Procarboxypeptidases. The changes in primary structure which occur upon activation of procarboxypeptidases have also been investigated.[148] Bovine procarboxypeptidase A exists in two forms with sedimentation coefficients of 5S ($PCPAS_5$) and 6S ($PCPAS_6$). In contrast to zymogens from other species (human, dogfish, lungfish) $PCPAS_5$ contains two subunits (I and II) whereas $PCPAS_6$ has three (I, II, and III). Subunit I is the immediate precursor of carboxypeptidase A and subunit II is the zymogen of an endopeptidase similar to chymotrypsin. The interest in the oligomeric zymogens stems from their extremely slow activation rates compared with those of the momomeric zymogens.

The intact subunits which have a high tendency for association have lysine (Subunit I) and half-cystine (Subunit II) as amino-termini. Trypsin at low concentrations activates Subunit II by cleavage at a Val-Arg bond, converting it into an enzyme like chymotrypsin. At this stage it is possible to dissociate the subunits by the addition of 1M-calcium ions. Further incubation leads to the generation of carboxypeptidase A from Subunit I. The nature of the product depends on the conditions used; mere incubation of the activation mixture (which contains the active Subunit II) as such, or with added chymotrypsin, yields carboxypeptidase A, which has the *N*-terminal sequence Ala-Arg-Ser. However, the presence of trypsin in the activation mixture also yields carboxypeptidase A in which the Ala-Arg peptide has been excised. An activation scheme has been proposed on the basis of sequence studies on the activation products. The isolation and characterization of carboxypeptidases A and B from activated bovine pancreatic juice have also been described.[149] Three forms of carboxy-peptidase B are produced during this endogenous activation, the two major forms being two-chain enzymes. Sequenator analysis showed that the light chains of the latter (molecular weight 10 000) correspond to the amino-terminal region of the intact enzyme.

[146] M. M. Weber and M. Sokolovsky, *Biochem. Biophys. Res. Comm.*, 1972, **48**, 384.
[147] G. R. Reeck and H. Neurath, *Biochemistry*, 1972, **11**, 3947.
[148] J. R. Uren and H. Neurath, *Biochemistry*, 1972, **11**, 4483.
[149] G. R. Reeck, K. A. Walsh, and H. Neurath, *Biochemistry*, 1971, **10**, 4690.

Other Proteases. Thermolysin. The elucidation of the full three-dimensional structure of thermolysin was the successful culmination of collaboration between Neurath's group who sequenced the protein and Matthews' group who constructed its three-dimensional structure. The series of publications [150–154] makes an impressive story. The sequencing of the protein was carried out by conventional means although more than half the sequence was confirmed by the automatic sequencer. Some unusual distributions of amino-acid residues are apparent in the sequence. The region between residues 101 and 182 is extremely acidic since it contains 10 carboxyl residues and no arginine or lysine, whereas the region from residues 227—316 contains 4 carboxy-groups and 10 positively charged residues. Comparison of the sequence of thermolysin with that of bovine carboxypeptidase, which it resembles in several respects, showed no significant structural homology and there was no evidence for internal duplication. Some of the residues involved in the active site have been identified from the three-dimensional structure. Thus the three zinc ligands were identified as His-142, His-146, and Glu-166. The same combination of zinc ligands is found in bovine carboxypeptidase A, and other possible similarities have been enumerated. Thus Glu-143 could be the counterpart of Glu-270 in bovine CPA and Arg-203 could correlate with Arg-145 in CPA. The only possible counterpart in thermolysin of Tyr-248 in CPA is Tyr-157, but the positions of the two do not correspond. The most striking difference between the two enzymes lies in the fact that the imidazole ring of His-231 is in front of the zinc atom in thermolysin, and forms a salt-link with Asp-226. Such an interaction has previously only been observed in the serine proteinases. Thus the possibility that thermolysin and carboxypeptidase A bear the same relationship to one another as subtilisin does to the mammalian serine proteinases has not been realized.

The main limitation of the above study is that it has not permitted the assignment of the catalytically functional residues at the active site. Some evidence for the involvement of tyrosyl residues in the function of the enzyme has been reported in a study of the effect of *N*-acetylimidazole on the enzyme.[155] However, the conformational stability of the enzyme was also affected by the extensive acetylation, so it is difficult to place much significance on these data.

Pepsin. The effect of chemical modification on the structure and function of pepsin has been studied with epoxy reagents which modify arginine

[150] K. Titani, M. A. Hermodson, L. H. Ericsson, K. A. Walsh, and H. Neurath, *Nature New Biol.*, 1972, **238**, 35.

[151] B. W. Matthews, J. N. Jansonius, P. M. Colman, B. P. Schoenborn, and D. Dupourque, *Nature*, 1972, **238**, 37.

[152] B. W. Matthews, P. M. Colman, J. N. Jansonius, K. Titani, K. A. Walsh, and H. Neurath, *Nature New Biol.*, 1972, **238**, 35.

[153] P. M. Colman, J. N. Jansonius, and B. W. Matthews, *J. Mol. Biol.*, 1972, **70**, 701.

[154] K. Titani, M. A. Hermodson, L. H. Ericsson, K. A. Walsh, and H. Neurath, *Biochemistry*, 1972, **11**, 2427.

[155] Y. Ohta, H. Nakamura, and T. Samejima, *J. Biochem.*, 1972, **72**, 521.

residues. In the first case porcine pepsin was treated with 1,2-epoxy-3-(*p*-nitrophenoxy)propane and the enzyme was inactivated when about 2 moles of epoxide were incorporated.[156] The epoxide-containing peptides obtained from peptic, subtilisin, thermolysin, and chymotryptic digests of the modified protein were isolated and used to determine the sequences around the reactive sites. Only two unique peptides could be identified in all cases and their sequences were

I Phe-Glu-Gly-**Met**-Asp-Val-Pro-Thr-Ser-Ser-Gly-Glu-Leu
II. Ile-Val-**Asp**-Thr-Gly-Ser-Asn

The modified residues are methionine in sequence I and aspartic acid in sequence II. Comparison with existing sequence data shows that the methionyl residue in sequence I is located 38 residues from the carboxy-terminus of the enzyme. It was not possible to locate the site of the modified aspartyl residue within any known sequence. Although the sequence around this residue showed remarkable homology with the active-centre aspartyl residue which is reactive towards diazo inactivators, it does not appear to be the same peptide since the epoxide-modified pepsin still reacts with the diazo reagent.

The above study has been extended to the selective modification of the aspartyl residue in sequence II to determine its function in the active centre of pepsin.[157] Use of a lower temperature led to the incorporation of one mole of epoxide per mole of pepsin when almost total inactivation had occurred. The sequence of the modified peptide corresponded to sequence II, and it was concluded that the aspartyl residue in this sequence is associated with the active centre of pepsin. It is interesting that there are at least two carboxyl residues in the active centre of pepsin since the possibility of this has been speculated upon previously.

The modification of arginyl residues in pepsin has been effected with 2,3-butanedione (biacetyl), the choice of this reagent being favoured by its reactivity under mild conditions in a near neutral solution.[158] Inactivation was slow but amino-acid analysis showed the loss of one arginine residue, and its modification paralleled the loss of activity. In contrast, pepsinogen treated with biacetyl at pH 7 was not modified significantly. The modified residue in pepsin was located in the sequence

Asp-Arg-Ala-Asn-Asn-Lys-Val-Gly-Leu-Ala-Pro-Val-Ala

Since this is the *C*-terminal sequence the modified arginine is located 12 residues from the *C*-terminus of the enzyme. The other arginine residue in the molecule remained unchanged. It is not likely that the modified arginine residue participates in the enzymic mechanism since the modified pepsin retains some activity.

156 K. C. S. Chen and J. Tang, *J. Biol. Chem.*, 1972, **247**, 2566.
157 J. Hartsuck and J. Tang, *J. Biol. Chem.*, 1972, **247**, 2576.
158 W. Y. Huang and J. Tang, *J. Biol. Chem.*, 1972, **247**, 2704.

Additional information on the sequence of the *N*-terminal cyanogen bromide fragment from pepsin has been obtained.[159] Three tryptic peptides were isolated from a digest of the aminoethylated fragment, one containing the *N*-terminal sequence and aminoethylcysteine whilst a second short peptide contained aminoethylcysteine at the *C*-terminus. The third large tryptic fragment contained homoserine and was assigned to the *C*-terminus of the cyanogen bromide fragment. The three tryptic fragments were aligned on the basis of this information although the sequence of the cyanogen bromide fragment is not yet complete. The phosphoserine residue was located in the third fragment so it appears to lie between the second half-cystine and the first methionine in pepsin.

The conversion of pepsinogen into pepsin has also been investigated by modification of the lysine residues in pepsinogen.[160] The reversible blocking agents maleic and citraconic anhydride were used and it was found that nine lysine residues and the α-amino-group in pepsinogen were blocked and that the pepsinogen derivatives were stable under neutral conditions. Exposure of the derivatives to acid resulted in the recovery of the pepsin activity. However, not all the potential activity could be recovered and it is argued that this implies a conformational change in the zymogen which precludes complete activation of the unblocked zymogen. The effect of gossypol (1,1′,6,6′,7,7′-hexahydroxy-3,3′-dimethyl-5,5′-di-isopropyl-2,2′-binaphthyl-8,8′-dialdehyde), which prevents activation of pepsinogen, has also been investigated.[161] The gossypol–pepsinogen, on digestion with Nagarse, yielded a decapeptide from the *N*-terminal portion of the molecule and a heptapeptide containing the *C*-terminus, both of which were associated with gossypol attached to the protein. It was also shown that the gossypol forms a predominantly intramolecular link between lysine-18 and lysine-358 in the zymogen. Therefore the modification could prevent the separation of the *N*- and *C*-terminal regions of pepsinogen, which is thought to be necessary for activation. It has also been reported that pure swine pepsin and pepsinogen contain less than 0.02 moles of carbohydrate per mole of protein.[162] This contradicts a suggestion that the acidolysis of glycoprotein bonds could be the triggering step in the activation of pepsinogen.

Papain. The state of the thiol group in cysteine-25 of papain, which becomes acylated during catalysis, has been investigated with 2,2′- and 4,4′-dipyridyl disulphide.[163, 164] The results are consistent with the view that the thiol group is in its unionized form, probably as a result of its interaction with the histidine-159–asparagine-175 system. This study has

[159] L. Moravek, *F.E.B.S. Letters*, 1972, **23**, 337.
[160] Y. Nakagawa and G. E. Perlmann, *Arch. Biochem. Biophys.*, 1972, **149**, 476.
[161] R. C. Wong, Y. Nakagawa, and G. E. Perlmann, *J. Biol. Chem.*, 1972, **247**, 1625.
[162] V. M. Stepanov, E. A. Timokhina, L. A. Baratova, L. P. Belyanova, V. P. Korzhenko, and I. G. Zhukova, *Biochem. Biophys. Res. Comm.*, 1971, **45**, 1482.
[163] K. Brocklehurst and G. Little, *Biochem. J.*, 1972, **128**, 471.
[164] G. Little and K. Brocklehurst, *Biochem. J.*, 1972, **128**, 475.

been extended to stem bromelain, which resembles papain in some respects.[165] However, it appears that the reaction of bromelain with 2,2′-dipyridyl disulphide depends on its method of preparation. Use of Amberlite CG-50 leads to preparations which are potentially capable of rapid reaction with the reagent. On the other hand, elimination of the Amberlite chromatography yields an unreactive species, indicating an alteration of the enzyme during that step.

Papain has also been modified with α-bromo-β-(5-imidazolyl)propionic acid (BIP) and the modification was shown to be the alkylation of cysteine-25, the catalytically important sulphydryl group.[166] The rationale behind the experiment is interesting. It has been known for some time that papain has one imidazole (histidine-159) at the active site. The reagent used in this study replaces a thiol group with an imidazole residue, thus producing a protein with two imidazole groups in close proximity to one another. Presumably this has been achieved, although the modified papain did not possess ribonuclease activity as hoped for.

The involvement of lysine residues in the function of the acid protease from *Mucor michei* has been suggested on the basis of experiments in which the protein was modified with cyanate.[167] However, although there was a correspondence between the carbamylation of the five lysine residues concerned and the loss of activity, it is unlikely that a specific lysine is present in the active site since the fully modified enzyme retains some activity. Furthermore, the number of lysine residues is too great to permit useful interpretation of the results.

The active site of clostripain, a sulphydryl protease from *Clostridium histolyticum*, has been investigated with α-*N*-tosyl-L-lysyl chloromethyl ketone (TLCK), since the enzyme only hydrolyses substrates containing arginine or lysine.[168] The enzyme was completely inhibited by a molar equivalent of the inhibitor and the kinetics of inactivation were coincident with those of TLCK incorporation. However, no attempt was made to map the site of modification.

Awamorin is an acid protease from *Aspergillus awamori* which resembles pepsin in many of its properties. Treatment with *N*-diazoacetyl-*N*′-2,4-dinitrophenylethylenediamine results in the specific modification of a single aspartyl residue in the protein.[169] The modified residue was located in a peptic tripeptide with the sequence

Ile-Ala-Asp

This sequence has also been found at the reactive sites of pepsin and penicillopepsin, showing that awamorin is probably homologous with the former two enzymes.

165 K. Brocklehurst, E. M. Crook, and M. Kierstan, *Biochem. J.*, 1972, **128**, 979.
166 C. J. Jolley and J. A. Yankeelov, *Biochemistry*, 1972, **11**, 164.
167 W. S. Rickert, *Biochim. Biophys. Acta*, 1972, **271**, 93.
168 W. H. Porter, L. W. Cunningham, and W. M. Mitchell, *J. Biol. Chem.*, 1971, **246**, 7675.
169 G. G. Kovaleva, M. P. Shimanskaya, and V. M. Stepanov, *Biochem. Biophys. Res., Comm.*, 1972, **49**, 1075.

Proteinase Inhibitors. Interest in the proteinase inhibitors has been gaining momentum with a vigour reminiscent of the era of proteinase structure determination. There are several reasons for this. Since they are involved in highly specific interactions with certain proteinases, the proteinase inhibitors are useful tools for the study of protein–protein interactions. The availability of three-dimensional structures for the proteinase components, the prospects of structures for the inhibitors, and a good understanding of the components of the active sites of the proteinases have increased the feasibility of such studies. A considerable advance in the understanding of these inhibitors came from the discovery that the cognate proteinases, usually trypsin or chymotrypsin, actually cleave bonds in the inhibitor.[170] Much of the protein chemistry done on these enzymes has therefore been concerned with analysing these splitting reactions.

Soybean trypsin inhibitor (*Kunitz*). Previous studies of this inhibitor have shown that it contains an area, the reactive site, which comes into close contact with the active site of trypsin in the stable trypsin–inhibitor complex.[170] The bond Arg-63—Ile-64 has been shown to be involved in the interaction since (i) incubation of virgin inhibitor with catalytic quantities of trypsin leads to specific hydrolysis of the Arg-63—Ile-64 bond in the inhibitor; both virgin and modified inhibitor are active but the former reacts more rapidly, (ii) removal of the newly formed *C*-terminal Arg-63 with carboxypeptidase B produces an inactive inhibitor, as does chemical modification of the newly formed *N*-terminal Ile-64, (iii) the complexes formed by trypsin with either virgin or modified inhibitor are indistinguishable, and (iv) controlled dissociation of these complexes leads to the formation of trypsin and predominantly virgin inhibitor.

The weak interaction of soybean trypsin inhibitor with chymotrypsin has now been investigated.[171] Tosyl-lysine chloromethyl ketone-treated bovine α-chymotrypsin was used and it was found that the virgin inhibitor was converted into a form similar to the trypsin-modified inhibitor. Several lines of evidence showed that contaminating trypsin was not the cause of this conversion, *e.g.* TPCK-treated α-chymotrypsin did not catalyse the conversion. However, re-association experiments with trypsin strongly suggested that the bond cleaved by chymotrypsin was the same as that normally cleaved by trypsin, *i.e.* Arg-63—Ile-64. This was also supported by the observation that carboxypeptidase B quantitatively removed a basic residue from chymotrypsin-modified inhibitor. A study was made of the complexes of chymotrypsin with virgin and modified inhibitor to show that the Arg-63—Ile-64 bond was not only hydrolysed by chymotrypsin but was also in the reactive-site peptide for chymotrypsin. The results are consistent with the formation of a 1 : 1 complex between chymotrypsin and modified inhibitor, but the weakness of the association

170 M. Laskowski and R. W. Sealock in 'The Enzymes', ed. P. D. Boyer, 1971, p. 376.
171 U. De Vonis Bidlingmeyer, T. R. Leary, and M. Laskowski, *Biochemistry*, 1972, **11**, 3303.

and the long incubation time required for complex formation make the quantitative interpretation of the data more difficult. A major consequence of this work is that it invalidates the approach of using binding to trypsin inhibitors as a criterion for trypsin-like enzymes.

Chemical modification has been used to study the reactive site of soybean trypsin inhibitor.[172] The reactive amino-groups in the virgin inhibitor were blocked by guanidination with *O*-methylisourea of the ε-amino-groups and carbamylation of the α-amino-group. The blocked inhibitor was fully active and yielded modified inhibitor (Arg-63—Ile-64 hydrolysed) on treatment with trypsin. The new amino-group in modified inhibitor was then blocked by citraconylation, amidination, or carbamylation. These procedures were used since they yield substituents containing a negative, positive, or no charge respectively. The results show that any of the substituents on the newly exposed α-amino-group of Ile-64 lead to quantitative loss of activity, demonstrating the need for a free α-amino-group in this position. This contradicts some previous reports that a free α-amino-group in the reactive site of modified inhibitors is not essential.

Bowman–Birk soybean proteinase inhibitor. This inhibitor differs functionally from the Kunitz inhibitor since it inhibits both trypsin and chymotrypsin. Its minimum molecular weight is about 8000 but it has a tendency to dimerize. Early studies showed that the protein had a high cysteine content (*ca.* 20%) and no glycine.[173] The full sequence of the protein has now been determined.[174] The approach used for assigning overlaps is interesting. First, the tryptic peptides from a digest of the *S*-methoxycarbonylated protein were purified and sequenced. The native protein was then subjected to a limited hydrolysis with chymotrypsin, which caused cleavage of a single peptide bond and yielded two fragments. The larger fragment contained one methionine residue and was cleaved with cyanogen bromide. A series of *N*- and *C*-terminal analyses on the fragments together with the sequence of the tryptic peptides was sufficient to determine unequivocally the sequence of the protein. In addition, it was evident that the limited hydrolysis by chymotrypsin had occurred between Leu-43 and Ser-44. It was also shown that trypsin cleaves the inhibitor between Lys-16 and Ser-17. The assignment of a Leu—Ser bond as the cleavage site by chymotrypsin has also been confirmed but the position of the bond cleaved is thought to be between Leu-48 and Ser-49.[175] A comparison of the two studies shows major discrepancies in the analyses of the larger chymotryptic fragment. Thus the analyses obtained by Odani and Ikenaka gave a total of 43 residues for the large fragment and 28 residues for the small one whereas those of Seidl and Liener show 48 residues in the larger and 23 in the smaller. Since the overall analyses seem the same in both cases it appears that two

[172] D. Kowalski and M. Laskowski, *Biochemistry*, 1972, **11**, 3451.
[173] S. Odani, T. Koide, and T. Ikenaka, *J. Biochem.*, 1972, **71**, 831.
[174] S. Odani and T. Ikenaka, *J. Biochem.*, 1972, **71**, 839.
[175] D. S. Seidl and J. E. Liener, *Biochim. Biophys. Acta*, 1972, **258**, 303.

residues of aspartic acid and one residue each of glutamic acid, proline, and lysine have been exchanged between the two fragments. It is not possible to decide on the available evidence just why the two results are in conflict but some doubt must remain about the sequence and the site of chymotryptic cleavage. It is also interesting to note that Odani and Ikenaka find the trypsin cleavage site between Lys-16 and Ser-17 whereas Seidl and Liener report it to be between Lys-17 and Ser-18.

In spite of the conflicting evidence described above, the two studies confirm that there is a high degree of homology between the Bowman–Birk inhibitor and that from lima bean. This is also reflected in the cleavage sites since the chymotrypsin-sensitive site of the latter is a Ser-55—Ser-56 bond and the trypsin-sensitive site is a Lys-28—Ser-28 bond. In addition there are other regions of extensive homology. As in the case of the lima bean inhibitor there is evidence for a repetition of a similar sequence in two separate portions of the molecule. Thus the distribution of half-cystine residues in positions 9—26 is quite similar to that in positions 36—53. It is interesting that these two regions contain the tryptic and chymotryptic sensitive sites respectively. It is possible therefore that the protein has evolved from a univalent precursor which underwent duplication and modification to accommodate the extra function. It has also been noted that an octapeptide sequence common to the two inhibitors has been found in peanut inhibitor whose overall sequence is quite distinct from the other two:

BBI 16—23	Lys-Ser-Asn-Pro-Pro-Glu-Cys-Arg
LBI IV	Lys-Ser-Ile -Pro-Pro-Glx-Cys-Arg
Peanut inhibitor	Arg-Ser-Asx-Pro-Pro-Glx-Cys-Arg

If this is a geniune case of convergent evolution the implication is that these octapeptide sequences play an essential role in anti-proteinase activity.

The positive identification of the antitrypsin site in lima bean protease inhibitor has been carried out.[176] It had previously been suggested on the basis of the homology between the repetitive regions 50—61 and 23—24, the former having been shown to contain the anti-chymotrypsin site, and the knowledge that the anti-trypsin site contained a Lys—X bond, that the bond cleaved was between Lys-28 and Ser-29. This has now been confirmed by end-group analysis of the fragments obtained from tryptic cleavage.

Some information on the mechanism of combination of the Bowman–Birk inhibitor with trypsin and chymotrypsin has been obtained with tetranitromethane-treated inhibitor.[177] Only one of the two tyrosines in the inhibitor is modified and there is no change in its inhibitory activity towards either trypsin or chymotrypsin. However, it was found that the p*K* of the nitrotyrosine was 7.4 for the free inhibitor and 6.8 in the presence of

[176] J. Krahn and F. C. Stevens, *Biochemistry*, 1972, **11**, 1804.
[177] R. F. Steiner, *European J. Biochem.*, 1972, **27**, 87.

excess trypsin or chymotrypsin. It is argued that the shift in p*K* is large enough to conclude that the tyrosine residue concerned is close to the zone of interaction between proteinase and inhibitor.

Pancreatic trypsin inhibitor. Interest in this protein is considerable since the publication of its structure.[178] A study on the role of the disulphide bridges in the interaction with trypsin has been carried out.[179] The Cys-14—Cys-38 bridge, which is adjacent to Lys-15 in the inhibitor structure, is very easily reduced in the native inhibitor but is masked in the trypsin–inhibitor complex. Selective reduction of the bridge with sodium borohydride does not prevent association with trypsin, but the dissociation constant of the complex is several orders of magnitude higher than that of the trypsin–inhibitor complex. After reduction of the disulphide bridge, chemical modification of the thiol groups with iodoacetamide, iodoacetic acid, or ethyleneimine yielded derivatives which had much weaker associations with trypsin than native inhibitors. Thus the integrity of the Cys-14-—Cys-38 bridge appears essential for a tight association with trypsin.

Another study of the effect of chemical modification on the conformational properties of the bovine pancreatic trypsin inhibitor has been reported.[180] The emphasis is on the unusual conformational stability of the inhibitor which remains folded both in 6M-guanidine–HCl and at 77 °C. However, selective reduction of the disulphide bridge between Cys-14 and Cys-38 reduced the transition temperature of the molecule by 21 °C at pH 2.1. Similar properties were observed when the reduced protein was alkylated and it was found that the alkylated inhibitor was completely unfolded in 6M-guanidine–HCl. Thus the loss of a disulphide bond appears to increase the flexibility of the protein considerably. The protein also undergoes a structural change which occurs between pH 5 and 2 and it appeared likely that the ionization of a carboxy-group was involved in the transition. Chemical modification was used to investigate this possibility. The carboxy-groups were converted into amides by treatment with glycine ethyl ester and 1-ethyl-3-(3-dimethylaminopropyl)carbodi-imide. The unmodified inhibitor incorporated 2.8 glycines in 6M-guanidine–HCl whereas the alkylated inhibitor incorporated 4.5 glycine residues under the same conditions. The modification of the 4.5 carboxylate groups resulted in a loss of inhibitory capacity since it required two moles of the modified inhibitor to inhibit one mole of trypsin. It is argued that this implies only one of the masked carboxy-groups to be important, both structurally and functionally, and that it is the less reactive of the two towards the soluble carbodi-imide. By means of a rather tortuous argument it is concluded that the most likely acidic residue to be involved in a structural role is Glu-7.

[178] R. Huber, D. Kukla, A. Ruhlmann, and W. Steigman, *Cold Spring Harbor Symp. Quant. Biol.*, 1971, **36**, 141.
[179] J. P. Vincent and M. Lazdunski, *Biochemistry*, 1972, **11**, 2967.
[180] J. P. Vincent, R. Chicheportiche, and M. Lazdunski, *European J. Biochem.*, 1971, **23**, 401.

It has also been shown that a trypsin inhibitor is present in bovine ovaries.[181] The protein was isolated by affinity chromatography through trypsin-Sepharose and found to have the same amino-acid composition as the basic pancreatic trypsin inhibitor. Tryptic peptide mapping gave the same nine peptides and the sequences of several tryptic peptides were found to be homologous with those from the pancreatic inhibitor. Furthermore, the first 17 amino-acids, determined on an automatic sequencer, were found to be the same as those in the pancreatic inhibitor. Not unexpectedly, it was concluded that the two inhibitors were identical.

Aminoacyl-tRNA Synthetases.—These enzymes are also called amino-acid:tRNA ligases, but the more commonly used nomenclature will be used in this discussion. It has been realized from the time of their discovery that the synthetases have considerable potential for studying protein–nucleic acid interactions. Unlike many other systems both components in the system can be isolated pure and studied separately in addition to being investigated as the complex. However, the main limitation has been the lack of availability of suitable quantities of the enzymes since they are not present in large amounts in cells. As a result several other approaches have been employed to circumvent the problem. The most obvious of these has been the use of large amounts of starting material, which in spite of its somewhat pedestrian nature has proved fairly successful.[182] A more useful approach has been employed with methionyl-tRNA synthetase from *E. coli*. A strain containing an episome was isolated and found to contain about four times as much methionyl-tRNA synthetase activity as wild-type *E. coli*, indicating that at least one extra copy of the structural gene for the enzyme exists on the episome.[183] It is particularly useful in this case, since excellent crystals have been obtained from a fragment of the enzyme and structural studies are imminent.[184] It has also been shown that the growth conditions have a significant effect on the amount of aminoacyl-tRNA synthetase present in cells.[185] Up to a four-fold increase in several synthetases was observed when the growth rate was relatively fast, and the amount of enzyme in the cell fell off rapidly in slow-growing cultures. It should be possible to exploit this observation and get better yields of these enzymes. The obvious appeal of affinity chromatography for the isolation of specific synthetases has not been properly exploited. Each enzyme has two specific ligands so that it should in principle be possible to use these for affinity binding. However, some properties of the enzymes themselves interfere with this strategy. The major problem is that in all cases investigated so far the amino-group of the amino-acid appears essential for binding and is therefore not available for modification.[186] Although tRNA can

181 J. Chauvet and R. Acher, *F.E.B.S. Letters*, 1972, **23**, 317.
182 J. G. Chirikjian, H. T. Wright, and J. R. Fresco, *Proc. Nat. Acad. Sci. U.S.A.*, 1972, **69**, 1638.
183 D. Cassio and J. P. Waller, *F.E.B.S. Letters*, 1971, **12**, 309.
184 J. P. Waller, J. L. Risler, C. Monteilhet, and C. Zelwer, *F.E.B.S. Letters*, 1971, **16**, 186.
185 J. Parker and F. C. Neidhart, *Biochem. Biophys. Res. Comm.*, 1972, **49**, 495.
186 T. S. Papas and A. H. Mehler, *J. Biol. Chem.*, 1970, **245**, 1588.

be bound to solid matrices such as Sepharose or polyacrylamide the nucleic acid is not very stable, especially in the presence of relatively crude extracts which could contain substantial amounts of nuclease activity.[187] An added complication with affinity chromatography is the recent discovery that the synthetases are not totally specific for either the cognate amino-acid or tRNA and non-specific binding is likely to occur.[188] However, affinity chromatography could prove to be very useful at the later stages in synthetase purification. A rapid method for isolating small amounts of synthetases has been developed which depends on the formation of synthetase–tRNA complexes.[189] The complexes have electrophoretic mobilities sufficiently unique to make them easy to separate. The main limitation is that the method is not amenable to the isolation of large amounts of pure enzyme.

Subunit Structures of Synthetases. Several synthetases have been purified to homogeneity and their subunit structures. The most interesting outcome of these studies is the wide variation in the size and distribution of the subunits. Broadly speaking, it is possible to identify four classes of enzymes. The enzymes of the A_1 class possess a single elongated protomer with a molecular weight of over 100 000, and appear to have only one binding site for amino-acid and tRNA.[190] The A_2 class is composed of proteins with two very similar protomers. The molecular weights of the protomers in this class of enzymes are usually in the range 35 000—45 000. Another class of synthetases, α_4, appears to contain enzymes with four very similar subunits. Only two enzymes have been characterized from this class and there are doubts about the validity of some of the data obtained. The fourth, and in some ways the most interesting, class of synthetases is the A_2B_2 class, in which the enzymes contain dissimilar units. There are usually two of each type of subunit.

It is noteworthy that most of the enzymes in the A_1 class are concerned with the activation of three similar amino-acids, valine, isoleucine, and leucine. It also appears that the enzymes from different sources have retained their elongated monomeric nature. The only exception to this is leucyl-tRNA synthetase from yeast which seems to contain two similar subunits with molecular weights of 55 000. The enzyme is of interest since it has been obtained in a crystalline form which appears suitable for *X*-ray diffraction studies.[182] However, the variation in subunit structure of various preparations and the contrast with leucyl-tRNA synthetase from other sources raise some doubt about the proposed subunits of the enzyme from yeast. The similarity between the three synthetases for valine, isoleucine, and leucine is also reflected in lack of specificity in the amino-acid-activating step. This was originally demonstrated with isoleucyl-tRNA synthetase

[187] S. Bartkowiak and J. Pawelkiewicz, *Biochim. Biophys. Acta*, 1972, **272**, 137.
[188] D. Kern, R. Giege, and J. P. Ebel, *European J. Biochem.*, 1972, **31**, 148.
[189] W. Siefert, G. Nass, and W. Zillig, *J. Mol. Biol.*, 1968, **33**, 507.
[190] M. Yarus and P. Berg, *J. Mol. Biol.*, 1967, **28**, 479.

from *E. coli*, which has the ability to form the aminoacyl-adenylates with both isoleucine and valine, but only transfers isoleucine to tRNA.[191, 192] The same enzyme from another strain of *E. coli* has the ability to activate both leucine and valine, showing that the discrimination between these amino-acids occurs at a later stage in the catalytic process.[193] The inability of the leucyl- and valyl-tRNA synthetases to activate isoleucine suggests that the incorrect activations only occur when the side-chain of the incorrect amino-acid is smaller than that of the correct one.

An interesting variant of valyl-tRNA synthetase has been isolated from *E. coli* which was infected by bacteriophage T_4.[194] Infection leads to the formation of a modified form of the enzyme which contains an additional component, T-factor, which has a molecular weight of about 10 000. However, the normal catalytic functions of the enzyme are not affected by attachment of the viral factor.

The majority of synthetases characterized so far contain two very similar protomers. Tyrosyl- and tryptophanyl-tRNA synthetases from several sources have been isolated and in all cases the enzymes have been dimers. The only exception was a report that tryptophanyl-tRNA synthetase from beef pancreas contained dissimilar subunits.[195] However, this has been disproved by a detailed study on the enzyme, which shows it to be composed of two similar protomers with molecular weights of about 55 000.[196] This is considerably larger than the corresponding enzymes from *E. coli.*[197] Prolyl-tRNA synthetase from *E. coli* which has protomers of about 47 000 molecular weight has the unusual property of reversibly dissociating at low temperatures with a concomitant loss of activity.[198]

The most unusual class of synthetases is composed of enzymes with significantly dissimilar subunits. The first of these to be characterized was glycyl-tRNA synthetase from *E. coli*, which has two types of subunit with molecular weights of 33 000 and 80 000 respectively.[199] Since then two other enzymes, phenylalanyl-tRNA synthetase from yeast [200] and glutamyl-tRNA synthetase from *E. coli*,[201, 202] have also been shown to contain dissimilar subunits. The latter is particularly interesting since it contains only one copy of each type of subunit, in contrast to the other two enzymes which contain two copies of each subunit. It has also been

[191] F. Bergmann, P. Berg, and M. Dieckmann, *J. Biol. Chem.*, 1961, **236**, 1735.
[192] A. T. Norris and P. Berg, *Proc. Nat. Acad. Sci. U.S.A.*, 1964, **52**, 330.
[193] M. Kondo and C. R. Woese, *Biochemistry*, 1969, **8**, 4177.
[194] G. L. Marchin, M. M. Comer, and F. C. Neidhardt, *J. Biol. Chem.*, 1972, **247**, 5132.
[195] E. C. Preddie, *J. Biol. Chem.*, 1969, **244**, 3958.
[196] C. Gros, G. Lemaire, R. van Rapenbusch, and B. Labouesse, *J. Biol. Chem.*, 1972, **247**, 2931.
[197] D. R. Joseph and K. H. Muench, *J. Biol. Chem.*, 1971, **246**, 7610.
[198] M. L. Lee and K. H. Muench, *J. Biol. Chem.*, 1969, **244**, 223.
[199] D. L. Ostrem and P. Berg, *Proc. Nat. Acad. Sci. U.S.A.*, 1970, **67**, 1967.
[200] F. Fasolo, N. Befort, Y. Boulanger, and J. P. Ebel, *Biochim. Biophys. Acta*, 1970, **217**, 305.
[201] J. Lapointe and D. Soll, *J. Biol. Chem.*, 1972, **247**, 4966.
[202] J. Lapointe and D. Soll, *J. Biol. Chem.*, 1972, **247**, 4975.

shown that only the larger subunit of glutamyl-tRNA synthetase has catalytic activity. Some modification of the affinity for the substrates was detected upon recombination of the subunits.

Proteolytic Fragmentation of Synthetases. A major complicating factor in studies on aminoacyl-tRNA synthetases is likely to be the sensitivity of these enzymes to proteolytic fragmentation, which in many cases yields fully active enzyme. A good example of this is methionyl-tRNA synthetase from *E. coli*, which has been studied extensively. The native enzyme has a molecular weight of 176 000 and it is thought to be tetrameric.[203] However, another form of the enzyme has also been purified and found to have a molecular weight of only 90 000.[204] It appears that the type of enzyme obtained depends on the manner in which it is purified. Introduction of autolysis as a purification step leads to the formation of the smaller form, indicating that it is at least partly due to proteolysis of the enzyme. An interesting observation is that both forms of the enzyme have the same number of binding sites for methionine and tRNA, showing that the active site is not located in the half which is removed by proteolysis. A further complication, albeit a useful one, is the existence of an alternative mode of proteolysis which can be effected by a variety of proteinases and which yields a smaller fragment of the enzyme with a molecular weight of only 66 000.[205] This fragment is fully active in aminoacylation and has been obtained in a crystalline form which promises to be very useful in *X*-ray diffraction studies.

Another synthetase which has been shown to exhibit sensitivity to proteolysis is leucyl-tRNA synthetase from *E. coli*. The native form of the enzyme is monomeric and has a molecular weight of about 110 000. However, a second form of the enzyme can carry out all the partial reactions catalysed by the native enzyme except for transfer of leucyl-adenylate to tRNA. It has also been shown that the native enzyme is converted into the second form by mild treatment with trypsin and that reconstitution of the native form from the latter can be effected.[206] It seems very likely that the second form is the product of limited proteolysis of the native enzyme, especially since the polypeptide chains in the former are about half the length of the latter.

As mentioned above, there is little doubt that the variety of sizes in the polypeptide chains of synthetases is considerably greater than might have been expected. However, the apparent sensitivity of these enzymes to proteolysis does raise the possibility that at least some of the enzymes which have been isolated so far do not represent the native forms of the enzymes. Limited proteolysis is a likely cause of the observation that several synthetases give multiple forms during purification.[207] Suggestions

[203] F. Lemoine, J. P. Waller, and R. van Rapenbusch, *European J. Biochem.*, 1968, **4**, 213.
[204] C. J. Bruton and B. S. Hartley, *Biochem. J.*, 1968, **108**, 281.
[205] D. Cassio and J. P. Waller, *European J. Biochem.*, 1971, **20**, 283.
[206] P. Rouget and F. Chapeville, *European J. Biochem.*, 1971, **23**, 459.
[207] L. L. Kisselev and I. D. Baturina, *F.E.B.S. Letters*, 1972, **22**, 231.

that there are several forms of a synthetase (isozymes) in a particular source should therefore be treated with extreme caution. Proteolysis could also explain the disproved claim that tryptophanyl-tRNA synthetase from beef pancreas contains dissimilar subunits, since the sum of the molecular weights of the latter is almost the same as the correct molecular weight of the protomer. There seems to be a case for re-evaluating the molecular weights of any synthetases which give multiple forms during purification or manipulation. The inclusion of proteinase inhibitors during purification, provided they do not affect the catalytic properties of the enzymes, could also be useful in this respect.

Other Enzymes.—The virtually complete sequence of chicken liver glutamate dehydrogenase has been reported.[208] Having 503 residues, this is the largest protein so far sequenced, and of the 485 residues definitely positioned only 26 differ from the bovine sequence published last year. Photo-oxidation of four histidine residues caused total loss of activity with this enzyme,[209] and at least one of these residues was considered to be essential. However, neither glutamate nor NAD protected the enzyme and in view of the lack of specificity of the method the results were not completely clear.

Procedures have been described for the purification of isocitrate dehydrogenase from rat heart, rat liver,[210] and yeast,[211] and also a method for its immobilization on Sepharose 4B.[212] The enzyme from yeast behaved very strangely on gel-filtration; Sephadex G-200 gave a molecular weight of 375 000 ± 10%, but passage through Sepharose 6B gave a figure of 25 000—30 000. Presumably the enzyme binds to the Sepharose and this is one case of affinity chromatography without the necessity of introducing functional groups. Yeast isocitrate dehydrogenase also appears to bind to the ampholytes used in isoelectric focusing experiments. This was concluded [213] because the multiple enzyme bands found using this technique varied according to the protein : ampholyte ratio and could not be detected in any system from which ampholytes were excluded. It is clear that caution must be used in invoking isoelectric focusing results as diagnostic of microheterogeneity.

The sequence around the active site phosphohistidine residue in succinyl CoA synthetase from *E. coli* has been determined:[214]

PO_3^{2-}
|
Met-Gly-His-Ala-Gly-Ala-Ile-Ile-Ala-Gly-Gly-Lys

[208] K. Moon, D. Piszkiewicz, and E. L. Smith, *Proc. Nat. Acad. Sci. U.S.A.*, 1972, **69**, 1380.
[209] N. Tudball, R. Bailey-Wood, and P. Thomas, *Biochem. J.*, 1972, **129**, 419.
[210] M. Islam, J. L. Bell, and D. N. Baron, *Biochem. J.*, 1972, **129**, 1003.
[211] J. A. Illingworth, *Biochem. J.*, 1972, **129**, 1119.
[212] A. E. Chung, *Arch. Biochem. Biophys.*, 1972, **152**, 125.
[213] J. A. Illingworth, *Biochem. J.*, 1972, **129**, 1125.
[214] T. Wang, L. Jurášek, and W. A. Bridger, *Biochemistry*, 1972, **11**, 2067.

These workers suggest that their purification scheme, chromatography on QAE-Sephadex, acid treatment to remove the phosphate group, and then re-chromatography on the QAE column, may be generally applicable to all phosphohistidine-containing peptides.

Although it is now generally agreed that the subunit structure of aspartate transcarbamylase from *E. coli* consists of two catalytic trimers and six regulatory units (see last year's Report), the exact arrangement of these subunits continues to attract attention. Now some very elegant electron-microscopy [215] and physicochemical studies and chemical cross-linking experiments [216] have led to a model where the regulatory subunits are arranged dimerically around the two groups of catalytic trimers (Figure 6). These two catalytic subunits are of slightly oblate structure having a triangular arrangement of the three polypeptide chains. Although they are superimposed above one another in an eclipsed configuration, according to the model they are not in direct physical contact. The three regulatory dimers, each a V-shaped structure consisting of two cylindrically shaped monomers, extend to the outside of the molecule and serve to interconnect the two catalytic chains displaced 120° in the lower trimer.

Tryptophanase has been purified from *E. coli* strains K-12 [217] and B.[218] In both cases the enzyme was found to be tetrameric with a subunit molecular weight of 55 000, thus correcting an earlier report [219] that the enzyme was an octamer with four pyridoxal phosphate binding-sites. 411 out of the 474 residues of the *E. coli* B enzyme have been sequenced,[220] including the *N*-terminal, *C*-terminal, and pyridoxal phosphate-binding peptides. The corresponding enzyme from *Bacillus alvei* has also been isolated [221] and it appears to be similar, having four identical subunits of molecular weight 50 000 ± 1800.

A revised structure of aspartokinase-1-homoserine dehydrogenase from *E. coli* has been published.[222, 223] The new subunit molecular weight is increased from 60 000 to 86 000, which implies that the enzyme is a tetramer rather than a hexamer. Comprehensive evidence for this view comes from SDS-gel electrophoresis, gel-filtration in 6M-guanidinium hydrochloride, isoelectric focusing in 6M-urea, and chemical cross-linking with dimethyl suberimidate, which gave bands whose mobilities corresponded to the monomer, dimer, trimer, and tetramer. The identical nature of the subunits was established by their electrophoretic homogeneity, and unique

[215] K. E. Richards and R. C. Williams, *Biochemistry*, 1972, **11**, 3393.
[216] J. A. Cohlberg, V. P. Pigiet, and H. K. Schachman, *Biochemistry*, 1972, **11**, 3396.
[217] J. London and M. E. Goldberg, *J. Biol. Chem.*, 1972, **247**, 1566.
[218] H. Kagamiyama, H. Wada, H. Matsubara, and E. E. Snell, *J. Biol. Chem.*, 1972, **247**, 1571.
[219] Y. Morino and E. E. Snell, *J. Biol. Chem.*, 1967, **242**, 5591.
[220] H. Kagamiyama, H. Matsubara, and E. E. Snell, *J. Biol. Chem.*, 1972, **247**, 1576.
[221] S. O. Hock and R. D. DeMoss, *J. Biol. Chem.*, 1972, **247**, 1750.
[222] F. Falcoz-Kelly, J. Janin, J. C. Saari, M. Véron, P. Truffa-Bachi, and G. N. Cohen, *European J. Biochem.*, 1972, **28**, 507.
[223] D. E. Wampler, *Biochemistry*, 1972, **11**, 4428.

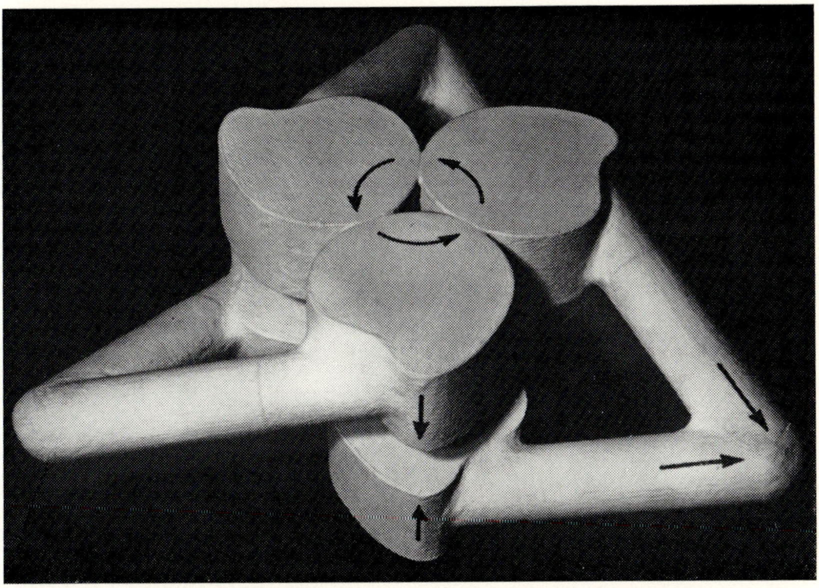

Figure 6 *Proposed arrangement of subunits in aspartate transcarbamylase*
(Reproduced by permission from *Biochemistry*, 1972, **11**, 3396)

N-terminal, *C*-terminal, cysteinyl, and tryptophanyl amino-acid sequences. Why was the previous molecular weight in error? This is not clear, but the main evidence came from sedimentation-equilibrium studies in 6M-guanidine, a system in which this protein is apparently not ideal. Mild proteolysis of the enzyme produces a dimeric fragment, molecular weight 110 000, which possesses only the homoserine dehydrogenase activity.[224] Carboxypeptidase experiments have indicated that this fragment originates from the *C*-terminal part of the enzyme. The same workers have also found a mutant, Gif 108, possessing only the aspartokinase activity, and fingerprints showed this fragment to be *N*-terminal in the native enzyme. Thus the two activities are localized in independent regions of the same polypeptide chain, possibly as a result of an earlier gene fusion.

The observation that certain L-asparaginases are capable of inhibiting and eliminating various types of tumour cells has stimulated research on the properties and large-scale purification of the enzyme. Batch production from *Erwinia carotovora* has been described[225] and the protein from this source appears more stable, and possibly therefore more suited to clinical use, than the enzyme from the alternative sources, *E. coli* or *Proteus vulgaris*.[226] Asparaginases from all three bacteria appear to be composed of four identical subunits, each of molecular weight 32 000—38 000.[227, 228] In an investigation[229] of the specificity of the enzyme only free asparagine or small peptides with *C*-terminal asparagine were found to be deamidated; thus the loss of amide groups from insulin cannot be responsible for the hyperglycaemia reputedly arising from clinical use of the enzyme.

The complete sequence of human erythrocyte carbonic anhydrase B has appeared.[230] Comparison with the known structure of carbonic anhydrase C showed that two of the zinc co-ordination ligands (His-94 and His-96) were conserved, but the third binding site (His-118 in the C enzyme) was replaced by isoleucine. Possibly the metal interacts with His-119 or Glu-117. The full *X*-ray crystallographic structure, reportedly now in progress, should clarify the situation. For a full review of the function of the zinc ion in carbonic anhydrase see reference 231.

Using a technique similar to that employed on aldolase (see above), interactions between native and chemically modified subunits of matrix-bound glycogen phosphorylase have been studied.[232] Binding to the support reduced the activity of the 'a' and the 'b' forms to 30% of the original

[224] M. Veron, F. Falcoz-Kelly, and G. N. Cohen, *European J. Biochem.*, 1972, **28**, 520.
[225] K. A. Cammack, D. I. Marlborough, and D. S. Miller, *Biochem. J.*, 1972, **126**, 361.
[226] T. Tosa, R. Sano, K. Yamamoto, M. Nakamura, and I. Chibata, *Biochemistry*, 1972, **11**, 217.
[227] A. C. Greenquist and J. C. Wriston, *Arch. Biochem. Biophys.*, 1972, **152**, 280.
[228] S. Shifrin and B. J. Grochowski, *J. Biol. Chem.*, 1972, **247**, 1048.
[229] J. B. Howard and F. H. Carpenter, *J. Biol. Chem.*, 1972, **247**, 1020.
[230] B. Anderson, P. O. Nyman, and L. Strid, *Biochem. Biophys. Res. Comm.*, 1972, **48**, 670.
[231] R. H. Prince and P. R. Woolley, *Angew. Chem. Internat. Edn.*, 1972, **11**, 408.
[232] K. Feldmann, H. Zeisel, and E. Helmreich, *Proc. Nat. Acad. Sci. U.S.A.*, 1972, **69**, 2278.

value, and treatment with SDS further reduced this figure to 1%. These figures imply that, unlike aldolase, the monomer has only about 3% of the native enzyme activity. Hybrids of the normal enzyme subunits and subunits with modified cofactor exhibited activity in one subunit only. However, the inactive subunit could elicit activity from the matrix-bound monomers, results which imply that the pyridoxal phosphate cofactor has a direct role in catalytic activity.[233, 234] Hybrids have also been found[235] between echinoderm arginine kinase and mammalian creatine kinase. The fact that these two phosphagen kinases, which have varying substrate specificity, may be hybridized suggests that their tertiary structures are similar, and this in turn implies a common evolutionary origin. The hybrid enzymes appeared to be active, but it is never clear in these cases whether activity simply stems from substrate-induced dissociation of the hybrid back to monomers and then reassociation to the native form.

Reconstitution of the two cyanogen bromide fragments of thioredoxin-C from *E. coli* has produced a form possessing 20% of the original enzyme activity with NADPH and thioredoxin reductase.[236] Chromatography on Sephadex G-50 gave three peaks, one of which represented 20% of the total mixture, and co-eluted with genuine thioredoxin-C. It appears that this enzyme can be added to the list of proteins, such as ribonuclease and staphylococcal nuclease, in which suitable fragments will spontaneously fold into an active conformation. The thioredoxin isolated from the cells of *E. coli* infected with the bacteriophage T_4 appears totally dissimilar to that of the native host.[237] The amino-acid sequences are totally unrelated[238] except for the S—S bridge between Cys-14 and Cys-17, which is essential for redox activity. A similar situation has been noted previously for T_4 and hen egg-white lysozymes.

Amyloid proteins are found in a complex proteinaceous substance present in tissues of patients suffering from amyloidosis. The physiological role and site of cellular synthesis of these proteins are unknown. Although a previous report found homologies with immunoglobulin light chains, the complete sequence of monkey amyloid protein A shows no such homology.[239] It is, however, very similar to a corresponding component of human amyloid fibrils.[240–242] Thus, the role and origin of these proteins remain unclear.

233 T. Pfeuffer, J. Ehrlich, and E. Helmreich, *Biochemistry*, 1972, **11**, 2125.
234 H. D. Weisshaar and D. Palm, *Biochemistry*, 1972, **11**, 2146.
235 D. C. Watts, B. Focant, B. M. Moreland, and R. L. Watts, *Nature New Biol.*, 1972, **237**, 51.
236 A. Holmgren, *F.E.B.S. Letters*, 1972, **24**, 351.
237 B.-M. Sjöberg, *J. Biol. Chem.*, 1972, **247**, 8058.
238 B.-M. Sjöberg and A. Holmgren, *J. Biol. Chem.*, 1972, **247**, 8063.
239 M. A. Hermodson, R. W. Kuhn, K. A. Walsh, H. Neurath, N. Eriksen, and E. P. Berditt, *Biochemistry*, 1972, **11**, 2934.
240 E. C. Franklin, M. Pras, M. Levin, and B. Frangione, *F.E.B.S. Letters*, 1972, **22**, 121.
241 D. Ein, S. Kimura, W. D. Terry, J. Magnotta, and G. G. Glenner, *J. Biol. Chem.*, 1972, **247**, 5653.
242 G.Husby, K. Sletten, T. E. Michaelsen, and J. B. Natvig, *Nature New Biol.*, 1972, **238**, 187.

Complete amino-acid sequences have also appeared this year for cytochromes *c* from the rust fungus *Ustilago sphaerogena*;[243] the camel, *Camelus dromedarius*;[244] the elephant seal, *Mirounga leonina*;[245] the protozoan *Crithidia oncopelti*;[246] tomato, *Lycopersicum esculentum* L.;[247] the bat *Miniopteris schreibersi*;[248] the garden snail, *Helix aspersa* Müller;[249] the thermophilic fungus *Humicola lanuginosa*;[250] spinach, *Spinacea oleracea* L.;[251] the niger, *Guizotia abyssinica* Cass.;[251] elder, *Sambucus nigra* L.;[251] and last but not least the cytochrome c_{553} from the bacterium *Desulphovibrio vulgaris*.[252] This brings the total number of sequences for the enzyme to 68 and no doubt next year's Report will contain a new crop from ever more imaginative sources. The validity of constructing phylogenetic trees from cytochrome *c* sequences has been discussed (see also the introduction to this section). A phylogeny of higher plants based on these primary structures has been presented and the biological implications have been discussed.[251]

Structures have also been reported for human [253] and guinea-pig [254] α-lactalbumins, guinea-hen egg-white lysozyme,[255] calf-lens γ-crystallin,[256, 257] and the anti-tumour protein from *Streptomyces carzinostaticus*, neocarzinostatin.[258]

4 Structural Proteins

Collagen.—Although collagen is probably the most abundant of cell proteins (10^{12} kg is the amount estimated to be distributed throughout the world) it attracts little interest from protein chemists other than those working directly in the field. This is a pity, since during the past few years a great deal of interesting work has been done and a number of useful methods, for instance the use of hydroxylamine to cleave Asn-Gly

243 K. G. Bitor, S. N. Vinogradov, C. Nolan, L. J. Weiss, and E. Margoliash, *Biochem. J.*, 1972, **129**, 561.
244 M. Sokolovsky and M. Moldovan, *Biochemistry*, 1972, **11**, 145.
245 R. C. Augusteyn, M. A. McDowall, E. C. Webb, and B. Zerner, *Biochim. Biophys. Acta*, 1972, **257**, 264.
246 G. W. Pettigrew, *F.E.B.S. Letters*, 1972, **22**, 64.
247 R. Scogin, M. Richardson, and D. Boulter, *Arch. Biochem. Biophys.*, 1972, **150**, 489.
248 D. J. Strydom, S. J. Vere der Walt, and D. P. Botes, *Comp. Biochem. Physiol.*, 1972, **43B**, 21.
249 R. H. Brown, M. Richardson, D. Boulter, J. A. M. Ramshaw, and R. P. S. Jefferies, *Biochem. J.*, 1972, **128**, 971.
250 W. T. Morgan, C. P. Hensley, and J. P. Riehm, *J. Biol. Chem.*, 1972, **247**, 6555.
251 D. Boulter, J. A. M. Ramshaw, E. W. Thompson, M. Richardson, and R. H. Brown, *Proc. Roy. Soc.*, 1972, **B181**, 441.
252 M. Bruschi and J. le Gall, *Biochim. Biophys. Acta*, 1972, **271**, 48.
253 J. B. C. Findlay and K. Brew, *European J. Biochem.*, 1972, **27**, 65.
254 K. Brew, *European J. Biochem.*, 1972, **27**, 341.
255 J. Jollès, E. Van Leemputten, A. Mouton, and P. Jollès, *Biochim. Biophys. Acta*, 1972, **257**, 497.
256 C. Slingsby and L. R. Croft, *Biochem. J.*, 1972, **127**, 609.
257 L. R. Croft, *Biochem. J.*, 1972, **128**, 961.
258 J. Meienhofer, H. Maeda, C. B. Glaser, J. Czombos, and K. Kuromizu, *Science*, 1972, **178**, 875.

G Y D E K S A G V S V P G P M G P S G P R G L hP G P hP G A hP
10 20 30
|——— CB-1 ———→←————————

G P Q G F Q G P hP G E hP G E hP G A S G P M G P R G P hP G P hP
40 50 60
——— CB-2 ———————→←————————

G K N G D D G E A G K P G R hP G Q R G P hP G P Q G A R G L hP
70 80 90
———————————— CB-4 ————————

O-Gal-Glc
|
G T A G L hP G M hK G H R G F S G L D G A K G N T G P A G P K
100 110 120
———————→←————————— CB-5 —

G E hP G S hP G Z B G A hP G Q M G P R E L hP G E R G R hP G P hP
130 140 150
———————————→←————————

G S A G A R G D D G A V G A A G P hP G P T G P T G P hP G F hP
160 170 180
————————————————————

G A A G A K G E A G P Q G A R G S E G P Q G V R G E hP G P hP
190 200 210
————————————————————

G P A G A A G P A G N hP G A D G Q hP G A K G A N G A hP G I A
220 230 240
————————————————————

G A hP G F hP G A R G P S G P Q G P S G A hP G P K G N S G E hP
250 260 270
——————— CB-8 ———————

G A hP G N K G D T G A K G E hP G P A G V Q G P hP G P A G E E
280 290 300
————————————————————

G K R G A R G E hP G P S G L hP G P hP G E R G G hP G S R G F hP
310 320 330
————————————————————

G A D G V A G P K G P A G E R G S hP G P A G P K G S hP G E A
340 350 360
————————————————————

G R hP G E A G L hP G A K G L T G S hP G S hP G P D G K T G P hP
370 380 390
————————————————————

G P A G Z B G R hP G P A G P hP G A R G Q A G V M G F hP G P K
400 410 420
————————————————→←——

G A A G E hP G K A G E R G V hP G P hP G A V E P A G K D G E A
430 440 450

G A Q G P hP G P A G P A G E R G E Q G P A G S hP G F Q G L hP
460 470 480

G P A G P hP G E A G K hP G E Q G V hP G D L G A hP G P S G A R
490 500 510

CB-3

G E R G F hP G E R G V E G P hP G P A G P R G A N G A hP G N D
520 530 540

G A K G D A G A hP G A hP G S Q G A hP G L Q G M
550 560

→ ←CB-7→

G P hP G L A G P hP G E S G R E G A hP G A E G S hP G R D G S hP
10 20 30

←

G A K G D R G E T G P A G A hP G P hP G A hP G A hP G P V G P A
40 50 60

G K S G D R G E T G P A G P I G P V G P A G A R G P A G P Q
70 80 90

G P R G B hK G Z T G Z Z G B R G I hK G H R G F S G L Q G P hP
100 110 120

CB-6

G P hP G S hP G E Q G P S G A S G P A G P R G P hP G S A G S hP
130 140 150

G K D G L N G L hP G P I G hP hP G P R G R T G D A G P A G P hP
160 170 180

G P hP G P hP G P hP G P P S G G Y
190

→

Figure 7 *The partial amino-acid sequence of the* α_1*-chain of collagen. The structure shown is from rat skin except for fragment* CB-6 *which is from calf skin. The* 273 *residue fragment*, CB-7, *is the only missing portion of the chain. The disaccharide residue* 2-*O*-α-D-*glucopyranosyl*-*O*-β-D-*galactopyranosyl is bound to the* δ-*hydroxy-group of the hydroxylysine residue at position* 99 *through an O-glycosidic linkage. The I.U.P.A.C.–I.U.B. one-letter notation* (*see Chapter* 7) *is used together with* hP *as hydroxyproline and* hK *as hydroxylysine*

sequences, have emerged. The general features of the structure, from the sequence to the fibril, are now established,[259] and, apart from biosynthesis and assembly, most of the activity is directed towards filling in the chemical detail, comparing the collagens of different species and tissues, and determining the basis of pathological defects.

Collagen appears to be synthesized as the precursor, procollagen, which is cleaved by a specific peptidase to form the native molecule.[260] Normal collagens consist of two types of chain, α_1 and α_2, in the mole ratio of 2 : 1, wound together to form a three-stranded rope, each strand having about 1050 amino-acids. With the completion of the sequences of the α_1-CB3 [261, 262] (the nomenclature refers to the elution order of the cyanogen bromide fragments from CM-cellulose columns), α_1-CB6,[263, 264] α_1-CB8,[265, 266] and α_2-CB4 [267] fragments, the complete primary structure of the α_1 and α_2 chains from rat and calf-skin collagen is almost complete. The sequence of most of the α_1-chain has been assembled in Figure 7. Throughout most of the sequence every third residue is glycine, followed commonly by proline, and often preceded by hydroxyproline. In CB-3 and CB-8 this pattern changes and leucine or phenylalanine occupies the second position. There is clear homology not only between collagens from different species, but also between the α_1 and α_2 chains [268] and, not surprisingly, between collagens derived from different tissues of the same animal.[269] In fact a comparison of the same fragment from the skin and teeth of rats has shown them to be identical in sequence and to differ only in the degree of hydroxylation of proline and lysine residues.

At the ends of the collagen chains the sequence regularities disappear and the chains become covalently cross-linked. An interesting kind of linkage has been found in cow-skin collagen.[270] Based on histidine, the 2,10-diamino-5-hydroxymethyl-6-(*N*-1-histidyl)undecanedioic acid (2) is presumably derived from addition of an imidazole to the known cross-linking species, diaminoformylundecanedioic acid. Other cross-links recently found include N^{ε}-hexosylhydroxylysine,[271] which may join the

[259] P. M. Gallop, O. O. Blumenfeld, and S. Seifter, *Ann. Rev. Biochem.*, 1972, **41**, 617.
[260] P. Bornstein, H. P. Ehrlich, and A. W. Wyke, *Science*, 1972, **175**, 544.
[261] P. Wendt, P. P. Fietzek, and K. Kühn, *F.E.B.S. Letters*, 1972, **26**, 69.
[262] P. P. Fietzek, P. Wendt, I. Kell, and K. Kühn, *F.E.B.S. Letters*, 1972, **26**, 74.
[263] P. P. Fietzek, F. W. Rexrodt, P. Wendt, M. Stark, and K. Kühn, *European J. Biochem.*, 1972, **30**, 163.
[264] P. Wendt, K. Von Der Mark, F. W. Rexrodt, and K. Kühn, *European J. Biochem.*, 1972, **30**, 169.
[265] J. Rauterberg, P. P. Fietzek, F. W. Rexrodt, U. Becker, M. Stark, and K. Kühn, *F.E.B.S. Letters*, 1972, **21**, 75.
[266] G. Balian, E. M. Click, M. A. Hermodson, and P. Bornstein, *Biochemistry*, 1972, **11**, 3798.
[267] P. P. Fietzek, I. Kell, and K. Kühn, *F.E.B.S. Letters*, 1972, **26**, 66.
[268] K. A. Piez, G. Balian, E. M. Click, and P. Bornstein, *Biochem. Biophys. Res. Comm.*, 1972, **48**, 990.
[269] W. T. Butler, *Biochem. Biophys. Res. Comm.*, 1972, **48**, 1540.
[270] R. B. Fairweather, M. L. Tanzer, and P. M. Gallop, *Biochem. Biophys. Res. Comm.*, 1972, **48**, 1311.
[271] M. L. Tanzer, R. Fairweather, and P. M. Gallop, *Arch. Biochem. Biophys.*, 1972, **151**, 137.

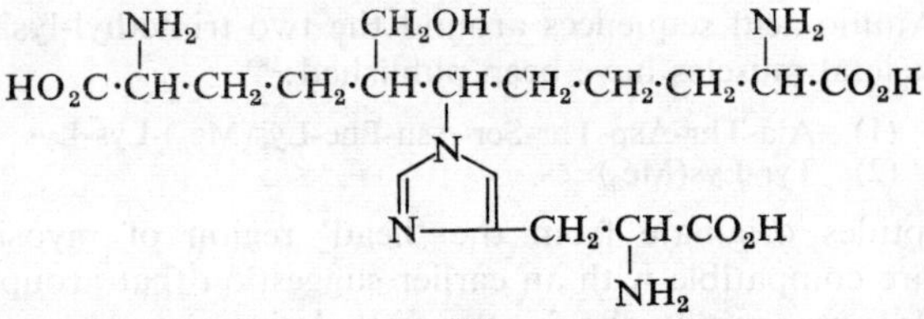

(2)

collagen to glycoproteins or proteoglycans in connective tissue, and ε-(γ-glutamyl)lysine,[272] which has been found in the medulla of hair and quill.

Actin.—Actin may no longer be considered simply as a component of muscle. Actin-like proteins have now been identified from a large number of sources and this ubiquitous protein seems to occur whenever the phenomenon of motility is encountered.

Following last year's demonstration of the presence of actin in *Acanthamoeba*, the peptides containing the N^{τ}-methylhistidine and N^{ε}-methyllysine residues have been isolated and their compositions determined.[273] These proved to be very similar to corresponding regions of the rabbit muscle protein. Actin has also been isolated [274] from amoebae of the slime-mould *Dictyostelium discoideum*. Polymerization with 0.1M-KCl produced filaments which could activate the Mg^{2+}-dependent ATPase of rabbit muscle myosin. Also, arrowhead structures could be seen in the electron microscope after combination with rabbit muscle heavy meromyosin (HMM), a test diagnostic of a true actin-like protein. Similar structures have been found with thrombosthenin A,[275, 276] the contractile protein isolated from blood platelets, and the actin isolated from mouse neuroblastoma cells.[277] Cultured chick embryo fibroblasts are yet another recently reported [278] source of this protein.

One can only speculate on the role of the actins from these diverse sources. Presumably they play a role in the implementation of intra- and extra-cellular movement and changes in cell shape, and for the neuroblastoma actin there is a possible involvement in axonal transport.

The three thiol groups on the surface of trout actin (see last year's Report) are, contrary to previous work, not involved in Ca^{2+} or ATP-binding, or the polymerization to F-actin.[279] A similar result has been obtained [280] by reaction of one of the surface thiol groups with Cu^{2+}.

[272] H. W. J. Harding and G. E. Rogers, *Biochim. Biophys. Acta*, 1972, **257**, 37.
[273] R. R. Weihing and E. D. Korn, *Biochemistry*, 1972, **11**, 1538.
[274] D. E. Woolley, *Arch. Biochem. Biophys.*, 1972, **150**, 519.
[275] M. Bettex-Galland, E. Probst, and O. Behnke, *J. Mol. Biol.*, 1972, **68**, 533.
[276] E. Probst and F. Lüscher, *Biochim. Biophys. Acta*, 1972, **278**, 577.
[277] P. R. Burton and W. L. Kirkland, *Nature New Biol*,, 1972, **239**, 244.
[278] Y.-Z. Yang and J. F. Perdue, *J. Biol. Chem.*, 1972, **247**, 4503.
[279] J. Bridgen, *Biochem. J.*, 1972, **126**, 21.
[280] S. S. Lehrer, B. Nagy, and J. Gergely, *Arch. Biochem. Biophys.*, 1972, **150**, 164.

Myosin.—Amino-acid sequences around the two trimethyl-lysine residues in rabbit skeletal muscles have been published:[281]

(1) Ala-Thr-Asp-Thr-Ser-Asn-Phe-Lys(Me_3)-Lys-Lys
(2) Tyr-Lys(Me_3)

Both peptides originate from the 'head' region of myosin and the sequences are compatible with an earlier suggestion that groups of lysines may provide a recognition site for the methylating enzyme.

The sequence around the methylated histidine in this skeletal myosin has been compared[282] with the corresponding non-methylated sequences from cardiac myosins:

	(τMe on His)
Rabbit skeletal myosin	Leu-Leu-Gly-Ser-Ile -Asp-Val-Asp-His-Gln-Thr-Tyr-Lys
Rabbit cardiac myosin	Leu-Leu-Ser -Ser-Leu-Asp-Ile -Asp-His-Gln-Asn-Tyr-Lys
Bovine cardiac myosin	Leu-Leu-Gly-Ser-Leu-Asp-Ile -Asp-His-Gln-Thr-Tyr-Lys

The important conclusion from this work is that cardiac and skeletal myosins are synthesized by different genes. Also, there is clearly a greater disparity between myosins from the same animal than between myosins from the same tissue. A correlation between histidine methylation and muscle development has been found[283] by studying the methylation of this peptide in adult and foetal muscle. In the case of foetal myosin two forms of the peptide are found:

	(τMe on His)
Rabbit skeletal myosin	Leu-Leu-Gly-Ser-Ile -Asp-Val-Asp-His-Gln-Thr-Tyr-Lys
Rabbit foetal myosin	Leu-Leu-Ala-Ser-Ile -Asp-Ile -Asp-His-Gln-Thr-Tyr-Lys
Rabbit foetal myosin	Leu-Leu-Ser -Ser-Leu-Asp-Ile -Asp-His-Gln-Asn-Tyr-Lys

Clearly at least three genes exist for myosin. The second foetal sequence is identical to that for rabbit cardiac myosin above, and the two forms of this peptide may result from a mixture of tissue-types. In all cases the sequences around the trimethyl-lysine residues were identical and the reason for the existence of a separate gene for foetal myosin is unknown, but a similar situation has been noted for haemoglobin. Presumably the differing forms of these proteins satisfy the differing requirements of the foetus for muscular movement and oxygen transport relative to its requirements later in the life cycle. The absence of histidine methylation, both in this form and in that from cardiac muscle, may be due either to a lack of the methylating enzyme system or to changes in the myosin tertiary structure. The significance of the methylation also remains unknown but the phenomenon appears to be associated with structural and motile proteins.

[281] G. Huszar, *J. Biol. Chem.*, 1972, **247**, 4057.
[282] G. Huszar and M. Elzinga, *J. Biol. Chem.*, 1972, **247**, 745.
[283] G. Huszar, *Nature New Biol.*, 1972, **240**, 260.

An actomyosin-like protein, possessing Ca^{2+} sensitivity in superprecipitation, has been isolated from equine leucocytes.[284] This protein resembled muscle actomyosin in both its superprecipitation and filament-formation properties. The calcium sensitivity presumably implies the presence of a tropomyosin–troponin system in leucocytes. Myosin, like actin, has now been isolated from cloned mouse fibroblasts [285] and again its properties are very similar to those of muscle myosin.

An interesting report has appeared on the separation of myosin functions.[286] It has for some time been known that blocking of the heavy meromyosin thiol group, SH_1, with *N*-ethylmaleimide completely inhibits the EDTA-ATPase activity whereas the Ca-ATPase is enhanced. It has now been shown that the actin activation of the blocked HMM-ATPase is only 1.5% of its normal value. However, in the presence of the relaxing system, troponin and tropomyosin, and in the absence of calcium, this actin activation is completely suppressed. This of course is the situation in normal actomyosin. The conclusion must be that SH_1 is required not only for EDTA-ATPase activity but also for activation of Mg^{2+} ATPase by actin, whereas it is not involved in the interaction with the troponin–tropomyosin complex.

The Troponin–Tropomyosin System.—The confusion surrounding the nature of troponin components at last appears to be resolved.[287–289] There are three components essential for the Ca^{2+}-dependent modification of actomyosin ATPase mediated by troponin and tropomyosin (Table 1). A

Table 1 *Components of rabbit skeletal muscle troponin and their properties*

Component	*Alternative names*	*Molecular weight*	*Properties*
I	Inhibitory factor, troponin I	23 000	Inhibits actomyosin ATPase
II	Troponin-T	37 000	Reverse inhibition regardless of whether calcium is bound or not
III	Ca^{2+}-binding protein, troponin-C, troponin-A	18 000	Binds 2Ca mol^{-1}. When combined with components I and II restores control of actomyosin ATPase by Ca^{2+}

[284] N. Shibata, N. Tatsumi, K. Tanaka, Y. Okamura, and N. Senda, *Biochim. Biophys. Acta*, 1972, **256**, 565.

[285] R. S. Adelstein, M. A. Conti, G. S. Johnson, I. Pastan, and T. D. Pollard, *Proc. Nat. Acad. Sci. U.S.A.*, 1972, **69**, 3693.

[286] R. Silverman, E. Eisenberg, and W. W. Keilley, *Nature New Biol.*, 1972, **240**, 207.

[287] M. L. Greaser, J. Gergely, M. H. Han, and E. S. Benson, *Biochem. Biophys. Res. Comm.*, 1972, **48**, 358.

[288] I. Staprans, H. Takihashi, M. P. Russell, and S. Watanabe, *J. Biochem.*, 1972, **72**, 723.

[289] S. Ebashi, *J. Biochem.*, 1972, **72**, 787.

fourth protein, tropocalcin, is not a constituent of the complex and its role in modifying the actin–myosin interaction is unknown. A previously reported component of molecular weight 14 000 is now considered to be a proteolytic cleavage product.[290]

Following the isolation of a tropomyosin-like protein from human platelets the role of this protein in the blood-clotting process has been discussed.[291] Owing to the constant presence of calcium ions in plasma the normal contraction process would be irreversible, and possibly the tropomyosin has an inhibitory role in this system. A tropomyosin, very similar to that from muscle, has also been isolated from plasmodia of the myxomycete *Physarum polycephalum*.[292] A Ca^{2+}-mediated regulatory role in protoplasmic streaming is postulated for this protein, particularly as actin has also been purified from this source.[293]

Previous studies have established that tropomyosin is an almost completely α-helical two-stranded coiled coil of molecular weight 63 000—68 000. Sequence studies [294–296] of the cyanogen bromide *C*-terminal fragment (molecular weight 17 000) now show a regular repeat of hydrophobic residues interrupted only occasionally by polar or charged groups (Figure 8). When these apolar residues are identified they can be seen to lie on the locus of a helix of long pitch around the surface of the α-helix. If two α-helices are tilted with respect to each other so that the long helices are in contact and the projections of one fit into the holes of the other, they may then coil around each other so that the long helices unwind and lie in parallel contact as the central core of a two-strand rope. Model building shows that the two strands can best be packed if one strand is staggered with respect to the other by fourteen residues. This would imply tail sections with one strand jutting out at each end of the molecule.

Protein Kinases in Muscle. There have been two simultaneous reports of the phosphorylation of troponin by endogenous enzymes. In one case [297] a cyclic-AMP-dependent protein kinase catalysed transfer of phosphate specifically from ATP to both actin and to the component of troponin of molecular weight 37 000. In the other case [298] the more specific enzyme, phosphorylase kinase, was used and this time the transfer of 0.4 mole phosphate per mole of troponin was to the inhibitory component (see Table 1). When, however, all endogenous phosphate was stripped with phosphorylase phosphatase and the resulting product was then rephosphorylated, uptake of one mole of phosphate per mole of protein was

[290] J. M. Wilkinson, S. V. Perry, H. A. Cole, and I. P. Trayer, *Biochem. J.*, 1972, **127**, 215.
[291] I. Cohen and C. Cohen, *J. Mol. Biol.*, 1972, **68**, 383.
[292] H. Tanaka and S. Hatano, *Biochim. Biophys. Acta*, 1972, **257**, 445.
[293] S. Hatano and F. Oosawa, *Biochim. Biophys. Acta*, 1966, **127**, 488.
[294] J. Sodek, R. S. Hodges, L. B. Smillie, and L. Jurasek, *Proc. Nat. Acad. Sci. U.S.A.*, 1972, **69**, 3800.
[295] R. S. Hodges and L. B. Smillie, *Canad. J. Biochem.*, 1972, **50**, 330.
[296] R. S. Hodges and L. B. Smillie, *Canad. J. Biochem.*, 1972, **50**, 312.
[297] E. Pratje and L. M. G. Heilmeyer, *F.E.B.S. Letters*, 1972, **27**, 89.
[298] J. T. Stull, C. O. Brostrom, and E. G. Krebs, *J. Biol. Chem.*, 1972, **247**, 5272.

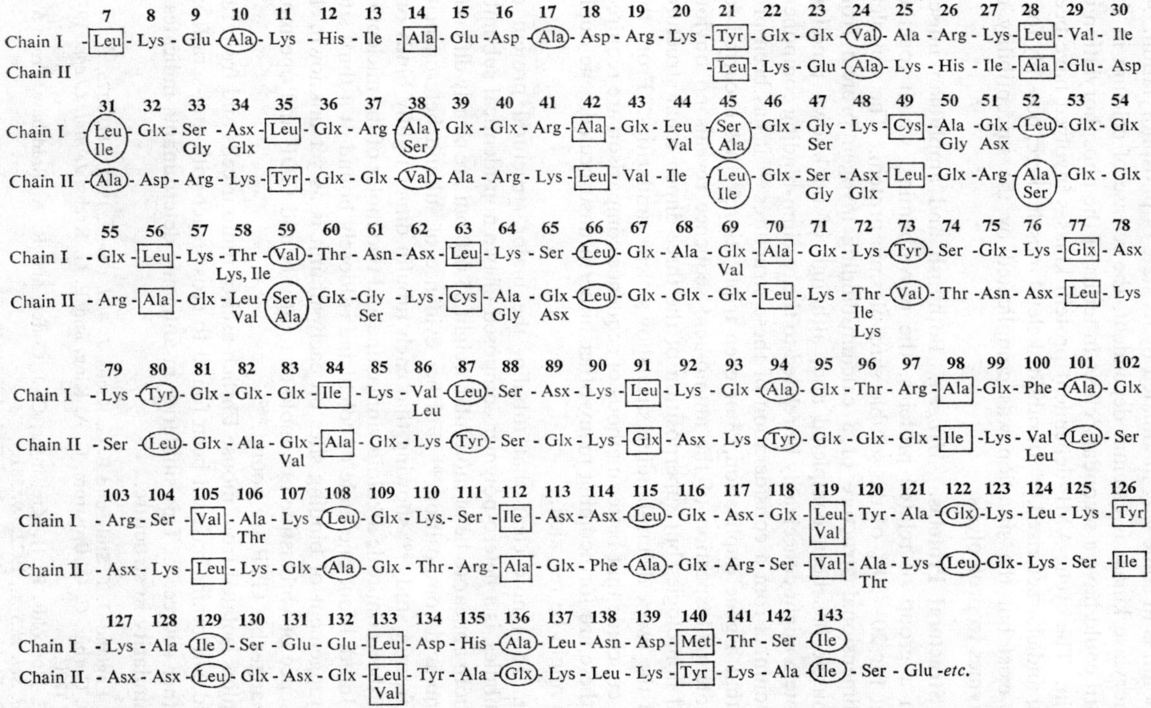

Figure 8 *Staggered arrangement of the two tropomyosin helices in a coiled coil where the two chains are out of register by* 14 *residues to maximize hydrophobic interaction and regularity of the coiled coil*
(Reproduced by permission from *Proc. Nat. Acad. Sci. U.S.A.*, 1972, **69**, 3802)

observed.[299] It seems possible that inhibition of actomyosin ATPase by troponin is in some way modulated by its state of phosphorylation. As phosphorylase kinase is calcium dependent, the release of calcium in the myofibril could have a secondary effect in terms of the phosphorylation of troponin. The cyclic-AMP-dependent protein kinases involved have been purified and characterized.[300] In rabbit skeletal muscle three, possibly four, species exist but the situation was complicated by the susceptibility of the enzymes to proteolysis.

Other Structural Proteins.—*Myogen.* Both fish and amphibian muscle contain a group of related proteins, the parvalbumins, which together account for 20—30% of the soluble sarcoplasmic protein. The complete three-dimensional structure of a calcium-binding myogen protein from carp muscle has been completed and, although of no known function, this protein might conceivably be related to the calcium-binding component of troponin. From a consideration of these structures [301] it has been proposed that a gene duplication,[302] or even triplication,[303] has occurred for this protein. Inspection of the amino-acid sequence reveals no obvious internal homologies but superposition of the three-dimensional structures around the two calcium-binding sites shows clear similarities. From this we can conclude that better methods of sequence comparison are required, and if these are forthcoming many other hitherto unsuspected gene duplications might be revealed.

Tubulin. It is now clear that tubulin, the major structural protein of microtubules, is a hetero-polymer composed of two non-identical subunits. These both appear to be of molecular weight 55 000 but are clearly different in amino-acid composition.[304] Further evidence of subunit heterogeneity comes from a study [305] showing that each tubulin dimer tightly binds one mole of the antimitotic alkaloid vinblastine and two moles of the nucleotide GTP. In addition, one mole of colchicine is loosely bound at a third site. The arrangement of binding sites on each subunit is as yet unknown. In the absence of vinblastine one mole of tightly bound GDP and one mole of exchangeable GTP are bound.[306]

Tubulins isolated from mouse Ehrlich ascites tumour cells and from rat, pig, and calf brains have all been found to cross-react with anti-sera from each other source.[307] This similarity in antigenic determinants indicates a close similarity in structure.

[299] P. J. England, J. T. Stull, and E. G. Krebs, *J. Biol. Chem.*, 1972, **247**, 5275.
[300] J. D. Corbin, C. O. Brostrom, C. A. King, and E. G. Krebs, *J. Biol. Chem.*, 1972, **247**, 7790.
[301] C. E. Nockolds, R. H. Kretsinger, C. J. Coffee, and R. A. Bradshaw, *Proc. Nat. Acad. Sci. U.S.A.*, 1972, **69**, 581.
[302] A. D. McLachlan, *Nature New Biol.*, 1972, **240**, 83.
[303] R. H. Kretsinger, *Nature New Biol.*, 1972, **240**, 85.
[304] J. Bryan, *J. Mol. Biol.*, 1972, **66**, 157.
[305] J. Bryan, *Biochemistry*, 1972, **11**, 2611.
[306] R. W. Berry and M. L. Shelanski, *J. Mol. Biol.*, 1972, **71**, 71.
[307] K. H. Dönges and E. Roth, *Naturwiss.*, 1972, **59**, 372.

Flagellar Proteins. There are two types of ATP-dependent contractile system, the actomyosin sliding-filament form and a second type, typified by protozoan cilia and sperm-tail flagella. Little is known about how this system contracts, but its ATPase, dynein, was thought to have very different enzymatic and physical properties from myosin. A recent report,[308] however, shows that similar intermediates are formed by both enzymes during ATP cleavage. This is in agreement with last year's work [309, 310] demonstrating an ATP-induced sliding-filament mechanism in the flagella of sea-urchin sperm. A more mathematical treatment of a sliding-filament model for flagellar movement has also been given.[311] Purification methods for the flagellar ATPase from sea-urchin spermatozoa have been presented.[312]

Preliminary sequence work has been reported on flagellin from *Proteus mirabilis* [313] and the phase-1 flagellar protein of *Salmonella typhimurium.*[314]

Viral Proteins. Preliminary sequence information has been published for the major structural protein of mammalian Type C RNA tumour viruses,[315] cucumber green mottle virus,[316] the cow pea strain of tobacco mosaic virus,[317] and alfalfa mosaic virus coat proteins.[318] The amino-terminal and carboxyl-terminal sequences of several Reovirus Type 3 capsid proteins have been determined,[319] and surprisingly found to be identical. The significance of this result remains obscure. The complete sequence of the β-protein of the bacteriophage ZJ-2 has been published.[320] Only two differences were found relative to the related phage fd β-protein but the paper is strongly recommended to anybody who has problems with insoluble peptides. The amino-acid sequence of turnip yellow mosaic virus coat protein has also been presented.[321] Many of the large number of proline residues present were found to be clustered in Pro-X-Y-Pro, Pro-X-Pro-Y-Pro, or Pro-Pro sequences, and the implications of these residues for the folding of the polypeptide chain are discussed.

Treatment of cow pea chlorotic mottle virus with chymotrypsin at pH 7.4 has been found [322] to release 18 amino-acids from the *N*-terminal

308 R. Barclay and R. G. Yount, *J. Biol. Chem.*, 1972, **247**, 4098.
309 K. E. Summers and I. R. Gibbons, *Proc. Nat. Acad. Sci. U.S.A.*, 1971, **68**, 3092.
310 C. J. Brokaw, *J. Exp. Biol.*, 1971, **55**, 289.
311 C. J. Brokaw, *Science*, 1972, **178**, 455.
312 K. Ogawa and H. Mohri, *Biochim. Biophys. Acta*, 1972, **256**, 142.
313 H. Glossmann and W. Bode, *Z. physiol. Chem.*, 1972, **353**, 298.
314 T. M. Joys and V. Rankis, *J. Biol. Chem.*, 1972, **247**, 5180.
315 S. Oroszlan, T. Copeland, M. Summers, and R. V. Gilden, *Biochem. Biophys. Res. Comm.*, 1972, **48**, 1549.
316 K. Kurachi, G. Funatsu, and M. Funatsu, *Agric. and Biol. Chem.* (*Japan*), 1972, **36**, 937.
317 B. Kraal, J. M. De Graaf, T. A. Bakker, G. M. A. Van Beynum, M. Goedhart, and L. Bosch, *European J. Biochem.*, 1972, **28**, 20.
318 M. W. Rees and M. N. Short, *Virology*, 1972, **50**, 772.
319 D. Roy, W. D. Graziadei, P. Lengyel, and W. Konigsberg, *Biochem. Biophys. Res. Comm.*, 1972, **46**, 1066.
320 D. T. Snell and R. E. Offord, *Biochem. J.*, 1972, **127**, 167.
321 R. Peter, D. Stehelin, J. Reinbolt, D. Collot, and H. Duranton, *Virology*, 1972, **49**, 615.
322 J. H. Tremaine, H. O. Agrawal, and J. Chidlow, *Virology*, 1972, **48**, 245.

region of the coat protein without alteration of the morphology of the virus particle. However, digestion with trypsin at the same pH caused the removal of 25 residues from the same region, but this time the cleavage was accompanied by disassembly of the virus into a 6S component. These workers have deduced the sequence of these 25 amino-acids and postulate that as residues 21—25 carry three positive charges they may be important for RNA binding and maintenance of the integrity of the virus.

5 Immunoglobulins

Complete Sequences.—The sequences of several light chains from immunoglobulins have been reported and not surprisingly many of these are Bence-Jones proteins. Three of the light chains are of the κ-type; two belong to subgroup I and one to subgroup III. In the case of Bence-Jones protein Au the sequence of the entire variable part (residues 1—108) was determined [323] but the sequence of the constant part (residues 109—214) was only partially characterized. However, the available sequence and the existence of the specific deletion at position 29 show that it belongs to subgroup II. It is suggested that comparisons of the sequence of this protein with other κ chains of subgroup I indicate that the amino-acid exchanges in a subgroup are not randomly distributed, but that linkage groups exist which allow further subdivisions of the subgroups. Some support for this view has been obtained from the sequence of the Bence-Jones protein SCW.[324] It is also a κ-type immunoglobulin light chain which has been assigned to subgroup I on the basis of its sequence and the characteristic deletion after position 29. Although the protein shares all the linked amino-acid exchanges characteristic of the K1/1 sub-subgroup, with the exception of positions 50 and 103, it has only about 80% homology with other members of the sub-subgroup, suggesting that a new linkage group might exist in subgroup I. Clearly more sequences of proteins in the subgroup will be necessary before such a possibility can be confirmed or denied.

The complete sequence of Bence-Jones proteins T_1, a κ-type immunoglobulin light chain, has also been determined,[325–328] and the variable region shows that the protein belongs to subgroup III of the κ-chains and is therefore the only protein of this subgroup with a completely known sequence.

The sequence of Bence-Jones protein NE1,[329] a λ-type light chain, shows the specific exchanges and the deletion at position 96 which are

[323] H. Schiechl and N. Hilschmann, *Z. physiol. Chem.*, 1972, **353**, 345.
[324] M. Eubitz, D. Gotze, and N. Hilschmann, *Z. physiol. Chem.*, 1972, **353**, 478.
[325] L. Suter, H. U. Barnikol, and N. Hilschmann, *Z. physiol. Chem.*, 1972, **353**, 143.
[326] L. Suter, H. U. Barnikol, and N. Hilschmann, *Z. physiol. Chem.*, 1972, **353**, 151.
[327] H. U. Barnikol, S. Watanabe, L. Suter, and N. Hilschmann, *Z. physiol. Chem.*, 1972, **353**, 160.
[328] L. Suter, H. U. Barnikol, S. Watanabe, and N. Hilschmann, *Z. physiol. Chem.*, 1972, **353**, 189.
[329] F. E. Garver and N. Hilschmann, *European J. Biochem.*, 1972, **26**, 10.

characteristic of subgroup II of the λ-chains. However, in contrast to all known λ-chains of this subgroup, the protein contains a sixth cysteine residue at position 88 which is adjacent to the invariant cysteine at position 87. It also contains carbohydrate which is joined to the protein at position 93 (Asx), one of the hypervariable areas of the variable part.

The existence of a large number of subgroups for the κ-chains of mouse immunoglobulins has often been cited as evidence for the germ-line theory of immunoglobulin variability. However, until the recent report on the mouse myeloma protein MOPC 21 no complete sequence had been determined for a mouse κ light chain [330] (Figure 9). A comparison of the sequences in the constant regions of mouse and human κ- and λ-chains indicates that the rate of evolution of the κ-chains has been faster than that of the λ-chains. Furthermore, the sequence of MOPC 21 does not seem to fit into any of the nine subgroups previously defined for the mouse κ-chains so it probably belongs to a new subgroup.

The sequence of β_2-microglobulin,[331] a protein appearing in increased amounts in humans with renal malfunction, has been determined. Interest in the protein resulted from the possibility that it might represent one of the putative domains in the whole immunoglobulin molecule since the protein contains about one hundred residues and a single disulphide bond. Furthermore, the sequence of the first 46 residues in a separate study [332] had shown that the region possesses considerable homology with parts of the constant region of the heavy chain in immunoglobulins Eu(IgG_1) and Vin(IgG_4). The total sequence shows substantial homology with the constant region of the Eu light chain and with the homology regions of the Eu heavy chain. Of the three homologous domains in the heavy chains β-microglobulin is most homologous with the C3 region of the heavy chain, *i.e.* the *C*-terminal region. Thus the protein could be an isolated domain from an immunoglobulin molecule although it is not yet possible to determine whether it is a proteolytic fragment or the product of a distinct gene. Most of the available evidence favours the latter. The β-microglobulin from dog [333] has also been isolated and its amino-terminal sequence determined. Of the 42 known positions only 7 are different from those in the human proteins, confirming the homology between the two proteins. The availability of an animal source of this protein should facilitate studies on the origin of the molecule.

Partial Sequences.—The availability of automatic sequencers has permitted studies on several problems which were previously almost insurmountable. Thus the question of the degree of heterogeneity in the light chains of rabbit antibodies directed against *p*-azophenyltrimethylammonium and

[330] J. Svasti and C. Milstein, *Biochem. J.*, 1972, **128**, 427.
[331] P. A. Peterson, B. A. Cunningham, I. Berggard, and G. M. Edelman, *Proc. Nat. Acad. Sci. U.S.A.*, 1972, **69**, 1697.
[332] O. Smithies and M. D. Poulik, *Science*, 1972, **175**, 187.
[333] O. Smithies and M. D. Poulik, *Proc. Nat. Acad. Sci. U.S.A.*, 1972, **69**, 2914.

```
              5           10           15           20           25           30
Asn-Ile-Val-Met-Thr-Gln-Ser-Pro-Lys-Ser-Met-Ser-Met-Ser-Val-Gly-Glu-Arg-Val-Thr-Leu-Thr-Cys-Lys-Ala-Ser-Glu-Asn-Val-Val-

             35           40           45           50           55           60
Thr-Tyr-Val-Ser-Trp-Tyr-Gln-Gln-Lys-Pro-Glu-Gln-Ser-Pro-Lys-Leu-Leu-Ile-Tyr-Gly-Ala-Ser-Asn-Arg-Tyr-Thr-Gly-Val-Pro-Asp-

             65           70           75           80           85           90
Arg-Phe-Thr-Gly-Ser-Gly-Ser-Ala-Thr-Asp-Phe-Thr-Leu-Thr-Ile-Ser-Ser-Val-Gln-Ala-Glu-Asp-Leu-Ala-Asp-Tyr-His-Cys-Gly-Gln-

             95          100          105          110          115          120
Gly-Tyr-Ser-Tyr-Pro-Tyr-Thr-Phe-Gly-Gly-Gly-Thr-Lys-Leu-Glu-Ile-Lys-Arg-Ala-Asp-Ala-Ala-Pro-Thr-Val-Ser-Ile-Phe-Pro-Pro-

            125          130          135          140          145          150
Ser-Ser-Glu-Gln-Leu-Thr-Ser-Gly-Gly-Ala-Ser-Val-Val-Cys-Phe-Leu-Asn-Asn-Phe-Tyr-Pro-Lys-Asp-Ile-Asn-Val-Lys--Trp-Lys-Ile-

            155          160          165          170          175          180
Asp-Gly-Ser-Glu-Arg-Gln-Asn-Gly-Val-Leu-Asx-Ser-Asx-Thr-Glx-Trp-Asx-Ser-Lys-Asp-Ser-Thr-Tyr-Ser-Met-Ser-Ser-Thr-Leu-Thr-

            185          190          195          200          205          210
Leu-Thr-Lys-Asp-Glu-Tyr-Glu-Arg-His-Asn-Ser-Tyr-Thr-Cys-Glu-Ala-Thr-His-Lys-Thr-Ser-Thr-Ser-Pro-Ile-Val-Lys-Ser-Phe-Asn-

Arg-Asn-Glu-Cys
```

Figure 9 *Amino-acid sequence of* MOPC 21 *light chain*

p-azobenzenearsonate haptens has been investigated.[334] The antibodies were first purified by immunoabsorption followed by preparative isoelectric focusing. Most of the electrofocused fractions had light chains with relatively unambiguous major sequences. This confirmed the belief that the methods used to separate the antibodies into fractions could yield essentially homogeneous light chains. It was also found that the light chains from a single antiserum are relatively homogeneous in overall sequence, but there was significant microheterogeneity within limited areas. The sequences were representative of either subgroup II or III of the κ-chain.

The amino-terminal region of the heavy chain from mouse immunoglobulin MOPC 173 has also been determined.[335] The region was isolated as a cyanogen bromide fragment which represents the first 104 residues of this heavy chain. Although this does not represent the entire length of the variable region it is argued on grounds of homology with immunoglobin Eu that the variable region could be about 116 residues. It is suggested that the sequence allows the definition of a third subgroup of variability for the heavy-chain pool.

Another hypothesis which has attracted some attention recently is the suggestion that 2-pyrrolidone-5-carboxylic acid, which has been found at the *N*-termini of immunoglobulins and some other proteins, is involved in the initiation of protein synthesis in eukaryotic cells. Evidence has been obtained that its presence in proteins is not an artifact and that it comes directly from glutamic acid rather than glutamine.[336] On the other hand, it has also been suggested that the pyrrolidonecarboxylic acid is not present at the initiation of protein synthesis in the cells concerned, but is incorporated after synthesis of the protein molecules.[337] However, this could not be possible if the claim that there is no glutamine or glutamine-producing system in the cells used in the above study is correct.

The availability of homogeneous rabbit antibody to type III pneumococci has permitted the determination of the amino-terminal sequence of the heavy chain from a functional antibody.[338] The *N*-terminal cyanogen bromide fragment was isolated and sequenced on an automatic sequencer. However, it was also necessary to use the more conventional techniques since the *N*-terminus is blocked. The high yield obtained from the sequencer confirmed the homogeneity of the preparation. In all, 69 residues at the *N*-terminus were identified. Two regions of the heavy chains (residues 35—46 and 62—69) are very similar to corresponding regions of heavy chains from other rabbit, mouse, guinea-pig, and human immunoglobulins. However, residues 47—62 appear to be variable.

[334] M. H. Freedman, R. B. Guyer, and W. D. Terry, *J. Biol. Chem.*, 1972, **247**, 7051.
[335] A. Bourgois, M. Fougereau, and C. De Preval, *European J. Biochem.*, 1972, **24**, 446.
[336] D. R. Twardzik and A. Peterkovsky, *Proc. Nat. Acad. Sci. U.S.A.*, 1972, **69**, 274.
[337] D. J. Stott and A. J. Munro, *Biochem. J.*, 1972, **128**, 1221.
[338] J. C. Jaton and D. G. Braun, *Biochem. J.*, 1972, **130**, 539.

A preliminary report has appeared on the partial amino-acid sequence of the λ-type chain of IgA(Ch).[339] Comparison of the sequence with four other human λ-Bence-Jones proteins shows that 33 residues out of the 74 positions compared are common to all five proteins. The *N*-terminus of IgA(Ch) is unblocked and there is an insertion of a tyrosine residue immediately after the *N*-terminal residue. It is suggested on the basis of comparison with other λ-type light chains that IgA(Ch) belongs to subgroup IV. There also is a substitution of glutamic acid or glutamine for lysine at position 130, *i.e.* in the constant region, and position 190 is arginine, indicating that it belongs to Oz(−).

The deletions in light and heavy chains of IgG(Sac) have now been characterized.[340] Previous studies had shown that the Fc region of the heavy chain was normal and that the deletions were in the Fab fragment. From comparison of the sequence of the light chain with that of Ig(1u) it appears that a deletion of 68 residues in the variable region has occurred and the discontinuity probably lies between residues 18 and 19 or 19 and 20. The sequence of the heavy chain from IgS(Sac) in the *N*-terminal region is unique among those reported for the heavy chains of myeloma proteins of all classes. On the basis of a comparison with other heavy chains it appears that the region is homologous with residues 1—102 of the heavy chain of IgG(Eu). In addition there appears to have been an insertion of two residues (14 and 15) corresponding to residues 116 and 117 of the latter, but there do not seem to be any alterations in the constant region of the heavy chain.

Disulphide Bridges.—The twelve intrachain disulphide bonds and the interchain disulphide bridges of the MOPC 173 (G2a) monoclonal immunoglobulin have been characterized in connection with the elucidation of the primary structure of this protein.[341] All cysteine-containing peptides were sequenced and positioned by homology with other mouse and human γ-chains. However, one of the intrachain bridges (HIV–HV) was inferred by alignment of the two separated branches around the bridge.

A peptide containing the inter-heavy-chain bridges ('hinge' region) of an IgA_2,Am2(+) myeloma protein has been isolated, sequenced, and compared with an IgA_1 myeloma protein.[342] The hinge peptide could be identified by its high proline content since there is a continuous sequence of five proline residues. There are three cysteine residues and diagonal electrophoresis has shown it to be bridged to two other peptides. The most striking difference between the IgA_2,Am2(+) proteins and the IgA_1 proteins is the absence of a region corresponding to the peptide containing the half-cysteine joining the heavy to the light chain present in the latter.

[339] Y. Okada, Y. Nozu, K. Titani, S. Watanabe, H. Hara, and M. Kitagawa, *Immunochemistry*, 1972, **9**, 207.
[340] D. M. Parr, M. E. Percy, and G. E. Connell, *Immunochemistry*, 1972, **9**, 51.
[341] C. De Preval and M. Fougereau, *European J. Biochem.*, 1972, **30**, 452.
[342] C. Wolfenstein-Todel, B. Frangione, and E. C. Franklin, *Biochemistry*, 1972, **11**, 3971.

This fact has been used to type immunoglobulins with these chains chemically.[343]

The intrachain disulphide bridges of rabbit immunoglobulin κ-type light chains have been partially characterized following the earlier discovery that there are three such bridges instead of the usual two.[344] The three bridges have been assigned to the approximate positions 23—88, 134—194, and 'somewhere in the V region'—171. Thus the extra disulphide bridge in rabbit κ-chains joins the variable and constant regions. The presence of this extra disulphide bridge in both the light chain and the Fd portion of the heavy chain is interesting since rabbits as a species probably evolved after the separation of H-chain and L-chain genes.

The disulphide bridges of the mouse immunoglobulin MOPC 21 (IfG1) have been characterized using [^{35}S]cysteine-labelled protein which was prepared from myeloma cells in tissue culture.[345] One peptide of 96 residues from peptic and tryptic digests contained both the heavy–light interchain disulphide bridge and all the inter-heavy-chain bridges. Thus all the heavy-chain half-cysteine residues form interchain bridges and no evidence was obtained for the labile intrachain disulphide bond described in rabbit IgG. In contrast with the large homology in the sequences defining the intrachain S—S bridges of heavy chains there is very little homology in the sequences containing the heavy-chain interchain bridges. It is suggested that it would be more appropriate to call this the 'bridge' region rather than the 'hinge' region. The arrangement of the disulphide bonds in this protein has also been determined.[346] A unique arrangement, out of 170 possibilities, was obtained and it shows that all the heavy–heavy interchain bonds are parallel. This is the first case in which the arrangement of all the interchain bonds of any of the human or mouse IgG subclasses has been established.

The interchain disulphide bonds in immunoglobulins have proved a useful handle in some studies on the addition of carbohydrate to the molecules.[347] The approach relies on the fact that the LHHL molecule is the most mature interchain-disulphide-linked structure, HH and HHL being more juvenile. The latter are found predominantly in the rough microsomal fraction whereas the completed molecules are the main species in the smooth microsome fraction, thus establishing the temporal relationship between the two compartments. It was also shown that the acquisition of galactose occurs at a much later stage than does that of mannose.

Partial Fragmentation.—Cleavage of immunoglobulins by limited proteolysis continues to yield valuable information on the structures of these proteins. Further incentive for this effort has been provided by the domain

343 B. Frangione and E. C. Franklin, *F.E.B.S. Letters*, 1972, **20**, 321.
344 M. E. Lamu and B. Frangione, *Biochem. J.*, 1972, **128**, 1357.
345 J. Svasti and C. Milstein, *Biochem. J.*, 1972, **126**, 837.
346 J. Svasti and C. Milstein, *European J. Biochem.*, 1972, **31**, 405.
347 E. W. Sutherland, E. H. Zimmerman, and M. Kerr, *Proc. Nat. Acad. Sci. U.S.A.*, 1972, **69**, 167.

hypothesis which proposes that there are several independently folded regions in the light and heavy chains.[348] Direct support for the hypothesis has been obtained from a study of the proteolytic fragmentation of light chains. A variety of proteinases was used and in all cases the light chains were split into two halves, one containing the variable region and one the constant region.[349] It also appears that each region on splitting acquires about five residues of the other, which has led to the conclusion that the region which is sensitive to proteolysis consists of less than ten residues. Physical characterization of the fragments indicates that whole light chains are longer than two half chains. It has also been found that the Fd fragment from rabbit IgG can be converted into two smaller fragments with papain, and the molecular weights of these fragments (11 500) are approximately those predicted for the variable and constant domains in the protein. However, the fragments have not been fully characterized.[350]

The existence of a single disulphide bond between the variable and constant regions of rabbit antibody light chains has also been demonstrated by limited proteolysis.[351] It was found that the size of the chain did not decrease after proteolysis unless subjected to full reduction and alkylation. This is interpreted to mean that the variable and constant regions are folded into separate tight domains and that proteolysis is confined to the switch regions between the two. Since the two domains are also joined by a disulphide bond they do not separate on proteolysis. This interdomain disulphide bond places an interesting constraint on the putative genes for the variable and constant regions since the code for one half-cysteine would have to be conserved in all genes covering variable regions.

The ability to split immunoglobulins into the various domains has also permitted the localization of the antibody combining sites within the variable portions of the heavy and light chains.[352] Pepsin splits the Fab′-fragment of mouse IgA myeloma (protein 315) into a smaller (Fv) fragment which was shown to contain the variable portions of the heavy and light chains held together by non-covalent bonds. Since Fv has about one binding site for dinitrophenol with the same binding constant as Fab′ it shows that the antibody site in this protein is contained entirely in the variable portion and is independent of the constant part of the molecule.

Cyanogen bromide cleavage has been used extensively for the isolation of fragments of the heavy chain which contain the variable part of the chain but are longer than the Fd fragment.[353] It appears that the remarkable similarity in the amino-acid analyses of various preparations of C_1 arises from the formation of aggregates in solvents such as 6M-urea or 1M-acetic

[348] G. M. Edelman, B. A. Cunningham, W. E. Gall, P. D. Gottlieb, V. Rutishauser, and M. J. Waxdal, *Proc. Nat. Acad. Sci. U.S.A.*, 1969, **63**, 78.
[349] F. A. Karlsson, P. A. Peterson, and J. Berggard, *J. Biol. Chem.*, 1972, **247**, 1065.
[350] J. S. Huston, I. Bjork, and C. Tanford, *Biochemistry*, 1972, **11**, 4256.
[351] K. Poulsen, K. J. Fraser, and G. Haber, *Proc. Nat. Acad. Sci. U.S.A.*, 1972, **69**, 2495.
[352] D. Inbar, J. Hochman, and D. Givol, *Proc. Nat. Acad. Sci. U.S.A.*, 1972, **69**, 2659.
[353] B. Friedenson, Y. Takeda, D. A. Roholt, and D. Pressman, *European J. Biochem.*, 1972, **27**, 31.

acid. This aggregation can be prevented by the use of 6M-guanidinium hydrochloride.

The cyanogen bromide fragments of the heavy chain from guinea pig immunoglobulin G_2 have been isolated and aligned.[354] The alignment was achieved by a combination of tryptic and chymotryptic overlaps around the methionine residues in the intact protein. The alignment achieved is probably correct since it accounts for the entire protein. The sequence of the 49-residue CNBr fragment from the *N*-terminal quarter has been determined.[355] The fragment which spans residues 35—83 of the heavy chain has two regions of almost constant sequence joined by a very variable region.

Chemical Modification.—As in the case of enzymes, specific modifications can be used to study both the combining sites and the conformations of immunoglobulins. In the case of the combining sites most of the work up to now has been a direct extension of the methods used in enzyme chemistry. The approach has mainly depended on a difference labelling technique, usually using iodination in the presence and absence of a hapten, and then comparing the sites of incorporation to deduce which ones are derived from the antibody site. In earlier studies the sequences of two heavy-chain tyrosyl peptides from the putative sites of the antibodies to 3-azopyridine were reported. More recently, the combining site of antibodies to *p*-azophenyltrimethylammonium ion have been investigated and also found to involve tyrosyl residues.[356] A peptide with the sequence Ile-Tyr-Asp was isolated and assigned to the heavy chain of the antibody.

The tyrosyl peptides from the combining sites of anti-3-azopyridine rabbit antibody which had been sequenced earlier have now been assigned to the variable region between residues 85 and 120 of the heavy chain.[357] The assignment relies on direct evidence that the appropriate tyrosyl residues are in a cyanogen bromide fragment representing residues 85—253 and the knowledge that the sequences within which they reside are not present in the constant region which begins at about residue 120.

Anti-dinitrophenyl antibodies from inbred guinea pigs have been modified with the affinity label *m*-nitrobenzenediazonium fluoroborate using the radioactive (^{14}C) reagent.[358] It was found that most of the label was incorporated into the heavy chains, mainly tyrosine residues in the amino-terminal region. Affinity labelling of two mouse myeloma proteins with anti-dinitrophenol activity has also been carried out with bromoacetyl derivatives of DNP ligands.[359] It had previously been shown that bromoacetyl-*N*-DNP-lysine (BADL) labels a lysyl residue in the heavy chain and

[354] D. C. Benjamin, Q. Z. Hussain, and J. J. Cebra, *Biochemistry*, 1972, **11**, 3641.
[355] B. K. Birshtein and J. J. Cebra, *Biochemistry*, 1971, **10**, 4930.
[356] Y. Takeda, B. Friedenson, O. A. Roholt, and D. Pressman, *J. Immunol.*, 1972, **108**, 1431.
[357] B. Friedenson, E. Appella, O. A. Roholt, and D. Pressman, *Biochem. Biophys. Res. Comm.*, 1972, **48**, 1208.
[358] A. Ray and J. J. Cebra, *Biochemistry*, 1972, **11**, 3647.
[359] J. Haimovich, H. N. Eisen, E. Hurwitz, and D. Givol, *Biochemistry*, 1972, **11**, 2389.

bromoacetyl-*N*-DNP-ethylenediamine (BADE) labels a tyrosyl residue in the light chain of protein-315. It now appears that BADE also labels a lysyl residue in the light chain of protein-460. The labelled residues in protein-315 are lysine-54 and tyrosine-34, whereas lysine-54 is labelled in protein-460. Since all these residues are located within restricted hypervariable segments of the variable regions it is suggested that the specific combining sites of the intact immunoglobulins are formed by amino-acids in this region. An interesting facet of this study was the discovery of a tryptophan residue in each of the affinity-labelled tryptic peptides and a red shift in the absorption spectrum of the DNP moiety of the peptides similar to that observed upon the binding of DNP ligands to anti-DNP antibody. The paired labelling technique has also been used to study the binding sites of anti-5-azoisophthalate antibodies.[360] The lysine residues were modified by maleylation and the arginine residues by glyoxalation, in the presence and absence of hapten. Although the results lack the clarity of some other data obtained by this approach, there are indications of the existence of a large population of antibody binding sites which contain arginine residues and a small population containing lysine in the combining site.

A comprehensive approach to the labelling of antibody binding sites is the use of reagents which might even attack C—H bonds since the main limitation of the methods used so far is that they only label chemically functional residues.[361] The generation of a highly reactive species such as a carbene or a nitrene at the active site could circumvent this problem. The main problem with this approach lies in the requirement for a chemically inert precursor, which automatically excludes the carbenes; however, the aryl azides, which are nitrene precursors, do have the required chemical stability. The hapten used was 4-azido-2-nitrophenyl (NAP) which was attached to bovine-γ-globulin (NAP-bovine γ-globulin) by reaction with 4-fluoro-3-nitrophenyl azide. Isolated anti-NAP antibody was usually labelled with ε-NAP-lysine by irradiation at wavelengths above 400 nm. It was found that the hapten was covalently attached to the antibody after irradiation and that the antibody binding site was blocked. There was a simultaneous labelling of the heavy and light chains in the ratio 2 : 5 : 1. Digests of the labelled heavy chain yielded two small peptides which showed that about 13% of the label in the antibody was attached to cysteine-92 and alanine-93, which are adjacent to the hypervariable region in rabbit heavy chain. It is clear that the approach has considerable potential, although a full evaluation of this will only be possible when a homogeneous protein with NAP-binding capacity is available.

The effect of maleylation on IgA has been studied and conditions have been obtained for the reversible blocking of amino-groups in the proteins.[362]

[360] G. L. Mayers, A. L. Grossberg, and D. Pressman, *Immunochemistry*, 1972, **9**, 169.
[361] G. W. J. Fleet, J. R. Knowles, and R. R. Porter, *Biochem. J.*, 1972, **128**, 499.
[362] A. Bezkorovainy and D. Grehlich, *Immunochemistry*, 1972, **9**, 137.

The interesting result from this study was that the removal of the maleyl groups was more difficult from the denatured than from the undenatured proteins.

The effect of modifying lysyl residues in human IgG on the conformational transitions of the protein has been investigated using the reversible blocking reagent citraconic anhydride.[363] Most of the lysine residues reacted and the modified protein underwent gross changes in conformation which were interpreted as expansion and disorganization of the molecule. The ability to form a precipitate with antiserum was also lost upon citraconylation, but it is argued that it is the loss of the native conformation rather than blocking of the lysine residues which causes this change since carbamylation does not cause analogous conformational changes or alter immunological precipitability. Citraconylation of this protein and some properties of the modified protein have also been the subject of another report.[364]

The nitrophenyl-binding human macroglobulin YMW ag has been shown to possess tyrosine residues with a very high susceptibility to tetranitromethane, and there is a close correlation between the degree of nitration and the loss of binding activity, in studies with the Fab′μ fragments.[365] It was also found that the major proportion of nitrotyrosine was present in the Fd′ piece rather than in the light chain. However, before concluding that the tyrosine residues concerned are directly involved in binding, an explanation will have to be found for the loss of binding activity when modification is carried out in the presence of a protecting ligand.

Concanavalin A.—A remarkable increase in the volume of research on this protein has followed the discovery that cells transformed by DNA tumour viruses or carcinogens are agglutinated by this lectin more readily than are normal cells. An excellent introduction to the entire subject of lectins can be found in a recent review.[366] The obvious place to start this discussion is at the description of the sequence (Figure 10) and three-dimensional structure of concanavalin A.[367] The details of the sequencing have not been published and three regions of the sequence were not determined chemically but were assigned on the basis of the interpretation of the election-density map. An interesting property of concanavalin A (Con A) is its existence in two forms, one with the intact subunit and the other with a subunit made up of two fragments. However, the conformation of the fragmented subunit must be very similar to that of the intact subunit since mixed crystals can be formed which are virtually indistinguishable from crystals of the intact molecule. This is probably due to the fact that

363 Y. Nakagawa, S. Capatello, and B. Jirgensons, *J. Biol. Chem.*, 1972, **247**, 5703.
364 F. S. A. Habeeb, R. E. Schrohenloher, and J. C. Bennett, *Biochim. Biophys. Acta*, 1972, **263**, 339.
365 N. A. Otchin and H. Metzger, *J. Biol. Chem.*, 1971, **246**, 7051.
366 N. Sharon and H. Lis, *Science*, 1972, **177**, 949.
367 G. M. Edelman, B. A. Cunningham, G. N. Reeke, J. W. Becker, M. J. Waxdal, and J. L. Wang, *Proc. Nat. Acad. Sci. U.S.A.*, 1972, **69**, 2580.

1 10 20 30
Ala-Asp-Thr-Ile-Val-Ala-Val-Glu-Leu-Asp-Thr-Tyr-Pro-Asn-Thr-Asp-Ile-Gly-Asp-Pro-Ser-Tyr-Pro-His-Ile-Gly-Ile-Asp-Ile-Lys-

40 50 60
Ser-Val-Arg-Ser-Lys-Lys-Thr-Ala-Lys-Trp-Asn-Met-Gln-Asp-Gly-Lys-Val-Gly-Thr-Ala-His-Ile-Ile-Tyr-Asn-Ser-Val-Asp-Lys-Arg-

70 80 90
Leu-Ser-Ala-Val-Val-Ser-Tyr-Pro-Asx-Ala-Asx-Asx-Ala-Thr-Ser-Val-Ser-Tyr-Asx-Val-Asx-Leu-Asx-Asx-Val-Leu-Pro-Gln-Trp-Val-

100 110 120
Arg-Val-Gly-Leu-Ser-Ala-Ser-Thr-Gly-Leu-Tyr-Lys-Glu-Thr-Asn-Thr-Leu-Ile-Ser-Phe-Ser-Trp-Thr-Ser-Lys-Leu-Lys-Ser-Asx-Ser-

130 140 150
Thr-His-Glx-Thr-Asx-Ala-Leu-His-Phe-Met-Phe-Asn-Gln-Phe-Ser-Lys-Asp-Gln-Lys-Asp-Leu-Ile-Leu-Gln-Gly-Asp-Ala-Thr-Thr-Gly-

160 170 180
Thr-Asp-Gly-Asn-Leu-Glu-Leu-Thr-Arg-Val-Ser-Ser-Asx-(Glx,Ser,Pro,Gly)-Gly-Ser-Ser-Val-Gly-Arg-Ala-Leu-Phe-Tyr-Ala-Pro-Val-

190 200 210
His-Ile-Trp-Glu-Ser-Ser-Ala-(Glx,Ala,Ser,Val)-Phe-Glu-Ala-Thr-Phe-(Thr,Leu,Val)-Ile-Lys-Ser-Pro-Asp-Ser-His-Pro-Ala-Asp-Gly-

220 230
Ile-Ala-Phe-Phe-Ile-Ser-Asn-Ile-Asx-Ser-Ser-Ile-Pro-Ser-Gly-Ser-Thr-Gly-Arg-Leu-Leu-Gly-Leu-Phe-Pro-Asp-Ala-Asn

Figure 10 *Tentative amino-acid sequence of concanavalin A.*[367]
(Reproduced by permission from *Proc. Nat. Acad. Sci. U.S.A.*, 1972, **69**, 2580)

the cleavage occurs between residues 119 and 120, which are located in a loop that extends well away from the main body of the molecule. However, the sites of proteolysis of the molecules by trypsin and chymotrypsin are different, indicating that the exposure of the loop is not the sole cause of its lability towards the naturally occurring proteolysis. The metal-ion binding sites have also been tentatively identified. The transition metal is liganded by Glu-8, Asp-10, Asp-19, and His-24 whereas the calcium-binding site is liganded by Tyr-12, Asn-14, Asp-10, and Asp-19. The report that amino-groups are involved in calcium binding has not been confirmed.

The position of the saccharide-binding site of Con A was deduced from difference projections of the β-IPG derivature. The iodine atom is surrounded by 11 hydrophobic side-chains and a serine residue. The presence of a tyrosine and a tryptophan residue in this region directly contradicts a report that these residues are not involved in the carbohydrate-binding sites of the molecule.[368] The latter study employed chemical modification of the protein by a wide variety of reagents but in no case were the modified residues identified, showing the necessity for such monitoring in chemical modification studies.

6 Membrane Proteins

Solubilization and Fractionation.—One of the main constraints on structural studies of membrane proteins has been the difficulty in obtaining sufficient quantities of these proteins. The need for good chemistry on membrane proteins has stimulated considerable interest in the fractionation of membranes and the starting point has usually been the study of Rosenberg and Guidotti,[369] which relies on serial extraction of membrane preparations with low-ionic-strength medium, high salt, and an organic solvent. The process has been used to distinguish between extrinsic proteins (adsorbed to membranes by ionic interactions) and intrinsic proteins (integrated into the structure of the membrane).[370] In beef erythrocyte ghosts 46% of the total membrane protein was solubilized by treatment with distilled water and 1.2M-NaCl, and the polypeptides removed were very large (molecular weight 200 000). The fraction containing the intrinsic proteins was separated into 10 polypeptides of which 4 were glycoproteins. One of the glycoproteins contained 50% of the total protein in the intrinsic fraction. However, the proteins were not isolated in preparative quantities since the final purification was based on electrophoretic mobility. A more comprehensive procedure for the isolation of proteins from human erythrocyte membranes which yields sufficient amounts of three major proteins for analytical studies has been described.[371] The sialoglyco-

368 G. S. Hassing and I. J. Goldstein, *Biochim. Biophys. Acta*, 1972, **271**, 378.
369 S. A. Rosenberg and G. Guidotti, *J. Biol. Chem.*, 1968, **243**, 1985.
370 R. A. Capaldi, *European J. Biochem.*, 1972, **29**, 74.
371 M. J. A. Tanner and D. H. Boxer, *Biochem. J.*, 1972, **129**, 333.

protein isolated from the pyridine supernatant had an apparent molecular weight of about 45 000 and contains about 54% of the protein. Neither the *C*-terminal nor the *N*-terminal analysis gave satisfactory results. The other two proteins were isolated from the pyridine precipitate but the ultimate separation relied on preparative electrophoresis so that the amounts of pure protein obtained were relatively small.

The value of pyridine extraction for solubilizing glycoproteins is apparent from a study on plasma membranes.[372] Stepwise extraction with alkali buffer and 33% pyridine gave only one glycoprotein in the alkali-soluble fraction whereas about 13 were evident in the pyridine-soluble fraction which could be fractionated further by gel-filtration.

Lithium di-iodosalicylate promises to become a most useful reagent for obtaining biologically active and water-soluble glycoproteins since they can be extracted from membranes by quite low concentrations of the reagent.[373] The solubility of the glycoproteins is retained even after removal of the reagent, so it is possible to handle them with relative ease. Thus it has been possible to isolate quite large amounts of a single glycoprotein from human erythrocyte membranes in this manner. Glycoproteins from this source have also been separated from other membrane proteins after extraction of the aqueous phase with chloroform–methanol.[374] The preparation contains three of the glycoproteins which are thought to extend through the erythrocyte membrane and one other unspecified protein.

The high affinity of the lectin concanavalin A for carbohydrates has been exploited in the development of a general procedure for the isolation of membrane glycoprotein.[375] The dissociation of the membrane was effected with deoxycholate and the product passed through a concanavalin–Sepharose column. Using methyl-α-D-glucopyranoside for elution 5% of the total membrane protein could be isolated. All this appeared to be glycoprotein. The main drawback to this promising technique was the loss of about 20% of the total membrane protein during chromatography, presumably as a result of irreversible binding to the column. An improvement in the yield of absorbed glycoprotein has been obtained by using phyto-haemagglutinin instead of concanavalin.[376] It is claimed that about twice as much glycoprotein was obtained with the latter procedure.

Isoelectric focusing has been used to obtain greater resolution of membrane proteins which have been dissolved in 8M-urea, and the effectiveness of the technique is apparent from the fact that about 40 species were detected.[377] This is substantially more than the number obtained using

372 J. W. Gurd, W. H. Evans, and H. R. Perkins, *Biochem. J.*, 1972, **126**, 459.
373 V. T. Marchesi and E. P. Andrews, *Science*, 1971, **174**, 1247.
374 H. Hamaguchi and H. Cleve, *Biochim. Biophys. Acta*, 1971, **278**, 271.
375 D. Allan, J. Auger, and M. J. Crumpton, *Nature New Biol.*, 1972, **236**, 23.
376 M. J. Hayna and M. J. Crumpton, *Biochem. Biophys. Res. Comm.*, 1972, **47**, 923.
377 D. C. Merz, R. A. Good, and H. W. Litman, *Biochem. Biophys. Res. Comm.*, 1972, **49**, 84.

SDS-gel electrophoresis alone, and a combination of the two techniques should prove very effective.

The value of lithium di-iodosalicylate in studies on membrane structure is apparent from a report on the structure of the major glycoprotein of human erythrocyte membranes.[378] The protein was isolated in sufficient quantities to permit characterization by tryptic digestion and cyanogen bromide cleavage. The studies on the peptides indicate that the protein component of the glycoprotein has a molecular weight of 55 000. The five cyanogen bromide fragments have been tentatively ordered and support the view that the protein only contains a single polypeptide chain. However, it now appears that two of the fragments are not obtained in good yield and may be artifacts.[379] The carbohydrate has been assigned to the *N*-terminal region of the protein and it is envisaged that oligosaccharide moieties are projected out of the membrane whereas the *C*-terminal region, which contains a large amount of non-polar amino-acids, is embedded within the membrane. Fifty-one residues from the *C*-terminal region have now been sequenced, part of the sequence coming from the *C*-terminal cyanogen bromide fragment and the rest from a tryptic peptide. Although details are lacking, there is support for the contention that the glycoprotein can be considered to have three regions, an *N*-terminal half containing all the carbohydrate receptor sites, an extremely hydrophobic region (23 'hydrophobic' amino-acids in a single stretch), and a *C*-terminal region which is relatively hydrophilic and rich in proline. It is argued that the hydrophobic region is the part of the molecule which actually spans the membrane.

Receptor Proteins.—The alternative approach of isolating membrane proteins by exploiting some facet of their function has been widely used to isolate various receptor proteins.[380] Notable amongst these is the acetylcholine receptor and, as with most receptor proteins, two basic approaches are employed. One is the 'affinity label technique' in which a radioactive reagent is bound irreversibly to the active site of the receptor. An extension of this method is the use of affinity chromatography. The second and more tedious method is to monitor the receptor through purification by a binding assay. However, the first priority in this field must be to decide whether a putative receptor is a protein, since the literature is riddled with reports of every imaginable type of compound as putative receptor. This decision is not easy when the approach relies on the formation of complexes with highly radioactive compounds. In a study using ^{3}H-acetylated α-bungarotoxin it is reported that binding of the toxin to potential binding sites in pig cerebral cortex homogenates is decreased

[378] V. T. Marchesi, T. W. Tillack, R. L. Jackson, J. P. Secrest, and R. E. Scott, *Proc. Nat. Acad. Sci. U.S.A.*, 1972, **69**, 1445.

[379] J. P. Secrest, R. L. Jackson, V. T. Marchesi, R. B. Guyer, and W. Terry, *Biochem. Biophys. Res. Comm.*, 1972, **49**, 964.

[380] R. D. O'Brien, M. E. Eldefrawi, and A. T. Eldefrawi, *Ann. Rev. Pharm.*, 1972, **12**, 19.

when the homogenate is treated with proteinases.[381] A complex of the toxin with a membrane component was purified and found to have a molecular weight of 94 000. Some properties of the complex were consistent with the view that the membrane component was a protein. Several studies with [^{125}I]-α-bungarotoxin have shown that complexes are formed with various membrane components, but the nature of the receptor has not been defined. The involvement of a protein in acetylcholine binding is also suggested by the observation that diazo-reagents and reagents which react with sulphydryl groups and disulphide bonds markedly reduce the binding of acetylcholine to particulate preparations of *Torpedo* electroplax.[382] Several other reports also provide direct evidence that the acetylcholine receptor is a protein. A five-thousand-fold purification of the 'receptor' for cobra toxin in the electric organ of *Electrophorus electricus* has been achieved and it has been concluded to be a protein on the basis of its sensitivity to trypsin and pronase.[383]

The use of affinity chromatography has led to the enrichment of several acetylcholine receptors and provided very strong evidence that they are proteins.[384] The technical barriers to the use of resins with covalently bound quaternary ammonium functions have been resolved and a resin with *N*-(ε-aminohexanoyl)-3-aminopropyltrimethylammonium bromide hydrobromide linked to agarose was used to purify a membrane protein preparation from the electroplax of *Narcine*. The final purification was about 14-fold and gel-electrophoresis shows one major component with a molecular weight of 28 000, which is about half that obtained for the putative receptor from *Torpedo marmorata*. Affinity chromatography has also been used to purify the receptor from *Electrophorus electricus* using another quaternary ammonium compound, CT 5263.[385] The purification achieved in a single step was about 150-fold, but the material gave several bands on SDS-gel electrophoresis. Affinity chromatography of membrane constituents from the electric organ of *Torpedo* have also been carried out using a column of *Naja* neurotoxin coupled to Sepharose.[386] The purification achieved was about 1000-fold, but in this case too the preparations were not homogeneous.

Considerable success has been achieved in the isolation of the insulin receptor through the use of affinity chromatography.[387] The final purification achieved using a column containing various insulin analogues was about 250 000-fold, which is thought to yield a protein of about 60% purity. It appears that the presence of the detergent necessary to solubilize the receptor does not interfere with the purification procedure. The main

[381] H. B. Bossmann, *J. Biol. Chem.*, 1972, **247**, 130.
[382] M. E. Eldefrawi and A. T. Eldefrawi, *Proc. Nat. Acad. Sci. U.S.A.*, 1972, **69**, 1776.
[383] B. Fulpuis, S. Cha, R. Klett, and E. Reich, *F.E.B.S. Letters*, 1972, **24**, 323.
[384] J. Schmidt and M. A. Raftery, *Biochem. Biophys. Res. Comm.*, 1972, **49**, 572.
[385] R. W. Olsen, J. C. Mennier, and J. P. Changeux, *F.E.B.S. Letters*, 1972, **28**, 96.
[386] E. Karlsson, E. Heilbronn, and J. Widlund, *F.E.B.S. Letters*, 1972, **28**, 107.
[387] P. Cuatracasas, *Proc. Soc. Nat. Acad. Sci. U.S.A.*, 1972, **69**, 1277.

limitation at the moment is the difficulty in scaling the procedure up to a level which will permit chemical studies on the receptor.

7 Peptides and Hormones

Hormones may act in two ways. After secretion of the hormone from the endocrine glands into the blood-stream, interaction with a specific receptor in the cell membrane causes activation of adenyl cyclase. This enzyme catalyses the formation of cyclic AMP from ATP, which in turn activates other enzymes, *i.e.* those involved in glycogen and fat metabolism. Alternatively, the hormone may interact directly with DNA, possibly by removal of repressor molecules, and thus initiate enzyme synthesis by direct gene activation. In this way glycogen synthetase, phosphofructokinase, and pyruvate kinase, key enzymes of the glycolytic pathway, may be induced with insulin. On the other hand induction of pyruvate carboxylase, phosphoenol pyruvate carboxykinase, and fructose-1,6-diphosphatase with glucocorticosteroids reverses the pathway, causing the synthesis rather than the degradation of glucose. A brief summary of the properties and biological activities of some hormones is given in Table 2. Structure–function relationships of some of these molecules are fully discussed in Chapter 5 of this Report.

Table 2 *Biological effects of some hormones*

Hormone	*Mode of action* *via Adenyl cyclase*	*Induction of enzyme synthesis*	*Biological effect*
Adrenocorticotropic hormone	+	+	Stimulation of steroid hormone synthesis
Thyroid-stimulating hormone	+	+	Stimulation of thyroid hormone synthesis
Luteinizing hormone	+	−	Stimulation of sex hormone synthesis
Vasopressin	+	−	Contraction of vascular muscle
Oxytocin	+	−	Contraction of smooth muscle
Catecholamines	+	−	Activation of phosphorylase, stimulation of cardiac, circulatory, and nervous systems
Insulin	+	+	Lowering of blood sugar level; promotion of glycogen, protein, and lipid synthesis
Glucagon	+	−	Activation of liver phosphorylase
Melanocyte-stimulating hormone	+	−	Regulation of pigment distribution in the skin
Parathyroid hormone	+	−	Mobilization of Ca^{2+}; increase in PO_4^{2-} excretion
Angiotensin	+	−	Increase in blood pressure; stimulation of aldosterone secretion

Table 2 (*cont.*)

Hormone	*Mode of action*: *via Adenyl cyclase*	*Induction of enzyme synthesis*	*Biological effect*
Prostaglandins	+	−	Smooth muscle contraction; inhibition of fatty tissue lipase
Histamine	+	−	Smooth muscle contraction; stimulation of HCl production in the stomach
Serotonin	+	−	Affects central nervous system
Glucocorticosteroids	−	+	Stimulation of gluconeogenesis; inhibition of inflammation; immunosuppresive action
Thyroid hormones	−	+	Induction of respiratory enzymes
Growth hormone	−	+	Growth; stimulation of protein synthesis and gluconeogenesis
Estradiol (estrogens)	−	+	Female sexual differentiation; stimulation of lipid synthesis
Testosterone	−	+	Male sexual differentiation; promotion of protein synthesis
Erythropoietin	−	+	Formation of erythrocytes

Pituitary Hormones.—The α-subunit of human chorionic gonadotropin, the hormone which prolongs the life of the corpus luteum during early pregnancy, has been found [388] to be identical to the corresponding subunit of human luteinizing hormone [389] except that there are three additional residues at the *N*-terminus. Although the sequence of the β-chain of human luteinizing hormone has not yet been reported there are obvious homologies between that from the ovine hormone and the β-subunit of the human chorionic gonadotropin.[390-392] The carbohydrate composition of both chains of chorionic gonadotropin is, however, completely different from that of the luteinizing hormones, and this might explain the difference in biological properties. Luteinizing hormones have previously been shown to be virtually identical with the thyroid-stimulating hormones, and it is interesting to note that in spite of the carbohydrate differences and sequence variations in the *N*-terminal regions, the α-subunits of all three types of hormone are interchangeable, with retention of biological activity.

[388] O. P. Bahl, R. B. Carlsen, R. Bellisario, and N. Swaminathan, *Biochem. Biophys. Res. Comm.*, 1972, **48**, 416.

[389] T. Inagami, K. Murakami, D. Puett, A. S. Hartree, and A. Nureddin, *Biochem. J.*, 1972, **126**, 441.

[390] M. R. Sairam, H. Papkoff, and C. H. Li, *Biochem. Biophys. Res. Comm.*, 1972, **48**, 530.

[391] W.-K. Liu, H. S. Nahm, C. M. Sweeney, G. N. Holcomb, and D. N. Ward, *J. Biol. Chem.*, 1972, **247**, 4365.

[392] W.-K. Liu, H. S. Nahm, C. M. Sweeney, W. M. Lamkin, H. N. Baker, and D. N. Ward, *J. Biol. Chem.*, 1972, **247**, 4351.

The releasing factor of ovine luteinizing hormone has been purified and sequenced:[393]

Glp-His-Trp-Ser-Tyr-Gly-Leu-Arg-Pro-Gly-NH_2

This is identical to the corresponding porcine releasing factor. A hypothalamic peptide which inhibits the release of melanocyte-stimulating hormone from the pituitary gland has been found[394] to have the sequence

Pro–His–Phe–Arg-Gly–NH_2

Although the authors used mass spectrometry to obtain this structure, apparently no particular care was taken to avoid quaternization of the histidine residue during the derivatization procedure (see the section on mass spectrometry). This is unusual as salt formation on histidine side-chains will generally render the derivative non-volatile.

Pancreatic Hormones.—In proinsulin, the insulin precursor molecule, the A- and B-chains are joined by an interconnecting C-peptide in the manner

B-chain–Arg-Arg–C-peptide–Lys-Arg–A-chain

The A- and B-chains are linked by S—S bridges, and after proteolytic cleavage the basic amino-acids and C-peptide are released, leaving the intact insulin molecule. Amino-acid sequences are now available for eight C-peptides (Figure 11) and it is clear that compared with native insulin there is considerable variability in both their composition and their length. Apart from maintaining the insulin in a non-active form a possible function of the C-peptide might simply be to position the A- and B-chains for correct disulphide-bridge formation, in which case considerable sequence variability can be tolerated.

A study has been made[399] of the effect of iodination on the hormonal activity of insulin, and the amino-acid sequence of this hormone from the mouse, *Mus musculus*, has been given.[400] Further information on all aspects of insulin biochemistry can be found in a recent review.[401]

[393] R. Burgus, M. Butcher, M. Amiss, W. Ling, M. Monahan, J. Rivier, R. Fellows, R. Blackwell, W. Vale, and R. Guillemin, *Proc. Nat. Acad. Sci. U.S.A.*, 1972, **69**, 278.
[394] R. M. G. Nair, A. J. Kastin, and A. V. Schally, *Biochem. Biophys. Res. Comm.*, 1972, **47**, 1420.
[395] A. S. C. Ko, D. G. Smyth, J. Markussen, and F. Sundby, *European J. Biochem.*, 1971, **20**, 190.
[396] J. D. Peterson, S. Nehrlich, P. E. Oyer, and D. F. Steiner, *J. Biol. Chem.*, 1972, **247**, 4866.
[397] H. S. Tager and D. F. Steiner, *J. Biol. Chem.*, 1972, **247**, 7936.
[398] J. Markussen and F. Sundby, *European J. Biochem.*, 1972, **25**, 153.
[399] C. J. Garratt, D. M. Harrison, and M. Wicks, *Biochem. J.*, 1972, **126**, 123.
[400] H. F. Bünzli, B. Glatthaar, P. Kunz, E. Mülhaupt, and R. E. Humbel, *Z. physiol. Chem.*, 1972, **353**, 451.
[401] T. Blundell, G. Dodson, D. Hodgkin, and D. Mercola, *Adv. Protein Chem.*, 1972, **26**, 279.

	1	2	3	4	5	6	7	8	9	10	11	12	13	14	15	16	17	18	19	20	21	22	23	24	25	26	27	28	29	30	31
Human [395]	Glu	Ala	Glu	Asp	Leu	Gln	Val	Gly	Gln	Val	Glu	Leu	Gly	Gly	Gly	Pro	Gly	Ala	Gly	Ser	Leu	Gln	Pro	Leu	Ala	Leu	Glu	Gly	Ser	Leu	Gln
Monkey [396]	Glu	Ala	Glu	Asp	Pro	Gln	Val	Gly	Gln	Val	Glu	Leu	Gly	Gly	Gly	Pro	Gly	Ala	Gly	Ser	Leu	Gln	Pro	Leu	Ala	Leu	Glu	Gly	Ser	Leu	Gln
Pig [396]	Glu	Ala	Glu	Asn	Pro	Gln	Ala	Gly	Ala	Val	Glu	Leu	Gly	Gly	Gly	Leu	Gly	—	Gly	—	Leu	Gln	Ala	Leu	Ala	Leu	Glu	Gly	Pro	Pro	Gln
Cow, [401] Sheep [396]	Glu	Val	Glu	Gly	Pro	Gln	Val	Gly	Ala	Leu	Glu	Leu	Ala	Gly	Gly	Pro	Gly	Ala	Gly	—	—	—	—	—	Gly	Leu	Glu	Gly	Pro	Pro	Gln
Horse [397]	Glu	Ala	Glu	Asp	Pro	Gln	Val	Gly	Glu	Val	Glu	Leu	Gly	Gly	Gly	Pro	Gly	Leu	Gly	Gly	Leu	Gln	Pro	Leu	Ala	Leu	Ala	Gly	Pro	Gln	Gln
Dog [396]	Asp	Val	Glu	—	—	—	—	—	—	—	—	Leu	Ala	Gly	Ala	Pro	Gly	Glu	Gly	Gly	Leu	Gln	Pro	Leu	Ala	Leu	Glu	Gly	Ala	Leu	Gln
Rat I [398]	Glu	Val	Glu	Asp	Pro	Gln	Val	Pro	Gln	Leu	Glu	Leu	Gly	Gly	Gly	Pro	Glu	Ala	Gly	Asp	Leu	Gln	Thr	Leu	Ala	Leu	Gln	Val	Ala	Arg	Gln
Rat II [398]	Glu	Val	Glu	Arg	Pro	Gln	Val	Ala	Gln	Leu	Glu	Leu	Gly	Gly	Gly	Pro	Gly	Ala	Gly	Asp	Leu	Gln	Thr	Leu	Ala	Leu	Gln	Val	Ala	Arg	Gln

Figure 11 *Amino-acid sequences of proinsulin connecting peptides*

Amino-acid sequences have been reported for glucagon, one of the hormones responsible for glycogen breakdown in the liver, from ducks [402] and humans.[403] These 29-residue peptides were completely degraded in the sequenator using the dilute-quadrol and variable-programme techniques pioneered by Niall.[404] The sequences determined proved to be identical to those from the porcine and bovine hormones reported earlier.

Growth Factors.—Amino-acid sequence studies on mouse nerve growth factor (NGF), a protein which has been shown to enhance the growth of sympathetic ganglia *in vitro* and *in vivo* and to elicit neurite outgrowth from explanted sympathetic and sensory ganglia *in vitro*, have shown [405] a clear similarity in primary structure to insulin, a hormone with which it shares some properties. If the sequences are aligned so that the *N*-terminus of the proinsulin β-chain corresponds to the *N*-terminus of the NGF (Figure 12) only four deletions need to be introduced into the latter sequence to obtain maximum homology. The NGF structure then corresponds to a complete proinsulin chain with an additional B-chain added to the *C*-terminus. Possibly NGF arose from a gene duplication of an ancestral proinsulin followed by either genetic deletion or proteolytic removal of the second A plus C region. This is an interesting theory as the NGF isolated from the venom of *Vipera ruselli*, although very similar in biological activity, appears totally unrelated in structure to insulin.[406] In fact it is a very basic glycoprotein of molecular weight 37 000 and contains 20% carbohydrate. Possibly there are several factors promoting neural growth or, alternatively, the proteins from different genera are unrelated in evolutionary terms. The primary structure of epidermal growth factor from mouse has also been presented; apparently the *C*-terminal hexapeptide is not essential for activity.[407, 408]

The growth hormones continue to attract attention mainly from the point of view of delineating those parts of the molecule important for biological activity, an essential prerequisite to synthetic studies. The primary structure of the ovine hormone has appeared from two groups [409, 410] along with a preliminary communication of the bovine sequence.[411] Limited tryptic digestion of the bovine hormone produces two fragments of

[402] F. Sundby, E. K. Frandsen, J. Thomsen, K. Kristiansen, and K. Brunfeldt, *F.E.B.S. Letters*, 1972, **26**, 289.
[403] J. Thomsen, K. Kristiansen, K. Brunfeldt, and F. Sundby, *F.E.B.S. Letters*, 1972, **21**, 315.
[404] H. D. Niall, *J. Agric. Food Chem.*, 1971, **19**, 638.
[405] W. A. Frazier, R. H. Angeletti, and R. A. Bradshaw, *Science*, 1972, **176**, 482.
[406] F. L. Pearce, B. E. C. Barks, D. V. Banthorpe, A. R. Berry, H. H. S. Davies, and C. A. Vernon, *European J. Biochem.*, 1972, **29**, 417.
[407] C. R. Savage, T. Inagami, and S. Cohen, *J. Biol. Chem.*, 1972, **247**, 7612.
[408] C. R. Savage and S. Cohen, *J. Biol. Chem.*, 1972, **247**, 7609.
[409] C. H. Li, J. S. Dixon, D. Gordon, and J. Knorr, *Internat. J. Peptide Protein Res.*, 1972, **4**, 151.
[410] J. T. Bellair, *Biochem. Biophys. Res. Comm.*, 1972, **46**, 1128.
[411] H. N. Fernández, C. Peña, E. Poskus, M. J. Biscoglio, A. C. Paladini, J. M. Dellacha, and J. A. Santomé, *F.E.B.S. Letters*, 1972, **25**, 265.

1 5 10 15

Mouse NGF Ser - —— - Ser - Thr - His - Pro - —— - Val - Phe - His - Met - Gly - Glu - —— - —— - Phe - Ser - Val - Cys - Asp - Ser -
Human proinsulin Phe - Val - Asn - Gln - His - Leu - Cys - Gly - Ser - His - Leu - Val - Glu - Ala - Leu - Tyr - Leu - Val - Cys - Gly - Glu -
Guinea pig insulin Phe - Val - Ser - Arg - His - Leu - Cys - Gly - Ser - Asn - Leu - Val - Glu - Thr - Leu - Tyr - Ser - Val - Cys - Gln - Asp -

B-1 B-5 B-10 B-15 B-20

20 25 30 35

Mouse NGF Val - Ser - Val - Trp - Val - Gly - Asp - Lys - Thr - Thr - Ala - Thr - Asn - Ile - Lys - Gly - Lys - Glu - Val - Thr - Val -
Human proinsulin Arg - Gly - Phe - Phe - Tyr - Thr - Pro - Lys - Thr - Arg - Arg - Glu - Ala - Glu - Asp - Leu - Gln - Val - Gly - Gln - Val -
Guinea pig insulin Asp - Gly - Phe - Phe - Tyr - Ile - Pro - Lys - Asp -

B-25 * B-30 C-1 C-5 C-10

40 45 50 55

Mouse NGF Leu - Ala - Glu - Val - Asn - Ile - Asn - Asn - Ser - Val - Phe - Arg - Gln - Tyr - Phe - Phe - Glu - Thr - Lys - Cys - Arg -
Human proinsulin Glu - Leu - Gly - Gly - Gly - Pro - Gly - Ala - Gly - Ser - Leu - Gln - Pro - Leu - Ala - Leu - Glu - Gly - Ser - Leu - Gln -
Guinea pig insulin *

C-15 C-20 C25 C-30

60 65 70 75

Mouse NGF Ala - Ser - Asn - Pro - Val - Glu - Ser - Gly - Cys - Arg - Gly - Ile - Asp - Ser - Lys - His - —— - Trp - Asn - Ser - Tyr -
Human proinsulin Lys - Arg - Gly - Ile - Val - Glu - Gln - Cys - Cys - Thr - Ser - Ile - Cys - Ser - Leu - Tyr - Gln - Leu - Glu - Asn - Tyr -
Guinea pig insulin Gly - Ile - Val - Asp - Gln - Cys - Cys - Ala - Gly - Thr - Cys - Thr - Arg - His - Gln - Leu - Glu - Ser - Tyr -

C-35 A-1 A-5 * A-10 * A-15 *

```
                    80                        85                     90                    95                 100
Mouse NGF           Cys- Thr- Thr- Thr- His- Thr- Phe- Val- Lys- Ala- Leu- Thr- Thr- Asp- Glu- Lys- Gln- Ala- Ala- Trp- Arg-
Human proinsulin    Cys- Asn- CO2H            NH2- Phe- Val- Asn- Gln- His- Leu- Cys- Gly- Ser- His- Leu- Val- Glu- Ala- Leu-
Guinea pig insulin  Cys- Asn- CO2H            NH2- Phe- Val- Ser- Arg- His- Leu- Cys- Gly- Ser- Asn- Leu- Val- Glu- Thr- Leu-
                    A-20                           B′-1      *    B′-5                 B′-10                       B′-15

                                   105                    110                  115        118
Mouse NGF           Phe- Ile- Arg- Ile- Asn- Thr- Ala- Cys- Val- Cys- Val- Leu- Ser- Arg- Lys- Ala- Thr- Arg- CO2H
Human proinsulin    Tyr- Leu- Val- Cys- Gly- Glu- Arg- Gly- Phe- Phe- Tyr- Thr- Pro- —— Lys- Thr- Arg- Arg- C-peptide
Guinea pig insulin  Tyr- Ser- Val- Cys- Gln- Asp- Asp- Gly- Phe- Phe- Tyr- Ile- Pro- —— Lys- Asp- CO2H
                                   B′-20                  B′-25      *              *
                                                                                    B′-30
```

Figure 12 *The alignment of the amino-acid sequence of the primary subunit of mouse NGF with those of human proinsulin and guinea-pig insulin. Numbers above the lines are those of the NGF residue positions and numbers below the lines indicate the positions of the proinsulin or insulin residues. Solid lines enclose sets of identical residues, and dashed lines enclose sets of residues considered to be favoured amino-acid substitutions, defined as those pairs of residues with an R_{ij} value (relatedness odds) greater than the random value of* **10**. *B′ denotes the repeated C-chain of insulin (see text). Asterisks indicate positions at which residues from other insulins and proinsulins increase the number of observed similarities.*

(Reproduced by permission from *Science*, 1972, **176**, 483)

molecular weights 5000 and 16 000, and a 37-residue sequence of the smaller fragment was found to be active in humans.[412] This is important since if a report [413] that the human hormone is of clinical value in cases of dwarfism is confirmed, a supply of synthetic material will become essential.

An earlier report proposing that the S—S bridge in equine growth hormone was displaced with respect to the proteins from other species has been disproved.[414] In fact the sequence around the S—S bridge proposed previously resembles the sequence around the S—S bridge in the *C*-terminal region of prolactin, and the error in the initial report may have arisen from prolactin contamination.

Miscellaneous.—Hormone amino-acid sequences, rather like those of histones, appear to suffer from a recurring need for revision. Two corrections have been published for ovine β-lipotropic hormone [415] and revised structures have appeared for porcine, human,[416] ovine, and bovine [417] adrenocorticotropins (ACTH). In all these cases amides have been repositioned and in human ACTH an Ala-Gly sequence has been reversed, thus improving the homology.

Yet another gene-duplication mechanism has been proposed for the evolution of neurophysins.[418] These proteins are not hormones but function as carrier-substances to which the neurohypophyseal hormones (vasopressin, oxytocin) are bound during transport and storage within the hypothalamo-neurohypophyseal system. A partial sequence of bovine neurophysin I shows obvious homologies with both bovine neurophysin II and porcine neurophysin I (Figure 13), and binding of the hormone may occur in the central region of the proteins where the sequences are virtually identical. Further evidence for the suggested gene-duplication mechanism, shown in Figure 14, comes from a paper describing the localization of the S—S bridges in bovine neurophysin II.[419]

The hormone responsible for inducing colour changes in the bodies of crustaceans has been isolated from the eye-stalks of the prawn *Pandalus borealis*; 100 nmol of peptide were obtained and mass spectrometry showed the sequence to be

$$\text{Glp-Leu-Asn-Phe-Ser-Pro-Gly-Trp-NH}_2$$

[412] N. Yamasaki, K. Kangawa, S. Kobayashi, M. Kikutani, and M. Sonenberg, *J. Biol. Chem.*, 1972, **247**, 3874.
[413] J. M. Tanner, *Nature*, 1972, **237**, 433.
[414] M. M. Zakin, E. Poskus, J. M. Dellacha, A. C. Paladini, and J. A. Santomé, *F.E.B.S. Letters*, 1972, **25**, 77.
[415] M. Chrétien, C. Gilardeau, and C. H. Li, *Internat. J. Peptide Protein Res.*, 1972, **4**, 263.
[416] B. Riniker, P. Sieber, W. Rittel, and H. Zuber, *Nature New Biol.*, 1972, **235**, 114.
[417] C. H. Li, *Biochem. Biophys. Res. Comm.*, 1972, **49**, 835.
[418] J. D. Capra, J. M. Kehoe, D. Kotelchuck, R. Walter, and E. Breslow, *Proc. Nat. Acad. Sci. U.S.A.*, 1972, **69**, 431.
[419] D. H. Schlesinger, B. Frangione, and R. Walter, *Proc. Nat. Acad. Sci. U.S.A.*, 1972, **69**, 3350.

BNP-II (12—31)	Pro-Cys-Gly-Pro-Gly-Gly-Lys-Gly-Arg-Cys-Phe-Gly-Pro-Ser-Ile-Cys-Cys-Gly-Asp-Glu
BNP-II (60—77)	Pro-Cys-Gly-Ser-Gly-[ ]-Gly-Arg-Cys-Ala-Ala-Ala-Thr-Ile-Cys-Cys-Ser-Asp-Glu
BNP-I (12—31)	Pro-Cys-Gly-Pro-Gly-Gly-Lys-Gly-Arg-Cys-Phe-Gly-Pro-Ser-Ile-Cys-Cys-Gly-Asp-Glu
PNP-I (12—31)	Pro-Cys-Gly-Pro-Gly-Gly-Lys-Gly-Arg-Cys-Phe-Gly-Pro-Ser-Ile-Cys-Cys-Gly-Asp-Glu
PNP-I (60—77)	Pro-Cys-Gly-Ser-Gly-[ ]-Gly-Arg-Cys-Ala-Ala-Ala-Gly-Ile-Cys-Cys-Asn-Asp-Glu

Figure 13 *Evidence for repeating sequence pattern in bovine neurophysin II. Residues 12—31 and 60—77 have been aligned, and the identities are outlined.*
(Reproduced by permission from *Proc. Nat. Acad. Sci. U.S.A.*, 1972, **69**, 433)

This material was synthesized and the synthetic product shown to be active.[420]

Finally, two analogous tetradecapeptides, named alytesin and bombesin, have been isolated from the skin of European discoglossid frogs.[421] The exact function of these peptides is unknown but they have been reported

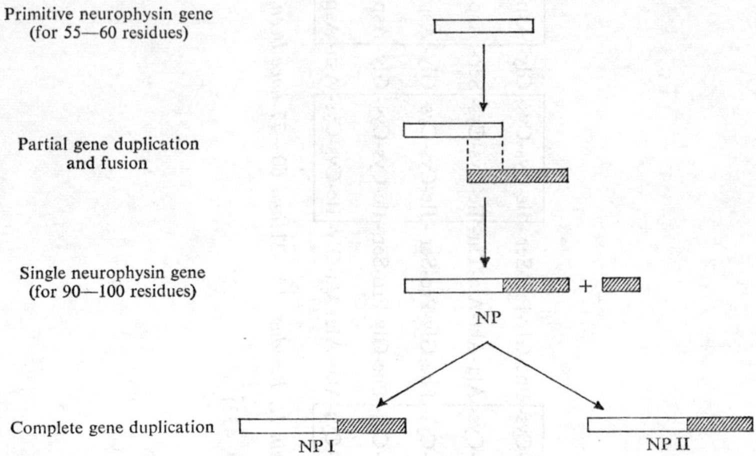

Figure 14 *Proposed evolutionary scheme for the neurophysins*
(Reproduced by permission from *Proc. Nat. Acad. Sci. U.S.A.*, 1972, **69**, 433)

to display a number of pharmacological actions on vascular and extravascular smooth muscle, on gastric secretion, and on venal circulation and function. The structures of alytesin and bombesin were found to be very similar to ranatensin, an undecapeptide of similar biological activity, isolated from the skin of the American frog *Rana pipiens*:[422]

Alytesin	Glp-Gly-Arg-Leu-Gly-Thr-Gln-Trp-Ala-Val-Gly-His-Leu-Met-NH_2
Bombesin	Glp-Gln-Arg-Leu-Gly-Asn-Gln-Trp-Ala-Val-Gly-His-Leu-Met-NH_2
Ranatensin	Glp-Val-Pro-Gln-Trp-Ala-Val-Gly-His-Phe-Met-NH_2

8 Toxins

The neurotoxins from several sources, particularly those from snake venoms, have attracted a considerable amount of attention over the past decade and the rate of progress has increased even further in the past few years. Thc increase in interest in the structures of these molecules undoubtedly stems from the discovery that some of them have curarizing pharmacological properties in spite of an overall similarity with the structures of other toxins which do not manifest neurotoxicity. Thus the

[420] P. Fernlund and L. Josefsson, *Science*, 1972, **177**, 173.
[421] A. Anastasi, V. Erspamer, and M. Bucci, *Arch. Biochem. Biophys.*, 1972, **148**, 443.
[422] V. Erspamer, G. Falconieri Erspamer, and M. Inselvini, *J. Pharm. Pharmacol.*, 1970, **22**, 85.

availability of a range of molecules which have considerable structural similarity but significant differences in neurotoxicity provides a convenient starting point for the elucidation of the structure–function relationships of neurotoxicity. Ultimately this should yield some insight into the nature of the toxin receptors.

The serological classification is a convenient starting point for any discussion on the snake neurotoxins. The first class are those toxins which display curarizing pharmacological properties, presumably through interaction with the cholinergic receptors (see above). The second class contains toxins with cardiotoxic pharmacological properties which act on muscular calcium metabolism. A third class of toxins has been defined on the basis of their failure to cross-react with antibodies to toxins of the first two classes.

Studies on the structures of a large number of toxins have been performed and meaningful comparisons are becoming possible. For the present purposes it will be more profitable to consider the details of the investigations which have been carried out over the past year before integrating these into the overall scheme.

Snake Toxins.—Two neurotoxins have been isolated from the common sea snake *Euhyduria schistosa*,[423, 424] the venom of which contains over sixteen different neurotoxins. The purified toxins, which are very similar in size (sixty residues each) and amino-acid composition, have four disulphide bonds and a single cysteine residue, the modification of which has no effect on activity. These proteins represent a new type of curariform toxin since all such toxins characterized so far only contain disulphide bonds. As in the case of many other toxins, the two major components isolated from this source are virtually identical, the only difference being the replacement of proline and serine at position 46. This was established from the amino-acid sequences, which also showed that the free sulphydryl group is at position 3.

The venom of the Indian cobra *Naja naja* contains a large number of toxins, several of which have been isolated and sequenced. A recent addition to the list is a basic protein Cytotoxin I which has a high cytotoxicity towards Ascites hepatoma cells.[425] The sequence and pharmacological properties of Cytotoxin I have led to the inference that Cytotoxins I and II are identical with Cobramines A and B, respectively. The sequence of Cytotoxin I and some of its physical properties show that it also resembles toxin A, the major neurotoxin from the Indian cobra. Reduction and methoxycarbonylation eliminate the cytotoxicity of Cytotoxin A, showing that the disulphide bridges are essential for the maintenance of activity.

Crotoxin, the neurotoxic component of the Brazilian rattlesnake *Crotalus durissus terrificus*, is unusual since it has been found to be a

[423] E. Karlsson, D. Eaker, L. Fryklund, and S. Kadin, *Biochemistry*, 1972, **11**, 4628.
[424] L. Fryklund, D. Eaker, and E. Karlsson, *Biochemistry*, 1972, **11**, 4633.
[425] K. Hayashi, M. Takechi, and T. Saski, *Biochem. Biophys. Res. Comm.*, 1971, **45**, 1357.

complex of two proteins, one of which is acidic and the other basic. The separated components possess no neurotoxicity but the basic component retains some haemolytic activity. Thus both proteins are necessary for the manifestation of full activity. These results have now been confirmed by two independent studies.[426, 427] It also appears that the acidic component previously assigned a molecular weight of 8000—10 000 is not a single polypeptide chain but is made up of several short chains held together by disulphide linkages.

The details of the purification of α-bungarotoxin and an extensive characterization of the protein have been published.[428] The results are in agreement with those obtained from previous studies, and with the amino-acid sequence (Figure 15), which shows that the protein contains 74 amino-

```
                5                 10                15               20
Ile-Val-Cys-His-Thr-Thr-Ala-Thr-Ile-Pro-Ser-Ser-Ala-Val-Thr-Cys-Pro-Pro-Gly-Glu-

                25                30                35               40
Asn-Leu-Cys-Tyr-Arg-Lys-Met-Trp-Cys-Asp-Ala-Phe-Cys-Ser-Ser-Arg-Gly-Lys-Val-Val-

                45                50                55               60
Glu-Leu-Gly-Cys-Ala-Ala-Thr-Cys-Pro-Ser-Lys-Lys-Pro-Tyr-Glu-Glu-Val-Thr-Cys-Cys-

                65                70
Ser-Thr-Asp-Lys-Cys-Asn-His-Pro-Pro-Lys-Arg-Gln-Pro-Gly
```

Figure 15 *The complete amino-acid sequence of reduced and* S-*aminoethylated α-bungarotoxin*

acid residues with five disulphide bridges.[429] Some difficulty was experienced with the carboxyl-terminus of the protein (Pro-Gly) and hydrazinolysis was the only effective method for the assignment of this region of the protein. The polypeptide chain of α-bungarotoxin is the longest yet obtained from the homologous toxins and it resembles the others in showing a high frequency of homodipeptides; ten pairs are present of which the serine and proline dipeptides occur twice. Ten homologous toxins of the curare-like type have been sequenced and α-bungarotoxin shows greatest homology with α-toxin from *Naja nivea*.

The elucidation of the sequence of Cytotoxin II from the Indian cobra *Naja naja* has been extended to the positioning of the four disulphide bridges.[430] The bridges link residues 3 with 21, 14 with 38, 42 with 53, and 54 with 59, and the close similarity between Cytotoxins I and II suggests

[426] J. Horst, R. A. Hendon, and H. Fraenkel-Conrat, *Biochem. Biophys. Res. Comm.*, 1972, **46**, 1042.

[427] K. H. Rubsamen, H. Breithaupt, and E. Haberman, *Nannyn-Shmiederbergs Arch. Pharmale*, 1971, **270**, 274.

[428] D. G. Clark, D. D. McMurchie, E. Elliot, R. G. Wolcott, A. M. Landel, and M. A. Raftery, *Biochemistry*, 1972, **11**, 1663.

[429] D. Mebs, K. Narita, S. Iwanaga, Y. Samejima, and C. Y. Lee, *Z. physiol. Chem.*, 1972, **353**, 243.

[430] M. Takechi and K. Hayashi, *Biochem. Biophys. Res. Comm.*, 1972, **49**, 584.

that a similar arrangement prevails in the former. It now appears that the positions of the homologous half-cysteine residues in neurotoxins with 61 or 62 amino-acid residues are identically cross-linked, implying that the differences in toxicity of these peptides are not due to variations in disulphide bridging (which seemed an obvious possibility) but a more subtle function of the amino-acid sequences.

The sequences of two neurotoxins from the genus *Dendroaspis* have been determined,[431] bringing the number of snake genera for which complete neurotoxin sequences are known to four. The two toxins have several unique features, notably a substantial increase in the amount of aromatic amino-acids. Toxin α, which is composed of 60 amino-acids, has four tyrosyl residues in contrast to the single tyrosyl residue usually found in snake neurotoxins. Toxin γ contains 72 amino-acids and is the only known snake neurotoxin with two tryptophanyl residues. It is argued that the presence of the second tryptophanyl residue lends support to the view that the 'long' toxins have evolved from the 'short' one as a result of unequal crossing-over of genes.

The two most abundant toxins from *Naja melanoleuca* venom have been sequenced and compared with other toxins.[432] The toxins closely resemble two toxins from *Naja nivea* venom in sequence and immunological properties. Toxin b, which is of the long type, has 71 residues, contains no methionine, and resembles toxin α from *Naja nivea* whereas toxin d contains 61 amino-acids, is devoid of alanine, phenylalanine, and leucine, and resembles toxin β from *Naja nivea* venom.

The possibility has been raised that the curariform toxins which block the nicotinic acetylcholine receptor might interact with the receptor through a cationic group in the toxin since the other ligands for the receptor, *e.g.* tubocurarine and acetylcholine, are cationic compounds. However, complete blocking of the amino-groups does not eliminate activity completely.[433]

The amino-acid sequences of about 18 snake venom neurotoxins have now been determined and it is clear that the serological classification needs to be matched with a classification based on the primary structures of these proteins. The homology between the proteins is high and there is no doubt that they are a closely related family (Figure 16). The most striking common feature of the proteins lies in the distribution of the half-cystine sequences which form the disulphide bonds. In all known cases eight of the half-cystines are in homologous positions and there is a high degree of homology in the surrounding sequences. Thus three of the half-cystines are always followed by tyrosine, proline, and asparagine. The only exceptions to the absence of cysteine residues in the proteins are the *Schistosa* toxins which have a free SH group at position 3. In most cases

[431] D. J. Strydom, *J. Biol. Chem.*, 1972, **247**, 4029.
[432] D. P. Botes, *J. Biol. Chem.*, 1972, **247**, 2866.
[433] E. Karlsson, D. Eaker, and G. Pouterius, *Biochim. Biophys. Acta*, 1972, **257**, 235.

```
      1                   10                  20                    30                  40
 1  L Q C H N Q Q S S Q P P T T K T C P - G E T N C Y K K R W R D H - - - - R G S I
 2  M I C H N Q Q S S Q R P T I K T C P - G E T N C Y K K R W R D H - - - - R G T I
 3  L E C H N Q Q S S Q T P T T Q T C P - G E T N C Y K K Q W S D H - - - - R G S R
 4  L E C H N Q Q S S Q P P T T K S C P - G D T N C Y N K R W R D H - - - - R G T I
 5  L E C H N Q Q S S Q P P T T K T C P - G E T N C Y K K V W R D H - - - - R G T I
 6  L E C H N Q Q S S Q T P T T T C C S G G E T N C Y K K R W R D H - - - - R G Y R
 7  R I C F N Q H S S Q P Q T T K T C P S G S E S C Y N K Q W S D F - - - - R G T I
 8  R I C F N Q H S S Q P Q T T K T C P S G S E S C Y H K Q W S D F - - - - R G T I
 9  I R C F - - - I T P D V T S Q A C P D G - H V C Y T K M W C D N F C G M R G K R
10  I R C F - - - I T P D V T S Q I C A D G - H V C Y T K T W C D N F C A S R G K R
11  M E C H N Q Q S S Q P P T T K T C P - G E T N C Y K K Q W S D H - - - - R G T I
12  R I C Y N H Q S T T R A T T K S C E - - E N S C Y K K Y W R D H - - - - R G T I
13  R T C N - - - K T F S D Q S K I C P P C E N I C Y T K T W C D A W C S Q R G K R

                          50                  60                      75
 1  T E R C C G - - C P S V K K G I E I N C C T T D K C N N - - - - - - -
 2  I E R C C G - - C P S V K K G V G I Y C C K T D K C N R - - - - - - -
 3  T E R C C G - - C P T V K P G I K L K C C T T D R C N K - - - - - - -
 4  I E R C C G - - C P T V K P G I N L K C C T T D R C N N - - - - - - -
 5  I E R G C G - - C P T V K P G I K L N C C T T D K C N N - - - - - - -
 6  T E R G C G - - C P S V K N G I E I N C C T T D R C N N - - - - - - -
 7  I E R G C G - - C P T V K P G I K L S C C E S E V C N N - - - - - - -
 8  I E R G C G - - C P T V K P G I K L S C C E S E V C N N - - - - - - -
 9  V D L G C A A T C P K V K P G V N I K C C S R D N C N P F P I R K R S
10  V D L G C A A T C P T V K P G V N I K C C S T D N C N P F P T R N R P
11  I E R G C G - - C P S V K K G V K I N C C T T D R C N N - - - - - - -
12  I E R G C G - - C P K V K P G V G I H C C Q S D K C N Y - - - - - - -
13  V E L G C A A T C P K V K A G V E I K C C S T D D C D K F Q F G K P R
```

1 = *Naja haje* α, *Naja nivea* δ; 2 = *Naja nivea* β; 3 = *H. haemachatus* IV; 4 = *H. haemachatus* II; 5 = *N. nigricollis* α; 6 = *N. n. atra* cobrotoxin; 7 = *L. semifasciata* 'a'; 8 = *L. semifasciata* 'b'; 9 = *Naja nivea* α; 10 = *N. melanoleuca* b; 11 = *N. melanoleuca* d; 12 = *D. polylepis* α; 13 = *D. polylepis* γ.

A—Ala; C—Cys; D—Asp; E—Glu; F—Phe; G—Gly; H—His; I—Ile; K—Lys; L—Leu; M—Met; N—Asn; P—Pro; Q—Gln; R—Arg; S—Ser; T—Thr; V—Val; W—Trp; Y—Tyr.

Figure 16 *Comparison of the amino-acid sequences of some snake toxins*[431]
(Reproduced by permission from *J. Biol. Chem.*, 1972, **247**, 4029)

the sequence between the third and fourth half-cystines from the *N*-terminus contains either four or five residues, the exceptions being the *Schistosa* toxins which have only four residues in this region. These two toxins and cobratoxin are the only ones in which proline does not follow the third half-cystine.

Eighteen residues occupy the same positions in all the snake neurotoxins considered and two of the most significant are the 'essential' tyrosine and tryptophan residues. Chemical modification studies indicate that these two residues might be required for the activity of the toxins. They are located in the large loop formed by two disulphide bridges and the feature of this region is the large amount of hydrophilic residues. It has been suggested that this loop could contain the 'active site' of the toxins by protruding outwards from the molecule. It is interesting that five out of the eight residues which are conserved in all known snake neurotoxins are confined to the 'active site' loop. The only exceptions to the rule of one tyrosine and one tryptophan residue per toxin are toxin α and toxin γ from *Dendroaspis* since the former contains four tyrosyl and the latter two tryptophanyl residues. The other general feature of the snake toxins is the presence of homodipeptides: in some cases there are as many as ten such pairs. Not unexpectedly, significant structural roles have been suggested for these arrangements.

The major subdivision on the basis of primary structure is between the 'short' and the 'long' neurotoxins. The former vary in length from 60 to 65 residues and the latter from 71 to 75 residues. In addition there are always four disulphide bridges in the short toxins and five in the long ones. The position of the second tryptophanyl residue in toxin γ of *Dendroaspis* has been construed as positive evidence that the long toxins evolved from the short ones and not *vice versa*. Thus toxin γ is considered to be an intermediate in the evolution of the long toxins since it has acquired an extra sequence and has also undergone a single-step mutation converting Ser-30 into Cys-30, an obligatory step for the formation of the extra disulphide bond in the long toxins. The reason for the existence of the long toxins is not clear since the pharmacological properties of the two types do not seem to depend on the length of the protein.

Other Toxins.—An attempt has been made to test the hypothesis that the neurotoxins from *Clostridium botulinum* have a common basic configuration. The suggestion is based on the observation that although the toxins are serologically different they have virtually identical pharmacological activity. Part of the difficulty in comparing the toxins lies in the fact that all of them do not undergo proteolytic activation during isolation. However, it has now been shown that the type E toxin progenitor is split by trypsin into two fragments with molecular weights of 50 000 and 102 000.[434] These sizes compare well with the sizes of the two components of type A

[434] B. R. Dasgupta and H. Sugiyama, *Biochem. Biophys. Res. Comm.*, 1972, **48**, 108.

and type B toxins, indicating the presence of a common, albeit gross, structural feature in the three types of proteins.

The scorpion (*Androctonus australis* Hector) contains four neurotoxins of which two have been sequenced previously.[435] The sequence of a third, neurotoxin II, has now been determined and the homology with neurotoxin I is about 50%.[436] An unusual feature of this protein is that it appears to contain a blocked *C*-terminus, histinamide. Amidated *C*-terminal histidine and glutamine have only been found previously in toxic peptides from bee venom.

9 Chromosomal Proteins

Histones.—The DNA–protein complex chromatin represents the sum of all the genes of any one cell type, and is contained in all genetic structures. The function of histones appears to be the preservation of the three-dimensional structure of chromatin and not the control of DNA transcription or replication. There are five major classes of histone and these, along with some of their properties, are shown in Table 3. Amino-acid sequences

Table 3 *Properties of calf-thymus histones*

Nomenclature	*Molecular weight*	*Net change*	*Minimum number of sub-fractions*
I (f1)	~21 000	+53	3–4
IIb_1 (f2a)	~14 500	+20	2
IIb_2 (f2b)	13 774	+22	0–2
III (f3)	15 324	+22	3
IV (f2a1)	11 282	+19	4

have now appeared [437] for proteins from all of these classes except histone 1, of which only the 108-residue *N*-terminal section has been published. The unusual distribution of amino-acids in histones is well known and the current position is summarized in Table 4.

The sub-fractions shown in Table 3 mainly arise from post-translational modification of histone side-chains by methylation, acetylation, or phosphorylation. Many attempts have been made to correlate these modifications with periods of the cell cycle and DNA transcription but as yet the reason for these alterations remains unclear. A recent study on the distribution and turnover of labelled methyl groups in histone fractions from mammalian cells concludes [438] that methylation is fraction-specific and, unlike acetylation or phosphorylation, is relatively permanent. Only histone IIb_2 was found to contain methylarginine. A correlation has been

[435] C. Rochat, F. Sampieri, H. Rochat, and F. Miranda, *Biochimie*, 1972, **54**, 445.
[436] H. Rochat, C. Rochat, F. Sampieri, and F. Miranda, *European J. Biochem.*, 1972, **28**, 381.
[437] R. J. De Lange and E. L. Smith, *Accounts Chem. Res.*, 1972, **5**, 368.
[438] P. Byvoet, G. R. Shepherd, J. M. Hardin, and B. J. Noland, *Arch. Biochem. Biophys.*, 1972, **148**, 558.

claimed [439, 440] between phosphorylation of histone I and cell replication. Exclusive modification of this protein was found during the exponential growth phase of hepatoma HTC cells and, furthermore, this phosphorylation was largely abolished when the cells moved into the stationary phase. This would imply an obligatory involvement of histone phosphorylation in DNA synthesis or cell replication, rather than specific gene activation.

Table 4 *Distribution of amino-acids in histones*

	Residue No.	Amino-acid type: Basic	Acidic	Aliphatic	Aromatic	Hydroxyl	Prolyl	Charge
Histone I	1—41	10	3	0	0	3	7	+7
(212 residues)	42—107	11	5	16	2	13	0	+6
	108—end	43	3	4	0	6	16	+40
Histone IIb$_1$	1—36	13	0	5	1	4	1	+13
(131 residues)	37—131	17	10	25	3	8	4	+7
Histone IIb$_2$	1—50	19	2	6	3	9	5	+17
(125 residues)	51—125	13	8	17	4	18	1	+5
Histone III	1—53	19	1	5	1	9	4	+18
(135 residues)	54—135	15	11	22	6	9	2	+4
Histone IV	1—45	17	1	8	0	2	1	+16
(102 residues)	46—102	10	7	16	6	11	0	+3

Histones compared are all from calf thymus except histone I which is from rabbit thymus. Only the first 107 residues of this histone have been sequenced.

Both phosphorylation and acetylation of histone IV during spermatogenesis in trout have been studied.[441, 442] The results suggest that an obligatory acetylation of newly synthesized histone IV produces an α-helical conformation suitable for binding in the major groove of the DNA helix. Deacetylation then liberates new cationic centres which increase the strength of the electronic interaction and lock the molecule in place. A similar model has also been proposed for binding of the trout testis histone IIb$_1$ to DNA.

The amino-acid sequences of histone III from calf thymus [443] and chicken erythrocytes [444] have been reported. As expected they are virtually identical, except that the chicken protein has an extra serine inserted at position 90. With such close homology over the rest of the chain perhaps this point should be reinvestigated. The sequence of histone IIb$_2$ from calf thymus

439 R. Balhorn, R. Chalkley, and D. Granner, *Biochemistry*, 1972, **11**, 1094.
440 R. Balhorn, J. Bordwell, L. Sellers, D. Granner, and R. Chalkley, *Biochem. Biophys. Res. Comm.*, 1972, **46**, 1326.
441 E. P. M. Candido and G. H. Dixon, *J. Biol. Chem.*, 1972, **247**, 3868.
442 A. Louie, P. Candido, and G. H. Dixon, *Fed. Proc.*, 1972, **31**, 417.
443 R. J. De Lange, J. A. Hooper, and E. L. Smith, *Proc. Nat. Acad. Sci. U.S.A.*, 1972, **69**, 882.
444 W. F. Brandt and C. Von Holt, *F.E.B.S. Letters*, 1972, **23**, 357.

has reappeared [445] with two corrections; the customary model of the histone–DNA complex was also presented. Corrections have also been made [446] to a proposed sequence of histone IIb_1, again from calf thymus, producing an identity with an alternative version published simultaneously.[447–449] Interestingly, the first nine residues of this histone are almost identical to the corresponding sequence in histone IV. The amino-terminal sequence of histone T, a histone found specifically in trout tissues, has been determined [450] and found to show no obvious similarities to the more common histones.

Protein kinases which specifically catalyse the phosphorylation of histones have been isolated from trout testis ribosomes,[451] rat thymus,[452] and human lymphocytes.[453] The enzymes are all cyclic-AMP-dependent and appear similar to the kinases found in muscle (see Section 4). In addition, three histone-specific acetyltransferases have been purified from rat thymus nuclei.[454] In these cases the donor molecule is acetyl-CoA and in one case the acceptor is histone I whereas in the other two cases histone IV is acetylated.

Acidic Nuclear Proteins.—The idea that acidic proteins may play a decisive role in the regulation of biological activity of chromatin was originally put forward thirty years ago.[455] The high concentration of these proteins within the nuclei and their exceptional metabolic activity compared with the nearly inert histones implies their possible involvement in gene regulation, but this cannot, as yet, be supported by conclusive experimental results. Why then is there so much known about histone structure and so little known about this, possibly more important, group of proteins? The answer is that the isolation of acidic proteins free of basic and non-chromosomal contaminants, and their subsequent purification to homogeneity, has proved extraordinarily difficult. No less than six papers [456–461] have recently appeared describing new fractionation procedures but even

[445] K. Iwai, H. Hayashi, and K. Ishikawa, *J. Biochem.*, 1972, **72**, 357.
[446] P. Sautiere, D. Tyrou, B. Laine, J. Mizon, M. D. Lambelin-Breynaert, P. Ruffin, and G. Biserte, *Compt. rend.*, 1972, **274**, *D*, 1422.
[447] M. O. J. Olson, N. Sugano, L. C. Yeoman, B. R. Johnson, J. Jordan, W. C. Starbuck, and H. Busch, *Physiol. Chem. Phys.*, 1972, **4**, 10.
[448] L. C. Yeoman, M. O. J. Olson, N. Sugano, J. J. Jordan, C. W. Taylor, W. C. Starbuck, and H. Busch, *J. Biol. Chem.*, 1972, **247**, 6018.
[449] N. Sugano, M. O. J. Olson, L. C. Yeoman, B. R. Johnson, C. W. Taylor, W. C. Starbuck, and H. Busch, *J. Biol. Chem.*, 1972, **247**, 3589.
[450] G. H. Huntley and G. H. Dixon, *J. Biol. Chem.*, 1972, **247**, 4916.
[451] B. Jergil, *European J. Biochem.*, 1972, **28**, 546.
[452] M. I. Klein and M. H. Makman, *J. Cell Physiol.*, 1972, **79**, 407.
[453] A. W. Murray, M. Froscio, and B. E. Kemp, *Biochem. J.*, 1972, **129**, 995.
[454] D. Gallwitz and I. Sures, *Biochim. Biophys. Acta*, 1972, **263**, 315.
[455] E. Stedman and E. Stedman, *Nature*, 1944, **153**, 500.
[456] K. H. Richter and C. E. Sekeris, *Arch. Biochem. Biophys.*, 1972, **148**, 44.
[457] S. C. R. Elgin and J. Bonner, *Biochemistry*, 1972, **11**, 772.
[458] S. Levy, R. T. Simpson, and H. A. Sober, *Biochemistry*, 1972, **11**, 1547.
[459] M. Yoshida and K. Shimura, *Biochim. Biophys. Acta*, 1972, **263**, 690.
[460] G. H. Goodwin and E. W. Johns, *F.E.B.S. Letters*, 1972, **21**, 103.
[461] J. S. Bhorjee and T. Pederson, *Proc. Nat. Acad. Sci. U.S.A.*, 1972, **69**, 3345.

now there is no clear idea of the number of these proteins or whether, like the histones, they can be grouped into specific classes. Obviously much work remains to be done in this important area of protein chemistry.

Histones have so far been found only in eukaryotic cells and there has been controversy over whether dinoflagellates, which in some respects are intermediate between prokaryotes and the more advanced eukaryotes, do in fact contain histones. A recent investigation [462] has shown that the chromosomal proteins of the dinoflagellate alga *Gyrodinium cohnii* are acid insoluble, and therefore almost certainly not histones.

Protamines.—The complete amino-acid sequence of thynnin Y2, the protamine of the tuna fish, has been reported.[463] Protamines, which are found in the mature sperm of some fish species, are a specific class of basic proteins, containing as much as 70% arginine. These proteins are obviously especially difficult to sequence and it was only by recourse to the much neglected limited pepsin digestion that suitable fragments could be obtained and the sequence deduced. A somewhat different approach was used to obtain the sequences of clupeines YII and YI.[464, 465] In the first case treatment with conc. H_2SO_4 for 96 h at 20 °C resulted in an $N \rightarrow O$ acyl rearrangement at serine and threonine bonds. After acetylation of the liberated amino-groups the resulting ester bonds were cleaved with dilute sodium bicarbonate. These peptides were then overlapped with conventionally produced tryptic peptides to give the full sequence. The structure (of clupeine YI) was deduced from the tryptic and thermolytic peptides only. The sequences of all three protamines are clearly homologous:

Thynnin YII	Pro-(Arg)$_4$-Gln——————Ala-Ser-Arg-Pro-Val-(Arg)$_5$-
Clupeine YII	Pro-(Arg)$_3$-Thr-(Arg)$_2$-Ala-Ser-Arg-Pro-Val-(Arg)$_4$-
Clupeine YI	Ala-(Arg)$_4$-Ser,————Ser-Ser-Arg-Pro-Ile -(Arg)$_4$-
Thynnin YII	Tyr-(Arg)$_2$-Ser -Thr-Ala-Ala-(Arg)$_5$-Val-Val-(Arg)$_4$
Clupeine YII	Pro-(Arg)$_2$-Val-Ser——————(Arg)$_4$-Ala- (Arg)$_4$
Clupeine YI	Pro-(Arg)$_3$-Thr-Thr——————(Arg)$_4$-Ala-Gly-(Arg)$_4$

A study of the biosynthesis of protamine in trout testis has shown [466] that at the late spermatid stage of development histone is completely replaced by protamine. Soon after synthesis the protamine is phosphorylated on all the serine residues and subsequently dephosphorylated before the sperm are completely mature. Although the basic nuclear protein of bull spermatozoa has 24 out of 47 residues as arginine it is not considered [467] to be a protamine since the arginine residues are all at the

[462] P. J. Rizzo and L. D. Noodén, *Science*, 1972, **176**, 796.
[463] G. Bretzel, *Z. physiol. Chem.*, 1972, **353**, 933.
[464] K. Suzuki and T. Ando, *J. Biochem.*, 1972, **72**, 1419.
[465] K. Suzuki and T. Ando, *J. Biochem.*, 1972, **72**, 1433.
[466] M. M. Sanders and G. H. Dixon, *J. Biol. Chem.*, 1972, **247**, 851.
[467] J. P. Coelingh, C. H. Monfoort, T. H. Rozijn, J. A. Gevers Leuven, R. Schiphof, E. P. Steyn-Parvé, G. Braunitzer, B. Schrank, and A. Ruhfus, *Biochim. Biophys. Acta*, 1972, **285**, 1.

centre of the molecule rather than, as is normally the case with protamines, at the ends. Also this protein contains six cysteine residues which, it is postulated, may be important for the cross-linking of DNA–protein complexes in sperm nuclei.

10 Ribosomal Proteins

The structure and function of the bacterial ribosome have been reviewed and a correlation of the various nomenclatures used for ribosomal proteins has been presented.[468] Throughout this Report the now widespread terminology of Wittmann *et al.*[469] will be used. Ribosomal proteins from the 30S subunit will be prefixed 'S' and those from the 50S subunit will be prefixed 'L'.

It is now apparent that in *E. coli* the smaller (30S) ribosomal subunit consists of a 16S RNA molecule and 21 proteins, whereas the larger (50S) subunit has both 23S and 5S RNA subunits and about 34 proteins (Table 5). These proteins can be resolved by two-dimensional polyacrylamide gel electrophoresis and this technique has recently been applied to characterize the ribosomal proteins from yeast,[470] *Drosophila*,[471] wheat leaf chloroplasts,[472] and rat liver.[473, 474]

In an attempt to obtain information about the topography of the ribosome the ribosomal proteins from rat liver were phosphorylated from γ-[^{32}P]ATP using a specific protein kinase.[475] In the whole ribosome S9 was predominantly labelled, the same result as was obtained when just the small subunit was phosphorylated. However, in this latter case the pattern was slightly more complex as S10, S14, S17, and S18 were also labelled but to a lesser extent. Phosphorylation of the large subunit gave a preponderance of the label in L4, L19, L27 + 29, L28, L33, and L34. These results differ somewhat from a similar experiment [476] using a rabbit skeletal muscle protein kinase and *E. coli* ribosomes. In the small subunit, S4, S9, S18, and S19 were all labelled whereas the major acceptor protein in the 50S subunit was L2. These workers found that the same proteins were phosphorylated in intact ribosomes as in the isolated subunits although the degree of phosphorylation did vary.

Considerable effort has gone into mapping ribosome topography by cross-linking adjacent pairs of proteins on the subunit using bifunctional

468 C. G. Kurland, *Ann. Rev. Biochem.*, 1972, **41**, 377.
469 H. G. Wittman, G. Stöffler, I. Hindennach, C. G. Kurland, L. Randall-Hazelbauer, E. A. Birge, M. Nomura, E. Kaltschmidt, S. Mizushima, R. R. Traut, and T. A. Bickle, *Molec. Gen. Genetics*, 1971, **111**, 327.
470 H. Schmitt, *F.E.B.S. Letters*, 1972, **26**, 215.
471 A. G. Lamberton, *Molec. Gen. Genetics*, 1972, **118**, 215.
472 B. L. Jones, N. Nagabhushan, A. Gulyas, and S. Zalik, *F.E.B.S. Letters*, 1972, **23**, 167.
473 H. Welfle, J. Stahl, and H. Bielka, *F.E.B.S. Letters*, 1972, **26**, 228.
474 C. C. Sherton and I. G. Wool, *J. Biol. Chem.*, 1972, **247**, 4460; T. Hultin and A. Sjöqvist, *Analyt. Biochem.*, 1972, **46**, 342.
475 J. Stahl, H. Welfle, and H. Bielka, *F.E.B.S. Letters*, 1972, **26**, 233.
476 J. A. Traugh and R. R. Traut, *Biochemistry*, 1972, **11**, 2503.

reagents. Two reports have appeared [477, 478] on the use of the imidoester dimethyl suberimidate, and it appears that cross-linking can occur between S9 and S6 or S7, between S9 and S5, and between S11 or S12 and S19 in *E. coli* 30S subunits. In contrast, no cross-linking was found between the

Table 5 *Some properties of* E. coli *ribosomal proteins*

Protein	*Molecular weight*	p*K value*	*Directly binding to RNA*	*Function*
S1	65 000	—	−	Stabilization of mRNA on 30S subunit
S2	27 000	6·7	−	Aminoacyl-tRNA binding
S3	28 000	12·0	−	Aminoacyl-tRNA binding
S4	26 000	10·4	+	Assembly
S5	21 000	9·9	−	Protein synthesis
S6	17 000	4·9	−	Interaction with initiation factor
S7	24 000	12·2	+	Assembly
S8	16 000	9·1	+	Assembly
S9	17 500	>12·0	−	Assembly
S10	16 000	7·8	−	Protein synthesis
S11	18 000	>12·0	−	mRNA–tRNA interaction
S12	17 500	>12·0	−	Protein synthesis
S13	14 000	>12·0	−	Protein synthesis (partial requirement)
S14	15 000	>11·0	−	Aminoacyl-tRNA binding
S15	13 000	>12·0	+	—
S16	13 000	11·6	−	Assembly
S17	10 000	9·7	−	Assembly
S18	12 000	>12·0	−	Protein synthesis (partial requirement)
S19	14 000	>12·0	−	Protein synthesis
S20	13 000	>12·0	+	Protein synthesis (partial requirement)
S21	13 000	>12·0	−	Interaction with initiation factor
L7	15 500	4·8	−	Translocation; G-factor binding
L12	15 500	4·9	−	Translocation; G-factor binding

proteins of the larger subunit. A similar experiment [479] using *NN'*-1,4-phenylenedimaleimide specifically linked S18 and S21, presumably *via* their exposed sulphydryl groups. Again no cross-linking of the 50S subunit was observed.

As a prelude to detailed fluorescence studies of ribosome structure an investigation of the reactivity of the 30S ribosomal proteins towards fluorescein isothiocyanate has been reported.[480] The major conclusions are

[477] T. A. Bickle, J. W. B. Hershey, and R. R. Traut, *Proc. Nat. Acad. Sci. U.S.A.*, 1972, **69**, 1327.

[478] L. I. Slobin, *J. Mol Biol.*, 1972, **64**, 297.

[479] F. N. Chang and J. G. Flaks, *J. Mol. Biol.*, 1972, **68**, 177.

[480] K. H. Huang and C. R. Cantor, *J. Mol. Biol.*, 1972, **67**, 265.

that S1, S3, S4, and S21 proteins are exposed both in the subunit and in the intact 70S ribosome, whereas S8 is buried in both forms. S5, S9, S12, S13, S14, and S19 all appear to be strongly shielded by the 50S particle but S21 is much more strongly labelled in the intact ribosome than in the free 30S particle. This is interpreted as being indicative of a conformational change in the smaller subunit on assembly to form a complete ribosome. An alternative approach is to unfold the ribosome with varying amounts of edta followed by enzymic digestion with either ribonuclease T_1 or ribonuclease I. Fractionation of the resultant ribonucleoproteins then allows electrophoretic identification of the ribosomal proteins attached to each RNA segment. These segments may then be overlapped, and if a sufficiently large number of different digestions are completed, a unique sequence of ribosomal proteins can be deduced. This method has now been attempted on the 50S subunit of *E. coli* ribosomes and the following tentative order for some of the proteins is proposed:[481]

$$\begin{pmatrix}1\\8\end{pmatrix}\text{-5-}\begin{pmatrix}2\\3\\6\end{pmatrix}\text{-}\begin{pmatrix}10\\11\end{pmatrix}\text{-}\begin{pmatrix}14\\16\end{pmatrix}\text{-20-}\begin{pmatrix}12\\13\end{pmatrix}\text{-15-9-}\begin{pmatrix}22\\26\end{pmatrix}\text{-24-}\begin{pmatrix}25\\21\end{pmatrix}$$

The sequence of the proteins in parentheses could not be determined from the results so far obtained.

The amino-acid replacements conferring spectinomycin and streptomycin resistance on mutant *E. coli* ribosomal proteins have been characterized. Resistance to spectinomycin, an aminoglycoside antibiotic which strongly inhibits protein synthesis, is conferred by exchanging serine for proline in protein S5.[482] The four streptomycin-resistant mutants examined were all found to have alterations in the S12 protein.[483] In three cases, a lysine residue was replaced by asparagine, threonine, or arginine, respectively, whereas in the fourth case, arginine replaced a different lysine. Thus, it appears that single amino-acid changes in one ribosomal protein can have a profound affect on biological activity. More extensive mutations in protein S4 have also been found to confer streptomycin resistance.[484, 485]

The only proteins in the larger 50S ribosomal subunit with defined function are the two acidic proteins L7 and L12. Several reports [486–491]

[481] H. Kagawa, L. Jishuken, and H. Tokimatsu, *Nature New Biol.*, 1972, **237**, 74.
[482] G. Funatsu, K. Nierhaus, and B. Wittmann-Liebold, *J. Mol. Biol.*, 1972, **64**, 201.
[483] G. Funatsu and H. G. Wittmann, *J. Mol. Biol.*, 1972, **68**, 547.
[484] D. Donner and C. G. Kurland, *Molec. Gen. Genetics*, 1972, **115**, 49.
[485] G. Funatsu, W. Puls, E. Schiltz, J. Reinbolt, and H. G. Wittmann, *Molec. Gen. Genetics*, 1972, **115**, 131.
[486] N. Brot, E. Yamasaki, B. Redfield, and H. Weissbach, *Arch. Biochem. Biophys.*, 1972, **148**, 148.
[487] W. Möller, *J. Electroanalyt. Chem. Interfacial Electrochem.*, 1972, **37**, 407.
[488] E. Hamel, M. Koka, and T. Nakamoto, *J. Biol. Chem.*, 1972, **247**, 805.
[489] M. L. Sopori and P. Lengyel, *Biochem. Biophys. Res. Comm.*, 1972, **46**, 238.
[490] H. Weissbach, B. Redfield, E. Yamasaki, R. C. Davis, S. Pestka, and N. Brot, *Arch. Biochem. Biophys.*, 1972, **149**, 110.
[491] G. Sander, R. C. Marsh, and A. Parmeggiani, *Biochem. Biophys. Res. Comm.*, 1972, **47**, 866.

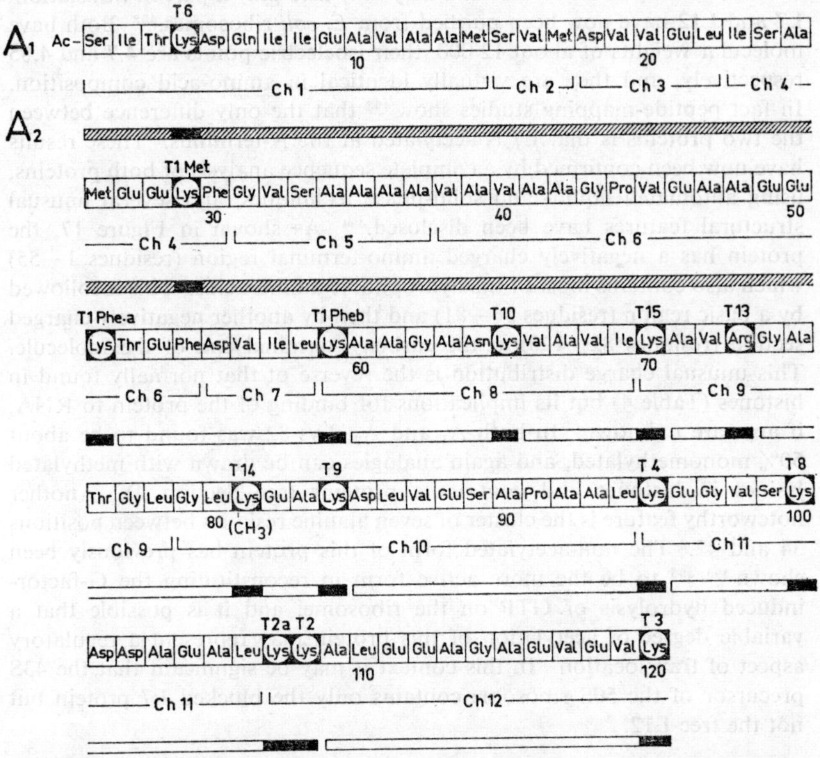

Figure 17 *The primary structure of A_1- and A_2-protein. A-protein consists of* 120 *amino-acids, A_1 starting with* N-*acetylserine and A_2-protein starting with serine at the* N-*terminal end. The tryptic peptides are indicated with T and the number placed above the lysine residues determines the* C-*terminal ends of each peptide. T_1-Met is a core peptide which contains the three methionine residues of the protein; T_1Phe is a core peptide which contains the two phenylalanine residues of A-protein. T_1Phe splits partially* (20%) *in T_1Phe a and T_1Phe b, each of which of the latter peptides contain one phenylalanine residue. Ch_1, Ch_2,* etc. *indicate the chymotryptic peptides of the protein; the numbering is consecutive from the* N-*terminus. The shaded bar, which extends till residue* 51, *represents the part of A_2-protein sequenced with the Beckman sequencer; a complete identity in sequence with A_1-protein for this stretch was found. The tryptic peptides of the remaining part of A_2 (residues* 52—120) *were sequenced with the aid of the automatic solid-phase Edman degradation technique (open bars) and found to be identical to the corresponding ones of A_1*

(Reproduced by permission from *F.E.B.S. Letters*, 1972, **28**, 326)

have appeared that these proteins are indispensable for G-factor-induced GTPase activity in ribosomes and they may also play a part in translation. L7 and L12 have now been purified from *E. coli* ribosomes.[492] Both have molecular weights of about 12 000, their isoelectric points are 4.7 and 4.85 respectively, and they are virtually identical in amino-acid composition. In fact peptide-mapping studies show [493] that the only difference between the two proteins is that L7 is acetylated at the *N*-terminus. These results have now been confirmed by a complete sequence analysis of both proteins, using automated liquid- and solid-phase techniques, and several unusual structural features have been disclosed.[494] As shown in Figure 17, the protein has a negatively charged amino-terminal region (residues 1—55) which also contains most of the hydrophobic amino-acids. This is followed by a basic region (residues 56—81) and then by another negatively charged section (residues 82—120) at the carboxyl-terminal end of the molecule. This unusual charge distribution is the reverse of that normally found in histones (Table 4) but its implications for binding of the protein to RNA, if any, are unknown. In both A_1 and A_2, Lys-82 was found to be about 50% monomethylated, and again analogies can be drawn with methylated lysines in histones and contractile proteins (see Section 4). Another noteworthy feature is the cluster of seven alanine residues between positions 34 and 42. The non-acetylated form of this protein has previously been shown [495, 496] to be the more active form in reconstituting the G-factor-induced hydrolysis of GTP on the ribosome, and it is possible that a variable degree of acetylation of this protein may represent a regulatory aspect of translocation. In this context it may be significant that the 43S precursor of the 50S ribosome contains only the blocked L7 protein but not the free L12.

11 Proteins of the Nervous System

Nervous systems are organized into three main structures: ganglia, nerves, and nerve nets. Ganglia are groups of nerve cells (neurons) surrounded by glial cells and enclosed by sheaths of connective tissue. Those ganglia which contain the terminations of sensory neurons and the cell bodies of motor fibres (which control the locomotion of the animal) and which are concerned with the elaboration of the overt behaviour of the animal are collectively called the central nervous system. For the purposes of this section this term will also be taken to include the cell processes (axons) which connect ganglia belonging to this system. Nerves may be defined as

[492] W. Möller, A. Groene, C. Terhorst, and R. Amons, *European J. Biochem.*, 1972, **25**, 5.
[493] C. Terhorst, B. Wittmann-Liebold, and W. Möller, *European J. Biochem.*, 1972, **25**, 13.
[494] C. Terhorst, W. Möller, R. Laursen, and B. Wittmann-Liebold, *F.E.B.S. Letters*, 1972, **28**, 325.
[495] K. Kischa, W. Möller, and G. Stöffler, *Nature New Biol.*, 1971, **233**, 62.
[496] H. E. Homann and K. H. Nierhaus, *European J. Biochem.*, 1971, **20**, 249.

tubular structures composed of long processes of neurons, the cell bodies of which may be situated in sense organs, tissues, or in ganglia. The total of all sensory, motor, and mixed nerves of an animal is referred to as the peripheral nervous system.

Myelin.—In vertebrates' peripheral and central nervous systems the peripheral axons are often surrounded by a specialized form of glial cell, the Schwann cell. Lipids, present in the plasma membrane of these Schwann cells, give rise to a sheath around the nerve fibres. This sheath is known as myelin.

Methods for the purification of proteins from rat brain myelin have been presented.[497] Differential and density-gradient ultracentrifugation resolved the basic encephalitogenic proteins from two proteolipid components. The two basic proteins from rat myelin have been characterized [498] and peptide-mapping studies show that the smaller component has a 40-residue deletion relative to the larger molecule. This deletion, corresponding to either residues 117—156 or residues 118—157 in the human sequence, removes a large part of the peptide reported to be encephalitogenic in the guinea pig and this correlates well with *in vivo* observations. Similar regions of the molecule have also been found to have encephalitogenic activity in rabbits. The larger of the two proteins, molecular weight 18 400, which is identical to the A1 protein previously isolated from bovine and human myelin, is known to be responsible for experimental allergic encephalomyelitis (EAE), an auto-immune demyelinating disease of the central nervous system. Sequence information now shows [499] that this protein is also identical to the proposed 'tissue-specific histone' previously isolated from pig brain.

Experimental allergic neuritis (EAN) is the peripheral nervous system equivalent of EAE. A basic protein, P2, has now been isolated [500] from rabbit sciatic nerve myelin which appears to be the counterpart of A1 in the induction of this disease. P2, molecular weight 11 000—12 000, represents about 18% of the total peripheral nerve myelin protein and does not correspond to the small basic myelin protein found in rats (see above) since it appears to have regions corresponding to A1 which had been deleted from the smaller protein. P1, a second, larger protein isolated from the same system as P2, has been found to be virtually identical in sequence to A1.[501] In particular, both proteins possess the reactive threonine residue which serves as an acceptor for glycosylation by *N*-acetylgalactosaminyltransferase, a triproline sequence, and the methylated arginine which has been localized previously at position 107 in the

497 T. V. Waehneldt and P. Mandel, *Brain Research*, 1972, **40**, 419.
498 R. E. Martenson, G. E. Diebler, M. W. Kies, S. S. McKneally, R. Shapira, and R. F. Kibler, *Biochim. Biophys. Acta*, 1972, **263**, 193.
499 S. E. Kornguth, L. R. Kozel, and O. Smithies, *Nature New Biol.*, 1972, **237**, 49.
500 S. W. Brostoff, P. Burnett, P. Lampert, and E. H. Eylar, *Nature New Biol.*, 1972, **235**, 210.
501 S. W. Brostoff and E. H. Eylar, *Arch. Biochem. Biophys.*, 1972, **153**, 590.

bovine sequence. A more recent report[502] indicates that the methylation of this arginine is heterogeneous and consists of a mixture of the mono-derivative: N^GN^G-dimethylarginine (3) and $N^GN'^G$-dimethylarginine (4). It will be interesting to see whether these derivatives also occur in P1.

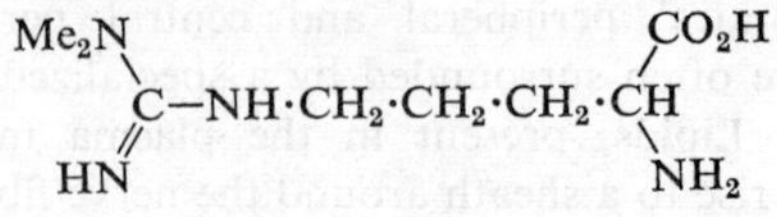

(3)

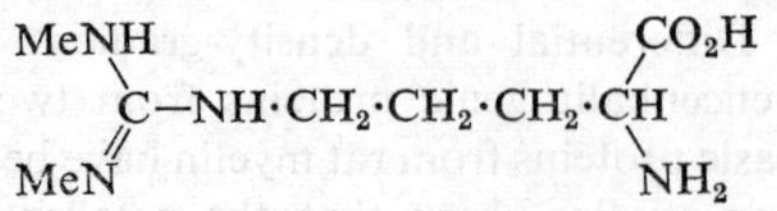

(4)

Brain Proteins.—The number and nature of the proteins specific to brain tissues, as well as the metabolic pathways of which some of them play a part, remain obscure. Isoelectric focusing procedures have been described for fractionation of both the soluble[503] and the insoluble[504] protein components from brain. Nineteen bands were formed from the soluble portion using a procedure which was capable of purifying up to 100 mg of protein. About 35—50 species were found in the insoluble part; here the procedure was on an analytical scale only.

The purification and properties of arylsulphatase A from chicken brain have been described.[505] Although the natural substrate of this enzyme is not known, it is possible that its role in brain tissue is the degradation of the glycolipid cerebroside-3-sulphate. The formation of the physiologically inactive form of serotonin, serotonin-*O*-sulphate, by a sulphotransferase of human brain has been demonstrated and a new radioisotopic assay for the enzyme described.[506] Using this assay, the highest enzyme activities were found in the hypothalamus, whereas the lowest were in the thalamus. It is possible that this enzyme is important in the regulation of serotonin (5-hydroxytryptamine), a putative neurotransmitter. Serotonin itself is synthesized in the brain from tryptophan *via* 5-hydroxytryptophan. The enzyme catalysing this hydroxylation, tryptophan hydroxylase, has now been purified from rabbit hindbrain and characterized.[507] For a review of the physiological action of serotonin see ref. 508.

[502] S. W. Brostoff, A. Rosegay, and W. J. A. Vandenheuvel, *Arch. Biochem. Biophys.*, 1972, **148**, 156.
[503] P. Milošević and L. Rakić, *J. Neurochem.*, 1972, **19**, 2459.
[504] G. D. Miner and L. L. Heston, *Analyt. Biochem.*, 1972, **50**, 313.
[505] A. A. Farooqui and B. K. Backhawat, *Biochem. J.*, 1972, **126**, 1025.
[506] H. Hidaka and J. Austin, *Biochim. Biophys. Acta*, 1972, **268**, 132.
[507] P. A. Friedman, A. H. Kappelman, and S. Kaufman, *J. Biol. Chem.*, 1972, **247**, 4165.
[508] A. Heller, *Fed. Proc.*, 1972, **31**, 81.

The reported isolation [509] of galactosylsphingosine galactosyl hydrolase opens up an alternative pathway for the degradation of the ganglioside galactosylacylsphingosine, a major constituent of brain and, in particular, of the myelin sheath. The normal degradation to the fatty acid ester of sphingosine, ceramide, by galactocerebrosidase may now be supplemented by an additional pathway *via* galactosylsphingosine.

The distribution of isoenzymes of monoamine oxidase in human brain has been mapped.[510] This enzyme degrades biologically active monoamines in the brain to aldehydes which may then be further oxidized to the corresponding carboxylic acid by an NAD-linked dehydrogenase. Alternatively, they may be reduced to the corresponding alcohol, and an NADPH-linked aldehyde reductase responsible for this has now been purified from pig brain.[511] The enzyme, which is located almost exclusively in the cytosol, can be resolved into two components on DE-cellulose,[512] a low-activity form (molecular weight 29 000) and a high-activity form (molecular weight 36 000). Specificity studies show that although it will reduce a wide range of aliphatic and aromatic aldehydes it has a marked preference for phenylglycoaldehydes.

A protein unique to the olfactory bulb (the site of primary synapse from the olfactory receptor cells) has been isolated from mouse brain.[513] Antiserum to this protein reacted only with olfactory bulb extracts and no other tissues or brain regions. The protein, minimum molecular weight 20 000, represents about 1% of the total olfactory protein and may be an example of selective genetic expression within the central nervous system.

The heterogeneity of the tissue-specific acidic protein S-100 (named after its solubility in 100% ammonium sulphate solution) has been investigated.[514] SDS–urea polyacrylamide gels indicate that the native protein, molecular weight 19 500, is composed of two α-subunits, molecular weight 4100, and two β-subunits, molecular weight 5300. There are four distinct types of α-subunit and two types of β-, an observation which may provide a basis for the observed [515] multiplicity of the native protein. The isolation of S-100 has also been reported from human acoustic neurinomas.[516] The function of this protein is unknown but a correlation has been claimed [517] between memory, in this case learning of transfer of handedness in rats, and S-100 synthesis. A similar report [518] now claims a correlation between these learning processes in rats and Ca^{2+}-induced conformational

509 T. Miyatake and K. Suzuki, *J. Biol. Chem.*, 1972, **247**, 5398.
510 M. B. H. Youdim, G. G. S. Collins, M. Sandler, A. B. B. Jones, C. M. B. Pare, and W. J. Nicholson, *Nature*, 1972, **236**, 225.
511 A. J. Turner and K. F. Tipton, *European J. Biochem.*, 1972, **30**, 361.
512 A. J. Turner and K. F. Tipton, *Biochem. J.*, 1972, **130**, 765.
513 F. L. Margolis, *Proc. Nat. Acad. Sci. U.S.A.*, 1972, **69**, 1221.
514 J. A. Stewart, *Biochem. Biophys. Res. Comm.*, 1972, **46**, 1405.
515 J. A. Stewart, *Biochim. Biophys. Acta*, 1972, **263**, 178.
516 S. E. P. Pfeiffer, P. L. Kornblith, H. L. Cares, J. Seals, and L. Levine, *Brain Research*, 1972, **41**, 187.
517 H. Hydén and P. W. Lange, *Exp. Cell Res.*, 1970, **62**, 125.
518 H. Haljamäe and P. W. Lange, *Brain Research*, 1972, **38**, 131.

changes in this protein. The difficulties inherent in the execution and interpretation of these experiments impose a need for great caution and these findings are unlikely to be generally accepted until improved experimental techniques become available.

The same general comments apply to the paper [519] on the isolation, identification, and synthesis of scotophobin, a pentadecapeptide isolated from the brains of rats trained to avoid the dark. Briefly, these authors attempted to show that rats trained in this way specifically synthesized a peptide, which, after isolation and injection into naïve rats, transferred dark avoidance to the untrained animals. The structure of the peptide was found to be: Ser-Asp-Asn-Asn-Gln-Gln-Gly-Lys-Ser-Ala-Gln-Gln-Gly-Gly-Tyr-NH_2, and this was confirmed by synthesis. The methods used for the structure determination have already been discussed [520] and only two extra points will be made. Firstly, the enzyme : substrate ratio used for the tryptic digestion of scotophobin (4 : 1) was 200 times greater than normal. Secondly, the statement 'in view of the small amount of material left at this point (130 μg), the only method of sequential analysis available to us was mass spectrometry' is erroneous. Conventional 'dansyl' techniques (see Section 2) are far more sensitive than mass spectrometry and are certainly applicable to these quantities of material. These criticisms apart, two reports have now appeared of dark-avoidance in mice [521] and goldfish [522] after injection of synthetic scotophobin. An independent synthesis of a structure originally proposed for this peptide (as above but with Glu-5 and Glu-11) possessed 10% of the biological activity of the natural product.[523] Unfortunately, the work was superseded by this new version of the structure, the sixth so far.[524–528]

A phosphoprotein which binds one mole of calcium per mole of protein has been isolated from pig brain.[529] This protein clearly differs in electrophoretic mobility and molecular weight (11 500) from the Ca^{2+}-binding S-100 protein (see above). Treatment with phosphatase removes the calcium-binding ability, and whether this protein is connected with the actomyosin-like proteins previously isolated from brain is unknown. The actin-like component of this complex, neurin, has now been characterized.[530]

519 G. Ungar, D. M. Desiderio, and W. Parr, *Nature*, 1972, **238**, 198.
520 W. W. Stewart, *Nature*, 1972, **238**, 202.
521 D. H. Malin and H. N. Guttman, *Science*, 1972, **178**, 1219.
522 H. N. Guttman, G. Matvyshyn, and G. H. Warriner, *Nature New Biol.*, 1972, **235**, 26.
523 A. Ali, J. H. R. Faesel, D. Sarantakis, D. Stevenson, and B. Weinstein, *Internat. J. Peptide Protein Res.*, 1972, **4**, 395.
524 G. Ungar and E. J. Fjerdingstad, in 'Biology of Memory', ed. G. Adam, Plenum Press, New York, 1971, p. 137.
525 G. Ungar, I. K. Ho, L. Galvan, and D. M. Desiderio, *Abstracts, Chem. Eng. News*, Feb. 9th, 1970, p. 11.
526 G. Ungar, quoted in ref. 523.
527 G. Ungar, in 'Methods in Pharmacology', ed. A. Schwartz, Appleton-Century-Crofts, New York, 1971, Vol. 1, p. 744.
528 D. M. Desiderio, G. Ungar, and P. A. White, *Chem. Comm.*, 1971, 432.
529 D. J. Wolff and F. L. Siegel, *J. Biol. Chem.*, 1972, **247**, 4180.
530 S. Puszkin and S. Berl, *Biochim. Biophys. Acta*, 1972, **256**, 695.

It appears to have properties very similar to muscle actin (see Section 4) and will stimulate the Mg^{2+}-ATPase activity of muscle myosin. It also contains one residue of τ-methylhistidine per mole of protein (50 000 g). The subcellular distribution of the whole contractile complex, neurostenin, has been investigated.[531] Although the function of this complex is unknown it has been suggested that it might occur only under particular conditions such as the junction of a synaptic vesicle and the membrane during release of a transmitter substance.

In addition to actomyosin, microtubule proteins may also be isolated from brain homogenates and a new procedure for the purification of these proteins from rat brain has been described.[532] Contrary to previous reports the microtubule protein contained no carbohydrate but 0.8 ml of phosphate per mole of protein dimer (110 000 g) was bound to the smaller, more acidic subunit. Whether the phosphorylation plays a role in the cellular control of either microtubule function of assembly is not yet known. Similar tubulins have also been isolated [533] from the adrenal medulla and from the nervous system (ganglia) of the marine mollusc *Aplysia californica*. Antiserum against brain tubulin was found to cross-react with adrenal tubulin [534] but, not surprisingly, it did not cross-react with actomyosin.

12 Chemical Modification

contributed by G. Allen

Some examples of chemical modification of proteins have been reported earlier in this chapter in connection with specific proteins, and this section will be concerned primarily with examples of more generally applicable methods.

A monograph has appeared on this topic [535] which gives a general coverage of the field. A new volume of 'Methods in Enzymology' [1] incorporates two chapters on chemical modification, updating and adding to the excellent Volume XI in the series.[536]

The number of reagents used for chemical modification of proteins continues to grow; however, most new reagents are related to existing ones. It is now widely recognized that care must be taken before the claim of specificity for particular functional groups in a modification reaction is made. For example, acylating reagents attack all nucleophilic groups in proteins (imidazole, thiol, phenolic hydroxy-, alkyl hydroxy-, and amine groups) although relative specificity may be conferred by certain features of the reagent. Specificity for amino-groups may be attained by reversal of

531 S. Puszkin, W. J. Nicklas, and S. Berl, *J. Neurochem.*, 1972, **19**, 1319.
532 B. A. Eipper, *Proc. Nat. Acad. Sci. U.S.A.*, 1972, **69**, 2283.
533 M. Jacobs, A. V. M. Brown, and G. L. Brown, *F.E.B.S. Letters*, 1972, **24**, 113.
534 D. A. Redburn, A. M. Poisner, and F. E. Samson, *Brain Research*, 1972, **44**, 615.
535 G. E. Means and R. E. Feeney, 'Chemical Modification of Proteins', Holden-Day, San Francisco, 1971.
536 'Methods in Enzymology', ed. C. H. W. Hirs, Volume 11, 'Enzyme Structure', Academic Press, New York, 1967.

acylation reactions at the other nucleophilic functions. Oxidizing reagents are also now recognized to be relatively non-specific, although varying the conditions of the reaction, notably pH, may result in specificity. Generally, thiol groups are the most reactive towards oxidizing agents. Selective modification of serine and threonine hydroxy-groups, carboxylic acid groups, and the guanidino-groups of arginine residues is still difficult to achieve under mild conditions unless their environment in the tertiary structure results in unusually high reactivity.

Since modification of proteins by a given reagent tends not to be specific for particular functional groups, this section is classified according to the type of reagent.

Alkylating Reagents.—The thiol group of cysteine residues is generally the most reactive towards alkylating reagents over the pH range 6—12 and, unless protected, cysteine residues exposed to the solvent are often modified. The most common type of alkylating reagents used have been activated alkyl halides.

Chloromethyl ketone derivatives of DL-leucine, DL-alanine, D-alanine, and glycine have been synthesized and tested for inhibition of leucine aminopeptidase.[537] The synthesis was designed to prevent polymerization of the reagents. The chloromethyl ketone derivatives of DL-leucine and DL-alanine were potent reversible inhibitors of the peptidase, but no covalent interaction was observed. The possibility that the enzyme catalyses the hydrolysis of the reagent was not investigated. Fluorescent alkylating reagents, N^{α}-dansyl-L-lysyl and N^{α}-dansyl-L-phenylalanyl chloromethyl ketones, have been prepared and used to label the active sites of trypsin and α-chymotrypsin, respectively.[538] The reagents are analogues of the N^{α}-tosyl derivatives, long known to react specifically with histidine residues in the active sites of these proteases. The use of the reagents as fluorescent probes of the active site was suggested.

Two new reagents suggested for the cross-linking of proteins, ethyl chloroacetimidate (5) [539] and 4-chloro-3,5-dinitrophenacyl bromide (6),[540] bear activated alkyl halide functions which react with thiol groups. Compound (5) was shown to form an internal cross-link between the sulphur and nitrogen atoms of cysteine,[539] and this amino-acid was also rapidly alkylated by the phenacyl bromide derivative. It was suggested that (5) cross-linked a lysyl and a histidyl residue in ribonuclease, whereas (6) may have cross-linked two histidine residues. The use of the dinitrophenacyl bromide as a spectral probe was also suggested.

An analogue of NAD, [3-(4-bromoacetylpyridino)propyl]adenosine pyrophosphate (7b), has been prepared and shown to inhibit lactate dehydrogenase, alcohol dehydrogenase, and glyceraldehyde-3-phosphate

[537] P. L. Birch, H. A. El-Obeid, and M. Akhtar, *Arch. Biochem. Biophys.*, 1972, **148**, 447.
[538] G. Schoellmann, *Internat. J. Protein Res.*, 1972, **4**, 221.
[539] M. Olomucki and J. Diopoh, *Biochim. Biophys. Acta*, 1972, **263**, 213,
[540] J. Diopoh and M. Olomucki, *Biochim. Biophys. Acta*, 1972, **263**, 220.

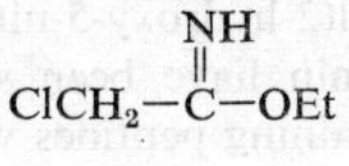

(5)

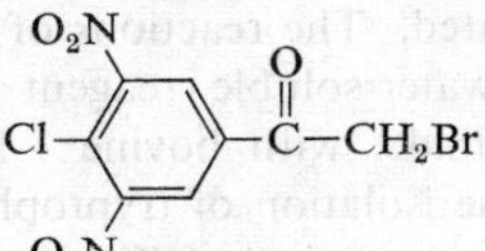

(6)

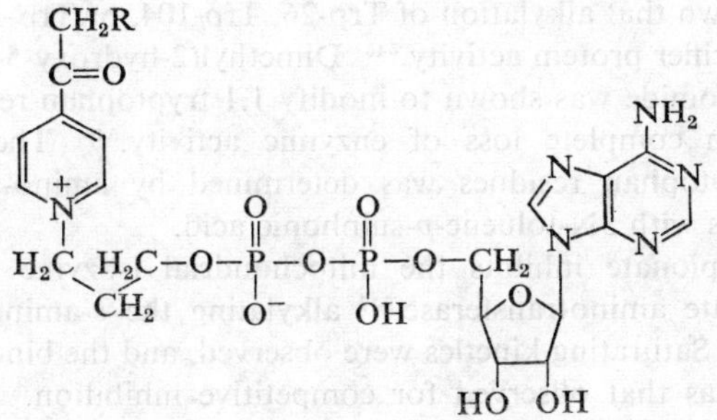

(7a) R = H (7b) R = Br

dehydrogenase.[541] NAD and NADH protected the enzymes against inhibition. One thiol group in glyceraldehyde-3-phosphate dehydrogenase was alkylated, with the loss of all enzymic activity. This result is not surprising, since most activated alkyl halides inhibit this enzyme. The analogous [3-(4-acetylpyridino)propyl]adenosine pyrophosphate (7a) was found to be a reversible inhibitor of the enzymes, competitive with respect to NAD and NADH, indicating that binding in the coenzyme binding sites does occur.

The reactions of the bromoacetyl analogue of histidine, α-bromo-β-(5-imidazolyl)propionic acid (see last year's Report) with cysteine[542] and papain[166] have been described. *N*-Bromoacetyl-β-D-galactosamine was shown to alkylate specifically a single methionine residue in β-galactosidase, with concomitant inhibition of the enzyme.[543] The inhibition reaction showed saturating kinetics, indicating the formation of an enzyme–inhibitor complex. The inhibited enzyme could be reactivated by thiolysis under mild conditions. Bromoacetamide and iodoacetamide were also found to inhibit the enzyme, but without forming a reversible enzyme–inhibitor complex and, in this case also, it was possible to regain enzyme activity. In an elegant experiment an active analogue of the enzyme, containing norleucine instead of methionine, was prepared and shown not to be inactivated by the alkylating reagents. Thus the methionine residue is not intimately involved in the catalysis.

Reactive benzyl halides, particularly 2-hydroxy-5-nitrobenzyl bromide which alkylates the indole ring of tryptophan residues at low pH, have also

541 C. Woenckhaus, E. Schättle, R. Jeck, and J. Berghäuser, *Z. physiol. Chem.*, 1972, **353**, 559.
542 J. A. Yankeelov and C. J. Jolley, *Biochemistry*, 1972, **11**, 159.
543 F. Naider, Z. Bohak, and J. Yariv, *Biochemistry*, 1972, **11**, 3202.

been investigated. The reactions of 2-hydroxy-5-nitrobenzyl bromide and the similar, water-soluble reagent dimethyl(2-hydroxy-5-nitrobenzyl)sulphonium bromide, with bovine α-lactalbumin have been studied.[544] A method for the isolation of tryptophan-containing peptides was suggested based on the adsorption of hydroxy-nitrobenzyl derivatives on Sephadex G-25. The mixture of modified α-lactalbumins was partially separated, and it was shown that alkylation of Trp-26, Trp-104, or Trp-118 led to the loss of the specifier protein activity.[545] Dimethyl(2-hydroxy-5-nitrobenzyl)-sulphonium bromide was shown to modify 1.1 tryptophan residues in hen lysozyme, with complete loss of enzymic activity.[546] The number of unreacted tryptophan residues was determined by amino-acid analysis after hydrolysis with 3N-toluene-*p*-sulphonic acid.

β-Bromopropionate inhibits the mitochondrial isozyme of pig heart muscle aspartate aminotransferase,[547] alkylating the ε-amino-group of a lysine residue. Saturating kinetics were observed, and the binding constant was the same as that observed for competitive inhibition. Succinic and maleic acids protected the enzyme. The pH-dependence of the rate of inhibition was associated with an apparent pK_a of 6.2, similar to the pK_a of the Schiff-base complex with the coenzyme determined spectrophotometrically. [1-^{14}C]-β-Bromopropionate labelled a lysine residue in the same sequence as that which was labelled after reduction of the enzyme–coenzyme complex with sodium borohydride.[548] Some alkylation of histidine[547] and some alkali-labile incorporation of radioactivity[548] were observed, but this did not correlate with the inactivation of the enzyme.

2-Bromoethyltrimethylammonium bromide reacts with cysteine to give 4-thialaminine (8), which can be easily identified by amino-acid analysis.[549]

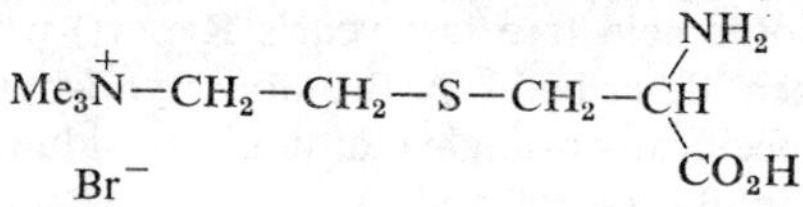

(8)

It was suggested that modification of cysteine residues in proteins with this alkylating reagent could be used for analysis of cysteine residues. However, high concentrations of the reagent and long reaction times are needed.

Cysteine residues may also be alkylated by activated alkenes. *N*-(*p*-2-benzimidazolyphenyl)maleimide, discussed in Volume 3 of these Reports, gives strongly fluorescent adducts with thiols (Scheme 1) and has been suggested for use in a sensitive determination of thiol concentration and as

544 T. E. Barman, *Biochim. Biophys. Acta*, 1972, **257**, 297.
545 T. E. Barman and W. Bagshaw, *Biochim. Biophys. Acta*, 1972, **278**, 491.
546 C. L. Borders, D. K. Jorkasky, and S. E. Pearson, *Biochem. Biophys. Res. Comm.*, 1972, **49**, 246.
547 M. Okamoto and Y. Morino, *Biochemistry*, 1972, **11**, 3188.
548 Y. Morino and M. Okamoto, *Biochemistry*, 1972, **11**, 3196.
549 H. A. Itano and E. A. Robinson, *J. Biol. Chem.*, 1972, **247**, 4819.

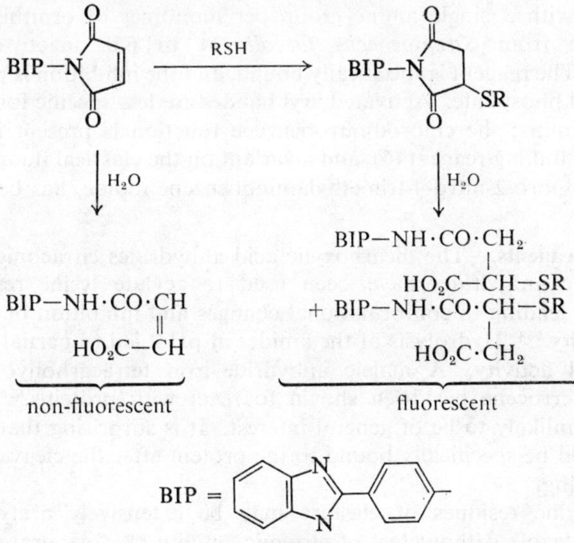

Scheme 1

a fluorescent label in proteins.[550] The reagent itself is hardly fluorescent, but the fluorescence intensity of the adducts increases upon hydrolysis, complicating an assay procedure. Thiol groups of proteins are alkylated by *p*-nitrostyrene: the adduct, *S*-β-(*p*-nitrophenethyl)-L-cysteine, was shown to be stable to acid hydrolysis.[551]

Ethyleneimine in liquid ammonia has also been reported to alkylate cysteine residues.[552] Triethyloxonium fluoroborate esterifies several carboxy-groups in pepsin, and a mathematical method relating the loss of enzymic activity with the incorporation of radioactivity, using [^{14}C]triethyloxonium fluoroborate, has shown that at least two carboxylic acid groups are essential to the activity of the enzyme.[553]

Reduction of Schiff bases, formed from aldehydes and amino-groups, with borohydride has led to the alkylation and identification of pyridoxal-phosphate-binding sites in haemoglobin[554] and to the methylation of the ε-amino-groups of lysine residues in ribonuclease.[50]

Arylating Reagents.—The reactivity of 2,4,6-trinitrobenzenesulphonate ion (TNBS) with primary amines and hydroxide ion has been investigated, and its reactivity with protein functional groups discussed.[555] TNBS reacts

[550] T. Sekine, K. Ando, M. Machida, and Y. Kanaoka, *Analyt. Biochem.*, 1972, **48**, 557.
[551] M. S. Masri, J. J. Windle, and M. Friedman, *Biochem. Biophys. Res. Comm.*, 1972, **47**, 1408.
[552] K. Kuromizu and J. Meienhofer, *J. Biol. Chem.*, 1972, **247**, 5646.
[553] A. K. Paterson and J. R. Knowles, *European J. Biochem.*, 1972, **31**, 510.
[554] R. E. Benesch, R. Benesch, R. D. Renthal, and N. Maeda, *Biochemistry*, 1972, **11**, 3576.
[555] G. E. Means, W. I. Congdon, and M. L. Bender, *Biochemistry*, 1972, **11**, 3564.

specifically with a single amino-group per monomer of ornithine transcarbamylase from *Streptococcus faecalis* at pH 6.0, inactivating the enzyme.[556] The reagent is specifically bound, and the inhibition is prevented by carbamyl phosphate. Activated aryl halides are less specific for reaction at amino-groups; the chlorodinitrobenzene function is present in a suggested cross-linking reagent (6), and a variant on the classical fluorodinitrobenzene, 1-fluoro-2-nitro-4-trimethylaminobenzene iodide, has been introduced.[557]

Acylating Reagents.—The dicarboxylic acid anhydrides citraconic, maleic, and itaconic anhydrides have been used to acylate lysine residues in pepsinogen, leading to conformational changes and inhibition of potential peptic activity.[558] Hydrolysis of the amides at pH 2 led to partial recovery of potential activity. A maleic anhydride–iron tetracarbonyl complex related to ferrocene has been shown to react with proteins,[559] but the reaction is unlikely to be of general interest. It is surprising that the iron atoms should be specifically bound to the protein after the cleavage of the anhydride ring.

The tyrosine residues of elastase may be extensively acetylated by *N*-acetylimidazole without loss of enzymic activity.[560] Cyanuration of the tyrosine residues of this protease[560] and of succinylated papain[561] did not lead to inhibition of the enzymes. *N*-Acetylimidazole was reported to acetylate several amino-groups in ornithine transcarbamylase without reaction at tyrosine or cysteine residues, causing inhibition of the enzyme.[556] Partial protection against inhibition was observed in the presence of carbamyl phosphate and norvaline.

The bifunctional active ester di(hydroxysuccinimide) succinate was shown to cross-link Gly-A1 and Lys-B29 of insulin, giving a product with 60% of the activity of native insulin.[562] The cross-linking of these two residues was first achieved in 1958 with 1,5-difluoro-2,4-dinitrobenzene.[563] *N*-Acyl-succinimides have been introduced as acylating agents for proteins,[564] and a study of the hydrolysis of the reagents and their reaction with n-butylamine has been reported.[565] Specificity for lysine residues in bovine serum albumin was claimed.[564] 3-Acetoxy-1-acetyl-5-methylpyrazole (9) is a less reactive acylating agent than is *N*-acetylimidazole, and suffers from the disadvantages of low solubility in water and strong u.v. absorption.[566]

[556] M. Marshall and P. P. Cohen, *J. Biol. Chem.*, 1972, **247**, 1669.
[557] D. A. Sutton, S. E. Drewes, and U. Welz, *Biochem. J.*, 1972, **130**, 589.
[558] Y. Nakagawa and G. E. Perlmann, *Arch. Biochem. Biophys.*, 1972, **149**, 476.
[559] R. W. Giese and B. L. Vallee, *J. Amer. Chem. Soc.*, 1972, **94**, 6199.
[560] M. J. Gorbunoff and S. N. Timasheff, *Arch. Biochem. Biophys.*, 1972, **152**, 413.
[561] L. A. Ae. Sluyterman and J. Wijdenes, *Biochim. Biophys. Acta*, 1972, **263**, 329.
[562] D. G. Lindsay, *F.E.B.S. Letters*, 1972, **21**, 105.
[563] H. Zahn and J. Meienhofer, *Makromol. Chem.*, 1958, **26**, 153.
[564] H. Boyd, S. J. Leach, and B. Milligan, *Internat. J. Protein Res.*, 1972, **4**, 117.
[565] H. Boyd, I. C. Calder, S. J. Leach, and B. Milligan, *Internat. J. Protein Res.*, 1972, **4**, 109.
[566] M. Irie, T. Miyasaka, and K. Arakawa, *J. Biochem.* (*Japan*), 1972, **72**, 65.

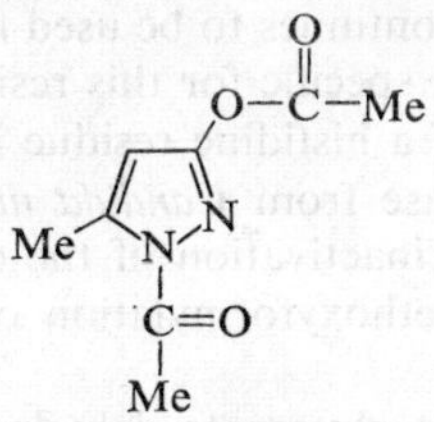

(9)

Homocysteine thiolactone may introduce a large number of thiol groups into proteins in a chain reaction catalysed by methylmercuric hydroxide and imidazole.[567] Extensive binding of methylmercury by the products may be useful for electron microscopy. The immunochemical properties of mercurated antibodies were investigated.[568]

Highly reactive acylating reagents are generally of low specificity. In trifluoroacetic acid, hydroxyamino-acids are acetylated by acetyl chloride,[569] cysteine is *S*-acetylated, and tryptophan residues are irreversibly acetylated;[570] the acetylation of amino-groups is suppressed. Hydrolysis of acylated tryptophan and tryptophan-containing peptides in 6N-HCl gave 1-substituted-3,4-dihydro-β-carboline-3-carboxylic acids (10) in about 70%

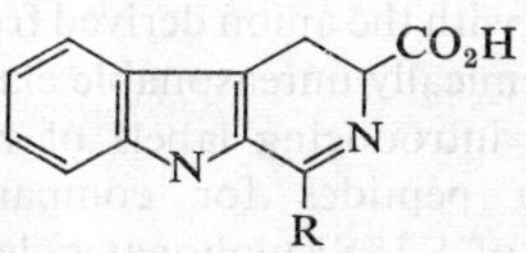

(10)

yield. Spectrophotometric analysis of this product was suggested as a method for the determination of tryptophan in proteins.[570] Treatment of thioredoxin with formic acid saturated with HCl resulted in the formylation of tryptophan residues and the loss of biological activity.[571] However, incubation of the product for 16 h at pH 9.5 in 6M-guanidinium chloride solution resulted in 50% recovery of biological activity and almost complete regeneration of the native u.v. absorbance spectrum. The conclusion that tryptophan residues are important for the biological activity[571] must be treated with caution, since other reactions, *viz*. esterification, cross-linking, and loss of amide groups, which are expected to occur under such conditions, were not investigated.

567 P. A. Kendall, *Biochim. Biophys. Acta*, 1972, **257**, 83.
568 P. A. Kendall, *Biochim. Biophys. Acta*, 1972, **257**, 101.
569 A. Previero, L.-G. Barry, and M.-A. Coletti-Previero, *Biochim. Biophys. Acta*, 1972, **263**, 7.
570 A. Previero, G. Prota, and M.-A. Coletti-Previero, *Biochim. Biophys. Acta*, 1972, **285**, 269.
571 A. Holmgren, *European J. Biochem.*, 1972, **26**, 528.

Diethyl pyrocarbonate continues to be used for modification of histidine residues, although it is not specific for this residue (see Volume 3 of these Reports). The presence of a histidine residue in the active site of 6-phosphogluconate dehydrogenase from *Candida utilis* has been suggested on the basis of the reversible inactivation of the enzyme by diethyl pyrocarbonate concomitant with ethoxyformylation of one histidine residue per subunit.[572]

The inhibition of bovine glutamate dehydrogenase by carbamyl phosphate has been shown to result from reaction of lysine-97 with cyanate, a breakdown product of carbamyl phosphate, rather than with the reagent itself.[573]

Oxidizing Reagents.—A great variety of protein modification reagents may be classed under this heading. Thiol groups may be specifically oxidized by weak oxidizing reagents such as disulphides, and the reaction of the thiol groups of ornithine transcarbamylase with disulphides has been investigated.[556] The rates of reaction of papain with 2,2′- and 4,4′-dipyridyl disulphides [163] and 5,5′-dithiobis-(2-nitrobenzoate) (DTNB) [164] have been studied, and unusual reactivity at low pH was investigated. Some preparations of bromelain also reacted quite rapidly with 2,2′-dipyridyl disulphide at low pH, but the observation was dependent upon the previous treatment of the protease in various ways.[165] A reinvestigation of the reactivity of protein disulphide bonds with the anion derived from reduction of DTNB [574] casts doubt upon the chemically unreasonable claim [575] that such bonds are cleaved. A method for introducing labels of high specific radioactivity into cysteine-containing peptides for comparative peptide mapping, involving the formation of *S*-[^{35}S]sulphonates, has been described.[576] For the labelling and isolation of a peptide containing the cysteine residue in cytoplasmic aspartate aminotransferase, which is syncatalytically modified by DTNB, the mixed disulphide formed was treated with [^{14}C]cyanide ion.[577] Rapidly reacting thiols not involved in the activity of the enzyme were blocked in the absence of substrates by *N*-ethylmaleimide.

Sulphenyl chlorides are much more powerful reagents than disulphides for the oxidative addition of sulphur derivatives to proteins, and at low pH both cysteine and tryptophan residues are modified. Separation by ion-exchange chromatography of the products obtained when lysozyme was treated with one equivalent of 2-nitro-4-carboxyphenylsulphenyl chloride gave a species in which Trp-108 only was modified, and which had 10% of the activity of the native enzyme.[578] Tryptophan residues may be converted into tryptophan-2-thiol by thiolysis of the adduct formed with

572 M. Rippa, M. Signorini, and S. Pontremoli, *Arch. Biochem. Biophys.*, 1972, **150**, 503.
573 F. M. Veronese, D. Piszkeiwicz, and E. L. Smith, *J. Biol. Chem.*, 1972, **247**, 754.
574 K. Brocklehurst, M. Kierstan, and G. Little, *Biochem. J.*, 1972, **128**, 811.
575 J. F. Robyt, R. J. Ackerman, and C. G. Chittenden, *Arch. Biochem. Biophys.*, 1971, **147**, 262.
576 G. R. Stark and L. V. Crawford, *Nature New Biol.*, 1972, **237**, 146.
577 W. Birchmeier, K. J. Wilson, and P. Christen, *F.E.B.S. Letters*, 1972, **26**, 113.
578 F. M. Veronese, E. Boccu, and A. Fontana, *F.E.B.S. Letters*, 1972, **21**, 277.

2,4-dinitrophenyl sulphenyl chloride.[579] The tryptophan-2-thiol residues regenerate 2,4-dinitrophenyl sulphenyl tryptophan upon treatment with fluorodinitrobenzene at pH 5.

Iodine, a classical protein-modification reagent, may be generated *in situ* by insolubilized lactoperoxidase[580] although, since the same species is formed as in iodinations with I_2 + KI, it seems unlikely that greater specificity can be obtained for iodination of tyrosine residues. Iodine gave no oxidative cross-linking in L-asparaginase from *E. coli* B, and 9.5 tyrosyl residues per tetramer were iodinated.[581] More reactive halogenating agents are generally less selective, but it is claimed that 2,4,6-tribromo-4-methylcyclohexadienone (11) can be used at pH 3 to cleave tryptophanyl peptide

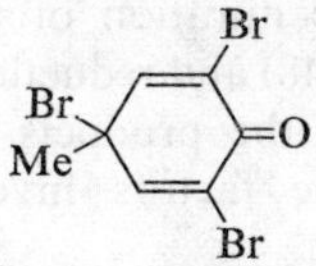

(11)

bonds selectively; 5—60% yields of cleavage products from lysozyme were reported.[582] Tyrosine residues were brominated and methionine and cysteine residues oxidized, but no cleavage at tyrosyl bonds was observed. The use of a large excess of *N*-bromosuccinimide to cleave tyrosyl but not *O*-acetyltyrosyl peptide bonds in ribonuclease[583] is unlikely to prove generally useful for the identification of exposed tyrosine residues.

Peptide bonds are chlorinated by t-butyl hypochlorite and an estimate of the extent of chlorination may be made iodometrically. Many side-chains also react. The method has been suggested as an alternative to tritium exchange for the estimation of α-helix or β-sheet structures in polypeptides.[584]

Tetranitromethane (TNM) has been widely used for the titration of tyrosine residues in proteins. However, oxidative side-reactions usually occur, and an investigation of the reactivity of TNM with pseudo-acids indicates that these may be of great complexity.[585] A 42-fold molar excess of TNM at 20 °C, pH 8, destroys the activity of bovine liver 2-keto-4-hydroxyglutarate aldolase, but only thiol groups were found to be oxidized.[586] TNM modified all the tyrosine residues in acid-soluble calf-skin tropocollagen,[587] but only 30% of the modified amino-acids were

579 M. Wilchek and T. Miron, *Biochem. Biophys. Res. Comm.*, 1972, **47**, 1015.
580 G. S. David, *Biochem. Biophys. Res. Comm.*, 1972, **48**, 464.
581 S. Shifrin and B. G. Solis, *J. Biol. Chem.*, 1972, **247**, 4121.
582 Y. Burstein and A. Patchornik, *Biochemistry*, 1972, **11**, 4641.
583 Y. Burstein and A. Patchornik, *Biochemistry*, 1972, **11**, 2939.
584 A. Matsushima, S. Yamazaki, K. Shibata, and Y. Inada, *Biochim. Biophys. Acta*, 1972, **271**, 243.
585 S. W. Jewett and T. C. Bruice, *Biochemistry*, 1972, **11**, 3338.
586 R. S. Lane and E. E. Dekker, *Biochemistry*, 1972, **11**, 3295.
587 M. K. Dabbous, M. Sief, and E. C. Brinkley, *Biochem. Biophys. Res. Comm.*, 1972, **48**, 1586.

present as 3-nitrotyrosine, and intramolecular cross-links were introduced. TNM in low molar excess at pH 8 caused aggregation of L-asparaginase from *E. coli* B, with the introduction of covalent cross-links.[581] The titration behaviour of the aggregates indicated that the cross-links were not exclusively derived from tyrosyl residues. TNM also caused cross-linking in thyroid-stimulating hormone; nitration of luteinizing hormone could, however, be accomplished in the absence of extensive cross-linking.[588]

In favourable circumstances, nitration of tyrosine residues may occur selectively. A nitrotyrosine-containing peptide sequence in aspartate aminotransferase inhibited syncatalytically by TNM was determined;[589] in the absence of substrates only low levels of nitration and little inactivation were observed. Specific nitration of the single tyrosine residue in staphylococcal nuclease-T-(6-48) and reduction to an aminotyrosine residue could be accomplished, and the products were used in the continuing intimate investigation of the factors involved in the conformational stability of this enzyme.[590]

Photo-oxidation of proteins by sunlight has been shown to yield *N'*-formylkynurenine residues derived from tryptophan.[591]

Other Reagents.—Diazonium salts have been known for several decades to react with proteins, and the modification of trypsin with a large excess of diazobenzenesulphonic acid [592] with little inhibition of the enzymic activity gave no new information. However, the specific diazo-coupling reaction of diazotized arsanilic acid with single tyrosine residues in the active sites of carboxypeptidases A (in the crystalline state) [593] and B [594] has been used to advantage in the isolation and sequence analysis of peptides containing these residues. The chromophore may also be useful as a spectral probe.

Spin-labelled analogues of di-isopropyl fluorophosphate have been used to label the active sites of serine proteases.[595] Doubly labelled derivatives were more sensitive indicators of conformation changes. 2-Chloro-1,3,2-dioxaphosphorinan 2-oxide inhibits eel acetylcholine-esterase, but spontaneous reactivation at pH 7 was observed, in contrast to the inhibition with diethylphosphoryl derivatives.[596]

A β-elimination reaction of substituted serine and threonine residues in phosphoproteins and glycoproteins, followed by addition of [^{35}S]sulphite, to give labelled cysteic acid, and 2-amino-3-sulphonylbutyric acid, respectively, has been described.[597] The reaction is accomplished at pH 11.5 in

[588] Kwong-Wah Cheng and J. G. Pierce, *J. Biol. Chem.*, 1972, **247**, 7163.
[589] O. L. Polyanovsky, T. V. Demidkina, and C. A. Egorov, *F.E.B.S. Letters*, 1972, **23**, 262.
[590] I. M. Chaiken, *J. Biol. Chem.*, 1972, **247**, 1999.
[591] A. Pirie, *Biochem. J.*, 1972, **128**, 1365.
[592] C. A. Bauer and G. Ehrensvärd, *Acta Chem. Scand.*, 1972, **26**, 1209.
[593] J. T. Johansen, D. M. Livingston, and B. L. Vallee, *Biochemistry*, 1972, **11**, 2584.
[594] M. Sokolovsky and L. Eisenbach, *European J. Biochem.*, 1972, **25**, 483.
[595] J. C. Hsia, D. J. Kosman, and L. H. Piette, *Arch. Biochem. Biophys.*, 1972, **149**, 441.
[596] Y. Ashani, S. L. Snyder, and I. B. Wilson, *Biochemistry*, 1972, **11**, 3518.
[597] D. L. Simpson, J. Hranisavljevic, and E. A. Davidson, *Biochemistry*, 1972, **11**, 1849.

the presence of 0.1—0.2M-Na_2SO_3. Cysteine and cystine residues may also undergo the reaction. The method is suggested for the localization and identification of substituted serine and threonine residues.

Glucose has been shown to react with peptides under mild conditions, and synthetic glucosyl-L-Val-L-His-OH has the properties of the substance released by proteolysis from the *N*-terminus of the β-chains of haemoglobin A_{IC}.[598] It is possible that the reaction of haemoglobin with glucose occurs spontaneously *in vivo*.

The use of photogenerated reagents for biological receptor-site labelling has been reviewed, with emphasis placed upon carbenes and nitrenes.[599] Proteins are labelled by radiolysis in the presence of high-specific-activity $^3HCCl_3$,[600] but it has been shown that the non-exchangeable tritium incorporated into L-alanyl-L-alanine was lost upon acid hydrolysis,[601] indicating that the major reaction with proteins is probably not, as previously suggested,[600] an exchange of carbon-bound hydrogen for tritium.

PART II: *X*-Ray Studies *by T. L. Blundell*

1 Introduction

From the small number of reported *X*-ray structures of proteins, 1972 would seem to be a year of retrenchment. Eleven new protein structures at high resolution were described in detail in 1971. In contrast only five high-resolution analyses were described in detail in 1972. These are ferredoxin, malate dehydrogenase, thermolysin, concanavalin, and a flavodoxin from *Desulphovibro vulgans*. Preliminary communications on some aspects of these high-resolution studies have previously been reported (see Vol. 4, p. 177).

There is a lesson to be learnt from this observation. The number of high-resolution *X*-ray studies successfully completed will not necessarily rise exponentially in a way we have come to expect of scientific developments. In fact the structures of the easily available, stable extracellular hydrolases and the small redox proteins have mostly been solved. For the future, the problems of successful analyses of large proteins such as phosphorylase or the nitrogenases are immense. These proteins may involve ten times as much *X*-ray data. However, the differences are not purely quantitative. There must be qualitative differences in technique, new breakthroughs in data collection and analysis before such structures are successfully elucidated.

598 H. B. F. Dixon, *Biochem. J.*, 1972, **129**, 203.
599 J. R. Knowles, *Accounts Chem. Res.*, 1972, **5**, 155.
600 M. Saunders, H. A. Jung, and W. L. Hamilton, *J. Amer. Chem. Soc.*, 1967, **89**, 472.
601 W. S. Brinigar and H. P. Rappaport, *Biochem. Biophys. Res. Comm.*, 1972, **48**, 1367.

2 Amino-acids and Peptides

One of the principal reasons for studying amino-acids and small peptides is to understand hydrogen bonding, which is so important to protein structure. Early *X*-ray studies using photographic data could not determine hydrogen positions precisely. However, modern techniques allow better analyses, and many of the recently reported studies have concerned the refinement of structures already published. From these studies the hydrogen atoms can usually be positioned easily.

An example is the precise analysis and refinement of the L-arginine phosphate structure[1] to a reliability value of 3.9%. The carboxy and guanidyl groups are essentially planar. The amino and guanidyl groups are protonated while the carboxyl and phosphate each carry one negative charge. The symmetrical guanidinium phosphate hydrogen bond geometry proposed for the interaction of basic proteins with DNA is not formed, and therefore this structure is not necessary to a complex hydrogen-bond network. The analysis of *N*-acetyl-L-histidine[2] also gives the position of the proton. In the crystal this amino-acid is a zwitterion with protonation of the imidazole ring.

X-Ray analyses of three forms of histidine have been reported.[3–5] Studies of L-methionine and L-norleucine show that they form layers held together by N—H···O bonds and packed in pairs so that the hydrocarbon side-chains are in contact.[6] Similar structures are also found with L-cysteine, L-valine, and L-isoleucine. They provide a model for the hydrogen bonding and hydrophobic packing found in globular proteins.

An alternative technique to find hydrogen positions is neutron diffraction. Neutron analyses of L-asparagine monohydrate,[7,8] L-glutamic acid,[9] and L-lysine monohydrochloride dihydrate[10,11] have been reported. Accurate hydrogen positions are given. In the study of lysine, an analysis of axial librations has led to the calculation of energy barriers to rotation. These are 7.0 kcal mol^{-1} for the α-NH_3^+ group and 2.9 kcal mol^{-1} for the ε-NH_3^+ group. Precision neutron diffraction studies of certain peptides have also been reported. In glycylglycine monohydrochloride monohydrate,[12] the C—H, N—H, and O—H bond lengths are 1.086, 1.028, and 1.003 Å. In

[1] W. Saenger and K. G. Wagner, *Acta Cryst.*, 1972, **B28**, 2237.
[2] T. J. Kistenmacher, D. J. Hunt, and R. E. Marsh, *Acta Cryst.*, 1972, **B28**, 3352.
[3] J. J. Madden, E. L. McGandy, and N. C. Seeman, *Acta Cryst.*, 1972, **B28**, 2377.
[4] J. J. Madden, E. L. McGandy, N. C. Seeman, M. M. Harding, and A. Hoy, *Acta Cryst.*, 1972, **B28**, 2382.
[5] K. Oda and H. Koyama, *Acta Cryst.*, 1972, **B28**, 639.
[6] K. Torii and Y. Iitaka, *Acta Cryst.*, 1972, **A28**, S39.
[7] M. Ramanadham, S. K. Sikka, and R. Chidambaram, *Acta Cryst.*, 1972, **B28**, 3000.
[8] J. J. Verbist, M. S. Lehmann, T. F. Koetzle, and W. C. Hamilton, *Acta Cryst.*, 1972, **B28**, 3006.
[9] A. Sequiera, H. Rajagopal, and R. Chidambaram, *Acta Cryst.*, 1972, **B28**, 2514.
[10] T. F. Koetzle, M. S. Lehmann, J. J. Verbist, and W. C. Hamilton, *Acta Cryst.*, 1972, **B28**, 3207.
[11] R. R. Bugayong, A. Sequiera, and R. Chidambaram, *Acta Cryst.*, 1972, **B28**, 3214.
[12] T. F. Koetzle, W. C. Hamilton, and R. Parthasarathy, *Acta Cryst.*, 1972, **B28**, 2083.

di-L-leucine hydrochloride,[13] the short hydrogen bond of 2.43 Å is of interest, but the position of the hydrogen is not unequivocal in this case.

The *X*-ray structures of some unusual amino-acids have been reported. Lanthionine, a diamino-acid analogous to cystine but with a single sulphur atom in place of the disulphide group, has a C—S—C angle of 103.9 (0.3)°

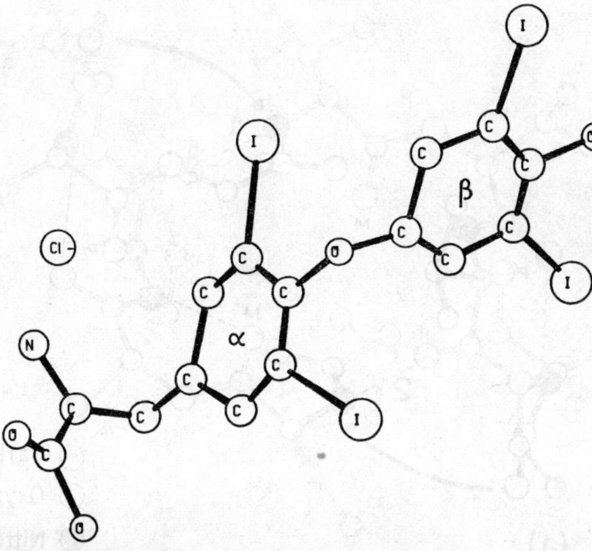

Figure 1 *The structure of thyroxin*
(Reproduced by permission from *Proc. Nat. Acad. Sci. U.S.A.*, 1972, **69**, 2130)

and S—C distances of 1.810 (3) Å.[14] The carboxyl carbons are antiperiplanar to the sulphur.

The structure of streptolidine,[15] a naturally occurring guanidine amino-acid (1), has been determined crystallographically.

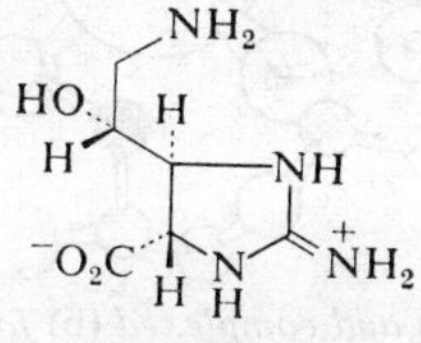

(1)

The mechanism of action of L-thyroxin, the principal hormone secreted by the thyroid gland, is largely unexplained. The crystal structure[16] as given by *X*-ray analysis is shown in Figure 1, but this does little to help our understanding of the biochemistry.

[13] L. Golič and W. C. Hamilton, *Acta Cryst.*, 1972, **B28**, 1265.
[14] R. E. Rosenfield, jun. and R. Parthasarathy, *Acta Cryst.*, 1972, **A28**, S39.
[15] B. W. Bycroft and T. J. King, *J.C.S. Chem. Comm.*, 1972, 652.
[16] N. Camerman and A. Camerman, *Proc. Nat. Acad. Sci. U.S.A.*, 1972, **69**, 2130.

The study of peptides is a useful way of analysing the geometry of the peptide bond. Although most peptide bonds are *trans*, there are certain situations where, for steric reasons, only *cis* conformations are allowed, *e.g.* in cyclic peptides. In any case, the *cis* conformation is similar in energy to

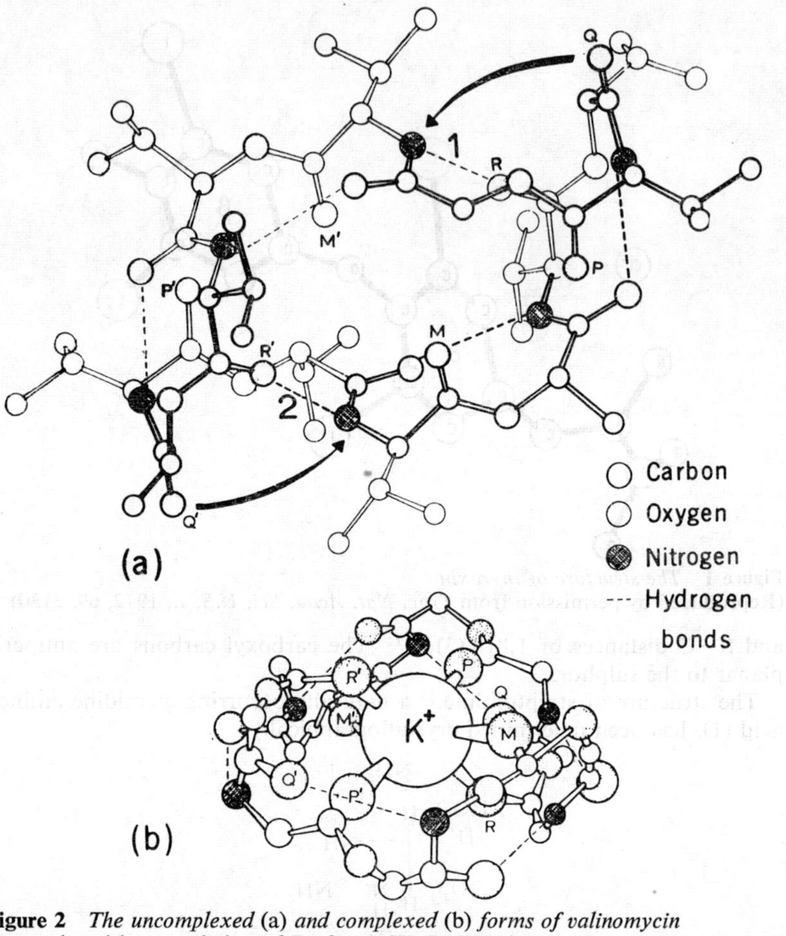

Figure 2 *The uncomplexed* (a) *and complexed* (b) *forms of valinomycin* (Reproduced by permission of Professor W. L. Duax)

the *trans* in the case of peptides containing imino-acid residues, *i.e.* those containing prolyl residues. It is therefore not surprising to find that *cyclo*-(tri-L-prolyl) and *cyclo*-(L-prolyl-L-prolyl-L-hydroxyprolyl) have all the peptide units in *cis* conformations.[17] The peptide bonds are close to planar but in one case show a deviation of 19°.

[17] G. Kartha and G. Ambady, *Acta Cryst.*, 1972, **A28**, S33.

There is also a significant non-planarity of 10.2° of the *trans* peptide bond in glycylalanine,[18] although in two derivatives of glycyl-L-phenylalanine [19] and in glycylglycine phosphate,[20] the peptide bonds are essentially planar.

The crystal structure of uncomplexed valinomycin has been determined by a novel direct method of Hauptman.[21] Four of the six intramolecular hydrogen bonds form β-turns (ten-membered rings) similar to those in the complexed molecule. The other two hydrogen bonds (1 and 2 in Figure 2) form thirteen-membered rings which flatten and open out the structure. The two free carbonyl groups (P and M) are sufficiently free to initiate complexing with the potassium ion. Once a complex is formed the two hydrogen bonds in the thirteen-membered rings may be broken, and eventually the fully complexed lipid-soluble structure is formed.

Small-angle *X*-ray diffraction studies of valinomycin are consistent with conformational changes in the peptide on binding K^+ ions.[22]

Further details of the crystallographic refinement of actinomycin have been presented.[23] Sobell and Jain [24] have also discussed the binding found in the actinomycin complex as a model for proteins which may utilize a two-fold axis in recognizing symmetrically arranged nucleotide sequences on the DNA helix.

The structure of viomycin (2), a tuberculostatic antibiotic, has been crystallographically determined. It contains a sixteen-membered ring with the familiar intra-ring hydrogen bond forming a ten-membered loop, or β-turn.[25]

A number of metal–amino-acid complexes have been described. These include bis(glycinato)bis(imidazole)nickel(II),[26] bis-(L-tyrosinato)copper-(II),[27] bis-(L-histidinato)zinc(II),[28] hydrogendi(bis-π-cyclopentadienyl)-L-cysteinatomolybdenum(IV) chloride [29] and hexafluorophosphate, and bis-π-cyclopentadienylglycinatomolybdenum(IV) chloride.[29] Di-μ-thio-n-butyl(bis-π-cyclopentadienyl)molybdenum, which is a model compound of the nitrogenase system, has a bent sulphur bridge and no metal–metal bond.[30]

The structure of the iron-containing peptide ferrichrysin [31] indicates that the hexapeptide sequence *cyclo*-(Gly-Ser-Ser-Orn-Orn-Orn) is stabilized by hydrogen bonding to give a pleated sheet structure.

18 P. S. Naganathan and K. Venkatesan, *Acta Cryst.*, 1972, **B28**, 552.
19 J. M. van der Veen and B. W. Low, *Acta Cryst.*, 1972, **B28**, 3548.
20 G. R. Freeman, R. A. Heam, and C. E. Bugg, *Acta Cryst.*, 1972, **B28**, 2906.
21 W. L. Duax, *Acta Cryst.*, 1972, **A28**, S46.
22 W. R. Krigbaum, F. R. Kuegler, and H. Oelschlaeger, *Biochemistry*, 1972, **11**, 4548.
23 C. S. Jain and H. M. Sobell, *J. Mol. Biol.*, 1972, **68**, 1.
24 H. M. Sobell and C. S. Jain, *J. Mol. Biol.*, 1972, **68**, 21.
25 B. W. Bycroft, *J.C.S. Chem. Comm.*, 1972, 660.
26 H. C. Freeman and J. M. Guss, *Acta Cryst.*, 1972, **B28**, 2090.
27 D. Van der Helm and C. E. Tatsch, *Acta Cryst.*, 1972, **B28**, 2307.
28 T. J. Kistenmacher, *Acta Cryst.*, 1972, **B28**, 1302.
29 C. K. Prout, G. B. Allison, L. T. J. Delbaere, and E. Gore, *Acta Cryst.*, 1972, **B28**, 3043.
30 C. K. Prout and T. S. Cameron, *Acta Cryst.*, 1972, **B28**, 453.
31 R. Norrestam and B. Stensland, *Acta Cryst.*, 1972, **A28**, S39.

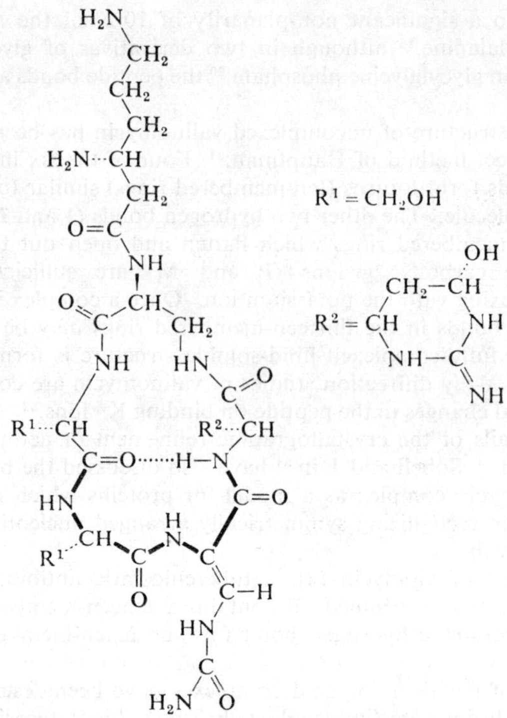

(2)

3 Methods of Protein Structure Analysis

Dobler and his colleagues have developed a microtechnique for batch crystallization using 3 μl of solution for each sample.[32] The solution (300 μl) at a spectrometrically determined concentration is placed in a small tube. A 30 μl aliquot of this solution is transferred to a further tube and replaced by 30 μl of water. This process is repeated to give ten tubes each containing 30 μl of solution with decreasing concentration. From each of these fractions, a 3 μl aliquot is removed and injected as a bead into a 1 mm melting-point capillary, which is then flame sealed. Each aliquot is replaced by 3 μl of 0.4M-ammonium sulphate. This process is repeated to give 10 capillary tubes from each 30 μl solution with successively increasing concentration of ammonium sulphate. The tubes are stored horizontally in rows of ten attached to transparent adhesive tape, so that they can be inspected easily. By this technique a wide range of protein and salting-out agent concentrations is explored in an attempt to find optional crystallization conditions. This method was successfully used for the crystallization of C-phycocyanin.

[32] M. Dobler, S. D. Dover, K. Laves, A. Binder, and H. Zuber, *J. Mol. Biol.*, 1972, **71**, 785.

A new microbalance procedure leads to a precise and reproducible measurement of the water content of protein crystals.[33] The method can be applied to small monocrystals (0.5—10 mg) or to masses of very small crystals (of the order of 0.1 mg each).

Glutaraldehyde is a useful agent for stabilizing protein crystal lattices, and has made certain *X*-ray studies possible under conditions which would normally destroy the crystals. The nature of this reagent and its homogeneity have been the subject of considerable discussion. Korn *et al.*[34] have demonstrated that aqueous solutions used for chemical modification consist of free glutaraldehyde (3), the cyclic hemiacetal of its hydrate (4),

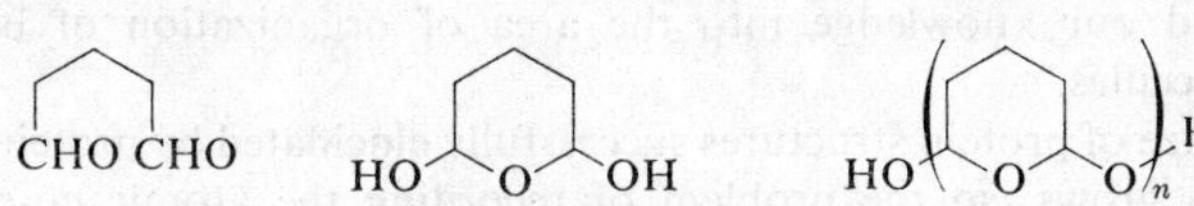

(3) (4) (5)

and oligomers of this (5) in equilibrium with each other. On dilution (3) and (4) are favoured, and all react as free glutaraldehyde and not as a condensation product as indicated by Richards and Knowles.[35] However, the reaction eventually involves four glutaraldehyde molecules per lysine residue, but the instability of the reaction products made identification impossible.

Although no complete structure of a protein has been determined by a technique other than the method of isomorphous replacement, an analysis of the native protein *X*-ray data using a Patterson or rotation function can give useful information about the symmetry of subunit structure in an oligomer. Rossmann[36] has now described a 'locked rotation function' which should give better contrast than his original rotation function. The new method only accepts peaks in the function which are consistent with all the assumed molecular symmetry. Lattman[37] has discussed optional sampling of the rotation function. He presented a method requiring fewer sampling points which associates equal volumes with all sample points and produces undistorted maps.

The problem of finding the arrangement of subunits in proteins may also be approached by other diffraction techniques, *e.g.* neutron or *X*-ray scattering in solution or image reconstruction of electron micrographs. Neutron scattering can be used to give the identity and relative special position of each component in a complex, provided the complex can be prepared in the deuteriated form and can be reconstituted in its separate parts. Engelman and Moore[38] have shown that if two components are rich in hydrogen and if the rest of the complex is deuteriated, the contrast

33 J. Berthou, F. Cesbron, and A. Laurent, *J. Mol. Biol.*, 1972, **71**, 809.
34 A. H. Korn, S. H. Feairheller, and D. E. M. Filachione, *J. Mol. Biol.*, 1972, **65**, 525.
35 F. M. Richards and J. R. Knowles, *J. Mol. Biol.*, 1968, **37**, 231.
36 M. G. Rossmann, *J. Mol. Biol.*, 1972, **64**, 246.
37 E. Lattman, *Acta Cryst.*, 1972, **B28**, 1065.
38 D. E. Engelman and P. B. Moore, *Proc. Nat. Acad. Sci. U.S.A.*, 1972, **69**, 1997.

in scattering power permits the measurement of separation of the pair from an interference cross-term in the scattering. The quaternary arrangement is obtained by triangulation from a set of measured pair separations. Hoppe[39] has suggested a similar technique using *X*-ray scattering of particles labelled specifically with heavy atoms.

Klug and Crowther[40] have reviewed the methods proposed for the derivation of a three-dimensional structure from a series of electron micrographs. They formulate the problem in a general way in the context of information theory. Lake[41] has suggested a method for helix structures. These techniques can add to the information gained from *X*-ray diffraction, and extend our knowledge into the area of organization of biological macromolecules.

As the size of protein structures successfully elucidated by protein crystallographers grows, so the problem of recording the atomic co-ordinates from a model becomes more tedious. Salemme and Fehr[42] have described an automatic device which can be used in conjunction with a Richards optical comparator, 'Fred's Folly', to measure rapidly and accurately the model molecular co-ordinates. A marker light may be automatically positioned by entering preselected co-ordinates from an electronic console. The device uses pulsed stepping motors, and incorporates modularized solid-state circuitry throughout.

Colman, Jansonius, and Matthews[43] have also published details of a 'Folly' in which the electron-density sections are stacked parallel to the half-silvered mirror. With this construction the alignment is simpler and there is easy access to both the model and the electron-density map. In addition, the vertical mirror is under no strain and does not bend, and one may work close to the mirror, thus improving accuracy.

The alternative technique of model building with the aid of computer graphics has also been discussed.[44]

4 Globular Proteins

General Features.—The structures of globular proteins reported during 1972 appear to pose as many fascinating general questions as they answer. For example, the striking similarity of the iron–sulphur complex in ferredoxin to that in high-potential iron protein gives no clue to their very different redox potentials. Further, the close resemblance of the active site in the thermolysin to that of carboxypeptidase extends to the arginine residue which, it has been proposed, binds to a polypeptide carboxy terminal. This makes carboxypeptidase a specific exopeptidase, but thermolysin is an endopeptidase!

[39] W. Hoppe, *J. Mol. Biol.*, 1973, **78**, 581.
[40] A. Klug and R. A. Crowther, *Nature*, 1972, **238**, 435.
[41] J. A. Lake, *J. Mol. Biol.*, 1972, **66**, 255.
[42] F. R. Salemme and D. G. Fehr, *J. Mol. Biol.*, 1972, **70**, 697.
[43] F. M. Colman, J. N. Jansonius, and B. W. Matthews, *J. Mol. Biol.*, 1972, **70**, 701.
[44] T. H. Jacobi, R. A. Ellis, and J. M. Fritsch, *J. Mol. Biol.*, 1972, **72**, 589.

Perhaps the most interesting general features concern quaternary structure. The antiparallel pleated sheet interaction binding two subunits in concanavalin is similar to that found in the insulin dimer. This secondary structure leads neatly to a two-fold axis relationship of the subunits, and it is very likely that this arrangement will be found in the self-association of many proteins. In concanavalin, in contrast to insulin, the two-fold axis of symmetry is precise. However, in concanavalin there are no bulky phenylalanine residues immediately adjacent to the symmetry axis as in insulin, although phenylalanine does occur in the pleated sheet structure.

A further example of a dimer with approximate rather than exact two-fold symmetry is given by malate dehydrogenase. Here the question arises as to whether the two-fold symmetry is, or is not, exact in solution. The single-crystal structure provides a tempting basis for speculation, but gives little evidence on this question. If the dimer were asymmetric in solution it would be likely to crystallize with an approximate symmetry axis, and not use crystallographic symmetry. However, the dimer could be symmetric in solution, and still occupy the crystallographic asymmetric unit. If this were so the two chemically equivalent units would have different crystallographic environments, and may adjust their structures in slightly different ways giving rise to distortion of two-fold symmetry. Many independent crystallographic studies of the same dimer in different crystallographic environments would shed some light on this difficult problem. Without resolving this aspect, it is obviously dangerous to extend the speculation of dimer lack of symmetry to the phenomenon of negative co-operativity.

Proteases.—The structure of thermolysin [45, 46] at 2.3 Å resolution provides a further fascinating example of the power of the *X*-ray method.

Thermolysin is an extracellular, thermostable, proteolytic enzyme isolated from *Bacillus thermoproteolyticus*. It is a neutral protease specific for peptide bonds on the imino side of hydrophobic residues, in particular isoleucine, leucine, and phenylalanine. The enzyme has a molecular weight of 34 600 and contains one catalytically active zinc and four calcium atoms.

Hexagonal crystals of space group $P6_122$ and cell dimensions a = 94.2 Å and c = 131.4 Å grow to 0.5 mm in each direction. Three heavy-atom derivatives were used in the analysis; these were dimercuryacetic acid, K_2PtI_6, and Hg–thermolysin derivative, with mercury replacing zinc in the last case. The electron-density map was first interpreted without reference to the chemically determined amino-acid sequence in order to estimate the degree of success which might be expected from an *X*-ray analysis of a protein of completely unknown sequence, and also to provide as far as possible a completely independent check on the chemical results.

[45] B. W. Matthews, P. M. Colman, J. N. Jansonius, K. Tibani, K. A. Walsh, and H. Neurath, *Nature New Biol.*, 1972, **238**, 41.

[46] B. W. Matthews, J. N. Jansonius, P. M. Colman, B. P. Schoenborn, and D. Dupourque, *Nature New Biol.*, 1972, **238**, 37.

Assuming that the *X*-ray method cannot distinguish Asp from Asn, or Glu from Gln, the overall fraction of residues correctly identified from the electron-density map was 53% (compared with 48% and 60% for myoglobin and carboxypeptidase A, respectively).

The 316 residues are folded into a three-dimensional structure, Figure 3, in which there are two distinct globular units linked through residues 153—158. The two halves are approximately spherical, and the internal hydrophobic regions of each are continuous across the junction. The catalytically

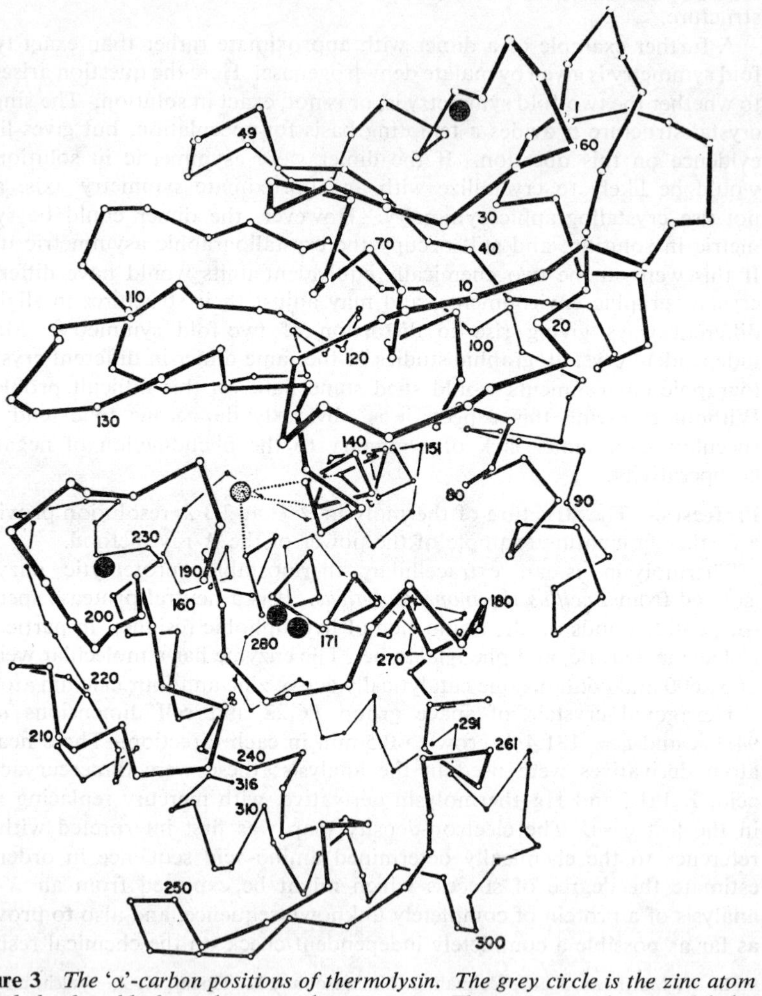

Figure 3 *The 'α'-carbon positions of thermolysin. The grey circle is the zinc atom and the four black circles are calcium atoms. There are two distinct globular units linked through residues* 153—158
(Reproduced by permission from *J. Mol. Biol.*, 1972, **70**, 701)

essential zinc atom lies in a deep cleft, and there is a hydrophobic pocket in the cleft about 8 Å from the zinc.

The helical regions comprising some 38% of the residues are located in seven sections each of approximately four turns of helix. Whereas the

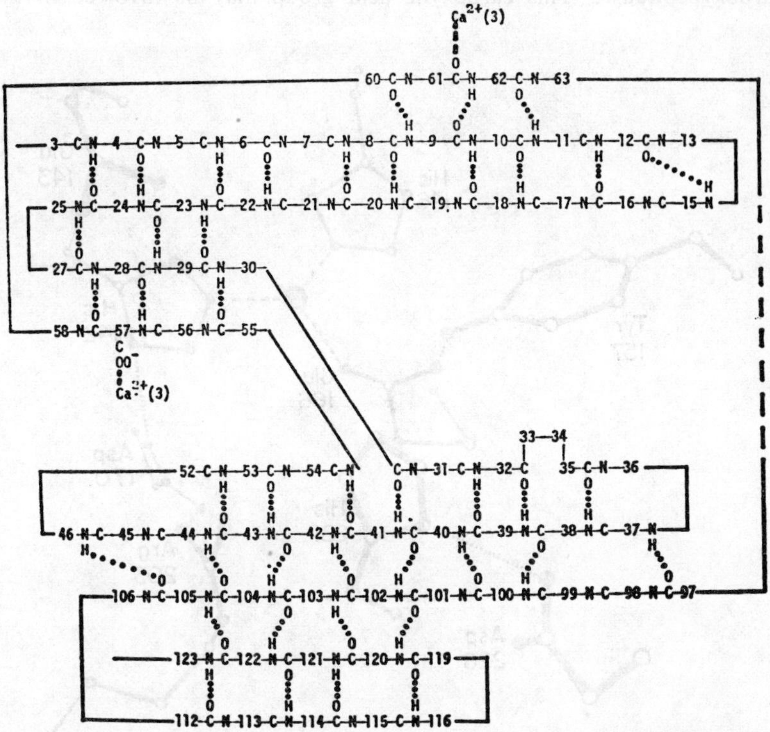

Figure 4 *The hydrogen-bonding scheme in the* N-*terminal half of thermolysin. The β-sheet structure is formed into a cylinder similar to those in α-chymotrypsin and its homologues*
(Reproduced by permission from *J. Mol. Biol.*, 1972, **70**, 701)

majority of the helix occurs in the *C*-terminal half, essentially all the β-structure in the molecule is in the amino-terminal lobe, shown schematically in Figure 4. This β-sheet structure is formed into a 'cylinder' similar to those in α-chymotrypsin and homologous proteins.

There are three clear and one less well-defined calcium binding sites. Two of the sites are only 3.8 Å apart. The sites were confirmed by replacement with strontium, barium, and tervalent lanthanide ions.[47] The interaction of these calcium ions with carboxylic acid side-chains is undoubtedly an important factor in the thermal stability of the enzyme.

[47] P. M. Colman, L. H. Weaver, and B. W. Matthews, *Biochem. Biophys. Res. Comm.*, 1972, **46**, 1999.

The zinc atom is bound by His-142, His-146, and Glu-166 (Figure 5), showing a striking resemblance to the co-ordination of the zinc in carboxypeptidase A. Furthermore, there is a glutamic acid residue, Glu-143, which has the same disposition relative to the zinc atom as Glu-270 in carboxypeptidase. This carboxylic acid group may be involved in the

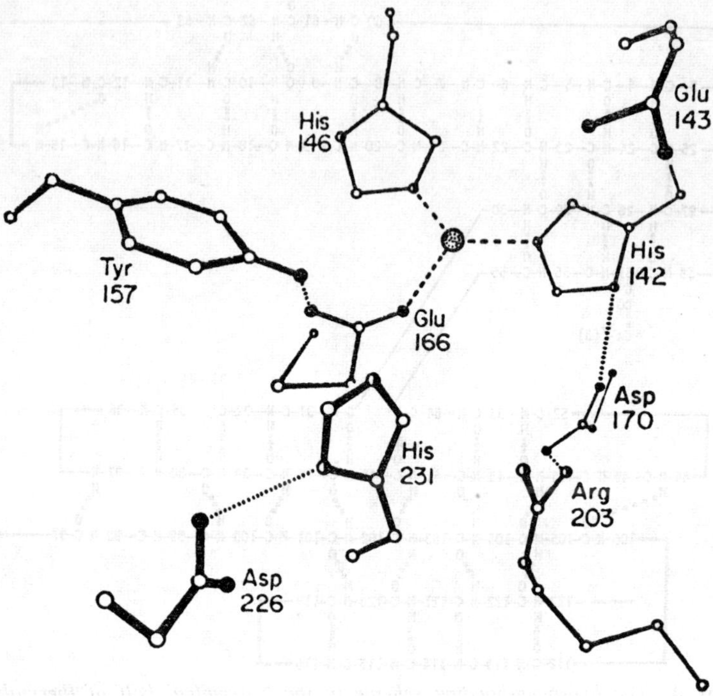

Figure 5 *The co-ordination of the zinc in thermolysin*
(Reproduced by permission from *J. Mol. Biol.*, 1972, 701)

mechanism of hydrolysis of the susceptible peptide bond. Further, similarities between the two proteases embrace a system of salt links from one of the histidine–zinc ligands through an aspartic acid to an arginine residue (His-69—Asp-142—Arg-145 in native carboxypeptidase A and His-142—Asp-170—Arg-203 in thermolysin) (Figure 5). The role of the arginine in the endopeptidase, thermolysin, must clearly be different from that of Arg-145 in carboxypeptidase, an exopeptidase. The occurrence of all the active-site residues within a span of about 90 amino-acids is unusual. Most enzymes, the serine proteases being examples, have the active-site residues in widely different parts of the primary structure, and usually on different globular structures within the enzyme tertiary structure.

Birktoft and Blow [48] have published a very detailed report on α-chymotrypsin at 2 Å resolution. This updates the preliminary reports of 1967 and 1968, and provides a wealth of structural information.

Crystalline methylchymotrypsin [49] has been studied in the hope that it would allow the study of enzyme–substrate complexes by *X*-ray diffraction. α-Chymotrypsin, methylated specifically at the N^{τ}-position of His-57, crystallizes isomorphously with the native enzyme, and so can be studied using phase information from tosylchymotrypsin. The difference electron-density maps indicate small movements of the active-site His-57 and Ser-195. The water molecule bound to His-57(N^{τ}) is displaced by the covalently bound methyl group, and this imidazole nitrogen is no longer a hydrogen-bond acceptor for the Ser-195 hydroxy-group. This is required in the charge-relay mechanism of action postulated by Blow and co-workers.

This structure analysis does not really explain the observed low inherent activity of this modified enzyme in solution.[50] The authors discuss the possibility that there is more freedom of movement of side-chains in the active molecule in solution. In particular in the active monomer, as opposed to the crystallographic dimer, the histidine may be several angstroms nearer the solvent so that a water molecule can still play a role in the mechanism.

A chemical and crystallographic study of carbamylchymotrypsin has also been reported.[51] In this case the carbamyl group occupies the same position as the carbonyl group in indoleacryloylchymotrypsin, which is bound to the active-site Ser-195. This position is proposed for the acyl group of the intermediate in the mechanism of action.

Kraut and his colleagues have reported a number of studies on binding of inhibitors to subtilisin BPN′.[52–54] The results of such difference Fourier studies of the binding geometry for various polypeptides are reviewed and new proposals for mechanism presented. Polypeptides corresponding to the acylating portion of good substrates bind in a fashion very similar to that seen previously for polypeptide chloromethyl ketone derivatives, which are assumed to represent a model for the Michaelis complex between the enzyme and actual substrates.[52] An important difference in the new series is that the hydrogen bond between the amido nitrogen of the specificity residue P_1 and the carbonyl oxygen of Ser-125 is not formed. Other areas on the surface are found to have affinity for the polypeptides tested.

Model-building studies have led to models for two other likely intermediates, a tetrahedral addition compound and the acyl enzyme. In the

[48] J. J. Birktoft and D. M. Blow, *J. Mol. Biol.*, 1972, **68**, 187.
[49] C. S. Wright, G. Hess, and D. M. Blow, *J. Mol. Biol.*, 1972, **63**, 295.
[50] R. Henderson, *Biochem. J.*, 1971, **124**, 13.
[51] G. T. Robillard, J. C. Powers, and P. E. Wilcox, *Biochemistry*, 1972, **11**, 1773.
[52] J. D. Robertus, R. A. Alden, J. J. Birktoft, J. Kraut, J. C. Powers, and P. E. Wilcox, *Biochemistry*, 1972, **11**, 2439.
[53] C. S. Wright, R. A. Alden, and J. Kraut, *J. Mol. Biol.*, 1972, **66**, 283.
[54] J. D. Robertus, J. Kraut, R. A. Alden, and J. J. Birktoft, *Biochemistry*, 1972, **11**, 4293.

tetrahedral addition compound, construction of a covalent linkage between Ser-221 and the carbonyl carbon of the susceptible bond results in a distortion of that carbon to a tetrahedral configuration, and formation of a hydrogen bond between the substrate specificity residue P_1 and the enzyme at Ser-125. There are also two further hydrogen bonds in the 'oxyanion hole' stabilizing the developed negative charge on the substrate. These together stabilize the tetrahedral intermediate and lower the activation energy for its formation. Further, model building shows that when the tetrahedral intermediate collapses to the acyl-enzyme, steric constraints prevent the ester carbonyl from remaining in the 'oxyanion hole'. There is a stereochemical similarity to the 'oxyanion hole' in chymotrypsin extending the degree of convergence in evolution beyond the charge-relay system and the mode of binding.

A comparison of the active-site stereochemistry and substrate conformation in α-chymotrypsin and subtilisin BPN′ has also been presented from the Cambridge laboratory.[55]

Studies of trypsin–trypsin inhibitor complexes are proceeding in a number of laboratories. Two further crystalline forms of the complex have been reported.[56] Blow and Huber and their co-workers have built a model which suggests a unique mode of binding for bovine pancreatic trypsin inhibitor to chymotrypsin or trypsin.[57] Seven hydrogen bonds and about 200 van der Waals contacts are made.

Crystals of a further rennin-like enzyme of fungal origin (from *Mucor Pusillus*) are reported.[58]

Nucleases.—Carlisle and his colleagues have computed a 2.5 Å resolution electron-density map of bovine pancreatic ribonuclease A.[59] The crystals are the same form as that studied by Kartha, Bello, and Harker.[60] The mean figure of merit is 0.64 falling off from 0.8 at 5 Å to 0.47 at 2.5 Å. The analysis clearly shows the main polypeptide folding and side-chain positions, and is in agreement with the results of Kartha *et al.* The amino-acid residues His-12, His-19, and Lys-41, which are involved in the activity of ribonuclease, and Lys-7, which lies close by, are clearly identified.

A low-resolution (6 Å) study[61] of ribonuclease S indicates that the enzyme in the monoclinic crystals may have a slightly different structure from that earlier reported for the trigonal form.[62]

Bovine seminal ribonuclease differs from other ribonucleases in that it is a dimer, of molecular weight 29 000. It crystallizes in space group $P2_12_12_1$

55 C. S. Wright, *J. Mol. Biol.*, 1972, **67**, 151.
56 F. E. Cole and R. Parthasarathy, *J. Mol. Biol.*, 1972, **71**, 105.
57 D. M. Blow, C. S. Wright, D. Kukla, A. Rühlmann, W. Steigemann, and R. Huber, *J. Mol. Biol.*, 1972, **69**, 137.
58 P. C. Moews and C. W. Bunn, *J. Mol. Biol.*, 1972, **68**, 389.
59 C. H. Carlisle, B. A. Gorinsky, S. K. Mazumdar, R. A. Palmer, and D. G. R. Yeates, *Acta Cryst.*, 1972, **A28**, S36.
60 G. Kartha, J. Bello, and D. Harker, *Nature*, 1967, **213**, 862.
61 Y. Mitsui, H. W. Wyckoff, and F. M. Richards, *Acta Cryst.*, 1972, **A28**, S36.
62 H. W. Wyckoff, D. Tsernoglou, A. W. Hanson, J. R. Knox, B. Lee, and F. M. Richards, *J. Biol. Chem.*, 1970, **245**, 305.

with cell dimensions a = 36.6, b = 66.6, and c = 107.4 Å.[63] The two molecules of the dimer are related by a two-fold axis.

Two fragments of staphylococcal nuclease containing residues 6—48 and residues 49—149, respectively, can be reconstituted to give an intact, non-covalently bound enzyme called nuclease T′ with approximately 8% biological activity. In the presence of deoxythymidine 3′,5′-diphosphate and Ca^{2+} ions, the nuclease T′ crystallizes in space group $P4_1$ with cell dimensions $a = b = 47.6 \pm 0.2$, $c = 63.3 \pm 0.02$ Å, compared with $a = b = 48.3$, $c = 63.3$ Å for crystalline native nuclease.[64] The crystals are highly isomorphous, indicating a close similarity of the three-dimensional structures of nuclease T′ and native nuclease in solution.

Glycoside Hydrolases.—Swan [65] has studied the inhibition of hen egg-white lysozyme by various imidazole and indole derivatives using kinetic and crystallographic methods. His results are consistent with the formation of charge-transfer complexes between imidazole derivatives and tryptophan residues of lysozyme. Comparison of the kinetic and crystallographic data suggests that direct interpretation of inhibition studies is open to question when the inhibitor is not a close analogue of the '*in vivo*' substrate.

Berthou, Laurent, and Jolles have crystallized duck egg-white lysozyme and are studying it at 4, 20, and 37 °C in an attempt to find the changes of structure at these different temperatures.[66]

Carbonic Anhydrases.—Further details of the 2 Å resolution analysis of human carbonic anhydrase C, originally published in 1971, are reported.[67] There is also a preliminary report on carbonic anhydrase B.[68] This includes a discussion of a technique of seeding which was used to grow crystals of 2 mm size, with space group $P2_12_12_1$ and cell dimensions a = 81.5, b = 73.6, and c = 37.1 Å. The position of the essential zinc ion has been established. This zinc can be replaced by mercury, giving a useful heavy-atom derivative.

Enzymes of the Glycolytic Pathway.—One of the most encouraging aspects of this work in 1972 has been the preliminary studies on phosphorylase, with a subunit molecular weight of 100 000. Tetrameric and dimeric forms have been crystallized.

Preliminary *X*-ray data for monoclinic phosphorylase were reported by Mathews in 1967.[69] This form has been under study by Madsen, Honikel,

63 L. Mazzarella and A. Ripamonti, *J. Mol. Biol.*, 1972, **64**, 311.
64 H. Taniuchi, D. R. Davies, and C. B. Anfinsen, *J. Biol. Chem.*, 1972, **247**, 3362.
65 I. D. A. Swan, *J. Mol. Biol.*, 1972, **65**, 59.
66 J. Bertou, A. Laurent, and P. Jolles, *J. Mol. Biol.*, 1972, **71**, 815.
67 A. Liljas, K. K. Kannan, P.-C. Bergste'n, I. Waara, K. Fridborg, B. Strandberg, U. Carlbom, L. Järup, S. Lövgren, and M. Petef, *Nature New Biol.*, 1972, **235**, 131.
68 K. K. Kannan, K. Fridborg, P.-C. Bergste'n, A. Liljas, S. Lovgren, M. Petef, B. Strandberg, I. Waara, L. Adler, S. O. Falkbring, P. O. Göthe, and P. O. Nyman, *J. Mol. Biol.*, 1972, **63**, 601.
69 F. S. Mathews, *Fed. Proc.*, 1967, **26**, 831.

and James in Edmonton,[70] by Johnson and co-workers in Oxford,[71] and by Huber and co-workers in Munich.[72] The monoclinic forms of the crystals of both phosphorylases a and b in the presence or absence of 10mM-AMP contain one tetramer per asymmetric unit and were found to exhibit very similar *X*-ray diffraction patterns.

A new crystal form of phosphorylase b containing dimers has been reported.[73] The crystals are grown in the presence of AMP and have space group $P4_12_12_1$ (or enantiomorph) and cell dimensions $a = b =$ 129.2 and $c =$ 116.2 Å. There is one subunit in the asymmetric unit; this offers considerable advantages for the *X*-ray analysis. It also implies that the subunits are identically related by a crystallographic two-fold axis and suggests that there is no gross structural rearrangement on addition of AMP to phosphorylase b. Data collection is underway on the native and two potential derivatives at 6 Å resolution. Crystals grown under the same conditions as the tetrahedral form have been studied in the electron microscope. The cells have similar dimensions $a = b =$ 125 Å, $\gamma = 107°$, but appear to contain phosphorylase tetramers in contrast to the dimers of the large crystals used for *X*-ray analysis.

A new crystal form of rabbit muscle aldolase with space group $P6_122$ and cell dimensions $a = 163.5$, $b = 335$ Å has been reported.[74]

The low-resolution structure of yeast phosphoglycerate kinase,[75] like the same enzyme from horse muscle (Vol. 4, p. 198),[76] has two distinct structural lobes or globular units. The importance of this arrangement is not understood, but it is suggested that it may have evolutionary significance. The 6 Å resolution structure of yeast phosphoglycerate mutase has also been reported by the Bristol group.[77] The yeast enzyme is a tetramer of 110 700 daltons, in contrast to the muscle enzymes which are dimers of about 65 000 daltons. Dimers in the tetramer are related by a crystallographic axis. The two subunits contained in the crystal asymmetric unit are very similar, suggesting that the small differences observed are probably due to errors in measurement rather than real differences in structure. These subunits are related by a local two-fold axis so that the tetramer has almost perfect 222 symmetry.

Preliminary *X*-ray studies of rabbit-muscle pyruvate kinase have shown that the spacegroup is $P6_122$ with $a = 187.2$ and $c = 170.0$ Å with one half of the tetrameric molecule in the asymmetric unit.[78]

[70] N. B. Madsen, K. O. Honikel, and N. M. G. James, 'Metabolic Conversion of Enzymes', Springer Verlag, New York, 1972.
[71] P. A. M. Eagles and L. N. Johnson, *J. Mol. Biol.*, 1972, **64**, 693.
[72] H. Fasold, F. Ortlanderl, R. Huber, K. Bartels, and P. Schwager, *F.E.B.S. Letters*, 1972, **12**, 229.
[73] P. A. M. Eagles, M. Iqbal, L. N. Johnson, J. Mosley, and K. S. Wilson, *J. Mol. Biol.*, 1972, **71**, 803.
[74] L. Sawyer, *J. Mol. Biol.*, 1972, **71**, 503.
[75] P. L. Wendell, T. N. Bryant, and H. C. Watson, *Nature New Biol.*, 1972, **240**, 134.
[76] C. C. F. Blake, P. R. Evans, and R. K. Scopes, *Nature New Biol.*, 1972, **235**, 195.
[77] J. W. Cambell, G. I. Hodgson, and H. C. Watson, *Nature New Biol.*, 1972, **240**, 137.
[78] A. McPherson and A. Rich, *J. Biol. Chem.*, 1972, **247**, 1334.

Dehydrogenases.—Glyceraldehyde-3-phosphate dehydrogenase (GPDH) catalyses the oxidation and subsequent phosphorylation of aldehyde substrates to their corresponding acyl phosphates.

X-Ray studies of human GPDH by Andreeva *et al.*[79] indicate tetrahedral symmetry for the tetramer of the holo-enzyme. The space group $C222_1$ demonstrates that one two-fold axis must be exact, but further 'quasi' systematic absences in the diffraction pattern suggest that the tetramer is also located at the intersection of local dyad axes. As this conclusion is based on 2.8 Å resolution, there may be some small distortion from perfect tetrahedral symmetry. However, the symmetry is clearly reduced in the apo-enzyme crystals. This is in contrast to the change in lactate dehydrogenase from 222 molecular symmetry to symmetry 2 during transformation of this enzyme from the apo- to the holo-form.

Human GPDH has also been studied by Watson and his colleagues,[80] and here progress in the past year has been very encouraging, a 6 Å resolution model being reported. The crystal form is very similar to that reported by Andreeva, but there are small but significant differences in the diffraction pattern indicating space group *C*2 with four equivalent positions in the cell. The tetramer therefore has at least one two-fold axis and the diffraction intensities indicate further local two-fold axes perpendicular to this. The crystal unit cell appears to be suitable for *X*-ray analysis but the crystals are often twinned. The 6 Å electron-density map clearly defines the molecular boundary and indicates an almost spherical hexamer with dimensions approximating 80, 80, and 75 Å along the crystallographic and local two-fold axes, respectively. There is no electron density on these two-fold axes as expected. Tentative assignments of the subunit boundaries have been made. The authors have not attempted to interpret the chain folding of the enzyme. However, they note that in the methyl mercury nitrate and gold chloride derivatives, in which the activity is lost, the metal positions are separated by 8.5 Å. These probably mark the active-site thiols. A tentative position for the coenzyme is identified. The authors also note that in some parts of the molecule there are deviations from local two-fold symmetry, which may explain the reported half-site reactivity.

The rotation function has been successfully exploited in the determination of the orientation of the tetramer of lobster glyceraldehyde-3-phosphate dehydrogenase in the crystal cell.[81] The results indicate 222 symmetry of the molecule.

Cytoplasmic malate dehydrogenase (MDH) catalyses the interconversion of oxaloacetate and L-malate and uses nicotinamide adenine dinucleotide as a coenzyme. The conformations of the two polypeptide chains of the dimer of MDH have been determined from an electron-density map at

79 A. I. Gorjunov, N. S. Andreeva, T. Baranowski, and M. Wolny, *J. Mol. Biol.*, 1972, **69**, 421.
80 H. C. Watson, E. Duée, and W. D. Mercer, *Nature New Biol.*, 1972, **240**, 130.
81 M. G. Rossmann, G. C. Ford, H. C. Watson, and L. J. Banaszak, *J. Mol. Biol.*, 1972, **64**, 237.

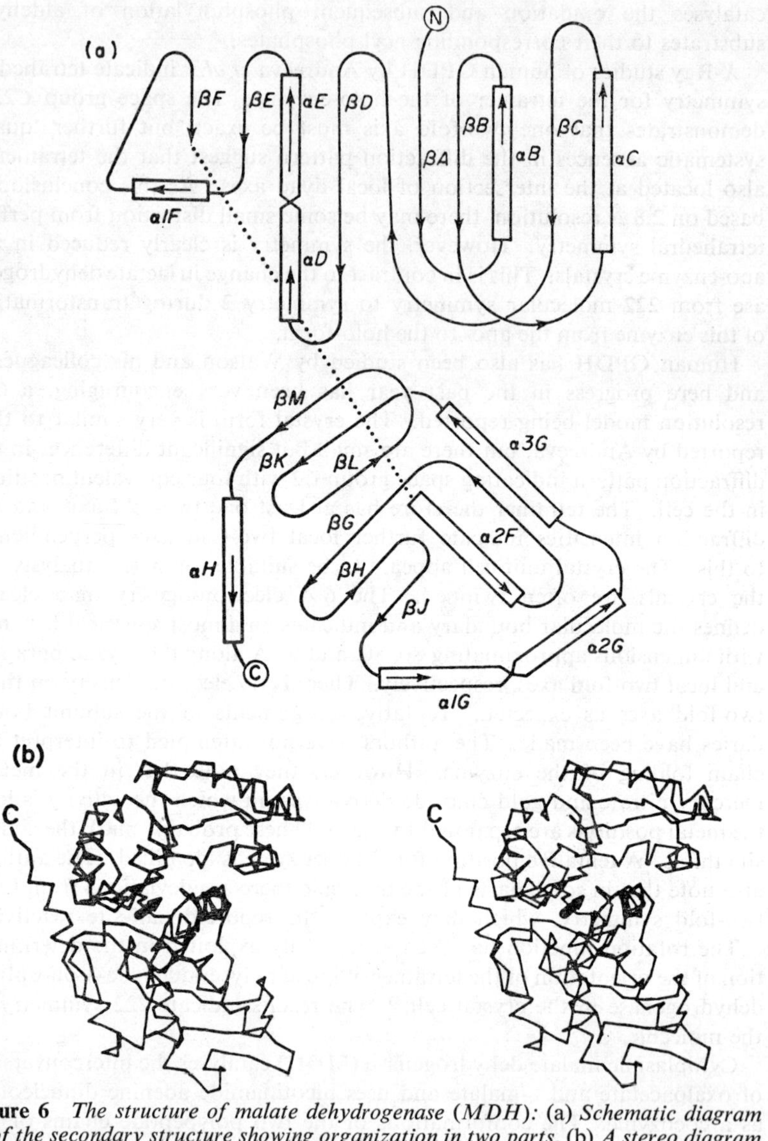

Figure 6 *The structure of malate dehydrogenase* (*MDH*)*:* (a) *Schematic diagram of the secondary structure showing organization in two parts*, (b) *A stereo diagram of the structure of MDH. The arrow shows the polypeptide connecting the two parts, each of which is a distinct globular unit*
(Reproduced by permission from *J. Mol. Biol.*, 1972, **72**, 577)

3 Å resolution (see Vol. 4, p. 199) and the structure has now been described in detail.[82] There has also been further discussion of the methods used in the 5 Å resolution analysis.[83] The conformation is defined in Figure 6a in terms of segments of chain in either 'α'-helices or 'β'-pleated sheets and connecting turns between them. The molecule is clearly folded in two halves as shown in Figure 6b.

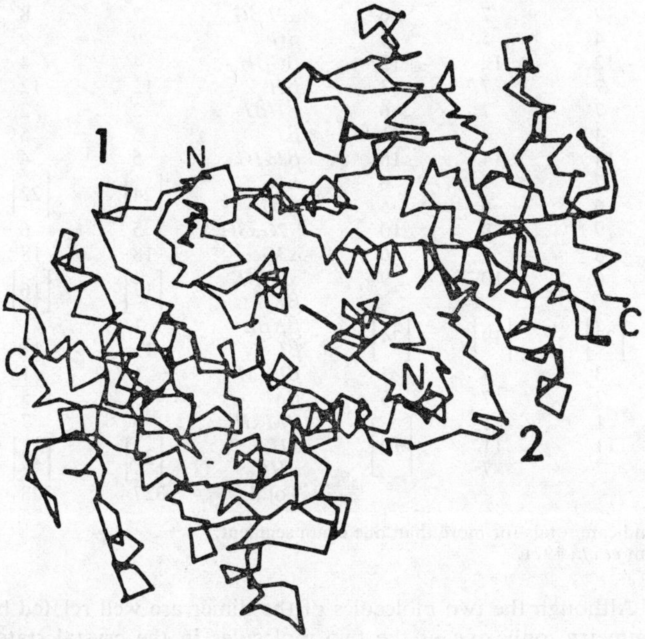

Figure 7 *The dimer of malate dehydrogenase viewed along the approximate two-fold axis*
(Reproduced by permission from *J. Mol. Biol.*, 1972, **72**, 577)

The association of the two molecules in the dimer is shown in Figure 7. The dimer is roughly spheroidal in shape with approximate dimensions 64 × 64 × 45 Å. The two polypeptide chains of the dimer are related by an approximate two-fold axis. Table 1 indicates the conformation of chain segments in the structure of MDH and LDH (lactate dehydrogenase). The two structures have remarkably similar conformations with the exception of an extra 21 residues at the *N*-terminal of LDH (see Figure 8). This is consistent with the role of these residues in LDH of strengthening association of dimers to tetramers, which is not observed in

[82] E. Hill, D. Tsernoglou, L. Webb, and L. J. Banaszak, *J. Mol. Biol.*, 1972, **72**, 577.
[83] D. Tsernoglou, E. Hill, and L. J. Banaszak, *J. Mol. Biol.*, 1972, **69**, 75.

Table 1 *Residues allocated to chain segments in malate dehydrogenase and lactate dehydrogenase*

Segment	s-MDH *Subunit* 1	s-MDH *Subunit* 2	LDH[a]	*Segment*	s-MDH *Subunit* 1	s-MDH *Subunit* 2	LDH
NαA	—	—	1				
αA	—	—	5	*βF*	4	4	3
αAβA	—	—	16	*βFα2F*	[16]	[15]	3
NβA	2	2	—	*α2F*			17
βA	7	7	6	*α2FβG*	7	8	6
βAαB	4	5	4	*βG*	9	9	5
αB	12	12	12	*βGβH*	4	4	7
αBβB	7	7	3	*βH*	12	12	8
βB	7	7	6	*βHβJ*	2	2	
βBαC	4	4	1	*βJ*	5	5	[19]
αC	14	14	16	*βJα1G*	5	4	
αCβC	7	7	6	*α1G*	[24]	[22]	[19]
βC	6	6	5	*α2G*			
βCβD	9	8	10	*α2Gα3G*	5	6	3
βD	8	8	6	*α3G*	18	18	15
βDαD	6	10	9	*α3GβK*	[17]	[16]	[3]
				βK			
αD	[23]	[19]	[24]	*βKβL*	1	1	1
αE				*βL*	15	15	15
αEβE	3	4	3	*βLβM*	1	1	2
βE	7	7	6	*βM*	5	5	6
βEα1F	1	1		*βMαH*	6	7	4
α1F	11	11	[19]	*αH*	[25]	[27]	20
α1FβF	7	7		*αHC*			2
				Total	327	328	331

Brackets indicate totals for more than one chain segment.
[a] Rossmann *et al.*, 1971.

MDH. Although the two molecules of the dimer are well related by two-fold symmetry, only one of the two molecules in the crystal state binds NAD^+ strongly. The binding resembles that of NAD^+ in LDH. There may be low NAD^+ occupancy at the equivalent site on the second subunit. The non-equivalence of the two subunits may be an example of negative co-operativity. In both subunits there is density corresponding to the 'essential' histidine of LDH. Further mechanistic information awaits the extension of the analysis to 2 Å and the sequence determination.

Regulatory Enzymes.—Crystals of glutamine synthetase from *E. coli* have been re-examined by *X*-ray diffraction techniques,[84, 85] and more symmetry than reported earlier has been revealed. The cell is hexagonal with $a = 142 \pm 2$ Å and $c = 148 \pm 2$ Å. The *X*-ray data extend to 15 Å. The most likely structure is now considered to be a square tetramer rather than an eclipsed two-storey ring dodecamer. Electron microscopy of

84 R. P. Bywater, C. H. Carlisle, R. B. Jackson, A. L. Mackay, and P. A. Timmins, *Acta Cryst.*, 1972, **A28**, S37.
85 R. P. Bywater, R. B. Jackson, and C. H. Carlisle, *J. Mol. Biol.*, 1969, **45**, 429.

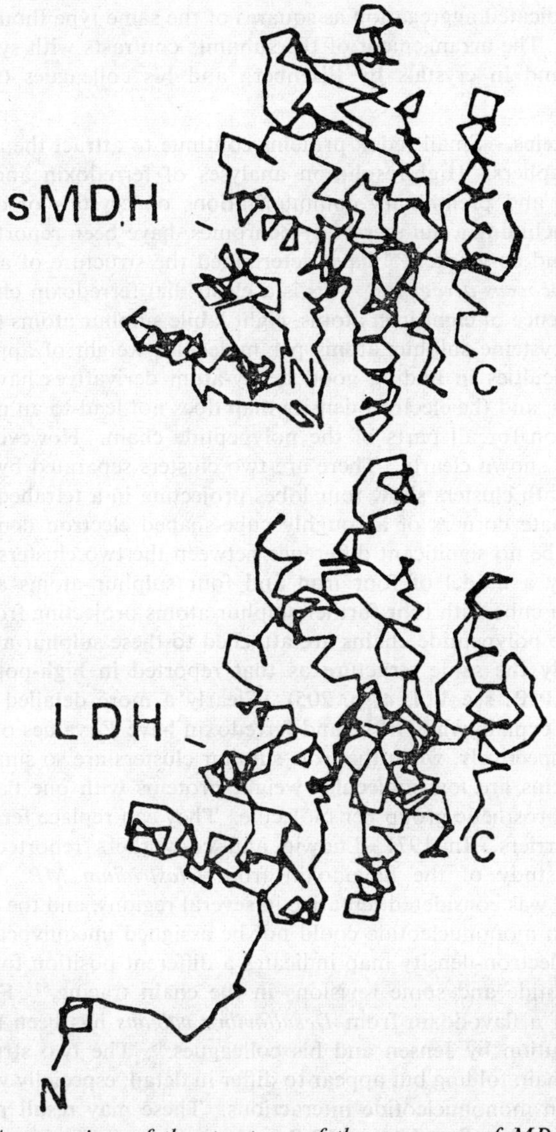

Figure 8 *A comparison of the structures of the monomers of MDH and LDH viewed from equivalent directions. Note the* N*-terminal arm of LDH which is partly responsible for tetramer formation; MDH only forms a dimer*
(Reproduced by permission from *J. Mol. Biol.*, 1972, **72**, 577)

solutions confirmed the presence of dodecamers both singly and as rods, but also indicated aggregation as squares of the same type thought to exist in crystals. The arrangement of the subunits contrasts with symmetries 6 or 622 found in crystals by Eisenberg and his colleagues (see Vol. 4, p. 201).

Redox Proteins.—Small redox proteins continue to attract the attention of crystallographers. High-resolution analyses of ferredoxin and a further flavodoxin, and preliminary communications on crystals of other redox proteins, including a number of cytochromes, have been reported.

Jensen and co-workers [86] have determined the structure of a ferredoxin from *Micrococcus aerogenes*. This is a clostridial ferredoxin characterized by the presence of eight iron atoms, eight labile sulphur atoms (inorganic), and eight cysteine sulphur atoms per molecular weight of approximately 6000. Difficulties in finding good heavy-atom derivatives have hindered the analysis, and the electron-density map does not lead to an unequivocal interpretation for all parts of the polypeptide chain. However, the iron clusters are shown clearly. There are two clusters separated by a distance of 12 Å. Both clusters show four lobes projecting in a tetrahedral fashion from alternate corners of a roughly cube-shaped electron density. There appears to be no significant difference between the two clusters which can be fitted by a model of four iron and four sulphur atoms at alternate corners of a cube with four further sulphur atoms projecting from the iron atoms. The polypeptide chains are attached to these sulphur atoms. This is essentially the same structure as that reported in high-potential iron protein (HiPIP; see Vol. 4, p. 205). Clearly a more detailed analysis is required to explain why HiPIP and ferredoxin have E_0 values of $+0.3$ and -0.4 V, respectively, when the iron–sulphur clusters are so similar.

Flavodoxins are low molecular weight proteins with one flavin mononucleotide prosthetic group per molecule. They can replace ferredoxins as electron carriers. In 1971, Ludwig and co-workers reported a 3.25 Å resolution study of the flavodoxin from *Clostridium MP*. The model constructed was considered tentative in several regions, and the orientation of the flavin mononucleotide could not be assigned unequivocally, but an improved electron-density map indicates a different position for the flavin mononucleotide and some revisions in the chain tracing.[87] Further, the structure of a flavodoxin from *Desulfovibro vulgans* has been reported at 2.5 Å resolution by Jensen and his colleagues.[88] The two structures are similar in chain folding but appear to differ in detail, especially with respect to the flavin mononucleotide interactions. These may result partly from the fact that the *Clostridium MP* flavodoxin is in the semiquinone form

[86] L. C. Sieker, E. Adman, and L. H. Jensen, *Nature*, 1972, **235**, 40.

[87] R. D. Anderson, P. A. Algar, R. M. Burnett, G. D. Darling, M. E. Lesquesne, S. G. Mayhew, and M. L. Ludwig, *Proc. Nat. Acad. Sci. U.S.A.*, 1972, **69**, 3189.

[88] K. D. Watenpaugh, L. C. Sieker, L. H. Jensen, J. Legall, and M. Dubourdieu, *Proc. Nat. Acad. Sci. U.S.A.*, 1972, **69**, 3185.

whereas that from *D. vulgans* is in the oxidized state. Figures 9—11 illustrate the flavodoxin structures viewed from equivalent directions. The central sheet is five-stranded with two long helices on either side.

In *D. vulgans* flavodoxin the methyl groups and one edge of the flavin, along with part of the ribitol group, are exposed at the surface. The flavin

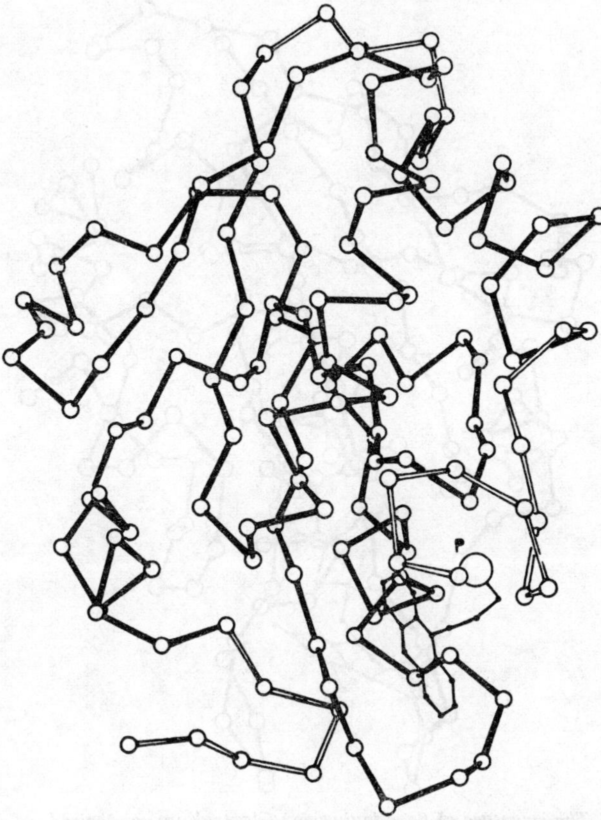

Figure 9 *The structure of flavodoxin from* Clostridium, *as postulated in* 1971 (Reproduced by permission from *Proc. Nat. Acad. Sci. U.S.A.*, 1972, **69**, 3189)

group appears to be planar and is packed between two, probably aromatic, side-chains. The two carbonyl oxygen atoms of the pyrimidine ring of the flavin are hydrogen-bonded to the protein. The ribitol group is clearly defined and the phosphate is the largest peak in the electron-density map. It is buried deeply in the protein and is bent away from the flavin rather than towards it.

In the *Clostridium* flavodoxin the reinterpretation indicates that the phosphate is bent away from the flavin group, but in a rather different way from that in *D. vulgans.* The flavin ring is less buried in *Clostridium*, and lies at an angle of about 30° relative to the position in *D. vulgans.*

An analysis of bonito ferrocytochrome *c* at 2.3 Å resolution [89] shows that this protein is similar in structure to horse ferricytochrome *c* reported in 1971. The crystals may be reversibly oxidized in the absence of a reducing

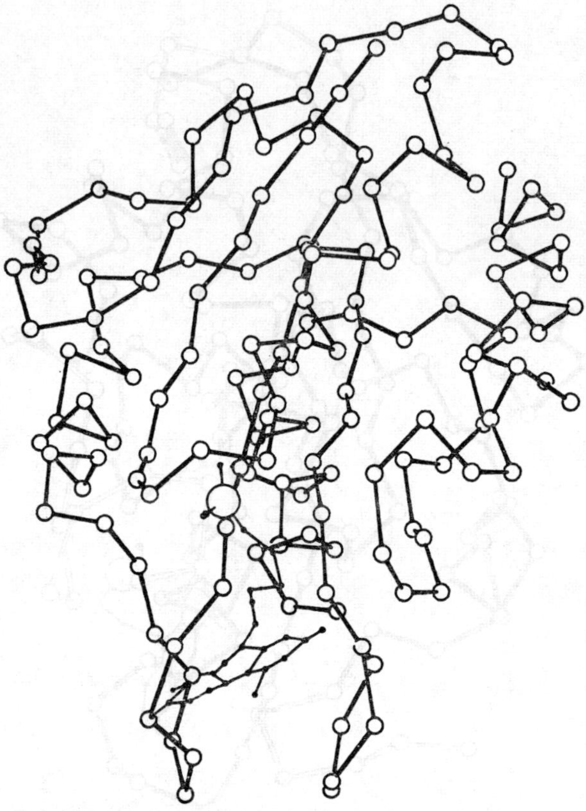

Figure 10 *The structure of flavodoxin from* Clostridium, *as revised* (Reproduced by permission from *Proc. Nat. Acad. Sci. U.S.A.*, 1972, **69**, 3189)

agent at pH 6 with no apparent deterioration of the crystal, and *X*-ray studies show that this oxidation–reduction reaction surprisingly does not bring about a large change in the main-chain folding. This is in strong contrast to the horse cytochrome *c* where the oxidized and reduced forms have very large conformational differences, and crystallize in quite different space groups.[90]

[89] M. Kakudo, T. Ashida, N. Tanaka, T. Tsukihara, and T. Yamane, *Acta Cryst.*, 1972, **A28**, S34.

[90] T. Takano, O. B. Kallai, and R. E. Dickerson, *Acta Cryst.*, 1972, **A28**, S35.

A preliminary study of crystals of rice cytochrome *c* shows that it has space group $P6_1$ or $P6_5$ with cell dimensions a = 43.8 and c = 110.6 Å and contains one molecule in the asymmetric unit.[91] Crystals of cytochrome *c* from *Micrococcus denitrificans* have also been studied. They have space group $P2_12_12_1$ with a = 42.7, b = 82.1, and c = 31.6 Å and one molecule

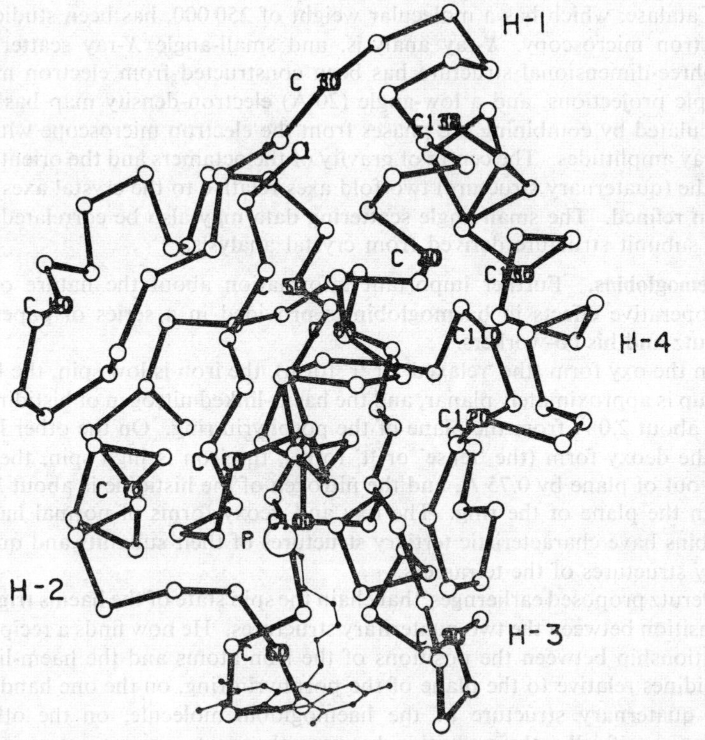

Figure 11 *The structure of flavodoxin from* D. vulgans
(Reproduced by permission from *Proc. Nat. Acad. Sci. U.S.A.*, 1972, **69**, 3189)

in the asymmetric unit.[92] Both of these studies are proceeding to high resolution and so detailed structural information on the very interesting evolution of cytochrome *c* should eventually be available.

Mathews *et al.* have published a full and very detailed discussion of the analysis of cytochrome b_5 at 2.8 Å resolution,[93] a summary of which was given in 1971. They have also reported preliminary studies of cytochrome b_{562} from *E. coli*.[94] This protein has three crystalline forms, one of which

[91] Y. Morita and S. Ida, *J. Mol. Biol.*, 1972, **71**, 807.
[92] R. Timkovich and R. E. Dickerson, *J. Mol. Biol.*, 1972, **72**, 199.
[93] F. S. Mathews, M. Levine, and P. Argos, *J. Mol. Biol.*, 1972, **64**, 449.
[94] E. W. Czerwinski, F. S. Mathews, P. Hollenberg, K. Drickamer, and L. P. Hager, *J. Mol. Biol.*, 1972, **71**, 819.

has data to 2.8 Å resolution and contains one molecule per asymmetric unit.

Preliminary *X*-ray studies of Japanese-radish peroxidase C[95] show that the enzyme crystallizes in space group *P*422 with cell dimensions $a = b =$ 111.2 Å and $c =$ 133.6 Å with sixteen molecules (molecular weight 32 000) in the unit cell.

Catalase, which has a molecular weight of 250 000, has been studied by electron microscopy, *X*-ray analysis, and small-angle *X*-ray scattering.[96] A three-dimensional structure has been constructed from electron microscopic projections, and a low-angle (20 Å) electron-density map has been calculated by combining the phases from the electron microscope with the *X*-ray amplitudes. The centre of gravity of the octamers and the orientation of the (quaternary structure) two-fold axes relative to the crystal axes have been refined. The small-angle scattering data may also be correlated with the subunit structure derived from crystal analysis.

Haemoglobins.—Further important information about the nature of the co-operative effects in haemoglobin is provided in a series of papers by Perutz and his co-workers.

In the oxy form (the 'relaxed' or 'r' form), the iron is low spin, the haem group is approximately planar, and the haem-linked nitrogen of histidine-F8 lies about 2.0 Å from the plane of the porphyrin ring. On the other hand, in the deoxy form (the 'tense' or 't' form), the iron is high spin, the iron lies out of plane by 0.75 Å, and the nitrogen of the histidine is about 2.9 Å from the plane of the ring. The oxy and deoxy forms of normal haemoglobins have characteristic tertiary structures of their subunits and quaternary structures of the tetramers.

Perutz proposed earliernges that chain the spin state of the haems trigger a transition between the two quaternary structures. He now finds a reciprocal relationship between the positions of the iron atoms and the haem-linked histidines relative to the plane of the porphyrin ring, on the one hand, and the quaternary structure of the haemoglobin molecule, on the other.[97] More specifically, the transition between the quaternary structures alters the spin state of the haems. In the deoxy state the quaternary structure is literally 'tense' and the haem is constrained. In the oxy state the structure is relaxed as in the free subunits.

Aquomethaemoglobin (metHb) and the closely related mutant methaemoglobin M Milwaukee are suitable for the study of haem–haem interaction.[98] In these forms the iron atoms occupy positions 0.3 Å from the porphyrin plane. This is intermediate between the position for the normal oxy and deoxy structures. Thus the tertiary structure of either oxy or deoxy form depends not only on the reaction at the iron atom, but also

[95] Y. Morita, *Acta Cryst.*, 1972, **A28**, S52.
[96] B. K. Vainshtein, G. V. Gurskaya, V. V. Barynin, and L. A. Feigin, *Acta Cryst.*, 1972, **A28**, S37.
[97] M. F. Perutz, *Nature*, 1972, **237**, 495.
[98] M. F. Perutz, P. D. Pulsinelli, and H. M. Ranney, *Nature New Biol.*, 1972, **237**, 259.

on the quaternary structure of the tetramer. This has been studied by crystallographic and optical absorption spectroscopic techniques.

In haemoglobin M Milwaukee, the valine residues at positions E11(67), which are close to the haem iron, have been replaced by glutamic acid. The latter are linked *via* their side-chain carboxy-groups to the ferric iron atoms. Crystals of the form in which the α-subunits are ferrous deoxy and the β-subunits are ferric are isomorphous with deoxyhaemoglobin. However, when all the subunits are ferric, the crystals are in the oxy crystal form. Thus this provides an example where iron atoms in the same spin and oxidation state can exist with different quaternary structures and almost certainly different tertiary structures. In agreement with Perutz's mechanism, the change to the oxy form does result in a shift of the iron atom and the haem-linked histidine towards the porphyrin plane. Similar conclusions are derived from spectroscopic study of mixed-state haemoglobins (containing ferric and ferrous iron). These experiments clearly show that there is a change of tension at the haem which accompanies a change of quaternary structure.

The study of the interaction of D-2,3-diphosphoglycerate (DPG) with haemoglobin is also important to understanding the relative stability of the oxy and deoxy forms.[99] DPG and haemoglobin are the two principal organic constituents inside the human erythrocyte; they both have 5mM *in vivo* concentration. The function of DPG is to lower the oxygen affinity of haemoglobin and thus facilitate the release of oxygen. It does this through binding one molecule of DPG to a deoxyhaemoglobin tetramer. The *X*-ray study [99] was achieved by first changing the solvent to 2-methyl-2,4-pentanediol, as even moderate concentrations of salt dissociate the DPG–deoxyhaemoglobin complex. The DPG was found to bind in a stereochemically complementary cavity on the two-fold axis. The acidic groups form salt bridges with seven cationic groups of the β-chains which include the valines 1, histidines 2 and 143, and one of the lysines 82. This complementarity would be lost in the oxy forms where the cavity contracts, and DPG thus shifts the allosteric equilibrium to the deoxy form. These results confirm Perutz's hypothesis.

Further confirmation of Perutz's hypotheses presented in 1970 comes from the study of the crystal structure of deoxyhaemoglobin Yakima.[100] This mutant has high oxygen affinity, no co-operativity, but a normal Bohr effect. It has Asp-Gl(99) replaced by histidine. In normal deoxyhaemoglobins this aspartic acid, which lies at the $\alpha_1\beta_2$ interface, forms a crucial hydrogen bond with an adjacent tyrosine of the α_1 subunit. The replacement of this aspartic acid by histidine in haemoglobin Yakima results in the removal of this single important hydrogen bond, and the bulky imidazole ring constrains the structure towards the oxy form. In fact, a hydrogen bond which usually characterizes the oxy structure is formed.

[99] A. Arnone, *Nature*, 1972, **237**, 146.
[100] P. D. Pulsinelli, *J. Mol. Biol.*, 1973, **74**, 57.

The disturbances in the stereochemistry affect both 'α' and 'β' haems. Thus this *X*-ray structure provides a stereochemical explanation for the abnormally high oxygen affinity and vastly dimished haem–haem interaction.

Sickle cell haemoglobin differs from normal haemoglobin by replacement of valine for glutamic acid at the sixth position of both β-chains. The increased hydrophobicity leads to aggregation under deoxygenation conditions which is responsible for 'sickling' of cells. Deoxygenated sickle cells are birefringent under crossed nicol prisms, and the electron microscope indicates rod-like structures with diameters 140—170 Å. Magdoff-Fairchild and co-workers [101a] have demonstrated that rods can be formed *in vitro* by rigorous exclusion of oxygen from concentrated solutions. *X*-Ray diffraction data on these preparations can be interpreted in terms of cylinders 170 Å in diameter with a 55 Å diameter central hole. Electron microscopy [101b] has shown that the rods consist of six parallel filamentous polymers of HbS molecules with the molecules in adjacent polymers in register along the rods.

Agglutinants.—Proteins capable of agglutinating cells are attracting considerable interest in the study of cell division and cancer. In 1971, two structure analyses of concanavalin A were reported at 4 Å resolution. The detail has now been improved to 2 Å resolution and the amino-acid sequence published simultaneously.[102, 103] The analysis of concanavalin A provides a fascinating picture of this tetrameric protein, which is capable of agglutinating various types of cell and is useful in the study of cell surfaces and the control of cell division. The nature of the quaternary interactions, and the binding sites for sugars, calcium, and manganese can now be described.

The study by Hardman and Ainsworth at 2.4 Å[103] was achieved by the method of isomorphous replacement using five heavy-atom derivatives and measuring *X*-ray data with an automated diffractometer. On the other hand, Edelman and co-workers [102] used photographic techniques in measuring the *X*-ray data to 2 Å with three principal derivatives. Edelman's group had the advantage of knowing the sequence when interpreting their electron density. Nevertheless, both electron-density maps appear to be of very good quality.

The electron-density maps reveal asymmetric units (of molecular weight 25 000 and 238 amino-acid residues) forming 'ellipsoidal domes', which are related by two-fold axes to give a roughly tetrahedral shape. Each unit contains regions of extended polypeptide chain arranged in the anti-

[101a] B. Magdoff-Fairchild, P. H. Swerdlow, and J. F. Bertles, *Nature*, 1972, **239**, 217.
[101b] J. T. Finch, M. F. Perutz, J. F. Bertles, and J. Döbler, *Proc. Nat. Acad. Sci., U.S.A.*, 1973, **70**, 718.
[102] G. M. Edelman, B. A. Cunningham, G. N. Reeke, J. W. Becker, M. J. Waxdal, and J. L. Wang, *Proc. Nat. Acad. Sci. U.S.A.*, 1972, **69**, 2580.
[103] K. O. Hardman and C. F. Ainsworth, *Nature*, 1972, **237**, 54; *Biochemistry*, 1972, **11**, 4910.

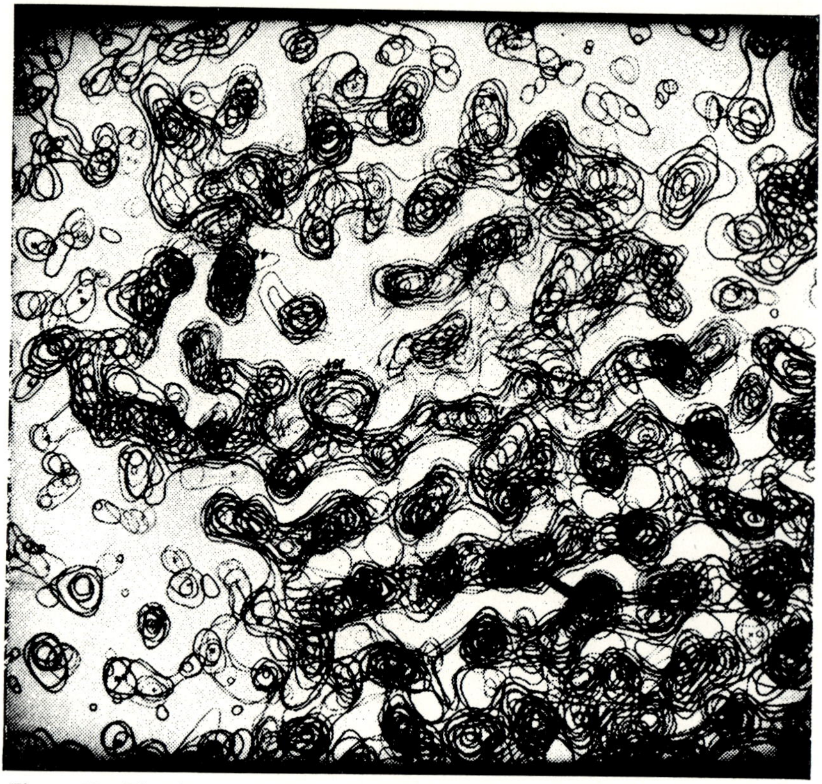

Figure 12 *The electron density in a pleated sheet region of concanavalin A* (Reproduced by permission from *Biochemistry*, 1972, **11**, 4910)

[*Facing p. 173*

parallel β-pleated sheets. The first contains 51 residues in seven hydrogen-bonded chains (residues 132—125, 106—116, 199—190, 48—55, 66—59, and 73—78) and nine residues in short connecting loops. The second sheet contains 34 residues arranged in six antiparallel chains (residues 144—140,

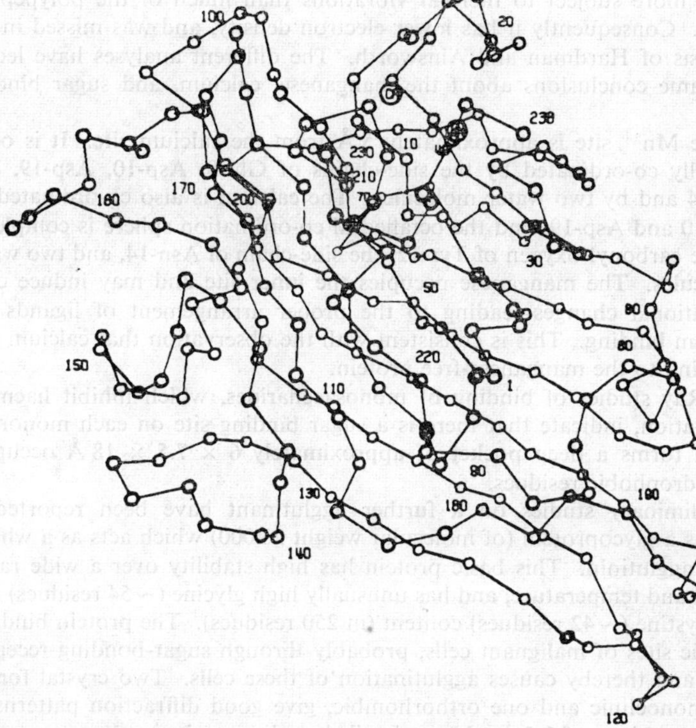

Figure 13 *The* α*-carbon positions in concanavalin A*
(Reproduced by permission from *Proc. Nat. Acad. Sci. U.S.A.*, 1972, **69**, 2580)

173—177, 97—92, 209—215, 9—4, and 25—29). The first sheet forms the back of the molecule and the second is perpendicular to it, dividing the molecule into two halves. One half contains a loosely arranged 'left-handed coil' while the other contains 'right-handed coils' of the amino and carboxy termini. There is only one turn of α-helix! Figure 12 shows the electron density of a pleated sheet region, and Figure 13 a skeletal model of the structure.

The first sheet structure is extended across the interface between two subunits, giving rise to an ellipsoidal dimer. The main-chain nitrogen and carbonyl oxygen atoms of residues 125—132 are hydrogen bonded to complementary atoms of the same residues in a molecule related by a two-fold axis. The dimers interact primarily through side-chains of the first sheet structure to give tetramers. There appear to be both hydrophilic and hydrophobic side-chain interactions.

Concanavalin A is cleaved naturally between residues 119 and 120, and Edelman and colleagues have observed that this chain extends well away from the main body of the structure. This part of the molecule appears to be more subject to thermal vibrations than much of the polypeptide chain. Consequently it has lower electron density, and was missed in the analysis of Hardman and Ainsworth. The different analyses have led to the same conclusions about the manganese, calcium, and sugar binding sites.

The Mn^{2+} site is approximately 5 Å from the calcium site. It is octahedrally co-ordinated by the side-chains of Glu-8, Asp-10, Asp-19, and His-24 and by two water molecules. The calcium is also co-ordinated by Asp-10 and Asp-19, and the octahedral co-ordination sphere is completed by the carbonyl oxygen of Tyr-12, the side-chain of Asn-14, and two water molecules. The manganese occupies the inner site and may induce conformational changes leading to the proper arrangement of ligands for calcium binding. This is consistent with the observation that calcium will not bind to the manganese-free protein.

X-Ray studies of binding of monosaccharides, which inhibit haemagglutination, indicate that there is a sugar binding site on each monomer, which forms a deep pocket of approximately $6 \times 7.5 \times 18$ Å occupied by hydrophobic residues.

Preliminary studies on a further agglutinant have been reported.[104] This is a glycoprotein (of molecular weight 23 000) which acts as a wheat-germ agglutinin. This basic protein has high stability over a wide range of pH and temperature, and has unusually high glycine (~54 residues) and half cystine (~42 residues) content (in 250 residues). The protein binds to specific sites of malignant cells, probably through sugar-bonding receptor sites, and thereby causes agglutination of these cells. Two crystal forms, one monoclinic and one orthorhombic, give good diffraction patterns to Bragg spacings of 2.5 Å. More detailed studies are in progress.

Toxins.—Preliminary *X*-ray studies of cobrotoxin, the main neurotoxin extracted and purified from Formosan cobra venom, indicate a space group of either $P4_12_1$ or $P4_32_1$ with cell dimensions $a = b = 40.2$ and $c = 71.0$ Å.[105] As in the case of the erabutoxin reported in 1971 the iodinated neurotoxin is a suitable derivative.

Protein Hormones.—The work on the structure of insulin has proceeded to higher resolution and has become a truly international exercise. Following the earlier determination of the structure at 2.8 Å by the Oxford group, a Japanese group report an electron-density calculation of 3.1 Å based on one heavy-atom derivative,[106] and a Chinese group have completed a very

[104] C. S. Wright, T. Sato, Y. Nagata, J. McMillan, R. Langridge, and M. M. Burger, *Acta Cryst.*, 1972, **A28**, S35.

[105] C. Wong, T. W. Chang, T. J. Lee, and C. C. Yang, *Acta Cryst.*, 1972, **A28**, S35.

[106] K. Sakabe and N. Sakabe, *Acta Cryst.*, 1972, **A28**, S34.

beautiful electron-density map at 2.5 Å.[107] This latter map is based on one derivative (Pb^{2+}) in common with the Oxford map, but their contribution has been to discover a new ethyl mercuric derivative. Although the electron-density map may not be as detailed as the recently computed one at 1.9 Å by Hodgkin and co-workers, it is clearly an improvement on their 2.8 Å map and it gives a more precise picture of the zinc co-ordination. It is a demonstration of the strength of Chinese crystallography, and a powerful refutation of those pundits who repeatedly assured us that all was lost in Chinese Science during the cultural revolution!

The structure of insulin and its relation to the chemistry and biology of this hormone have been the subjects of a comprehensive review.[108]

Calcium Binding Protein.—Kretsinger and co-workers have presented an improved and considerably revised model for carp calcium-binding protein.[109, 110] This protein closely resembles mammalian muscle troponin A. The electron-density map at 1.8 Å resolution gives a main-chain configuration into which all the known homologous amino-acid sequences can readily be built. The tightly packed hydrophobic core contains ten phenylalanine residues packed in a herringbone arrangement. The calcium ions are each co-ordinated by groups close to each other in the primary structure. One site has co-ordination through the side-chains of Asp-90, Asp-92, Asp-94, and Glu-101, as well as by the carbonyl oxygen of Lys-96. The second site is co-ordinated through carboxylates of Asp-51, Asp-53, Glu-59, and Glu-62, in addition to the hydroxy-group of Ser-55 and the carbonyl oxygen of Ala-57. The two structures co-ordinating the calcium ions have similar topologies and occur between analogously placed helices.

Immunoglobulins.—No further studies have been reported on complete immunoglobulins but there are some interesting results on Bence-Jones proteins and Fab fragments prepared by pepsin digestion. These give a clear demonstration of the existence of domain structure in immunoglobulins.

Huber and co-workers[111] have studied the Bence-Jones (the light chain of the immunoglobulin) Rei of the antigenic type K. This is monomeric in solution and crystallizes with one molecule per asymmetric unit. The space group is $P6_1$ with lattice constants: $a = b = 75.8$, $c = 98.2$ Å. Patterson search calculations using 5 Å data indicate that the monomeric Bence-Jones protein has a local pseudo two-fold axis. This indicates that the general folding of the two halves, the constant and variable parts, is very similar.

[107] Peking *X*-ray Group, personal communication.

[108] T. L. Blundell, G. G. Dodson, D. C. Hodgkin, and D. Mercola, *Adv. Protein Chem.*, 1972, **26**, 279.

[109] R. H. Kretsinger, *Acta Cryst.*, 1972, **A28**, S52.

[110] C. E. Nockolds, R. H. Kretsinger, C. J. Coffee, and R. A. Bradshaw, *Proc. Nat. Acad. Sci. U.S.A.*, 1972, **69**, 681.

[111] O. Epp, W. Palm, H. Fehlhammer, A. Rühlmann, W. Steigemann, P. Schwager, and R. Huber, *J. Mol. Biol.*, 1972, **69**, 315.

A 4 Å resolution electron-density map based on phases from four heavy-atom derivatives (details not given) shows globular features of high electron-density lying on the three-fold screw axes. These globules are related together in pairs by local two-fold axes perpendicular to the three-fold axes in the way predicted by the Patterson analysis. There is electron density joining the two globules which may be polypeptide connecting the constant and variable halves of the molecule.

Edmundson and co-workers [112] have published further details of an analogous Bence-Jones protein of λ type.

IgG class immunoglobulins can be digested by pepsin to give a fragment consisting of a complete light chain and the *N*-terminal half of the heavy chain. This is the Fab′ fragment (a Fab′ fragment contains ten more amino-acid residues in the heavy chain than the Fab fragment). Both heavy and light chains have a constant and a variable region in the Fab′ fragment. Its structure is of considerable interest in that it retains the antibody activity and specificity of the parent molecule, which resides in the variable portion.

Poljak *et al.*[113] have studied Fab′ New crystallized in space group *C*2, lattice parameters: $a = 111.4$, $b = 56.7$, $c = 90.3$ Å, $\beta = 116.5°$ with one molecule in the asymmetric unit. They have computed a 6 Å electron-density map with a mean figure of merit of 0.85. This shows two structural features of globular shape, relatively compact and nearly equal in size. Arguments are presented for the identification of these as constant and variable regions and for the further differentiation of light and heavy chains. The overall symmetry of the Fab′ model is that of a distorted tetrahedron, approximately described by the point group 222. At the beginning of the variable subunit the chains define together an easily accessible cavity compatible with the dimensional requirements of the antigen binding site of antibodies.

The antigen binding specificity of most Fab fragments from human myeloma patients such as those described above is not known. However, a few homogeneous immunoglobulins of human and mouse origin have antigen binding activity, as shown by screening against test antigens. Davies and co-workers [114] have reported a crystallization and a preliminary *X*-ray investigation of the Fab fragments from two myeloma proteins that bind phosphocholine. These Fab fragments give large crystals from ammonium sulphate solutions. They crystallize in hexagonal space groups *P*622 and *P*63, and one diffracts to about 2.7 Å resolution. The crystals are shown to bind one molecule of hapten per molecule of Fab fragment, but with a binding constant significantly lower than observed in solution.

[112] A. B. Edmundson, M. Schiffer, K. R. Ely, and M. K. Wood, *Biochemistry*, 1972, **11**, 1822.

[113] R. J. Poljak, L. M. Amzel, H. P. Avey, L. N. Becka, and A. Nisonoff, *Nature New Biol.*, 1972, **235**, 137.

[114] S. Rudikoff, M. Potter, D. M. Segal, E. A. Padlan, and D. R. Davies, *Proc. Nat. Acad. Sci. U.S.A.*, 1972, **69**, 3689.

Other Globular Proteins.—*C*-Phycocyanin,[115] a protein pigment involved in photosynthesis from thermophilic algae, has been crystallized using a microbatch technique. Crystals of dimensions 0.2 × 0.2 × 0.5 mm are obtained which have space group *P*63 and cell dimensions $a = 154$, $c = 40.5$ Å. There appear to be two molecules of molecular weight 28 000 per asymmetric unit. Intensity changes with several mecurial derivatives have been observed.

Small-angle *X*-ray scattering techniques indicate that human serum low-density lipoprotein has an approximately spherical symmetry.[116] There is a spherical lipid bilayer of average radius 65 Å, and the outer surface is covered in a two-dimensional network of protein subunits, probably 60 in number, with icosahedral symmetry. The subunits have a molecular weight of about 8000. The phospholipid and cholesterol esters are uniformly distributed on each side of the bilayer, and the cholesterol moiety appears to interact with the protein subunits. The centre of the molecule is probably a protein core.

Details of the preparation of crystals and crystallographic data for different α-lactalbumins have been published.[117, 118]

5 Fibrous Proteins

Collagens and Synthetic Analogues.—Collagen is widely distributed in the animal kingdom, although samples from different species differ appreciably in amino-acid composition and size of polypeptide chain. Traub and Salem[119] have examined high- and low-angle diffraction patterns to determine whether the molecular conformations or modes of assembly of molecules to form fibrils are the same. *X*-Ray patterns indicate that a specific mode of assembly as well as conformation was established at an early stage of evolution.

Crystallographic studies of a synthetic collagen analogue $(\text{Pro-Pro-Gly})_{10}$ indicate a new structure which has some helical parameters in common with those of collagen as proposed by Ramachandran. (See Table 2.)[120]

Fibrinogen.—Fibrinogen is a large molecule (molecular weight 340 000) with a complex structure. Electron microscopy indicates that it has a length of about 475 Å and comprises three globular units connected by thin rods. Blood clots are formed when thrombin cleaves negatively charged peptides from fibrinogen, which then aggregates to form the

[115] M. Dobler, S. D. Dover, K. Laves, A. Binder, and H. Zuber, *J. Mol. Biol.*, 1972, **71**, 785.

[116] L. Mateu, A. Tardieu, V. Luzzati, L. Aggerbeck, and A. M. Scanu, *J. Mol. Biol.*, 1972, **70**, 105.

[117] R. Aschaffenburg, R. E. Fenna, B. O. Handford, and D. C. Phillips, *J. Mol. Biol.*, 1972, **67**, 525.

[118] R. Aschaffenburg, R. E. Fenna, and D. C. Phillips, *J. Mol. Biol.*, 1972, **67**, 529.

[119] W. Traub and G. Salem, *Acta Cryst.*, 1972, **A28**, S38.

[120] K. Okuyama, N. Tanaka, T. Ashida, and M. Kakudo, *Acta Cryst.*, 1972, **A28**, S38; K. Okuyama, N. Tanaka, T. Ashida, M. Kakudo, S. Sakakibara, and Y. Kishida, *J. Mol. Biol.*, 1972, **72**, 571.

Table 2 *Parameters of triple-helical conformation proposed for* (Pro-Pro-Gly)$_{10}$ *and for the collagen model*

Parameter	(Pro-Pro-Gly)$_{10}$	*Collagen model*
Distance between centres of individual chains	5.0 Å	4.4 Å[a]
Major helix radius	2.8 Å	—
Major helix pitch	60.24 Å	104.8 Å
Vertical distance between residues	2.87 Å	2.91 Å
Number of residues in the repeat distance	21	36
Number of turns of minor helix in the repeat distance	7	12

[a] For structure with two hydrogen bonds. This value is 5.0 Å for the structure with one hydrogen bond.

fibrin clot. Fibrin has an axial repeat of only 225 Å, and various mechanisms including 'shrinkage' and 'staggering' have been proposed to account for this small axial period. Fibrinogen has been difficult to study by electron microscopy and *X*-ray diffraction as it has not previously been prepared in a proper crystalline form. Tooney and Cohen[121] now report the preparation of microcrystals of a fibrinogen modified by a protease from *Pseudomonas*. The modified fibrinogen appears to be 15% smaller in molecular weight, but optical diffraction patterns of electron micrographs suggest that no large structural changes have taken place. The A-fibrinopeptides are still attached, and the fibrinogen still clots when treated with thrombin. Preparation of crystals large enough for *X*-ray analysis is under study.

A small-angle *X*-ray scattering study of bovine fibrinogen has been reported, and the results have been compared with various molecular models from the electron microscope.[122]

Synthetic Polyamino-acid Structures.—Physicochemical studies on synthetic polymers of L-proline from dimer to octamer indicate that oligomers containing five or more residues assume a left-handed helical structure identical with that of poly-L-proline II.[123]

Matsuzaki[123] has crystallized and determined the structure of t-butyloxycarbonyltetra-L-proline benzyl ester. The molecule forms one turn of a poly-L-proline-like helix. However, the ϕ and ψ angles and the average translation per residue along the helix axis appear to be closer to collagen models than to those of poly-L-proline II.

A study of poly-L-ornithine hydrobromide[124] shows that fibres with up to four molecules of water per ornithine residue have a parallel pleated sheet structure which does not vary significantly with hydration. However, structures with more than four water molecules per residue appear to

[121] N. M. Tooney and C. Cohen, *Nature*, 1972, **237**, 23.
[122] K. Lederer, *J. Mol. Biol.*, **63**, 315.
[123] T. Matsuzaki, *Acta Cryst.*, 1972, **A28**, S38.
[124] M. Suwalsky, *Acta Cryst.*, 1972, **A28**, S39.

have an α-helical structure which reverts to the pleated sheet arrangement on drying.

Detailed *X*-ray analysis structures of dry fibres and of liquid-crystalline solutions of racemic poly-γ-benzyl glutamate have been described.[125] Regular side-chain conformations occur in these systems, probably as a result of strong interaction between the benzyl groups on the side-chains. The structures have been refined using least-squares analysis of the reciprocal space co-ordinates of reflections. The axial repeat of 64 Å suggests the presence of α-helical molecules with 43 residues in 12 turns of a helix, and this is confirmed by equatorial Patterson and Fourier syntheses.

Muscle.—There are a number of studies on muscle which are concerned with the mechanism of transition from the rigor to the relaxed condition. Rome [126] has reported low-angle *X*-ray studies of relaxed glycinerated muscle, which she concludes has a similar structure to living resting muscle. Miller and Tregear [127] have discussed the structure of insect fibullar flight muscle in the presence and absence of ATP. Vibert *et al.*[128] have considered structural changes in actin-containing filaments of muscle, and Matsubara and Elliot [129] have reported studies on skinned single fibres of frog skeletal muscle. Small and Squire [130] have discussed the structural basis of contraction in vertebrate smooth muscle. Their study confirms and extends the evidence for ribbon-shaped elements being the *in vivo* form of the myosin component of vertebrate smooth muscle.

PART III: Conformation and Interaction of Peptides and Proteins in Solution

edited by R. H. Pain, with contributions by P. M. Bayley, C. E. Johnson, G. L. Kellett, P. Knowles, G. R. Penzer, H. W. E. Rattle, B. Robson, R. D. Ryder, and R. M. Stephens

1 Introduction

by R. H. Pain

The two most important phases as far as proteins are concerned are the aqueous and the membrane phases. A large proportion of the work on proteins during the past year has involved the use of physical techniques to probe their structure in aqueous solution, and some of these techniques are also being used to a lesser extent to examine proteins in the membrane environment.

There is little doubt today that the structure of individual protein molecules as revealed by *X*-ray crystallography reflects in large part the

[125] J. M. Squire and A. Elliot, *J. Mol. Biol.*, 1972, **65**, 291.
[126] E. Rome, *J. Mol. Biol.*, 1972, **65**, 331.
[127] A. Miller and R. T. Tregear, *J. Mol. Biol.*, 1972, **70**, 85.
[128] P. J. Vibert, J. C. Haselgrove, J. Lowy, and F. R. Poulsen, *J. Mol. Biol.*, 1972, **71**, 757.
[129] I. Matsubara and G. F. Elliott, *J. Mol. Biol.*, 1972, **72**, 657.
[130] J. V. Small and J. M. Squire, *J. Mol. Biol.*, 1972, **67**, 117.

solution conformation, but it will still be necessary to rely on solution and non-crystalline-phase methods of probing structure for a great many proteins and enzymes. N.m.r. and c.d. techniques, for example, continue to be developed and are beginning to provide detailed information which can be validated by using known structures, and in this way confidence is steadily being built up for their use on unknown structures. Certain dynamic features of protein structure are not accessible to *X*-ray crystallography and while, for example, this technique has had remarkable success in stimulating much informed inference about catalytic mechanisms, combination of such information with results from kinetic, spin-label, fluorescence, and n.m.r. techniques amongst others is still essential if a complete picture is to be built up. The interaction of subunits in multimeric proteins is important in biological control and in the organization of certain metabolic pathways. More-sophisticated theoretical treatments are now available for studying the equilibria of such interactions and the continued development and use of gel-filtration as a technique should provide a powerful stimulus to further understanding, at least of the thermodynamics of these systems.

Besides the mechanism of action of enzymes, the other major outstanding problem is the means and the pathway by which a linear polypeptide chain folds up to form the unique three-dimensional conformation of the biologically active protein. This field has stimulated a fresh look at the kinetics of folding and unfolding reactions, together with experimental and theoretical approaches to the whole problem of 'nucleation' which is seen conceptually as being necessary to enable the folding process to take place in a realistic period of time. Theoretical studies of polypeptide and simple protein structures, originally carried out on molecules '*in vacuo*', are now beginning to take into account the role of a solvent environment. Combined with c.d. and n.m.r. studies, energy calculations have made significant steps towards our understanding of what were, not so long ago, formidably complex polypeptide structures.

2 Theoretical Aspects of Protein Structure

contributed by B. Robson

Since the last discussion of this topic in Volume 4,[1] further progress has been made towards prediction of the native, biologically active conformation of a protein from its primary sequence. Here the problem is treated as one of understanding the code relating the amino-acid sequence to the values adopted by the 'soft' variables[2] (the internal rotations around single bonds) in the native conformation. Although emphasizing applications to carbohydrate conformation, a recent review by Brant[3] is of interest.

[1] B. Robson, in 'Amino-acids, Peptides, and Proteins', ed. G. T. Young (Specialist Periodical Reports), The Chemical Society, London, 1973, vol. 4, p. 224.
[2] H. A. Scheraga, *Chem. Rev.*, 1971, **71**, 195.
[3] D. A. Brant, *Ann. Rev. Biophys. Bioengineering*, 1972, **1**, 369.

The Analytic Approach.—In the past there has been great interest in making predictions of backbone configuration based on statistical analysis of the available data from proteins of known sequence and conformation. This analytic approach usually makes no assumptions except that interactions between atoms close together in the primary sequence have a major role in deciding the configuration of the backbone in that region. Although it has met with some success,[4–6] progress in 1972 has been slower. Nevertheless, a technique[4] for quantifying the information provided by the analytic approach has recently been used to characterize the directional nature of the information concerning the α-helix.[7] This work shows that many amino-acids direct a helix-forming or -breaking influence preferentially in either the *N*- or *C*-terminal direction, in some cases with a strength approaching that of proline.

A number of publications have recently appeared which show that the scope of the analytic approach is by no means exhausted and which promise well for further development in the near future. Similar conclusions to those using directional information have been obtained by Crawford, Lipscomb, and Schellman,[8] but their analysis extends to backbone turns and loops in general. This interest in loops is also reflected in the analyses of Nagano.[9] However, the relatively extended pleated-sheet conformation has not been forgotten.[8–10] In addition, Wu and Kabat have extended their analysis of the occurrence of tripeptides[11] in proteins of known conformation to attempt a prediction of all the Φ, Ψ angles in immunoglobulin[12] and cytochrome *c*.[13] The nature of their method restricted it to proteins for which a considerable number of homologous variants have been sequenced.

Finally, the spirit of the analytic approach has spread to encompass the concept of locally folded and spatially separate intermediate regions within globular proteins.[14] These may well be the intermediate stage not only in the hierarchy of structural organization but in the actual folding process of the protein molecule, thus providing further justification for studying local interactions.

The Theoretical Background to the Use of Conformational Energy Calculations.—The past two years have seen the consolidation of the theoretical justification for the use of energy calculations.

[4] R. H. Pain and B. Robson, *J. Mol. Biol.*, 1971, **58**, 237.
[5] P. N. Lewis, H. Momany, and H. A. Scheraga, *Proc. Nat. Acad. Sci. U.S.A.*, 1971, **68**, 2293.
[6] P. N. Lewis, N. Gō, D. Kotelchuck, and H. A. Scheraga, *Proc. Nat. Acad. Sci. U.S.A.*, 1970, **65**, 810.
[7] B. Robson and R. H. Pain, *Nature New Biol.*, 1972, **238**, 107.
[8] J. L. Crawford, W. N. Lipscomb, and C. G. Schellman, *Proc. Nat. Acad. Sci. U.S.A.*, 1973, **70**, 538.
[9] K. Nagano, *J. Mol. Biol.*, 1973, **75**, 401.
[10] C. Chothia, *J. Mol. Biol.*, 1973, **75**, 295.
[11] T. T. Wu and E. A. Kabat, *Proc. Nat. Acad. Sci. U.S.A.*, 1971, **68**, 1501.
[12] E. A. Kabat and T. T. Wu, *Proc. Nat. Acad. Sci. U.S.A.*, 1972, **69**, 960.
[13] T. T. Wu and E. A. Kabat, *J. Mol. Biol.*, 1973, **75**, 13.
[14] D. B. Weltaufer, *Proc. Nat. Acad. Sci. U.S.A.*, 1973, **70**, 697.

Firstly, a fundamental hypothesis continues to be accepted as the central dogma of those using conformational energy calculations.[15] This hypothesis is that the native conformation of a protein is the conformation of lowest free energy that is accessible to the system comprising the protein molecule and its solvent, providing the protein has not been covalently modified since biosynthesis. Certainly no new experimental evidence has refuted this essentially experimentally based[16–18] concept. In accord with this hypothesis, the co-ordinates of the atoms of the protein are determined as recently described by Ooi *et al.*[19] from the soft variables, the conformational free energy is calculated from those co-ordinates and parameters of the atoms associated with them, and finally the conformational free-energy surface (which results from the free energies corresponding to all possible values of the soft variables) is examined for the deepest minimum. Unfortunately, the deepest minimum is, by definition, only discovered when it is shown that there is none deeper, and this requires an exhaustive search of the free-energy surface. Current computing facilities render this an impossible task for a molecule as complex as a protein because of the immensity and complexity of the free-energy surface.[1,2] This explains the continued popularity of simple cyclic polypeptides as an object of study[20,21] (cyclic molecules have considerably more restricted degrees of freedom). Moreover, attention has been drawn[22] to the possibility that the native conformation of a protein may correspond to a deep local, rather than global, minimum if the global minimum cannot be attained during the lifetime of the molecule as a biologically active and experimentally observed entity. Hence the hypothesis continues to be a conceptual directive rather than a working rule, and in the immediate future minimization procedures are likely to be applied in the spirit of discovering possible local structures or metastable intermediates rather than the global minimum.

Secondly, emphasis has been placed[3] on a statistical-mechanical description of the protein in solution, following an earlier discussion of Gō and Scheraga.[23] The classical partition function is a function of the classical Hamiltonian of the protein–solvent system, integrated over the space of soft variables and their associated conjugate momenta. As shown by Lifson and Openheim,[24] when only the soft variables are considered, an effective potential function of the soft variables alone may be used to

[15] W. R. Krigbaum and B. H. Rubin, *Biochim. Biophys. Acta*, 1971, **229**, 368.
[16] C. B. Anfinsen, *Brookhaven Symp. Biol.*, 1962, **15**, 184.
[17] C. B. Anfinsen, *Harvey Lectures*, 1967, **61**, 95.
[18] C. Tanford, *Adv. Protein Chem.*, 1970, **24**, 1.
[19] T. Ooi, R. A. Scott, G. Vanderkooi, and H. A. Scheraga, *J. Chem. Phys.*, 1967, **46**, 4410.
[20] N. Gō and H. A. Scheraga, *Macromolecules*, 1970, **3**, 188.
[21] K. P. Serathy and C. Ramakristna, *Internat. J. Protein Res.*, 1971, **3**, 209.
[22] D. A. Brant and P. R. Schimmel, *Proc. Nat. Acad. Sci. U.S.A.*, 1967, **58**, 428.
[23] N. Gō and H. A. Scheraga, *J. Chem. Phys.*, 1969, **51**, 4751.
[24] S. Lifson and I. Openheim, *J. Chem. Phys.*, 1960, **33**, 109.

evaluate the partition function. To this effective potential function is contributed a term obtained by averaging over the space of solvent conformations for each protein conformation. The practical evaluation of this solvent term is problematic. Further, as described elsewhere,[22] the above rationalization of the statistical-mechanical framework of the problem highlights the involvement of an equally problematic conformation-dependent matrix comprising elements from a kinetic energy term which should, in principle, contribute to the Hamiltonian. However, in spite of these observations, the practices continue of severely approximating the solvent term, of neglecting the conformation-dependent matrix, and even of avoiding the integration of the space of soft variables (except for simplifications in certain special cases [20, 25–28]). Again, what emerges is not a working rule but a more general awareness of the deficiencies.

Quantum-mechanical Calculations.—The most active investigation of peptide conformation by quantum mechanics continues to be that of Pullman and co-workers using their PCILO method.[29–31] One problem in the use of the quantum-mechanical approach is in its extension to larger structures, although the CNDO/11 procedure has recently been applied to polypeptides.[32] Another problem is that the details of the approximate quantum-mechanical energy surface may be misleading.[33] Further, it is relatively difficult to modify these calculations in the light of observational experience since the free-energy surface is in this case approximately derived from a limited set of *ab initio* postulates. This has, however, been done by Pople [34] in the case of predictions of charge distribution by the CNDO/11 method.

Empirical Energy Calculations.—Because of the difficulties in the quantum-mechanical method, the approach of Lifson and Warshel,[35] which requires that the free-energy surface be constructed from the addition of arbitrarily but meaningfully partitioned sources, has continued to be popular. These partitioned sources may be identified with the classical energy contributions, namely the distortions of valence lengths, valence angles, torsion angles, and non-bonded interactions from their equilibrium values; the non-bonded interactions are composed of van der Waals, electrostatic, and hydrogen-bonding interactions. They differ, however, from the classical

25 N. Gō, M. Gō, and H. A. Scheraga, *Proc. Nat. Acad. Sci. U.S.A.*, 1968, **59**, 1030.
26 M. Gō, N. Gō, and H. A. Scheraga, *J. Chem. Phys.*, 1970, **52**, 2060.
27 M. Gō, N. Gō, and H. A. Scheraga, *J. Chem. Phys.*, 1971, **54**, 4489.
28 P. J. Flory, 'Principles of Polymer Chemistry', Cornell University Press, Ithaca, 1953, p. 672.
29 B. Maigret, P. Pullman, and D. Perahia, *J. Theoret. Biol.*, 1971, **31**, 269.
30 B. Pullman, *Internat. J. Quantum Chem.*, 1971, **4**, 319.
31 B. Pullman, in 'Aspects de la Chimie Quantique Contemporaine', ed. R. Dandel and A. Pullman, Colloque International du CNRS, Paris, 1971.
32 K. Morokuma, *Chem. Phys. Letters*, 1971, **9**, 129.
33 A. E. Tonelli, *Macromolecules*, 1971, **4**, 618.
34 J. A. Pople and D. L. Beveridge, 'Approximate Molecular Orbital Theory', McGraw-Hill, New York, 1970, p. 214.
35 S. Lifson and A. Warshel, *J. Chem. Phys.*, 1968, **49**, 5116.

energy contributions because the emphasis is on improved parametrization by reference to experimental systems at the expense of being able to attach any physical significance to the partitioned contributions examined individually. This is particularly true when high levels of refinement require the introduction of additional contributions which bear little analogy to any classical energy term.[3] Of course, providing the parameters can be carried over from one system to another, there can be no argument against applying them to protein molecules when calculation of the conformational energy *in toto* is the sole objective.

Assignment of Parameters to Empirical Functions.—In recent years considerable attention has been directed to the parametrization of empirical functions for hydrocarbons by reference to a large body of structural, spectral, and thermodynamic data, and these efforts continue.[36] These findings are applicable to protein conformational energy calculations but by no means exhaust the greater variety of interactions which occur in proteins. Rather than to understand this variety of interactions by investigation of numerous simple model compounds, attention is naturally directed to polypeptide systems. The case for refinement by reference to crystal homopolymer systems has recently been argued by Hopfinger,[37] while that for reference to known protein structures has been argued by Levitt and Lifson.[38] Abe and Krimm [39] have preferred the use of crystalline polyglycine I in a refinement of the valence force of the polypeptide backbone. Some studies on artificial polypeptide crystals emphasize intramolecular interactions.[40]

Recent studies by the Scheraga school [41] on enzyme–substrate intermolecular forces open up the possibility of parametrization, not only by *X*-ray crystallography of enzyme–substrate analogue complexes, but by the kinetic elucidation of binding constants. Solution studies of the helix–coil transition of polypeptides provide a potential means of obtaining useful parameters. A further measurement of the enthalpy for the transition [42] is of interest to the assignment of limits to the difference in energy between peptide–peptide and peptide–water hydrogen-bonding linkages and hence to the problem of a stabilizing role, as well as a structure-forming role, for the globular conformation.[1] Snell and Fasman [43] have determined the Zimm–Rice parameter σ from helix–coil transitions, and this may be directly related to theoretical energy calculations in the manner

[36] A. Warshel and S. Lifson, *J. Chem. Phys.*, 1970, **53**, 582.
[37] A. J. Hopfinger, *Biopolymers*, 1971, **10**, 1299.
[38] M. Levitt and S. Lifson, *J. Mol. Biol.*, 1969, **46**, 269.
[39] Y. Abe and S. Krimm, *Biopolymers*, 1972, **11**, 1817.
[40] R. F. McGuire, G. Vanderkooi, F. A. Momany, R. T. Ingwall, G. M. Crippen, N. Lotan, R. W. Tuttle, K. L. Kashuba, and H. A. Scheraga, *Macromolecules*, 1971, **4**, 112.
[41] K. E. B. Platzer, F. A. Momany, and H. A. Scheraga, *Internat. J. Protein Res.*, 1972, **4**, 187, 201.
[42] A. Teranato and T. Narisny, *Biopolymers*, 1972, **11**, 1693.
[43] C. R. Snell and G. D. Fasman, *Biopolymers*, 1972, **11**, 1723.

of Gō, Gō, and Scheraga.[44] It is also possible that a recent study[45] on frequency of helix–coil fluctuations may also be relevant to the theoretical free-energy surface from the point of view of activation energy for the process. Generally speaking, however, the relationship between the experimental observation and the theoretical energy functions is relatively complex in investigations of this type.

A more direct way of obtaining function refinement from solution studies involves adjusting energy parameters to account for the experimentally observed time-averaged dimensions of random-coil polypeptides.[46] A related concept is the notion of symmetry in the distribution of low-energy conformations of each residue, so that the co-operative use of optical methods may be of value in the future. In this context, a recent theoretical calculation of the circular dichroism of disordered polypeptides may be of interest.[47]

These examples illustrate that ample opportunities exist for providing parameters in the spirit of the empirical energy method. Although the empirical estimate of a conformational energy is only as reliable as its least reliable term, an appeal can always be made to quantum mechanics to bolster insubstantial information. For example, Hopfinger and Walton[48] represent Pitzer strain terms on the basis of the extended Hückel method. Similarly, the problematic treatment of hydrogen bonding continues to attract quantum-mechanical estimates.[49]

Hard Variables.—The soft variables are not the only variables which determine overall molecular geometry, and hence the dimensionality of the free-energy surface. There are also the 'hard' variables, the valence lengths and valence angles which have relatively high energies associated with their distortion from equilibrium values.[50] The peptide link between the carbonyl carbon and amide nitrogen atom may also be considered as a hard variable because of the high energy required to distort the bond from planarity.[50] The hard variables are often neglected from consideration of the code relating sequence to conformation, each conformation defined by soft variables alone being considered as a whole continuum of conformations associated with the oscillations of the hard variables. This is true even of conformations described by *X*-ray crystallographic analysis in which atoms are represented by thermal ellipsoids.[51] Even with this simplification, however, the fact that the vibrational degree of freedom contributes to the free energy should,[1] could,[52] and recently has been taken

[44] N. Gō, M. Gō, and H. A. Scheraga, *Proc. Nat. Acad. Sci. U.S.A.*, 1968, **59**, 1030.
[45] M. Nahanshi, M. Tsuboi, A. Ikegami, and M. Kamehisa, *J. Mol. Biol.*, 1972, **64**, 363.
[46] P. J. Flory, 'Statistical Mechanics of Chain Molecules', Wiley, New York, 1969, p. 432.
[47] E. W. Konish and S. Krimm, *Biopolymers*, 1972, **11**, 1919.
[48] A. J. Hopfinger and A. E. Walton, *Biopolymers*, 1970, **9**, 29.
[49] S. Takashima, *Biopolymers*, 1972, **11**, 1903.
[50] C. Tanford, *Adv. Protein Chem.*, 1968, **23**, 121.
[51] C. K. Johnson, in 'Or Tep: A Fortran Thermal-Ellipsoid Plot Program for Crystal Structure Illustrations', Oak Ridge National Laboratory, 1970.
[52] H. H. Mills and J. C. Speakman, *Progr. Stereochem.*, 1969, **4**, 273.

into account in the cases of a small cyclic polypeptide[20, 53] and larger polypeptides.[25, 27]

The division of variables into two categories is arbitrary because there are degrees of 'softness'. It is therefore possible that the softest of the hard variables contribute to the dimensionality of the free-energy surface as it is considered in practice. In proteins, this may involve the valence angle of the C_α-atom and the rotation angle of the peptide link. The effect of these extra dimensions on the allowed conformations[54] and the energy surface[55] of single residues and the associated polypeptide backbone has recently been investigated. However, caution should be used in drawing conclusions about the distortions of these variables from data provided by *X*-ray crystallographic analysis of proteins. Of the four variables (namely dihedral angles ϕ around $N-C_\alpha$, ψ around $C_\alpha-CO$, ω around the peptide bond, and valence angle $N-C_\alpha-CO$) which the backbone of each residue contributes to the dimensionality of the energy surface, any three are sufficient to fit the backbone adequately to the electron-density maps (R. Diamond, personal communication).

Solvent Effects.—The role of the solvent in the stabilization of macromolecular structure has recently been reviewed by Dandliker and de Saussure.[56] These workers favour the description of water structure due to Pople,[57] and extend the model to meet the problem of hydrophobic bonding. Water is envisaged as a hydrogen-bonded network with a continuum of bond energies and geometries varying over a wide range. The alternative type of model is one in which water is considered as a mixture of molecular species in which hydrogen bonds are essentially either maximally stable or completely dissociated. It has not always been self-evident in past literature as to what extent these two types of model are mutually exclusive, and to what extent they are merely different kinds of treatment. To the extent, however, that they represent incompatible models, Franks has concluded[58] that the data are currently insufficient to decide between them. More recently, experimental work on the solvation of proteins[59] has revealed an i.r. absorption suggestive of bound water with ice-like properties.

However, it should be remembered that proteins are complex entities with hydrogen-bonding as well as non-polar groups, and that the effects of these two kinds of group on the structure of water may be rather different. In the computer treatment of the solvated polypeptide exemplified by the work of Hopfinger,[60] the accent is on the solvent molecules immediately

[53] N. Gō, P. N. Lewis, and H. A. Scheraga, *Macromolecules*, 1970, **3**, 628.
[54] C. Ramakrishnan and R. Asubramanian, *Internat. J. Protein Res.*, 1972, **4**, 79.
[55] R. Asubramanian and C. Ramakrishnan, *Internat. J. Protein Res.*, 1972, **4**, 91.
[56] W. B. Dandliker and V. A. de Saussure, in 'Chemistry of Biosurfaces', ed. M. L. Hair, Dekker, New York, 1971, vol. 1, p. 1.
[57] J. A. Pople, *Proc. Roy. Soc.*, 1951, **A205**, 168.
[58] D. H. S. Frank, *Science*, 1970, **169**, 635.
[59] U. Buontempo, G. Caveri, and P. Fasella, *Biopolymers*, 1972, **11**, 519.
[60] A. J. Hopfinger, *Macromolecules*, 1971, **4**, 731.

bound to the different atoms and groups and the free energy required to displace them. In this case, a clash involving a protein group having solvent molecules results in the displacement of a calculated number of the solvent molecules and hence a change in the free energy of the system. Conformational energy calculations on this basis have recently been applied to oxytocin in solution.[61] There have been no recent improvements in solvent treatment, and it seems reasonable to assume that the involvement of the solvent is one of the least reliable factors in the application of empirical energy calculations to polypeptides in solution. This is mainly because uncertainty about water structure and microscopic properties results in uncertainty not only as to the set of parameters but also in the choice of overall algorithm for handling the consequence of approaching protein atoms. This difficulty is also reflected in the choice of dielectric constant for the immediate environment of charged moieties. This problem is discussed by Brant.[3] Typically, the dielectric constant is treated either as independent of the distance of the charged moieties [62] or as a simple step function.[63]

General Conclusions.—The quest for an understanding of the code relating sequence to conformation in globular proteins is proceeding rapidly, although recent work has in general represented a consolidation of theory and data. The refinement of energy functions is not the sole determinant of active progress in this field: the fact that a typical protein has approximately one thousand soft variables and an energy surface with many minima means that some very elaborate minimization procedures will have to be developed in order to locate and recognize the native conformation in reasonable computer time. The practice of supplementing computer reasoning by human decisions is continuing, and one of the most entertaining ways of developing man–machine interaction for this purpose is to use computer-generated stereoscopic pictures of the protein in which the soft variables may readily be manipulated.[64, 65]

3 Folding Mechanisms in Globular Proteins

contributed by R. D. Ryder

Introduction.—The amino-acid sequence of a protein contains all the information required for the folding of the polypeptide chain into its biologically active conformation.[66] It is unlikely that a process of random wandering of the backbone can form the native structure within a reasonable time-scale so it can safely be assumed that there exist some folding

[61] D. Kotelchuck, H. A. Scheraga, and R. Walter, *Proc. Nat. Acad. Sci. U.S.A.*, 1972, **69**, 3629.
[62] D. A. Brant, W. Miller, and G. Flory, *J. Mol. Biol.*, 1967, **23**, 47.
[63] K. D. Gibson and H. A. Scheraga, *Proc. Nat. Acad. Sci. U.S.A.*, 1967, **58**, 420.
[64] C. D. Barry and A. C. T. North, *Cold Spring Harbor Symp. Quant. Biol.*, 1971, **36**, 577.
[65] L. Katz and C. Levinthal, *Ann. Rev. Biophys. Bioengineering*, 1972, **1**, 465.
[66] C. J. Epstein, R. F. Goldberger, and C. B. Anfinsen, *Cold Spring Harbor Symp. Quant. Biol.*, 1963, **28**, 439; C. B. Anfinsen, *Harvey Lectures*, 1967, **61**, 95.

pathways specified by the primary sequence. Because the native conformation represents the lowest accessible free-energy state under (near) physiological conditions, an important approach to the understanding of the relationship between the three-dimensional structure and the primary sequence is afforded by kinetic and thermodynamic studies of the denaturation and renaturation processes.[67]

The reversible unfolding of several small proteins shows apparent two-state behaviour,[67, 68] an approximation which has aroused considerable controversy concerning its validity. Certainly the 'all or none' mechanism, in which only completely native and completely denatured forms are envisaged, is inferior to a less restrictive view in which the two states present need not be in the extremes of folding, but in fact may be two dynamic states whose average properties will change with temperature, pressure, and solvent composition due to thermodynamic redistribution.[68, 69] However, other protein-folding reactions are known in which intermediate states have been detected,[70–73] so the problem is less the validity of the two-state approximation but rather more the resolution of the pathways of protein folding.[72]

One probable mechanism for folding involves the formation of structural nuclei, such as a short section of α-helix, followed by the formation of hydrophobic interactions to give the native structure.[71] The kinetics of many reversible denaturation reactions are complex, but criteria for the evaluation of several possible pathways have been proposed[74] and tested,[75, 76] and sequential models exist that are based on experimental evidence for the existence of the nucleation reaction.[71, 77–83] However, the kinetics of the nucleation reaction are still somewhat ambiguous, and some workers envisage a rapid nucleation step,[71] whereas others are of the opinion that the step is slow and rate-limiting.[77–82]

67 C. Tanford, *Adv. Protein Chem.*, 1968, **23**, 1; 1970, **24**, 259.
68 J. F. Brandts, in 'Structure and Stability of Biological Macromolecules', ed. S. N. Timasheff and G. D. Fasman, Marcel Dekker Inc., New York, 1969, p. 213.
69 R. Lumry, R. Biltonen, and J. F. Brandts, *Biopolymers*, 1966, **4**, 917.
70 A. N. Schechter, R. F. Chen, and C. B. Anfinsen, *Science*, 1970, **167**, 886.
71 H. F. Epstein, A. N. Schechter, R. F. Chen, and C. B. Anfinsen, *J. Mol. Biol.*, 1971, **60**, 499.
72 O. Jardetzky, H. Thielmann, Y. Arata, J. L. Markley, and M. N. Williams, *Cold Spring Harbor Symp. Quant. Biol.*, 1971, **36**, 257.
73 A. Ikai and C. Tanford, *Nature*, 1971, **230**, 100.
74 A. Ikai and C. Tanford, *J. Mol. Biol.*, 1973, **73**, 145.
75 A. Ikai, W. W. Fish, and C. Tanford, *J. Mol. Biol.*, 1973, **73**, 165.
76 C. Tanford, K. C. Aune, and A. Ikai, *J. Mol. Biol.*, 1973, **73**, 185.
77 T. Y. Tsong, R. L. Baldwin, P. McPhie, and E. L. Elson, *J. Mol. Biol.*, 1972, **63**, 453.
78 T. Y. Tsong, R. L. Baldwin, and E. L. Elson, *Proc. Nat. Acad. Sci. U.S.A.*, 1971, **68**, 2712.
79 E. L. Elson, *Biopolymers*, 1972, **11**, 1499.
80 T. Y. Tsong and R. L. Baldwin, *J. Mol. Biol.*, 1972, **69**, 145.
81 T. Y. Tsong and R. L. Baldwin, *J. Mol. Biol.*, 1972, **69**, 149.
82 T. Y. Tsong, R. L. Baldwin, and E. L. Elson, *Proc. Nat. Acad. Sci. U.S.A.*, 1972, **69**, 1809.
83 M. R. Summers and P. McPhie, *Biochem. Biophys. Res. Comm.*, 1972, **47**, 831.

Staphylococcal Nuclease.—The folding of acidified staphylococcal nuclease has been followed by measuring changes in tryptophan fluorescence upon neutralization.[71] The process is highly co-operative and is reversible in the range pH 3—7, although there is a small perturbation of tyrosyl fluorescence above the transition pH. However, the major fluorescence change follows two first-order rate processes, the faster rate being essentially independent of temperature but the slower being strongly temperature-dependent. This suggests a sequence involving nucleation of ordered structure followed by the formation of hydrophobic interactions. On the basis of the known crystallographic structure,[84] the formation of helices surrounding the tryptophan residue would be consistent with the observed changes of the fast process, since the growth of helices has been demonstrated to be a fast process and not rate-limiting.[85]

In contrast to the co-operativity observed in the folding from acid solution as followed by tryptophan fluorescence, the 220 MHz ^{1}H n.m.r. spectrum of one of the four histidine C-2 protons shows a biphasic transition although the other three histidine proton resonances do appear to follow the same single equilibrium of the tryptophan fluorescence.[86] This can be interpreted as signifying a local conformational change which may be related to the asymmetry of the tryptophan fluorescence change above the transition pH. Nevertheless, the remaining histidine residues appear to be involved in the major conformational change associated with changes in tryptophan fluorescence.

In addition to this, there is evidence for several intermediate conforformational states being present in the re-folding from high pH in the range 8—11.[72] The use of selectively deuteriated analogues of staphylococcal nuclease has made it possible to identify the spectral lines originating from individual aromatic residues. The changes in chemical shifts observed on titration, combined with the known crystallographic structure, have made possible a tentative sequence of unfolding in terms of segments of the polypeptide chain. In this case a folding sequence has not been proposed, although it is probably pictured as the reverse of the unfolding sequence, and no mention of the possibility of incorrectly folded intermediate states is made.

Lysozyme and Cytochrome *c*: Mathematical Analysis.—Kinetic studies on the reversible denaturation of proteins have now shown that metastable intermediates exist which are not on the direct pathway between the native and denatured states.[73] Analysis of the observed kinetic data for cytochrome *c*, β-lactoglobulin, ovalbumin, and staphylococcal nuclease in various denaturants was attempted in terms of simple three- or four-state models. The conclusion was reached that incorrectly folded intermediate

[84] A. Arnone, C. J. Bier, F. A. Cotton, E. E. Hazen, D. C. Richardson, J. S. Richardson, and A. Yonath, *J. Biol. Chem.*, 1971, **246**, 3202.

[85] A. N. Schechter, *Science*, 1970, **170**, 273.

[86] H. F. Epstein, A. N. Schechter, and J. S. Cohen, *Proc. Nat. Acad. Sci. U.S.A.*, 1971, **68**, 2042.

states could be a general feature of the denaturation process, and it was suggested that initially the folding steps are rapidly reversible and without influence on the final result. This is consistent with the idea that *in vitro* folding can be initiated at any part of the extended chain, whereas the *in vivo* folding presumably begins near the amino-terminal end of the polypeptide chain.[73]

The simple models used could not be fitted by all the experimental data, and so a more rigorous mathematical approach to the analysis of unimolecular reaction kinetics has been developed.[74] Equations are presented to relate unknown parameters, such as microscopic rate constants, to the observable kinetic data. In some cases all unknown parameters can be evaluated for simple mechanisms provided that kinetic measurements are made for both the folding and unfolding reactions. This enables criteria for the rejection of certain unlikely mechanisms to be established.

Using these procedures, the reversible denaturation of horse heart ferricytochrome *c*[75] and of lysozyme[76] by guanidinium chloride (GuCl) has been analysed. In the case of cytochrome *c*[75] the most satisfactory mechanism appears to be one in which four species are present:

$$\mathrm{N} \rightleftharpoons \mathrm{X_1} \rightleftharpoons \mathrm{D} \rightleftharpoons \mathrm{X_2}$$

in which X_1 is an intermediate on the pathway between the native (N) and denatured (D) states, and X_2 is a highly ordered state on a dead-end pathway. The conclusion reached is that X_2 is an incorrectly folded form of the polypeptide chain which is the first state formed by the rapid folding of the denatured protein. The ultimate conversion of all the protein into the native state occurs more slowly, and the rate-limiting step in the transition zone is envisaged as the reaction $D \rightarrow X_1$. The magnitude of the change in physical properties observed in the slow process is consistent with the idea that it represents a steady-state process in highly co-operative folding.

In contrast with the results observed for cytochrome *c*, the reversible denaturation of lysozyme follows strictly first-order kinetics under most conditions, showing that no intermediates accumulate to a significant extent during the reaction. However, at pH 2.6, under extremes of denaturant (guanidinium chloride) conditions, kinetic intermediates have been observed.[76] Analysis as before[74] shows that the principal intermediate observed at low denaturant concentration is different from that at high denaturant concentration, and that both are on the main pathway between states N and D. All observed results can be accounted for by the mechanism:

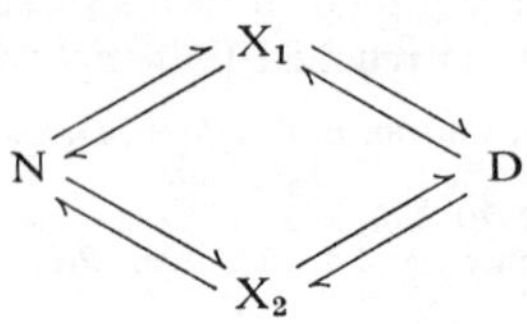

where the state X_1 is that observed at low denaturant concentration. X_1 is spectrally very similar to the native state and is probably highly ordered. The fact that the lysozyme disulphide bridges were intact during these experiments may account for the differences between lysozyme and cytochrome *c*. The presence of the cross-links is assumed to limit the number of possible pathways for folding and will make incorrectly folded 'dead-end pathway' intermediates less accessible.

Ribonuclease.—The thermal unfolding of bovine pancreatic ribonuclease A (RNase A) below pH 2 is apparently a highly co-operative two-state process. Calorimetric measurements of the transition offer a criterion for the estimation of closeness to the two-state approximation since the enthalpy change calculated from the van't Hoff equation (ΔH_{vH}) is equal to the calorimetric value (ΔH_{cal}) for a two-state process under given conditions. Application of this criterion has shown that deviations from two-state behaviour become increasingly evident in the thermal unfolding of RNase A as the pH is raised above 2.[87] However, the analysis of the thermal data obtained (*H vs. T*) and the subsequent conclusions have been subjected to some criticism,[88] which will be discussed later with reference to chymotrypsinogen. Nevertheless, the finding of deviation from two-state behaviour has been verified using other techniques.[77–82]

A sequential model for protein folding has been proposed[77, 79] which includes a nucleation reaction as the first step based on the differences between the activation energies for unfolding (large) and folding (small, in the midpoint of the transition). The model suggests that nucleation is the rate-limiting step and that the unfolding can approximate to a two-state process even in the presence of sizeable concentrations of stable intermediates. It also predicts that the kinetics of unfolding can be divided into a rapid transient phase followed by a slow steady-state unfolding.

The kinetics of the slow thermal unfolding of RNase A were followed by the change in exposure to solvent of the tyrosine residues at both neutral and low pH, and the results obtained were compared with the sequential model.[77, 78] At pH 7.0 it was found that the kinetics of unfolding were closer to two-state kinetics than at pH 1.3.[77] This is in contrast with equilibrium studies,[87] but can be explained by the sequential model in the light of the finding of a rapid initial phase in the unfolding[78] as predicted by the model.

To guard against the possibility that these kinetics were indicative of two distinct groups of tyrosine residues, one exposed more rapidly than the other, the kinetics of unfolding were studied by following the perturbation of the dinitrophenyl group of DNP-Lys(41)-RNase A.[81] The results obtained were identical with those for the tyrosyl exposure,[77, 78] thus giving futher support for the biphasic nature of the kinetics.

[87] T. Y. Tsong, R. P. Hearn, D. P. Wrathall, and J. M. Sturtevant, *Biochemistry*, 1970, **9**, 2666.

[88] W. M. Jackson and J. F. Brandts, *Biochemistry*, 1970, **9**, 2294.

Studying the kinetics of both unfolding and re-folding reactions following pH-jump experiments [82] apparently confirms two basic postulates of the sequential model,[77] in that the rate of re-folding is limited by a slow step thought to be nucleation, and that unfolding proceeds *via* a series of fast steps before the rate-limiting step is reached. Under certain conditions most of the absorbance change in the unfolding process has been shown to occur in the fast phase. The pH- and temperature-dependence of the amplitude has been interpreted as being indicative of a rapid equilibrium between nucleated forms, which is then followed by a slow reversal of the nucleation reaction.[83] The simplicity of this model need not preclude the sort of mechanism involving incorrectly folded intermediates [73] since, if an early step in normal folding occurs only slowly, then some rapidly formed, incorrectly folded, metastable intermediates may exist. This may be consistent with findings from the potassium thiocyanate denaturation of RNase A,[89] where the unfolding has been found to be incomplete. Re-folding of the incompletely denatured form leads to a non-native ordered form, which may be evidence for a branching process on the main pathway.

Chymotrypsinogen.—Bovine pancreatic chymotrypsinogen A, like RNase A, under certain conditions shows apparent two-state behaviour as judged by spectroscopic and calorimetric studies of the denaturation reaction.[67, 68, 88] In the range pH 2—3, good agreement between ΔH_{cal} and ΔH_{vH} has been found for the unfolding reaction at the transition mid-point,[88] indicating a close approximation to two-state behaviour.[87] Although the conclusions regarding the denaturation of RNase A [87] and chymotrypsinogen are the same (with respect to the different conditions employed), the treatment of the experimental results is different and warrants mention. A plot of *H vs. T* is obtained in both cases which shows upward curvature with increasing temperature below the transition zone. In the case of RNase A [87] the curvature was ignored and the base line was taken as a line of zero slope intersecting the experimental data at the lowest temperature measured. The enthalpy at the transition midpoint was measured with respect to this line. In the case of chymotrypsinogen the upward curvature was extrapolated beyond the transition midpoint and was used at this point in the calculation of the enthalpy of unfolding. If the data for RNase A had been treated in this way it would have been necessary to conclude that the transition did not approximate to two-state behaviour.[88] The difference in the treatment of the data stems from the difference in the definition of the native state of the enzyme. In the case of chymotrypsinogen the native state is thus defined as that macroscopic state existing exclusively below the major unfolding transition as determined by other physical methods.[68, 69] The temperature dependence of the heat capacity in this region is then explained in terms of small conformational changes induced by thermal perturbation, changes in solvent–

[89] J. H. Bradbury and N. L. R. King, *Austral. J. Chem.*, 1972, **25**, 209.

solvent and solvent–protein interactions, and so on. The application of this definition to both high- and low-temperature states is thus self-consistent, and the analysis of the transition between these states, as defined in terms of two-state behaviour, becomes a more valid treatment.

Fast kinetic measurements of the change in exposure to solvent of buried tryptophan groups in the thermal unfolding of chymotrypsinogen, however, have shown the presence of a fast phase not detected by equilibrium measurements.[79] The biphasic kinetics are also observed in pH-jump experiments, but it is not known whether or not the observed intermediates are on the direct pathway of unfolding. The kinetics are interpreted as being consistent with the sequential model proposed earlier [77] in that the slow phase is thought to be an almost steady-state reaction and typical of a nucleation-limited re-folding.

Myoglobin.—The unfolding and re-folding of sperm-whale myoglobin following pH-jumps have been studied by stopped-flow spectrophotometry.[90] In the transition range the unfolding reaction is apparently two-state, although below pH 3.8 the kinetics are biphasic (*cf.* ribonuclease and chymotrypsinogen), showing a rapid phase and a slow phase. Re-folding in the transition range also shows a single phase, whilst at pH 5—6 the kinetics are again biphasic. If the denatured solution is left long after the first observable changes have occurred, renaturation becomes more difficult. These results are thought to indicate a scheme involving four states: $N \rightleftharpoons X \rightleftharpoons D \rightleftharpoons D^*$, where D^* is a modified denatured form which is formed slowly and is difficult to renature. The intermediate X is not observed in the transition range, and renaturation is thought to follow the reaction $D \rightarrow N$. The states D and D^* are spectroscopically similar but renaturation of D near the transition range is accompanied by parallel changes in optical rotation and u.v. absorption, whereas the renaturation of D^* leads to a more rapid recovery of helical structure than regain of u.v. absorption. D^* is thought to be a reversible aggregate of haem and protein which slowly dissociates and is followed by a regain of structure and recombination to form myoglobin molecules in a rate-limiting reaction.

Ovalbumin.—The denaturation of ovalbumin by sodium dodecyl sulphate and GuCl has been followed by sedimentation velocity, optical rotation, and viscometry.[91] Full denaturation could be achieved only with GuCl, irrespective of whether or not the single disulphide bond was reduced. Complete recovery of physical parameters indicative of the native protein could not be obtained upon renaturation owing to aggregation and precipitation, which led to the conclusion that only partial renaturation was achieved.

Although no information about the kinetics of protein folding is obtained, some important points are raised. Aggregation during attempted renaturation occurred at protein concentrations at which other proteins have been

[90] L. L. Shen and J. Hermans, jun., *Biochemistry*, 1972, **11**, 1836.
[91] J. C. Holt and J. M. Creeth, *Biochem. J.*, 1972, **129**, 665.

renatured successfully. This is interpreted as supporting the necessity of a specific pathway for correct folding, since the failure to renature appears to be due to causes intrinsic to the protein. The implication is that in the case of ovalbumin kinetic barriers are raised to render the thermodynamically stable state inaccessible. However, the presence of the carbohydrate moiety and of the phosphate groups on the serine residues are recognized as special features of ovalbumin which may play an important part in the re-folding reaction.

Carbonic Anhydrase.—More information about the importance of extra-thermodynamic considerations in protein folding is presented in the case of carbonic anhydrase, which is a metalloenzyme containing one equivalent of zinc as Zn^{II}. The denaturation and renaturation have been studied [92] and found to be essentially reversible, with greater than 95% recovery of activity. The transition between native and denatured forms appears to be thermodynamically reversible with or without Zn^{II}, although in the latter case it occurs at lower denaturant concentrations, showing the greater structural stability in the presence of Zn^{II}.

Removal of zinc from the native enzyme has little structural effect, but re-folding in the absence of Zn^{II} occurs extremely slowly. In the presence of Zn^{II} the re-folding occurs readily, suggesting that Zn^{II} is bound during the early stages of folding in a form of nucleation step which influences the pathway of folding but not the final conformational state.

Multimeric Proteins.—The renaturation of denatured subunit proteins may theoretically be divided into two processes:

(i) re-folding of the polypeptide chains, and

(ii) reassociation of the chains to give full quaternary structure.

The enzymes fumarase, aldolase, enolase, lactate dehydrogenase, glyceraldehyde phosphate dehydrogenase, and malate dehydrogenase have been denatured in 6M-guanidinium chloride, and the re-folding and reassociation have been followed kinetically.[93] The rates and final extents of reactivation are increased in the presence of cofactors or substrates, but altering the protein concentration had varying effects. The rate of equilibration between different conformational species formed in renaturation was relatively rapid for fumarase, enolase, and aldolase, but slow for glyceraldehyde phosphate dehydrogenase and lactate dehydrogenase. It is suggested that in the former group the re-folding gives states solely determined by thermodynamic considerations, whereas in the latter group the folding forms metastable states specified by kinetic factors.

The renaturation as followed by optical rotation and fluorescence showed a major regain of structure within one minute, followed by minor structural alterations associated with regain of enzymic acitivity over a

[92] A. Yazgan and R. W. Henkens, *Biochemistry*, 1972, **11**, 1314.
[93] J. W. Teipel and D. E. Koshland, jun., *Biochemistry*, 1971, **10**, 792, 798.

longer period of time. The implication of this is that renaturation proceeds *via* stable precursor conformations which subsequently rearrange to give either active or inactive species.

The major questions to be answered thus appear to be (i) does re-folding precede, accompany, or follow reassociation?, and (ii) does regain of enzymic activity precede, accompany, or follow reassociation?

Aldolase. In an attempt to answer these questions, the kinetics of re-folding, reassociation, and reactivation of rabbit muscle aldolase have been investigated.[94] The renaturation is thought to follow a mechanism in which the dissociated unfolded chains rapidly re-fold and partially reassociate to give a mixture of highly structured but inactive monomers and dimers. This is followed by a slow first-order minor conformational change and then rapid association to tetramers with simultaneous regain of enzymic activity. Using the techniques employed here, it is not possible to distinguish between regain of activity which precedes association and that which follows association, although there is some evidence for the existence of a folded active monomer.

Lactate Dehydrogenase. The course of the unfolding of dogfish muscle lactate dehydrogenase induced by guanidinium chloride has been followed by the exposure of free thiol groups and by u.v. difference spectroscopy.[95] Equilibrium data for the two methods show good agreement, and the unfolding has a high degree of co-operativity as judged by these methods. The exposure of the thiol groups apparently follows first-order kinetics at various GuCl concentrations in the transition range. The apparent two-state nature of the reaction is of interest, since dissociation of the tetramer must accompany denaturation. Presumably fast-reaction techniques would be able to give more valuable data on the denaturation process of this enzyme (*cf.* RNase A, *etc.*).

RNA Polymerase. The DNA-dependent RNA polymerase of *E. coli* has been reversibly dissociated into subunits by *p*-chloromercuribenzoate [96] and urea.[97] Heterologous individual subunits designated α, β, β', and σ have been isolated, and recombination of various binary complexes has been attempted.[97] This was only possible in the case of the α and β subunits, which formed an $\alpha_2\beta$ complex. This complex could then combine with the β' subunit upon its addition, giving a core enzyme of the structure $\alpha_2\beta\beta'$ which possesses full polymerase activity but cannot initiate transcription from DNA. The addition of the σ subunit restores full enzymic activity. Thus the assembly appears to be a sequential process, with the formation of the $\alpha_2\beta$ complex an important step.

[94] J. W. Teipel, *Biochemistry*, 1972, **11**, 4100.
[95] P. M. Wassarman and J. W. Burgner, *J. Mol. Biol.*, 1972, **67**, 537.
[96] A. Ishihama, *Biochemistry*, 1972, **11**, 1250.
[97] A. Ishihama and K. Ito, *J. Mol. Biol.*, 1972, **72**, 111.

4 Spin Labels

contributed by P. Knowles

Spin-label studies into protein structure and function are becoming more sophisticated. Both *direct* study (by e.p.r.) of the bound spin label and *indirect* study (by n.m.r.) through perturbation of neighbouring nuclear spins by the bound spin label have been applied with considerable success. This Report attempts to survey the literature in this field to the end of 1972. The format used in last year's Report will be followed for continuity.

Haemoglobin.—Ogata and McConnell [98] have proposed a new model (the Generalized Concerted Transition or G.C.T. model) for haemoglobin in solution. This model is derived from the Monod, Wyman, and Changeux allosteric model [99] by (i) removing the restriction that partially liganded molecules should preserve symmetry and (ii) taking account of the non-identical α and β subunits present in haemoglobin.

The results of studies with triphosphate spin labels, whose binding only to deoxyhaemoglobin mimics that of the effector 2,3-diphosphoglycerate, were discussed in terms of the G.C.T. model. Thus haemoglobin Chesapeake (α92, Arg $\rightarrow$ Lys) and Hb $\alpha_2^{CN}\beta$ show similarities in their binding of triphosphate spin label;[100] this is equally true of haemoglobin Kempsey (β99, Asp $\rightarrow$ Asn) and Hb $\alpha_2\beta_2^{CN}$. (Haemoglobin Chesapeake and haemoglobin Kempsey both show low co-operativity in oxygen binding.) These results can be successfully explained by the G.C.T. model. McConnell and co-workers speculate [101] further that the mutation α92, Arg $\rightarrow$ Lys in haemoglobin Chesapeake locks the α subunit in the oxygenated (R) state while the mutation β99, Asp $\rightarrow$ Asn in haemoglobin Kempsey locks the β subunit in the R state. However, this attractive idea is not borne out by *X*-ray evidence on haemoglobin Chesapeake.[102]

Further spin-label studies on the haemoglobin–haptoglobin complex have been made by Smith and co-workers.[103] It was concluded from previous studies [104] using haemoglobin spin-labelled at cysteine β93 with *N*-(1-oxyl-2,2,6,6-tetramethyl-4-piperidinyl)iodoacetamide (IAM 6) that the conformation of the haemoglobin–haptoglobin complex is very similar to that of deoxyhaemoglobin. However, when cysteine β93 was spin-labelled with *N*-(1-oxyl-2,2,6,6-tetramethyl-3-pyrrolidinyl)iodoacetamide (IAM 5), the haptoglobin complex had an e.p.r. spectrum intermediate between those of oxy- and deoxy-haemoglobin. This clearly illustrates the caution which should be exercised in interpreting spin-label results in terms of protein conformations. Similarly, McConnell noted isosbestic

[98] R. T. Ogata and H. M. McConnell, *Biochemistry*, 1972, **11**, 4792.
[99] J. Monod, J. Wyman, and J.-P. Changeux, *J. Mol. Biol.*, 1965, **12**, 88.
[100] R. T. Ogata and H. M. McConnell, *Proc. Nat. Acad. Sci. U.S.A.*, 1972, **69**, 335.
[101] R. T. Ogata, H. M. McConnell, and R. J. Jones, *Biochem. Biophys. Res. Comm.*, 1972, **47**, 157.
[102] J. Green, *J. Mol. Biol.*, 1971, **62**, 241.
[103] B. Malchy, H. Dugas, F. Ofosu, and I. C. P. Smith, *Biochemistry*, 1972, **11**, 1669.
[104] M. W. Makinen and H. Kon, *Biochemistry*, 1971, **10**, 43.

points in the e.p.r. spectrum as oxygen was added to haemoglobin spin-labelled with IAM 5 at cysteine β93 but *not* with IAM 6.[105] The first result favours the Monod, Wyman, and Changeux allosteric model [99] whilst the latter favours the sequential model of Koshland.[106]

Shiga *et al.*[107] have studied the conformation of haemoglobin as bulky ligands, for example nitrobenzene, are co-ordinated to the haem 6-position. The conformational change was monitored through the e.p.r. spectra of spin labels attached to cysteine β93. It could be concluded that steric crowding at the iron centre causes conformational changes distinct from the T $\rightarrow$ R changes.

Enzyme Binding Studies.—*Carbonic Anhydrase.* Mushak and Coleman [108] have studied binding of a spin-labelled sulphonamide derivative to carbonic anhydrase. The enzyme was used both in its native form and also with the Zn^{2+} replaced by the paramagnetic ion Co^{2+}, which still leaves the enzyme with most of its activity. The e.p.r. spectra were similar for these two forms of the enzyme and indicated that the spin label was appreciably immobilized. The authors concluded that binding of the sulphonamide grouping to the metal had occurred, with the nitroxide being sufficiently remote to prevent any spin–spin interaction with the cobalt. In other studies, the rate of release of spin-labelled sulphonamide from the inhibited enzyme following treatment with urea and guanidinium chloride suggested different modes of action for these two denaturants. Hower, Henkens, and Chesnut [109] have attempted a detailed analysis of the e.p.r. spectrum resulting from binding of 4-(*p*-sulphonamido)benzoyloxyl-2,2,6,6-tetramethylpiperidin-1-oxyl to the enzyme. These authors considered *X*-ray crystallographic, fluorescence polarization, and triplet–triplet exchange evidence in arriving at the conclusion that the aromatic ring was firmly bound but that the nitroxide could undergo motion about the bond between a carboxylate oxygen and C-4 of the piperidinyl ring. The nitroxide hyperfine spectra suggested that this grouping on the inhibitor was located in a highly polar environment.

Lysozyme. Attempts are being made to compare distances calculated from magnetic resonance data with those from *X*-ray diffraction studies on the enzyme. McConnell and co-workers [110] have studied dipolar broadening by various spin-labelled compounds of specific protons on the enzyme or substrate. Calculation of distances is only possible when one species is exchanging between a bound and an unbound state. Thus when histidine-15 of lysozyme was covalently linked to bromoacetamide spin label, the n.m.r. spectrum from the enzyme was almost uniformly broadened and no distances could be calculated. However, the effect of

[105] H. M. McConnell and B. B. McFarland, *Quart. Rev. Biophys.*, 1970, **3**, 91.
[106] D. E. Koshland and K. E. Neet, *Ann. Rev. Biochem.*, 1968, **37**, 359.
[107] T. Shiga, Y. Ueda, and I. Tyuma, *J. Biochem. (Japan)*, 1972, **72**, 849.
[108] P. Mushak and J. E. Coleman, *J. Biol. Chem.*, 1972, **247**, 373.
[109] J. F. Hower, R. W. Henkens, and D. B. Chesnut, *J. Amer. Chem. Soc.*, 1971, **93**, 6665.
[110] R. W. Wien, J. D. Morrisett, and H. M. McConnell, *Biochemistry*, 1972, **11**, 3707.

this spin label on *N*-acetylglucosamine (NAG) could be studied since this substrate analogue exchanges on and off subsite C of the enzyme. From linewidth measurements on the methyl protons of the acetamido-group, a distance from the nitroxide of 14 Å was estimated. The e.p.r. spectrum of the bound spin label indicated a correlation time of less than 5×10^{-9} s; this meant that the attached nitroxide had motion independent of tumbling of the whole enzyme molecule, for which a correlation time of 10^{-8} s was calculated. From molecular models representing the two extreme positions for the nitroxide grouping, a distance of 18 ± 7 Å could be measured. McConnell pointed out that the distance was more probably between 11 and 18 Å, which agreed quite well with the value of 14 Å that had been reported. The converse studies where spin-labelled substrate analogues perturbed the C-2 protons on histidine-15 of the enzyme were claimed to yield more precise distance estimates. However, the mobilities of the spin labels in these cases were not discussed. The anticipation that *N*-(1-oxyl-2,2,6,6-tetramethyl-4-piperidinyl)acetamide would bind at a specific site on lysozyme was disproved by *X*-ray diffraction studies on the enzyme–inhibitor complex.[111] Binding of (1-oxyl-2,2,5,5-tetramethyl-3-pyrrolidinyl) methyl-β-chitobiose and (1-oxyl-2,2,5,5-tetramethyl-3-pyrrolidinyl) methyl-β-*N*-acetyl-D-glucosamide to the enzyme was more specific, and distances to histidine-15 C-2 protons of ~20 Å were calculated, which agreed with measurements from molecular models of 21 Å. It was claimed that, in favourable situations, distances up to 30 Å might be possible by this spin-label method.

Other studies on lysozyme have been presented at a conference on 'Magnetic Resonance in Biology and Medicine', held in New York, December 1972.[112] Very-high-resolution ^{1}H n.m.r. spectra of the enzyme have been obtained through use of a 270 MHz Fourier-transform spectrometer on samples to which paramagnetic ions had been specifically bound. For example, Gd^{III} was shown to bind in the active site of lysozyme[113] and broadened the resonances of protons in its vicinity. The difference spectrum between free lysozyme and the Gd^{III}–lysozyme complex then consisted only of resonances from the active-site region. Once these resonances had been assigned to specific residues, the tertiary structure of the active-site region could be determined by use of computer-aided conformational analysis.[114]

Phosphorylase. Campbell *et al.*[115] have extended their spin-label studies on phosphorylase *b*. A single SH group in each subunit reacted with the

[111] L. J. Berliner, *J. Mol. Biol.*, 1971, **61**, 189.
[112] I. D. Campbell, C. M. Dobson, R. J. P. Williams, and A. V. Xavier, *Ann. N.Y. Acad. Sci.*, 1973, in the press.
[113] K. G. Morallee, E. Nieboer, F. J. C. Rossotti, R. J. P. Williams, A. V. Xavier, and R. A. Dwek, *Chem. Comm.*, 1970, 1132.
[114] C. D. Barry, A. C. T. North, J. A. Glasel, R. J. P. Williams, and A. V. Xavier, *Nature*, 1971, **232**, 236.
[115] I. D. Campbell, R. A. Dwek, N. C. Price, and G. K. Radda, *European J. Biochem.*, 1972, **30**, 339.

iodoacetamide spin label IAM 6 (see p. 196) without affecting either the enzymic activity or the co-operativity between AMP sites. The authors emphasized that the attached spin label was not completely immobilized, having a rotational correlation time of 2.7 ns compared to 100 ns for the whole protein molecule. As pointed out on p. 198, this mobility introduces considerable uncertainty into distance calculations. However, the distance data were self-consistent with those obtained using Mn^{2+} as an additional paramagnetic probe. The main conclusions from the paper were (i) that at least four different conformations of the enzyme exist, corresponding to free enzyme and to enzyme bound to AMP, glucose-1-phosphate, and AMP plus glucose-1-phosphate, and (ii) that the allosteric effector (AMP) and substrate (glucose-1-phosphate) occupy sites on the enzyme which are very close together. Dwek *et al.*[116] have used this same spin-labelled phosphorylase *b* to study the activation of the enzyme by conversion into phosphorylase *a*. Again a conformational change was involved.

Chymotrypsin and Other Active-site Serine Enzymes. Shimshick and McConnell[117] described a spin-label method for measuring the rotational correlation time of chymotrypsin. This method could have wide applicability. Morrisett and Broomfield[118] have studied conformational changes in chymotrypsin during guanidinium chloride denaturation by monitoring the c.d. and by e.p.r. studies on the enzyme spin-labelled at the active-site serine. It was concluded that the two physical techniques reflected structural changes at different parts of the protein molecule. Hsia, Kosman, and Piette[119] have described a doubly spin-labelled fluorophosphate inhibitor that is specific for the active site in chymotrypsin and other serine enzymes. The spin–spin interaction between the two nitroxide groups is very sensitive to conformational changes in the protein. Clear differences between the e.p.r. of these labelled enzymes were apparent, reflecting differences in their active sites. The spin label was useful for monitoring denaturation of chymotrypsin by urea and by guanidinium chloride. Kosman and Piette[120] used the same spin label to study why chemical modification of methionine-192 affects enzyme activity. It could be concluded that this was due to steric blocking and not to a conformational change. Methionine-192 still moved, even when alkylated, in response to ionization of isoleucine-16 and aspartate-194.

Phosphofructokinase. Jones, Dwek, and Walker[121] have specifically spin-labelled one SH group per subunit with IAM 6 (see p. 196). The spin label was not completely immobilized. Isosbestic points in the e.p.r.

[116] R. A. Dwek, J. R. Griffiths, G. K. Radda, and U. Strauss, *F.E.B.S. Letters*, 1972, **28**, 161.
[117] E. J. Shimshick and H. M. McConnell, *Biochem. Biophys. Res. Comm.*, 1972, **46**, 321.
[118] J. D. Morrisett and C. A. Broomfield, *J. Amer. Chem. Soc.*, 1971, **93**, 7297.
[119] J. C. Hsia, D. J. Kosman, and L. H. Piette, *Arch. Biochem. Biophys.*, 1972, **149**, 441.
[120] D. J. Kosman and L. H. Piette, *Arch. Biochem. Biophys.*, 1972, **149**, 452.
[121] R. Jones, R. A. Dwek, and I. O. Walker, *F.E.B.S. Letters*, 1972, **26**, 92.

spectrum following titration with Mg-ATP suggested that only two conformations of the enzyme were involved in this allosteric transition. Similar experiments with Mn-ATP led to broadening of the spin-label spectrum and allowed a rough estimate of the Mn–nitroxide distance (12 Å). The authors admitted that they were unable to say yet whether the ATP was binding at an allosteric *or* an active site.

Dehydrogenases. Spallholz and Piette[122] have studied binding of an *o*-phenanthroline spin label to horse liver alcohol dehydrogenase. This compound was a competitive inhibitor to NAD^+ binding and probably reacted with active-site zinc atoms. Two SH groups per mole of enzyme reacted with IAM 6 (see p. 196). Blocking of these two SH groups reduced the amount of *o*-phenanthroline which could be bound, which suggested that the zinc and SH sites are close together.

Elek *et al.*[123] have used a series of iodoacetamide spin labels to block the reactive SH group of glyceraldehyde-3-phosphate dehydrogenase. The results suggested that there was a cleft on the surface of the enzyme near to the reactive sulphydryl.

Grande *et al.*[124] have bound *N*-(1-oxyl-2,2,5,5-tetramethyl-3-pyrrolidinyl)maleimide to SH groups in the flavoproteins lipoamide dehydrogenase and D-amino-acid oxidase. In both cases the e.p.r. spectrum indicated two classes of bound spin label differing in their mobility. It has further been shown that a spin-labelled substrate analogue could be bound to D-amino-acid oxidase. Removal of FAD from the enzyme caused considerable immobilization of this spin label.

β-Galactosidase and β-Galactoside Permease. A spin-labelled β-galactoside, tritiated in position 4 of the piperidinyl ring, has been synthesized.[125] This compound was a substrate for β-galactosidase. Uptake of the spin-labelled galactoside by *E. coli* cells, which could conveniently be monitored by scintillation counting, was *inhibited* by amytal and by lactate but it *increased* following induction of the cells with isopropylthiogalactoside. These results support active transport of the nitroxide in *E. coli* membranes by β-galactoside permease. The low concentrations of the permease in the membrane have so far prevented e.p.r. detection of the bound galactoside.

New Spin Labels.—A new method for nitroxide radical synthesis has been reported.[126] This involves treatment of an appropriately substituted nitroso-olefin with iodine. Yields of *ca.* 30% in this conversion were obtained. Keana and Dinerstein[127] reported the synthesis of a highly

[122] J. E. Spallholz and L. H. Piette, *Arch. Biochem. Biophys.*, 1972, **148**, 596.
[123] G. Elek, M. Sajgo, G. L. Grigorian, V. M. Chibrikin, and T. Keleti, *Acta Biochim. Biophys. Acad. Sci. Hung.*, 1972, **7**, 119.
[124] H. J. Grande, A. J. W. G. Visser, J. L. de Wit, F. Muller, and C. Veeger, *Z. Naturforsch.*, 1972, **27b**, 1058.
[125] W. G. Struve and H. M. McConnell, *Biochem. Biophys. Res. Comm.*, 1972, **49**, 1631.
[126] W. B. Motherwell and J. S. Roberts, *J.C.S. Chem. Comm.*, 1972, 328.
[127] J. F. W. Keana and R. J. Dinerstein, *J. Amer. Chem. Soc.*, 1971, **93**, 2808.

anisotropic dinitroxide keten spin label for use as a probe of membrane structure. Suitable adaptation of this synthetic procedure could yield a range of useful spin-labelled compounds for protein studies similar to those reported by Hsia *et al.*[119]

Membranes.—During 1972 there have been approximately as many papers reporting spin-label studies on phospholipid membranes as on proteins. Since this research area is outside the scope of the present literature survey, only a selection of the more important papers will be given.

There is an excellent review on spin labelling and membrane structure.[128] This covers both model membranes (multibilayers and vesicles) and biological membranes.

Lateral-diffusion Studies. Kornberg and McConnell[129] have incorporated a polar-head-group, spin-labelled phospholipid into phosphatidylcholine bilayer vesicles and studied the broadening of the *N*-methyl protons. It was concluded that the rate of the elementary step for diffusion must be greater than $3 \times 10^3\ s^{-1}$. Devaux and McConnell[130] noted that the e.p.r. spectrum of a highly concentrated spot of this same spin label, deposited on an oriented phosphatidylcholine multibilayer, changed with time as the spin label diffused into the surrounding lipid. The rate for diffusion was approximately $10^7\ s^{-1}$, which corresponds to a diffusion coefficient of $1.8 \pm 0.6 \times 10^{-8}\ cm^2\ s^{-1}$ at 25 °C. Since a similar value for the diffusion coefficient was found[131] for rabbit sarcoplasmic reticulum, lateral diffusion could have biological significance. Lee, Birdsall, and Metcalfe[132] have calculated a value for the rate of lateral diffusion in sarcoplasmic reticulum from proton n.m.r. linewidths that is in good agreement with the value reported by Devaux and McConnell.

In an important series of papers, Sackmann and Traüble[133] described their spin-label studies on dipalmitoyl-lecithin monolayer vesicles and aqueous suspensions. The spin label (an *N*-oxyl-4′,4′-dimethyloxazolidine derivative of 5α-androstan-3-on-17β-ol) was incorporated at both low and high concentrations. Interpretation of the e.p.r. spectra on the basis of dipolar and exchange-broadening terms allowed an estimate of the lateral diffusion coefficient to be made ($1 \times 10^{-8}\ cm^2\ s^{-1}$).

Marsh and Smith[134] have calculated the distance between adjacent spin-labelled molecules incorporated into a number of biological and synthetic oriented bilayer films. Dipole–dipole interaction between pairs of spin-labelled molecules contributed to broadening of the central lines of the

128 P. Jost, A. S. Waggoner, and O. H. Griffith, in 'Structure and Function of Biological Membranes', ed. L. Rothfield, Academic Press, New York, 1971, p. 83.
129 R. D. Kornberg and H. M. McConnell, *Proc. Nat. Acad. Sci. U.S.A.*, 1971, **68**, 2564.
130 P. Devaux and H. M. McConnell, *J. Amer. Chem. Soc.*, 1972, **94**, 4475.
131 C. J. Scandella, P. Devaux, and H. M. McConnell, *Proc. Nat. Acad. Sci. U.S.A.*, 1972, **69**, 2056.
132 A. G. Lee, N. J. M. Birdsall, and J. C. Metcalfe, *Biochemistry*, 1973, **12**, 1650.
133 E. Sackmann and H. Traüble, *J. Amer. Chem. Soc.*, 1972, **94**, 4482, 4492, 4499.
134 D. Marsh and I. C. P. Smith, *Biochem. Biophys. Res. Comm.*, 1972, **49**, 916.

spin-label spectrum but also produced new lines at the high- and low-field extremities of the spectrum. These new lines only appeared when the magnetic field was applied perpendicular to the bilayer. It was suggested that the method provides a convenient approach to the study of bilayer lateral structure.

Membrane Fluidity. There is a growing interest in how other membrane components, particularly proteins, affect the properties of biological membranes. McConnell *et al.*[135] have suggested a simple spin-label assay to determine the fraction of phospholipid in a biological membrane which is in a fluid state. They found, for example, that 84% of lipid in rabbit muscle sarcoplasmic reticulum was fluid at 25 °C. This result is consistent with the studies of lateral diffusion. The addition of the protein rhodopsin to spin-labelled phosphatidylcholine bilayers inhibited segmental motion of the hydrocarbon chains,[136] an effect similar to that of cholesterol. From other studies [137, 138] it is known that the rhodopsin is rotating about an axis perpendicular to the membrane surface. Extended studies of this system should aid our understanding of the photoreceptor membrane.

Hydrocarbon Chain Mobility. A detailed spin-label analysis of mobility at different positions along the fatty acid chain of phospholipids in oriented multibilayers has been carried out by Griffith *et al.*[139] The e.p.r. spectra have been computer-simulated assuming a restricted random-walk model. Seelig [140] has investigated the temperature dependence of alkyl chain flexibility using a series of spin labels and has calculated an apparent configurational entropy per methylene group of 2 e.u.

Eletr and Keith [141] have examined how the presence of double bonds in the alkyl chains of the phospholipid affects their mobility. Fatty acids with the double bond at the 6, 9, or 11 position from the carboxylate were added to the culture medium of yeast mutants deficient in their ability to synthesize fatty acids. Fatty acids with a spin-labelled grouping at various positions down the chain were incorporated into the membranes, and order parameters [142] were calculated from the e.p.r. spectra. One important conclusion was that greater alkyl chain mobility was present below the double bond, moving towards the centre of the bilayer.

Marsh *et al.*[143] concluded, from studies involving the incorporation of the spin label 5′-spiro-[2′-(*N*-oxyl-4′,4′-dimethyloxazolidine)]stearic acid into sonicated lecithin dispersions, that single bilayer vesicles allowed

135 H. M. McConnell, K. L. Wright, and B. G. McFarland, *Biochem. Biophys. Res. Comm.*, 1972, **47**, 273.
136 K. Hong and W. L. Hubbell, *Proc. Nat. Acad. Sci. U.S.A.*, 1972, **69**, 2617.
137 R. A. Cone, *Nature New Biol.*, 1972, **236**, 39.
138 P. K. Brown, *Nature New Biol.*, 1972, **236**, 35.
139 P. Jost, L. J. Libertini, V. C. Herbert, and O. H. Griffith, *J. Mol. Biol.*, 1971, **59**, 77.
140 J. Seelig, *J. Amer. Chem. Soc.*, 1971, **93**, 5017.
141 S. Eletr and A. D. Keith, *Proc. Nat. Acad. Sci. U.S.A.*, 1972, **69**, 1353.
142 W. L. Hubbell and H. M. McConnell, *J. Amer. Chem. Soc.*, 1971, **93**, 314.
143 D. Marsh, A. D. Phillips, A. Watts, and P. F. Knowles, *Biochem. Biophys. Res. Comm.*, 1972, **49**, 641.

significantly greater alkyl chain mobility than was possible in multibilayer vesicles. This conclusion is supported by the results of Sheetz and Chan.[144]

5 Fluorescence

contributed by G. R. Penzer

The use of fluorescence spectroscopy to study proteins continued to expand in 1972. The techniques have become so routine that they have ceased automatically to merit special mention in the titles (and occasionally even the abstracts) of all relevant articles. For this reason the aim has been to illustrate, with specific examples, the main areas in which fluorescence measurements are proving useful. The range of the applications and the sophistication and precision of the measurements have grown during the year, but no major innovatory methods have come to the attention of this Reporter. The Report is confined to work with purified proteins. Two large areas are omitted – immunofluorescence and studies of membrane preparations and subcellular particles (recently reviewed).[145] Other reviews have been published.[146–149]

Theory.—It has been recognized for many years that the time dependence of fluorescence emission anisotropy after pulsed excitation yields more information about the rotational diffusion of the emitter than does fluorescence polarization measured under conditions of static excitation and emission. However, there has been disagreement over the correct theory. The recent increase in the availability of instruments capable of providing decay anisotropy measurements made resolution of the disagreement urgent.[150–152] The situation with regard to a single rigid absorber/emitter on a macromolecule is resolved ('all parties are now in agreement' [151]). The final expression [151] for the time dependence of the polarization anisotropy [153] shows that the decay of fluorescence polarization comprises contributions from five exponential terms (no more than three of them independent). In most circumstances, however, only two decay rates will be distinct within realistic experimental error.[151] A computer simulation of fluorescence anisotropy decay due to Brownian motion has yielded results in agreement with the theoretical expression.[154] The situation in which the macromolecule contains a randomized distribution of transition moment orientation has also been considered.[150]

144 M. P. Sheetz and S. I. Chan, *Biochemistry*, 1972, **11**, 4573.
145 G. K. Radda and J. Vanderkooi, *Biochim. Biophys. Acta*, 1972, **265**, 509.
146 J. R. Brocklehurst, in 'Amino-acids, Peptides, and Proteins', ed. G. T. Young (Specialist Periodical Reports), The Chemical Society, London, 1972, vol. 4, p. 236.
147 R. F. Chen, *Methods Pharmacol.*, 1972, **2**, 1.
148 C. F. Chignell, *Methods Pharmacol.*, 1972, **2**, 33.
149 L. Brand and J. R. Gohlke, *Ann. Rev. Biochem.*, 1972, **41**, 843.
150 M. Ehrenberg and R. Rigler, *Chem. Phys. Letters*, 1972, **14**, 539.
151 G. C. Belford, R. L. Belford, and G. Weber, *Proc. Nat. Acad. Sci. U.S.A.*, 1972, **69**, 1392.
152 T. J. Chuang and K. B. Eisenthal, *J. Chem. Phys.*, 1972, **57**, 5094.
153 T. Tao, *Biopolymers*, 1969, **8**, 609.
154 S. C. Harvey and H. C. Cheung, *Proc. Nat. Acad. Sci. U.S.A.*, 1972, **69**, 3670.

Two samples, both emitting quanta at identical rates, will yield different signals in spectrofluorometers with the usual 90° geometry if for one sample the polarization is zero whilst for the other it is high.[155] It has now been shown that this result of the anisotropy of the fluorescence, which occurs even with unpolarized excitation, depends on instrument geometry according to the expression:[156]

$$I(\alpha)/I_0 = 1 - \tfrac{1}{4}r(3\cos^2\alpha - 1)$$

where $I(\alpha)$ is the emission intensity detected at an angle α to excitation and I_0 is the emission detected if the sample has emission anisotropy (r) of zero. Thus $I(\alpha) = I_0$ (and is independent of r) in instruments for which $3\cos^2\alpha = 1$ ($\alpha \approx 90° \pm 35°$).

Other theoretical treatments have considered the applicability of the theory of resonance energy transfer,[157] self-depolarization of rigid isotropic solutions as a result of energy transfer,[158] and the temperature dependence of non-radiative excited singlet deactivation processes.[159, 160]

Measurements and their Interpretation.—A number of developments in fluorescence instrumentation were reported during 1972. The popular method for measuring fluorescence lifetimes remains study of decay curves after pulsed excitation, and it has been shown that the time resolution of the method can be improved to the sub-nanosecond range by using photon counting and deconvolution of the signal by a method of moments.[161] A new simple source for pulsed excitation has also been described.[162] A simple modification of the alternative phase-shift lifetime measurement has been reported in which the time delay between the modulated emission and excitation beams is detected.[163] The time resolution in this instance was no better than 5 ns but could certainly be improved. Various instruments for studying small samples (both steady-state fluorescence[164, 165] and emission decay curves[166]) have been developed, as has a simple, small volume, windowless cell for front surface fluorometry in standard instruments.[167] Simple devices for the measurement of polarization of fluorescence have been described.[168] They are based on the fact that all the information necessary to obtain fluorescence polarization can be calculated from the successive maxima and minima of (total) fluorescence emitted at 90° to the

155 J. Paoletti and J.-B. Le Pecq, *Analyt. Biochem.*, 1969, **31**, 33.
156 M. Shinitzky, *J. Chem. Phys.*, 1972, **56**, 5979.
157 A. S. Agabekyan and A. O. Melikyan, *Optics and Spectroscopy*, 1972, **32**, 153.
158 A. Jablonski, *Acta Phys. Polon., A*, 1972, **41**, 85.
159 S. H. Lin, *J. Chem. Phys.*, 1972, **56**, 2648.
160 E. R. Pantke and H. Labhart, *Chem. Phys. Letters*, 1972, **16**, 255.
161 R. Schuyler, I. Isenberg, and R. D. Dyson, *Photochem. and Photobiol.*, 1972, **15**, 395.
162 J. Zynger and S. R. Crouch, *Appl. Spectroscopy*, 1972, **26**, 631.
163 J. Jessop, R. P. Wayne, and T. J. Wayne, *J. Phys.* (*E*), 1972, **5**, 638.
164 M. Sernetz and A. Thaer, *Analyt. Biochem.*, 1972, **50**, 98.
165 E. Kohen, C. Kohen, and B. Theorell, *Biochim. Biophys. Acta*, 1972, **286**, 189.
166 C. N. Loeser, E. Clark, and H. Tarkmeel, *Exp. Cell. Res.*, 1972, **72**, 480.
167 J. A. McHard and J. D. Winefordner, *Analyt. Chem.*, 1972, **44**, 1922.
168 J. Lavorel, C. Vernolt, B. Arrio, and F. Rodier, *Biochimie*, 1972, **54**, 161.

exciting beam when a single rotating polarizer is placed in it. When suitable compensation for the intrinsic polarization of the instrument is provided it is shown that the technique gives good results for some dyes with intense emission and high polarization.

Two new types of measurement involving light emission have been developed. The first studies the circular polarization of fluorescence from excited states in asymmetric environments. An instrument with greatly improved sensitivity has been described [169] and some fluorescence probes bound to chymotrypsin have been studied by the technique.[170] The second method is the optical detection of magnetic resonance in which phosphorescence (at low temperature) is studied for samples subjected to microwave radiation which affects the relative populations of the three spin sub-levels of the triplet state. The microwave frequencies at which effects are observed give a direct measure of the zero-field splittings of the triplet. The zero-field splitting parameters of free tryptophan and tryptophan in bovine serum albumin are different,[171] and it is suggested that the technique may have applications in the study of conformational changes. A similar suggestion was made in a report of the detection of delayed fluorescence emission from some proteins,[172] but in both cases it is likely that the very low sample temperatures necessary for the measurements will greatly limit the use of the technique in the study of biological samples.

Most standard fluorescence instruments provide spectra uncorrected for the spectral responses of monochromators and photomultipliers and for the emission spectrum of the lamp. Corrections are possible and a set of correction curves for Hitachi MPF-2A instruments has been published.[173] Automatic correction is naturally preferable, and some computerized data-acquisition and -handling systems for fluorescence instruments have been described.[174, 175]

Two papers sounding warnings for fluorescence spectroscopists have appeared. One describes the fluorescent impurities in polythene containers [176] and the other the photochemical transformation of tryptophan residues into *N*-formylkinurenine when irradiated with visible light.[177]

A number of new ways of applying fluorescence measurements (using conventional instruments) have been reported. It has been common to estimate inter- and intra-molecular distances by studying the efficiency of resonance energy transfer between pairs of chromophores located, if possible, at specific sites on the protein. However, one can go to the other

[169] I. Z. Steinberg and A. Gafni, *Rev. Sci. Instr.*, 1972, **43**, 409.
[170] J. Schlessinger and I. Z. Steinberg, *Proc. Nat. Acad. Sci. U.S.A.*, 1972, **69**, 769.
[171] J. Zuclich, D. Schweitzer, and A. H. Maki, *Biochem. Biophys. Res. Comm.*, 1972, **46**, 1764.
[172] M. Bazin, M. Aubailly, and R. Santus, *Chem. Phys. Letters*, 1972, **13**, 310.
[173] F. Sundholm, H. Konschin, and H. Tylli, *Suomen Kem.*, 1972, **45**, *A*, 77.
[174] R. R. Schehl and R. Friedel, *J. Phys. (E)*, 1972, **5**, 1038.
[175] T. M. Shepherd and C. A. Vincent, *Chem. in Britain*, 1973, **9**, 66.
[176] R. F. Chen, *Analyt. Letters*, 1972, **5**, 663.
[177] A. Pirie, *Biochem. J.*, 1972, **128**, 1365.

extreme and label the surface of the protein randomly and extensively with fluorescers.[178,179] The theoretical treatment of the results (from which distances are calculated) involves some assumptions,[178] but application of the method to known systems (*e.g.* the trypsin–trypsin-inhibitor interaction) yields good results.[179] It is suggested that the technique will be useful both for measuring the separation between two macromolecules, and for measuring anhydrous volumes (hence, perhaps, detecting conformation changes).[179]

Another development has been the direct measurement of the rates of excited-state proton-transfer reactions by study of fluorescence lifetimes. These rates depend on the environment of the emitter, and it is suggested that the method will be useful for detecting the acidity of the binding site of the probes (*e.g.* naphtholsulphonic acids) – and perhaps also of tyrosine – which undergo these reactions.[180]

There have been a number of discussions of fluorescence titrations in which the binding of a ligand to a protein is studied. The problem of inner-filter effects can be circumvented in several ways, depending on the system. When the emitter is being titrated with increasing amounts of a chromophore which absorbs at the wavelength of excitation (frequently the case when protein fluorescence is studied in the presence of nucleotides), compensation for the variable inner filtering can be achieved either by comparison with the effects on a similar titration with a model compound to which the chromophore does not bind,[181] or by calculation knowing the absorbance of the chromophore at the wavelength of excitation.[182,183] The latter technique requires assumption of the effective path length of the fluorescence cell, but is adequate when corrections are small. A technique avoiding this assumption has been described that can be applied when the inner-filter effect is due to the emitter itself.[184] It involves comparison of the emissions of a given concentration of emitter in the presence and absence of the compound to which it binds. Titrations are usually performed an aliquot at a time, but a double-beam instrument capable of automatic fluorescence titrations has been described, and shown to provide good results.[182] The binding of proteins to polymers (*e.g.* DNA) can be studied by fluorescence anisotropy.[185] Under these circumstances inner-filter effects are unimportant.

The interpretation of fluorescence measurements is helped by an understanding of the behaviour of the individual emitting groups. It has been shown that the yield of phenylalanine fluorescence is greatest at neutral

178 R. B. Gennis and C. R. Cantor, *Biochemistry*, 1972, **11**, 2509.
179 L. S. Gennis, R. B. Gennis, and C. R. Cantor, *Biochemistry*, 1972, **11**, 2517.
180 M. R. Loken, J. W. Hays, J. R. Gohlke, and L. Brand, *Biochemistry*, 1972, **11**, 4779.
181 N. C. Price, *F.E.B.S. Letters*, 1972, **24**, 21.
182 J. J. Holbrook, *Biochem. J.*, 1972, **128**, 921.
183 G. Engel, H. Heider, A. Maelicke, F. von der Haar, and F. Cramer, *European J. Biochem.*, 1972, **29**, 257.
184 J. S. Franzen, I. Kuo, and A. E. Chung, *Analyt. Biochem.*, 1972, **47**, 426.
185 R. Wickett, H. J. Li, and I. Isenberg, *Analyt. Biochem.*, 1972, **50**, 608.

pH, and falls when either charged group becomes neutral.[186] The interactions between peptides and nucleic acids have also been studied, and it has been deduced that although both tryptophan and tyrosine residues interact with nucleic acids, only the former does so by intercalation.[187]

Fluorescence Probes.—The use of extrinsic fluorescence probes in the study of proteins is growing, and was extensively reviewed in 1972.[149] Both covalent and non-covalent interactions are useful, and in both classes some specific probes which interact strongly with a single known site on the protein have been developed (Table 1). Probes of this type are harder

Table 1 *Fluorescence probes and their applications*

*Probe**	*Application(s)*	*Ref.*
Anilinonaphthalene sulphonates (ANS)	$(Na^+ + K^+)$-ATPase	198
	Bovine serum albumin	199
	κ-Casein	200
	Glyceraldehyde-3-phosphate dehydrogenase	201
	Glycogen phosphorylase	202
	Phycocyanin	203
	Thyroxine-binding globulin	204, 205
Toluidinonaphthalene sulphonates (TNS)	Chymotrypsin	170
	Isoleucine:tRNA ligase	206
	Various linear and cyclic peptides	207, 208
5-Dimethylaminonaphthalene-1-sulphonyl (Dns) [C]	Adrenocorticotropic hormone	221
	Bovine serum high-density lipoprotein	245
	α-Lactalbumin	213
	Lysozyme	213
	Poly-L-glutamic acid	239
	Poly-L-lysine	239, 240
1-Dns-3-*NN*-dimethylamino-propane [S] and 1-Dns-3-trimethylammonium iodide [S]	Horse serum cholinesterase	*a*
N-α-Dns-L-phenylalanylchloro-methane [C, S]	Chymotrypsin	*b*
N-α-Dns-L-lysylbromomethane [C, S]	Trypsin	*b*
14,*N*-terminal Dns peptides [S]	Pepsin	*c*
Dns-$(Gly)_n$-Phe [S] and Dns-$(Gly)_n$-Trp [S]	Carboxypeptidase	220, 247, *d*
Pyrenebutyric acid [C]	Bovine serum high-density lipoprotein	245
	L-Glutamate dehydrogenase	224

[a] R. T. Mayer and C. H. Himel, *Biochemistry*, 1972, **11**, 2082.
[b] G. Schoellmann, *Internat. J. Protein Res.*, 1972, **4**, 221.
[c] G. P. Sachdev, M. A. Johnston, and J. S. Fruton, *Biochemistry*, 1972, **11**, 1080.
[d] S. A. Latt, D. S. Auld, and B. L. Vallee, *Analyt. Biochem.*, 1972, **50**, 56.

* Probes marked [C] form covalent compounds with proteins: those marked [S] interact specifically with a site on the protein which is directly involved in its biological activity.

[186] J. Tournon, E. Kuntz, and M. A. El-Bayoumi, *Photochem. and Photobiol.*, 1972, **16**, 425.
[187] C. Helene and J. L. Dimicoli, *F.E.B.S. Letters*, 1972, **26**, 6.

Table 1 (*cont.*)

*Probe**	*Application(s)*	*Ref.*
Fluorescein	Bovine serum albumin	164
Fluorescein isothiocyanate [C]	Tropomyosin	229
	Trypsin, trypsin inhibitors	179
Rhodamine and rosamine derivative isothiocyanates [C]	Synthesis	*e,f*
5′-(*N*-isomaleimido)-rhodamine B [C]	Haemoglobin	*g*
1,4-Sulphonaphthylhydrazone of hellebrigenin [S]	($Na^+ + K^+$)-ATPase	198
Tetracyclines	Bovine serum albumin	195
	Mg^{2+} and Ca^{2+}	196
	Sarcoplasmic reticulum	197
Naphthol and naphthol-sulphonic acid	Bovine serum albumin	180
p-Nitrophenylanthranilate [C]	Chymotrypsin	170
N-Methylacridinium ion [S], bis-(3-aminopyridinium)-1,10-decane [S] and 7-dimethyl-carbonyl-*N*-methylquinolinium iodide [S]	Acetylcholinesterase	*h*
1,N^6-ethenoadenosine and its nucleotides [S]	Synthesis of ethenoadenosine, etheno-AMP, etheno-ADP, and etheno-ATP, and their interactions with pyruvate, adenylate, and hexo- and phospho-fructokinases	188—190
	Phosphorylase *b*	191
	Synthesis of cyclic etheno-AMP	192
	Synthesis of etheno-NAD	193
	Synthesis of etheno-FAD	194

[e] Y. E. Sklyar, A. V. Afanas'eva, and G. I. Mikhailov, *Metody Poluch. Khim. Reaktiv. Prep.*, 1970, No. 22, p. 167 (*Chem. Abs.*, 1972, **77**, 50 111).

[f] Y. E. Sklyar, A. V. Afanas'eva, N. N. Shcherbakova, and G. I. Mikhailov, *Metody Poluch. Khim. Reaktiv. Prep.*, 1970, No. 22, p. 186 (*Chem. Abs.*, 1972, **77**, 50 112).

[g] E. Y. Alfimova and G. I. Likhtenshtein, *Biofizika*, 1972, **17**, 49.

[h] G. Mooser, H. Schulman, and D. S. Sigman, *Biochemistry*, 1972, **11**, 1595.

* Probes marked [C] form covalent compounds with proteins: those marked [S] interact specifically with a site on the protein which is directly involved in its biological activity.

to devise and synthesize than the non-specific variety, but their use can lead to very precisely interpretable information. One new class, although specific, may be widely useful because its members are derivatives of adenosine and its nucleotides. It is based on 1,N^6-ethenoadenosine.[188—194]

[188] J. A. Secrist, J. R. Barrio, and N. J. Leonard, *Science*, 1972, **175**, 646.

[189] J. R. Barrio, J. A. Secrist, and N. J. Leonard, *Biochem. Biophys. Res. Comm.*, 1972, **46**, 597.

[190] J. A. Secrist, J. R. Barrio, N. J. Leonard, and G. Weber, *Biochemistry*, 1972, **11**, 3499.

[191] R. F. Steiner, *F.E.B.S. Letters*, 1972, **23**, 139.

[192] J. A. Secrist, J. R. Barrio, and N. J. Leonard, *Science*, 1972, **177**, 4045.

[193] J. R. Barrio, J. A. Secrist, and N. J. Leonard, *Proc. Nat. Acad. Sci. U.S.A.*, 1972, **69**, 2039.

[194] R. A. Harvey and S. Damle, *F.E.B.S. Letters*, 1972, **26**, 341.

Among the attractions of these derivatives are their simple synthesis from the parent compounds and the fact that they retain many of the biochemical properties of the unmodified nucleotides. The fluorescence emissions of 1,N^6-etheno-derivatives of adenosine, AMP, ADP, and ATP are very similar and are not very sensitive to environmental changes.[190] The emission of etheno-NAD is quenched, and a ten-fold increase in fluorescence follows hydrolysis.[193] Another useful probe is chlorotetracycline, which shows environment-sensitive fluorescence on binding to proteins [195] and also in its interactions with bivalent metal ions.[196, 197] Other probes which have featured in previous Reports of this series (*e.g.* pyridoxal phosphate, auramine *O*, 7-chloro-4-nitrobenzo-2,1,3-oxadiazole, and iodoacetamidosalicylic acid), on the other hand, had a quiet year (at least in the literature).

In spite of these developments, the front runners of the environment-sensitive fluorescence probes remain anilino- and toluidino-naphthalene-sulphonates.[170, 198–208] The molecular basis of their environmental sensitivity remains puzzling, though it is now clear that binding to proteins does not necessarily cause fluorescence enhancement,[207, 208] and that enhancements can be produced in polar (hydrophilic) environments.[209] The importance of molecular flexibility for efficient fluorescence quenching in water has been stressed,[209] and it seems that if fluorescence enhancement is to accompany binding of a probe to a protein there must be both polar and hydrophobic regions in the binding site.[207, 208]

Structure Mapping.—A common use of fluorometry is in plotting the relative positions of fluorescent groups in, or bound to, a protein. The molecular resolution with which the results can be interpreted varies very widely. At one extreme are studies of the tryptophan emissions of lysozyme.[210–212] The fluorescence decays of both native and oxidized hen

[195] P. G. Popov, K. I. Vaptsarova, G. P. Kosekova, and T. K. Nikolov, *Biochem. Pharmacol.*, 1972, **21**, 2363.
[196] A. H. Caswell, *J. Membrane Biol.*, 1972, **7**, 345.
[197] A. H. Caswell and S. Warren, *Biochem. Biophys. Res. Comm.*, 1972, **46**, 1757.
[198] A. Yoda and L. E. Hokin, *Mol. Pharmacol.*, 1972, **8**, 30.
[199] E. E. Santos and A. A. Spector, *Biochemistry*, 1972, **11**, 2299.
[200] R. F. L. Clark and S. Nakai, *Biochim. Biophys. Acta*, 1972, **257**, 61.
[201] N. K. Nagradova, R. A. Asriyants, and M. V. Ivanov, *Biokhimiya*, 1972, **37**, 299.
[202] V. L. Seery and S. R. Anderson, *Biochemistry*, 1972, **11**, 707.
[203] A. Binder, D. A. Deranleau, and H. Zuber, *F.E.B.S. Letters*, 1972, **23**, 185.
[204] A. M. Green, J. S. Marshall, J. Pensky, and J. B. Stanbury, *Science*, 1972, **175**, 1378.
[205] A. M. Green, J. S. Marshall, J. Pensky, and J. B. Stanbury, *Biochim. Biophys. Acta*, 1972, **278**, 305.
[206] E. Holler and M. Calvin, *Biochemistry*, 1972, **11**, 3741.
[207] C. F. Beyer, L. C. Craig, and W. A. Gibbons, *Biochemistry*, 1972, **11**, 4920.
[208] C. F. Beyer, L. C. Craig, and W. A. Gibbons, *Nature New Biol.*, 1973, **241**, 78.
[209] G. R. Penzer, *European J. Biochem.*, 1972, **25**, 218.
[210] G. Y. Yashinsky, *F.E.B.S. Letters*, 1972, **26**, 123.
[211] T. Imoto, L. S. Forster, J. A. Rupley, and F. Tanaka, *Proc. Nat. Acad. Sci. U.S.A.*, 1972, **69**, 1151.
[212] V. I. Teichberg, T. Plasse, S. Sorell, and N. Sharon, *Biochim. Biophys. Acta*, 1972, **278**, 250.

egg-white lysozyme are both best interpreted as the sum of the same two exponentials, in spite of the fact that the overall quantum yield drops to about 40% when Trp-108 is oxidized.[210] When both Trp-108 and Trp-62 are oxidized the fluorescence drops further, and it is concluded that energy transfer from Trp-108 to Trp-62 (perhaps by way of Trp-63) is important in the native enzyme.[210] The effects of pH, quenchers, and tri-*N*-acetylglucosamine are all consistent with this conclusion. Human lysozyme has also been studied, and the fluorescence changes that are observed when substrates bind have been interpreted in terms of changes in Trp-64–Asp-101 and Trp-108–Glu-35 interactions.[212] All the fluorescence results suggest that the structure and properties of the human and hen-egg-white enzymes are similar. In contrast, although the molecular weights and amino-acid sequences of hen egg-white lysozyme and α-lactalbumin are very similar, the rotational relaxation times of their 5-dimethylaminonaphthalene-1-sulphonyl (Dns) conjugates are different (25 ns and 35 ns, respectively).[213]

Usually interpretations have to be at a more general level. Human [214] and bovine [215] growth hormones have been purified and their fluorescence emissions compared. The emission of the single tryptophan in the human hormone is more readily perturbed by both solvents and quenchers than the single tryptophan of the bovine hormone. Energy transfer from tyrosine to tryptophan is also more efficient in the former. In the case of alkaline phosphatase (from *E. coli*), however, the emission spectrum is interpreted (by comparison with models) as showing that most of the tryptophan residues are inside the protein (three-quarters of them close to groups inducing static quenching of fluorescence).[216] The excitation spectrum indicates negligible tyrosine to tryptophan energy transfer.[217] The variation of the fluorescence of asparaginase with pH indicates the presence of a quenching group with pK_a 6.0 close to the emitting tryptophan.[218] Addition of substrate leads to a time-dependent quenching, and it is deduced that histidine and tryptophan are adjacent in the substrate-binding site.

The fluorescence emission of adrenodoxin implicates the presence of tyrosine in an unusual environment.[219] Its emission is red-shifted to about 330 nm, and most artifactual reasons for this observation have been excluded. The authors suggest an explanation involving exciplex formation, but until the second component of the complex is identified this cannot be considered to have been established.

Distances in protein topography can be calculated from the results of fluorescence energy transfer measurements. Förster's resonance energy

[213] A. B. Rawich, *Arch. Biochem. Biophys.*, 1972, **151**, 22.
[214] V. T. Maddaiah, P. J. Collip, R. K. Sharma, S. Y. Chen, and J. Thomas, *Biochim. Biophys. Acta*, 1972, **263**, 133.
[215] V. T. Maddaiah and P. J. Collip, *F.E.B.S. Letters*, 1972, **23**, 208.
[216] D. Gerard, G. Laustriat, and H. Lami, *Biochim. Biophys. Acta*, 1972, **263**, 482.
[217] D. Gerard, H. Lami, and G. Laustriat, *Biochim. Biophys. Acta*, 1972, **263**, 496.
[218] R. Homer, *Biochim. Biophys. Acta*, 1972, **278**, 395.
[219] T. Kimura, J. J. Ting, and J. J. Huang, *J. Biol. Chem.*, 1972, **247**, 4476.

transfer theory is satisfactory providing that reasonable assumptions about donor:acceptor orientations and the refractive index of the medium can be made. The method has been used in a study of the active site of carboxypeptidase A during catalysis.[220] Cobalt (rather than native zinc) enzyme was studied with Dns-(Gly)$_n$-Phe (n = 1, 2, 3, or 4) and Dns-(Gly)$_n$-Trp (n = 1, 2, or 3) as the substrates. Energy transfers from enzyme tryptophan to Dns and Dns to cobalt both occur. The distances calculated for Dns–cobalt separation using the different substrates are all consistent with the assumption that an extended peptide (substrate) chain interacts with cobalt at the carbonyl oxygen of the bond to be split. Another study utilizing resonance energy transfer was on the conformation of adrenocorticotropic hormone-(1—24)-tetrakosipeptide.[221] The peptide was labelled with Dns at Lys-21 and the efficiency of energy transfer from Trp-9 to the Dns group was measured. The distance separating these two groups (2.3 ± 0.3 nm) is almost independent of solvent. It was deduced that the hormone conformation is close to random-coil in all solvents studied.

The environment of the pyridoxal phosphate coenzyme of phosphorylase *b* has been investigated by comparing the absorption and emission spectra of enzyme-bound coenzyme with those of pyridoxal in simpler model systems.[222] It is concluded that the most probable pyridoxal environment in phosphorylase *b* is non-polar. In the ground state there is carbinolamine formation between the aldehyde of pyridoxal phosphate and a lysine ε-amino-group in the protein. The excited state of the carbinolamine expels water to give a Schiff base which is the fluorescence emitter. It is rehydrated back to carbinolamine in the ground state. A non-aqueous environment for the coenzyme is consistent with fluorometric studies of the sodium-borohydride-reduced coenzyme–enzyme compound.[223] These measurements also suggest that below pH 4.8 the enzyme adopts a different shape in which the cofactor site is more exposed to solvent.

Aggregation and Shape Changes.—Probably the most common use of fluorescence spectroscopy is the detection and study of changes in the states of aggregation and the conformations of proteins. Sometimes interpretation is helped by combining these studies with the application of other physical techniques (*e.g.* c.d. and, increasingly, n.m.r.). Often the measurements utilize fluorescence anisotropy (either decay anisotropy or steady-state polarization).

The aggregation of L-glutamate dehydrogenase has been investigated by measuring the fluorescence anisotropy of enzyme labelled with pyrenebutyric acid as a function of protein concentration.[224] The results were tested against the alternative hypotheses that there is continuous association of monomers (with a constant association constant) to give either

[220] S. A. Latt, D. S. Auld, and B. L. Vallee, *Biochemistry*, 1972, **11**, 3015.
[221] P. W. Schiller, *Proc. Nat. Acad. Sci. U.S.A.*, 1972, **69**, 975.
[222] K. O. Honikel and N. B. Madsen, *J. Biol. Chem.*, 1972, **247**, 1057.
[223] M. Cortijo and S. Shaltiel, *European J. Biochem.*, 1972, **29**, 134.
[224] D. A. Malencik and S. R. Anderson, *Biochemistry*, 1972, **11**, 3022.

spherical or linear aggregates. The results favour linear aggregation. A second globular protein whose state of aggregation in solution has been investigated (along with its conformation) is κ-casein.[200] A new way to measure protein association by studying the fluorescence of labelled protein when it changes in a non-linear way with the state of aggregation has also been suggested.[225]

The molecular states of muscle proteins have been studied by fluorescence. Changes in intrinsic (tryptophan) fluorescence associated with the G–F transformation of actin and when either form is denatured have been reported.[226, 227] Both the intrinsic (tyrosine) fluorescence [228] and that of the fluorescein conjugate [229] of tropomyosin indicate that the molecule undergoes a major conformational change close to physiological temperatures (306—307 K). The tryptophan fluorescence of heavy meromyosin is sensitive to the binding of ATP.[230] The results are interpreted in terms of a conformational change associated with the formation of an enzyme–product complex after the ATP has been hydrolysed. No doubt an advantage seen for the fluorescence method in the study of muscle proteins is its suitability for the study of intact samples. The sensitivity of the polarization of tryptophan fluorescence to the extent of contraction in single muscle fibres has already been demonstrated.[231, 232]

The sensitivities of fluorescence parameters to protein conformation changes depend on the system studied. Sometimes quantum yields vary substantially. For example, the intrinsic fluorescence of pyruvate kinase increases by 22% in the presence of the allosteric modifier phenylalanine.[233] The different effects of Mg^{2+} and Mn^{2+} on the allosteric effect are simply demonstrated fluorometrically. N.m.r. measurements have enabled an estimate (1.2 nm) of the separation between Mn^{2+} and phenylalanine sites. In this instance intrinsic fluorescence is much more sensitive to the conformational change than arc the magnetic resonance parameters. Other studies in which conformational changes have been monitored by changes in quantum yield of fluorescence involved Bence-Jones proteins,[234] thioredoxin,[235] the 55 000 dalton protein component of sarcoplasmic reticulum,[236]

225 J. J. Holbrook, D. W. Yates, S. J. Reynolds, R. W. Evans, C. Greenwood, and M. G. Gore, *Biochem. J.*, 1972, **128**, 933.
226 J. K. Weltman, R. P. Szaro, A. R. Franckelton, and R. M. Dowken, *F.E.B.S. Letters*, 1972, **22**, 61.
227 S. S. Lehrer and G. Kerwar, *Biochemistry*, 1972, **11**, 1211.
228 A. Satoh and K. Mihashi, *J. Biochem. (Japan)*, 1972, **71**, 597.
229 K. Mihashi, *J. Biochem. (Japan)*, 1972, **71**, 607.
230 M. Werber, A. G. Szent-Györgyi, and G. D. Fasman, *Biochemistry*, 1972, **11**, 2872.
231 C. G. DosRemedios, R. G. C. Millikan, and M. F. Morales, *J. Gen. Physiol.*, 1972, **59**, 103.
232 U. S. Borovikov, U. M. Rosanov, L. A. Shiffers, and M. S. Shudel, *Tsitologiya*, 1972, **14**, 1342.
233 F. J. Kayne and N. C. Price, *Biochemistry*, 1972, **11**, 4415.
234 R. Pollet, G. Rossi, and H. Edelhoch, *J. Biol. Chem.*, 1972, **247**, 5921.
235 A. Holmgren, *J. Biol. Chem.*, 1972, **247**, 1992.
236 N. Ikemoto, G. M. Bhatnagar, B. Nagy, and J. Gergely, *J. Biol. Chem.*, 1972, **247**, 7835.

rhodopsin,[237] and a sweet-sensitive protein.[238] In this last instance it was found that glucose and sucrose produce opposite responses – the former decreased quantum yield of protein fluorescence, while the latter increased it.

In some systems fluorescence anisotropy is sensitive to protein conformation. Helix–coil transitions of synthetic polypeptides have been studied this way.[239, 240] So have κ-casein and some of the muscle proteins (see above, p. 212). The fluorescence polarization of carp 15S immunoglobulin shows it to be much less flexible than mammalian immunoglobulins.[241] There have been several papers on histones. Both fluorescence (tyrosine emission anisotropy) and circular dichroism detect phosphate-induced conformation changes in histone IV,[242, 243] but similar time-dependent changes were not observed for histone II b 2 because they are too fast.[244] Bovine serum high-density lipoprotein has been shown to be a rigid spherical particle with local flexibility at the level of amino-acid side-chains in water. Octanol, however, disrupts the structure.[245]

Sometimes the time courses for changes in different fluorescence parameters are different. The interconversion of one electrophoretic form of alcohol dehydrogenase (from *Drosophila*) to another occurs with concomitant changes in both quantum yield of fluorescence and the wavelength of the emission maximum.[246] The rate of quantum-yield change is fast and correlates with the amount of starting protein which remains. The emission shifts more slowly, at a rate which correlates with the rate of change of enzyme activity. Presumably two distinct changes of protein conformation or environment are being detected in these measurements.

Interactions with Small Ligands.—A number of fluorescence studies in which the interactions between proteins and ligands were investigated have already been mentioned in this Report. Sometimes the interactions are observed by intrinsic protein fluorescence, sometimes by the emission of a fluorescent (or labelled) substrate, and sometimes by use of a reporter group. The speed with which fluorescence can be detected makes it suitable for monitoring the kinetics of protein–ligand interactions by fast reaction techniques – both stopped-flow [206, 220, 247] and temperature-jump.[248]

237 T. G. Ebrey, *Photochem. and Photobiol.*, 1972, **15**, 585.
238 F. R. Dastoli, *Experientia*, 1972, **28**, 389.
239 T. J. Gill, C. T. Ladoulis, H. W. Kunz, and M. F. King, *Biochemistry*, 1972, **11**, 2644.
240 T. Iio, Y. Iwashita, and H. Watanake, *Bull. Chem. Soc. Japan*, 1972, **45**, 2206.
241 R. Richter, P. Nuhn, H. Ambrosius, Y. A. Zagyansky, L. A. Tumerman, and R. S. Nezlin, *F.E.B.S. Letters*, 1972, **27**, 184.
242 R. R. Wickett, H. J. Li, and I. Isenberg, *Biochemistry*, 1972, **11**, 2952.
243 H. J. Li, R. Wickett, A. M. Craig, and I. Isenberg, *Biopolymers*, 1972, **11**, 375.
244 J. A. D'Anna and I. Isenberg, *Biochemistry*, 1972, **11**, 4017.
245 A. Jonas, *J. Biol. Chem.*, 1972, **247**, 7773.
246 J. A. Knopp and K. B. Jacobson, *Arch. Biochem. Biophys.*, 1972, **149**, 36.
247 D. S. Auld, S. A. Latt, and B. L. Vallee, *Biochemistry*, 1972, **11**, 4994.
248 A. D. B. Malcolm, *European J. Biochem.*, 1972, **27**, 453.

Measurement of equilibrium binding constants is usually performed by means of some kind of fluorescence titration. The reasons why fluorescence is sensitive to ligand binding can vary. When protein emission is studied, possible causes are conformation changes, formation of a non-fluorescent complex, and energy transfer from protein to ligand. The third of these possibilities has been considered in detail for enzymes which contain a number of indistinguishable, independent, and intrinsically identical binding sites, where there is a non-linear decrease in protein fluorescence with

Table 2 *Quantitative fluorescence binding studies**

Protein	*Ligand(s)*	*Emitter(s)*	*Ref.*
Acetylcholinesterase	*N*-Methylacridinium ion, bis-(3-aminopyridinium)-1,10-decane, and 7-dimethylcarbamyl-*N*-methylquinolinium iodide	Ligands	*a*
Carboxypeptidase	Phenylacetate and phenylalanine	Dns-peptide	*b*
Pepsin	Various Dns-peptides	Ligands	*c*
Lysozyme	Various chitin oligo-saccharides	Protein	212
Ribonuclease	DNA	Protein	185
Creatine kinase	ADP and ATP	Protein	181
Pyruvate kinase	Phenylalanine	Protein	233
	ADP and ATP	Protein	181
Isoleucine:tRNA ligase	Isoleucine, ATP, and pyrophosphate	TNS	206
Serine:tRNA ligase	ATP and $tRNA^{Ser}$	Protein	183
Ribonucleotide reductase	dATP	Protein	184
Isocitrate dehydrogenase	NADPH	Ligand	184
Lactate dehydrogenase	NADH	Protein and ligand	225
Alcohol dehydrogenase	NADH	Protein and ligand	225
Glycerol-3-phosphate dehydrogenase	NADH	Protein and ligand	225
Malate dehydrogenase	NAD, NADH, hydroxy-malonate, and D-malate	NADH	*d*
Neurospora crassa glutamate dehydrogenase	NADPH	Protein and ligand	225
Ox liver glutamate dehydrogenase	NADH and GTP	Protein and NADH	225
	NADH and NADPH	Ligand	248
Retinol-binding protein	Retinol	Protein and ligand	*e*
Flavodoxin apoproteins	FMN	Proteins and ligand	249
Glucose oxidase	FAD	Protein and ligand	249

[a] G. Mooser, H. Schulman, and D. S. Sigman, *Biochemistry*, 1972, **11**, 1595.
[b] S. A. Latt, D. S. Auld, and B. L. Vallee, *Analyt. Biochem.*, 1972, **50**, 56.
[c] G. P. Sachdev, M. A. Johnston, and J. S. Fruton, *Biochemistry*, 1972, **11**, 1080.
[d] J. J. Holbrook and R. G. Wolfe, *Biochemistry*, 1972, **11**, 2499.
[e] D. S. Goodman and R. B. Leslie, *Biochim. Biophys. Acta*, 1972, **260**, 670.

* This table is not intended to be comprehensive. Studies measuring the binding of non-specific probes (*e.g.* ANS) have been omitted, as have reports of binding interactions detected fluorometrically but not studied quantitatively.

the fraction of the total number of coenzyme binding sites occupied.[182] It was found that $F = [1 - \alpha(1 - x)]^n$ (where F is protein fluorescence intensity, α is the fraction of sites occupied by ligand, n is the number of sites, $x = F_s^{1/n}$, $F_s = F$ when $\alpha = 1$) fits the results obtained for the binding of NADH to pig heart lactate dehydrogenase. Geometric quenching of this kind requires that for each additional ligand binding, the protein emission falls by the same fraction, irrespective of the number of sites already occupied. The interactions between nucleotides and other dehydrogenases were studied, and in two cases (binding of GTP to ox liver glutamate dehydrogenase and of NADH to *Neurospora crassa* glutamate dehydrogenase) quenching of protein fluorescence was linear.[225] This situation is predicted either when $n = 1$ or when all n sites are completely co-operative. The *Neurospora crassa* enzyme results are attributed to the former situation, the ox liver enzyme results to the latter.

When extrinsic probes are used to monitor ligand binding, the fluorescence changes may reflect changes in the environment of the bound probe or, alternatively, changes in the number (usually because of displacement) of bound probes. In several systems it has been well demonstrated that the latter is the case.[202, 204]

A nice example of the way protein and probe emissions may reflect different changes is provided by the binding of the coenzyme FAD (considered in this case as the probe) to glucose oxidase.[249] Protein fluorescence is quenched much more rapidly than the flavin emission, so binding to form holoenzyme is not a one-step process.

A list of some of the binding studies reported in 1972 is given in Table 2 to illustrate the wide range of systems to which the fluorescence techniques have been applied.

6 Mössbauer Spectroscopy

contributed by C. E. Johnson

Introduction.—Research has continued in several laboratories on the investigation of the state of iron in biological molecules using the ^{57}Fe Mössbauer effect. The work may be classified as (*a*) haem proteins, (*b*) iron–sulphur proteins, and (*c*) other molecules.

The literature in this field is listed in 'Index of Publications in Mössbauer Spectroscopy of Biological Materials', compiled by Leopold May, Department of Chemistry, The Catholic University of America, Washington D.C. 20017, U.S.A. Additions to the Index are issued in January and June. In 1972 about 40 entries were noted. There is also a section on 'Biological Compounds' in the 'Mössbauer Effect Data Index' by John G. Stevens and Virginia G. Stevens, published by Hilger, London. Volumes covering publications appearing in the years 1969, 1970, and 1971 have been issued; the 1972 volume is expected later in 1973.

[249] J. A. D'Anna and G. Tollin, *Biochemistry*, 1972, **11**, 1073.

Haem Proteins.—Further studies of iron(III) haemoglobin (methaemoglobin) derivatives have been carried out by Winter *et al.*[250] In addition to measurements on the fluoride, cyanide, and azide (which had previously been investigated by Lang and Marshall[251]), data on the imidazole, cyanate, and methanethiol derivatives were presented. The measurements were carried out at temperatures between 4 and 195 K, both in the presence and absence of magnetic fields. In some cases (*e.g.* Met Hb CN, which is pure low-spin Fe^{III}) measurements at the higher temperatures allow the spin state to be determined. Generally, however, the changes in isomer shift from compound to compound and from one spin state to another were small, and the spin state is most directly determined from observations of magnetic hyperfine interactions in Mössbauer spectra measured at 4.2 K in a magnetic field. It is shown that the Mössbauer spectra of methaemoglobin derivatives of intermediate magnetic moment (*e.g.* Met Hb NCO and Met Hb–methanethiol complex) may be approximated by the addition of appropriate amounts of the spectra of a purely high-spin and a purely low-spin derivative, *i.e.* the Mössbauer spectra enable the high-spin–low-spin equilibrium to be observed directly.

The effects of ligand nuclear magnetic moments on the Mössbauer spectra of some iron–porphyrin complexes have been studied by Oosterhuis and Viccaro.[252]

The magnetic hyperfine splitting observed at low temperatures in the Mössbauer spectra of haemoglobins may be highly complex in zero applied magnetic field. The splitting is observable because the electron spins of the iron relax slowly, but they interact with the nuclear magnetic moments of the near neighbours (presumably the nitrogens of the haem group), as well as their own ^{57}Fe nucleus which is being observed with the Mössbauer effect. In acid metmyoglobin at pH 6 this interaction may be well represented by an effective randomly oriented field of *ca.* 9 G at the ^{57}Fe nuclei. When an external magnetic field larger than this is applied (as in the measurements of Winter *et al.*[250]) the effects of the ligand nuclei become negligible, and a relatively simple and symmetrical hyperfine spectrum results.

Winterhalter *et al.*[253] have studied the electronic structure of iron in haemoglobin Zurich β (63 His $\rightarrow$ Arg) and they showed by Mössbauer spectroscopy that, even in the deoxy form, the two haem irons on the abnormal β-chains were low-spin.

A series of measurements have been made on haemin derivatives.[254] These include α-, β-, γ-, and δ-tetraphenylporphine, oxo-bridged Fe^{III} porphines, and α-, β-, γ-, and δ-tetra-arylporphines.

250 M. R. C. Winter, C. E. Johnson, G. Lang, and R. J. P. Williams, *Biochim. Biophys. Acta*, 1972, **263**, 515.
251 G. Lang and W. Marshall, *Proc. Phys. Soc.*, 1966, **87**, 3.
252 W. T. Oosterhuis and P. J. Viccaro, *Biochim. Biophys. Acta*, 1972, **264**, 11.
253 K. H. Winterhalter, E. E. Di Iorio, J. G. Beetlestone, J. B. Kushimo, H. Uebelhack, H. Eicher, and A. Mayer, *J. Mol. Biol.*, 1972, **70**, 665.
254 C. Maricondi, D. K. Straub, and L. M. Epstein, *J. Amer. Chem. Soc.*, 1972, **94**, 4157; M. A. Torrens, D. K. Straub, and L. M. Epstein, *ibid.*, pp. 4160, 4162.

Iron–Sulphur Proteins.—Münck *et al.*[255] have measured the ^{57}Fe Mössbauer spectra of putidaredoxin, a 2Fe–2S protein from *Pseudomonas putida*, and also of its selenium analogue. The spectra of the reduced proteins measured in strong magnetic fields showed that the spins on the two iron atoms (one Fe^{3+} with $S_a = \frac{5}{2}$ and one Fe^{2+} with $S_b = 2$) couple antiferromagnetically to a total spin $S = \frac{1}{2}$. This is similar to results obtained on other two-iron iron–sulphur proteins (plant ferredoxins, adrenodoxin, *etc.*) and confirms the model proposed for the two-iron centre by Gibson *et al.*[256] Values of the magnetic and electric hyperfine tensors are given.

A MO calculation of the electron distribution, magnetic moment, and electric field gradient of the active-site complex of two-iron ferredoxins has been carried out by Loew and Steinberg.[257] They claim that the antiferromagnetic coupling between the pairs of Fe^{3+} atoms arises naturally from electron delocalization.

Mössbauer spectroscopy has been used to characterize the ferredoxin from the blue-green alga *Microcystis flos-aquae.*[258] The spectra were similar to those of other two-iron ferredoxins.

Mössbauer-effect measurements of ^{57}Fe in rubredoxin from *Chloropseudomonas ethylica* and from *Clostridium pasteurianum* have been made.[259] Rubredoxins are proteins containing one iron atom in a tetrahedral sulphur environment whereas the plant-type ferredoxins contain two such units antiferromagnetically coupled together. The study of the hyperfine interaction tensors both in the oxidized (Fe^{3+}) and reduced (Fe^{2+}) state is, therefore, valuable in trying to understand the reduced two-iron proteins which contain a coupled Fe^{3+}–Fe^{2+} active centre. In the oxidized state the Fe^{3+} has an isotropic hyperfine field of (-370 ± 3) kG. In the reduced state the Fe^{2+} hyperfine interaction was measured from the Mössbauer spectrum in an applied magnetic field; the hyperfine field is anisotropic with a component perpendicular to the symmetry axis of the ion of -200 kG. The sign of the electric field gradient in the reduced protein is negative, *i.e.* the ground state of the Fe^{2+} is a d_{z^2} orbital. There is a large non-cubic ligand-field splitting ($\Delta/k \approx 900$ K), and a small spin–orbit splitting (DS_{z^2} where $D \approx +4.4$ cm^{-1}) of the Fe^{2+} levels. The contributions from core polarization to the hyperfine field in the Fe^{3+} and Fe^{2+} atoms were estimated to be -370 and -300 kG, respectively.

Other Molecules.—As well as being used to study the electronic state in biological molecules, the Mössbauer effect is becoming increasingly used to study biochemical reactions in which iron is involved.

255 E. Münck, P. G. Debrunner, J. C. M. Tsibris, and I. C. Gunsalus, *Biochemistry*, 1972, **11**, 855.
256 J. F. Gibson, D. O. Hall, J. H. M. Thornley, and F. R. Whatley, *Proc. Nat. Acad. Sci. U.S.A.*, 1966, **56**, 987.
257 G. H. Loew and D. A. Steinberg, *Theor. Chim. Acta*, 1972, **26**, 107.
258 K. K. Rao, R. V. Smith, R. Cammack, M. C. W. Evans, D. O. Hall, and C. E. Johnson, *Biochem. J.*, 1972, **129**, 1159.
259 K. K. Rao, M. C. W. Evans, R. Cammack, D. O. Hall, C. L. Thompson, P. J. Jackson, and C. E. Johnson, *Biochem. J.*, 1972, **129**, 1063.

Bock and Lang [260] made measurements of iron chelated by deferoxamine. Zabinski *et al.*[261] have made kinetic and Mössbauer studies on the mechanisms of protocatechuic acid 4,5-oxygenase.

7 Nuclear Magnetic Resonance

contributed by H. W. E. Rattle

New Techniques.—An increasing flow of papers on the biological applications of n.m.r. now bears witness to the power of the techniques already in use; among advances just beginning to make an impact, the new generation of Fourier Transform (FT) machines holds out the promise of still greater things to come. Apart from its vastly increased effective sensitivity, the pulsed spectrometer, with its associated on-line computer, offers opportunities for the storage and manipulation of data in many ways hitherto impossible, and in addition usually has a full range of pulse-sequence methods for measuring the relaxation times T_1 and T_2. Several of the papers mentioned in this brief review describe work done on FT machines.

Among the most interesting of the developments in technique described in 1972 is the application of homonuclear inter-nuclear double resonance (INDOR) to amino-acids and peptides.[262, 263] This technique appears similar to the well-used spin-decoupling method of detecting spin-coupled nuclei, but with the important difference that instead of holding the irradiating (decoupling) frequency constant and sweeping the measuring frequency it is the measuring frequency that is held constant, monitoring the absorption at a given point, while the decoupling frequency is swept through the spectrum. The result is a spectrum which contains peaks (positive or negative) corresponding only to transitions having an energy level in common with the monitored transition. As an example,[263] the ^{1}H n.m.r. spectrum of gramicidin S-A contains the overlapping resonances of five different amino-acids, but by monitoring the frequency of the valine C_α proton, the C_β proton multiplets of the valine residues appear as the only signals in the INDOR spectrum, permitting a complete analysis of the coupling constants and a precise determination of the chemical shift of the valine C_β resonance, valuable information which would otherwise have been completely inaccessible for conformational analysis of that residue. The implications for the analysis of protein spectra are obvious. Another decoupling method, this time for the αC-protons hidden under the ubiquitous water solvent peak, has been described.[264]

260 J. L. Bock and G. Lang, *Biochim. Biophys. Acta*, 1972, **264**, 245.
261 R. Zabinski, E. Münck, P. M. Champion, and J. M. Wood, *Biochemistry*, 1972, **11**, 3212.
262 W. A. Gibbons, H. Alms, J. Sogn, and H. R. Wyssbrod, *Proc. Nat. Acad. Sci. U.S.A.*, 1972, **69**, 1261.
263 W. A. Gibbons, H. Alms, R. S. Bockman, and H. R. Wyssbrod, *Biochemistry*, 1972, **11**, 1721.
264 J. Dadok, P. H. Von Dreele, and H. A. Scheraga, *J.C.S. Chem. Comm.*, 1972, 1055.

Amino-acids and Small Synthetic Peptides.—Particular interest has been shown this year in the unusual proline residue and peptides containing it. Studies on *N*-acyl-prolines,[265] on the effect of proline on the ^{13}C chemical shift of its neighbours,[266] and on the *cis–trans* isomerism of proline-containing dipeptides [267] have been reported, while cyclic peptides containing proline have also come in for attention. Papers on the cyclic pentapeptides *cyclo*(-Gly-L-Ala-Gly-Gly-L-Pro-),[268] related pentapeptides,[269] and on some cyclic hexapeptides [270–272] have all shown how n.m.r. can detect and characterize the multiple conformational states of such systems. The effect of a glycine residue on the spectrum of a neighbour in a dipeptide [273] should also add to the knowledge needed for fuller analysis of natural peptides, while data have also been published on the correlation times of glycine and glycylglycine molecules.[274]

Another amino-acid of particular interest to the investigator of protein structure and function is histidine. A paper [275] which gives accurate values for the ionization constants of histidine in various molecular situations and which uses computer curve-fitting and a mathematical model to improve the analysis of the data should be useful here. Also of interest is work on the binding of cobalt(III) to histidine [276] and the binding of Gly-His-Gly to DNA.[277] The dissociation of cysteine with changes in pH has been discussed.[278]

Synthetic Polypeptides.—Studies have been reported on the conformation or conformational transitions of poly Pro-Gly and poly Gly-Gly-Pro-Gly,[279] of block copoly DL-Lys-L-Ala-DL-Lys,[280] poly-sarcosine,[281] poly-(*N*-δ-carbobenzoxy-L-ornithine),[282] poly-L-tyrosine,[283] poly-γ-benzyl-L-glu-

[265] W. A. Thomas and M. K. Williams, *J.C.S. Chem. Comm.*, 1972, 788.
[266] M. Christl and J. D. Roberts, *J. Amer. Chem. Soc.*, 1972, **94**, 4565.
[267] W. A. Thomas and M. K. Williams, *J.C.S. Chem. Comm.*, 1972, 994.
[268] K. Wuthrich, A. Tun-Kyi, and R. Schwyzer, *F.E.B.S. Letters*, 1972, **25**, 104.
[269] J. P. Meraldi, R. Schwyzer, A. Tun-Kyi, and K. Wuthrich, *Helv. Chim. Acta*, 1972, **55**, 1962.
[270] D. A. Torchia, A. Di Corato, S. C. K. Wong, C. M. Deber, and E. R. Blout, *J. Amer. Chem. Soc.*, 1972, **94**, 609.
[271] D. A. Torchia, S. C. K. Wong, C. M. Deber, and E. R. Blout, *J. Amer. Chem. Soc.*, 1972, **94**, 616.
[272] F. A. Bovey, A. I. Brewster, D. J. Patel, A. E. Tonelli, and D. A. Torchia, *Accounts Chem. Res.*, 1972, **5**, 193.
[273] M. Scheinblatt, *J. Magn. Resonance*, 1972, **8**, 55.
[274] J. P. Behr and J. M. Lehn, *J.C.S. Perkin II*, 1972, 1488.
[275] R. I. Schrager, J. S. Cohen, S. R. Heller, D. H. Sachs, and A. N. Schechter, *Biochemistry*, 1972, **11**, 541.
[276] S. Bagger, K. Gibson, and C. S. Sørensen, *Acta Chem. Scand.*, 1972, **26**, 2503.
[277] H. Fritzsche, *F.E.B.S. Letters*, 1972, **23**, 105.
[278] L. Flohe, E. Breitmaier, W. A. Guenzler, W. Voelter, and G. Jung, *Z. physiol. Chem.*, 1972, **353**, 1159.
[279] D. A. Torchia, *Biochemistry*, 1972, **12**, 1462.
[280] J. C. Howard and H. A. Scheraga, *Macromolecules*, 1972, **5**, 328.
[281] M. Sisido, Y. Imanishi, and T. Higashimura, *Biopolymers*, 1972, **11**, 399.
[282] G. Boccalon, A. S. Verdini, and G. Giacometti, *J. Amer. Chem. Soc.*, 1972, **94**, 3639.
[283] E. M. Bradbury, C. Crane-Robinson, V. Giancotti, and R. M. Stephens, *Polymer*, 1972, **13**, 33.

tamate by ^{13}C resonance,[284] and on the binding of hexachlorophene to poly-γ-benzyl-L-glutamate and poly-methionine.[285]

Small Natural Peptides.—The conformations of small naturally occurring peptide molecules in solution provide a rich field for n.m.r. analysis; the structures are complex, but their n.m.r. spectra lack some of the peak overlap which makes the analysis of larger protein spectra so difficult. An example of the variety of n.m.r. methods which may be used is found in the papers published this year on the gramicidins. These include chemical-shift changes induced by solvent change,[286] the use of a probe group (in this case a nitroxide radical, which can help distinguish internal from solvent-exposed peptide protons),[287] and the use of ring-current shifts, NH–αCH couplings, and hydrogen–deuterium exchange rates to support the presence of $\pi_{L,D}$ helices in the gramicidin structure.[288, 289] Similar approaches have been applied to the structure and the effect of structural modifications of lysine vasopressin,[290] a hormone whose resonances have been assigned by techniques including spin-decoupling.[291] Other similar molecules which have been investigated by n.m.r. include oxytocin,[292] valinomycin,[293] ferrichromes,[294] a pentapeptide of gastrin,[295] viomycin,[296] glutathione,[297] and some interesting studies employing the Overhauser effect on bovine neurophysin II.[298, 299]

Iron-containing Proteins.—The presence of the iron atom, with its variable oxidation and spin states, coupled with the considerable biological importance of molecules such as haemoglobin (Hb) and the cytochromes, have resulted in a considerable number of n.m.r. studies. An additional attraction is, of course, the ready availability of molecules differing at various known sites. Again, approaches have been made from several directions;

284 L. Paolillo, T. Tancredi, P. A. Temussi, E. Trivellone, E. M. Bradbury, and C. Crane-Robinson, *J.C.S. Chem. Comm.*, 1972, 335.
285 R. Hague and D. R. Buhler, *J. Amer. Chem. Soc.*, 1972, **94**, 1824.
286 T. P. Pitner and D. W. Urry, *J. Amer. Chem. Soc.*, 1972, **94**, 1399.
287 K. D. Kopple and T. J. Schamper, *J. Amer. Chem. Soc.*, 1972, **94**, 3644.
288 J. D. Glickson, D. F. Mayers, J. M. Settine, and D. W. Urry, *Biochemistry*, 1972, **11**, 477.
289 D. W. Urry, *Biochemistry*, 1972, **11**, 487.
290 J. D. Glickson, D. W. Urry, R. T. Havran, and R. Walter, *Proc. Nat. Acad. Sci. U.S.A.*, 1972, **69**, 2136.
291 P. H. Von Dreele, A. I. Brewster, J. Dadok, H. A. Scheraga, F. A. Bovey, M. F. Forger, and V. Du Vigneaud, *Proc. Nat. Acad. Sci. U.S.A.*, 1972, **69**, 2169.
292 R. Deslauriers, R. Walter, and I. C. P. Smith, *Biochem. Biophys. Res. Comm.*, 1972, **48**, 854.
293 M. Ohnishi, M. C. Fedarko, J. D. Baldeschwieler, and L. F. Johnson, *Biochem. Biophys. Res. Comm.*, 1972, **46**, 312.
294 M. Llinas, M. P. Klein, and J. B. Neilands, *J. Mol. Biol.*, 1972, **68**, 265.
295 J. Feeney, G. C. K. Roberts, J. P. Brown, A. S. V. Burgen, and H. Gregory, *J.C.S. Perkin II*, 1972, 601.
296 P. Viglino, C. Franconi, A. Lai, E. Brosio, and F. Conti, *Org, Magn. Resonance*, 1972, **4**, 237.
297 G. Jung, E. Breitmaier, and W. Voelter, *European J. Biochem.*, 1972, **24**, 438.
298 P. Balaram, A. A. Bothner-By, and E. Breslow, *J. Amer. Chem. Soc.*, 1972, **94**, 4017.
299 P. Balaram, A. A. Bothner-By, and J. Dadok, *J. Amer. Chem. Soc.*, 1972, **94**, 4015.

the titration of histidine residues in haemoglobin has been followed for different oxidation states of the molecule,[300, 301] and an accurate value for the pK_a of His β 146 has been obtained by following the ^{19}F resonance of a specific label in deoxyHb.[302] The same probe technique has also given information on tertiary structure changes in Hb.[303] The linkage between the α and β chains of Hb is another phenomenon of general interest which may be investigated by n.m.r., and reports of the use of ^{13}C resonances,[304] ligand binding at the α-haems,[305] ring-current shifts,[306] and the differences between the spectra of different forms of Hb[307] all contribute to the discussion on the origin and significance of this linkage.

A similar variety of approaches is found in recent work on the cytochromes. Results obtained include evidence that the redox carrier ability of modified mammalian cytochromes *c* is retained only if a methionine is bound to one of the axial positions of the Fe,[308] and the conclusion that the exchange of oxidation states by electron transfer in mixed solutions of the two oxidation states of cytochrome *c* is governed simply by the binary collisions of the molecules.[309] Other papers indicate the presence of high-spin iron in both mammalian[310] and bacterial[311] cytochromes, and the presence of a highly polar binding site for cytochrome on mitochondrial membranes.[312]

Determination of the correlation time τ for ferredoxin molecules[313] has allowed a calculation of the dipolar contribution to the linewidth of contact-shifted resonances. Once this parameter is known, the distance of the contact-shifted protons from the paramagnetic centre in the molecule may be estimated.

Other Proteins.—The ribonuclease molecule has been the subject of studies involving its denaturation,[314, 315] the binding of Cu^{2+} to the whole molecule and its S-peptide,[316] the binding of the S-peptide to the S-protein,[317]

300 H. Sick, K. Gersonde, J. C. Thompson, W. Maurer, W. Haar, and H. Rueterjans, *European J. Biochem.*, 1972, **29**, 217.
301 N. J. Greenfield and M. N. Williams, *Biochim. Biophys. Acta*, 1972, **257**, 187.
302 W. H. Huestis and M. A. Raftery, *Proc. Nat. Acad. Sci. U.S.A.*, 1972, **69**, 1887.
303 W. H. Huestis and M. A. Raftery, *Biochemistry*, 1972, **11**, 1648.
304 R. B. Moon and J. H. Richards, *Proc. Nat. Acad. Sci. U.S.A.*, 1972, **69**, 2193.
305 T. R. Lindstrom, C. Ho, and A. V. Pisciotta, *Nature New Biol.*, 1972, **237**, 263.
306 T. R. Lindstrom, I. B. E. Noren, S. Charache, H. Lehmann, and C. Ho, *Biochemistry*, 1972, **11**, 1677.
307 K. Winterhalter and K. Wuthrich, *J. Mol. Biol.*, 1972, **63**, 477.
308 K. Wuthrich, A. Aviram, and A. Schejter, *Biochim. Biophys. Acta*, 1971, **253**, 98.
309 R. K. Gupta, S. H. Koenig, and A. G. Redfield, *J. Magn. Resonance*, 1972, **7**, 66.
310 R. M. Keller, I. Aviram, A. Schejter, and K. Wuthrich, *F.E.B.S. Letters*, 1972, **20**, 90.
311 R. M. Keller, K. Wuthrich, and P. G. Debrunner, *Proc. Nat. Acad. Sci. U.S.A.*, 1972, **69**, 2073.
312 A. Azzi, A. M. Tamburro, G. Frania, and E. Gobbi, *Biochim. Biophys. Acta*, 1972, **256**, 619.
313 I. Salmeen and G. Palmer, *Arch. Biochim. Biophys.*, 1972, **150**, 767.
314 J. H. Bradbury, N. L. R. King, and J. M. O'Shea, *Internat. J. Peptide Protein Res.*, 1972, **4**, 257.
315 J. H. Bradbury and N. L. R. King, *Austral. J. Chem.*, 1972, **25**, 209.
316 M. Ihnat, *Biochemistry*, 1972, **11**, 3483.
317 F. M. Finn, J. Dadok, and A. A. Bothner-By, *Biochemistry*, 1972, **11**, 455.

deuterium exchange,[318] and some assignments of its ^{13}C spectrum.[319] Indications of a very fruitful future are to be found in studies on lysozyme[320, 321] in which the use of a ^{19}F-labelled inhibitor is combined with that of a lanthanide shift reagent to compare the structure in solution with that in the crystal form. Again, subtlety of technique holds the key to overcoming the problems of peak overlap in n.m.r. spectroscopy. Another interesting use of a ^{19}F probe molecule in lysozyme,[322] utilizing a comparison of ^{19}F and ^{1}H chemical shifts in a probe ring containing a $-CH_2F$ group, enables the separation of ring-current effects from electrostatic and van der Waals shifts and the subsequent detection of structural changes of about 1 Å. Other studies which must perforce be simply listed have been carried out on substrate binding to lysozyme,[323] the stereospecificity of 20β-hydroxy-steroid dehydrogenase,[324] interaction between the subunits of horse liver alcohol dehydrogenase,[325] the histidine titration of human and bovine carbonic anhydrases,[326] the relaxation times of bovine carbonic anhydrase and its complexes,[327, 328] and on aspartate transcarbamylase,[329] aspartate transaminase,[330] pyruvate kinase,[331, 332] myosin,[333] and α-chymotrypsin.[334–336]

The histones continue to fulfil their promise of interesting results from n.m.r. studies. The interactions of whole and cleaved histones have been reported[337, 338] – the latter paper using a simple computer-aided method to generate analogue spectra for use in analysing spectral changes observed when histone–histone interactions take place. Another particularly interesting result has been the recording and analysis of the spectrum of

[318] D. N. Silverman, D. Kotelchuck, G. T. Taylor, and H. A. Scheraga, *Arch. Biochem. Biophys.*, 1972, **150**, 757.
[319] V. Glushko, P. J. Lawson, and F. R. N. Gurd, *J. Biol. Chem.*, 1972, **247**, 3176.
[320] C. G. Butchard, R. A. Dwek, P. W. Kent, R. J. P. Williams, and A. V. Xavier, *European J. Biochem.*, 1972, **27**, 548.
[321] C. G. Butchard, R. A. Dwek, S. J. Ferguson, P. W. Kent, R. J. P. Williams, and A. V. Xavier, *F.E.B.S. Letters*, 1972, **25**, 91.
[322] F. Millett M. and A. Raftery, *Biochem. Biophys. Res. Comm.*, 1972, **47**, 625.
[323] F. Millett and M. A. Raftery, *Biochemistry*, 1972, **11**, 1639.
[324] E. S. Szymanski, C. S. Furfine, and C. F. Hammer, *Steroids*, 1972, **19**, 243.
[325] B. Lindman, M. Zeppezauer, and A. Akeson, *Biochim. Biophys. Acta*, 1972, **257**, 173.
[326] J. S. Cohen, C. T. Yim, M. Kandel, A. G. Gornall, S. I. Kandel, and M. H. Freedman, *Biochemistry*, 1972, **11**, 327.
[327] A. Lanir and G. Navon, *Biochemistry*, 1972, **11**, 3536.
[328] G. Navon and A. Lanir, *J. Magn. Resonance*, 1972, **8**, 14.
[329] C. H. McMurray, D. R. Evans, and B. D. Sykes, *Biochem. Biophys. Res. Comm.*, 1972, **48**, 572.
[330] M. W. Dybel, S. Cheng, and M. Martinez-Carrion, *Arch. Biochem. Biophys.*, 1972, **148**, 320.
[331] T. Nowak and A. S. Mildvan, *Biochemistry*, 1972, **11**, 2813.
[332] A. S. Mildvan, *Biochemistry*, 1972, **11**, 2819.
[333] R. G. Bryant, Y. Legler, and H. Moon Han, *Biochemistry*, 1972, **11**, 3846.
[334] J. T. Gerig and R. A. Rimerman, *J. Amer. Chem. Soc.*, 1972, **94**, 7549.
[335] J. T. Gerig and R. A. Rimerman, *J. Amer. Chem. Soc.*, 1972, **94**, 7558.
[336] J. T. Gerig and R. A. Rimerman, *J. Amer. Chem. Soc.*, 1972, **94**, 7565.
[337] E. M. Bradbury, P. D. Cary, C. Crane-Robinson, P. L. Riches, and E. W. Johns, *European J. Biochem.*, 1972, **26**, 482.
[338] E. M. Bradbury and H. W. Rattle, *European J. Biochem.*, 1972, **27**, 270.

one histone fraction in deoxyribonucleoprotein gel,[339] which despite its apparent high viscosity has sufficient mobility to give good protein spectra.

Water in Biological Systems.—One of the earliest uses for n.m.r. with biological systems continues to appear in the literature. The states of water in muscle,[340–342] in ribosomes,[343] in collagen,[344] and in whole nerve [345] have all been reported in 1972.

8 Infrared and Raman Spectroscopy

contributed by R. M. Stephens

Model Compounds.—The influence of substituents upon the amide I and amide A frequencies has been analysed for ten amides substituted by aliphatic side-chains.[346] The behaviour of model amides in different solvents shows that the nature of the substituents has no influence upon the proton-donor or proton-acceptor power of the amido-group. Information is, however, provided about very local interactions and bond conformation which bears on polypeptide conformation. According to the i.r. spectra obtained from the heated forms of nine cyclic hexapeptides with potassium bromide, their conformations were all considered to be a single folded-leaf structure, compared with a mixture of different conformations at room temperature.[347] Several differences were observed in the i.r. spectra obtained from Ac-L-Ala-L-Phe-OMe, Ac-L-Ala-D-Phe-OMe, and Ac-(LAla)$_n$-OMe, suggesting that two conformations could be formed.[348] These spectral differences are related to configuration, sequence, and chain length in the peptides.

From the polarized i.r. spectra of oriented single crystals of L-alanine, the dichroic behaviour of bands between 1615 and 320 cm^{-1} was analysed and correlated with a C_2 molecular point-group.[349] The majority of the assignments previously made were consistent with this model, but eight reassignments based on band dichroisms were suggested.

Polypeptide Conformation.—The conformational behaviour of poly-L-tyrosine in dimethyl sulphoxide–water and dimethyl sulphoxide–trimethyl phosphate solvents has been studied using i.r. and n.m.r. spectroscopy.[350]

339 E. M. Bradbury, B. G. Carpenter, and H. W. Rattle, *Nature*, 1972, **241**, 123.
340 W. Derbyshire and J. L. Parsons, *J. Magn. Resonance*, 1972, **6**, 344.
341 F. W. Cope, *Nature New Biol.*, 1972, **237**, 215.
342 M. M. Civan and M. Shporer, *Biophys. J.*, 1972, **12**, 404.
343 J. P. White, I. D. Kuntz, and C. R. Cantor, *J. Mol. Biol.*, 1972, **64**, 511.
344 Z. T. Lalowiz and M. Remin, *Inst. Nucl. Phys. Cracow Rep.*, 1972, INP-790, IPS.
345 P. Dea, S. I. Chan, and F. J. Dea, *Science*, 1972, **157**, 206.
346 F. Fillaux and C. de Lozé, *Biopolymers*, 1972, **11**, 2063.
347 G. A. Kogan, V. M. Tul'chinskii, V. V. Shilin, and V. T. Ivanov, *Khim. prirod. Soedinenii*, 1972, 361.
348 V. M. Tul'chinskii, G. A. Kogan, and A. T. Miroshnikov, *Khim. prirod. Soedinenii*, 1973, 353.
349 R. Adamowicz and E. Fishman, *Spectrochim. Acta*, 1972, **A28**, 889.
350 E. M. Bradbury, C. Crane-Robinson, V. Giancotti, and R. M. Stephens, *Polymer*, 1972, **13**, 33.

The results suggest that in dimethyl sulphoxide the polymer has a random-coil conformation and develops a partially right-handed α-helical conformation when 20% deuterium oxide or 50% trimethyl phosphate is added to the solution.

Infrared spectra have been obtained from poly-L-proline in aqueous calcium chloride solutions.[351] As the salt concentration is increased the carbonyl stretching band develops a component at the frequency found in the solid state, while the CH_2 bending bands broaden. These results are correlated with c.d. measurements which indicate progressive disordering of the chain with increasing salt concentration. The disordering is associated primarily with an increase in the range of C_α—C=O rotation angles rather than with the random introduction of *cis* amide bonds in the chain. However, other near-i.r. studies of poly-L-proline in concentrated aqueous salt solutions suggest that further absorption bands associated with the carbonyl and methylene absorptions indicate that both *cis* and *trans* amide bonds are present. Further, an increase in the temperature shifts the equilibrium in favour of more *cis* amide bonds.[352]

When polarized i.r. spectra are obtained from oriented helical polypeptides it is possible to calculate the angle between the transition moment associated with the vibration and the helix axis, thus giving detailed information about molecular arrangements. The side-chain orientations in poly-L-aspartate esters have been investigated using polarized i.r. spectroscopy.[353] The results have shown that in poly-β-benzyl-L-aspartate, prepared in either the right-handed or left-handed α-helical form, the side-chain conformations are different and dependent upon the helix sense. All other right-handed poly-L-aspartate esters have similar side-chain conformations to that in right-handed poly-β-benzyl-L-aspartate. However, the conformation of side-chains in left-handed α-helical poly-L-aspartate esters is dependent upon the length of the side-chain.

Hydration.—The state of water adsorbed on to proteins is a matter of considerable interest, but the interpretation of the data in complex systems has often proved difficult. Polarized i.r. techniques have been used to examine the water adsorbed on to α-helical poly-L-alanine.[354] The results indicate that the adsorbed water has a preferred orientation which is independent of the relative humidity over the range 0—100%. An attempt has been made to show that water is adsorbed through a hydrogen-bond mechanism *via* the carbonyl oxygen and by non-bonded dipole–dipole interaction. Other studies on how bound water affects the peptide group of globular proteins and their state of aggregation are reported by Buontempo *et al.*[355]

[351] N. Johnston and S. Krimm, *Biopolymers*, 1971, **10**, 2597.
[352] C. A. Swenson, *Biopolymers*, 1971, **10**, 2591.
[353] E. M. Bradbury, G. B. Carpenter, and R. M. Stephens, *Macromolecules*, 1972, **5**, 8.
[354] C. B. Baddiel, M. M. Breuer, and R. M. Stephens, *Colloid Interface Sci.*, 1972, **40**, 429.
[355] U. Buontempo, G. Coreri, and P. Fasella, *Biopolymers*, 1972, **11**, 519.

Conformation.—Theoretical calculations of the vibrational frequencies of groups often aid the analysis of an i.r. spectrum of a protein, and a normal co-ordinate treatment for cyclic tetra-L-alanine has allowed a comparison of the amide modes and certain low-frequency vibrations that are sensitive to conformational changes to be made.[356] Intramolecular coupling coefficients have been calculated using a perturbation treatment. A complete normal co-ordinate analysis of glycylglycine and its *N*-deuteriated analogue has been made using a Urey–Bradley force field. This, together with i.r. absorption spectroscopy and inelastic neutron scattering studies, has made assignments in the low-frequency region possible.[357] A different theoretical approach is to use a computer for curve-fitting analyses of i.r. spectra. This procedure has been used for the analysis of the amide I band in the i.r. spectra of isolated or membrane-bound proteins dissolved or suspended in deuterium oxide.[358] It yields the percentage of β-structure and the sum of α-helical and unordered structures. The basic information given to the computer was calculated from data obtained from lysozyme, ribonuclease, myoglobin, and serum albumin.

For a long time the frequency and dichroism of the main amide bands have been used to determine the conformations of polypeptides, and the knowledge gained from this work has been used to interpret the spectra obtained from proteins. The presence of the antiparallel β-pleated-sheet conformation in isolated human amyloid protein fibrils was confirmed using i.r. spectroscopy.[359] In most amyloid samples this conformation was enhanced under acidic conditions. While the amide I and amide V bands observed from a film of immunoglobulin indicated little sign of β-structure being present, Bence-Jones proteins exhibited some β-pleated-sheet structure upon acid or heat treatment. Whole light or heavy chains can assume β-structure under conditions of acid or heat, but it is the variable regions which are most easily capable of assuming a solid-state, antiparallel β-pleated-sheet conformation. This evidence was taken to be consistent with the suggestion that amyloid fibrils originated from the variable regions of the immunoglobulin molecule.

The far-i.r. spectra, between 40 and 400 cm^{-1}, of films of globular proteins (lysozyme, myoglobin, serum albumin, ribonuclease, chymotrypsinogen, and subtilisin), nylon 66, and poly-γ-benzyl-L-glutamate have been investigated using a Michelson interferometer.[360] While polypeptides are known to present several peaks in this region which can be assigned mostly to hydrogen-bonding modes, all the globular proteins investigated displayed only one broad intense band between 100 and 200 cm^{-1} which remains after denaturation or partial digestion. The hydrogen–deuterium exchange in aqueous solutions of defatted human serum albumin can easily

[356] V. D. Gupta and R. D. Singh, *Biopolymers*, 1972, **11**, 2099.
[357] A. M. Dioiverdi and V. D. Gupta, *Biopolymers*, 1972, **11**, 2091.
[358] R. Grosse, J. Malur, and K. R. H. Repke, *Acta Biol. Med.*, 1972, **28**, 11.
[359] J. D. Termine, E. D. Eanes, D. Ein, and G. G. Glenner, *Biopolymers*, 1972, **11**, 1103.
[360] U. Buontempo, G. Careri, P. Fasella, and A. Ferraro, *Biopolymers*, 1971, **10**, 2377.

be followed using i.r. techniques. At pH 7 and 25 °C, where maximum stability of defatted human serum albumin occurred, less than 20% of the peptide groups were protected from solvent exposure.[361] Attenuated total reflection spectra have been obtained from human skin *in situ*.[362] The reproducibility of the technique to study normal skin from one individual was better than 5% for those absorptions which were not influenced by moisture and by hydrogen-bonding, and differences between individuals, particularly of different ages, could be detected. Structural studies on insulin fibrils heated in acid, using i.r., *X*-ray, and electron microscopy techniques, have shown that the fibres form uniform cross β-structures of cross-section 29 Å × 47 Å.[363] A detailed structure was suggested in which individual insulin molecules were stacked in layers 4.7 Å thick in the direction of the fibril axis with one molecule per layer. Interaction of the extended chains in adjacent layers was believed to be of the parallel pleated-sheet type.

Raman Spectra.—Although a great deal of structural information can be obtained from the i.r. spectra of proteins, especially when polarized i.r. radiation is used with oriented samples, certain information may be lacking because some molecular vibrations do not have a changing dipole moment and are thus not i.r.-active. However, the molecular vibration may be Raman-active instead, if a change in electrical polarizability occurs during the vibration, and Raman spectroscopy is often used as a complementary technique to i.r. spectroscopy of proteins.

The Raman spectra of aqueous bovine serum albumin and β-lactoglobulin have been recorded in the range 1600—500 cm^{-1}. Each spectrum contains about thirty lines and nearly half of these were assigned to functional groups of the constituent amino-acids. From the intensities and positions of the amide III lines near 1250 cm^{-1}, it appears that β-lactoglobulin is predominantly random-coil. The amide III lines of bovine serum albumin are less clear-cut but are consistent with a higher proportion of ordered structure than is found in β-lactoglobulin.[364] The Raman and i.r. spectra of *N*-acetylglycine have shown that there are 37 active Raman lines and only 33 i.r. absorptions in the fundamental region.[365] Assignment of the observed frequencies was made by comparing them with the spectra of other similar compounds, and the origin of two broad bands in the O—H stretching region is discussed. Similar spectra have been recorded from single crystals of L- and DL-glutamic acid hydrochloride.[366] In the L-compound 56 Raman lines were identified and assigned. The corresponding DL-compound had 49 bands.

[361] A. Hvidt and K. Wallervik, *J. Biol. Chem.*, 1972, **247**, 1530.
[362] N. A. Puttnam, *J. Soc. Cosmetic Chemists*, 1972, **23**, 209.
[363] M. J. Burke and M. A. Rougise, *Biochemistry*, 1972, **11**, 2435.
[364] A. M. Bollocq, R. C. Lord, and R. Mendelsohn, *Biochim. Biophys. Acta*, 1972, **257**, 280.
[365] V. N. Sankaranarayanan and R. S. Krishnan, *Indian J. Pure Appl. Phys.*, 1972, **10**, 378.
[366] V. N. Sankaranarayanan and R. S. Krishnan, *Indian J. Pure Appl. Phys.*, 1972, **10**, 382.

Raman spectroscopy is a low-sensitivity technique, so that its advantage over i.r. of relative insensitivity to solvent is often more than outweighed by the high protein concentration required. Resonance Raman spectroscopy,[367] on the other hand, is a technique of much higher sensitivity which may prove a useful structural probe for chromophore-containing proteins in solution. When the exciting wavelength falls within an absorption envelope there follows an interaction of electronic and vibrational transition, with the consequence that certain vibrational modes of the chromophore are greatly enhanced in the spectrum. The nature of any covalent or non-covalent interaction between the chromophore and the protein will be expected to modify the spectrum.

The FeS_4 tetrahedron shown to exist in rubredoxin by *X*-ray crystallography, and which is the optically active chromophore, has been shown by resonance Raman spectroscopy to be present also in aqueous solutions of the protein.[368] Structural modifications, such as the co-ordination of water to iron, are probably ruled out. The spectra for oxyhaemoglobin and methaemoglobin are similar, minor differences being correlated with the out-of-plane distance of the iron atom.[369] The spectrum of ferrocytochrome *c* is also similar but that of ferricytochrome *c* is very different.[370] The basis of this difference, clearly not dependent solely on the spin states of the iron, is at present not completely understood. These spectra are obtained in the concentration range 10^{-5}—10^{-3} mol l^{-1} and the technique is thus potentially useful for proteins with intrinsic or, perhaps, extrinsic chromophores.

9 Dissociation and Association of Proteins

contributed by G. L. Kellett

Analytical Ultracentrifugation Techniques.—*Interferometry*. Interferometry has long been accepted as being the most precise method for measuring the distributions of macromolecules in the ultracentrifuge cell. Significant developments in the design of interference systems are, therefore, to be welcomed, particularly by those interested in the quantitative analysis of interacting protein systems. 1972 saw a major step forward in the realization of the full potential of interferometry with the introduction of the laser.[371, 372]

The immediate advantages of a laser light source are clear:

(1) The coherence length of a laser beam is about three orders of magnitude greater than that of light from the standard AH6 mercury lamp, so

[367] J. Behringer, in 'Raman Spectroscopy', ed. H. A. Szymanski, Plenum Press, New York, 1967, Ch. 6.
[368] T. V. Long, T. M. Loehr, J. R. Alkins, and W. Lovenberg, *J. Amer. Chem. Soc.*, 1971, **92**, 1809.
[369] T. C. Strekas and T. G. Spiro, *Biochim. Biophys. Acta*, 1972, **263**, 830.
[370] T. C. Strekas and T. G. Spiro, *Biochim. Biophys. Acta*, 1972, **278**, 188.
[371] R. C. Williams, jun., *Analyt. Biochem.*, 1972, **48**, 164.
[372] C. H. Paul and D. A. Yphantis, *Analyt. Biochem.*, 1972, **48**, 588, 605.

that the resolution of interference fringes and observable path differences is greatly enhanced. Higher solute concentrations and a wider range of gradients can therefore be studied without compensation for higher refractive index differences.

(2) The diameter of a laser beam is generally *ca.* 1—2 mm, and this may readily be focused to about 50 μm to form an almost ideal point source. The fringe blurring caused normally by finite slit length is thus reduced.

(3) The greatly increased intensities of the laser source permit the use of finer-grain emulsions, which are much slower than the usual IIG plate.

Williams [371] has realized these advantages simply by replacing the AH6 lamp with a 3 mW helium–neon laser. The helium–neon laser emits light of 6328 Å and may therefore be used in the study of haem proteins. In a study of cyanomethaemoglobin, Williams found that the observable concentration range with the AH6 source was seven fringes whereas it was 40 fringes with the laser source.

When an ultracentrifuge cell is held stationary in the light path, the interference fringes obtained using laser and AH6 sources are very similar. When the cell is spinning, however, major differences from the stationary state are seen. Some of these differences arise from the changes in alignment of the slits with respect to the two sectors of the centrepiece during rotation. Thus, there are three orientations in which light may pass through the system, namely two asymmetrical configurations in which one slit is aligned with the opposite channel and one symmetrical configuration in which both slits are aligned with the corresponding channels. Fringes are, of course, formed only in the latter situation. Unfortunately, asymmetrical configurations cause a background darkening of the photographic plate, resulting in a reduction in fringe contrast. Paul and Yphantis [372] have overcome this problem by using Pulsed Laser Interferometry (PLI). They have incorporated a pulsed argon-ion laser, filtered at 5154 Å. The pulse frequency was 60 Hz maximum with a peak power of 0.8 W. The effective aperture of the system is determined by the laser pulse timing and the speed of rotation. For example, a pulse of 6 μs at 60 000 r.p.m. corresponds to a rotational distance of 2—4 mm, which is less than the distance of most centrepiece channels. With appropriate programming, Paul and Yphantis have arranged for the pulse to coincide with the correct symmetrical configuration of slit and channels only, resulting in vastly improved fringes. The use of a pulsed laser system also assists greatly in overcoming problems associated with variation in effective slit width and precession, reduces the need for a cell mask, and permits direct multiplexing without the use of wedge windows.

Narrowing the slits would reduce the problems of the conventional system, but is precluded because of the increased exposure times necessary. This will be possible with a laser system. The design, construction, and placement of a Rayleigh mask have been considered in great detail by

Richards *et al.*,[373] who have described two mask assemblies, one very flexible for investigating the effects of different slit widths and centre-to-centre separations, and one very precise for difference sedimentation experiments.[374] The latter may be aligned laterally to 25 μm and rotationally to 0.1° in a single adjustment by the use of equations based on the geometry of the cells and the mask.[373, 375] De Rosier *et al.* have described the automatic measurement of interference fringes.[376]

Theory. Several investigators have dealt with the complication of non-ideality in the study of interacting systems by sedimentation equilibrium and have described methods for the evaluation of the second virial coefficient.[377–380] In most of these cases a single coefficient is assumed to be applicable to all species. Although consideration of the effects of non-ideality at high protein concentrations is clearly essential, such treatments will be important in unusual solvent conditions at lower concentrations. Munk and Cox have, for example, re-emphasized that the virial coefficients of proteins in concentrated guanidine hydrochloride solutions are about one order of magnitude greater than in dilute buffers.[379]

Cann and Goad[381] have extended their theories[382] on the transport of macromolecular systems interacting with buffer components by taking account of the hydrodynamic dependence of sedimentation coefficient of each macromolecular species in the case of ligand-induced dimerization. Howlett and Nichol have considered an equivalent case for sedimentation equilibrium, the interaction of *N*-acetylglucosamine with lysozyme.[383] Kegeles has shown that the sedimentation coefficient of a differential boundary[384] defines the *Z*-average sedimentation coefficient of the system.[385] If the weight-average coefficient is known, then for a monomer–single higher polymer system, the equilibrium constant may be readily obtained.

Binding Stoicheiometry. A simple method for determining the stoicheiometry of an interacting system has been described by Klee and Klee.[386] The interaction of the A protein of lactose synthetase with α-lactalbumin is

[373] E. G. Richards, J. Bell-Clark, M. Kirschner, A. Rosenthal, and H. K. Schachman, *Analyt. Biochem.*, 1972, **46**, 295.
[374] M. W. Kirschner and H. K. Schachman, *Biochemistry*, 1971, **10**, 1900, 1919.
[375] E. G. Richards, D. C. Teller, V. D. Hoagland, jun., R. H. Haschemeyer, and H. K. Schachman, *Analyt. Biochem.*, 1971, **41**, 215.
[376] D. J. De Rosier, P. Munk, and D. J. Cox, *Analyt. Biochem.*, 1972, **50**, 139.
[377] P. W. Chun, S. J. Kim, J. D. Williams, W. T. Cope, L. H. Fing, and E. T. Adams, jun., *Biopolymers*, 1972, **11**, 197.
[378] L. A. Holladay and A. J. Sophianopoulos, *J. Biol. Chem.*, 1972, **247**, 427.
[379] P. Munk and D. J. Cox, *Biochemistry*, 1972, **11**, 687.
[380] J. Visser, R. Deonier, E. T. Adams, jun., and J. W. Williams, *Biochemistry*, 1972, **11**, 2635.
[381] J. R. Cann and W. B. Goad, *Arch. Biochem. Biophys.*, 1972, **153**, 603.
[382] J. R. Cann, 'Interacting Macromolecules', Academic Press, New York, 1970.
[383] G. J. Howlett and L. W. Nichol. *J. Biol. Chem.*, 1972, **247**, 5681.
[384] G. Kegeles, *Proc. Nat. Acad. Sci. U.S.A.*, 1972, **69**, 2577.
[385] G. L. Kellett, in 'Amino-acids, Peptides, and Proteins', ed. G. T. Young (Specialist Periodical Reports), The Chemical Society, London, 1973, vol. 4, p. 287.
[386] W. A. Klee and C. B. Klee, *J. Biol. Chem.*, 1972, **247**, 2336.

very weak, and it has been difficult to establish directly by physical methods the existence of the complex.[387] In the technique as described by Klee and Klee,[386] the Type I band-forming centrepiece developed by Vinograd[388] was used to layer a small volume of 5—15 μl of the more rapidly sedimenting A protein as a narrow band at the top of a solution of the more slowly sedimenting α-lactalbumin. The experiment was performed in a double-sector cell with both sides filled with the same solution of α-lactalbumin, but with the A protein layered on one side only and an equal volume of buffer on the other side to serve as the blank. Photoelectric scanning absorption optics were employed and the absorption of the α-lactalbumin in both sectors was automatically subtracted by the scanner so that only the band of A protein and any excess α-lactalbumin complexed with it was visible. The sedimentation coefficient of the band increased with complex formation and the area provided a measure of the number of moles of A initially applied and the number of moles of α-lactalbumin complexed. The experiment was repeated as a function of α-lactalbumin concentration at constant A protein concentration and the resulting Scatchard plot indicated that the stoicheiometry was 1 : 1. All band-forming experiments are very economical in their use of the protein layered in the band, which is particularly fortunate in this case, as it is the A protein which is difficult to obtain.[389] Many of the considerations taken into account in the design of active enzyme sedimentation experiments[390] clearly must apply to this experiment. One point which is perhaps worth particular notice, however, is that the band must be stabilized from convective disturbance by a density gradient. In the lactose synthetase study, this was achieved by making the bulk solvent 85% in D_2O. There have been several reports recently that D_2O may markedly increase the extent of association of an interacting system.[391—396] It may, therefore, be advantageous to use D_2O to emphasize interaction where possible, when working with systems where the interaction is very weak.

Gel Chromatography.—It is conventional practice in the study of interacting protein systems by gel chromatography to utilize frontal-analysis elution techniques.[397—400] A major reason for this is that commercial spectro-

[387] P. Andrews, *F.E.B.S. Letters*, 1970, **9**, 297.
[388] J. Vinograd, R. Radloff, and R. Bruner, *Biopolymers*, 1965, **3**, 481.
[389] I. P. Trayer and R. L. Hill, *J. Biol. Chem.*, 1971, **246**, 6666.
[390] R. Cohen and M. Mire, *European J. Biochem.*, 1971, **23**, 267, 276.
[391] J. J. Lee and D. S. Berns, *Biochem. J.*, 1968, **110**, 465.
[392] S. Paglini and M. A. Lauffer, *Biochemistry*, 1968, **7**, 1827.
[393] R. F. Henderson, T. R. Henderson, and B. M. Woodfin, *J. Biol. Chem.*, 1970, **245**, 3733.
[394] K. C. Aune, L. C. Goldsmith, and S. N. Timasheff, *Biochemistry*, 1971, **10**, 1617.
[395] J. Osborne and R. F. Steiner, *Arch. Biochem. Biophys.*, 1972, **152**, 849.
[396] P. A. Baghurst, L. W. Nichol, and W. H. Sawyer, *J. Biol. Chem.*, 1972, **247**, 3199.
[397] D. J. Winzor and H. A. Scheraga, *J. Amer. Chem. Soc.*, 1963, **2**, 1263.
[398] E. Chiancone, L. M. Gilbert, G. A. Gilbert, and G. L. Kellett, *J. Biol. Chem.*, 1968, **243**, 1212.
[399] G. A. Gilbert and G. L. Kellett, *J. Biol. Chem.*, 1971, **246**, 6079.
[400] G. K. Ackers, *Adv. Protein Chem.*, 1970, **24**, 343.

photometers may be used directly to monitor the column effluent. Moreover, the record of both a leading and tailing profile often shows the non-enantiography that is characteristic of a system in which the rate of equilibration between species is greater than that due to their separation by transport.[399, 401, 402] Measurement of the weight-average partition coefficient as a function of concentration then permits the determination of the equilibrium constant(s) of the system by suitable statistical methods.[398] For quantitative purposes, however, such procedures have the drawback that only two data points are provided in each experiment. In principle this limitation may be overcome by scanning the column to measure directly the degree of partitioning at many different levels within the column. Thus the statistical accuracy of a single partition coefficient may be improved enormously. The technical problems in achieving direct scanning are potentially formidable. Nevertheless, three papers[403–405] from Ackers' laboratory describe how they have been surmounted successfully. In the original scanning system, a 1 cm diameter quartz column was driven horizontally through a horizontally collimated beam of monochromatic light and the absorbance values were recorded as a function of distance along the column by an X–Y recorder.[403, 404] At 220 nm, the apparent absorbance of the gel relative to air is 1.8. This absorbance is apparently caused by internal light scattering within the column. However, Ackers and co-workers were able to demonstrate that no spectral broadening or shifts occur and that Beer's Law holds to a very close approximation. Partition coefficients may be measured by saturating the column with protein solution, which is then everywhere at mechanical and thermodynamic equilibrium. The background absorbance of the gel varies slightly as a result of packing variations. To measure the absorbance of the protein in the gel, the column must first be scanned in the absence of protein and the readings subtracted at each level from the corresponding value in the presence of protein. The partition coefficient, σ, is given by the equation

$$A_b/A_a = \alpha + \beta\sigma$$

where A_b and A_a are the absorbances of the protein in the gel and in the solution above the gel, respectively, and α and β are the void and internal column volume fractions. Ackers and co-workers have now replaced the X–Y plotter with a digital data-acquisition system. With a 17 cm column, this system collects 382 points during a three minute scan. The partition coefficients determined in this extremely precise manner have an accuracy of a few tenths of a percent.[405] The system also readily lends itself to the design of a new kind of experiment in which gels of increasing porosity may be stacked on each other in a single column. It is then possible in a

401 G. A. Gilbert, *Proc. Roy. Soc.*, 1959, **A250**, 377.
402 G. A. Gilbert and R. C. Ll. Jenkins, *Proc. Roy. Soc.*, 1959, **A253**, 420.
403 E. E. Brumbaugh and G. K. Ackers, *J. Biol. Chem.*, 1968, **243**, 6815.
404 E. E. Brumbaugh and G. K. Ackers, *Analyt. Biochem.*, 1971, **41**, 543.
405 H. S. Warshaw and G. K. Ackers, *Analyt. Biochem.*, 1971, **42**, 405.

single experiment to determine the partition coefficients of a given protein on each of the gels. Potentially, this is of particular importance in the study of interacting protein systems, since ways exist in theory to determine equilibrium constants from such data in a manner independent of relatively imprecise absolute molecular weight calibrations.[405]

The association–dissociation equilibrium of haemerythrin has been studied by frontal-analysis techniques.[406] The elution patterns show a sharp leading and a diffuse trailing profile. The gradient profile of the latter is bimodal, the position and size of the slow peak being independent of concentration above a limiting value of concentration, as predicted by Gilbert theory. The haemerythrin data are consistent with a monomer–octamer equilibrium, in accord with previous work.

Kinetic Studies.—Relatively few kinetic studies of association–dissociation reactions on the millisecond time-scale have been reported in the past. This is largely due to the fact that there are few systems in which convenient signals are available. Studies have therefore been limited primarily to those systems in which an absorption change in the accessible region of the spectrum occurs, although such changes are usually very small,[407] or to those in which molecular weights are sufficiently large (often in excess of 10^6 dalton) to produce a change in light scattering that is detectable by conventional techniques.[408] Although these technical requirements have restricted the scope of systems available for study, there are nevertheless several reports which are particularly interesting for the light they throw upon the mechanism of ligand-induced association–dissociation reactions.

Huang and Frieden[409] have now extended their earlier stopped-flow studies[408] to produce a most detailed report on the mechanism of depolymerization induced in bovine liver glutamate dehydrogenase by coenzyme and guanine nucleotides. Under the conditions chosen for this study, the molecular weight of the enzyme in the absence of effectors is so large that the depolymerization induced by them may be followed by the changes in turbidity observed at 300 nm. The reaction may also be observed at 365 nm, where the signal arises primarily from changes in coenzyme binding together with a small change in turbidity. The effects of coenzyme and nucleotide in this system are concerted; neither alone will produce substantial depolymerization. However, Huang and Frieden observed that the time course for the depolymerization reaction observed when TPNH was mixed with enzyme–GTP complex was quite different from that when GTP was mixed with enzyme–TPNH complex. The data are consistent with an ordered sequence of events, and Huang and Frieden have proposed a mechanism which assumes that the ligand-binding steps are in rapid equilibrium and that after ligands bind to the enzyme there are at least two

[406] A. L. Rao and S. Keresztes-Nagy, *Arch. Biochem. Biophys.*, 1972, **150**, 493.
[407] G. L. Kellett and H. Gutfreund, *Nature*, 1971, **227**, 921.
[408] C. Y. Huang and C. Frieden, *Proc. Nat. Acad. Sci. U.S.A.*, 1969, **64**, 338.
[409] C. Y. Huang and C. Frieden, *J. Biol. Chem.*, 1972, **247**, 3638.

conformational changes, with the one induced by GTP occurring after that induced by TPNH. The effects of TPNH–GDP, DPNH–GTP, and DPNH–GDP pairs were also examined.

Chiancone *et al.*[410] have described a simple stopped-flow light-scattering apparatus consisting of modified Brice–Phoenix apparatus in which the conventional light source is replaced by a 1 mW helium–neon laser. The flow system incorporates a simple two-jet mixer and has a dead-time of about 20 ms. Chiancone *et al.* have used the instrument to study the pH-induced dissociation of earthworm erythrocruorin. The protein has a molecular weight of about 3.4×10^6 dalton at pH 7 and dissociates in the range pH 8.9—10.0 to produce several species of different sizes down to 4.6×10^4 dalton. The extent and rate of the dissociation increase with increase of pH and the kinetics are characterized by an initial rapid phase ($t_{\frac{1}{2}}$ between 50 and 500 ms) followed by a much slower phase ($t_{\frac{1}{2}}$ equal to several minutes). The changes are not completely reversible.

The addition of magnesium ion to enolase (brewers' yeast) causes a small change in tryptophan absorbance with a difference maximum at 296 nm. Brewer and DeSa [411] have used this change to follow the increase in association induced by magnesium ion and the decrease induced by magnesium ion removal with ethylenediaminetetra-acetic acid (H_4edta). The addition of magnesium to partially dissociated enolase causes a rapid absorbance change in the dead-time of the stopped-flow instrument, a protein-concentration-dependent change representing association from monomer to dimer, followed finally by a first-order change. The latter conformational change in the dimer parallels the recovery of activity as determined in the stopped-flow instrument by observation at 245 nm of the production of phosphoenol pyruvate from 2-phosphoglycerate. The removal of endogenous magnesium ion by the addition of H_4edta is characterized by an absorbance change within the apparatus dead-time followed by two consecutive first-order changes. Since these reactions occur even if the enzyme is partially dissociated, they do not correspond to the changes seen on association. Enolase is therefore a 'hysteretic enzyme' as defined by Frieden.[412] Measurements in the far-u.v. with stopped-flow machines are difficult, and Brewer and DeSa found it necessary to modify substantially the commercial Gibson–Durham machine employed.

Tiepel has studied the reassembly of tetrameric aldolase from enzyme denatured in 6M-guanidine hydrochloride.[413] When the denatured enzyme is diluted one-hundred-fold into the renaturation buffer at 0 °C, the regain of secondary structure as judged by optical rotation was found to be complete by the first measurement time of 30 s. The rate of reassociation

[410] E. Chiancone, P. Vecchini, M. R. Rossi-Fanelli, and E. Antonini, *J. Mol. Biol.*, 1972, **70**, 73.
[411] J. M. Brewer and R. J. DeSa, *J. Biol. Chem.*, 1972, **247**, 7941.
[412] C. Frieden, *Ann. Rev. Biochem.*, 1971, **40**, 653.
[413] J. W. Tiepel, *Biochemistry*, 1972, **11**, 4100.

was followed by light-scattering in a conventional Brice–Phoenix apparatus. Within this 30 s, a rapid reassociation to dimer occurred which was followed by a very slow ($t_{\frac{1}{2}}$ greater than one hour) reassociation to tetramer with simultaneous regain of activity. The reassociation rate was first-order, implying that the dimeric species underwent a conformational change which was rate-limiting.

In certain favourable cases, association–dissociation equilibria and reactions can be conveniently investigated when the subunits and oligomer possess different kinetic properties, as is the case with haemoglobin.[407] Human haemoglobin A tetramer may exist in two different conformations, oxygenated and deoxygenated, which correspond to the allosteric R and T states. The ratio of the dimer to tetramer association constant for deoxy- to oxy-haemoglobin has been shown to be not less than 10^5, so that under normal conditions the degree of dissociation of deoxy A tetramer is negligible.[414] However, substantial dissociation to dimer may be achieved in 2M-NaI solutions at pH 7.0 [415] or at extremely alkaline pH in dilute buffer,[416] when it is found that the rate of reaction of the dimer with carbon monoxide is about forty-fold greater than that with the tetramer (T state). Haemoglobin chains and the R state are characterized by the same fast rate. Perutz [417] has proposed that the integrity of deoxy A tetramer is due to the formation between dimers of salt bridges which serve to hold the penultimate tyrosines of the α and β chains in pockets between the F and H helices. When ligand is bound, the tyrosines are ejected and the salt bridges broken in the conversion into the R state. In haemoglobin Bethesda the penultimate tyrosine of the β chain is replaced by histidine, which would be expected to modify the normal deoxy T structure. Indeed, Olson and Gibson [418] have shown that deoxyhaemoglobin Bethesda binds carbon monoxide at the fast rate characteristic of the R structure. However, when the haemoglobin is incubated for a few seconds with the allosteric effector inositol hexaphosphate (IHP), which binds preferentially to the T state in normal haemoglobin A, only the slow rate is observed. The rate of the conformational change from R to T may be studied by mixing deoxyhaemoglobin Bethesda with carbon monoxide solution containing IHP. The resulting reaction is markedly biphasic, with the relative amplitude of the slow phase dependent on concentration and showing a second-order time course. Olson and Gibson [418] have also shown that in the absence of IHP, deoxyhaemoglobin Bethesda is partially dissociated to dimer, but not in the presence of excess IHP. The biphasic curve may therefore be attributed to a rate-limiting self-association of dimers prior to a more rapid conformational change from R to T states in the tetramer. The latter is accompanied by a small spectral change which may serve as an indicator for following the self-association process directly.[407, 418]

[414] G. L. Kellett, *J. Mol. Biol.*, 1971, **59**, 401.
[415] G. L. Kellett, *Nature New Biol.*, 1971, **234**, 189.
[416] M. E. Anderson, J. K. Moffat, and Q. H. Gibson, *J. Biol. Chem.*, 1971, **246**, 2796.
[417] M. F. Perutz, *Nature*, 1970, **228**, 726.
[418] J. Olson and Q. H. Gibson, *J. Biol. Chem.*, 1972, **247**, 3662.

Electron Microscopy.—Perhaps the most novel way to measure the kinetics of ligand-induced association–dissociation reactions is to count directly, as a function of time after the addition of ligand, the changes in the numbers of subunits and oligomer seen in the electron microscope. This unusual feat has been achieved by Wrigley *et al.*,[419] who have studied the interconversion of the tetramers, dimers, and monomers of human erythrocyte glucose-6-phosphate dehydrogenase (G6PDH). The monomers have dimensions of approximately 68 Å × 34 Å and a molecular weight close to 50 000 dalton. The monomers apparently become shorter and fatter on association to the dimer. The latter associate to a tetrahedron which may readily be identified in the electron microscope. The distinction between dimers and monomers is less easy, but may be achieved with practice. The extent of dissociation of G6PDH depends on pH and on $NADPH_2$ and G6P concentrations. Both association and dissociation may readily be induced, and Wrigley *et al.*[419] have described a simple, rapid technique for applying samples to the grids so that these processes may be followed at two-second intervals. With a single technique, therefore, they can determine, in this case, the number of subunits in the molecule, their quaternary structure, the equilibrium constant of the association–dissociation equilibria, and the rates of reactions comprising these equilibria.

The quaternary structure of aspartate transcarbamylase (ATCase) has been investigated with the electron microscope by Richards and Williams.[420] Two different methods of negative staining with phosphotungstate result in orthogonal orientations of the native enzyme, the first showing three-fold and the second two-fold symmetry with axes normal to the plane of the support. In the first orientation, one catalytic subunit lies directly above the other and appears as a clearly demarcated triangle with distinct trimeric division, as is also observed in the isolated catalytic subunit. Each side of the triangle has a small arm which is identified as the dimeric regulatory subunit. In the second orientation, the catalytic subunits appear as two prominent and identical portions with the regulatory subunits as indistinct extensions beyond. In the following paper, Cohlberg *et al.*[421] show conclusively by the cross-linking technique of Davies and Stark [422] that the regulatory subunit does indeed exist in the native enzyme as a dimer. The enzyme has a requirement for zinc.[423] The apo regulatory subunit exists in a monomer–dimer equilibrium. The addition of zinc, which binds to the extent of one mole per polypeptide chain, results in changes in circular dichroism, and the regulatory subunit then exists as a homogeneous dimer. The feedback inhibitor CTP causes a change in the increase

419 N. G. Wrigley, J. V. Heather, A. Bonsigrove, and A. De Flora, *J. Mol. Biol.*, 1972, **68**, 483.
420 K. E. Richards and R. C. Williams, *Biochemistry*, 1972, **11**, 3393.
421 J. A. Cohlberg, V. P. Pigiet, and H. K. Schachman, *Biochemistry*, 1972, **11**, 3396.
422 G. E. Davies and G. R. Stark, *Proc. Nat. Acad. Sci. U.S.A.*, 1970, **66**, 651.
423 M. E. Nelbach, V. P. Pigiet, J. C. Gerhart, and H. K. Schachman, *Biochemistry*, 1972, **11**, 315.

in association of the apo subunits. Moreover, CTP was found to bind with greater affinity to the Zn than to the apo regulatory subunits. From a consideration of the available physical evidence and in particular the electron-microscope evidence, Cohlberg *et al.* have proposed a model for the quaternary structure of ATCase, in which two catalytic subunits having a triangular arrangement of polypeptide chains are superimposed above each other in an eclipsed configuration without being in direct physical contact. The three regulatory dimers extend outside the molecule, interconnecting one catalytic chain in the upper trimer with one in the lower trimer, displaced by 120°. This model is quite different from that proposed by Rosenbusch and Weber solely on the basis of physical evidence.[424]

The quaternary structure of transcarboxylase [425] and the self-assembly of glutamate dehydrogenase into ordered superstructures [426] have also been studied.

Subunit Structure of Proteins.—The customary list on the subunit constitution of proteins has been omitted in deference to the comprehensive list of some 230 proteins compiled by Darnall and Klotz [427] in the latest version of their earlier tables.[428]

SDS-Gel Electrophoresis. One of the most attractive techniques for the determination of the number of subunits in an oligomer is the cross-linking technique of Davies and Stark.[422] In this method, the oligomeric protein is incubated under appropriate conditions with dimethyl suberimidate, which reacts with the ε-amino-groups of lysyl residues to form cross-links between the subunits. Subsequent electrophoresis in SDS-gels results in a series of bands whose molecular weights are integral values of that of the monomer. One of the problems in such studies is the choice of suitable molecular-weight standards. Carpenter and Harrington have investigated this problem by the intermolecular cross-linking of monomeric proteins, and recommend the use of cross-linked ovalbumin as a reliable standard over the range 45 000—360 000 dalton.[429] Carpenter and Harrington also emphasize that the technique has the potential for use as a probe of the spatial relationships between subunits. They observe that of the six bands produced from hexameric leucine aminopeptidase, those corresponding to dimers, tetramers, and hexamers were much stronger than those for trimers and pentamers. They conclude that, providing the ability of the subunits to form cross-links reflects their spatial arrangement within the oligomer, the latter consists of a trimer of dimers. Several hexameric structures are consistent with these findings. However, a planar hexamer consisting of all heterologous interactions is ruled out.

[424] J. P. Rosenbusch and K. Weber, *J. Biol. Chem.*, 1971, **246**, 1644.
[425] N. M. Green, R. C. Valentine, N. G. Wrigley, F. Ahmad, B. Jacobson, and H. G. Wood, *J. Biol. Chem.*, 1972, **247**, 6284.
[426] R. Josephs and G. Borisy, *J. Mol. Biol.*, 1972, **65**, 127.
[427] D. W. Darnall and I. M. Klotz, *Arch. Biochem. Biophys.*, 1972, **149**, 1.
[428] D. W. Darnall and I. M. Klotz, *Ann. Rev. Biochem.*, 1970, **39**, 25.
[429] F. H. Carpenter and K. T. Harrington, *J. Biol. Chem.*, 1972, **247**, 5580.

Glutaraldehyde has also been used as a cross-linking reagent.[430, 431]

Associating–Dissociating Systems.—The degree of association or dissociation of a number of interacting proteins in systems is influenced by effectors which bind preferentially to one particular polymeric form. These phenomena are often referred to as 'ligand-induced' association or dissociation, and may provide a mechanism for the control of activity for those enzymes where association and dissociation occur readily at physiological concentrations.[412] Alternatively, where association–dissociation reactions occur at lower concentrations, they provide a useful tool for the exploration of protein design.[432] Cytidine triphosphate synthetase displays ligand-induced association.[433] In the presence of either ATP or UTP, the dimeric enzyme associates to form a tetramer. However, ATP and UTP are synergistic, for the total effect of the two ligands together is greater than the sum of their independent effects. A similar effect is seen with the ligand-induced dissociation of glutamate dehydrogenase by TPNH and GTP.[409] Levitzki and Koshland have compiled a short list of several regulatory enzymes in which ligand-induced association–dissociation reactions occur.[433] Some other proteins in which similar effects occur are insulin,[434] enolase,[411] haemoglobin,[414] glucose-6-phosphate dehydrogenase,[419] the regulatory subunit of aspartate transcarbamylase,[421, 423] hexokinase,[435, 436] lysozyme,[378] lysine-sensitive aspartokinase,[437] and phosphofructokinase.[438–440]

The theory of interacting systems has been reviewed at length by Nichol and Winzor.[441]

10 Circular Dichroism and Optical Rotatory Dispersion

contributed by P. M. Bayley

Introduction.—A large volume of work with these techniques continues to be published; circular dichroism is now overwhelmingly preferred to optical rotatory dispersion. Extensions of the technique to cope with extreme experimental conditions (low temperatures, films, particulate samples) are becoming more frequent, and magnetic circular dichroism

430 I. P. Griffith, *Biochem. J.*, 1972, **126**, 553.
431 P. Halleux, C. Legrain, V. A. Stalon, A. Pierard, and J.-M. Wiame, *European J. Biochem.*, 1972, **31**, 386.
432 A. J. Cornish-Bowden and D. E. Koshland, jun., *J. Biol. Chem.*, 1970, **245**, 6241.
433 A. Levitzki and D. E. Koshland, jun., *Biochemistry*, 1972, **11**, 247.
434 P. T. Grant, T. L. Coombs, and B. H. Frank, *Biochem. J.*, 1972, **126**, 433.
435 J. S. Easterby and M. A. Rosemeyer, *European J. Biochem.*, 1972, **28**, 241.
436 M. Derechin, Y. M. Rustum, and E. A. Barnard, *Biochemistry*, 1972, **11**, 1793.
437 P. J. von Dippe, A. Abraham, C. A. Nelson, and W. G. Smith, *J. Biol. Chem.*, 1972, **247**, 2433.
438 A. Parmegianni, J. H. Luft, D. S. Love, and E. G. Krebs, *J. Biol. Chem.*, 1966, **241**, 4625.
439 K. R. Leonard and I. O. Walker, *European J. Biochem.*, 1972, **26**, 442.
440 P. R. Aaronson and C. Frieden, *J. Biol. Chem.*, 1972, **247**, 7502.
441 L. W. Nichol and D. J. Winzor, 'Migration of Interacting Systems', Oxford University Press, Oxford, 1972.

(m.c.d.) is finding more biological applications. The combination of c.d. with n.m.r. and simultaneous evaluation of theoretical c.d. and conformational energy is most promising for small peptide systems.

There is a large literature on protein c.d. (see Volume 4), dealing with (i) aromatic and disulphide chromophores, (ii) peptide chromophores, (iii) chromophoric proteins, (iv) added chromophores, (v) chemical modification. In this Report consideration of protein c.d. is restricted to papers involving new principles. The section dealing with peptide antibiotics, peptide hormones, and biologically active peptides is retained.

General.—*Reviews.* Several articles on the elucidation of molecular conformation in solution cite c.d. results as of primary diagnostic importance. In transition-metal complexes of amino-acids and peptides, rotational strength derives from (*a*) ligand transitions, possibly modified by complex formation, (*b*) charge-transfer interactions between metal and ligand, and (*c*) the *d–d* transitions of metals.[442] The net sign of the *d–d* manifold is affected by conformation of multidentate ligands, and also by vicinal effects of local configurations. For small side-chains these effects are additive, and conform to Sector Rules. Conformational properties of polypeptides and antibiotics [443] and of hormones [444] are reviewed with respect to their primary structures and biological activity: a major problem is the flexibility of structures, as judged by solvent-sensitivity of the c.d. properties.

In the use of c.d. (and dipole-moment and n.m.r. properties) for considering the conformation and conformational transitions of synthetic poly-α-amino-acids, examples of homopolymers and repeating peptide sequences are considered.[445] Further synthetic work is indicated for these ordered polypeptides, and for purely synthetic sequences, to extend comparisons of observed properties with those computed for theoretically predicted structures. To understand the establishment of a specific conformation for a specific sequence, methods are required for prediction of structure, and computations of energies and optical properties of such structures. At the other extreme, optical methods are ideal for monitoring conformational changes in globular proteins and co-operative conformational changes are of the greatest importance in non-linear and regulatory phenomena,[446] although the optical parameters may not as yet give a clear indication of the nature of the conformational change involved. The use of spectrochemical probes, in particular the extrinsic c.d. effects of chromophores extra to the peptide backbone, has also been reviewed.[447] Applica-

442 R. B. Martin, in 'Metal Ions in Biological Systems', ed. H. Siegel, M. Dekker, New York, 1973, Vol. 1.
443 C. Toniolo, *Farmaco* (*Pavia*) *Ed. Sci.*, 1971, **26**, 741.
444 C. Toniolo, *Farmaco* (*Pavia*) *Ed. Sci.*, 1972, **27**, 156.
445 N. Lotan, A. Berger, and E. Katchalski, *Ann. Rev. Biochem.*, 1972, **41**, 869.
446 F. M. Pohl, *Angew. Chem. Internat. Edn.*, 1972, **11**, 894.
447 B. L. Vallee, J. F. Riordan, and J. T. Johansen, *Cold Spring Harbor Symp. Quant. Biol.*, 1971, **36**, 517.

tions are described to metallo-enzymes, zymogen activation, and enzyme modification by proteolysis.

Finally, general considerations of the use of c.d. in studying protein conformation in biomembranes have been presented:[448] the existence of artifacts in measuring c.d. of particulate systems is now well documented (see Volumes 3 and 4 and p. 255 of this volume). The use of the dissymmetry factor ($g = \Delta\varepsilon/\varepsilon$) is proposed [448] for considering different membrane systems. Membrane proteins are not in general α-helical or β-structure: evidence may be adduced for the presence (in part) of such conformations, but determination of their relative abundances is restricted by the same considerations as in the case of globular proteins, even assuming that all artifactual contributions have been eliminated.

Theory. Further derivation of symmetry rules for optical rotation from group-theoretical considerations suggests that polarizability rather than polarity effects should dominate the mechanisms of the Octant Rule for saturated ketones.[449] In considering the implications of the Quadrant Rule for peptides, the nature of the perturbant and its proximity to the chromophore must be considered specifically. Individual cases require specific evaluation, and in general different mechanisms are weighted differently by several authors.

In an extension of the methods for evaluating dipeptide optical properties, the optical activity has been evaluated for polypeptides of regular structures and for proteins of known (*X*-ray) conformation.[450] The α-helix, β-structures (parallel and antiparallel) polyproline I and II and poly-*N*-methyl-L-alanine structures are accounted for adequately in polymeric systems. Short α-helical regions in proteins are best assessed by the 222 nm c.d. band (n–π^*); both 207 nm and 191 nm bands are strongly dependent on chain length, and this will cause deviations in a three-component analysis. Similarly, β-structures in proteins are found to deviate from ideal; the irregular, non-periodic regions are even less predictable and are certainly not accounted for by the experimentally produced random-coil structure.

The c.d. properties of 'random' polypeptides are still controversial. Various theoretical attempts to assess these properties have been made (see Volume 4 of these Reports). Problems attend the specific definition of randomness, the different absolute values of points in the conformational energy surfaces used, the method of weighting of different regions of conformational space, as well as different methods of performing the purely optical part of the calculation. Assuming such a calculation could be done uncontroversially, the question remains as to the experimental spectrum to be taken as characteristic of a random polypeptide. The charged polypeptides may have short-range order, and indeed a specific conformation

[448] D. W. Urry, *Biochim. Biophys. Acta*, 1972, **265**, 115.
[449] P. J. Stiles, *Nature Phys. Sci.*, 1971, **232**, 107.
[450] V. Madison and J. A. Schellman, *Biopolymers*, 1972, **11**, 1041.

has been proposed from conformational energy calculations.[451] Proline-containing polypeptides are particularly subject to *cis–trans* isomerization, frequently promoted by cations: such additional randomization is not generally considered in theoretical assessments based upon all-*trans*-systems. Proteins in denaturing solvents may suffer similar effects, plus specific complexing of the denaturants. Homopolymers may, by their very nature, be biased against forming random conformations to be compared with unfolded proteins.

The irregular regions of proteins offer the following guidelines (while still not representing a random population): (*a*) c.d. properties vary widely for the different conformations,[450] (*b*) calculations at the oligopeptide level are not adequately simulated by the sum of dipeptides (supported by data in ref. 452, although a different conclusion is drawn), (*c*) *cis*- and *trans*-forms affect the computations significantly.[450, 488] In addition, (*d*) energy surfaces are evaluated for *in vacuo* conditions and (*e*) solvent interactions may show complex ionic-strength and temperature effects. This reviewer therefore feels that a new experimental initiative is required, possibly using synthetic peptides.

Computation of c.d. properties of proposed peptide structures is clearly optimal for medium-sized molecules (n = 10—20 peptides) since variations of individual parameters are not excessively overlaid with the remaining structure, and the interaction of contiguous residues is optimal. Comparison of proposed structures for gramicidin S showed that (i) the total structure must be considered, summation of nearest-neighbour interactions being inadequate; (ii) proline residues are particularly useful in diagnosis; (iii) relatively similar structures could be distinguished on the basis of proline properties; and (iv) the c.d. is consistent with the cross-chain hydrogen-bonded structure inferred from synthetic, n.m.r., and deuterium-exchange work.[453]

Further development is reported of computation of the c.d. of polypeptides containing aromatic chromophores extrinsic to the polypeptide backbone chromophore.[454] The theoretical c.d. of poly-L-phenylalanine, and poly-L-*p*-aminophenylalanine is consistent with a right-handed α-helix (as is that of poly-L-tyrosine). Notable effects are that the ^1L_a transition is a better conformational indicator than the ^1L_b, and that the c.d. of n–π^* transitions for a given chirality may be inverted in sign owing to interactions with ^1L_a transitions. Inversion of the chirality of the helical backbone does not automatically simply invert the sign of all transitions, but magnitudes are also changed owing to a combination of conformational and configurational effects.

[451] W. A. Hiltner, A. J. Hopfinger, and A. G. Walton, *J. Amer. Chem. Soc.*, 1972, **94**, 4324.
[452] M. L. Tiffany and S. Krimm, *Biopolymers*, 1972, **11**, 2309.
[453] P. M. Bayley, *Biochem. J.*, 1971, **125**, 90P.
[454] R. W. Woody, *Biopolymers*, 1972, **11**, 1149.

Circular polarization of emitted radiation has recently attracted interest. New instrumentation for the determination of the circular polarization of emission of fluorescence from a chiral chromophore[455] allows comparison of the dissymmetry ratio g_{abs} for absorption (*i.e.* c.d.) and $g_{emission}$, that for emission processes. The significant differences found for the complex of chymotrypsin with an extrinsic chromophore are interpreted as being due to a change in asymmetric environment upon electronic excitation.[456] This would argue for caution in the interpretation of fluorescence-probe properties. Some theoretical developments are noted:[457] the case where g varies through the transitions of a vibrational manifold, and even changes sign, may present special problems, and the ability to elicit circularly polarized emission from a racemic system[457] also suggests caution in interpreting differences in g_{abs} and $g_{emission}$.

Analysis. Attempts to fit o.r.d. and c.d. spectra of proteins with a linear combination of the spectra of a small number of components continue to be made. Three 'Reference Spectra' have been evaluated using the known molar fractions of α-, β-, and irregular regions of five proteins studied by *X*-ray crystallography.[458] The spectral references for β- and irregular structure differ widely from those of standard β- and disordered states. Reasonable simulations of experimental spectra at $\lambda > 205$ nm were achieved, but some fundamental criticisms arise. The reference states themselves are dependent upon the initial choice of proteins. With five proteins for three spectra the system is overdetermined and the evidence from curve-fitting and theory is that proteins cannot be considered as the sum of three components only. Estimates of helical content are possible, because of the high rotational strength of the n–π^* transition at 225 nm,[450] but will be unreliable once this structure is a minor component.

In multicomponent analysis, the necessity of including four or more components has been further shown by rank-analysis from the matrix-reduction of 30 sets of c.d. data on histones and well-characterized proteins.[459] This was found to be most important when data below 205 nm were included, consistent with the theoretical prediction.[450] A general matrix-formulation of the problem has been presented.[460]

C.d. parameters may readily be obtained from an analogue device which simulates gaussian or lorentzian curves (for absorption c.d. or e.s.r.), and which allows the simulated spectrum and its components to be recorded on the experimental trace.[461] The ability to simulate overlapping effects by differentiation or convolute functions would be particularly useful for c.d.

For measurements in the far-u.v., data on refractive indices of solutions of urea (to 8 mol l^{-1}) and guanidinium chloride (to 6 mol l^{-1}) down to

455 I. Z. Steinberg and A. Gafni, *Rev. Sci. Instr.*, 1972, **43**, 409.
456 J. Schlessinger and I. Z. Steinberg, *Proc. Nat. Acad. Sci. U.S.A.*, 1972, **69**, 769.
457 C. A. Emeis and L. J. Oosterhoff, *J. Chem. Phys.*, 1972, **54**, 4809.
458 Y. H. Chen, J. T. Yang, and H. M. Martinez, *Biochemistry*, 1972, **11**, 4120.
459 D. G. Dalgleish, *F.E.B.S. Letters*, 1972, **24**, 134.
460 M. E. Magar, *J. Theoret. Biol.*, 1971, **33**, 105.
461 E. Pedersen, *J. Phys.* (*E*), 1972, **5**, 492.

200 nm have been tabulated.[462] Approximations by Sellmeyer-type equations in simple form facilitate correcting spectroscopic measurements.

Instrumental. A comparison has been made of commercial circular dichroic equipment for measuring in the normal and difference modes.[463] Comparable accuracy was found for measurements of standard samples for the Cary 61, Durrum–Jasco J-10, and Jouan CD 185 systems, when properly calibrated. Signal:noise ratios (S/N), indicative of instrument sensitivity, are unfortunately not quoted. Calibration with D-10-camphorsulphonic acid is preferred; use of epiandrosterone can give values 5% low. In the difference mode, serious limitations are imposed by strongly absorbing samples; at total absorbance (both samples) 1.8, $S/N < 1$ for $[\theta] = 40°$ cm^2 decimole^{-1} for the system lysozyme + N-acetylglucosamine. The difference of two independent measurements is currently more reliable than the use of difference attachments when high absorbances are encountered. A new c.d. spectrophotometer has been described and the virtues of different modulation devices are discussed. The instrument uses an elasto-optic modulator and photomultiplier detection.[464]

Low-temperature techniques are used to arrest transient phenomena in the enzyme–substrate interactions of ferriperoxidase.[465] Similarly, labile conformational equilibria in the phytochrome system [466] are stabilized at −45 °C: the bleached chromophore has a compact folded form, achieved by photoisomerization, while other forms are extended – a result with important consequences in the holoprotein system.

Measurements in the vacuum-u.v. have been performed on polypeptides in aqueous solution (to 167 nm) and in hexafluoropropan-2-ol (to 140 nm), revealing the presence of a variety of additional c.d. bands of opposite sign that are as yet unassigned.[467] Problems in photolysis of samples during far-u.v. measurements have been described in detail:[468] helical polypeptides are more susceptible than globular proteins, and 225 nm radiation promotes bond cleavage.

Magnetic Circular Dichroism (m.c.d.). The sensitivity of m.c.d. to chemical constitution and its relative insensitivity to conformation allows its application for the analysis of peptide and protein systems and to liganded chromophoric metals. The ratio of tyrosine:tryptophan residues may be assessed [469] from the intensity of the m.c.d. at 292 nm, ΔA_n for proteins in 6M-GuCl at neutral pH (6.0) and ΔA_{alk} at alkaline pH (11.7), by

$$\text{Tyr/Trp} = 3.20\ (\Delta A_n - \Delta A_{alk}/\Delta A_n)$$

The denaturant is used to remove natural c.d. and the alkaline shift allows expression of the tyrosine ionization. Using accurate values of the tyrosine

[462] J. R. Krivacic and D. W. Urry, *Analyt. Chem.*, 1971, **43**, 1508.
[463] K. H. Chau and J. T. Yang, *Analyt. Biochem.*, 1972, **46**, 616.
[464] R. H. Breeze and B. Ke, *Analyt. Biochem.*, 1972, **50**, 281.
[465] F. Travers and P. Douzou, *Compt. rend.*, 1972, **274**, *D*, 1403.
[466] M. J. Burke, D. C. Pratt, and A. Moscowitz, *Biochemistry*, 1972, **11**, 4025.
[467] W. C. Johnson, jun. and I. Tinoco, jun., *J. Amer. Chem. Soc.*, 1972, **94**, 4389.
[468] W. D. Wilson and J. F. Foster, *Biophys. J.*, 1972, **12**, 609.
[469] G. Barth, E. Bunnenberg, and C. Djerassi, *Analyt. Biochem.*, 1972, **48**, 471.

content (from amino-acid analysis), the m.c.d. method does not require accurate knowledge of protein concentration. Values are in good agreement with the literature values for well-established proteins, and further resolve certain ambiguous results. The method is a valuable analytical tool for tryptophan, which is uncertainly estimated by amino-acid analysis or absorption spectroscopy. Related measurements on indole alkaloids may assist with analysis of chemically modified proteins.[470] Tryptophan residues of azurin and carbonic anhydrase appear in m.c.d. with characteristic intense positive extrema at 292—293 nm.[471] The spectra of azurin correlate closely with that of *N*-acetyl-L-tryptophan; m.c.d. spectra of carbonic anhydrase are more complicated, but bear some resemblances in positions of peaks, fine structure, *etc.*, to the natural c.d. The distinctive band frequently extending from 295 to 305 nm in the natural c.d. of tryptophan derivatives and proteins is not separately observed in m.c.d.

M.c.d. measurements on extrinsic chromophores of xanthine oxidase show only diffuse bands in the region of iron and molybdenum complex absorptions.[472] With Co^{II}–carbonic anhydrase, comparisons with m.c.d. of model compounds of Co^{II} in tetrahedral, distorted tetrahedral, and octahedral co-ordination suggest that the formation of a number of anion complexes of the enzyme occurs without change of co-ordination geometry. This corresponds to a nearly tetrahedral form, trigonally distorted.[473] The similarity of the alkaline Co^{II}–carbonic anhydrase to the pentaco-ordinated tris-(2-dimethylaminoethyl)aminecobalt(II) has been noted;[474] a change in geometry with liganding by acetazolamide was observed.

Small Molecules, Model Compounds, and Synthetic Polymers.—*Amino-acids and Derivatives.* The synthesis and resolution of *cis*- and *trans*-5-methylproline is reported:[475] the isomers are separated by a differential reaction with toluene-*p*-sulphonyl chloride and the absolute configurations are related to L-proline. The kinetics of racemization of amino-acids show that the rate of interconversion of isomers is independent of pH in the neutral range;[476] some correlation with the electron-withdrawing capacity of the R substituent is found.

Chromophoric derivatives of amino-acids reflecting chirality are given by the reaction with 2-fluoro-3-nitropyridine:[477] differential reactivity is shown with thiol group and primary amino-groups in reduced glutathione, and the selective determination of cystine has been proposed.[478]

[470] G. Barth, R. E. Linder, and E. Bunnenberg, *Helv. Chim. Acta*, 1972, **55**, 2168.
[471] T. M. McFarland and J. E. Coleman, *European J. Biochem.*, 1972, **29**, 521.
[472] E. Bayer, A. Bacher, P. Krauss, W. Voelter, G. Barth, E. Bunnenberg, and C. Djerassi, *European J. Biochem.*, 1972, **22**, 580.
[473] J. E. Coleman and R. V. Coleman, *J. Biol. Chem.*, 1972, **247**, 4718.
[474] T. A. Kaden, B. Holmquist, and B. L. Vallee, *Biochem. Biophys. Res. Comm.*, 1972, **46**, 1654.
[475] C. G. Overberger, K. H. David, and J. A. Moore, *Macromolecules*, 1972, **5**, 368.
[476] J. L. Bada, *J. Amer. Chem. Soc.*, 1972, **94**, 1371.
[477] C. Toniolo, D. Nisato, L. Biondi, and A. Signor, *J.C.S., Perkin I*, 1972, 1179.
[478] C. Toniolo, D. Nisato, L. Biondi, and A. Signor, *J.C.S., Perkin I*, 1972, 1182.

Amino-acid and peptide ligands of transition metals are readily studied by c.d. because of the high dissymmetry ratios obtaining with weak absorption bands associated with *d–d* transitions.[442, 479–484] While the complexity of effects involved has been noted,[442] relatively simple Sector Rules are sought extensively. Such rules are strongly dependent upon the physical mechanism generating rotational strength, and hence upon the specific orbitals involved in the transitions of the chromophore. For reliable symmetry rules, the electronic characteristics must be unequivocally represented in theoretical treatments, and this is rarely achieved. Recent experimental observations include: complexes of *N*-methyl-L-alanine (2 : 1) with Cu^{II} and Pd^{II} showing a combination of substituent and conformational effects;[479] tetramer formation of glycyl-L-histidine (1 : 1) with Ni^{II} at pH 9.6;[480] additive and independent effects with *N*-methyl-L-amino-acids and peptides (2 : 1) with Co^{II} and Co^{III};[481] change of liganding properties with high pH for glycylglycyl-alanine (1 : 1) with Pd^{II} and Cu^{II}, and ethylenediamine and L-histidine (1 : 1 : 1) with Pd^{II};[482] solvent effects in the chelation of *NN*-dialkyl-L-alanine (2 : 1) with Cu^{II};[483] and the correlation of c.d. intensity with Taft constant σ^* for amino-acid complexes (2 : 1) with Cu^{II}.[484] Sector Rules predicting net rotational strength of the *d–d* transition manifold must be compared with data over the full *d–d* range: this may involve transitions at $\lambda > 1.0\ \mu m$ (*e.g.* Co^{II}).[481] These studies also relate to the liganding in metalloproteins, for which m.c.d. is providing additional important information.[472–474]

Dipeptides and Oligopeptides. *N*-acetyl-amino-acid amines and *N*-methylamides, and the peptides Gly-Tyr-Gly and Gly-Phe-Gly in dioxan, show strong negative dichroism characteristic of the effect of intramolecular hydrogen-bonding on the n–π^* rotational strength,[485] as previously observed with proline derivatives[486, 487] and verified by theoretical calculations.[488] The results for *N*-acetylalanine amide, in agreement with o.r.d. measurements,[489] support the assignment of (ϕ, ψ) values for the C_α within the ring of approximately (120, 240) rather than (240, 120). The latter, which would have a strong positive n—π^* rotational strength, had been assigned tentatively from n.m.r.[490] and used in support of conformational energy calculations.[491]

479 P. J. Morris and R. B. Martin, *Inorg. Chem.*, 1971, **10**, 964.
480 P. J. Morris and R. B. Martin, *J. Inorg. Nuclear Chem.*, 1971, **33**, 2913.
481 E. W. Wilson and R. B. Martin, *Inorg. Chem.*, 1971, **10**, 1197.
482 T. P. Pitner, E. W. Wilson, and R. B. Martin, *Inorg. Chem.*, 1972, **11**, 738.
483 C. P. Nash and C. A. Jacks, *J. Amer. Chem. Soc.*, 1972, **94**, 1767.
484 B. C. Verma and Y. P. Myer, *Bioinorg. Chem.*, 1972, **1**, 141.
485 J. R. Cann, *Biochemistry*, 1972, **11**, 2654.
486 V. Madison and J. A. Schellman, *Biopolymers*, 1970, **9**, 511.
487 V. Madison and J. A. Schellman, *Biopolymers*, 1970, **9**, 569.
488 P. M. Bayley, E. B. Neilson, and J. A. Schellman, *J. Phys. Chem.*, 1969, **73**, 228.
489 E. B. Neilson and J. A. Schellman, *Biopolymers*, 1971, **10**, 1559.
490 V. F. Bystrov, S. L. Portnova, V. I. Isetlin, V. T. Ivanov, and Yu. A. Ovchinnikov, *Tetrahedron*, 1969, **25**, 493.
491 B. Maigret, B. Pullman, and D. Perahia, *Biopolymers*, 1971, **10**, 107.

Conformational analysis of cyclohexapeptides from n.m.r. gives structures for mono- and 1,4-di-substituted derivatives, with two internal solvent-shielded peptide protons, presumed hydrogen-bonded; the substituents assume forms of minimum conformational energy.[492] The c.d. properties of the non-aromatic cyclohexapeptides are compared with the aromatic substituted derivatives[514] on the assumption that the same backbone persists in each.

Polypeptides. As well as continued work on homopolypeptides, block copolypeptides, and random-sequence copolypeptides, the use of ordered sequential copolypeptides allows the conformational effects of specific substitutions to be studied in considerable detail. In this work, which has increased in volume owing to the adoption of the appropriate synthetic methods, polymer molecular weight and physical conditions (film, solid, solution; solvent, pH, temperature) are important variables to be considered in comparing various results.

For homopolypeptides, the optical and hydrodynamic properties of poly-L-alanine in dichloroacetic acid have been compared.[493] The optical properties are relatively invariant with increasing molecular weight, and indicate that the conformation is α-helical: the molecular weight dependence of viscosity suggests that the helices are interrupted. These studies show the relatively short-range order which is reflected by the optical properties. The solution properties of poly-(*cis*-5-methylproline) show the interconversion (*cis–trans*) characteristic of poly ProI/II.[494] Disordering with concentrated salts or with acid shows only small changes in c.d., relative to large changes in hydrodynamic properties, suggesting that short-range order is maintained.

The conformational properties of poly-L-lysine are known to depend strongly upon the solution conditions; this polymer is therefore a useful one for studying the effects on conformational equilibria of other residues included in the sequence. Random copolymers of L-lysine and L-leucine up to 0.41 mole fraction of leucine have been shown to have α-helical structure.[495] The presence of the leucine residues stabilizes the helix and causes a progressive decrease in $[\theta]_{222}$ relative to $[\theta]_{208}$, the proportionality of $[\theta]_{208}$ to $[\theta]_{192}$ being constant. This is interpreted as an environmental rather than a conformational effect, the n—π^* rotational strength being sensitive to local solvation and molecular polarizability. A more dramatic change in properties occurs with poly-(*N*-benzyl-L-lysine) which resembles the parent poly-lysine in showing a helix–coil transition with 8% dichloroacetic acid in dioxan, and at pH 7.0.[496] No transition to the β-structure was observed on heating at pH > 8.0. Poly-lysine itself was found to

[492] K. D. Kopple, A. Go, R. H. Logan, jun., and J. Savrda, *J. Amer. Chem. Soc.*, 1972, **94**, 973.
[493] A. Nakajima and M. Murakami, *Biopolymers*, 1972, **11**, 1295.
[494] C. G. Overberger and K. H. David, *Macromolecules*, 1972, **5**, 373.
[495] C. R. Snell and G. D. Fasman, *Biopolymers*, 1972, **11**, 1723.
[496] H. Yamamoto and T. Hayakawa, *Biopolymers*, 1972, **11**, 1259.

adopt the helical form when chondroitin sulphate was added in 1 : 1 ratio at pH 7.0 to the random conformation of the polymer.[497] This is an ionic interaction, dependent upon the presence of the sulphate groups, and is enhanced by the addition of dioxan.

An extensive study has been made of the effect of including L-leucine in random copolymers with *N*-hydroxypropyl-L-glutamine.[498] Again the thermal stability of the random copolymer is enhanced by the presence of leucine. Thermodynamic analysis allows an extrapolation to be made of the properties of the poly-L-leucine helix, which is consequently found to be even more stable than that of poly-L-alanine. This result has been compared with the frequency of occurrence of various residues in the internal helical regions of globular proteins, when it is indeed found that leucine is the most frequent. These studies indicate the predictive value of analysis of the properties of synthetic polypeptides and suggest that studies with sequence-specific substitution in such systems will be of considerable value. Thus in poly Lys-Ala-Ala,[499] the distribution of the included lysine residue is known exactly. This polymer is α-helical when unionized (*cf.* poly-lysine), with T_m = 90 °C; even at pH 6.5 (*i.e.* when charged) it can exist as a helical form below 12 °C (water) or 29 °C (20% methane). The helix-stabilizing effect of lysine is inferred to be comparable with that of alanine.

In two studies on random copolymers, the conformational transitions of different regions of the polymer chain can be distinguished, indicating a differential stability for the two components and showing that relatively simple synthetic polymers can undergo limited conformational changes. The stability of random copoly(benzyl-L-glutamate: benzyl-L-aspartate) in chloroform with increasing concentrations of trifluoroacetic acid shows different helix–coil transition characteristics from the n.m.r. properties of aspartate and glutamate residues.[500] (The occurrence of significant stretches of each residue is inferred, *i.e.* a departure from true random copolymerization.) This distinction cannot be made when the analogous poly-(benzyl-D-glutamate:benzyl-L-aspartate) is examined: this polymer is entirely left-handed α-helix, in contrast to the former which includes helices of both handedness. Compositions of more than 10% L-glutamate induce a right-handed helix, the stability of which increases with glutamate content.

In the random copolymer of L-glutamate and L-tyrosine (23 : 1), mol. wt. 17 000, in water–dioxan (2 : 1), the aromatic c.d. of tyrosine residues included in an α-helix is negative, and reflects the asymmetry of individual residues.[501] The pH profile of this effect is different from that of the

[497] R. A. Gelman, W. B. Rippon, and J. Blackwell, *Biochem. Biophys. Res. Comm.*, 1972, **48**, 708.
[498] P. Y. Chou, M. Wells, and G. D. Fasman, *Biochemistry*, 1972, **11**, 3028.
[499] A. Yaron, N. Tal, and A. Berger, *Biopolymers*, 1972, **11**, 2461.
[500] L. Paolillo, P. A. Temussi, and E. M. Bradbury, *Biopolymers*, 1972, **11**, 2043.
[501] W. B. Gratzer and G. H. Beaven, *Biopolymers*, 1972, **11**, 689.

polypeptide chromophore; the relative instability of the α-helix close to tyrosine is inferred.

Ordered sequential polypeptides have been synthesized as models for the fibrous proteins,[502, 503] collagen,[504–506] and more complex systems,[507, 508] as well as for studying helix-forming capabilities of individual residues.[505, 506]

Periodic copolymers of general formula poly Ala$_x$-Gly$_y$ have been studied by c.d. as films or in suspensions.[502] Polymers predominating in alanine are obtained as α-helices in films; the species with $x = 2$, $y = 1$, 2, or 3 and $x = 3$, $y = 3$ may adopt β-structure. The most interesting forms are with (Ala-Gly-Gly)$_n$ and (Ala-Gly-Gly-Gly)$_n$, which adopt the poly-glycine II form in films, with c.d. resembling that of collagen: poly Ala-Gly in aqueous suspension has c.d. of a form previously unobserved, with positive dichroism at 223 and 200 nm. The solvents and substrates used in casting films are important factors in determining the molecular conformation adopted.

The effect of increasing the size of the side-chain in the series poly Ala-Gly, poly Ser-Gly, and poly Glu(OEt)-Gly (studied as films) is to increase the inter-sheet spacing in the fibrous cross β-form.[503] I.r. dichroism indicates that the molecular axis is perpendicular to the film, rather than parallel to it, as in the higher molecular weight *Bombyx mori* silk fibroin: c.d. shows the Iβ conformation in the film.

Sequential poly-tripeptides have established several factors contributing to the stability of the collagen helix, in which repeating tripeptide sequences have a structural role. Thus the sequence poly Ala-Pro-Gly, though ordered in ethylene glycol or hexafluoro-isopropanol, gives an unordered conformation in aqueous solution, where the isomer poly Pro-Ala-Gly is ordered.[504] Comparison of poly Pro-Ser-Gly, poly Pro-Ala-Gly, and poly Pro-Gly-Pro shows that the Pro-Pro sequence stabilizes the triple helix. Also, Pro-X-Gly confers more stability than X-Pro-Gly, the extra conformational freedom of the Gly having a greater destabilization effect when it follows rather than precedes Pro.[505] Likewise, the sequence poly Ala-Gly-Gly in dilute aqueous solution can adopt the 3_1-helix (as in one strand of the collagen triple helix) as well as a cross-β and a disordered form in other solvents[506] (*cf.* ref. 502). Replacement of one glycine by a residue other than Pro in poly Gly-Gly-X causes destabilization. Similar conclusions have been drawn from the sequences poly Glu(OEt)-Glu(OEt)-Gly and poly Glu(OEt)-Gly-Gly.[507] Two polymers of (Leu-Gly-Lys-Ala-Glu-X-Gly), with X = Ser or Ala, as models of paramyosin

502 A. Brack and G. Spach, *Biopolymers*, 1972, **11**, 563.
503 J. M. Anderson, H. H. Chen, W. B. Rippon, and A. G. Walton, *J. Mol. Biol.*, 1972, **67**, 459.
504 B. B. Doyle, W. Traub, G. P. Lorenzi, and E. R. Blout, *Biochemistry*, 1971, **10**, 3052.
505 F. R. Brown, A. J. Hopfinger, and E. R. Blout, *J. Mol. Biol.*, 1972, **63**, 101.
506 W. B. Rippon and A. G. Walton, *J. Amer. Chem. Soc.*, 1972, **94**, 4319.
507 W. B. Rippon, H. H. Chen, J. M. Anderson, and A. G. Walton, *Biopolymers*, 1972, **11**, 1411.

have been reported.[508] Relatively low helicity is found. However, the results indicate the power of present synthetic methods.

Proteins.—Consideration is limited in this section to aromatic and disulphide chromophores and their use as chiro-optical probes of protein structure.

Model Aromatic and Disulphide Chromophores. Assignment of electronic transitions and a knowledge of their behaviour in model systems is clearly a prerequisite for their use as probes of the more complex conformational properties of globular proteins. Aromatic chromophores are consequently referred to benzene as a spectroscopic prototype and this allows a clear distinction between the ^1L_a and ^1L_b transitions of phenylalanine and tyrosine. For tryptophan, the two transitions overlap; changes of solvent affect them similarly, but appropriately placed charges can affect them differently.[509] The broader ^1L_a bandwidth can be differentially perturbed by solvent, and small solvent shifts of ^1L_b to the red (in 80% methanol relative to water) can be observed in model compounds, allowing the assignment of the negative c.d. above 288 nm to this source. Both solvent (*i.e.* polarizability) and charge effects are to be expected with the tryptophan residue in globular proteins.

In α-aryl α-amino-acids that are analogues of phenylalanine, ^1L_b transitions are characterized by fine-structured c.d., of some considerable intensity when the chromophore is in the α-position. In the region of 220 nm, ^1L_a transitions plus carboxylate or peptide n—π^* transitions may contribute rotational strength;[510] the enhanced c.d. of the aryl relative to the saturated cyclohexyl derivatives argues in favour of the ^1L_a assignment in these model compounds.

Model compounds related to tyrosine have been studied extensively in nonpolar media at normal and low temperatures.[511, 512] Vibrational assignments within the ^1L_b transition have been made with *p*-cresol and *p*-methylanisole. For *N*-stearyl-L-tyrosine n-hexyl ester in cyclohexane the 0–0 band is at 283 nm: the red shift (1—4 nm) occurring with dioxan, dimethylacetamide, n-butanol, and methanol is attributed to hydrogen-bonding. For the first two solvents, the c.d. correlates with the absorption, and no change in symmetry is indicated, but for the alcohols there is a 50% loss of rotational strength consequent upon specific complex formation. These shifts are not found with the *O*-methylated derivatives, whose rotational strength is also unaffected by change of solvent. A striking ten-fold intensification of rotational strength occurs on cooling *N*-acetyl-*O*-methyl-L-tyrosine ethyl ester in EPA (ether–isopentane–ethanol mixture)

508 R. D. Cowell and J. H. Jones, *J.C.S., Perkin I*, 1972, 1814.
509 L. J. Andrews and L. S. Forster, *Biochemistry*, 1972, **11**, 1875.
510 W. Klyne, P. M. Scopes, R. N. Thomas, and H. Dahn, *Helv. Chim. Acta*, 1971, **54**, 2420.
511 E. H. Strickland, M. Wilchek, J. Horwitz, and C. Billups, *J. Biol. Chem.*, 1972, **247**, 572.
512 J. Horwitz and E. H. Strickland, *J. Biol. Chem.*, 1971, **246**, 3749.

from 297 to 140 K. This is currently one of the most powerful demonstrations of the removal of conformational motility in amino-acid side-chains, and these observations correlate directly with previous work on tryptophan and phenylalanine. The possibility of using these methods for assessing side-chain mobility in proteins has been demonstrated with carboxypeptidase.[513]

Near-u.v. c.d. bands due to tyrosine transitions have been assigned to specific conformations in cyclohexapeptides containing tyrosine either as a 'corner' residue or as an 'extended' residue in an intramolecularly hydrogen-bonded system.[514] It is not clear, though, that the aromatic chromophore necessarily adopts a single conformation relative to the neighbouring peptide system, and low-temperature data on these compounds would be of great interest. The general view that the ^1L_a transitions of aromatics are more definitive as conformational indicators (see *e.g.* ref. 452) may also be associated with the ^1L_b transitions' being more susceptible to the effects of conformational motility.

The chiro-optical properties of the disulphide chromophore have been shown theoretically to follow a Quadrant Rule. This implies degenerate solutions for the rotational strength as a function of dihedral angle ϕ_{ss} obtained at ϕ_{ss} and $\phi_{ss} + 180°$, and that additional information is necessary in making an assignment. In the case of 2,7-cystinyl gramicidin S [515] this derives from n.m.r., where shielding effects indicate *P* chirality consistent with $\phi_{ss} = 120°$. Theory predicts negative c.d. for the longest-wavelength transition and this is indeed found ($R_{271} = -12.3 \times 10^{-40}$, $R_{230} = 58.6 \times 10^{-40}$). In the case of *cyclo*-cystine, the diketopiperazine with an intramolecular disulphide link, the absence of c.d. at $\lambda > 250$ nm and the location of the absorption at $\lambda = 250$ nm argue for $\phi_{ss} = 90°$ or $270°$.[516] Evidence from n.m.r. suggests the presence of only one isomer, and perturbation arguments (to account for the relatively strong n—π^* effects) would imply *P* chirality once more. The presence of a single isomer at normal temperatures might appear unlikely in view of the previous arguments of conformational motility. The system is spatially restrained by the ring systems, and this could confer a unique conformation. Conformational mobility is clearly indicated in variable-temperature studies of L-cystine itself, judged by the enhancement of negative $[\theta]_{255}$ on cooling.[517] The picture of a mixed population of rotational isomers at normal temperatures is enforced by a combined n.m.r. and c.d. study of L-cystine derivatives. The predominant contributor would be of *M* chirality, $\phi_{ss} < 90°$, to give the observed negative $[\theta]_{280}$.[518] Crystal structures of disulphide analogues

513 L. Fretto and E. H. Strickland, *Biochim. Biophys. Acta*, 1971, **235**, 473.
514 C. A. Bush and D. E. Gibbs, *Biochemistry*, 1972, **11**, 2421.
515 R. Schwyzer, *Angew. Chem. Internat. Edn.*, 1972, **11**, 854.
516 B. Donzel, B. Kamber, K. Wuthrich, and R. Schwyzer, *Helv. Chim. Acta*, 1972, **55**, 947.
517 T. Takagi and N. Ito, *Biochim. Biophys. Acta*, 1972, **257**, 1.
518 J. P. Casey and R. B. Martin, *J. Amer. Chem. Soc.*, 1972, **94**, 6141.

of penicillin are found to have $\phi_{SS} = 60°$ and *M* chirality, consistent with the broad negative c.d. band ($\lambda = 293$ nm) observed in solution.[519]

Aromatic and Disulphide Chromophores in Proteins. The properties of these chromophores, as illustrated above, reflect local asymmetry with some sensitivity. This has led to their widespread use in monitoring the conformational properties of proteins undergoing the perturbation of ligand binding, changes in external conditions, and interaction with other macromolecular systems.

In general, comparisons are made with relatively simple model systems, and some theoretical treatments have been advanced. Thus for the tyrosine residues of ribonuclease it has been shown that considerable rotational strength can derive from non-degenerate interactions between Tyr-73 and Tyr-115.[520] The pairwise interactions of Tyr-92–Tyr-97 and Tyr-25–Tyr-97, and coupling with peptide transitions, also contribute. The properties of Tyr-25 show a low rotational strength which could readily be enhanced by a small conformational change to allow interaction with Phe-46 and His-48. This mechanism could account for the increased c.d. at 288.5 nm in RNase A.

Computations of residue–residue interactions have also been used to rationalize the differences observed in c.d. between lysozymes and the α-lactalbumins, for which there is extensive sequence homology.[521] The strong characteristic positive 295 nm c.d. of the avian lysozymes could derive from Trp–Trp interactions at residues 62—63 (or the analogous Tyr–Trp in the human leukaemic urinary lysozyme). This sequence is missing in the lactalbumins. However, the properties of this band and of a weak negative region at 305 nm show a p*K* of 6, and have been assigned to the interaction of Trp-108 and Glu-35.[522] Both positive and negative extrema are intensified on binding oligomers of *N*-acetylglucosamine, an interaction which is known to involve these residues. Similar effects are shown by the human lysozyme in pH dependence and the effect of oligosaccharides.[523] Differential effects at 295 and 305 nm with different substrate analogues suggest that while both Trp-62 and Trp-108 appear to contribute to $[\theta]_{295}$, effects in $[\theta]_{305}$ are due principally to Trp-108. It is of interest that reversible concentration-dependent c.d. effects (principally at 295 nm), showing a p*K* of 6.2 and attributed to dimerization, are effectively eliminated by *N*-acetylglucosamine oligomers.[524] Again, the active-site Trp residues are implicated.

Such interpretations depend strongly on the background knowledge of the *X*-ray crystallographic structure of the protein. While this may detract from the independence of the methods, it provides valuable corroboration

519 S. Kukolja, P. V. Demarco, and N. D. Jones, *J. Amer. Chem. Soc.*, 1972, **94**, 7592.
520 E. H. Strickland, *Biochemistry*, 1972, **11**, 3465.
521 D. A. Cowburn, K. Brew, and W. B. Gratzer, *Biochemistry*, 1972, **11**, 1228.
522 K. Ikeda and K. Hamaguchi, *J. Biochem.* (*Japan*), 1972, **71**, 265.
523 K. Ikeda, K. Hamaguchi, S. Miwa, and T. Nishina, *J. Biochem.* (*Japan*), 1972, **71**, 371.
524 L. A. Holladay and A. J. Sophianopoulos, *J. Biol. Chem.*, 1972, **247**, 1976.

of the validity of the crystalline conformation for studies in solution. However, a clear example of the usefulness of aromatic c.d. in monitoring conformational changes has been presented for the allosteric enzyme aspartate transcarbamylase in its interactions with substrates.[525] The perturbation of the aromatic c.d. of the native enzyme (E) and catalytic subunit (C) contains effects due to tyrosine and tryptophan. The tryptophan effects are weak and extend to 310 nm; they are effectively the same for both E and C and possibly involve tryptophan in the active-site region. The tyrosine perturbations (286 and 279 nm) are stronger, and those for E significantly more intense than those for C. A change in asymmetry is indicated for tyrosine residues involved in the homotropic interactions between catalytic and regulatory subunits of the native enzyme. The effects are shown for the binding of either carbamyl phosphate or succinate. The aromatic residues may therefore be used as probes for conformational events in two distinct regions of the protein: the fact that one region involves tyrosine residues buried in a subunit contact region which undergoes conformational changes associated with homotropic activation is particularly interesting.

The state of near-u.v. c.d. of proteins is therefore most promising. Extensive studies of model compounds, particularly at low temperatures, have characterized vibrational fine-structure and environmental factors such as solvent polarizability and hydrogen-bonding. Theoretical computations allow semi-quantitative predictions for known geometries and experimental techniques allow the aromatic c.d. to be recorded with sufficient precision to preserve fine-structural details, following subtractive processes. Finally, variable-temperature techniques, in conjunction with absorption and solvent-perturbation spectra, may allow residues to be distinguished which are conformationally mobile in the native structure. These features suggest that much useful information is likely to emerge which is of direct relevance to the conformation of globular proteins in solution.

Peptide Antibiotics, Hormones, *etc.*—This section includes non-enzymic peptides with antibiotic activity or biological activity, hormonal or otherwise.

Antibiotics and Ionophores. The cyclic decapeptide antamanide, *cyclo*-(-Val-Pro-Pro-Ala-Phe-Phe-Pro-Pro-Phe-Phe-), is known to complex univalent cations and the proposed mode of complex formation, analogous to that of the ionophore valinomycin, requires chelation by six peptide carbonyl groups.[526] Model structures with all-*trans* peptides have been proposed for this form (with four intramolecular hydrogen bonds) and for a conformationally distinct form found in non-polar solvents. Progressive shifts in c.d. spectra are found from the strongly negative trough either

525 J. H. Griffin, J. P. Rosenbusch, and K. K. Weber, *J. Biol. Chem.*, 1972, **247**, 6482.

526 V. T. Ivanov, A. I. Mioshnikov, N. D. Abdullaev, L. B. Senyavina, S. F. Arkhipova, N. N. Uvarova, K. K. Khalilulina, V. F. Bystrov, and Yu. A. Ovchinnikov, *Biochem. Biophys. Res. Comm.*, 1971, **42**, 654.

with increasing polarity of the solvent, or on complexing with Na^+. A different interpretation is placed on these shifts, namely the effect of solvent on an essentially rigid chromophoric array, by workers who argue against the presence of intramolecular hydrogen-bonding.[527] There is some conflict of data here but the overall effects are not disputed. These effects, involving inversion of sign of the n–π^* rotational strength, would appear to be too large to be accounted for without conformational mobility in the peptide backbone. The absence of high-energy π–π^* transitions of Phe is indicated by studies of the perhydro-derivatives:[526, 528] arguments for the non-planarity of the peptide[528] do not appear necessary from the intensity of the spectra. The polarity of the medium evidently affects the binding constants for all anions, but does not alter the relative affinities, which decrease in the order $Na^+ > Ca^{2+} > Tl^+$.[529] Again, these studies show the benefit of combined n.m.r. and c.d. studies.

In contrast to this carrier-type of ionophore, the properties of the linear peptide gramicidin A require a relatively massive flux of ions per molecule of ionophore, consistent with a membrane-pore model.[530] The Π_{L-D} model for gramicidin A has undergone further refinement to include conducting and non-conducting forms,[531] involving a proposed conformational transition from the conducting-channel structure. There is at present no experimental evidence on the conformation of gramicidin A included in a lipid bilayer at the low concentrations at which conductance is mediated. N.m.r. evidence from bulk solution in trifluoroethanol and dimethyl sulphoxide has been interpreted in favour of the Π_{L-D} structure.[532] However, the c.d. of gramicidin A in trifluoroethanol,[532] also observed on the purified isomers,[533] is of very low intensity; the propensity of gramicidin A to form β-structure in anhydrous n-propanol (the structure is destabilized by water) has been noted.[533] Interpretation of the far-u.v. c.d. is complicated by the presence of intense transitions associated with tryptophan; the properties of hydrogenated gramicidin A,[532] as yet not fully characterized, will be of the greatest interest.

An extensive study of the phallotoxins, based on a cyclic structure bridged by a modified tryptophan residue, shows intense rotational strengths, as is often found with conformationally restricted structures.[534] A strong correlation has been established between the cleavage of one ring, the loss of toxicity, and the reduction in rotational strength associated

527 A. E. Tonelli, D. J. Patel, M. Goodman, F. Naider, H. Faulstich, and Y. Wieland, *Biochemistry*, 1971, **10**, 3211.
528 H. Faulstich, W. Burgermeister, and T. Wieland, *Biochem. Biophys. Res. Comm.*, 1972, **47**, 975.
529 T. Wieland, H. Faulstich, and W. Burgermeister, *Biochem. Biophys. Res. Comm.*, 1972, **47**, 984.
530 S. B. Hladky and D. A. Haydon, *Nature*, 1970, **225**, 451.
531 D. W. Urry, *Proc. Nat. Acad. Sci. U.S.A.*, 1972, **69**, 1610.
532 D. W. Urry, J. D. Glockson, D. F. Mayers, and J. Haider, *Biochemistry*, 1972, **11**, 487.
533 B. E. Isbell, C. Rice-Evans, and G. H. Beaven, *F.E.B.S. Letters*, 1972, **25**, 192.
534 H. Faulstich and T. Wieland, *European J. Biochem.*, 1971, **22**, 79.

with the modified indole residue. Evidently small conformational variations (in the non-toxic derivatives) cannot be compensated for at the point of action, unlike the binding of flexible molecules, for example during hormone action. The site of action of the toxins is not known: the possibility exists of potential activation of some chemical grouping in the full cyclic bridged structure, and this would be irreversibly lost on ring opening.

Hormones. It is not possible to generalize across the wide range of peptide hormones which have been studied: no single conformational feature emerges. One characteristic of the smaller linear peptides, evidently too small to have a strongly determined secondary or tertiary conformation, is a conformational mobility which can be induced by a change in solvent conditions. Thus glucagon (29 residues) and a fragment (27 residues), while adopting in aqueous solution a compact globular structure still devoid of pronounced secondary characteristics, both change to helix in chloroethanol.[535] Similar, but less pronounced, transformations are shown by the related molecule secretin and its *C*-terminal peptide components.[536] Perhaps of more biological significance, a more ordered structure can also be induced in glucagon by binding to lysolecithin micelles.[537] In contrast to this, departures from native structure in insulin assessed by both peptide and tyrosine c.d. correlate strongly with loss of potency (*in vitro* with fat cells) of chemically modified derivatives.[538] The deviations from native properties increase in the order ε-B29-acetyl < α-A1-acetyl < ε-B29,α-A1-diacetyl < ε-B29,α-A1,α-B1-triacetyl < tetra-acetyl (involving *O*-acetylation). In contrast A1,B29 (intramolecularly linked) adipoyl-insulin shows no marked c.d. changes, but is of low activity.[539] Either conformational mobility is not allowable in these regions of insulin, or vital residues in a static structure have been modified.

A well-characterized reversible conformational change of insulin gives rise (following a brief heating at 80—100 °C at pH 2.0) to insulin fibrils. The optical properties are characteristic of β-structure, and i.r. and *X*-ray studies indicate the formation of a cross-β structure with extensive and regular intermolecular hydrogen-bonding.[540]

The common genetic origin of human, bovine, and ovine growth and lactogenic hormones, exemplified by the degree of sequence homology, allows some comparisons.[541–545] With mol. wt. = 20 000, these molecules

[535] R. M. Epand, *J. Biol. Chem.*, 1972, **247**, 2132.
[536] A. Bodanszky, M. A. Ondetti, and M. Bodanszky, *J. Amer. Chem. Soc.*, 1972, **94**, 3600.
[537] A. B. Schneider and H. Edelhoch, *J. Biol. Chem.*, 1972, **247**, 4992.
[538] D. Brandenburg, H. G. Gattner, and A. Wollmer, *Z. physiol. Chem.*, 1972, **353**, 599.
[539] D. Brandenburg, *Z. physiol. Chem.*, 1972, **353**, 869.
[540] M. J. Burke and M. A. Rougvie, *Biochemistry*, 1972, **11**, 2435.
[541] S. Aloj and H. Edelhoch, *J. Biol. Chem.*, 1972, **247**, 1146.
[542] T. A. Bewley and C. H. Li, *Biochemistry*, 1972, **11**, 884.
[543] T. A. Bewley and C. H. Li, *Biochemistry*, 1972, **11**, 927.
[544] T. A. Bewley, H. Kawauchi, and C. H. Li, *Biochemistry*, 1972, **11**, 4179.
[545] T. A. Bewley, M. R. Sairam, and C. H. Li, *Biochemistry*, 1972, **11**, 932.

possess some extensive secondary and tertiary structure. The properties at neutrality are relatively similar, and the structures are relatively stable to extremes of pH and denaturants. Thus human growth hormone, the most stable, shows little conformational variation (by peptide c.d.) from pH 2 to pH 12: reversible (non-conformational) effects due to tyrosine ionization are observed in the near-u.v. Also, the molecule is resistant to 9M-urea: conformational effects of 5M guanidinium chloride and 50% acetic acid are reversed on removal of the perturbant.[542] Bovine growth hormone and ovine prolactin are less stable: the dimer (mol. wt. 42 000) dissociates at pH < 3.6 and pH > 11.5, with some tertiary structural changes evident in the aromatic c.d.[543] Human chorionic somatomammotropin is of comparable stability: studies of the nitrophenylsulphenyl derivative of the single tryptophan residue show it to be more exposed than in the human growth hormone.[544] Again, conformational mobility does not appear to be a feature of the mode of action of this group of hormones.

Data on parathyroid hormone[546] (mol. wt. 10 000) and epidermal growth factor[547] (mol. wt. 6100) show c.d. minima at 200 nm ($[\theta]_{200} = -7.5 \times 10^3$ and -12.7×10^3, respectively), indicating an indeterminate but low amount of secondary structure. Likewise human chorionic gonadotrophin (both native and the form from which sialic acid has been removed) is a compact globular molecule low in secondary structure.[548] C.d. minima at 210 and 195 nm are found for the hormone and its two subunits:[549] the spectra of the components are non-additive, indicating conformational changes on forming the associated native species.

Other Peptides and Non-enzymic Systems. The interest in neurotoxins is stimulated in part by parallel studies of receptors for neuronal transmitters with which certain toxins compete effectively and fatally. Structure–function relationships in toxins from *Naja haje* venom show that modification of Tyr-24 eliminates toxicity.[550] This group of molecules, 60—70 residues long, has a high disulphide content; the structure is resistant to denaturation at neutral pH, exposure to pH 2 being required to affect the peptide conformation.

Concanavalin A, the protein with specific affinity for carbohydrate residues on the external surface of cells and with transforming potency for lymphocytes, has been further studied by c.d. Subunits of Con-A and naturally occurring fragments have similar tertiary structure, possibly including considerable β-structure; both forms show a time-dependent change in peptide c.d. in dilute alkali, that with the intact subunit being slower, and both forms are stabilized by binding of α-methyl-D-mannoside.[551]

546 S. Aloj and H. Edelhoch, *Arch. Biochem. Biophys.*, 1972, **150**, 782.
547 J. M. Taylor, W. M. Mitchel, and S. Cohen, *J. Biol. Chem.*, 1972, **247**, 5928.
548 K. F. Mori and T. R. Hollands, *J. Biol. Chem.*, 1971, **246**, 7223.
549 V. Hilgenfeld, W. E. Merz, and R. Brossmer, *F.E.B.S. Letters*, 1972, **26**, 267.
550 R. Chicheportiche, C. Rochat, F. Sampieri, and M. Lazdunski, *Biochemistry*, 1972, **11**, 1681.
551 B. A. Cunningham, J. L. Wang, M. N. Pflumm, and G. M. Edelman, *Biochemistry*, 1972, **11**, 3233.

Chemical modification of tyrosine residues suggests that the ellipticity of the native protein derives from buried rather than exposed residues, which may therefore act as probes of internal structure.[552]

Another extremely topical protein system which has now been isolated in quantities sufficient for near- and far-u.v. spectroscopic characterization is the lac repressor from *E. coli*.[553–555] Far-u.v. c.d. indicates an unexceptional protein with moderate secondary structure, which is not grossly perturbed by binding various inducers, *e.g.* isopropyl-β-D-thiogalactosides.[553, 554] Stopped-flow fluorescence shows the kinetics of binding to proceed in at least two stages;[553] u.v. difference spectra show that both tyrosine and tryptophan residues are affected.[555] Subunit dissociation is effected by 0.03% sodium dodecylsulphate and complete unfolding by urea:[554] there is a small hydrodynamic change on binding inducer.[553]

Membranes.—*Artifacts*. The distortion and flattening of c.d. spectra of particulate samples are particularly serious in the far-u.v. Characterization of the effects (in particular the enhancement of c.d. following sonication) and theoretical treatments of the factors involved have been discussed previously (Volume 4 of these Reports). Certain aspects of this work have been reviewed[448] and the classical scattering from particulate, optically active suspensions has been discussed further.[556] The application of correction factors to observed spectra is described in a paper on plasma membranes and sarcotubular vesicles.[557] The corrections depend upon the simultaneous measurements of ellipticity and optical density (*i.e.* transmission) of the sample, and an iterative procedure is used to compute the attenuation of the observed signals due to the scattering and shadowing effects. Following correction, significantly lower absorbance and ellipticity were found for the sarcoplasmic vesicles.

In studies of the plasma membrane of *Streptococcus faecalis*, the usual artefacts were observed associated with light scattering, and could be reduced by sonication.[558] $[\theta]_{222}$ was found to be less sensitive to distortion than $[\theta]_{208}$ and could be used uncorrected in assessing secondary structure. These relative sensitivities are undoubtedly a function of particle size and will therefore differ significantly with membranes from other sources.

Membranes and Membrane Proteins. The vogue for studying membrane systems by c.d. continues in spite of the considerable difficulties in correcting the observed spectra and interpreting the results. The main contribution made by c.d. has been to establish that membrane proteins are not simply

552 W. D. McCubbin, K. Oikawa, and C. M. Kay, *F.E.B.S. Letters*, 1972, **23**, 100.
553 S. L. Laiken, C. A. Gross, and P. H. von Hippel, *J. Mol. Biol.*, 1972, **66**, 145.
554 M. Matsuura, Y. Oshima, and T. Horiuchi, *Biochem. Biophys. Res. Comm.*, 1972, **47**, 1438.
555 Y. Oshima, M. Matsuura, and T. Horiuchi, *Biochem. Biophys. Res. Comm.*, 1972, **47**, 1444.
556 D. Gordon, *Ann. New York Acad. Sci.*, 1972, **195**, 147.
557 L. Masotti, D. W. Urry, J. R. Krivacic, and M. M. Long, *Biochim. Biophys. Acta*, 1972, **266**, 7.
558 B. J. Litman, *Biochemistry*, 1972, **11**, 3243.

β-structure, as required by early models, but show many characteristics of globular proteins, including a moderate amount of helicity. Further, membranes from different sources have widely different properties of size, lipid and protein composition, solubilization of proteins into aqueous solution, and requirement for cations. The degree of association, the relative importance of artefacts, and structural variation will all contribute to the observed spectra. A general review of the field is given[559] in a volume devoted to membrane structure, in which the results of physical methods, including c.d., are also presented.[560, 561]

Studies on erythrocyte membranes[561, 562] show that a large fraction of the protein may be solubilized into aqueous solution by removal of cations with chelating agents. This dependence on inorganic cations for integrity of the membrane–protein complex is not necessarily shared by all membrane systems: the variation in lipid composition and the possible determining role of minor lipid components have been noted.[561]

Succinylation of the erythrocyte membrane solubilizes the proteins into a state retaining a loose association with lipid.[563] Some solubilization or fragmentation of the membrane is indicated by increase in $[\theta]_{208}$ at pH > 9.5, correlating with a decrease in the amount of scattered light.[564]

Aggregation of microsomal membranes on addition of Mg^{2+} is indicated by a sharp increase in turbidity.[565] Changes in the c.d. of smooth-surfaced vesicles from the microsomal fraction of acid-secreting cells of the gastric mucosa (oxyntic cells) have been observed on addition of Mg–ATP but not Mg–AMP.[566] Conformational changes in membrane protein, as well as aggregation, have been inferred. An effect of Mg–ATP on erythrocyte ghosts and inhibition of the effect by preventing ATP utilization has previously been noted,[567] with evidence (including i.r.) in favour of an increase in β-structure. A similar interpretation is placed on the effects of Na^+ or K^+ on proteins in excitable membranes from brain.[568] The effect of Ca^{2+} on the purified protein from sarcoplasmic reticulum (mol. wt. 55 000; K_{diss} for $Ca^{2+} \approx 1$ mmol l^{-1}) has been studied by c.d.[569] The enhancement of $[\theta]_{222}$ by Ca^{2+} can be produced by 0.1M-KCl; in the presence of chloride Ca^{2+} causes little additional c.d. effect, but calcium-specific fluorescence changes have been observed.

Studies of mitochondrial membranes subjected to various extraction processes suggest that a large amount of phospholipid can be removed

559 O. Hechter, *Ann. New York Acad. Sci.*, 1972, **195**, 506.
560 D. W. Urry, *Ann. New York Acad. Sci.*, 1972, **195**, 108.
561 J. A. Reynolds, *Ann. New York Acad. Sci.*, 1972, **195**, 75.
562 J. A. Reynolds and H. Trayer, *J. Biol. Chem.*, 1971, **246**, 7337.
563 C. F. Moldow, D. Zucker-Franklin, A. Gordon, V. Hospelhorn, and R. Silber, *Biochim. Biophys. Acta*, 1972, **255**, 133.
564 R. Strom and B. Mondovi, *Biochemistry*, 1972, **11**, 1908.
565 J. C. Reinert and J. L. Davis, *Biochim. Biophys. Acta*, 1971, **241**, 921.
566 L. Masotti, M. M. Long, G. Sachs, and D. W. Urry, *Biochim. Biophys. Acta*, 1972, **255**, 420.
567 J. M. Graham and D. F. H. Wallach, *Biochim. Biophys.Acta*, 1971, **241**, 180.
568 G. Papakostidis, G. Zundel, and E. Mehl, *Biochim. Biophys. Acta*, 1972, **288**, 277.
569 N. Ikemoto, G. M. Bhatnagar, and B. Nagy, *J. Biol. Chem.*, 1972, **247**, 7835.

from submitochondrial vesicles without notably affecting the conformation of the remaining protein.[570] The intrinsic membrane proteins were found to be relatively stable to 8M-urea. A coarse fractionation of mitochondrial membrane proteins was obtained on methylated columns using chloroform–methanol as solvent:[571] the proteins exist in a disaggregated state, and show enhanced secondary and tertiary characteristics in this solvent system.

Finally, the role of water in membrane structure has been studied by measurement of the c.d. of films of erythrocyte ghosts as a function of hydration.[572] C.d. profiles similar to those of suspensions are obtained, but with less scattered light, and the profiles show little change from 92 to 0% relative humidity. It may be noted that *X*-ray work indicates that a dispersion of the lipids occurs at $<20\%$ relative humidity. The c.d. results therefore indicate that the conformation of the membrane proteins is essentially independent of water content and the state of the lipid phase. This may indicate that since there is only a low proportion of bimodal protein, *i.e.* embedded within and across the bilayer,[573] most of the protein being extractable,[561, 562] the films may represent the additional 'extrinsic' protein which covers the basic bilayer but which may be dissociated from it. The forces stabilizing the film would then resemble those acting in the two-dimensional array at the membrane surface.

Lipoproteins. The study of well-defined lipoprotein systems is clearly of direct relevance to the membrane properties described above, as well as being of intrinsic importance as a class of discrete, heterogeneous macromolecules. Studies continue of the plasma high-density lipoprotein, which has a characteristically high helical content of $\sim 70\%$.[574] Removal of lipid gives an apolipoprotein of lower helicity which may be dissociated into two further protein components. Reconstitution with phosphatidylcholine and cholesteryl oleate effectively fully restores the helicity of the lipoprotein, and the near-u.v. c.d. of the complex is likewise restored. Phosphotidylcholine alone is less effective, providing further evidence for a specific multicomponent complex. The phospholipid binding site has been located as being in the *C*-terminal portion of the subunit.[575]

Finally, a conformational variation has been inferred in a plasma lipoprotein isolated at different times from 30 subjects (following electrophoretic identification and isolation of the single α_2-globulin). Measurements were made at unusually low concentrations and no correlation between helical content and concentration was found; the possibility remains of dissociations from a multicomponent complex.[576]

[570] W. L. Zahler, D. Puett, and S. Fleischer, *Biochim. Biophys. Acta*, 1972, **255**, 365.
[571] P. J. Curtis, *Biochim. Biophys. Acta*, 1972, **255**, 833.
[572] M. J. Schneider and A. S. Schneider, *J. Membrane Biol.*, 1972, **9**, 127.
[573] D. E. Green, *Ann. New York Acad. Sci.*, 1972, **195**, 150.
[574] S. E. Lux, R. Hirz, R. I. Shrager, and A. M. Gotto, *J. Biol. Chem.*, 1972, **247**, 2598.
[575] S. E. Lux, K. M. John, and S. Fleischer, *Biochem. Biophys. Res. Comm.*, 1972, **49**, 23.
[576] C. R. Harmison and C. E. Frohman, *Biochemistry*, 1972, **11**, 4985.

3
Peptide Synthesis

BY J. H. JONES AND B. RIDGE

1 Introduction

The general arrangement of this chapter follows that used in previous volumes, but whereas in the last three volumes some attempt has been made here to comment on a few points of biological interest, such matters are this year more fully dealt with elsewhere (Chapter 5).

2 Methods

The year under review has seen the development of new protecting groups which are removable under such widely diverse conditions as chemical oxidation, electrolytic reduction, and photolysis. Progress in coupling methods has been steady rather than spectacular. A review has appeared on 'Peptide Synthesis in Medical Research'.[1] A handy guide[2] to the literature on protected amino-acid derivatives which are useful in peptide synthesis has appeared. The guide contains a list of the melting points, specific rotations, and literature citations for approximately one thousand compounds.

Protective Groups.—*Established Methods of Amino-group Protection.* The preparation of *N*-formyl-amino-acids, the use of the formyl group in peptide synthesis, its removal, and the action of formic acid on proteins have been reviewed.[3] The *N*-formyl group can be removed by treatment with hydrazine diacetate in methanol, phenylhydrazine, semicarbazide hydrochloride, t-butyl carbazate, aniline, hydroxylamine hydrochloride, or *p*-anisidine.[4]

Reaction of *N*-methylpyrrolidine and tosyl chloride, in the presence of silver perchlorate, affords *N*-methyl-*N*-tosylpyrrolidinium perchlorate (1), a stable crystalline salt, which can be used for selectively tosylating amino-groups in the presence of hydroxy-groups.[5] The reagent has not yet been used for the tosylation of amino-acid or peptide derivatives, but commends itself on account of its selectivity and the fact that it is reasonably stable in aqueous solution.

[1] M. Bodanszky, *Chim. Ther.*, 1972, **7**, 145 (*Chem. Abs.*, 1972, **77**, 102 150f).
[2] G. A. Fletcher and J. H. Jones, *Internat. J. Peptide and Protein Res.*, 1972, **4**, 347.
[3] P. U. Lakshmi and L. K. Ramachandran, *J. Sci. Ind. Res., India*, 1971, **30**, 680 (*Chem. Abs.*, 1972, **76**, 154 108x).
[4] Farbwerke Hoechst A.-G. B.P. 1 182 450 (*Chem. Abs.*, 1972, **76**, 72 794d).
[5] T. Oishi, K. Kamata, S. Kosuda, and Y. Ban, *J.C.S. Chem. Comm.*, 1972, 1148.

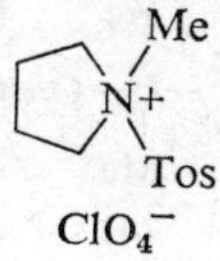

(1)

N-α-Benzyloxycarbonyl-lysine (2) is usually prepared, albeit in low yield, by the Schötten–Baumann acylation of *N*-ε-benzylidenelysine. A modification of this method involves the direct acylation of lysine in the presence of benzaldehyde, and subsequent chromatographic separation of the reaction products (Scheme 1).[6] The product (2) is obtained in only moderate yield

$$\text{Lys} \xrightarrow{\text{i, ii}} \underset{(2)}{\text{Z-Lys}} \xrightarrow{\text{iii}} \text{Z-Lys-OMe}$$

Conditions: i, PhCHO (25% excess)–ZCl–NaOH; ii, chromatography on Al_2O_3–$CuCO_3$; iii, BF_3–MeOH

Scheme 1

and is contaminated with inorganic materials, but can be used in further steps.

The most commonly used method for the introduction of the t-butoxycarbonyl group involves the acylation of the amino-acid with t-butyl azidoformate (3), at a carefully controlled alkaline pH. In a modification

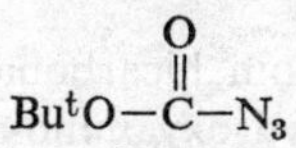

(3)

of this procedure the amino-acid 1,1,3,3-tetramethylguanidinium salt is acylated in DMF in the presence of excess tetramethylguanidine.[7] The reaction proceeds at room temperature, and the yields of t-butoxycarbonylamino-acids are excellent, except in the cases of alanine, aspartic acid, asparagine, glutamic acid, and tyrosine, when they are only moderate. The method used by Sieber and Iselin for the introduction of the 1-(*p*-biphenylyl)-1-methylethoxycarbonyl group [*via* the mixed phenyl carbonate (4)] has now been applied by others for the synthesis of t-butoxycarbonyl, *p*-methoxybenzyloxycarbonyl, benzhydryloxycarbonyl, and t-amyloxycarbonyl derivatives.[8] The mixed carbonate (5) is used to acylate the amino-acid tetramethylguanidinium salt in DMF, DMSO, or aqueous dioxan, at 20—40 °C. In general, good yields of the acylated products are obtained. This procedure has the advantage of simplicity, plus the fact that the

[6] N. L. Benoiton, R. E. Demayo, G. J. Moore, and J. R. Coggins, *Canad. J. Biochem.*, 1971, **49**, 1292.
[7] A. Ali, F. Fahrenholz, and B. Weinstein, *Angew. Chem. Internat. Edn.*, 1972, **11**, 289.
[8] U. Ragnarsson, S. M. Karlsson, and B. E. Sandberg, *Acta Chem. Scand.*, 1972, **26**, 2550.

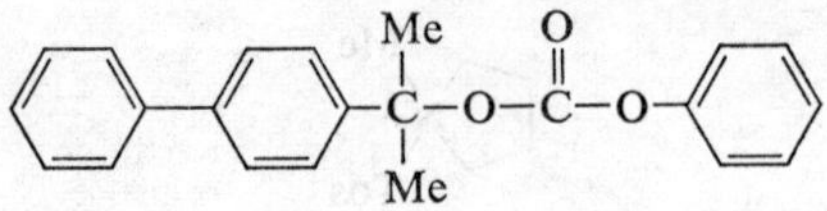

(4)

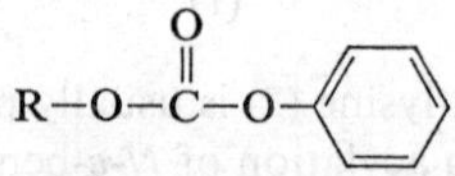

(5) R = Me_3C, *p*-$MeOC_6H_4CH_2$, Ph_2CH, or $EtCMe_2$

acylating agents can be prepared in a one-step process, and that excess base is not present for the acylation. A new acylating reagent has been used for the introduction of the t-butoxycarbonyl group into amino-esters. Di-t-butyl dicarbonate (6), which is prepared from potassium t-butoxide according to Scheme 2, on refluxing in an organic solvent with an amino-

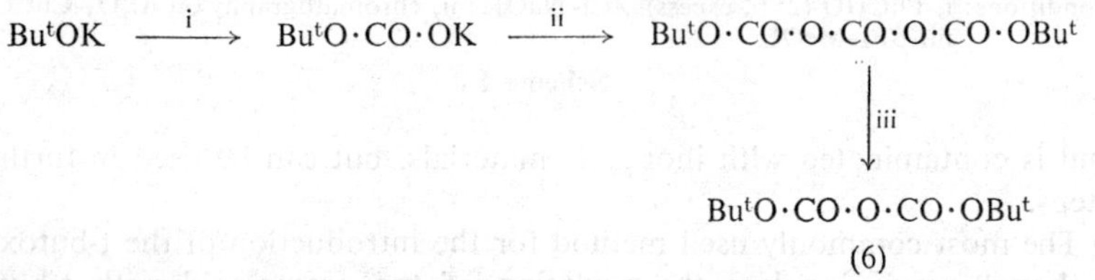

Conditions: i, CO_2; ii, $COCl_2$; iii, 1,4-diazabicyclo[2,2,2]octane

Scheme 2

ester hydrochloride and sodium bicarbonate gives a good yield of the protected derivative.[9] When hydroxyamino-acid esters are treated with one molar equivalent of the reagent, no diacylated materials are obtained. Comparison of the optical rotations of tyrosine methyl ester, before the acylation and after acidic cleavage of the product, showed that no appreciable racemization had taken place. The synthesis of tritiated t-butyl azidoformate (3) has been reported.[10] t-Butyl alcohol was tritiated using the Wilzbach technique, and converted *via* the mixed phenyl carbonate into (3) of activity approximately 366 mC $mol.^{-1}$.

Kessler and Iselin had rejected thiocyanate ions as a means of cleaving the *o*-nitrophenylsulphenyl N-protecting group, on the grounds that only partial cleavage occurred.[11] It has now been shown that if the reaction is carried out in the presence of an agent capable of trapping the *o*-nitrophenylsulphenyl thiocyanate produced, then the reaction proceeds smoothly to completion.[12] The cleavage occurs within five minutes when

[9] D. S. Tarbell, Y. Yamamoto, and B. M. Pope, *Proc. Nat. Acad. Sci. U.S.A.*, 1972, **69**, 730.

[10] J. F. Kennedy, C. J. Gray, S. A. Barker, and S. Ramonvongse, *J. Labelled Compounds*, 1972, **8**, 99.

[11] W. Kessler and B. Iselin, *Helv. Chim. Acta*, 1966, **49**, 1330.

[12] E. Wünsch and R. Spangenberg, *Chem. Ber.*, 1972, **105**, 740.

the substrate is treated with ammonium thiocyanate in an organic solvent containing acetic acid and 2-methylindole (Scheme 3). The products are,

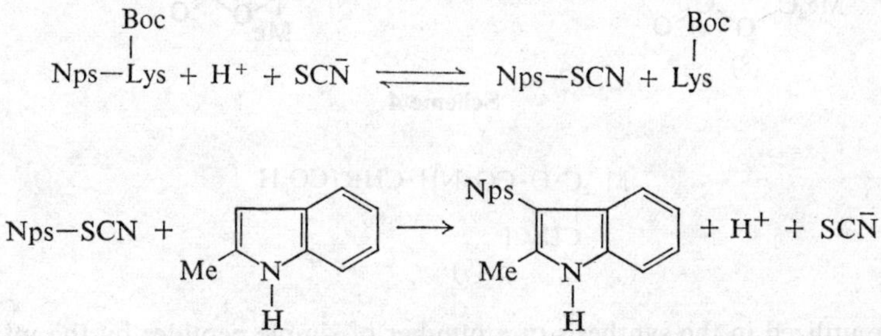

Scheme 3

in general, clean; t-butoxycarbonyl groups are stable under these conditions and, providing excess trapping agent is used, tryptophan derivatives are also cleaved satisfactorily. The method does not engender racemization.

The tritylation of some amino-acids and peptides in anhydrous pyridine has been studied.[13] The method may be useful for protecting peptides for use in semisynthetic studies, providing that the naked peptides are soluble in pyridine.

New Methods of Amino-group Protection. The idea of a protecting moiety which is stable under a wide variety of conditions but subject to rapid cleavage simply by heating the derivative in an appropriate solvent is now a reality. The preparation of 2-bromo- and 2-chloro-1,1-dimethylethoxycarbonyl-amino-acid derivatives* can be accomplished by reacting the corresponding chloroformate (7) with the amino-acid in aqueous solution,

$$\begin{array}{l} Me_2C\cdot O\cdot COCl \\ \,| \\ CH_2X \end{array}$$

(7) X = Br or Cl

or with the amino-ester in a mixture of water, chloroform, and tetrahydrofuran.[14] This circumvents the problem of urea formation which occurs when the acylation is carried out under anhydrous conditions (see Vol. 4 of these Reports, p. 317). Prolonged exposure of 2-bromo-1,1-dimethylethoxycarbonyl-amino-acids (8) to aqueous base results in the formation of 2-oxazolidinone derivatives (Scheme 4). Several bromo-derivatives (8) have been prepared in moderate yields, but the chloro-derivatives (9) were obtained in good yields. The two new protecting groups

[13] J. Halstrøm and K. Brunfeldt, *Z. physiol. Chem.*, 1972, **353**, 1204.
[14] T. Ohnishi, H. Sugano, and M. Miyoshi, *Bull. Chem. Soc. Japan*, 1972, **45**, 2603.

* 2-Bromo-1,1-dimethylethoxycarbonyl is preferred to α-bromo-t-butoxycarbonyl, which is inconsistent with IUPAC recommendations.

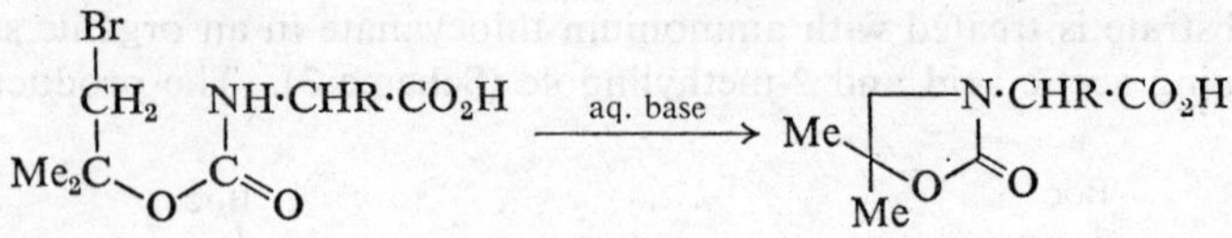

(8)

Scheme 4

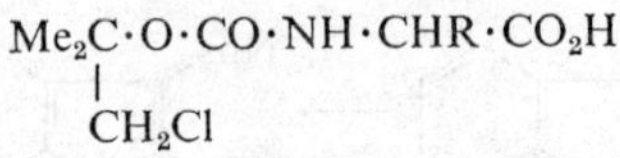

(9)

were utilized in the synthesis of a number of simple peptides by the mixed anhydride method. As judged by the criterion of optical rotation, these peptides are optically pure. The bromo-protecting group can be removed simply by heating the derivative in ethanol (Scheme 5), whereas removal

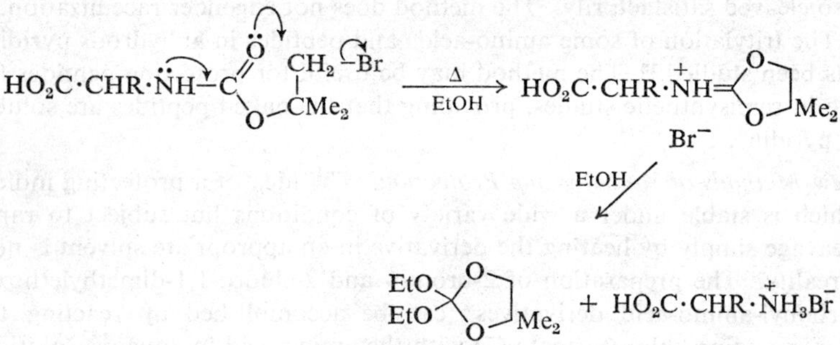

Scheme 5

of the 2-chloro-1,1-dimethylethoxycarbonyl protecting group is accomplished in a more conventional manner with hydrogen bromide in acetic acid. The *N*-2-bromo-1,1-dimethylethoxycarbonyl group is selectively removable in the presence of an *N*-ε-benzyloxycarbonyl group.

The use of the isobornyloxycarbonyl moiety for amino-group protection has been investigated.[15] Stable oily isobornyloxycarbonyl chloride (10) can be prepared from (+)-isoborneol and phosgene in almost quantitative yield. Isobornyloxycarbonyl-amino-acids (11), most of which are crystalline, can be prepared in very good yield by reaction with the chloroformate (10) under Schötten–Baumann conditions. The protecting group is stable to hydrogenolysis, hydrazinolysis, or alkaline saponification but is removed by treatment with trifluoroacetic acid, hydrogen bromide in acetic acid, or anhydrous hydrogen fluoride – presumably *via* a non-classical carbonium

[15] M. Fujino, S. Shinagawa, O. Nishimura, and T. Fukuda, *Chem. and Pharm. Bull. (Japan)*, 1972, **20**, 1017.

ion (12) (Scheme 6). The protecting group was evaluated in a successful synthesis of bradykinin.[16]

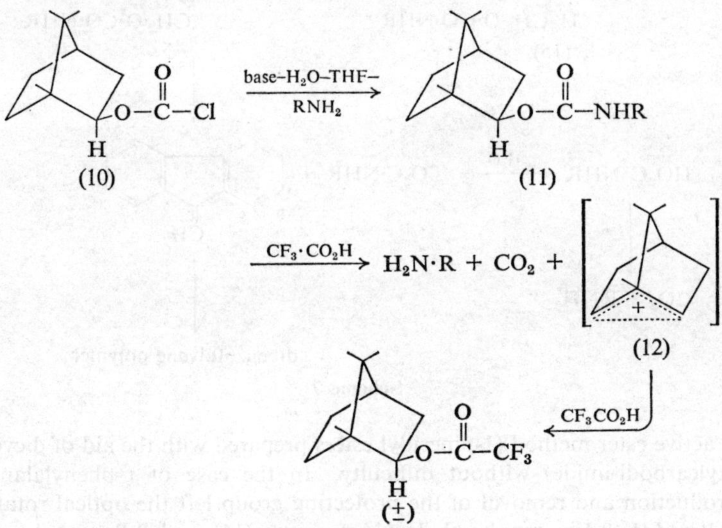

Scheme 6

A full paper has now appeared on the 9-fluorenylmethoxycarbonyl amino-protecting group (13).[17] 9-Fluorenylmethyl chloroformate (14) is a stable crystalline material, which can be prepared from 9-fluorenylmethanol and phosgene. It reacts normally with amino-compounds under basic

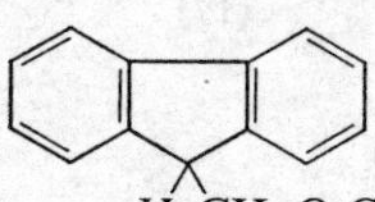

H CH_2·O·CO~

(13)

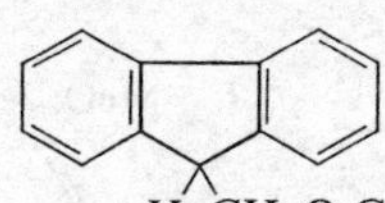

H CH_2·O·COCl

(14)

conditions to give the corresponding crystalline carbamates (15) in very good yield. The group is stable towards trifluoroacetic acid, hydrogen bromide in acetic acid, catalytic hydrogenation, and hydrazine in ethanol, but is readily cleaved under mildly basic, non-hydrolytic conditions. The preferred conditions require leaving the substrate in contact with liquid ammonia for several hours, work-up with water leading to the direct formation of the amine. This is probably an *E*1cb-type elimination process (Scheme 7). Cleavage can also be effected with secondary amines such as morpholine or piperidine. A number of simple peptides were prepared using

[16] M. Fujino and S. Shinagawa, *Chem. and Pharm. Bull. (Japan)*, 1972, **20**, 1021.
[17] L. A. Carpino and G. Y. Han, *J. Org. Chem.*, 1972, **37**, 3404.

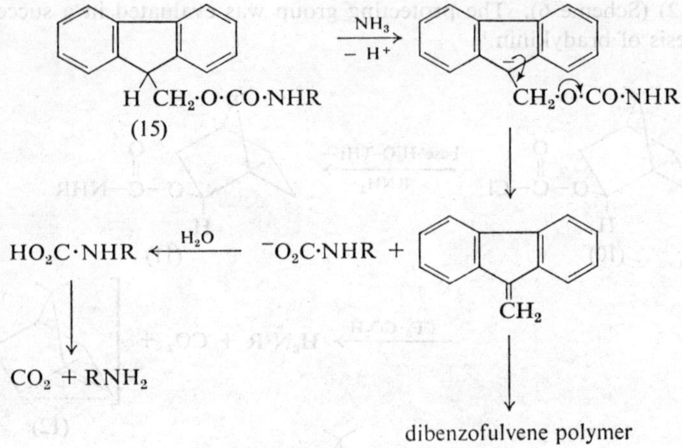

Scheme 7

the active ester method (1-piperidyl esters prepared with the aid of dicyclohexylcarbodi-imide) without difficulty. In the case of L-phenylalanine, introduction and removal of the protecting group left the optical rotation unaltered.[17] 9-Fluorenylmethyl azidoformate (16) and 9-fluorenylmethyl carbazate (17) were also prepared.

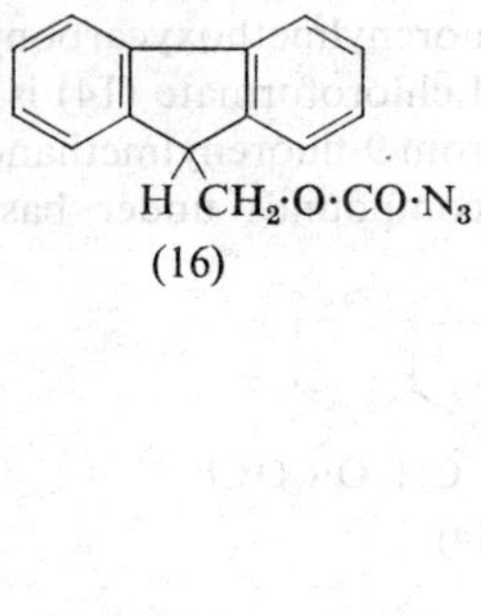

(16)

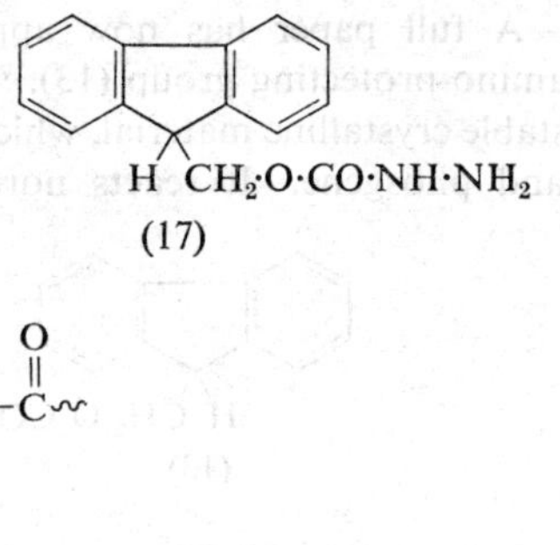

(17)

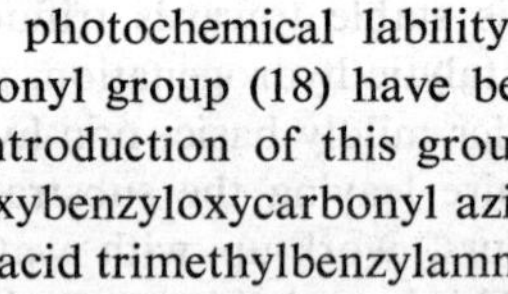

(18)

The introduction and photochemical lability of the αα-dimethyl-3,5-dimethoxybenzyloxycarbonyl group (18) have been discussed.[18] The preferred reagent for the introduction of this group is the stable crystalline αα-dimethyl-3,5-dimethoxybenzyloxycarbonyl azide (19), prepared according to Scheme 8. Amino-acid trimethylbenzylammonium salts are smoothly acylated by the azide in pyridine solution to give very good yields of crystalline derivatives (20). This new protecting group was shown to possess an acid lability intermediate between that of the t-butoxycarbonyl and 1-(*p*-biphenylyl)-1-methylethoxycarbonyl groups, *e.g.* complete cleavage

[18] C. Birr, W. Lochinger, G. Stahnke, and P. Lang, *Annalen*, 1972, **763**, 162.

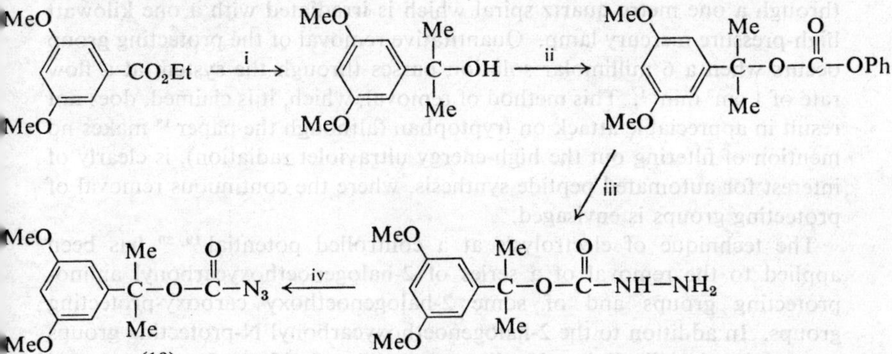

(19)

Conditions: i, MeMgI; ii, $ClCO_2Ph$; iii, N_2H_4; iv, HOAc–aq. $NaNO_2$

Scheme 8

occurs in three hours in 80% aqueous acetic acid or in eight minutes in 5% trifluoroacetic acid in dichloromethane (Scheme 9). Photochemical deacylation is carried out by passing a tetrahydrofuran solution of the substrate

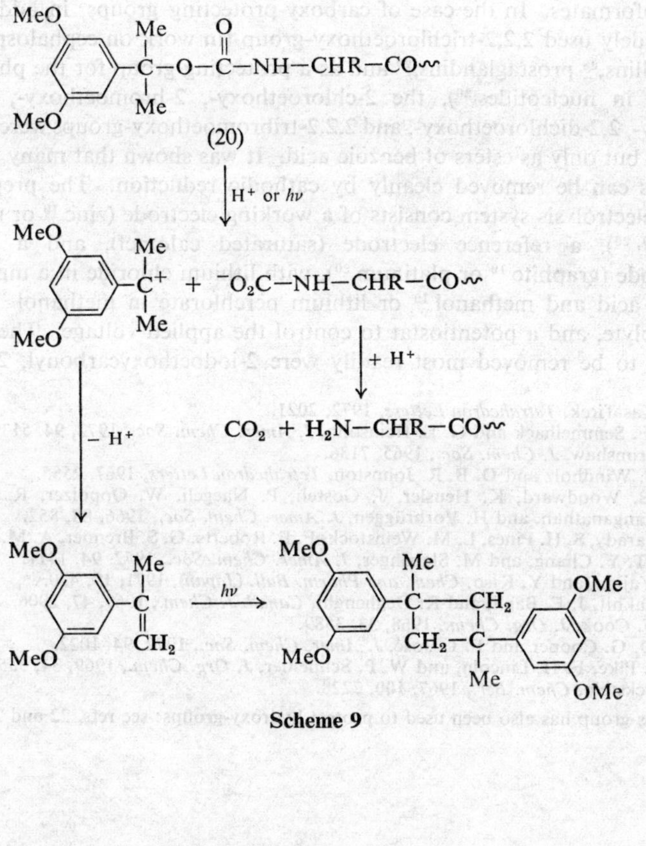

(20)

Scheme 9

through a one metre quartz spiral which is irradiated with a one kilowatt high-pressure mercury lamp. Quantitative removal of the protecting group occurs when a 6 millimolar solution passes through the system at a flow rate of 1 $cm^3 min^{-1}$. This method of removal, which, it is claimed, does not result in appreciable attack on tryptophan (although the paper [18] makes no mention of filtering out the high-energy ultraviolet radiation), is clearly of interest for automated peptide synthesis, where the continuous removal of protecting groups is envisaged.

The technique of electrolysis at a controlled potential [19, 20] has been applied to the removal of a series of 2-halogenoethoxycarbonyl amino-protecting groups and of some 2-halogenoethoxy carboxy-protecting groups. In addition to the 2-halogenoethoxycarbonyl N-protecting groups previously described in the literature (*viz.* 2-chloroethoxycarbonyl,[21] 2-bromoethoxycarbonyl,[21] 2-iodoethoxycarbonyl,[21] and 2,2,2-trichloroethoxycarbonyl,[22–25]* which were removed by zinc dust in methanol or aqueous acetic acid, or by a zinc–copper couple), the following new groups have been investigated: 2,2-dichloroethoxycarbonyl,[20] 2,2,2-trifluoroethoxycarbonyl,[19] and 2,2,2-tribromoethoxycarbonyl [19] (previously used to protect hydroxy-functions in nucleosides [27]). Amino-acids were converted into the 2-halogenoethoxycarbonyl derivatives using the corresponding chloroformates. In the case of carboxy-protecting groups, in addition to the widely used 2,2,2-trichloroethoxy-group (in work on cephalosporins,[23] penicillins,[28] prostaglandins,[29] and as a protecting group for the phosphate group in nucleotides [30]), the 2-chloroethoxy-, 2-bromoethoxy-, 2-iodoethoxy-, 2,2-dichloroethoxy-, and 2,2,2-tribromoethoxy-groups were investigated, but only as esters of benzoic acid. It was shown that many of these groups can be removed cleanly by cathodic reduction. The preparative scale electrolysis system consists of a working electrode (zinc [19] or mercury pool [19, 20]), a reference electrode (saturated calomel), and a counter electrode (graphite [19] or platinum [20]), with lithium chloride in a mixture of acetic acid and methanol [19] or lithium perchlorate in methanol [20] as the electrolyte, and a potentiostat to control the applied voltage. The groups found to be removed most readily were 2-iodoethoxycarbonyl, 2,2,2-tri-

19 E. Kasafírek, *Tetrahedron Letters*, 1972, 2021.
20 M. F. Semmelhack and G. E. Heinsohn, *J. Amer. Chem. Soc.*, 1972, **94**, 5139.
21 J. Grimshaw, *J. Chem. Soc.*, 1965, 7136.
22 T. B. Windholz and D. B. R. Johnston, *Tetrahedron Letters*, 1967, 2555.
23 R. B. Woodward, K. Heusler, J. Gosteli, P. Naegeli, W. Oppolzer, R. Ramage, S. Ranganathan, and H. Vorbrüggen, *J. Amer. Chem. Soc.*, 1966, **88**, 852.
24 S. Karady, S. H. Pines, L. M. Weinstock, F. E. Roberts, G. S. Brenner, A. M. Hoinowski, T. Y. Chang, and M. Sletzinger, *J. Amer. Chem. Soc.*, 1972, **94**, 1411.
25 H. Yajima and Y. Kiso, *Chem. and Pharm. Bull.* (*Japan*), 1971, **19**, 420.
26 S. Rakhit, J. F. Bagli, and R. Deghenghi, *Canad. J. Chem.*, 1969, **47**, 2906.
27 A. F. Cook, *J. Org. Chem.*, 1968, **33**, 3589.
28 R. D. G. Cooper and F. L. José, *J. Amer. Chem. Soc.*, 1972, **94**, 1022.
29 J. E. Pike, F. H. Lincoln, and W. P. Schneider, *J. Org. Chem.*, 1969, **34**, 3552.
30 F. Eckstein, *Chem. Ber.*, 1967, **100**, 2228.

* This group has also been used to protect hydroxy-groups: see refs. 22 and 26.

chloroethoxycarbonyl, 2,2,2-tribromoethoxycarbonyl, 2,2,2-trichloroethoxy, and 2,2,2-tribromoethoxy; in addition the 2-chloroethoxycarbonyl group could be removed in the presence of sodium iodide. However, under certain conditions both the 2,2,2-trichloroethyl protecting groups undergo degradation to the corresponding 2,2-dichloroethyl derivatives (Scheme 10). The *N*-benzyloxycarbonyl and *N*-tosyl groups are claimed[19] to be

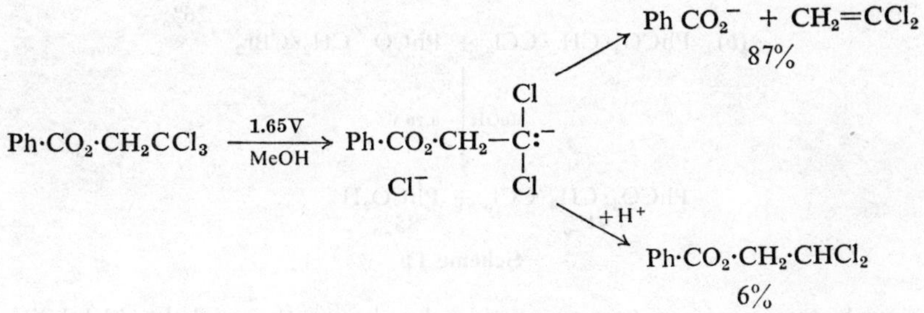

Scheme 10

stable under the conditions of electrolysis which were used (see, however, Vol. 4 of these Reports, p. 316). The application of the 2-iodoethoxycarbonyl group in peptide synthesis was tested by the synthesis of a few peptides by the mixed anhydride, active ester, or dicyclohexylcarbodiimide methods (cathodic reduction of a methionine peptide proceeds uneventfully), but no study of racemization has been reported.[19] The most important development, however, may follow from the determination of the half-wave potentials for the reduction of these halogeno-protecting groups in the analytical polarograph.[19, 20] It was argued that if the half-wave potentials for a given pair of groups differed by at least 0.3 V, then the selective reduction of a particular 2-halogenoethoxy-group in the presence of a less easily reduced group should be possible (Table 1). It must be

Table 1

Amino-protecting group	$E_{\frac{1}{2}}$/V
$ICH_2 \cdot CH_2 \cdot O \cdot CO$~~~	-1.28^a
$Cl_3C \cdot CH_2 \cdot O \cdot CO$~~~	-0.89^a
$Br_3C \cdot CH_2 \cdot O \cdot CO$~~~	-0.09^a

[a] Half-wave potentials ($E_{\frac{1}{2}}$) are in volts, presumably relative to the saturated calomel electrode, although not explicitly stated.[19]

stressed that this work is preliminary in nature and the examples of such selective removals are sparse (see Scheme 11). Although this kind of protecting group is chemically ingenious, much work remains to be done before peptide chemists can accept these methods into their armoury. However, controlled electrolytic reduction of protecting groups could in

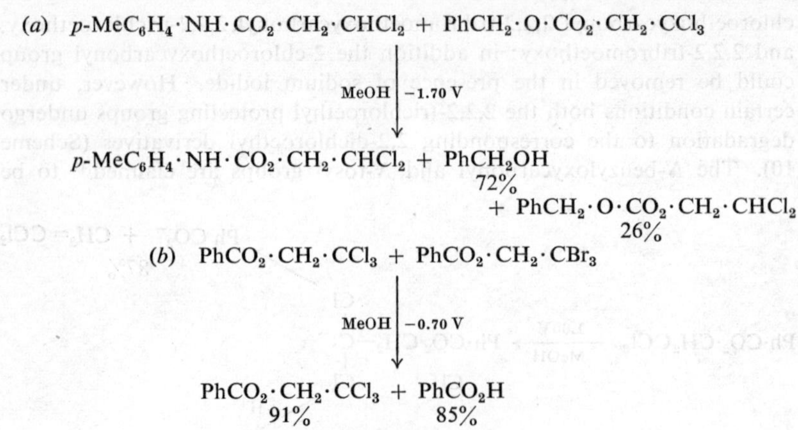

Scheme 11

time become as important to peptide chemists as the graded acid lability of different protecting groups is at present.

The stability of a number of N-protected derivatives to polarographic reduction in aqueous solution has been studied.[31] Although the 2-nitrophenylsulphenyl group is cleaved and the nitro-group is cleanly removed from nitroarginine, formyl, acetoacetyl, xanthyl, benzyloxycarbonyl, and t-butoxycarbonyl groups are stable: phthaloyl derivatives are converted into phthalamic acids.

A crafty protecting group which replaces both hydrogens of a primary amine, and resembles an *N*-benzyloxycarbonyl-*N*-benzylamino-group (21),

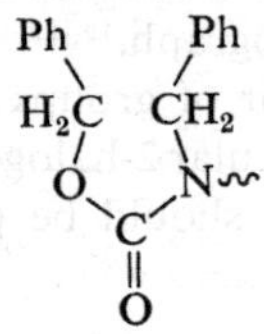

(21)

has been devised. It incorporates the primary amine nitrogen into the stable, unreactive, 4,5-diphenyl-4-oxazolin-2-one ring system (22).[32] The preparation of these derivatives is given in Scheme 12, together with the preparation of the acylating agent, benzoin cyclic carbonate (23). The highly crystalline amino-acid oxazolinones (22) are stable to aqueous base, refluxing ethanolic hydrazine, ethanolic hydrogen chloride, hydrogen bromide in acetic acid, refluxing trifluoroacetic acid, and anhydrous hydrogen fluoride. However, the protecting group may be removed quantitatively by low-pressure catalytic hydrogenation, or by sodium in liquid

[31] G. Capobianco and G. Vivaldi, *Internat. J. Protein Res.*, 1971, 3, 325.
[32] J. C. Sheehan and F. S. Guziec, jun., *J. Amer. Chem. Soc.*, 1972, **94**, 6561.

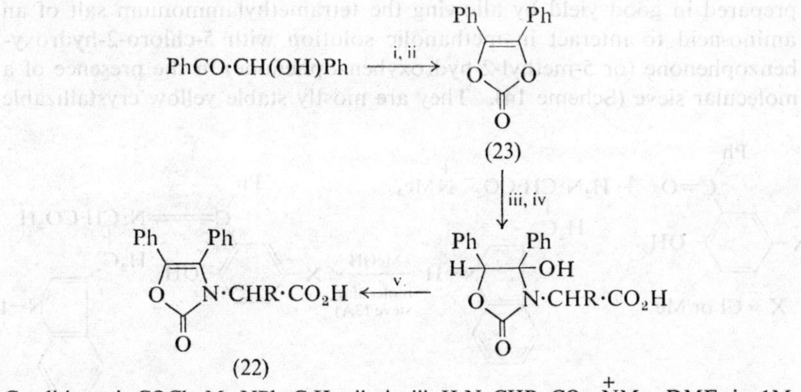

Conditions: i, $COCl_2$–Me_2NPh–C_6H_6; ii, Δ; iii, $H_2N \cdot CHR \cdot CO_2^- \overset{+}{N}Me_4$–DMF; iv, 1M-HCl; v, CF_3CO_2H

Scheme 12

ammonia. Alternatively, the group may be cleaved under oxidative conditions, with a peracid, followed by mild hydrolysis (these points are summarized in Scheme 13). In this preliminary communication [32] only a few

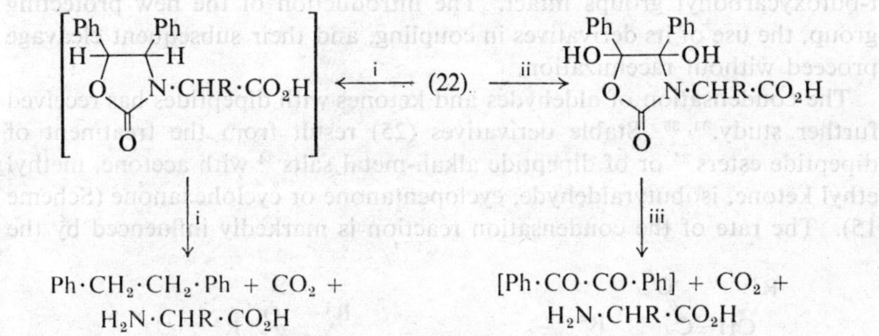

Conditions: i, H_2–Pd/C, organic solvent + aq. acid, 24 h; ii, *m*-$ClC_6H_4 \cdot CO_3H$–CF_3CO_2H; iii, hydrolysis

Scheme 13

protected amino-acids were described, together with a few dipeptide syntheses carried out using 1-ethyl-3-(3-dimethylaminopropyl)carbodiimide hydrochloride as condensing agent. The deblocked dipeptides were obtained in high yield, free of racemization. The fact that this protecting moiety is fluorescent and stable under solvolytic conditions suggests that it might also be useful for *N*-terminal residue analysis. Further developments of this new concept in amino-group protection are awaited with interest.

Schiff bases have also been employed as protecting groups capable of replacing both hydrogen atoms of a primary amine. Ketimines [33] can be

[33] A. Hope and B. Halpern, *Tetrahedron Letters*, 1972, 2261.

prepared in good yield by allowing the tetramethylammonium salt of an amino-acid to interact in methanolic solution with 5-chloro-2-hydroxybenzophenone (or 5-methyl-2-hydroxybenzophenone) in the presence of a molecular sieve (Scheme 14). They are mostly stable yellow crystallizable

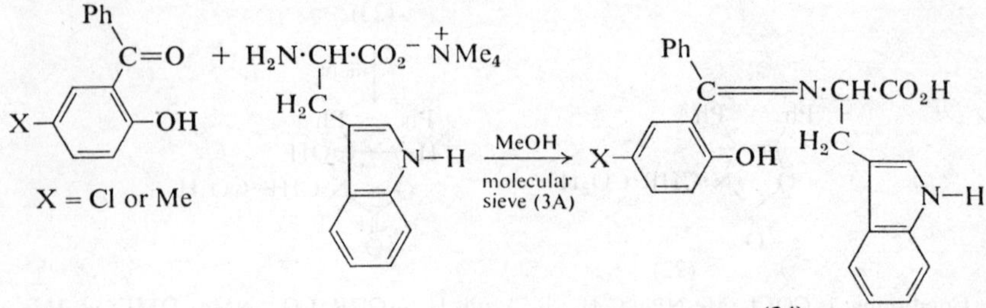

(24)

Scheme 14

solids [*e.g.* (24)] and can be utilized in peptide synthesis using dicyclohexylcarbodi-imide as coupling agent. The new protecting group is removed by 80% acetic acid (20 min at 80 °C, or 10 h at 25 °C), which conditions leave t-butoxycarbonyl groups intact. The introduction of the new protecting group, the use of its derivatives in coupling, and their subsequent cleavage proceed without racemization.

The condensation of aldehydes and ketones with dipeptides has received further study.[34, 35] Stable derivatives (25) result from the treatment of dipeptide esters [35] or of dipeptide alkali-metal salts [34] with acetone, methyl ethyl ketone, isobutyraldehyde, cyclopentanone or cyclohexanone (Scheme 15). The rate of the condensation reaction is markedly influenced by the

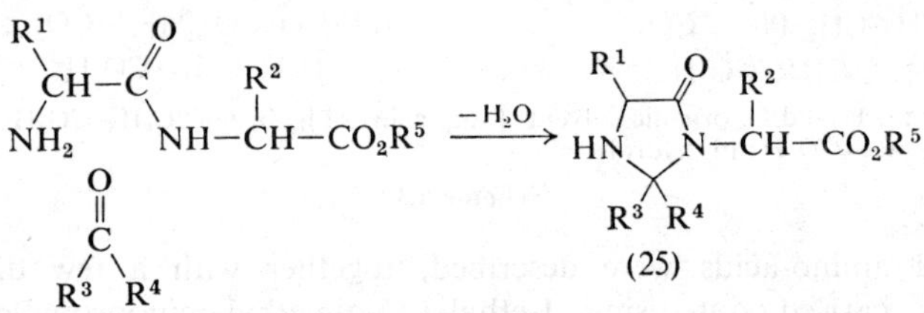

(25)

R^5 = Na, Me, or Et

Scheme 15

nature of the groups R^1 and R^2. In general, if R^1 is the same size or larger than R^2, one product results, namely (25). The spectroscopic properties of these 4-imidazolidinones have been documented.[34, 35] The reaction has some synthetic utility for separating mixtures of α- and β-aspartyl peptide

[34] C. A. Panetta and M. Pesh-Imam, *J. Org. Chem.*, 1972, **37**, 302.
[35] Y. Ariyoshi and N. Sato, *Bull. Chem. Soc. Japan*, 1972, **45**, 2015.

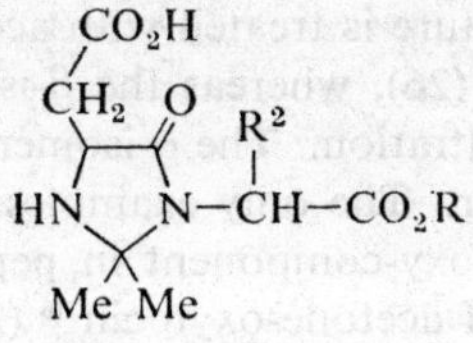

(26)

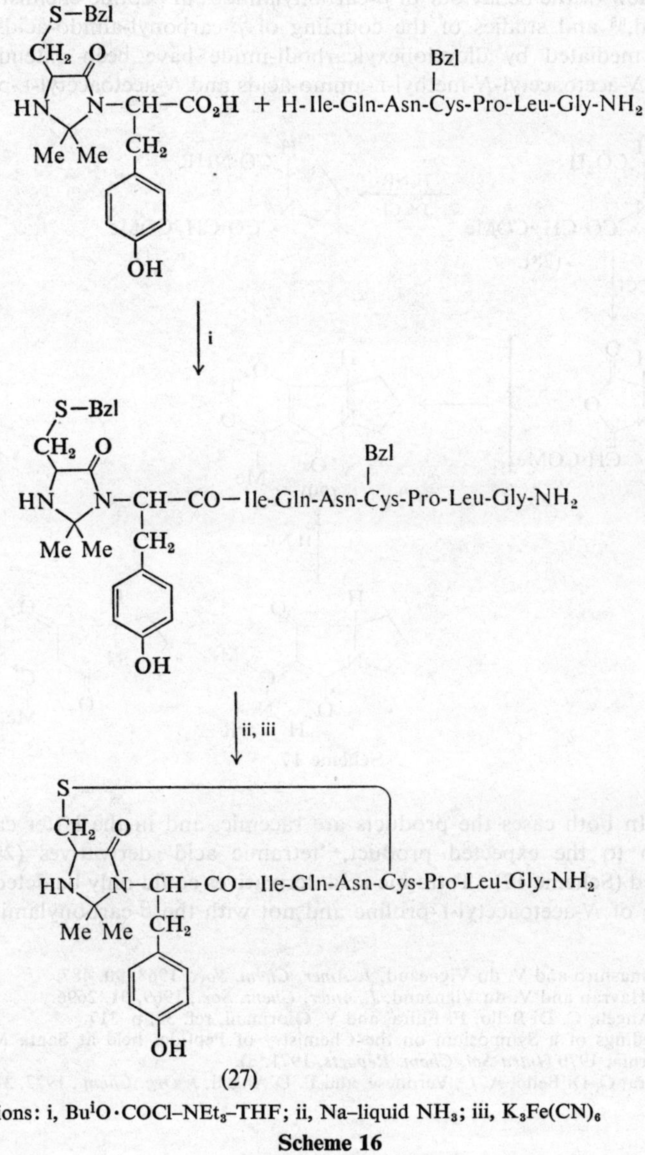

(27)

Conditions: i, $Bu^iO\cdot COCl$–NEt_3–THF; ii, Na–liquid NH_3; iii, $K_3Fe(CN)_6$

Scheme 16

esters.[35] When such a mixture is treated with acetone, the α-isomer forms a soluble imidazolidinone (26), whereas the β-isomer remains unchanged and can be recovered by filtration. The α-isomer is recovered after hydrolysis of (26) with hot water. The only examples of the use of a dipeptide imidazolidinone as a carboxy-component in peptide synthesis are in the preparation (Scheme 16) of acetone-oxytocin[36] (27) and of acetone-lysine-vasopressin.[37]

A review of the behaviour of β-carbonylamides in peptide chemistry has appeared,[38] and studies of the coupling of β-carbonyl-amido-acids with amines mediated by dicyclohexylcarbodi-imide have been extended to include *N*-acetoacetyl-*N*-methyl-L-amino-acids and *N*-acetoacetyl-L-proline

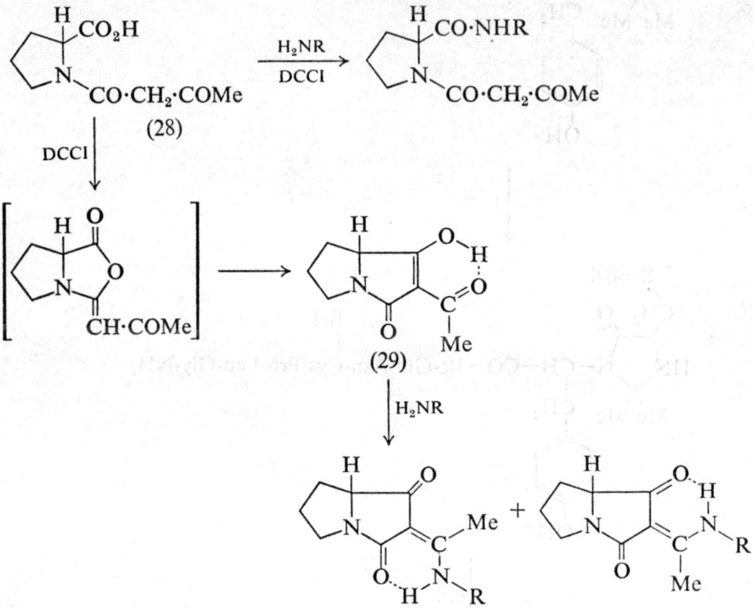

Scheme 17

(28).[40] In both cases the products are racemic, and in the latter case, in addition to the expected product, 'tetramic acid' derivatives (29) are produced (Scheme 17). Tetramic acid derivatives could only be detected in the case of *N*-acetoacetyl-L-proline and not with the β-carbonylamides of

[36] D. Yamashiro and V. du Vigneaud, *J. Amer. Chem. Soc.*, 1968, **90**, 487.
[37] R. T. Havran and V. du Vigneaud, *J. Amer. Chem. Soc.*, 1969, **91**, 2696.
[38] F. D'Angeli, C. Di Bello, F. Filira, and V. Giormani, ref. 39, p. 317.
[39] Proceedings of a Symposium on the Chemistry of Peptides held at Santa Monica, California, 1970 (*Intra-Sci. Chem. Reports*, 1971, **5**).
[40] F. Filira, C. Di Bello, A. C. Veronese, and F. D'Angeli, *J. Org. Chem.*, 1972, **37**, 3265.

other amino-acids. Further studies on the β-carbonylamides of serine have been reported.[41]

Di-t-butyl dithiodicarbonate (30), which is prepared according to Scheme 18, reacts with amino-esters and amino-acids under moderate

$$\mathrm{Bu^tSH \xrightarrow{i,\,ii} Bu^tS\cdot CO_2^-\,Na^+ \xrightarrow{iii,\,iv} Bu^tS\cdot CO\cdot O\cdot CO\cdot SBu^t}$$

(30)

Conditions: i, NaH; ii, CO_2; iii, $COCl_2$; iv, $\Delta(-CO_2)$

Scheme 18

conditions to yield the corresponding *N*-t-butylthiocarbonylamino-acid derivatives (31).[9] These thio-analogues of t-butoxycarbonyl derivatives are stable to acidic conditions (*e.g.* formic or trifluoroacetic acids) but can be cleaved with peracids. No racemization is detectable by optical rotation criteria. In the formation of the tyrosine derivative, no *O*-acylation was apparent. The corresponding serine derivative, however, is unstable and so is converted into the corresponding trimethylsilyl ether (32).

$\mathrm{Bu^tS\cdot CO\cdot NH\cdot CHR\cdot CO_2H}$ (31)

$\mathrm{Bu^tS\cdot CO\cdot NH\cdot CH(CH_2\cdot OSiMe_3)\cdot CO_2H}$ (32)

An example has appeared of the use of a metal as an *N*-terminal protecting group.[42] (*trans*-Dichlorobisamino-acid)platinum(II) (33) can be prepared by heating potassium tetrachloroplatinum(II) with excess amino-acid in aqueous solution (Scheme 19). Treatment of the complex (33) with

$$\mathrm{K_2PtCl_4 + NH_2\cdot CHR^1\cdot CO_2H \xrightarrow{i} Cl{-}Pt(NH_2\cdot CHR^1\cdot CO_2H)_2{-}Cl}$$

(33)

$\downarrow$ ii

$$\mathrm{Cl{-}Pt(NH_2\cdot CHR^1\cdot CO\cdot Gly\cdot OR^2)_2{-}Cl}$$

(34)

R^1 = H, Me, Pr^i, or CH_2OH
R^2 = Me or Et

Conditions: i, H_2O, 90 °C; ii, Gly·OR^2–DCCI–DMF or Me_2CO

Scheme 19

[41] A. Balog, D. Breazu, C. Daicoviciu, E. Vargha, L. Beu, and F. Gonczy, *Rev. Roumaine Chem.*, 1971, **16**, 1601 (*Chem. Abs.*, 1972, **76**, 34 561t).
[42] B. Purucker and W. Beck, *Z. Naturforsch.*, 1972, **27b**, 1140.

glycine ester and dicyclohexylcarbodi-imide yields the corresponding dipeptide complex (34). A number of simple dipeptides of the type (34) were prepared, but as yet the liberation of the dipeptide from the platinum has not been studied, although it is likely to require rather vigorous conditions.

Protection of Carboxy-groups. Peptide chemists have, whenever possible, avoided alkaline saponification as a method of removing ester protecting groups, on account of the side reactions which often result. This year has seen the publication of figures showing the extent of racemization which can attend alkaline hydrolysis of unactivated esters.[43,44] The products of model saponifications were resolved into their diastereoisomers and estimated on an amino-acid analyser after suitable deprotection. Scheme 20

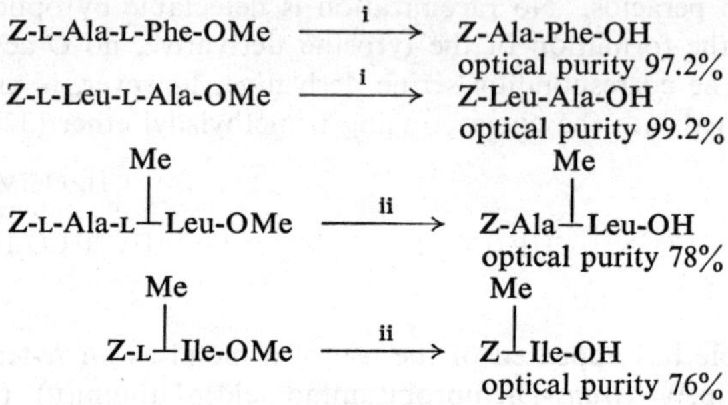

Conditions: i, 1 equivalent 0.25M-NaOH + 3 volumes acetone, 1 h, 20 °C; ii, saponification, conditions not specified

Scheme 20

shows that the racemization is by no means insignificant: particularly noteworthy are the results with *N*-methylamino-acid derivatives, which have generally been presumed to be resistant to racemization.[44] The warning is clear.

A convenient method for the preparation of certain amino-esters which are labile to acid involves protecting the amino-group as an enamine by reacting the potassium salt of the amino-acid with ethyl acetoacetate. These derivatives are readily converted into the required esters by alkylation with benzyl chloride, 4-methoxybenzyl chloride, or 2,4,6-trimethylbenzyl chloride. Mild acidolysis (HCl–MeOH) removes the N-protecting moiety to yield the salt of the required amino-ester. The three-step synthesis can be carried out without isolation of the intermediate compounds and gives very good yields with most amino-acids. It is convenient for the preparation of benzyl, 4-methoxybenzyl, 2,4,6-trimethylbenzyl, 4-nitrobenzyl, and

[43] G. W. Kenner and J. H. Seely, *J. Amer. Chem. Soc.*, 1972, **94**, 3259.

[44] J. R. McDermott and N. L. Benoiton, in 'Chemistry and Biology of Peptides', Proceedings of the Third American Peptide Symposium, Ann Arbor Science Publishers, 1972, p. 369.

4-picolyl esters (in the latter case, tetramethylguanidine is used as the base in the alkylation step). The products are free of racemate.[45]

Further papers have appeared illustrating the usefulness of the picolyl ester technique[46, 47] of peptide synthesis. Examples are recorded of the use of the azide method of coupling in conjunction with this ester group.

Conditions have been described for the synthesis of oligopeptide phenyl esters, which appear to crystallize readily, and for the removal of the phenyl group under very mild conditions.[43] In general, hydrolysis of *C*-terminal phenyl esters can be accomplished rapidly and efficiently at pH 10.5 in the presence of 0.8 equivalents of hydrogen peroxide in mixtures of water and an organic solvent such as acetone, dioxan, or dimethylformamide. Presumably a peptide peracid is the initial product, which is rapidly converted into the carboxylic acid. Racemization could not be detected using an amino-acid analyser after the hydrolysis of (35) or (36) (*cf.* Scheme 20). However, in the absence of peroxide, racemization was

Z-L-Ala-L-Phe-OPh	Z-L-Leu-L-Ala-OPh	Z-L-Asp(OBu^t)-Gly-OPh
(35)	(36)	(37)

observed. Model studies confirmed the absence of an $\alpha \rightarrow \beta$ aspartyl shift during the deprotection of (37). Indole and sulphur-containing side-chains (*e.g.* methionine or *S*-acetamidomethyl-cysteine) could be protected from unwanted oxidation by carrying out the peroxide-catalysed hydrolysis in the presence of excess dimethyl sulphide. The authors[43] also point out that acylation, coupled with deacylation by peroxide anion, provides a potentially general method for protecting phenolic hydroxy-groups.

NN'-Di-isopropylhydrazine has been introduced as a new reagent for the protection of carboxy-groups. It can be introduced by coupling (Scheme 21) the mixed anhydride of the carboxy-component with the disubstituted hydrazine, which is itself readily available by catalytic reduction of acetone azine. The resulting *N*-acyl-*NN'*-di-isopropylhydrazide (38) is stable to both acidic and basic conditions, but is rapidly removable under mild conditions by selective oxidation.[48] The preferred oxidant is lead tetra-acetate (a probable mechanism for this reaction is shown in Scheme 21), although sodium periodate, *N*-bromosuccinimide, and chromium trioxide in acetic acid have been used. Since the oxidation of the hydrazide fragment produces materials which are gaseous under the conditions of work-up, the products are particularly clean. The new protecting group has been used for carboxy-protection in the penicillin series but since the intermediate species formed during the oxidative

[45] J. A. Maclaren, *Austral. J. Chem.*, 1972, **25**, 1293.
[46] D. J. Schafer, *J.C.S. Perkin I*, 1972, 1452.
[47] G. A. Fletcher and G. T. Young, *J.C.S. Perkin I*, 1972, 1867.
[48] D. H. R. Barton, M. Girijavallabhan, and P. G. Sammes, *J.C.S. Perkin I*, 1972, 929.

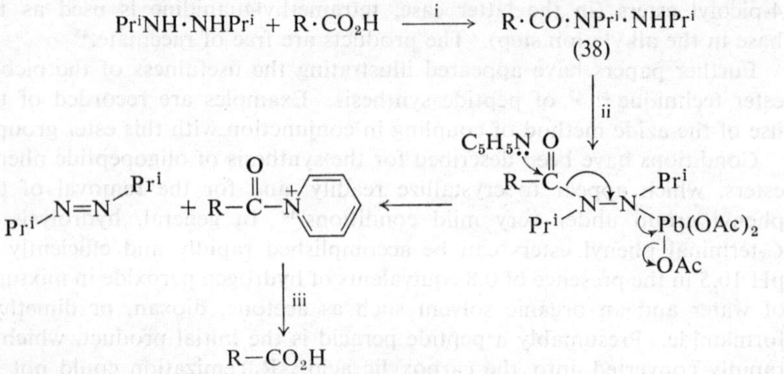

Conditions: i, EtOCOCl–NEt_3; ii, $Pb(OAc)_4$–C_5H_5N; iii, H_2O

Scheme 21

cleavage reaction may themselves be acylating agents (*e.g.* the proposed acylpyridinium salt in Scheme 21), there is risk of racemization and other side reactions with more typical linear peptides.

The utility of the benzoin moiety as a photosensitive carboxy-protecting group has been investigated.[49] A careful study of the photolysis of a series of substituted benzoin acetates indicated the superiority of 3′,5′-dimethoxybenzoin acetate (39), which on irradiation (medium-pressure mercury arc) smoothly yields 5,7-dimethoxy-2-phenylbenzofuran (40) as shown in Scheme 22. Phthaloylglycine 3′,5′-dimethoxybenzoin ester (41), prepared

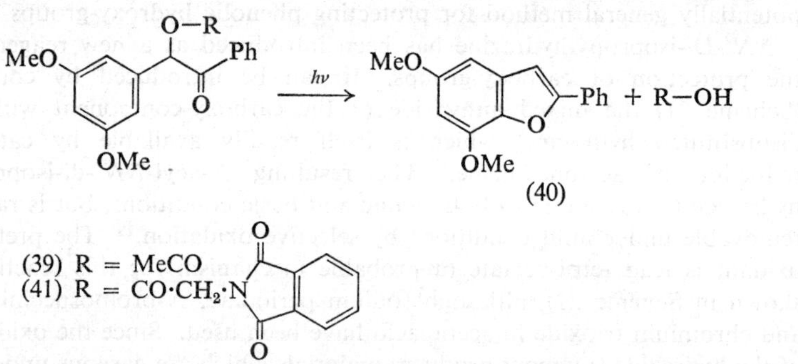

Scheme 22

from phthaloylglycyl chloride, releases phthaloylglycine in very high yield on photolysis in benzene. This new protecting group possesses the advantages that the cleavage reaction proceeds in very high yield, the blocking-group photoproduct is readily separated from the required carboxylic

[49] J. C. Sheehan, R. M. Wilson, and A. W. Oxford, *J. Amer. Chem. Soc.*, 1971, **93**, 7222.

acid, the excited state is short lived (thus avoiding undesirable quenching processes), and the wavelength of the exciting radiation is between 320 and 400 nm (thus ensuring that amino-acid side-chains will not absorb radiation intended for the blocking group, and that the peptide synthesis does not have to be carried out in the dark!). However, since the blocking moiety contains an asymmetric centre, difficulties are to be anticipated in the purification of peptide diastereoisomers.

Carboxylic acids can be converted into 2-oxazolines by acid-catalysed rearrangement of the corresponding acylaziridines (Scheme 23).[50] The

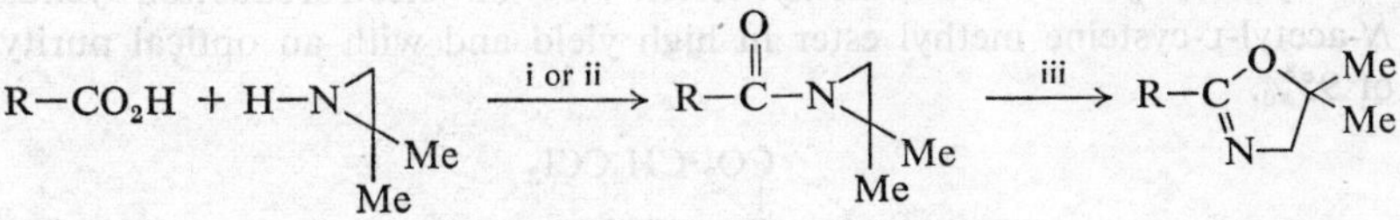

Conditions: i, DCCI; ii, acid chloride route–$SOCl_2$; iii, H_2SO_4 (catalytic amount) in Et_2O or CH_2Cl_2

Scheme 23

protecting group is stable to lithium aluminium hydride and is converted into the corresponding ethyl ester on treatment with ethanolic sulphuric acid. This method may prove useful for the selective reduction of functional groups in amino-acid derivatives.

The synthesis of terpene esters of amino-acids has been reported.[51]

Protection of Thiol Groups and the Synthesis of Cystine Peptides. The 4-picolyl group is suitable for the protection of the thiol group of cysteine.[52] *S*-4-Picolyl-L-cysteine (42) can be prepared by the reaction of purified

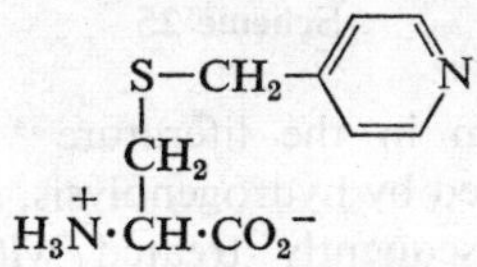

(42)

4-picolyl chloride with the solution obtained by the reduction of L-cystine by sodium in liquid ammonia. It is stable in the presence of trifluoroacetic acid or hydrogen bromide in acetic acid, but can be removed by electrolytic reduction of a solution in 0.25M sulphuric acid. Scheme 24 illustrates the use of this group in the synthesis of L-cystinyl-bis-glycine. The *S*-2,2,2-trichloroethoxycarbonyl group can also be removed from cysteine derivatives by electrolytic reduction.[20] For example, *N*-acetyl-*S*-2,2,2-trichloro-

[50] D. Haidukewych and A. I. Meyers, *Tetrahedron Letters*, 1972, 3031.
[51] E. G. Titkova and S. A. Kozhin, *Zhur. obshchei Khim.*, 1972, **42**, 1175.
[52] A. Gosden, D. Stevenson, and G. T. Young, *J.C.S. Chem. Comm.*, 1972, 1123.

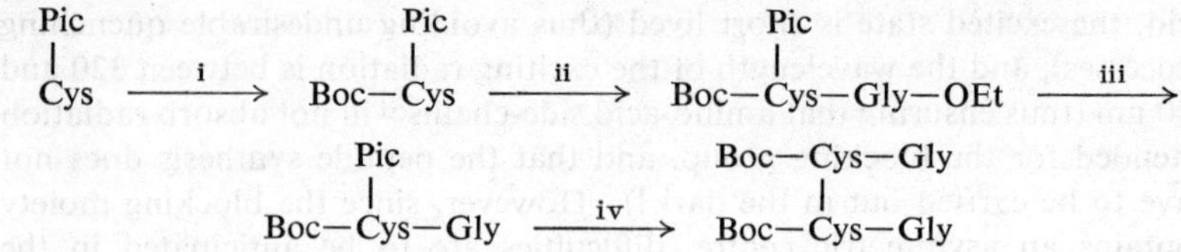

Conditions: i, Boc—N_3; ii, Gly—OEt and DCCI; iii, NaOH-dioxan; iv, electrolytic reduction and autoxidation

Scheme 24

ethoxycarbonyl-L-cysteine methyl ester (43) on electroreduction yeilds *N*-acetyl-L-cysteine methyl ester in high yield and with an optical purity of 95%.

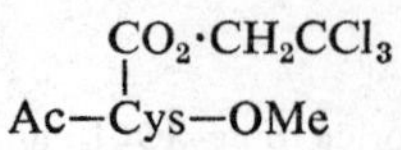

(43)

S-p-Nitrobenzyl-L-cysteine (44) affords *S-p*-aminobenzyl-L-cysteine on catalytic hydrogenation, and this, on treatment with the Hopkins reagent and subsequent decomposition of the mercury mercaptide followed by oxidation, yields cystine (Scheme 25).[53] These observations offer an

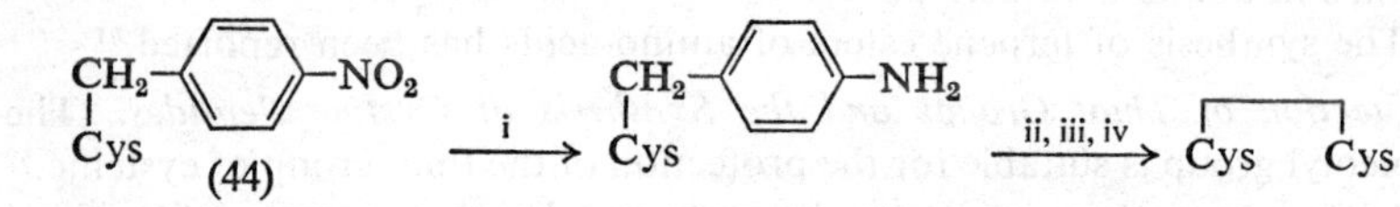

Conditions: i, $3H_2$–Pd/C; ii, $HgSO_4$–aq. H_2SO_4; iii, H_2S; iv, air

Scheme 25

explanation for the claim in the literature[54] that the *S-p*-nitrobenzyl protecting group is removed by hydrogenolysis, since in the work cited the supposed thiol was subsequently treated with the Hopkins reagent. *S*-Benzyl- and *S-p*-nitrobenzyl-L-cysteine are unaffected by this reagent.

A detailed report[55] discusses the behaviour of the *S*-ethylcarbamoyl group (45) in the synthesis of a ribonuclease T_1 fragment (46), which contains three cysteine residues. The stability of the *S*-ethylcarbamoyl moiety to trifluoroacetic acid and hydrogen bromide in trifluoroacetic acid and its lability to alkali (pH 8.5 or higher) were confirmed. In addition, however, this protecting group is readily cleaved by silver or mercuric acetate with the formation of a mercaptide. *p*-Chloromercuribenzoate reacts rapidly with *S*-ethylcarbamoyl derivatives at pH 7.0 but not at pH 4.6; this observation

[53] M. D. Bachi and K. J. Ross-Petersen, *J. Org. Chem.*, 1972, **37**, 3550.
[54] C. Berse, R. Boucher, and L. Piche, *J. Org. Chem.*, 1957, **22**, 805.
[55] H. T. Storey, J. Beacham, S. F. Cernosek, F. M. Finn, C. Yanaihara, and K. Hofmann, *J. Amer. Chem. Soc.*, 1972, **94**, 6170.

forms the basis of an analytical method for determining the *S*-ethylcarbamoyl content of synthetic fragments. In the synthesis of (46), cysteine residues were introduced using the active ester (47), and in order to minimize thiol-deblocking during the stepwise elongation of the peptide chain,

—S—CO·NHEt
(45) Ec

Ec
|
Boc—Cys—OTcp
(47)

Ec | Ec | Ec
HCO·Ala-Cys-Asp-Tyr-Thr-Cys-Gly-Ser-Asn-Cys-Tyr-NH·NH·Z
(46)

only 90% of the required molar quantity of triethylamine was used for neutralization.

A detailed paper on the *S*-acetamidomethyl protecting group has also appeared.[56] *S*-Acetamidomethyl-cysteine (48) is prepared by acid-catalysed *S*-alkylation of cysteine with *N*-hydroxymethylacetamide (49). *S*-Alkyl-

CH_2·NH·COMe
|
Cys
(48)

MeCO·NH·CH_2OH
(49)

ation of proteins with the reagent (49) is best carried out in anhydrous hydrogen fluoride. The protecting group is stable to alkali (pH 13), cold 1M hydrochloric acid, 15% ammonia, hydrazine, zinc dust in acetic acid, trifluoroacetic acid, and 6M-hydrogen chloride in ethyl acetate. It is not stable in the presence of 1M-hydrochloric acid at 100 °C, 30% hydrogen bromide in acetic acid, or anhydrous hydrogen fluoride. Although *S*-acetamidomethyl-cysteine can be oxidized with hydrogen peroxide, sulphoxide formation has not been detected under the conditions of peptide synthesis. Added dimethyl sulphide acts as a scavenger and protects the *S*-acetamidomethyl group from unwanted alkylation during deblocking of the growing peptide chain. The protecting group can be removed by the action of mercuric ions in water, aqueous acetic acid, or urea solution at pH 4. The derivatives (50)—(53) have proved useful in synthesis: racemization does not appear to be a problem.

Acm	Acm	Acm	Acm
Boc—Cys—OH	Boc—Cys—ONSu	Z—Cys—OH	Z—Cys—ONSu
(50)	(51)	(52)	(53)

[56] D. F. Veber, J. D. Milkowski, S. L. Varga, R. G. Denkewalter, and R. Hirschmann, *J. Amer. Chem. Soc.*, 1972, **94**, 5456.

The synthesis of some unsymmetrical disulphides containing cysteine have been reported, using the reaction of thiols with *S*-phthalimido-derivatives.[57] *N*-Trifluoroacetyl-*S*-phthalimido-L-cysteine methyl ester (54) was prepared according to Scheme 26. Thiolysis with benzyl thiol,

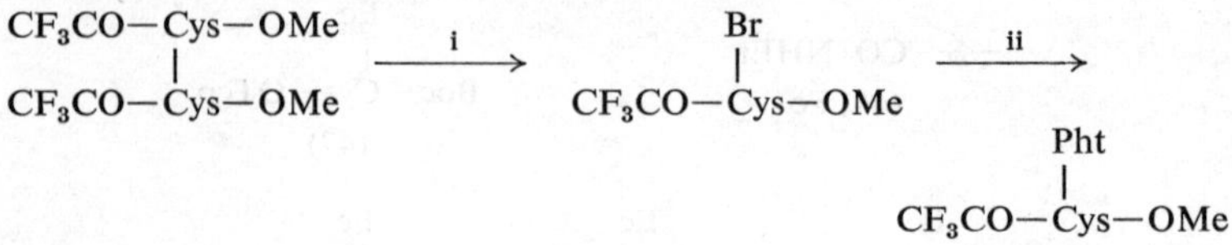

(54)

Conditions: i, Br_2; ii, potassium phthalimide

Scheme 26

cysteine hydrochloride monohydrate, or glutathione gives excellent yields of the corresponding unsymmetrical disulphides [*e.g.* (55)]. Only in the last case was a trace of symmetrical disulphide detected. The thiolysis

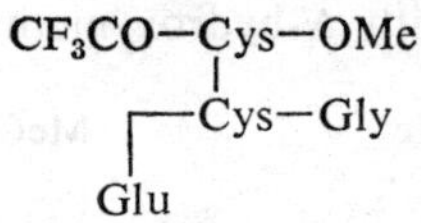

(55)

reaction appears to be suitable for the construction of disulphide bridges, providing selectively removable amino- and carboxy-protecting groups can be found which are stable under the conditions required for the synthesis of the *S*-phthalimido-derivatives. Alkyl phenyl disulphides, and unsymmetrical diaryl disulphides, can be prepared conveniently in very good yield by the interaction of methyl benzenesulphenate and a thiol (Scheme 27).[58]

$$\text{PhS·OMe} + \text{RSH} \longrightarrow \text{PhS·SR} + \text{MeOH}$$

Scheme 27

The synthesis of symmetrical cystine peptides comprising the sequences A14—A21 and A15—A21 of the insulin A-chain has been reported. This was achieved by reacting excess of a peptide azide [*e.g.* (56)] with a bifunctional cystine peptide (57) as shown in Scheme 28.[59] Further papers have appeared dealing with the fragment condensation of peptides containing a preformed disulphide bridge. Oxytocin has been prepared by coupling a hexapeptide disulphide (58) with a tripeptide amide (59) using dicyclohexylcarbodi-imide in the presence of 1-hydroxybenzotriazole

[57] D. N. Harpp and T. G. Back, *J. Org. Chem.*, 1971, **36**, 3828.
[58] D. A. Armitage, M. J. Clark, and C. C. Tso, *J.C.S. Perkin I*, 1972, 680.
[59] P. Fehrenbach and H. Zahn, *Chem. Ber.*, 1972, **105**, 1749.

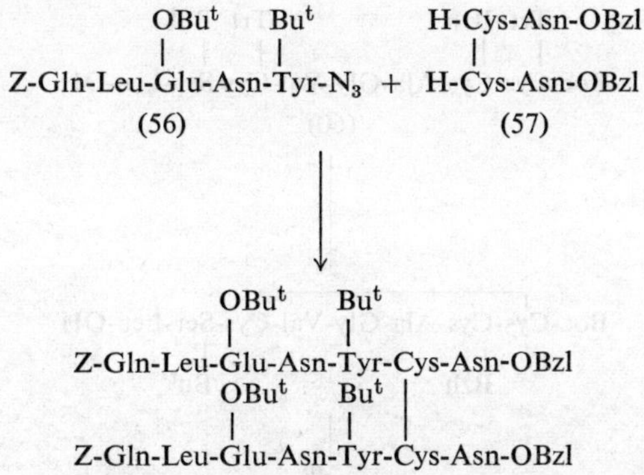

Scheme 28

(Scheme 29).[60] A synthesis of the protected A6—A13 segment of ovine insulin has been accomplished by a route using acid-labile protecting groups and coupling procedures involving *N*-hydroxysuccinimide. The resulting octapeptide (60) was cyclized with thiocyanogen to preform the intrachain disulphide bridge and then selectively deprotected, thus providing a peptide disulphide capable of chain-elongation at its amino- or

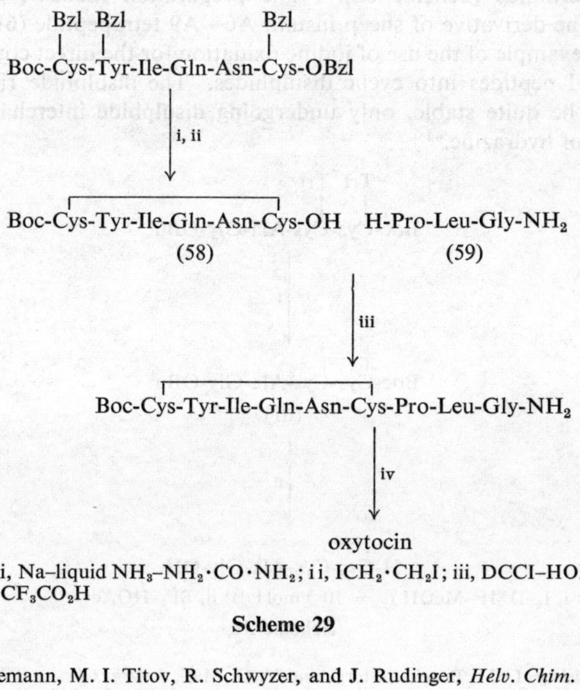

Conditions: i, Na–liquid NH_3–$NH_2 \cdot CO \cdot NH_2$; ii, $ICH_2 \cdot CH_2I$; iii, DCCI–HOBt; iv, 90% CF_3CO_2H

Scheme 29

[60] M. Mühlemann, M. I. Titov, R. Schwyzer, and J. Rudinger, *Helv. Chim. Acta*, 1972, **55**, 2854.

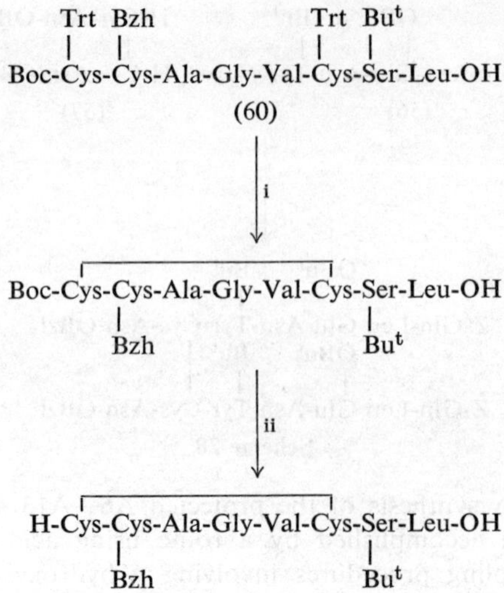

Conditions: i, $(SCN)_2$–$CHCl_3$–HOAc; ii, $CF_3\cdot CO_2H$–$CHCl_3$ at −20 °C

Scheme 30

carboxy-terminus (Scheme 30).[61] The preparation (Scheme 31) of the cyclocystine derivative of sheep insulin A6—A9 tetrapeptide (61) provides a further example of the use of iodine oxidation for the direct conversion of bis-*S*-trityl peptides into cyclic disulphides. The disulphide ring of (61) seems to be quite stable, only undergoing disulphide interchange in the presence of hydrazine.[62]

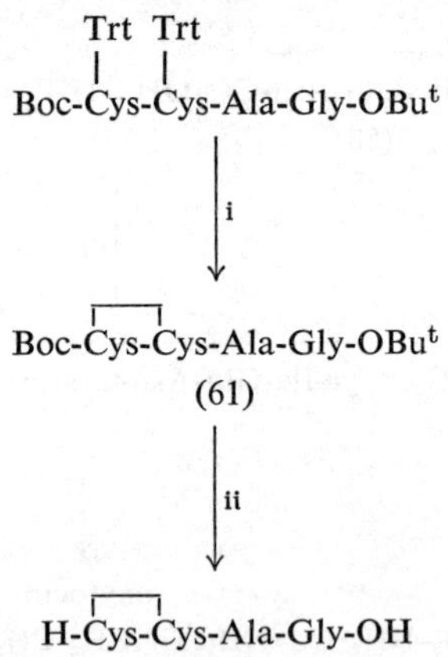

Conditions: i, I_2–DMF–MeOH ($c = 10^{-3}$ mol l^{-1}); ii, BF_3–HOAc

Scheme 31

[61] R. G. Hiskey, L. M. Beacham, tert., and V. G. Matl, *J. Org. Chem.*, 1972, **37**, 2472.
[62] H. Berndt, H. Klostermeyer, and H. Zahn, *Annalen*, 1972, **759**, 114.

The rates of cleavage of a series of aralkylthioethers of *N*-acetyl-L-cysteine-methylamide have been investigated under conditions commonly employed in peptide synthesis.[63] In the case of the *S*-trityl, *S*-(α-methylbenzhydryl), and *S*-(αα-dimethylbenzyl) substrates, an equilibrium is rapidly established in hydrogen bromide in acetic acid, which results in approximately 80% cleavage. All of the groups studied are cleaved slowly and incompletely in trifluoroacetic acid. In addition to the *S*-trityl-group, the *S*-(4,4′-dimethoxybenzhydryl) and *S*-(α,4,4′-trimethylbenzhydryl) groups are rapidly removed by iodine in methanol.

Protection of Primary Amide Groups and the Synthesis of Pyroglutamyl Peptides. The use of the 2,4-dimethoxybenzyl group for amide protection has been reported.[64] The reagent 2,4-dimethoxybenzylamine (62) is obtained according to Scheme 32. The dimethoxybenzyl derivatives of asparagine

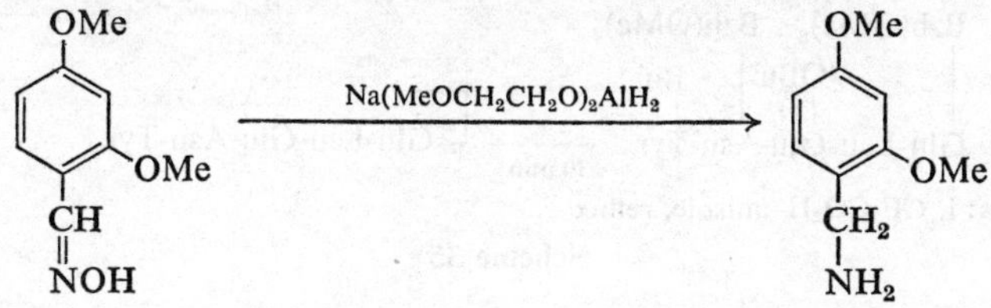

(62)

Scheme 32

and glutamine are obtained by coupling this reagent with the appropriate *N*-protected aspartic or glutamic acid α-esters, using either dicyclohexylcarbodi-imide or 1-diethylaminoprop-1-yne (*e.g.* Scheme 33). The deriva-

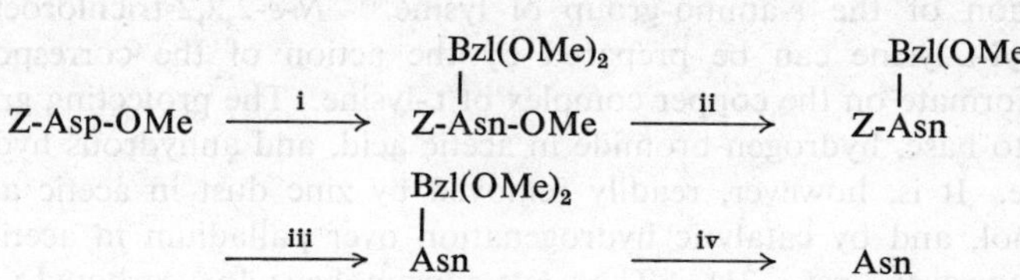

Conditions: i, (62)–$Me_2N{\cdot}C{:}C{\cdot}Me$; ii, NaOH; iii, H_2–Pd/C; iv, CF_3CO_2H or HF

Scheme 33

tives are crystalline, and the primary amide can be exposed by treating the derivative with trifluoroacetic acid or anhydrous hydrogen fluoride. Dimethoxybenzyl derivatives of glutaminyl peptides are stable to hydrogenolysis and treatment with base. The corresponding derivatives of asparaginyl peptides are stable to hydrogenolysis, but unstable in the presence of aqueous alkali. The authors state that the formation of pyroglutamyl derivatives was not observed during the removal of protecting

[63] G. Losse and T. Stölzel, *Tetrahedron*, 1972, **28**, 3049.

[64] P. G. Pietta, P. Cavallo, and G. R. Marshall, *J. Org. Chem.*, 1971, **36**, 3966.

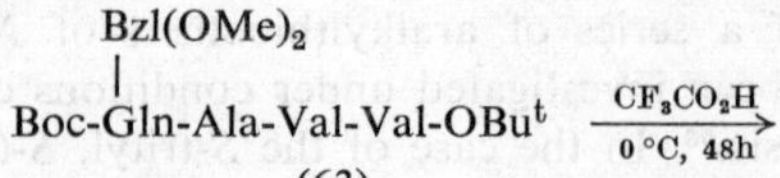

(63)

Scheme 34

groups from peptides such as (63) (Scheme 34). However, it has been shown that *N*-α-benzyloxycarbonyl-*N*-γ-(4,4′-dimethoxybenzhydryl)glutaminyl peptides are converted cleanly into pyroglutamyl peptides on refluxing in trifluoroacetic acid containing anisole[65] (Scheme 35). The synthesis of t-butyl

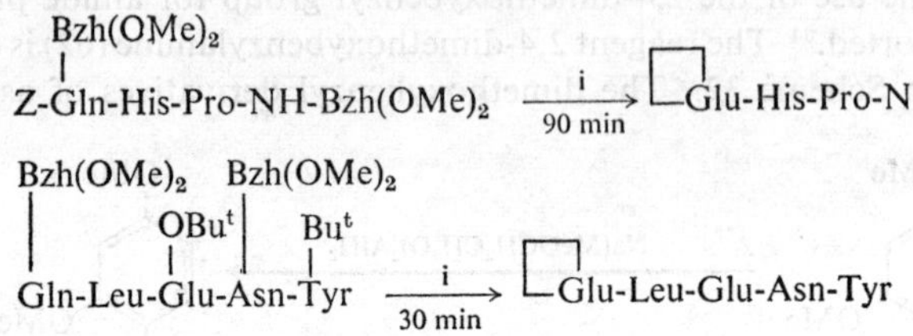

Conditions: i, CF_3CO_2H–anisole, reflux

Scheme 35

pyroglutamate[66] has been described. Treatment of pyroglutamyl peptides with hydrogen chloride in methanol leads to opening of the pyrrolidone ring.[67]

Miscellaneous Matters relating to Protective Groups and Functional Side-chains. The 2,2,2-trichloroethoxycarbonyl group has been used for the protection of the ε-amino-group of lysine.[68] *N*-ε-2,2,2-trichloroethoxy-carbonyl-L-lysine can be prepared by the action of the corresponding chloroformate on the copper complex of L-lysine. The protecting group is stable to base, hydrogen bromide in acetic acid, and anhydrous hydrogen fluoride. It is, however, readily removed by zinc dust in acetic acid or methanol, and by catalytic hydrogenation over palladium in acetic acid (see, however, ref. 21). The *N*-ε-*p*-bromobenzyloxycarbonyl-L-lysine derivative (64) has been used in solid-phase synthesis.[69] This protecting

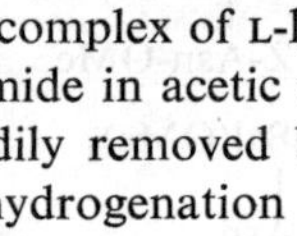

(64)

[65] W. König and R. Geiger, *Chem. Ber.*, 1972, **105**, 2872.
[66] M. Hollosi, M. Kajtar, Z. Rathonyi, and J. Tomasz, *Acta Chim. Acad. Sci. Hung.*, 1972, **71**, 104 (*Chem. Abs.*, 1972, **76**, 72 765v).
[67] I. Kawasaik and H. A. Itano, *Analyt. Biochem.*, 1972, **48**, 546.
[68] H. Yajima, H. Watanabe, and M. Okamoto, *Chem. and Pharm. Bull.* (*Japan*), 1971, **19**, 2185.
[69] D. Yamashiro and C. H. Li, *Internat. J. Peptide and Protein Res.*, 1972, **4**, 181.

group is more stable to acidolysis than is the benzyloxycarbonyl group, but is cleaved by anhydrous hydrogen fluoride.

A simplified synthesis of histidine peptides has been suggested in which the imidazole ring is used as a basic handle to simplify the isolation of products (*cf.* the picolyl ester method).[70]

It has been reported that the *N-im*-dinitrophenyl group is labile in the presence of aqueous alcoholic caustic soda (10 min, 0.2M reagent), ammonia in methanol (24 h), or hydrazine in ethanol (6 h), but that it is stable to triethylamine in methanol. Hydrazinolysis of a peptidyl polymer leads to simultaneous loss of the *im*-dinitrophenyl protecting group.[71]

The instability of γ-benzyl protecting groups to hydrogen bromide in acetic acid has been reiterated.[72] The direct preparation of the L-aspartic anhydride salts (65) has been reported (Scheme 36).[73] Preparations

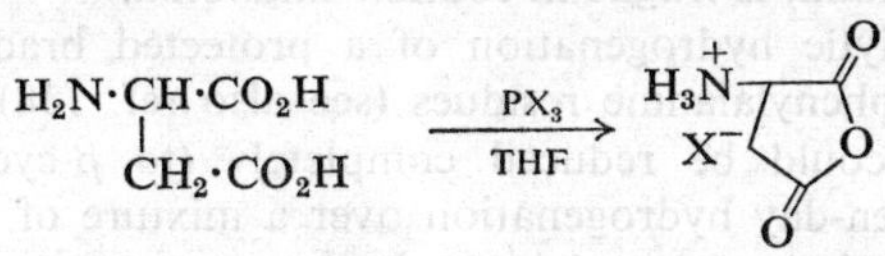

X = Cl or Br

(65)

Scheme 36

of t-butoxycarbonyl-L-aspartic acid α-benzyl ester, t-butoxycarbonyl-L-aspartic acid α-t-butyl ester, and the corresponding glutamic acid derivative have been described.[74]

The 4-picolyl group has been proposed for masking the hydroxy-function of tyrosine. Alkylation of the nickel complex of tyrosine yields *O*-4-picolyl-L-tyrosine. The protecting group is stable to trifluoroacetic acid, but electrolytic reduction liberates free tyrosine.[52]

Use of hydrogen bromide in trifluoroacetic acid to cleave the benzyl ether group from *O*-benzyltyrosyl peptides results in an irreversible modification of the tyrosyl residue[69, 75] (even when a carbonium ion scavenger is used). This problem can be obviated by using hydrogen bromide in acetic acid; this results in the phenolic moiety being acetylated, which can be reversed by alkaline treatment.[75] The use of the *m*-bromobenzyl ether of tyrosine (66) in solid-phase synthesis has been described.[69] This group is much more stable to acidolysis than the parent benzyl ether, but is removed with anhydrous hydrogen fluoride.

[70] D. J. Schafer and L. Carlsson, *J.C.S. Chem. Comm.*, 1972, 276.
[71] Yu. P. Shvachin, M. N. Ryabtsev, and A. P. Krymov, *J. Org. Chem.* (*U.S.S.R.*), 1972, **8**, 665.
[72] F. H. C. Stewart, *Austral. J. Chem.*, 1971, **24**, 2739.
[73] Y. Ariyoshi, T. Yamatani, N. Uchiyama, and N. Sato, *Bull. Chem. Soc. Japan*, 1972, **45**, 2208.
[74] J. Tomasz, *Acta Chim. Acad. Sci. Hung.*, 1971, **70**, 255 (*Chem. Abs.*, 1972, **76**, 14 871r).
[75] Y. Trudelle and G. Spach, *Tetrahedron Letters*, 1972, 3475.

Bzl(*m*-Br)
|
Boc-L-Tyr
(66)

Preparations of *O*-benzyl-L-serine,[76] and of t-butoxycarbonyl-L-threonine benzyl ester,[77] have been recorded.

Final deprotection of benzyl-protected peptides with sodium in liquid ammonia has been carried out in the presence of urea, which acts as an acid in the ammonia system and avoids the formation of the strong base sodium amide.[60] A study of the factors which influence acyl–proline cleavage by sodium in liquid ammonia has shown that the cleavage can be minimized by limiting both the amount of sodium used and the time of exposure of the peptide to it.[78] It has been shown that the seryl–proline bond, like the threonyl–proline bond, is fragile in sodium–ammonia.

Prolonged catalytic hydrogenation of a protected bradykinin caused reduction of the phenylalanine residues (see also ref. 138). The phenylalanine residues could be reduced completely (to β-cyclohexylalanine residues) with a ten-day hydrogenation over a mixture of palladium and platinum catalysts.[79]

Formation of the Peptide Bond.—A comprehensive review of reagents and reactions for the formation of peptide bonds has appeared.[80]

Activated Esters. A study of the kinetics of the amine-catalysed aminolysis and hydrolysis of 2-pyridyl thio-esters in water shows that the reactivity of primary and secondary amines is greater than that of tertiary amines, which react by a different mechanism. The greater reactivity of primary and secondary amines can be attributed to intramolecular general base catalysis (67).[81] The rate equation for the silver-ion promoted aminolysis of thioesters is consistent with a transition state which involves an approximately trigonal arrangement about the silver atom which ties the two reactants together and permits the synchronous transfer of electrons, as in (68).[82] A study of the kinetics and mechanism of aminolysis of esters in aprotic solvents has appeared,[83] and a further mention has been made of the catalysis of active ester coupling by acetic acid.[84]

The use of esters (69) of pyridine-4-aldoxime for peptide synthesis has been investigated.[85] Pentachlorophenyl and pentafluorophenyl esters of

[76] M. Masaki, T. Kato, and N. Izumiya, *Mem. Fac. Sci., Kyushu Univ., Ser. C*, 1972, **8**, 89 (*Chem. Abs.*, 1972, **76**, 14198j).
[77] C. W. Mosher and L. Goodman, *J. Org. Chem.*, 1972, **37**, 2928.
[78] A. Marglin, *Internat. J. Peptide and Protein Res.*, 1972, **4**, 47.
[79] D. J. Schafer, G. T. Young, D. F. Elliot, and R. Wade, *J. Chem. Soc.* (*C*), 1971, 46.
[80] Y. S. Klausner and M. Bodanszky, *Synthesis*, 1972, 453.
[81] G. J. Kasperek and T. C. Bruice, *J. Org. Chem.*, 1972, **37**, 1456.
[82] B. Boopsingh and D. P. N. Satchell, *Chem. and Ind.*, 1972, 426.
[83] F. M. Menger and J. H. Smith, *J. Amer. Chem. Soc.*, 1972, **94**, 3824.
[84] L. M. Litvinenko, Yu. A. Sharanin, V. V. Kosmynin, and L. P. Dridzh, *Doklady Akad. Nauk S.S.S.R.*, 1971, **200**, 854 (*Chem. Abs.*, 1972, **76**, 14911d).
[85] P. Y. Romanovskii, V. E. Muizhnieks, and G. I. Chipens, *Latv. P.S.R. Zinat. Akad. Vestis, Kim. Ser.*, 1972, 227 (*Chem. Abs.*, 1972, **77**, 75438x).

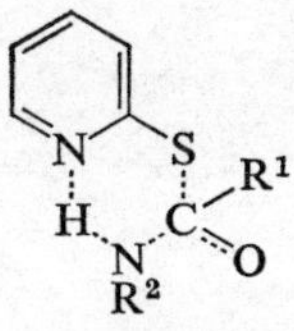

(67)

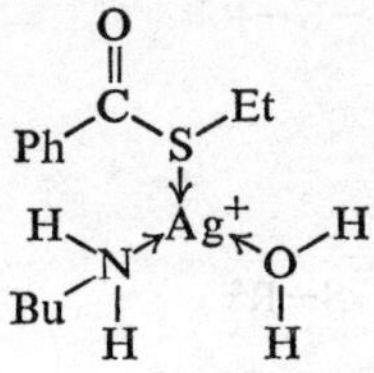

(68)

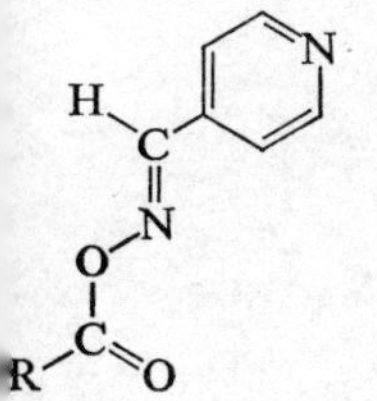

(69)

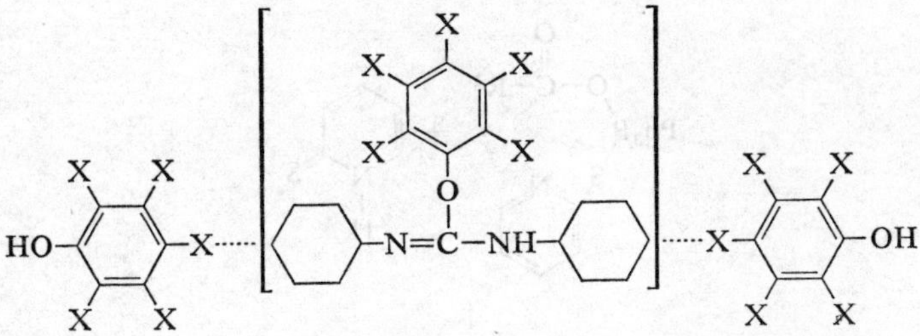

(70) X = F or Cl

acylpeptides are prepared with high optical purity using complexes (70) of dicyclohexylcarbodi-imide and the phenol.[86] The solid complexes are prepared by reacting dicyclohexylcarbodi-imide with three moles of the phenol. They dissociate in solution, the excess phenol being responsible for the low level of racemization encountered, since it reacts rapidly with any oxazolinone formed before it can racemize to any marked extent.[87] An example of the use of the complex in preparing a pentachlorophenyl active ester of the *N*-terminal pentapeptide sequence of cholecystokinin-pancreozymin is given in Scheme 37.[88]

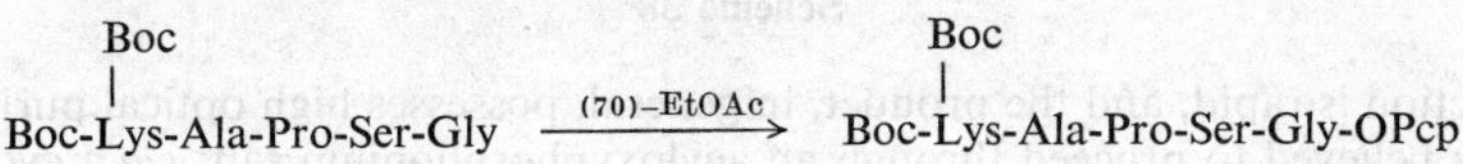

Scheme 37

Coupling Methods involving Acyloxyphosphonium Salts. A detailed review [89] has appeared dealing with the development of the 'oxidation–reduction' method of condensation.[90] The preferred coupling technique involves the interaction of the amino- and carboxy-component with triphenylphosphine and 2,2′-dipyridyl disulphide in methylene chloride (Scheme 38). The

[86] J. Kovacs, L. Kisfaludy, and M. Q. Ceprini, *J. Amer. Chem. Soc.*, 1967, **89**, 183.
[87] J. Kovacs, L. Kisfaludy, M. Q. Ceprini, and R. H. Johnson, *Tetrahedron*, 1969, **25**, 2555.
[88] M. Bodanszky, N. Chaturvedi, D. Hudson, and M. Itoh, *J. Org. Chem.*, 1972, **37**, 2303.
[89] T. Mukaiyama, *Synthetic Comm.*, 1972, **2**, 243.
[90] T. Mukaiyama and R. Matsueda, *Kobunshi*, 1972, **21**, 150 (*Chem. Abs.*, 1972, **76**, 127 387f).

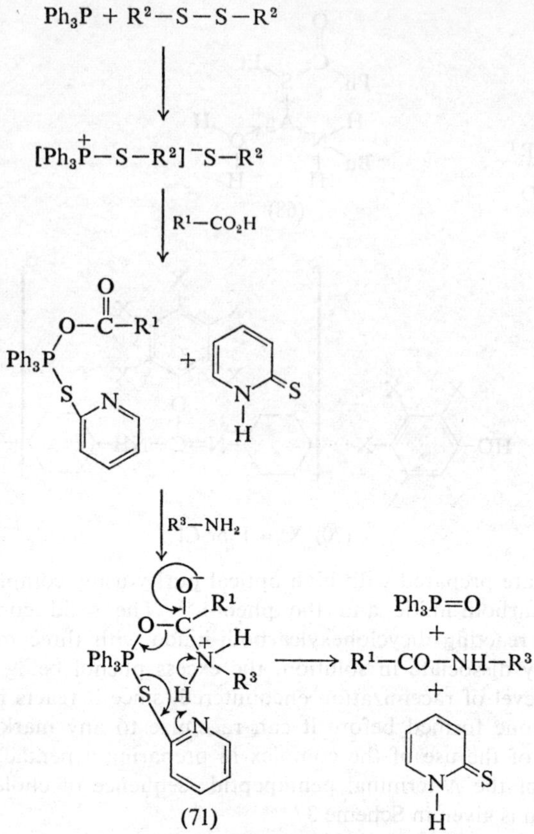

(71)

Scheme 38

reaction is rapid, and the product, in general, possesses high optical purity. It is believed to proceed through an acyloxyphosphonium salt, *via* a cyclic transition state (71), which obviates oxazolinone formation and hence racemization by this mechanism.

A further example of the use of triphenylphosphine plus carbon tetrachloride as a coupling reagent in peptide synthesis confirms that complete racemization occurs in the case of peptide carboxy-components (Scheme 39).[91]

Chlorotrisdimethylaminophosphonium perchlorate (72) is a stable salt, soluble in organic solvents, which readily forms anhydrides from carboxylic acids, and amides from amino- and carboxy-components.[92] It is prepared by treating an emulsion of carbon tetrachloride, ether, and aqueous ammonium perchlorate with trisdimethylaminophosphine (Scheme 40).

[91] Th. Wieland and A. Seeliger, *Chem. Ber.*, 1971, **104**, 3992.
[92] B. Castro and J. R. Dormoy, *Tetrahedron Letters*, 1972, 4747.

$$\text{Boc-L-Ala} + \text{HCl,H-L-Phe-OMe} \xrightarrow{\text{i}} \text{Boc-L-Ala-L-Phe-OMe}$$

$$\text{Boc-L-Ala-L-Phe-OMe} \xrightarrow{\text{ii}} \text{Boc-L-Ala-L-Phe}$$

$$\text{Boc-L-Ala-DL-Phe-L-Pro-OMe} \xleftarrow{\text{iii}} \text{Boc-L-Ala-L-Phe}$$

Conditions: i, NEt_3–Ph_3P–CCl_4; ii, aq. NaOH; iii, HCl,H-L-Pro-OMe–NEt_3–Ph_3P–CCl

Scheme 39

$$P(NMe_2)_3 + CCl_4 + NH_4^+\,ClO_4^- \longrightarrow Cl{-}\overset{+}{P}(NMe_2)_3\,ClO_4^-$$

(72)

Scheme 40

A number of simple peptides were prepared in moderate to good yields (Scheme 41), but complete racemization occurs in the Young test. The authors suggest [92] that the actual acylating species is a symmetrical anhydride, formed immediately by attack of the highly reactive acyloxyphosphonium salt by carboxylate ion.

$$R^1{-}CO_2H + Cl^-,H_3\overset{+}{N}{-}R^3 + (72) + 3NEt_3$$

$$\downarrow$$

$$R^1{-}CO{-}NH{-}R^3 + O{=}P(NMe_2)_3 + H\overset{+}{N}Et_3,ClO_4^- + 2H\overset{+}{N}Et_3,Cl^-$$

Scheme 41

The toluene-*p*-sulphonic anhydride–hexamethylphosphoramide method (see Vol. 2 of these Reports, p. 156, and Vol. 4, p. 344) has been thoroughly reinvestigated.[93] The reaction between toluene-*p*-sulphonic anhydride and hexamethylphosphoramide under modified experimental conditions has enabled the previously postulated dicationic intermediate to be isolated and converted into its stable crystalline tetrafluoroborate salt (73). Unfortunately, extensive racemization has been observed in the Izumiya test using (73) as the coupling agent, as well as under the original reaction conditions. Evidence from ^{31}P n.m.r. studies suggests that when the carboxyl component is benzyloxycarbonyl-valine the acylating species is the symmetrical anhydride rather than an acyloxyphosphonium salt.

Phosphorous acid and its esters, on oxidation with mercuric chloride in pyridine, yield *N*-phosphonium salts of pyridine (74) which react with carboxylic acids, with alcohols, and with amines, and can be used for preparing esters and simple peptides.[94] An extension of this study showed

[93] A. J. Bates, G. W. Kenner, R. Ramage, and R. C. Sheppard, Proceedings of the 12th European Peptide Symposium, 'Peptides, 1972', ed. H. Hanson and H.-D. Jakubke, North-Holland, Amsterdam, 1973.

[94] N. Yamazaki and F. Higashi, *Tetrahedron Letters*, 1972, 415.

$$(Me_2N)_3\overset{+}{P}{-}O{-}\overset{+}{P}(NMe_2)_3 \quad 2BF_4^-$$

(73)

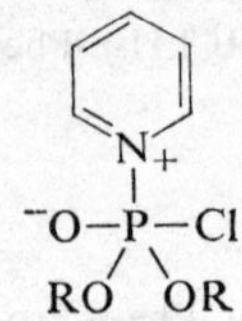

(74)

that the oxidant could be omitted from the reaction mixture. Thus esters or amides are formed when the reactants are left in pyridine in contact with diphenyl phosphite.[95] A few simple peptides were prepared by this method. Presumably the reaction proceeds as outlined in Scheme 42.

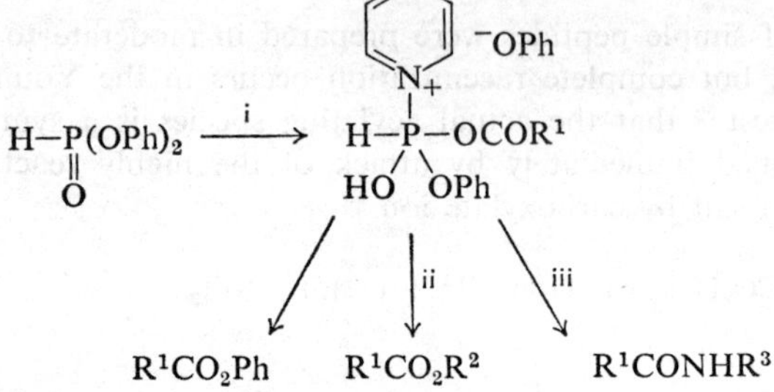

Conditions: i, R^1CO_2H–pyridine; ii, R^2OH; iii, R^3NH_2

Scheme 42

Active esters can be prepared from the carboxylic acid, diphenyl phosphite, and *p*-nitrophenol. This is highly reminiscent of the formation of active esters using triaryl phosphites.[96]

The use of diphenylphosphoryl azide (75) in peptide synthesis has been investigated.[97] The azide is a stable, non-explosive liquid, prepared according to Scheme 43. The coupling of acylamino-acids or acylpeptides

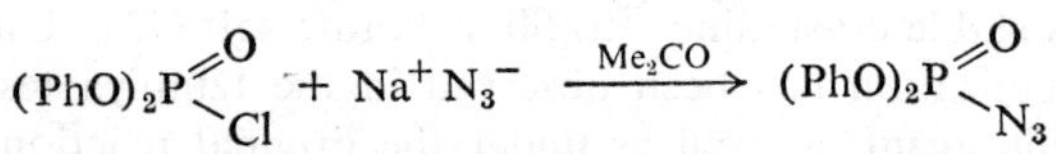

(75)

Scheme 43

with amino-esters or peptide esters can be accomplished smoothly by using the reagent (75) at 0 °C in dimethylformamide containing two equivalents of triethylamine. The optical purity of benzoyl-leucylglycine ethyl ester prepared by this method was 89%. Oxazolinone formation during the

[95] N. Yamazaki and F. Higashi, *Tetrahedron Letters*, 1972, 5047.

[96] J. A. Farrington, P. J. Hextall, G. W. Kenner, and J. M. Turner, *J. Chem. Soc.*, 1957, 1407; B. Iselin, W. Rittel, P. Sieber, and R. Schwyzer, *Helv. Chim. Acta*, 1957, **40**, 373.

[97] T. Shioiri, K. Ninomiya, and S. Yamada, *J. Amer. Chem. Soc.*, 1972, **94**, 6203.

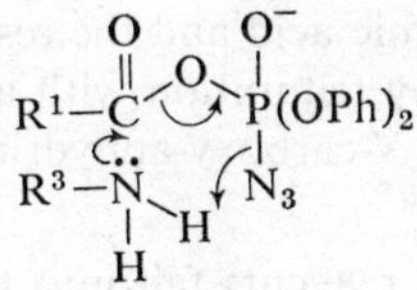

(76)

coupling could not be detected, and the 'chloride ion effect' was not noticed. No complications were observed in coupling derivatives with functional side-chains such as asparagine, glutamine, serine, nitroarginine, threonine, tyrosine, methionine, tryptophan, and histidine. The mechanism of coupling is not known with certainty; it could proceed *via* an acyl azide or an anhydride or by a concerted process [see (76)]. The reagent also converts carboxylic acids into urethanes (Scheme 44).

$$R^1{-}CO_2H + NEt_3 + (74) + R^2{-}OH \xrightarrow{\Delta} R^1{-}NH{-}CO_2R^2$$

Scheme 44

N-*Carboxy-anhydrides*. The use of amino-acid *N*-carboxy-anhydrides in the controlled stepwise synthesis of oligopeptides in aqueous solution, and the subsequent fragment condensations in the synthesis of ribonuclease S-protein, have been critically reviewed.[98] The polymerization of amino-acid *N*-carboxy-anhydrides, polymerization catalysts, and the co-polymerization of *N*-carboxy-anhydrides have been reviewed.[99] The influence of organo-aluminium compounds on the polymerization of *N*-carboxy-anhydrides and the stereochemistry of the resulting polymers have been studied.[100]

The controlled reaction of *N*-carboxy-anhydrides in the acetonitrile–water heterogeneous reaction medium has been extended to include coupling of the anhydride with amino-acids containing functional side-chains.[101] Use of this system enables coupling with glycine *N*-carboxy-anhydride to be carried out without encountering problems due to the formation of hydantoic acids,[102] which was observed in the earlier studies using an aqueous medium.[103]

In the synthesis of oligopeptides in aqueous solution, the isolation of free peptide from a reaction mixture which contains buffer components can present problems. The use of an organic base (triethylamine) to establish and maintain the pH during the reaction of an *N*-carboxy-anhydride with another amino-acid possesses the advantage that the base

[98] R. Hirschmann, ref. 39, p. 203.

[99] I. Yukio, *Kobunshi*, 1972, **21**, 92 (*Chem. Abs.*, 1972, **76**, 100 000n).

[100] T. Makino, S. Inoue, and T. Tsuruta, *Makromol. Chem.*, 1971, **150**, 137 (*Chem. Abs.*, 1972, **76**, 59 997c).

[101] R. Kataki, M. Oya, K. Uno, and Y. Iwakura, *Biopolymers*, 1971, **10**, 2199.

[102] R. Kataki, M. Oya, K. Uno, and Y. Iwakura, *J. Org. Chem.*, 1972, **37**, 327.

[103] R. Hirschmann, R. G. Strachan, H. Schwam, E. F. Schoenewaldt, H. Joshua, H. Barkmeyer, D. F. Veber, W. J. Paleveda, jun., T. A. Jacob, T. E. Beesley, and R. G. Denkewalter, *J. Org. Chem.*, 1967, **32**, 3415.

can be neutralized with formic acid and the resulting organic salt removed from the reaction residue by trituration with an organic solvent.[104]

The preparation of the *N*-carboxy-anhydride of γ-methyl-L-glutamate has been mentioned.[105]

Polymeric Reagents. These reagents fall into two categories: (i) the polymeric active esters (Scheme 45), and (ii) polymeric coupling reagents (Scheme 46).

Table 2 *Polymeric reagents used to form polymeric active esters*

Polymeric reagents[a]	*Notes, use, loading, literature*
(i) Based on nitrophenols	
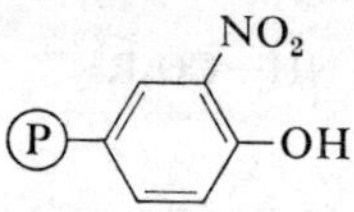	Polymeric active esters of N-protected amino-acids and N-protected peptides prepared by DCCI method. Cyclic[b] and linear peptides,[c,d] including bradykinin[e] (1.0—2.0 mmol g^{-1} available reagent sites)
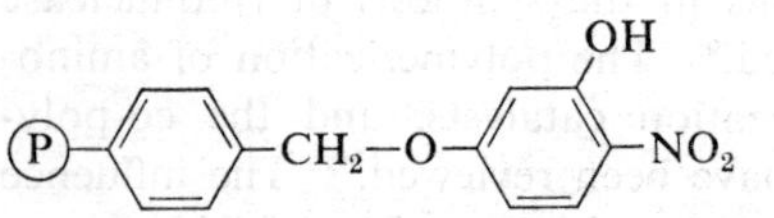	Active esters of N-protected amino-acids prepared by DCCI method. Simple linear peptides[f]
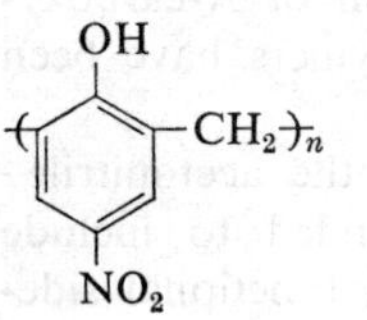	Active esters of N-protected amino-acids prepared by DCCI and mixed anhydride methods. Simple peptides[d] (0.05—0.5 mmol g^{-1} available reagent sites)
(ii) Based on N-*hydroxysuccinimide*	
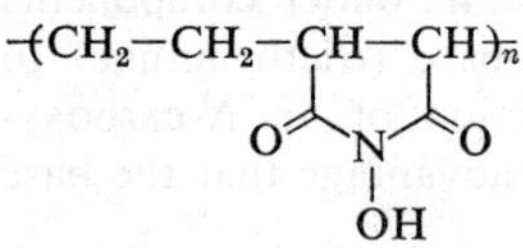	Active esters of t-butoxycarbonyl amino-acids prepared by DCCI or mixed anhydride method.[g,h] Polymer cross-linked by high-energy electron irradiation[g] or with bases such as hydrazine, spermine, or spermidine.[h] Series of oligopeptides.[g,h] Acylation reactions enhanced at elevated temperature (1.0—1.7 mmol g^{-1} available reagent sites)

[104] R. J. Gait, J. R. Langlois, and R. E. Williams, *Canad. J. Chem.*, 1972, **50**, 299.
[105] M. Iwatsuki, S. Mori, and S. Sakwai, *Kobunshi Kagaku* (*Chem. High Polymers*), 1972, **29**, 163 (*Chem. Abs.*, 1972, **76**, 100 122d).

Table 2 (*cont.*)

Polymeric reagents[a]	*Notes, use, loading, literature*
(*iii*) *Based on* 8-*hydroxyquinoline*	
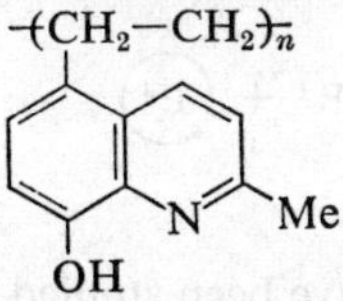	Active ester of N-protected amino-acid prepared by the anhydride method. Simple peptide.[i]
(*iv*) *Miscellaneous*	
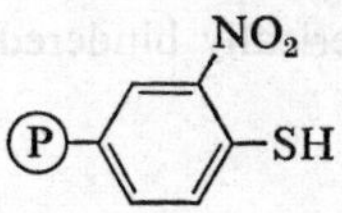	Active esters of N-protected amino-acids prepared by DCCI and mixed anhydride methods. Simple peptides (0.05—0.5 mmol g^{-1} available reagent sites)[d]
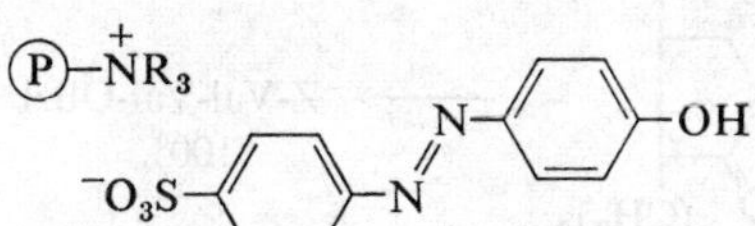	Simple peptides[j] (1.5 mmol g^{-1} of acyl active ester bound to Dowex 1 resin)
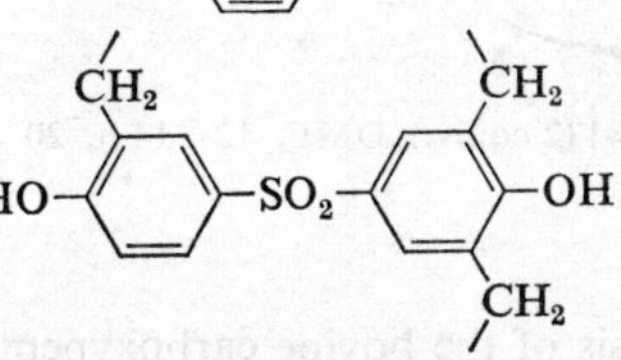	Active esters prepared by DCCI method (0.8—1.0 mmol g^{-1}), mixed anhydride (1.2 mmol g^{-1}), and thionyl chloride (0.9—1.1 mmol g^{-1})[k]
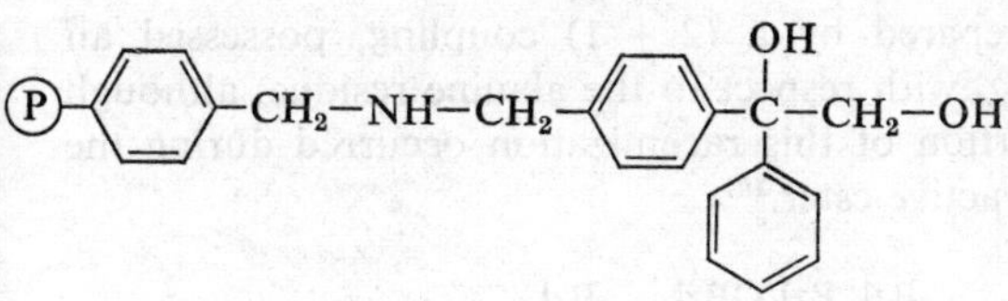	The N-protected amino-acid is esterified to the polymer's primary hydroxy-groups; treatment of the polymer with trifluoroacetic acid causes dehydration to the polymer diphenylvinyl active ester (1.3 mmol g^{-1} of glycol derivative)[l]

[a] Ⓟ- = styrene–divinylbenzene copolymer; [b] ref. 106*a*; [c] ref. 106*b*; [d] ref. 106*c*; [e] ref. 106*d*; [f] ref. 106*e*; [g] ref. 106*f*; [h] ref. 106*g*; [i] ref. 106*h*; [j] ref. 106*i*; [k] ref. 106*j*; [l] ref. 106*k*.

[106] (*a*) M. Fridkin, A. Patchornik, and E. Katchalski, *J. Amer. Chem. Soc.*, 1965, **87**, 4646; (*b*) M. Fridkin, A. Patchornik, and E. Katchalski, *J. Amer. Chem. Soc.*, 1966, **88**, 3164; (*c*) L. Yu. Sklyarov, V. I. Gorbunov, and L. A. Shchukina, *J. Gen. Chem.* (*U.S.S.R.*), 1966, **36**, 2217; (*d*) M. Fridkin, A. Patchornik, and E. Katchalski, *J. Amer. Chem. Soc.*, 1968, **90**, 2953; (*e*) M. Fridkin, A. Patchornik, and E. Katchalski, Proceedings of the Tenth European Peptide Symposium, Abano, Italy, 1969, 'Peptides 1969', ed. E. Scoffone, North Holland, Amsterdam, 1971, p. 166; (*f*) D. A. Laufer, T. M. Chapman, D. I. Marlborough, V. M. Vaidya, and E. R. Blout, *J. Amer. Chem. Soc.*, 1968, **90**, 2696; (*g*) M. Fridkin, A. Patchornik, and E. Katchalski, *Biochemistry*, 1972, **11**, 466; (*h*) G. Manecke and E. Haake, *Naturwiss.*, 1968, **55**, 343; (*i*) Th. Wieland and C. Birr, *Chimia* (*Switz.*), 1967, **21**, 582; (*j*) Th. Wieland and C. Birr, *Angew. Chem. Internat. Edn.*, 1966, **5**, 310; (*k*) Th. Wieland, C. Birr, and P. Fleckenstein, *Annalen*, 1972, **756**, 14.

$$R{-}CO_2{-}\textcircled{P^1} + H_2N{-}R^1 \longrightarrow R{-}CO{-}NH{-}R^1 + HO{-}\textcircled{P^1}$$

Scheme 45

$$R{-}CO_2H + \textcircled{P^2} + H_2N{-}R^1 \longrightarrow R{-}CO{-}NH{-}R^1 + \textcircled{P^3}$$

Scheme 46

Table 2 summarizes the polymeric active esters which have been studied. Fridkin and colleagues have also investigated the use of esters of copoly-(ethylene-*N*-hydroxymaleimide) cross-linked with hydrazine, spermine, or spermidine.[106g] The method, which is illustrated for a sterically hindered

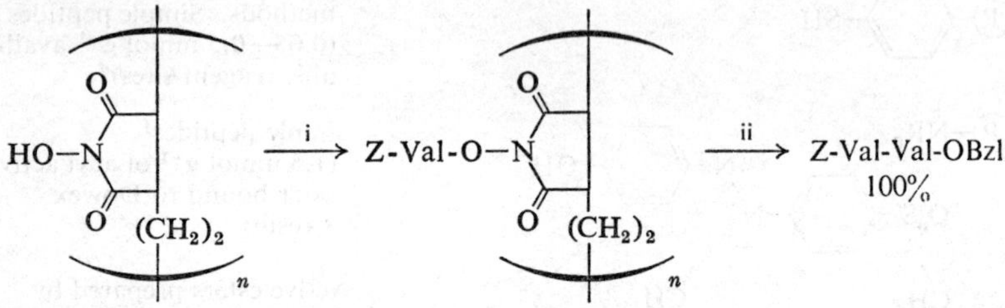

Conditions: i, DCCI–Z-Val-OH–DMF; ii, Val–OBzl (2 equiv.)–DMF, 12—14 h, 20 °C (or 1 h, 70 °C)

Scheme 47

case in Scheme 47, was used for the synthesis of the bovine carboxypeptidase A peptide (77). However, in the Izumiya racemization test the tripeptide Gly-Ala-Leu prepared by a (2 + 1) coupling, possessed an optical purity of only 87.2% with respect to the alanine residue, although it is not clear what proportion of this racemization occurred during the formation of the dipeptide active ester.[107]

```
     Bzl      Bzl Bzl OBzl      Bzl
      |        |   |   |         |
Boc-Ser-Pro-Cys-Ser-Glu-Thr-Tyr-OBzl
```

(77)

Polymeric coupling reagents which have been prepared are listed in Table 3. A further polymeric carbodi-imide has been prepared (Scheme 48) but not yet used in peptide synthesis.[108b] A polymeric arene sulphonyl chloride has been utilized for the synthesis of internucleotide phosphate bonds.[109] The reagent EEDQ has been prepared in a polymeric form

[107] D. R. Lauren and R. E. Williams, *Tetrahedron Letters*, 1972, 2665.

[108] (*a*) Y. Wolman, S. Kivity, and M. Frankel, *Chem. Comm.*, 1967, 629; (*b*) N. M. Weinshenker and C.-M. Shen, *Tetrahedron Letters*, 1972, 3281.

[109] M. Rubinstein and A. Patchornik, *Tetrahedron Letters*, 1972, 2881.

Table 3 *Reagents useful in peptide synthesis*

Polymeric reagents	*Notes, use, loading, literature*
(i) Based on carbodi-imides	
$-[(CH_2)_6-N=C=N]_n-$	Synthesis of simple peptides[a]
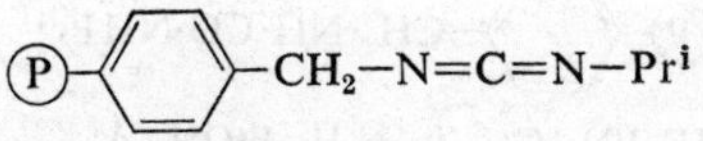	Synthesis of carboxylic acid anhydrides[b] (1.4—2.4 mmol g^{-1} available reagent sites)
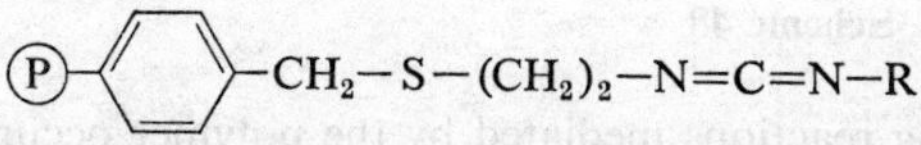	Synthesis of simple peptides[c]
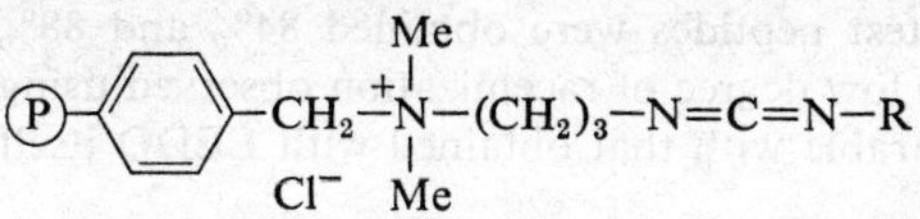	*c*
$-(CH_2-CH)_n-$ with pendant $N=C=N-R$	*c*
(ii) Based on arenesulphonyl halides	
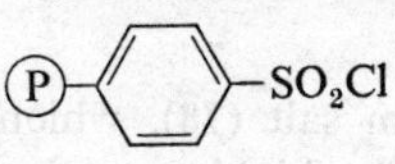	Synthesis of simple peptides by mixed carboxylic–sulphonic anhydride[c]
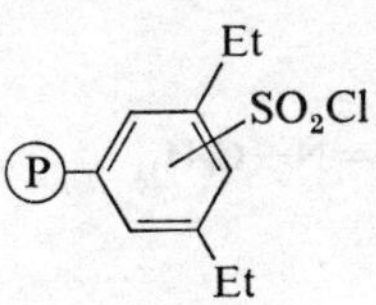	Synthesis of nucleotides[d]
(iii) Based on 'EEDQ'	
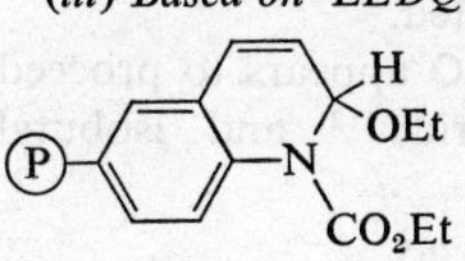	Synthesis of simple peptides (0.2—0.3 mmol g^{-1} available reagent sites)[e]

[a] Ref. 108*a*; [b] ref. 108*b*; [c] ref. 106*e*; [d] ref. 109; [e] ref. 110.

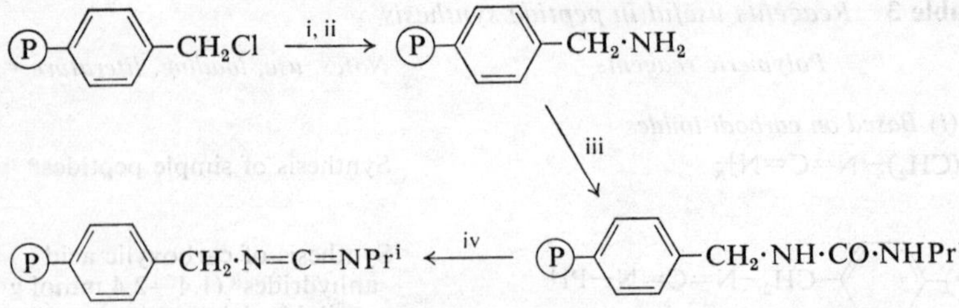

Conditions: i, potassium phthalimide–DMF 100 °C; ii, N_2H_4–EtOH, Δ; iii, Pr^i—N=C=O–tetrahydrofuran; iv, TosCl–Et_3N–CH_2Cl_2, Δ.

Scheme 48

(Scheme 49).[110] Model coupling reactions mediated by the polymer occur cleanly in high yield, and the coupling agent can be readily regenerated. The Young test and Izumiya test peptides were obtained 84% and 88% optically pure, respectively: the low degree of racemization observed using the polymeric reagent is comparable with that obtained with EEDQ itself under certain conditions.[107]

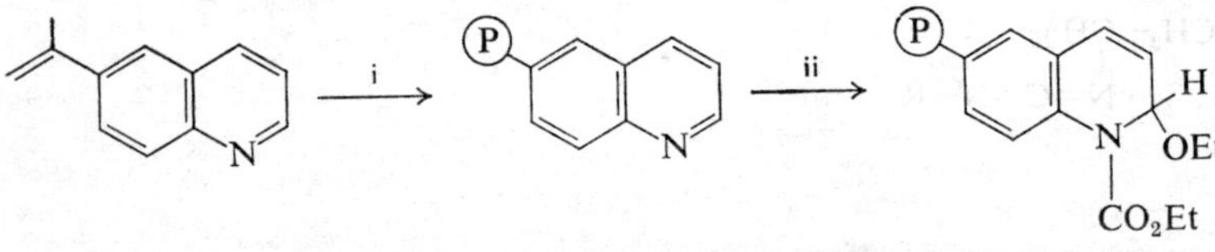

Conditions: i, $PhCH{=}CH_2$, $C_6H_4(CH{=}CH_2)_2$, *hν*; ii, Cl·CO_2Et–EtOH–NEt–CH_2Cl_2

Scheme 49

Other Methods. The preparation of a carbodi-imidium salt (78), which may be dimeric, has been reported (Scheme 50).[111] The highly reactive

$$C_6H_{11}{-}N{=}C{=}N{-}C_6H_{11} \xrightarrow[\text{MeI}]{\Delta} C_6H_{11}{-}\overset{\text{Me}}{\overset{|}{\underset{+}{N}}}{=}C{=}N{-}C_6H_{11} \quad I^-$$

(78)

Scheme 50

carbodi-imidium system should be able to activate carboxy-groups towards aminolysis, perhaps with a decreased tendency to *N*-acylurea formation, but its use in peptide synthesis has not yet been reported.

Since the formation of peptide bonds using EEDQ appears to proceed *via* the intermediate carboxylic–carbonic anhydride[112] and isobutyl

110 J. Brown and R. E. Williams, *Canad. J. Chem.*, 1971, **49**, 3765.
111 R. Scheffold and E. Saladin, *Angew. Chem. Internat. Edn.*, 1972, **11**, 229.
112 B. Belleau and G. Malek, *J. Amer. Chem. Soc.*, 1968, **90**, 1651.

chloroformate is sometimes preferable to ethyl chloroformate for generating such anhydrides, a study has been made of the use of 2-isobutoxy-1-isobutoxycarbonyl-1,2-dihydroquinoline (79) as a coupling reagent.[113] This compound and some analogues have been prepared according to Scheme 51. In test coupling reactions the reagent (79) gave significantly higher yields than the others.

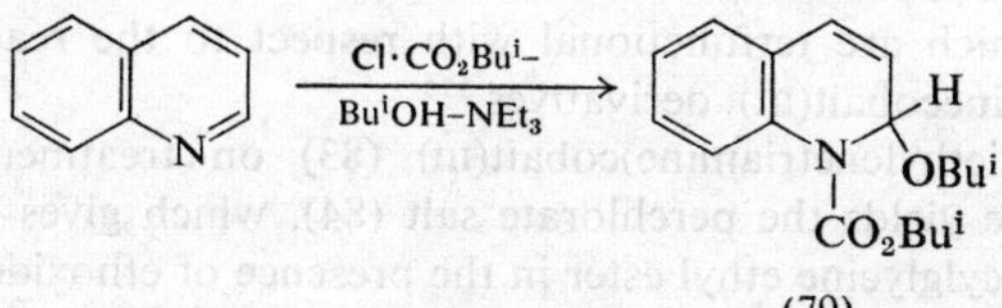

(79)

Scheme 51

NN′-Carbonyldi-(3-phenyl-1,2,4-oxadiazol-5-one) and the corresponding 3-benzyl derivative [(80) and (81), respectively; prepared as indicated in Scheme 52] have been shown to be highly reactive coupling reagents in

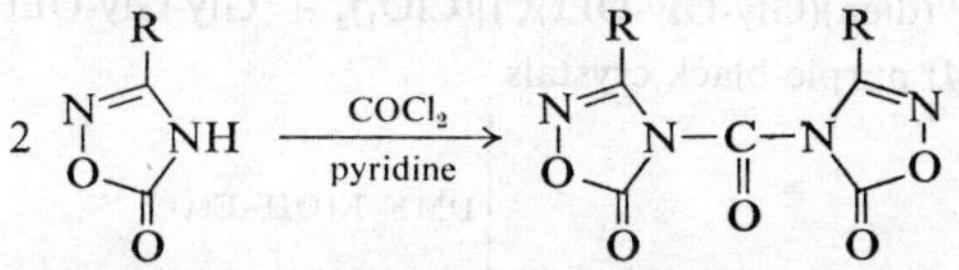

(80) R = Ph
(81) R = Bzl

Scheme 52

preparations of a number of dipeptides, but complete racemization occurs in the Anderson test.[114]

3-Substituted 1-benzyloxycarbonylaziridin-2-ones [*e.g.* (82)] can be obtained from benzyloxycarbonylamino-acids as shown in Scheme 53.[115]

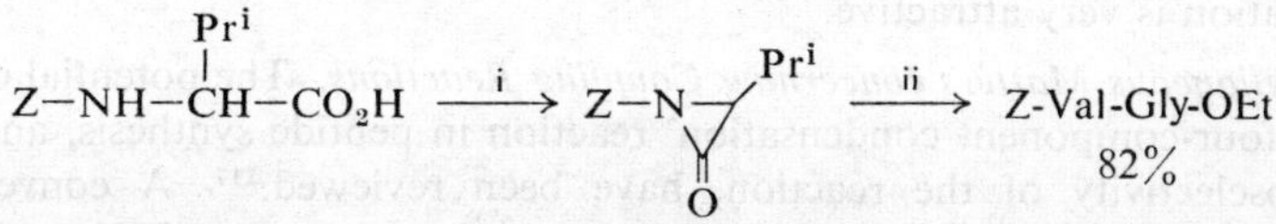

(82)

Conditions: i, THF–$COCl_2$–Et_2O–NEt_3, 2 h, −20 °C; ii, HCl,Gly-OEt–NEt_3–$CHCl_3$, 2 h, −20 °C

Scheme 53

[113] Y. Kiso and H. Yajima, *J.C.S. Chem. Comm.*, 1972, 942.
[114] T. Fujii and M. Miyoshi, *Chem. Letters*, 1972, 949.
[115] M. Miyoshi and H. Tamura, Ger. Offen. 2 132 957 (*Chem. Abs.*, 1972, **76**, 72 796f); see also M. Miyoshi, *Bull. Chem. Soc. Japan*, 1973, **46**, 212.

These compounds couple readily with amino-esters at low temperature to give high yields of dipeptides.

Previous studies on the formation of peptide bonds within the co-ordination sphere of cobalt(III) were concerned with complexes having only two adjacent co-ordination sites available for accommodation of the reactant, *viz.* bis(ethylenediamine)cobalt(III) and triethylenetetraminecobalt(III) derivatives. A new paper discusses amide bond formation mediated by cobalt(III) complexes which are terfunctional with respect to the reactant, *viz.* diethylenetriaminecobalt(III) derivatives.[116]

Trichloro(diethylenetriamine)cobalt(III) (83) on treatment with ethyl glycylglycinate yields the perchlorate salt (84), which gives (85) on treatment with glycylglycine ethyl ester in the presence of ethoxide ion (Scheme 54).* The peptide bond formation was interpreted in the terms outlined in

$[Co^{III}(dien)Cl_3]$ + Gly-Gly-OEt
(83)

↓

$[Co^{III}(dien)(Gly\text{-}Gly\text{-}OEt)Cl](ClO_4)_2$ + Gly-Gly-OEt
(84) purple-black crystals

↓ DMF–EtOH–EtO^-

$[Co^{III}(dien)(Gly\text{-}Gly\text{-}Gly\text{-}Gly\text{-}OEt)]Cl_2$
(85) orange-red crystals

Scheme 54

Scheme 55.* The stepwise condensation of a polypeptide within the co-ordination sphere of a metal ion is clearly in its infancy as a practicable peptide synthesis. However, the idea of a reagent which could co-ordinate amino- and carboxy-compounds in close proximity for amide bond formation is very attractive.

Miscellaneous Matters concerning Coupling Reactions. The potential of the Ugi 'four-component condensation' reaction in peptide synthesis, and the stereoselectivity of the reaction, have been reviewed.[117] A convenient synthesis of isocyanides has been devised, in which a dimethylformamide solution of *NN*-dimethylchloromethyleneammonium chloride, prepared *in situ* from thionyl chloride and dimethylformamide, is used to dehydrate formamides (Scheme 56).[118]

[116] Y. Wu and D. H. Busch, *J. Amer. Chem. Soc.*, 1972, **94**, 4115.
[117] I. Ugi, ref. 39, p. 203.
[118] H. M. Walborsky and G. E. Niznik, *J. Org. Chem.*, 1972, **37**, 187.

* dien = diethylenetriamine, also abbreviated N N N.

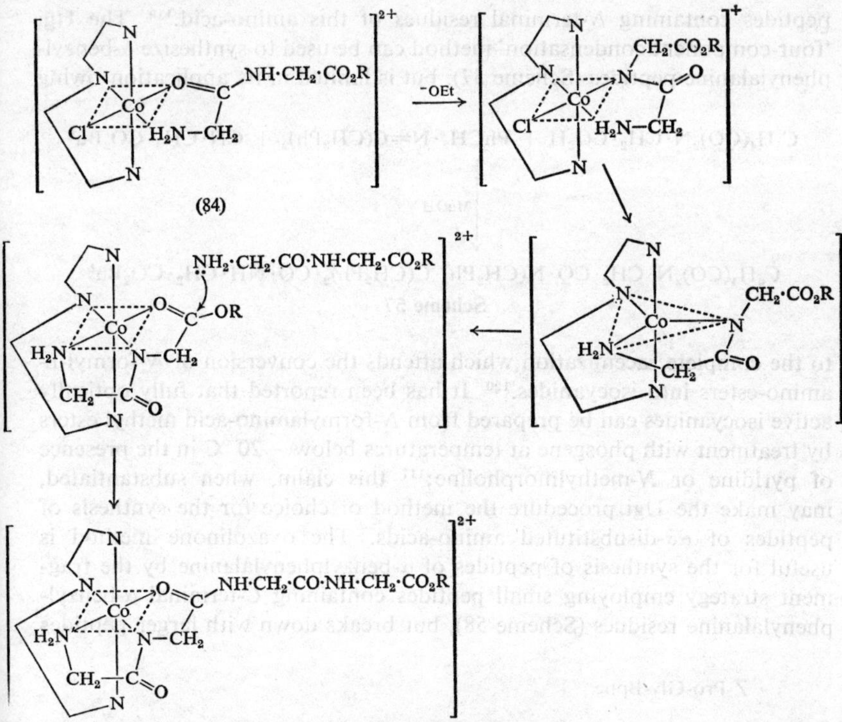

Scheme 55

A difficult $\alpha\alpha$-disubstituted amino-acid to introduce into peptides is the α-benzylphenylalanine residue (abbreviated Bphe). Dicyclohexylcarbodi-imide is not generally applicable as a coupling reagent for the synthesis of peptides of α-benzylphenylalanine, but can, however, be used successfully for the synthesis of dipeptides of α-benzylphenylalanine and of larger

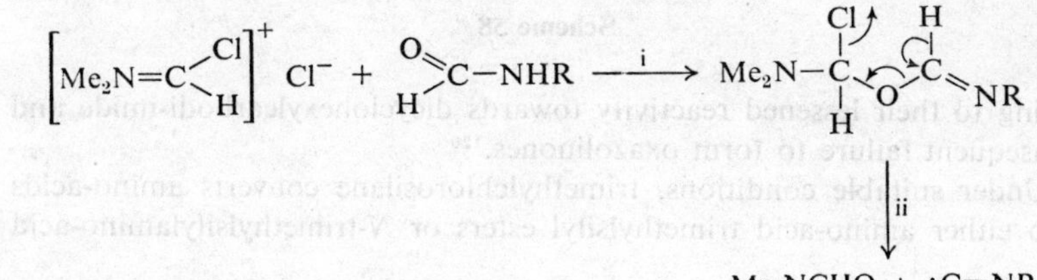

Conditions: i, -50 °C; ii, Na_2CO_3, 20 °C

Scheme 56

peptides containing *N*-terminal residues of this amino-acid.[119] The Ugi 'four-component condensation' method can be used to synthesize α-benzylphenylalanine peptides (Scheme 57), but is limited in its application owing

$$C_6H_4(CO)_2N\cdot CH_2\cdot CO_2H + PhCH_2\cdot N{=}C(CH_2Ph)_2 + CN\cdot CH_2\cdot CO_2Bu^t$$

$$\downarrow \text{MeOH}$$

$$C_6H_4(CO)_2N\cdot CH_2\cdot CO\cdot N(CH_2Ph)\cdot C(CH_2Ph)_2\cdot CO\cdot NH\cdot CH_2\cdot CO_2Bu^t$$

Scheme 57

to the complete racemization which attends the conversion of *N*-formyl-L-amino-esters into isocyanides.[120] It has been reported that fully optically active isocyanides can be prepared from *N*-formylamino-acid methyl esters by treatment with phosgene at temperatures below −20 °C in the presence of pyridine or *N*-methylmorpholine;[117] this claim, when substantiated, may make the Ugi procedure the method of choice for the synthesis of peptides of αα-disubstituted amino-acids. The oxazolinone method is useful for the synthesis of peptides of α-benzylphenylalanine by the fragment strategy employing small peptides containing *C*-terminal α-benzylphenylalanine residues (Scheme 58), but breaks down with larger peptides

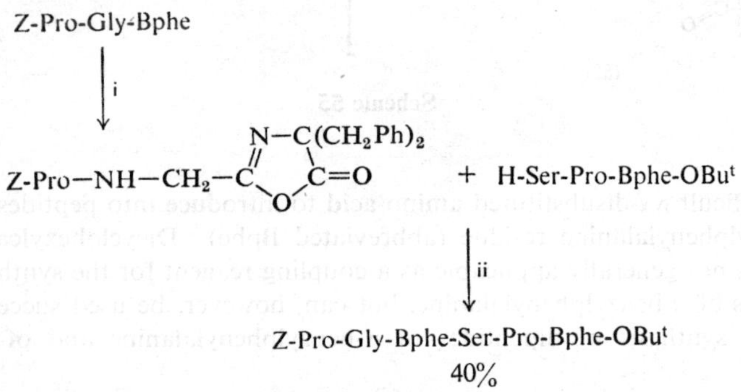

Conditions: i, DCCI–Et_2O; ii, MeCN, Δ, 14 h

Scheme 58

owing to their lessened reactivity towards dicyclohexylcarbodi-imide and consequent failure to form oxazolinones.[120]

Under suitable conditions, trimethylchlorosilane converts amino-acids into either amino-acid trimethylsilyl esters or *N*-trimethylsilylamino-acid

[119] G. C. Barrett, P. M. Hardy, T. A. Harrow, and H. N. Rydon, *J.C.S. Perkin I*, 1972, 2634.
[120] H. L. Maia, B. Ridge, and H. N. Rydon, *J.C.S. Perkin I*, 1973, 98.

trimethylsilyl esters. These compounds have been used, without isolation, for peptide synthesis using a variety of coupling agents.[121] The reaction of N-protected amino-acids with reagents such as dimethyldichlorosilane leads to the formation of the corresponding acyloxysilanes. However, when the carboxy-component is treated with excess dimethyldichlorosilane and triethylamine, a 2-silaoxazolidinone results (Scheme 59).[122]

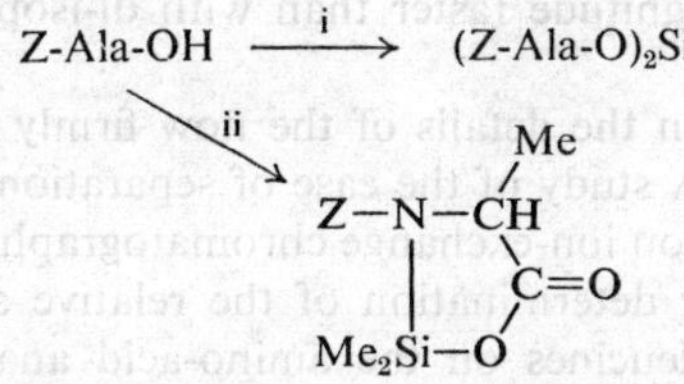

Conditions: i, Me_2SiCl_2–2 NEt_3; ii, excess Me_2SiCl_2–excess NEt_3

Scheme 59

The claim has been made that the aminolysis of N-protected amino-acid active esters can be accelerated [123] when the amino-component is first converted into the corresponding triethylammonium carbamate salt by reaction with carbon dioxide in the presence of base. *N*-Tosyl peptide active esters in pyridine solution undergo cyclization with the formation of dioxopiperazines.[124]

Racemization.—The optical purity of amino-acids can be determined by gas–liquid chromatography of the corresponding *N*-trifluoroacetylamino-acid isopropyl esters using cyclohexyl *N*-trifluoroacetyl-L-α-aminobutyryl-L-α-aminobutyrate as the optically active stationary phase.[125] Gas–liquid radio-chromatography has been applied to the determination of the optical purity of ^{14}C-labelled amino-acids in the form of the *N*-trifluoro-acetyl-L-prolylamino-acid methyl esters.[126] Both of these methods are accurate to 1%. Gas–liquid chromatography has also been applied to the separation of racemic 2-trifluoromethyloxazolin-5-ones.[127]

The kinetics of the racemization of amino-acids [128] as a function of pH have been studied at elevated temperatures. The kinetics of racemization of 4-isobutyl-2-phenyl-L-oxazolin-5-one (86) have been studied polarimetrically.[129] With triethylamine as base, compound (86) racemizes

[121] H. R. Kricheldorf, *Annalen*, 1972, **763**, 17.
[122] H. Buchwald, Ch. Seidel, H. Kunzek, P. Ludwig, B. Frölich, and K. Rühlmann, *J. Organometallic Chem.*, 1972, **37**, C1.
[123] M. Itoh, *Chem. and Pharm. Bull.* (*Japan*), 1972, **20**, 664.
[124] G. Lucente and P. Frattesi, *Tetrahedron Letters*, 1972, 4283.
[125] W. Parr and P. Y. Howard, *Angew. Chem. Internat. Edn.*, 1972, **11**, 529.
[126] A. V. Barooshian, M. J. Lautenschleger, and W. G. Harris, *Analyt. Biochem.*, 1971, **44**, 543.
[127] O. Grahl-Nielsen and E. Solheim, *J.C.S. Chem. Comm.*, 1972, 1092.
[128] J. L. Bada, *J. Amer. Chem. Soc.*, 1972, **94**, 1371.
[129] O. Grahl-Nielsen, *Chem. Comm.*, 1971, 1588.

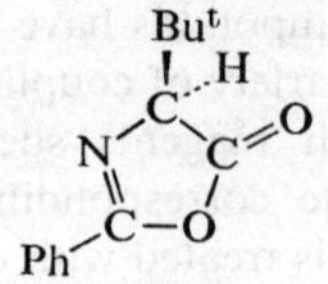

(86)

approximately an order of magnitude faster than with di-isopropylethylamine.

A full paper has appeared on the details of the now firmly established Izumiya racemization test.[130] A study of the ease of separation of a series of diastereoisomeric tripeptides on ion-exchange chromatography led to the elaboration of a procedure for determination of the relative amounts of diastereoisomeric glycylalanyl-leucines on the amino-acid analyser. The procedure was used to examine the racemization caused by various coupling agents in the synthesis of the peptide mentioned (Scheme 60). The sensi-

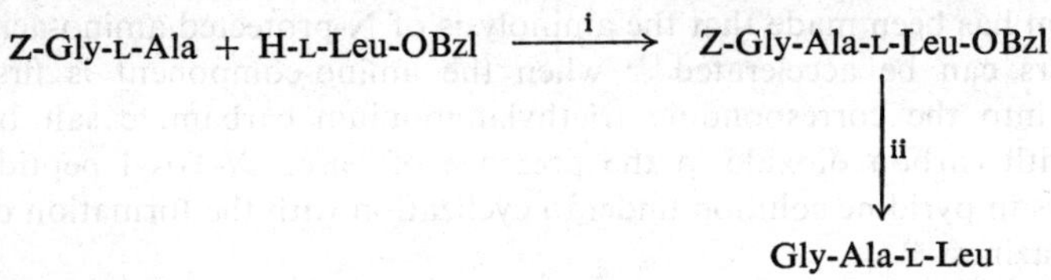

Conditions: i, coupling reagent; ii, H_2–Pd/C

Scheme 60

tivity of the method is 0.1% L–D-isomer in 5 μmol L–L-peptide. Low levels of racemization were found to occur, for example, with the modified dicyclohexylcarbodi-imide procedure (in THF in the presence of *N*-hydroxysuccinimide – optical purity 100.0%) and with EEDQ (in THF in the presence of triethylamine – optical purity 99.6%). Another recent application [107] of the Izumiya test, on the other hand, gave an optical purity of 94.2% for the peptide prepared by the modified dicyclohexylcarbodi-imide procedure (in dimethoxyethane in the presence of *N*-hydroxysuccinimide and triethylammonium toluene-*p*-sulphonate) and 88.0% for the peptide prepared with EEDQ (in dimethoxyethane in the presence of triethylammonium toluene-*p*-sulphonate). This underlines the necessity of precisely defining the conditions of coupling when carrying out a racemization test, and illustrates the care which must be exercised in making comparisons of racemization data from diverse studies.

The degree of racemization of amino-acids can be determined by reacting them with an optically active *N*-carboxy-anhydride and separating the diastereoisomers on an amino-acid analyser. The procedure has been

[130] N. Izumiya, M. Muraoka, and H. Aoyagi, *Bull. Chem. Soc. Japan*, 1971, **44**, 3391.

simplified, and separation of the diastereoisomers can be accomplished by thin-layer chromatography on silica-gel or cellulose.[131]

N-Acetylphenylalanylalanine methyl ester is a suitable test peptide for the determination of racemization by nuclear magnetic resonance spectroscopy.[132] The extent of racemization in the carboxy-component can be estimated by examination of the two abutting doublets, due to the alanyl methyl protons of each diastereoisomer. The method is convenient and rapid, with a sensitivity of $\pm 3\%$.

A study of the kinetics of the triethylamine-induced racemization of some *N*-(*p*-nitrobenzylidene)amino-acid esters (87) in various solvents

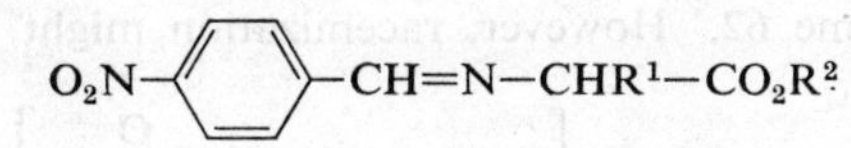

(87)

shows that the second-order rate constant for racemization is several orders of magnitude less than that for oxazolinones derived from the corresponding *N*-acylamino-acids. Further, the Schiff bases of amino-acid t-butyl esters are more resistant to racemization than are those of the corresponding methyl esters.[133]

Some evidence has been presented to show that *N*-methyl-amino-acid derivatives racemize as readily, and in some cases more readily, than the corresponding primary amino-acid derivatives under conditions of deprotection and peptide coupling.[44] A representative selection of results is shown in Scheme 61: under similar conditions the corresponding derivatives of

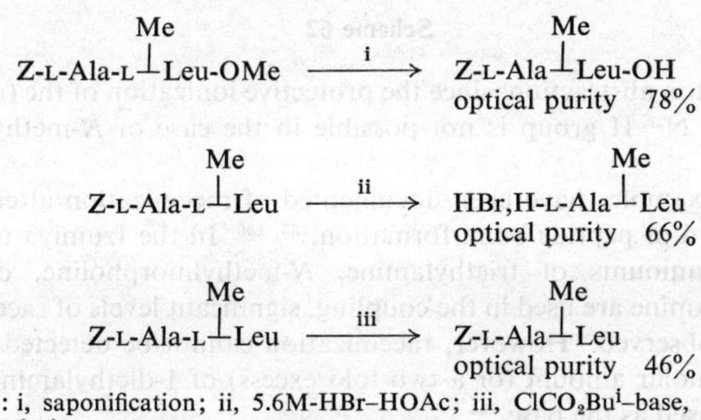

Conditions: i, saponification; ii, 5.6M-HBr–HOAc; iii, $ClCO_2Bu^i$–base, 5 min, hydrolysis

Scheme 61

[131] A. V. Barooshian, M. J. Lautenschleger, J. M. Greenwood, and W. G. Harris, *Analyt. Biochem.*, 1972, **49**, 602.
[132] B. Weinstein and A. E. Pritchard, *J.C.S. Perkin I*, 1972, 1015.
[133] I. Z. Siemion and L. Wilschowitz, *Z. Naturforsch.*, 1971, **26b**, 762.

alanyl-leucine and -isoleucine give products with an optical purity greater than 98%. The extent of racemization observed on coupling N-protected L-alanyl-L-*N*-methyl-leucine with glycine benzyl ester is minimal in the case of the *N*-hydroxysuccinimide ester, with dicyclohexylcarbodi-imide in the presence of *N*-hydroxysuccinimide, and using EEDQ (coupling solvent – THF in the absence of salt). However, in most of these couplings a pronounced salt effect is observed, in contradistinction to the corresponding couplings with N-protected alanyl-leucine derivatives as carboxy-components. Evidence that a 5-oxo-oxazolinium derivative [134] can be formed under peptide coupling conditions has been obtained by trapping it with methyl propiolate in a 1,3-dipolar cycloaddition reaction, as indicated in Scheme 62. However, racemization might also proceed by

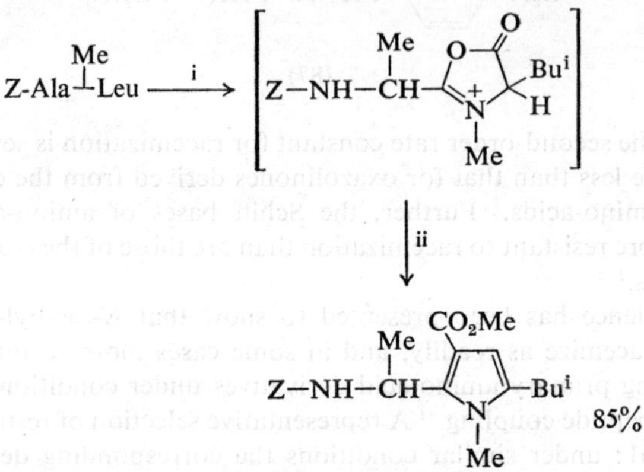

Conditions: i, DCCI–tetrahydrofuran, 10 min; ii, HC≡CCO$_2$Me

Scheme 62

direct α-proton abstraction, since the protective ionization of the (generally) more acidic N—H group is not possible in the case of *N*-methyl-amino-acids.

Further examples have been documented of racemization attending the azide method of peptide bond formation.[135, 136] In the Izumiya test, when equivalent amounts of triethylamine, *N*-methylmorpholine, or di-isopropylethylamine are used in the coupling, significant levels of racemization have been observed. However, racemization cannot be detected when an equivalent molar amount (or a two-fold excess) of 1-diethylaminopropan-2-ol (88) is used as the base.[135]

$$Et_2N\cdot CH_2\cdot CHOH\cdot Me$$

(88)

[134] G. V. Boyd and P. H. Wright, *J.C.S. Perkin I*, 1972, 909.
[135] L. Kisfaludy and O. Nyéki, *Acta Chim. Acad. Sci. Hung.*, 1972, **72**, 75.
[136] A. Metallidis and D. Theodoropoulos, *Chem. Chron.*, 1972, **1**, 154 (*Chem. Abs.*, 1972, **77**, 102 193x).

The model peptide histidylglutamic acid has been used for testing the lability of various protected histidine derivatives to racemization,[137] using different coupling agents and additives (Scheme 63). The method depends

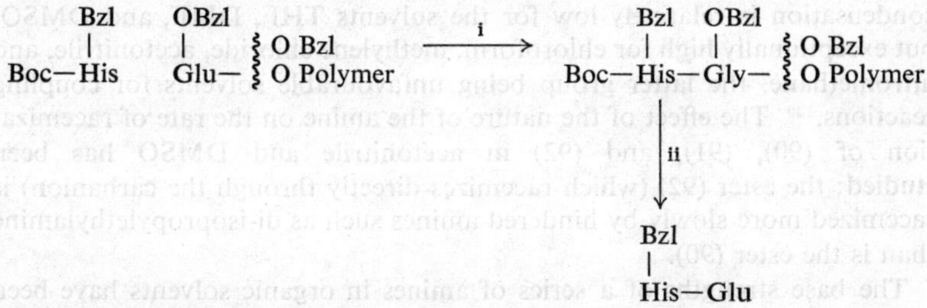

(89)

Conditions: i, coupling agent; ii, HBr–CF_3CO_2H

Scheme 63

on the determination of the diastereoisomeric *im*-benzyl-histidylglutamic acids (89) and is sensitive to 0.1%. When the coupling is mediated by dicyclohexylcarbodi-imide (in various solvents such as DMF, pyridine, or methylene chloride, in the presence of tributylammonium toluene-*p*-sulphonate) the product (89) is obtained with an optical purity, with respect to the histidyl residue, of greater than 94.2%. However, in the presence of *N*-hydroxysuccinimide the optical purity rises to at least 99.8%, although the yield drops owing to the formation of *β*-alanine derivatives from the additive. Use of *N*-hydroxybenzotriazole, on the other hand, enables the model peptide to be prepared in good yield and high optical purity (at least 99.8%). *im*-Benzyl-protected histidine residues appear to be more prone to racemization in solid-phase synthesis than in homogeneous solution work. Dicyclohexylcarbodi-imide-mediated coupling gives the model dipeptide with an optical purity of 78% with respect to the histidine, whereas in the presence of the additive *N*-hydroxybenzotriazole racemization is reduced and the optical purity of the dipeptide is 99.4%.[138]

A detailed study has been made of the factors which influence the racemization of some N-protected amino-acid and peptide active esters.[139] The pseudo first-order rate coefficients for the racemization in various solvents of benzyloxycarbonylglycyl-L-phenylalanine *p*-nitrophenyl ester (90), benzoyl-L-leucine *p*-nitrophenyl ester (91) (both models for racemization by the oxazolinone mechanism), and of benzyloxycarbonyl-L-phenylalanine *p*-nitrophenyl ester (92) (model for racemization by direct α-proton

Z-Gly-L-Phe-ONp	Bz-L-Leu-ONp	Z-L-Phe-ONp
(90)	(91)	(92)

[137] G. C. Windridge and E. C. Jorgensen, ref. 39, p. 375.
[138] G. C. Windridge and E. C. Jorgensen, *J. Amer. Chem. Soc.*, 1971, **93**, 6318.
[139] A. W. Williams and G. T. Young, *J.C.S. Perkin I*, 1972, 1194.

abstraction) were determined polarimetrically. The second-order rate coefficient for the condensation of the ester (92) with glycine ethyl ester was determined spectrophotometrically in the same series of solvents. The ratio of the rate coefficient for racemization to the rate coefficient for condensation is relatively low for the solvents THF, DMF, and DMSO, but exceptionally high for chloroform, methylene chloride, acetonitrile, and nitromethane, the latter group being unfavourable solvents for coupling reactions.[139] The effect of the nature of the amine on the rate of racemization of (90), (91), and (92) in acetonitrile and DMSO has been studied; the ester (92) (which racemizes directly through the carbanion) is racemized more slowly by hindered amines such as di-isopropylethylamine than is the ester (90).

The base strengths of a series of amines in organic solvents have been determined using 2,4-dinitrophenol as a spectrophotometric indicator. In THF, *N*-methylmorpholine is a weaker base than glycine ethyl ester, so it should be borne in mind that in THF the liberation of an ester from its salt may be incomplete if *N*-methylmorpholine is used. The change in base strength of glycine ethyl ester with solvent roughly parallels the change in its nucleophilicity, as measured by the rate of its reaction with ester (92). The increased racemization which is observed with coupling reactions carried out in the presence of a tertiary amine hydrochloride was shown in the case of (91) not to be due solely to the increase in ionic strength of the solution. On the other hand, addition of a salt can accelerate the racemization by bases, so it is concluded that the effect is due both to the basicity of the anion and to the increase in ionic strength of the solution.[139]

A method for determining the acid or base content of organic compounds using n.m.r. spectroscopy[140] may be useful to peptide chemists dealing with the exact neutralization of salts in a non-aqueous medium prior to a coupling reaction. The method employs titrants possessing a characteristic n.m.r. spectrum (*e.g.* methanesulphonic acid, 2,4,6-trinitrobenzenesulphonic acid, or tetramethylammonium hydroxide).

Repetitive Methods of Peptide Synthesis.—*Solid-phase Synthesis.* A list of *ca.* 300 peptides prepared by the solid-phase technique up to 1971 has been published.[141] Reviews of the principles of solid-phase synthesis,[142, 143] a discussion of the solid-phase synthesis of ribonuclease A,[144] and a general review of the use of solid supports and insoluble reagents in peptide chemistry[145] have appeared.

General Improvements to the Merrifield Technique. It is a current viewpoint that solid-phase peptide synthesis is best adapted for the synthesis of

140 Y. Dagani and A. Patchornik, *Analyt. Chem.*, 1972, **44**, 2170.
141 G. R. Marshall and R. B. Merrifield, 'Handbook of Biochemistry', 2nd edn., ed. H. A. Sober, The Chemical Rubber Co., Cleveland, 1970, p. C-145.
142 R. B. Merrifield, *Beckmann Rep.*, 1972, (1), 3 (*Chem. Abs.*, 1972, 77, 102 155m).
143 'Biochemical Aspects of Reactions on Solid Supports', ed. G. R. Stark, Academic Press, New York and London, 1971.
144 R. B. Merrifield, ref. 39, p. 283.
145 G. R. Marshall and R. B. Merrifield, ref. 144, chapter 3, pp. 111–169.

oligopeptides, which can subsequently be linked together in homogeneous solution using a fragment strategy. The application of two previously described polymers for the solid-phase synthesis of fragment peptides has been discussed.[146] A t-alkyloxycarbonylhydrazine–copoly(styrene–divinylbenzene) resin can be used to prepare carrier-bound peptide hydrazides, which are subsequently released by acidolysis. This method is illustrated by the synthesis of a decapeptide hydrazide; it utilizes a three-stage selectivity in acid-labile protecting groups, depicted in Scheme 64. The other

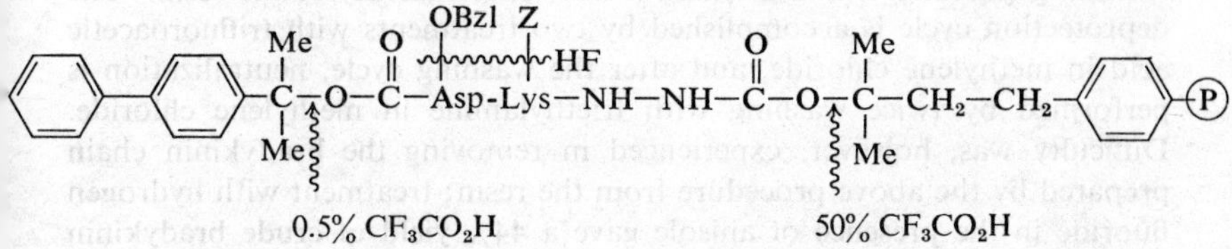

Relative sensitivity to anhydrous acid, R^1CO_2Bzl or R^2NHZ : $R^3NH{\cdot}NH{\cdot}CO{\cdot}O{\cdot}CMe_2{\cdot}CH_2{\cdot}CH_2{\cdot}C_6H_4{\cdot}$Ⓟ : $R^4NH{\cdot}$Bpoc (1 : 3000 : 1 000 000)

Scheme 64

matrix is based on a polymeric derivative of t-butyl alcohol, which can be esterified with phthaloylglycine using the benzenesulphonyl chloride mixed-anhydride method, although, in the case of optically active carboxy-components such as phthaloyl-L-phenylalanine, esterification to the resin by this method is accompanied by racemization. The phthaloylglycyl resin was N-deprotected with hydrazine acetate (Scheme 65) and the glycyl

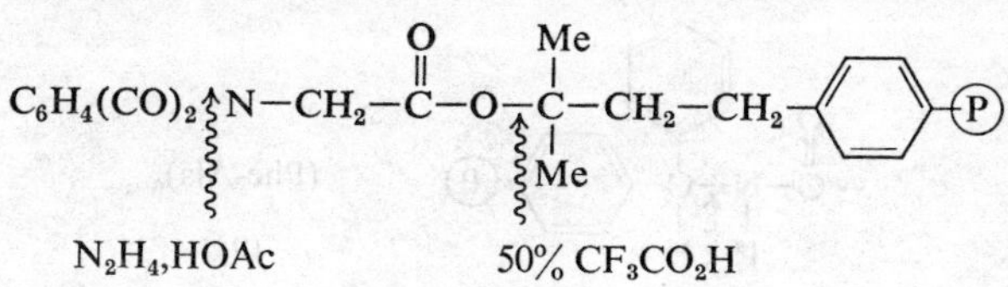

Scheme 65

resin was employed for the synthesis of a tetrapeptide which was removed from the carrier and subsequently coupled with a dipeptide amino-component in homogeneous solution.

By modifying the general Merrifield technique it is possible to synthesize bradykinin in less than five hours.[147] The modifications are based on the observations that multiple short periods of coupling and deprotection are more effective than a single exposure to the reagents; that alternate swelling and shrinking of the polymeric carrier enable entrained reagents and solvents to be removed more effectively and more rapidly; and that the

[146] S.-S. Wang and R. B. Merrifield, *Internat. J. Peptide and Protein Res.*, 1972, **4**, 309.
[147] L. Corley, D. H. Sachs, and C. B. Anfinsen, *Biochem. Biophys. Res. Comm.*, 1972, **47**, 1353.

dicyclohexylcarbodi-imide coupling step is essentially complete in one to two minutes (this may enable asparagine and glutamine derivatives to be introduced by the dicyclohexylcarbodi-imide method rather than by the active ester method). The preferred procedure involves adding a ten-fold excess of the coupling agent to a similar excess of the carboxy-component prior to the coupling step; coupling is allowed to proceed for three minutes and the coupling cycle repeated (the volume of solvent being kept as low as is consistent with good mixing). The carrier phase is washed by alternate shrinking (methanol) and swelling (methylene chloride) of the resin. The deprotection cycle is accomplished by two treatments with trifluoroacetic acid in methylene chloride, and after the washing cycle, neutralization is performed by twice washing with triethylamine in methylene chloride. Difficulty was, however, experienced in removing the bradykinin chain prepared by the above procedure from the resin; treatment with hydrogen fluoride in the presence of anisole gave a 44% yield of crude bradykinin (34% after chromatographic purification).[147]

Polymeric Carriers. The preparation of cross-linked polystyrenes for use as solid supports or insoluble reagents has been reviewed.[148]

Several further examples of the use of benzhydrylamine resins (93) in solid-phase synthesis have appeared. This resin possesses the advantage that final deprotection liberates the peptide primary amide under relatively mild conditions. It has therefore been used to prepare various biologically active amides such as human calcitonin,[149] luteinizing-hormone releasing factor,[150] some analogues,[151] and related peptides.[152] Further experimental details for the preparation of the resin have been recorded.[150]

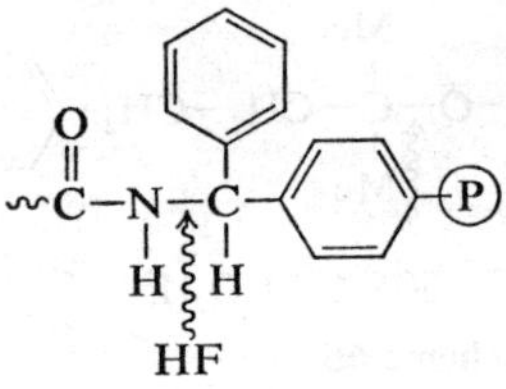

(93)

$(Phe\text{-}Ala)_6$

(94)

The recently introduced inorganic-based pellicular matrix for solid-phase synthesis possesses the disadvantage that it is difficult to remove the ultimate peptide from the silicate polymer.[153] This difficulty can be circumvented by inserting a short aliphatic chain between the three-dimensional silicate network and the benzyl anchor groups. The synthesis of this matrix

148 J. A. Patterson, ref. 144, chapter 5, p. 189.
149 P. Rivaille and G. Milhaud, *Helv. Chim. Acta*, 1972, **55**, 1617.
150 M. W. Monahan and J. Rivier, *Biochem. Biophys. Res. Comm.*, 1972, **48**, 1100.
151 M. W. Monahan, J. Rivier, W. Vale, R. Guillemin, and R. Burgus, *Biochem. Biophys. Res. Comm.*, 1972, **47**, 551.
152 R. Guillemin, M. Amoss, R. Blackwell, J. Rivier, N. Ling, and W. Vale, *Biochem. Biophys. Res. Comm.*, 1972, **48**, 1093.
153 W. Parr and K. Grohmann, *Tetrahedron Letters*, 1971, 2633.

is outlined in Scheme 66 (available reaction sites 0.03—0.08 mmol g^{-1}). The resin was used for the synthesis of the model peptide (94), which was removed from the carrier with hydrogen bromide in trifluoroacetic acid in *ca.* 90% overall yield. No short failure sequences could be detected.[154]

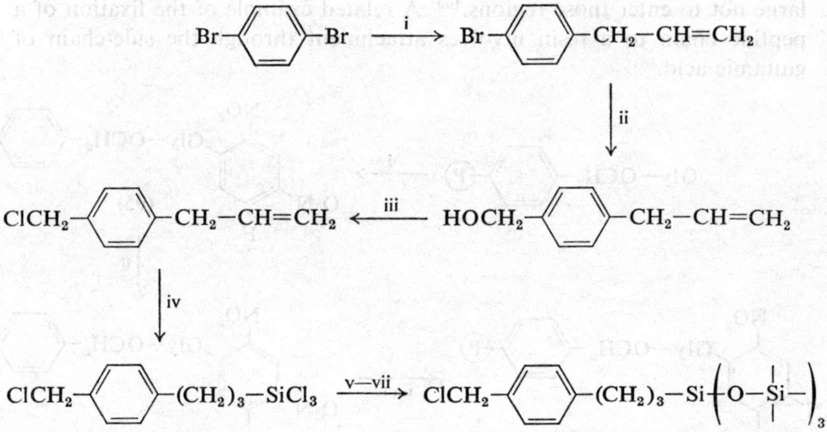

Conditions: i, Mg–CH_2=$CHCH_2Br$; ii, Mg–HCHO; iii, HCl–anhydrous Na_2SO_4; iv, $HSiCl_3$–H_2PtCl_6; v, porous glass beads; vi, hydrolysis; vii, polymerization at 100 °C

Scheme 66

Linkage to the Polymeric Carrier. The normal method of esterification to a chloromethylated resin proceeds unsatisfactorily in the case of t-butoxycarbonylmethionine, probably owing to formation of a sulphonium derivative. A method has been developed for introducing methionine as its sulphoxide and then reducing the sulphoxide to a thioether with sodium iodide and acetyl chloride in DMF.[155] One commercial chloromethylated polystyrene polymer on esterification with t-butoxycarbonylmethionine gives an extremely low incorporation of the amino-acid.[156] A low degree of esterification is obtained when t-butoxycarbonylnitroarginine triethylammonium salt is refluxed in ethanol with chloromethylpolystyrene, but a satisfactory incorporation can be achieved using the corresponding dimethylsulphonium–methylpolystyrene resin.[157] The oxidation–reduction method of condensation can be used for attaching amino-acid and peptide derivatives to hydroxymethylpolystyrene under mild conditions.[158] The course of esterification of t-butoxycarbonyl-

154 W. Parr and K. Grohmann, *Angew. Chem. Internat. Edn.*, 1972, **11**, 314.
155 K. Norris, J. Halstrøm, and K. Brunfeldt, *Acta Chem. Scand.*, 1971, **25**, 945.
156 W. Voelter, K. Zech, G. Jung, and K.-F. Senring, *Tetrahedron*, 1972, **28**, 2649.
157 D. F. Elliot, P. Moritz, and R. Wade, *J.C.S. Perkin I*, 1972, 1862.
158 R. Matsueda, E. Kitazawa, H. Maruyama, H. Takahagi, and T. Mukaiyama, *Chem. Letters*, 1972, 379.

phenylalanine to chloromethylated styrene–divinylbenzene copolymers of different structure has been studied.[159]

The difficulty arising from the esterification of amino-acid derivatives at sites on the resin which are inaccessible to further chain elongation can be overcome by directly linking to the resin peptides which are sufficiently large not to enter those regions.[160] A related example of the fixation of a peptide chain to a resin involves attachment through the side-chain of glutamic acid.[161]

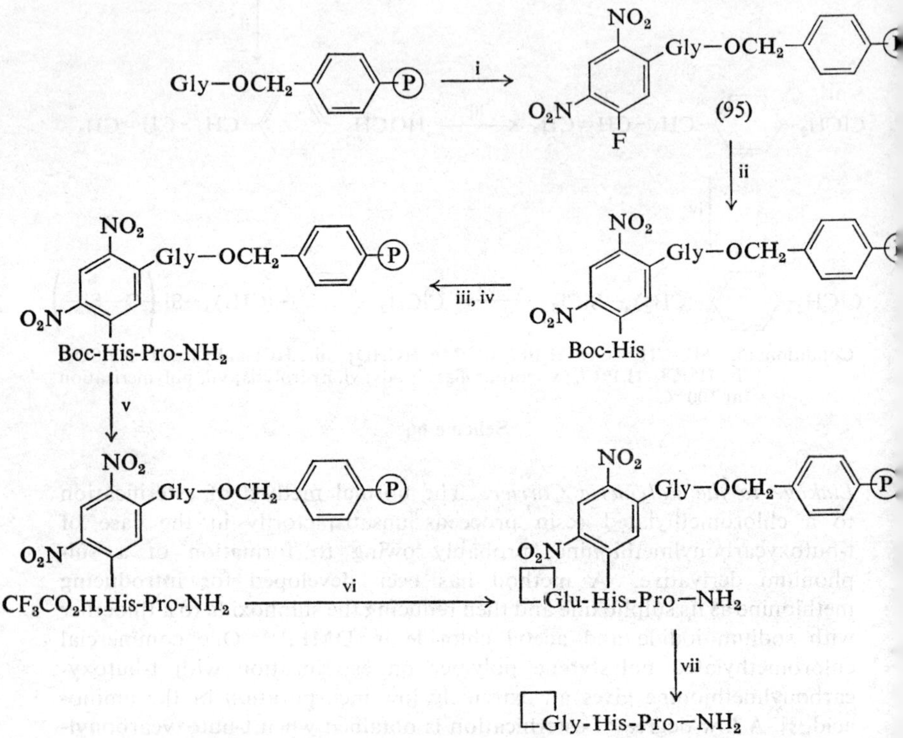

Conditions: i, 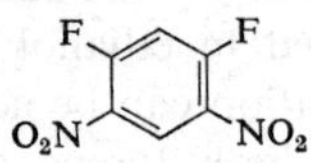 $-NEt_3$; ii, Boc-His-O$^-$ $\overset{+}{H}NEt_3$; iii, F_3CCO_2Np–py; iv, Pro-NH_2,HCl–NEt_3; v, F_3CCO_2H; vi, 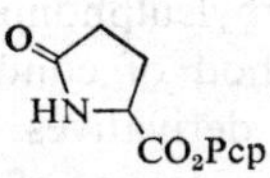 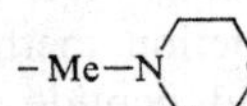; vii, $HSCH_2$-CH_2OH–NEt_3

Scheme 67

[159] A. Losse, *Z. Chem.*, 1971, **11**, 386 (*Chem. Abs.*, 1972, **76**, 34 570v).

[160] G. A. Korshunova, G. P. Mishin, Yu. A. Semiletov, and Yu. P. Shvachkin, *Zhur. obshchei Khim.*, 1972, **42**, 482 (*Chem. Abs.*, 1972, **77**, 34 918k).

[161] G. P. Mishin and Yu. P. Shvachkin, *Zhur. obshchei Khim.*, 1971, **41**, 234 (*Chem. Abs.*, 1972, **76**, 72 787d).

An ingenious synthesis of thyrotropin-releasing hormone has been described which illustrates the bidirectional extension of peptide chains from a starting amino-acid which is anchored to the resin by its side-chain.[162] Glycyl resin is treated with 1,5-difluoro-2,4-dinitrobenzene and the product (95) is treated with *N*-α-t-butoxycarbonylhistidine to form an *N-im*-dinitrophenylene bridge to the glycyl polymer (Scheme 67). The pendant protected histidine residue can be used for chain elongation at its carboxy-terminus or at the amino-group after suitable deprotection. Crude thyrotropin-releasing hormone was obtained after thiolysis together with two trace contaminants, and after chromatographic purification gave a 49% yield of pure material based on the starting glycyl resin. The fluorodinitrophenyleneglycyl resin (95) is also suitable for bridging to the side-chains of cysteine and tyrosine. Especially noteworthy are the mild conditions required to cleave the finished peptides from the resin.

N-Deprotection. A new procedure for the cleavage of the t-butoxycarbonyl group in the solid-phase synthesis of tryptophan-containing peptides without oxidation of the indole side-chain has been developed.[163] It arises from the observation that treatment of tryptophan with hydrogen chloride in formic acid results in the formation of *N-in*-formyltryptophan, which effectively protects the indole nucleus against oxidation. Further, *N*-α-t-butoxycarbonyl groups are rapidly cleaved by 0.1—1.0M-HCl in formic acid. The *in*-formyl group can be removed by treating the substrate with a reagent such as 0.1M aqueous piperidine, or hydrazine hydrate in DMF, but is resistant to anhydrous hydrogen fluoride, and triethylamine in DMF. In the application of this new reagent to solid-phase synthesis of the tryptophan-containing heptapeptide (96), oxidative destruction of tryptophan could not be detected.[163, 164] These observations suggest that (97) may

Lys-Ala-Gly-Leu-Gly-Trp-Leu
(96)

CHO
|
Boc—Trp
(97)

be a suitable reagent for the introduction of tryptophan residues into peptides, both by solid-phase and by classical techniques.

The Coupling Step. Improved coupling results can be obtained by treating a solution of the carboxy-component (six equivalents) with dicyclohexylcarbodi-imide (three equivalents), removing the precipitated dicyclohexylurea, and using the resulting activated amino-acid derivative directly in the solid-phase method. A nonapeptide was prepared in 95% overall yield by this method, which has the advantage that very concentrated

[162] J. D. Glass, I. L. Schwartz, and R. Walter, *J. Amer. Chem. Soc.*, 1972, **94**, 6209.
[163] M. Ohno, S. Tsukamoto, and N. Izumiya, *J.C.S. Chem. Comm.*, 1972, 663.
[164] M. Ohno, S. Tsukamoto, S. Makisumi, and N. Izumiya, *Bull. Chem. Soc. Japan*, 1972, **45**, 2852.

solutions of the anhydride can be used and avoids precipitation of the urea in the resin matrix.[165]

An investigation has been carried out of the kinetics of the coupling process using different types of resin [166] (Merrifield resin, some macroporous resins with different degrees of cross-linking, and a pellicular resin), and of the effect on the rate of coupling when sterically hindered components are involved, or when the free amino-group is located close to the Merrifield matrix, or is separated from it by a hydrocarbon chain.[167]

The oxidation–reduction method of condensation has been applied to the solid-phase synthesis of some oligopeptides [158] (both by stepwise and fragment condensation on the resin), oxytocin, and [2-phenylalanine]-lysine-vasopressin.[168] 2-Isobutoxy-1-isobutoxycarbonyl-1,2-dihydroquinoline (79) has been used to couple a hexapeptide to a dipeptide resin.[113]

In the solid-phase synthesis of the heptapeptide (98), the histidine residues were introduced either as *N*-α-*N*-*im*-di-t-butoxycarbonyl-L-

Ala-His-Arg-Leu-His-Gln-Leu

(98)

histidine or *N*-α-1-(*p*-biphenylyl)-1-methylethoxycarbonyl-*N*-*im*-t-butoxycarbonyl-L-histidine.[169] In both syntheses some loss of the *im*-t-butoxycarbonyl protecting group during the acidolytic deprotection cycle was evident. This results in subsequent coupling of N-protected amino-acids to the imidazole group, as well as to the α-amino-group, and accounts for a proportion of the newly introduced amino-acid appearing in the filtrate when later deprotection and neutralization steps are performed.

The free amino-groups of the polymer-supported growing peptide chain can act as anion-exchange sites (this is additional to the ion-exchange capacity of the resin-bound quaternary ammonium or sulphonium groups, which may result from the esterification step). Esko and Karlsson [170] demonstrated that acylamino-acids bind to the peptide chain and only the excess carboxy-component can be removed from the resin by washing with methylene chloride. This fact has now been turned to advantage in synthesis of some bradykinin analogues.[157] The procedure involves shaking the resin with a three-fold excess of the carboxy-component (the excess of which is removed by filtration and can be re-used), and bringing about condensation with a three-fold excess of dicyclohexylcarbodi-imide. The modified coupling procedure enabled the products of final deprotection to be obtained in high yield and of greater purity than those obtained by the conventional procedure. The side-products in this synthesis were shown to be largely of the 'truncated sequence' type.

[165] H. Hagenmaier and H. Frank, *Z. physiol. Chem.*, 1972, **353**, 1973.
[166] A. Losse, *Tetrahedron Letters*, 1971, 4989.
[167] G. Losse and R. Ulbrich, *Tetrahedron*, 1972, **28**, 5823.
[168] T. Mukaiyama, *Yuki Gosei Kagaku Kyokai Shi*, 1971, **29**, 848 (*Chem. Abs.*, 1972, **76**, 100 001p).
[169] D. Yamashiro, J. Blake, and C. H. Li, *J. Amer. Chem. Soc.*, 1972, **94**, 2855.
[170] K. Esko and S. Karlsson, *Acta Chem. Scand.*, 1970, **24**, 1415.

Removal of the Finished Peptide from the Resin. The thiol-mediated cleavage of the *im*-dinitrophenyl protecting group from histidine peptides also results in the conversion of any methionine sulphoxide residues present into methionine.[171] Some *C*-terminal primary amide gastrin analogues have been generated by direct ammonolysis of a peptidyl-nitrobenzyl ester resin.[172] Further reports have appeared dealing with the ammonolysis of peptidyl benzyl ester resins in liquid ammonia and DMF under pressure. The method is illustrated by the preparations of a number of t-butoxycarbonyl-amino-acid amides,[173] the *C*-terminal hexapeptide of secretin,[173] some eledoisin analogues,[156] and a scotophobin analogue.[173]

The synthesis of 4′-aminomethyl-2,2-diphenylethanediol (99) (Scheme 68) has been reported.[174] It is possible to attach (99) to a chloromethylated polymer *via* its aminomethyl function. Preliminary experiments show that N-protected amino-acids can be esterified to the pendant primary hydroxy-groups of the resin-bound glycol derivative. After a conventional solid-

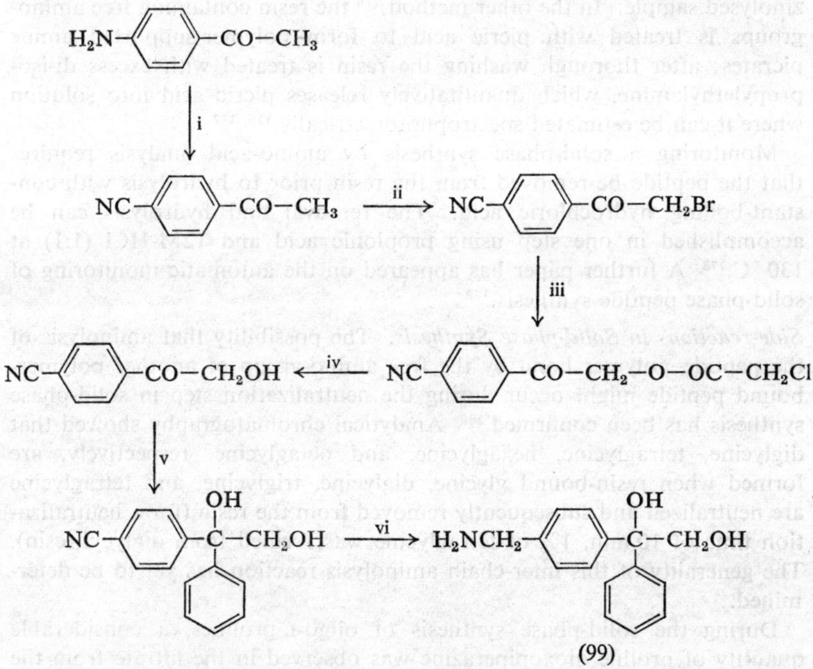

(99)

Conditions: i, Sandmeyer reaction; ii, Br_2; iii, $ClCH_2CO_2^-$; iv, aq. $KHCO_3$; v, PhMgBr; vi, $NaBH_4$ (or Raney Ni)

Scheme 68

[171] K. P. Polzhofer, *Tetrahedron*, 1972, **28**, 855.
[172] M. M. Badawi, *J. Drug Res.*, 1972, **3**, 223 (*Chem. Abs.*, 1972, **77**, 102 209g).
[173] W. Parr, C. Yang, and G. Holzer, *Tetrahedron Letters*, 1972, 101.
[174] Th. Wieland, C. Birr, and P. Fleckenstein, *Annalen*, 1972, **756**, 14.

phase synthesis, it is envisaged that treatment of the glycol ester with trifluoroacetic acid will yield an activated diphenylvinyl ester, which will enable the peptide to be removed from its carrier using very mild hydrolysis, solvolysis, hydrazinolysis, or ammonlysis. Furthermore, it should be possible to carry out fragment condensations or cyclization reactions with the polymeric active ester. The experimental details of this application of the 'safety-catch' principle are awaited with interest, although racemization of the *C*-terminal residue is clearly a danger.

Monitoring the Progress of Solid-phase Synthesis. Two analytical methods have been reported for following the incorporation of amino-acids into resin-bound peptides, and for following the deblocking of resin bound peptides. In one method,[175] the amino-groups of the polymer are dansylated; after thorough washing with methylamine solution, the resin is subjected to hydrazinolysis, and the dansyl-containing material released is estimated fluorimetrically. The useful detection limit is 1.5 nmol mg^{-1} of hydrazinolysed sample. In the other method,[176] the resin containing free amino-groups is treated with picric acid to form polymer-supported amine picrates; after thorough washing the resin is treated with excess di-isopropylethylamine, which quantitatively releases picric acid into solution where it can be estimated spectrophotometrically.[176, 177]

Monitoring a solid-phase synthesis by amino-acid analysis requires that the peptide be removed from the resin prior to hydrolysis with constant-boiling hydrochloric acid. The removal and hydrolysis can be accomplished in one step using propionic acid and 12M-HCl (1:1) at 130 °C.[178] A further paper has appeared on the automatic monitoring of solid-phase peptide synthesis.[179]

Side-reactions in Solid-phase Synthesis. The possibility that aminolysis of the peptide–polymer bond by the free amino-group of another polymer-bound peptide might occur during the neutralization step in solid-phase synthesis has been confirmed.[180] Analytical chromatography showed that diglycine, tetraglycine, hexaglycine, and octaglycine, respectively, are formed when resin-bound glycine, diglycine, triglycine, and tetraglycine are neutralized and subsequently removed from the resin (for a neutralization time of 10 min, 1% of tetraglycine was formed from diglycyl-resin). The generality of this inter-chain aminolysis reaction has yet to be determined.

During the solid-phase synthesis of oligo-L-prolines, a considerable quantity of proline dioxopiperazine was observed in the filtrate from the

175 J. Garden, jun. and A. M. Tometsko, *Analyt. Biochem.*, 1972, **46**, 216.
176 B. F. Gisin, *Analyt. Chim. Acta*, 1972, **58**, 248.
177 B. F. Gisin and R. B. Merrifield, *J. Amer. Chem. Soc.*, 1972, **94**, 3102.
178 F. C. Westall, J. Scotchler, and A. B. Robinson, *J. Org. Chem.*, 1972, **37**, 3363.
179 K. Brunfeldt, T. Christensen, and P. Villemoes, *F.E.B.S. Letters*, 1972, **22**, 238.
180 H. C. Beyerman, E. W. B. de Leer, and W. van Vossen, *J.C.S. Chem. Comm.*, 1972, 929.

neutralization step of the diprolylresin.[179, 181, 182] The intra-chain aminolysis results in the liberation of hydroxymethyl groups to which a newly introduced amino-acid could be esterified, thus leading to contamination of the final product with a peptide lacking two residues at the carboxy-terminus. This has been observed in the attempted synthesis of [8-(*N*-methyl-L-phenylalanine)]angiotensin II (100) where the major product

Asp-Arg-Val-Tyr-Ile-His-Pro-MePhe (100)   Asp-Arg-Val-Tyr-Ile-His (101)

was the hexapeptide (101);[183] in addition, cyclo-L-prolyl-*N*-methyl-L-phenylalanine was isolated from the appropriate filtrate. It has been found that although intra-chain aminolysis is slow during the neutralization step, it is markedly accelerated by presence of carboxylic acids,[177] *e.g.* H-D-Val-L-Pro-O-Resin yields 98% diketopiperazine after treatment with 0.08M acetic acid–methylene chloride with $t_{\frac{1}{2}} = 8.1$ min. A study of a series of peptidyl resins shows that losses of the order of 1—5% can be expected for a normal coupling procedure to a primary amino-dipeptide resin providing its peptide bond is not a tertiary amide; in the latter case much greater losses are to be expected. The problem can be minimized by adding the coupling reagent to the dipeptide resin prior to the carboxy-component.[177]

It has been shown that acetic acid (commonly used as the solvent for hydrogen chloride in the deprotection cycle) is absorbed by teflon and released at later stages, particularly the coupling steps, of the solid-phase synthesis cycles.[179] A careful study of the synthesis of an antamanide sequence (102) revealed incomplete sequences due to acetylation by acetic acid leached from the teflon at intermediate coupling stages.[184]

Phe-Pro-Pro-Phe-Phe-Val-Pro-Pro-Ala-Phe
(102)

Boc-Ala-Phe-Ile-Gly-Leu-Met-NH_2
(103)

An analytical method based on an iodometric titration has been developed to determine the sulphoxide content of peptidyl resins.[155] It was used to show that, in the synthesis of a protected hexapeptide amide (103), no oxidation of methionyl residues to methionine sulphoxide residues could be detected, either during the synthesis on the resin or during cleavage from the resin. In the solid-phase synthesis of human calcitonin the methionine residue was protected against possible oxidation during N-α-deprotection by including 1% ethanethiol in the trifluoroacetic acid.

181 M. Rothe and J. Mazánek, *Angew. Chem. Internat. Edn.*, 1972, **11**, 293.
182 M. Rothe and J. Mazánek, *Tetrahedron Letters*, 1972, 3795.
183 M. C. Khosla, R. R. Smeby, and F. M. Bumpus, *J. Amer. Chem. Soc.*, 1972, **94**, 4721.
184 K. Brunfeldt, T. Christensen, and P. Roeptsorff, *F.E.B.S. Letters*, 1972, **25**, 184.

It appears that this additive does not interfere with the *im*-dinitrophenyl protecting group.[149] In the synthesis of the peptide (98) the arginine residue was protected by nitration; after the removal of the peptide from the resin with anhydrous hydrogen fluoride or hydrogen bromide in trifluoroacetic acid, some ornithine peptides were detected in the product mixture.[169] This side reaction may be sequence dependent.

Apparatus and Instrumentation. The design and construction of two systems for automated solid-phase peptide synthesis have been described. One machine, controlled by a programmer, utilizes a pellicular resin packed in a chromatographic column, all synthetic operations being carried out on the column.[185] The other machine is more conventional, using a reaction vessel which is shaken.[186] A computer programme has been described[187, 188] which is capable of generating control tapes to direct automatic solid-phase synthesis, solely from the input of the desired amino-acid sequence. A parity check circuit has been used to monitor the translation of control tapes during automatic synthesis.[189] A modified reaction vessel for manual solid-phase synthesis has been described.[190] It consists of a round flask with a side-arm through which a gas dispersion tube can be inserted for use in deprotection cycles. Agitation of the resin is achieved by a wrist-action shaker.

Other Repetitive Methods. A solid-phase synthesis has been developed in which *N*-carboxyanhydrides are coupled with a resin-bound *N*-trimethylsilylamino-ester (Scheme 69).[191] The resin is not cross-linked. The func-

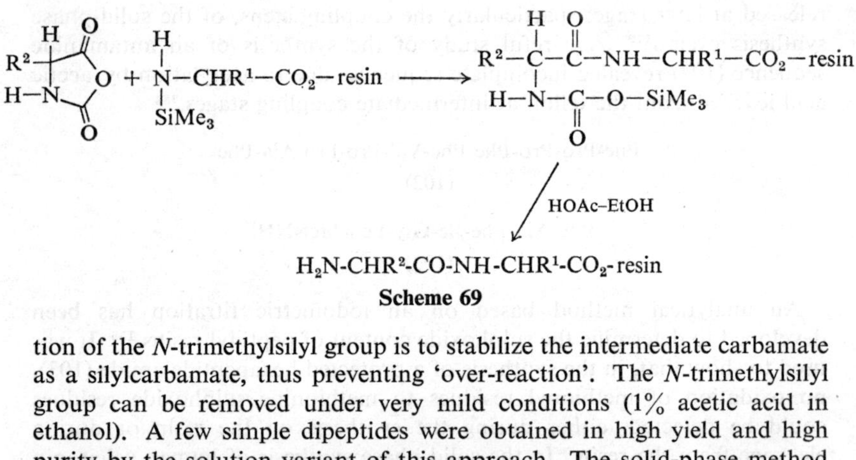

Scheme 69

tion of the *N*-trimethylsilyl group is to stabilize the intermediate carbamate as a silylcarbamate, thus preventing 'over-reaction'. The *N*-trimethylsilyl group can be removed under very mild conditions (1% acetic acid in ethanol). A few simple dipeptides were obtained in high yield and high purity by the solution variant of this approach. The solid-phase method

[185] P. R. W. Scott, S. Zolty, and K. K. Chan, *J. Chromatog. Sci.*, 1972, **10**, 384.
[186] V. J. Hruby, L. E. Barstow, and T. Linhart, *Analyt. Chem.*, 1972, **44**, 343.
[187] A. M. Tometsko, *Comput. Biomed. Res.*, 1972, **5**, 156.
[188] A. M. Tometsko, *Comput. Biomed. Res.*, 1972, **4**, 101.
[189] A. M. Tometsko, *Chem. Instrum.*, 1971, **3**, 235.
[190] W. K. Park and D. Regoli, *Canad. J. Biochem.*, 1972, **50**, 755.
[191] J. J. Maher, M. E. Furey, and L. J. Greenberg, *Tetrahedron Letters*, 1972, 1581.

has been applied to the synthesis of oxytocin but details of the work are not yet available.

A further paper has appeared dealing with the liquid-phase synthesis of peptides on a soluble polymeric support.[192] In this method, t-butoxycarbonyl-amino-acids are coupled in homogeneous solution by the mixed anhydride procedure with an amino-ester supported on poly(ethylene glycol); impurities are removed by ultrafiltration. In the synthesis of a few simple peptides no racemization could be detected, and failure sequences were not recognized. Transesterification with sodium methoxide in methanol was used for the removal of the peptide chain from the polymer.

Synthesis of Polymeric Models for Studies in Protein Chemistry.—*Poly-(amino-acids) and Sequential Polypeptides.* Poly-(*N*-ε-menthyloxycarbonyl-L-lysine) has been prepared by the *N*-carboxy-anhydride method.[193] Poly-(α-benzyl-L-aspartate), $M_n = 17\,500$, has been prepared[194] by polymerization of a succinimido-ester and converted into poly-(β-L-aspartic acid), $M_n = 9600$, $M_w = 11\,600$, by hydrogenolysis; treatment of the poly(benzyl ester) with hydrogen bromide in acetic acid leads to a polymer (104) containing a substantial proportion of aspartimide residues. It is

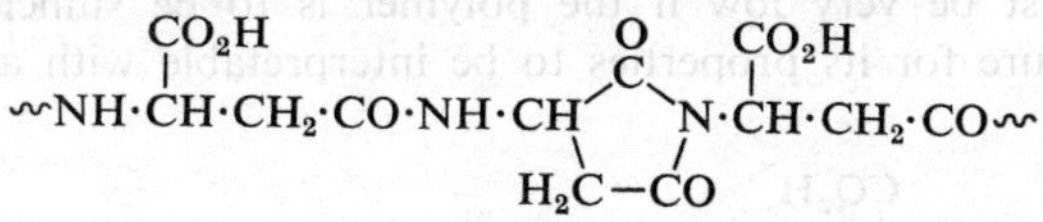

(104)

quite likely that samples of poly-(β-L-aspartic acid), previously reported in the literature,[195] contain a substantial proportion of aspartimide residues. Poly-(α-benzyl L-glutamate) has been prepared as shown in Scheme 70.[194] Poly-(γ-L-glutamic acid), produced by the action of hydrogen bromide in acetic acid on the poly-(α-benzyl ester), had properties in satisfactory agreement with the enantiomeric natural product from *Bacillus anthracis*. The related sequential polypeptides (105)[196] and (106)[197] have been prepared (by the pentachlorophenyl ester method) to investigate the importance of the distance between carboxy side-chain groups for serological activity with anti-serum against *B. anthracis* polypeptide.

The occurrence of a small degree of racemization of the *C*-terminal residue during polymerization of an oligopeptide monomer can result in serious configurational inhomogeneity in the polymer product. The Figure illustrates the results of calculations of the fraction of the desired stereochemically homogeneous product, for various degrees of polymeriza-

[192] E. Bayer and M. Mutter, *Nature*, 1972, **237**, 512.
[193] H. Yamamoto and T. Hayakawa, *Bull. Chem. Soc. Japan*, 1971, **44**, 1990.
[194] P. M. Hardy, J. C. Haylock, and H. N. Rydon, *J.C.S. Perkin I*, 1972, 605.
[195] J. Kovacs, R. Ballina, R. L. Rodin, D. Balasubramanian, and J. Applequist, *J. Amer. Chem. Soc.*, 1965, **87**, 119.
[196] J. Kovacs, A. Kapoor, U. R. Ghatak, G. L. Mayers, V. R. Giannasio, R. Giannotti, G. Senyk, D. E. Nitecki, and J. W. Goodman, *Biochemistry*, 1972, **11**, 1953.
[197] R. L. Rodin and J. Kovacs, *Bioorg. Chem.*, 1972, **2**, 65.

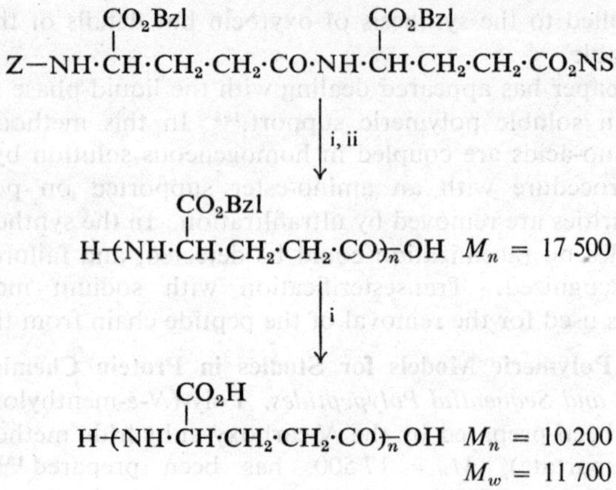

Conditions: i, HBr–CH_3CO_2H; ii, DMF–NEt_3

Scheme 70

tion, at different levels of racemization. It is clear that the level of racemization must be very low if the polymer is to be sufficiently stereochemically pure for its properties to be interpretable with any real confidence.[198]

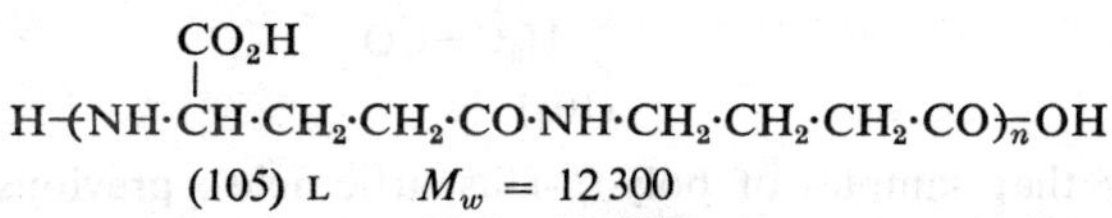

(105) L $M_w = 12\,300$; D $M_w = 12\,500$

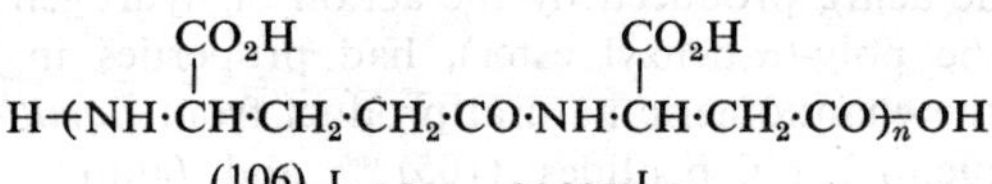

(106) L/D $M_w = 28000$ L/D

The synthesis of a series of diastereoisomeric poly-(γ-t-butyl glutamate)s and the derived poly-(glutamic acid)s has been described.[199] The monomers were polymerized using dicyclohexylcarbodi-imide in acetonitrile (Scheme 71). Racemization of *C*-terminal residues during polymerization was determined by comparing the optical rotations of the appropriate hydrolysis mixture of the polymer with that of an amino-acid standard. However, in the special case of the polymers (107; $x = 1$, LD or DL) the optical rotation in a helix-disrupting solvent is a direct measure of the racemization, assuming that end effects can be neglected. Considerable racemization was indicated by these measurements and the dicyclohexylcarbodi-imide method of polymerization, although simple and convenient, is clearly not satis-

[198] P. M. Hardy, H. N. Rydon, and R. C. Thompson, *J.C.S. Perkin I*, 1972, 5.
[199] D. I. Marlborough and H. N. Rydon, *J.C.S. Perkin I*, 1972, 1.

factory. It has been shown that the polymerization of oligopeptide succinimido-esters is accompanied by little racemization.[198] The succinimido-esters (108; $x = 1$—4) have been prepared from the corresponding acids (Scheme 72), which were synthesized by the dicyclohexylcarbodi-imide

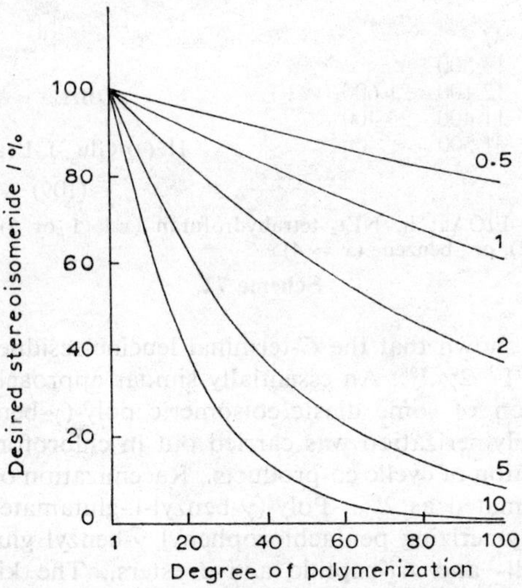

Figure *The effect of racemization of the* C*-terminal residue on the amount of the derived stereochemically homogeneous polypeptide formed in the polymerization of an oligopeptide. The number against each curve is the % racemization at each polymerization stage*
(Reproduced from *J.C.S. Perkin I*, 1972, 5)

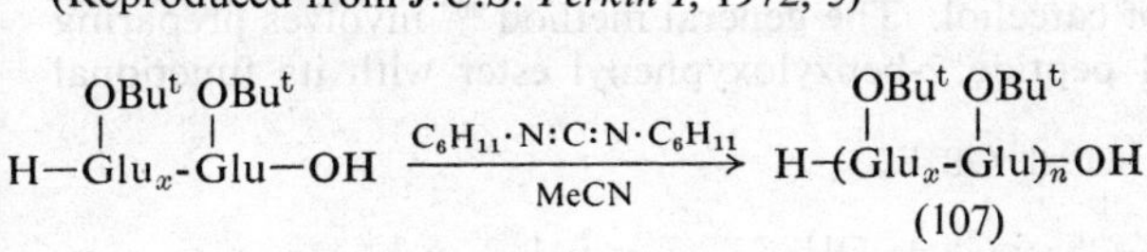

(107)

			M_n
L$_x$	D	$x = 1, 2, 3$	11 500, 6 500, 14 000
L	L	$x = 1$	42 000
D	L	$x = 1$	14 000

Scheme 71

method with the t-butoxycarbonyl group for amino-protection and the 2-methylthioethyl safety-catch ester for carboxy-protection. The latter group is removed by mild alkaline treatment after molybdate-catalysed oxidation to the corresponding 2-methylsulphonylethyl ester. The compounds (108; $x = 1$—4) on polymerization gave moderate yields of the sequential polypeptides (109; $x = 1$—4) which were accompanied in the appropriate cases by cyclo-di-, -hexa-, -octa-, and -deca-peptides. The optical purity of the sequential polymers was checked by digestion of the leucine (produced by hydrolysis of the polymer with constant boiling hydrochloric acid and removal of D-glutamic acid) with L-amino-acid

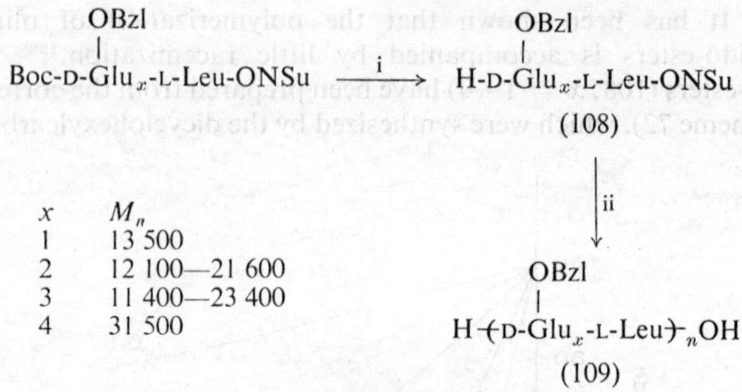

x	M_n
1	13 500
2	12 100—21 600
3	11 400—23 400
4	31 500

Conditions: i, HCl–EtOAc; ii, NEt_3–tetrahydrofuran (x = 1 or 3), –light petroleum (x = 2), or –benzene (x = 4).

Scheme 72

oxidase. It was shown that the *C*-terminal leucine residues are racemized to the extent of 1—2%.[198] An essentially similar approach has been used for the formation of some diastereoisomeric poly-(γ-benzyl glutamate)s (110).[200] The polymerization was carried out in chloroform, which minimized the formation of cyclic co-products. Racemization of the *C*-terminal residue was estimated as 2%. Poly-(γ-benzyl-L-glutamate) has also been prepared by polymerizing pentachlorophenyl γ-benzyl glutamate and the corresponding di- and tri-peptide active esters. The kinetics of polymerization of these monomers in benzene has been studied.[201] Polymerization of succinimido-esters has also been used to prepare a series of sequential polypeptides of β-methyl aspartate and γ-methyl glutamate (111).[202]

The other major method of racemization-free active ester polymerization utilizes monoesters of catechol. The general method [203] involves preparing a benzyloxycarbonyl peptide 2-benzyloxyphenyl ester with its functional

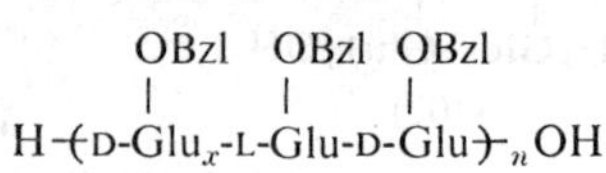

(110)

x = 0, 1, 2, or 3
n = 17 ~ 53
M_n = 15 000 ~ 45 000

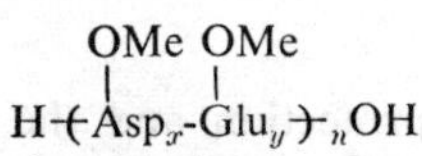

(111)

				M_n	M_w
$x = y = 1$	L	L		11 300	15 400
$x = 2, y = 1$	L	L	L	12 200	15 500
	L	L	D	9 900	12 200
$x = 1, y = 2$	L	D	D	9 900	12 200

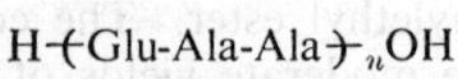

M_w = 2 500 ~ 6 000

(112)

[200] P. M. Hardy, H. N. Rydon, and H. T. Storey, *J.C.S. Perkin I*, 1972, 1523.
[201] (*a*) A. Brack and G. Spach, *Bull. Soc. chim. France*, 1971, 4481; (*b*) A. Brack and G. Spach, *ibid.*, p. 4485; (*c*) Y. Trudelle and G. Spach, *ibid.*, p. 4489.
[202] A. Ali, P. M. Hardy, and H. N. Rydon, *J.C.S. Perkin I*, 1972, 1070.
[203] R. D. Cowell and J. H. Jones, *J.C.S. Perkin I*, 1972, 2236.

side-chains masked by t-butyl-based protection. Deprotection to liberate the monomer is readily achieved by hydrogenolysis (provided sulphur-containing reagents are rigorously avoided in previous synthetic steps), and the resulting acetate salts are not isolated but polymerized directly with triethylamine in DMSO. This method has been illustrated with the synthesis of poly-(L-glutamyl-L-alanyl-L-alanine) (112) and poly-(L-lysyl-L-alanyl-L-alanine) (Scheme 73). The latter sequential polymer has also been

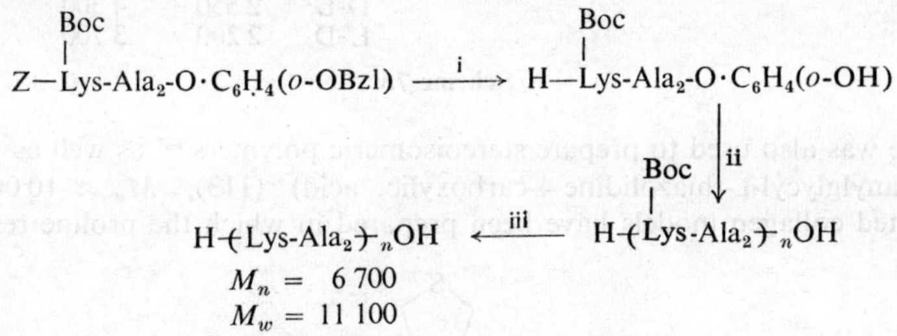

Conditions: i, H_2–Pd/C; ii, Et_3N–DMSO; iii, 90% CF_3CO_2H

Scheme 73

prepared, with comparable molecular weight, *via* the corresponding succinimido-ester (Scheme 74).[204] Enzymic digestion of the polymer indicated racemization at the *C*-terminal alanine residue of the monomer of less than 0.5%. Half-esters of catechol have also been used to synthesize

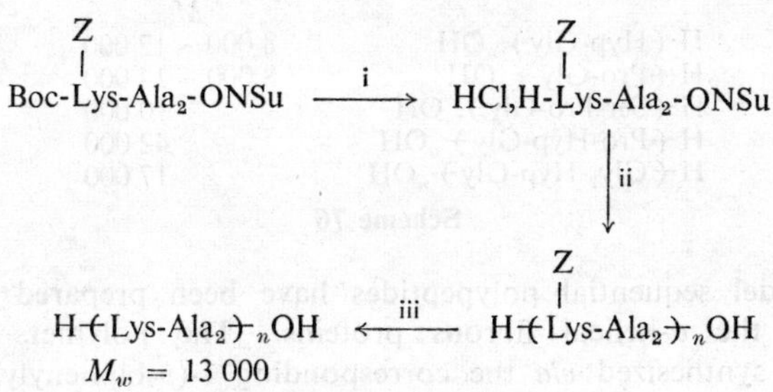

Conditions: i, HCl–EtOAc; ii, NEt_3–DMF; iii, HBr–AcOH

Scheme 74

poly-(L-glutaminyl-L-alanine), M_w = 5000—10 000, a model for the silks produced by sawflies of the family Argidae.[205]

Although poly-(L-prolyl-L-serylglycine), M = 10 000—18 000, and poly-(L-prolyl-L-alanylglycine), M = 7200—14 000, were prepared without difficulty by polymerization of the corresponding *p*-nitrophenyl esters,[206]

[204] A. Yaron, N. Tal (Turkeltaub), and A. Berger, *Biopolymers*, 1972, **11**, 2461.
[205] J. H. Jones and J. Walker, *J.C.S. Perkin I*, 1972, 2923.
[206] F. R. Brown, tert., A. di Corato, G. P. Lorenzi, and E. R. Blout, *J. Mol. Biol.*, 1972, **63**, 85.

with the latter polymer another group of workers [207] found the active ester approach with this particular sequence to be unsatisfactory and so prepared it from alanylglycylproline pentachlorophenyl ester (Scheme 75). The

HBr,H-Ala-Gly-Pro-OPcp $\xrightarrow{\text{DMSO}}$ H-(Ala-Gly-Pro)$_n$OH

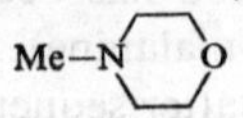

	M_n	M_w
L-L	9 000	12 000—14 000
D-D	8 900	12 700
D-L	2 550	4 300
L-D	2 200	3 700

Scheme 75

route was also used to prepare stereoisomeric polymers [207] as well as poly-(L-alanylglycyl-L-thiazolidine-4-carboxylic acid) (113), $M_w \approx 10\,000$.[208] Related collagen models have been prepared in which the proline residue

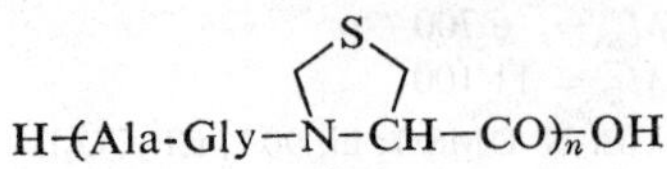

(113)

has been replaced with either L-piperidine-2-carboxylic acid or L-azetidine-2-carboxylic acid, using the catechol ester method.[208] Further collagen models which have been prepared by the *p*-nitrophenyl ester method of polymerization are listed in Scheme 76.[209]

	M_w
H-(Hyp-Gly)$_n$OH	8 000 ~ 12 000
H-(Pro-Gly)$_n$OH	8 000 ~ 14 000
H-(Ser-Pro-Gly)$_n$OH	10 000
H-(Pro-Hyp-Gly)$_n$OH	42 000
H-(Gly$_2$-Hyp-Gly)$_n$OH	17 000

Scheme 76

Two model sequential polypeptides have been prepared for studies related to the α-helical fibrous proteins. The polymers (114) and (115) were synthesized *via* the corresponding 1-(*p*-biphenylyl)-1-methylethoxycarbonyl heptapeptide acids with t-butyl-based side-chain protection. The *C*-terminal carboxy-group was activated by reaction with bis-*p*-nitrophenyl sulphite; selective acidolysis of the *N*-α-protecting group followed by treatment with triethylamine in DMSO gave polymers which yielded (114) and (115) on treatment with trifluoroacetic acid.[210, 211]

A spectroscopic study of the conformations and conformational stabilities of poly(leucyl-leucyl-β-benzyl aspartate) (116) [212] and the derived

[207] R. Fairweather and J. H. Jones, *J.C.S. Perkin I*, 1972, 1908.
[208] R. Fairweather and J. H. Jones, *J.C.S. Perkin I*, 1972, 2475.
[209] De los F. DeTar, R. J. Albers, and F. Gilmore, *J. Org. Chem.*, 1972, **37**, 4377.
[210] R. D. Cowell and J. H. Jones, *J.C.S. Perkin I*, 1972, 1809.
[211] R. D. Cowell and J. H. Jones, *J.C.S. Perkin I*, 1972, 1814.
[212] M. D'Alagni, P. Bemporad, and A. Garofolo, *Polymer*, 1972, **13**, 419.

H$\leftarrow$Leu-Glu-Lys-Ala-Glu-Ser-Gly$\rightarrow_n$OH $M_n = 6\,000$, $M_w = 11\,000$
(114)

H$\leftarrow$Leu-Glu-Lys-Ala-Glu-Ala-Gly$\rightarrow_n$OH $M_n = 3\,000$, $M_w = 6\,000$
(115)

H$\leftarrow$Leu$_2$-Asp(OBzl)$\rightarrow_n$OH
(116)

H$\leftarrow$Ala$_x$-Gly$_y\rightarrow_n$OH $x = 1, y = 1, 2$ or 3; $x = 2, y = 1, 2$ or 3; $x = 3, y = 3$
(117)

poly-acid[213] have been carried out. The polymer (116) was prepared *via* the tripeptide *p*-nitrophenyl active ester. A series of alanylglycine polymers (117) has been prepared from oligopeptide pentachlorophenyl esters, for conformational study.[214]

Further applications have appeared of the polymerization of oligopeptide pentachlorophenyl esters on a preformed monomer in connection with antigenicity studies.[215–218]

Polyglutathione, polyaspartathione, polyisoglutathione, and polyisoaspartathione (Scheme 77) have been prepared by the self-condensation

H$\leftarrow$Glu(–Cys-Gly$\rightarrow_n$OH)-OH $M_w = 9\,000$

H$\leftarrow$Asp(–Cys-Gly$\rightarrow_n$OH)-OH $M_w = 7\,000$

H$\leftarrow$Glu(OH)-Cys-Gly$\rightarrow_n$OH $M_w = 16\,000$

H$\leftarrow$Asp(OH)-Cys-Gly$\rightarrow_n$OH $M_w = 6\,000$

Scheme 77

of the corresponding tripeptide pentachlorophenyl esters.[219] The optical pure tripeptide monomers were prepared by a 'backing off' procedure from glycine active ester using t-butoxycarbonyl groups for *N*-α-protection and benzyl groups for side-chain protection. The polymers were stripped of their protecting groups with sodium in liquid ammonia, since anhydrous hydrogen fluoride gave less pure products of lower molecular weight.

A report has appeared on the synthesis of a series of sequential macromolecular polypeptolides[220] of L-leucine and L-2-hydroxy-4-methylpentanoic acid. Block copolymers of α-amino-acids and α-hydroxy-acids

213 M. Carità Morelli and M. D'Alagni, *Polymer*, 1972, **13**, 515.
214 A. Brack and G. Spach, *Biopolymers*, 1972, **11**, 563.
215 B. J. Johnson, C. Cheng, and N. Tsang, *J. Medicin. Chem.*, 1972, **15**, 95.
216 B. J. Johnson, *J. Medicin. Chem.*, 1972, **15**, 423.
217 B. J. Johnson and N. Tsang, *J. Medicin. Chem.*, 1972, **15**, 488.
218 B. J. Johnson, *J. Pharm. Sci.*, 1972, **61**, 1990.
219 J. Kovacs, C. Kalita, and U. R. Ghatak, *J. Org. Chem.*, 1972, **37**, 30.
220 B. Ridge, H. N. Rydon, and C. R. Snell, *J.C.S. Perkin I*, 1972, 2041.

have been prepared by treating mixtures of amino-acid *N*-carboxy-anhydrides and hydroxy-acid anhydro-sulphites with triethylamine in benzene.[221]

Synthetic Operations with Peptides of Biological Origins.—The semi-synthetic conversion of porcine insulin into human insulin is discussed later in these reports (see p. 330). De-[pentapeptide-(B26-B30)]-bovine insulin has been prepared by pepsin cleavage of the insulin.[222] It seems a likely candidate for semi-synthetic modification. N.m.r. studies have been aided by the semi-synthetic replacement of phenylalanine at position B1 in insulin by trifluoroacetyl-[1-^{13}C]- or -[2-^{13}C]-glycine probes.[223]

Offord[224a] has outlined the possible use of peptides derived from cyanogen bromide cleavage at methionyl peptide bonds in proteins as precursors for semi-synthesis. Either the terminal lactone ring is hydrolysed, and the resulting protected peptide acid coupled with amino-components using dicyclohexylcarbodi-imide in the presence of 1-hydroxybenzo-triazole, or it is hydrazinolysed, the resulting protected peptide hydrazide being used in an azide synthesis. The latter procedure achieves the selectivity from side-chain carboxy-groups necessary for general application but, in both methods of activation, re-formation of the lactone ring may be a seriously competing side-reaction. On the other hand, the lactone ring itself may possess sufficient intrinsic reactivity to undergo direct aminolysis under favourable conditions. Such conditions may be achieved when the coupling reaction becomes intramolecular, as when the amino- and carboxy-components are linked by prior disulphide bond formation.[224b]

3 Syntheses Achieved

A comprehensive list of the year's achievements is given in Appendix I (p. 338); the discussion below is highly selective. Biological aspects are this year dealt with in Chapter 5.

Adrenocorticotropic Hormone (ACTH).—Following the demonstration that the formerly accepted amino-acid sequence of human ACTH is erroneous in the segment (25—30),[225] a total synthesis of the corrected sequence (118)

Ser-Tyr-Ser-Met-Glu-His-Phe-Arg-Trp-Gly-Lys-Pro-Val-Gly-Lys-Lys-Arg-
1 **10** **11** **16** **17**
-Arg-Pro-Val-Lys-Val-Tyr-Pro-**Asn-Gly-Ala-Glu-Asp-Glu**-Ser-Ala-Glu-Ala-
24 **25**
Phe-Pro-Leu-Glu-Phe
39

(118)

[221] M. Goodman, C. Gilon, G. S. Kirshenbaum, and Y. Knobler, *Israel J. Chem.*, 1972, **10**, 867.
[222] Yu. P. Shvachkin, G. A. Shmeleva, V. F. Krivtsov, V. P. Fedotov, and A. I. Ivanova, *Biokhimiya*, 1972, **37**, 966.
[223] D. J. Saunders and R. E. Offord, *F.E.B.S. Letters*, 1972, **26**, 286.
[224a] R. E. Offord, *Biochem. J.*, 1972, **129**, 499.
[224b] D. F. Dyckes, T. Creighton, and R. C. Sheppard, *Nature*, in the press.
[225] B. Riniker, P. Sieber, W. Rittel, and H. Zuber, *Nature New Biol.*, 1972, **235**, 114.

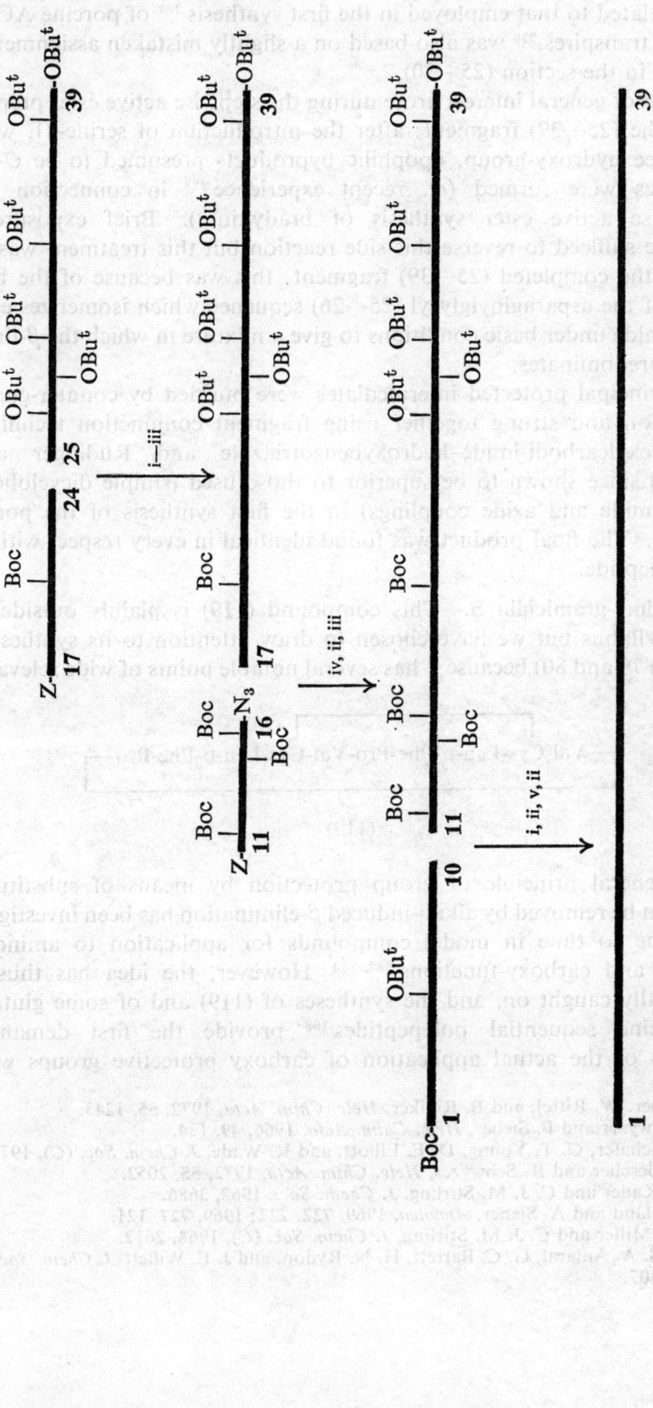

Scheme 78

Conditions: i, DCCI–HOBt; ii, c.c.d.; iii, H_2–catalyst; iv, DMF–*N*-methylmorpholine; v, CF_3CO_2H

has been reported.[226] The overall strategy of the synthesis (Scheme 78) was closely related to that employed in the first synthesis [227] of porcine ACTH, which, it transpires,[225] was also based on a slightly mistaken assignment of sequence in the section (25—30).

A point of general interest arose during the stepwise active ester preparation of the (25—39) fragment: after the introduction of serine-31, which had a free hydroxy-group, lipophilic byproducts presumed to be *O*-acyl derivatives were formed (*cf.* recent experience [228] in connection with a stepwise active ester synthesis of bradykinin). Brief exposure to hydrazine sufficed to reverse this side reaction but this treatment was not used on the completed (25—39) fragment: this was because of the base-lability of the asparaginylglycyl (25—26) sequence which isomerizes easily *via* the imide under basic conditions to give a mixture in which the β-linked peptide predominates.

The principal protected intermediates were purified by counter-current distribution and strung together using fragment conjunction techniques (dicyclohexylcarbodi-imide–hydroxybenzotriazole and Rudinger azide methods) since shown to be superior to those used (simple dicyclohexylcarbodi-imide and azide couplings) in the first synthesis of the porcine hormone. The final product was found identical in every respect with the natural peptide.

[2,7-Cystine]-gramicidin S.—This compound (119) is plainly outside our proper syllabus but we have chosen to draw attention to its synthesis [229] (Schemes 79 and 80) because it has several notable points of wide relevance.

Val-Cys-Leu-D-Phe-Pro-Val-Cys-Leu-D-Phe-Pro

(119)

The general principle of group protection by means of substituents which can be removed by alkali-induced β-elimination has been investigated from time to time in model compounds for application to amino-,[230] thiol-,[231] and carboxy-functions.[232, 233] However, the idea has thus far never really caught on, and the syntheses of (119) and of some glutamic acid–leucine sequential polypeptides [198] provide the first demanding examples of the actual application of carboxy protective groups which

[226] P. Sieber, W. Rittel, and B. Riniker, *Helv. Chim. Acta*, 1972, **55**, 1243.
[227] R. Schwyzer and P. Sieber, *Helv. Chim. Acta*, 1966, **49**, 134.
[228] D. J. Schafer, G. T. Young, D. F. Elliott, and R. Wade, *J. Chem. Soc.* (*C*), 1971, 46.
[229] U. Ludescher and R. Schwyzer, *Helv. Chim. Acta*, 1972, **55**, 2052.
[230] A. T. Kader and C. J. M. Stirling, *J. Chem. Soc.*, 1962, 3686.
[231] T. Wieland and A. Sieber, *Annalen*, 1969, **722**, 222; 1969, **727**, 121.
[232] A. W. Miller and C. J. M. Stirling, *J. Chem. Soc.* (*C*), 1968, 2612.
[233] M. J. S. A. Amaral, G. C. Barrett, H. N. Rydon, and J. E. Willett, *J. Chem. Soc.* (*C*), 1966, 807.

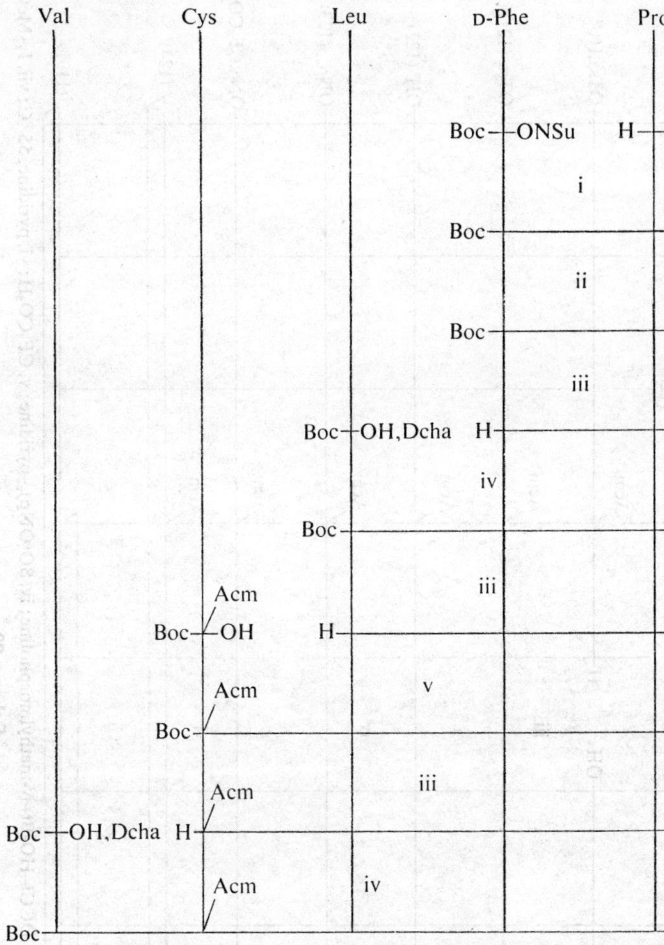

(120)

Conditions: i, *N*-methylmorpholine–pyridine; ii, DCCI–TsaOH; iii, HCl–EtOAc; iv, DCCI–HOBt; v, DCCI–HOBt–*N*-methylmorpholine.

Scheme 79

depend on the principle. In the case of (119) the β-(toluene-*p*-sulphonyl)-ethyl ester group served its purpose very well, being quite stable under the conditions required for stepwise construction of the fully protected pentapeptide (120), as shown in Scheme 79. The extensive use of dicyclohexylcarbodi-imide–hydroxybenzotriazole is noteworthy, and the demonstration that acylamino-acid dicyclohexylammonium salts can be coupled with peptide ester salts by this means directly without added tertiary base is valuable. The same method of coupling was used (Scheme 80) to join the two pentapeptide intermediates obtained from (120) by selective deprotec-

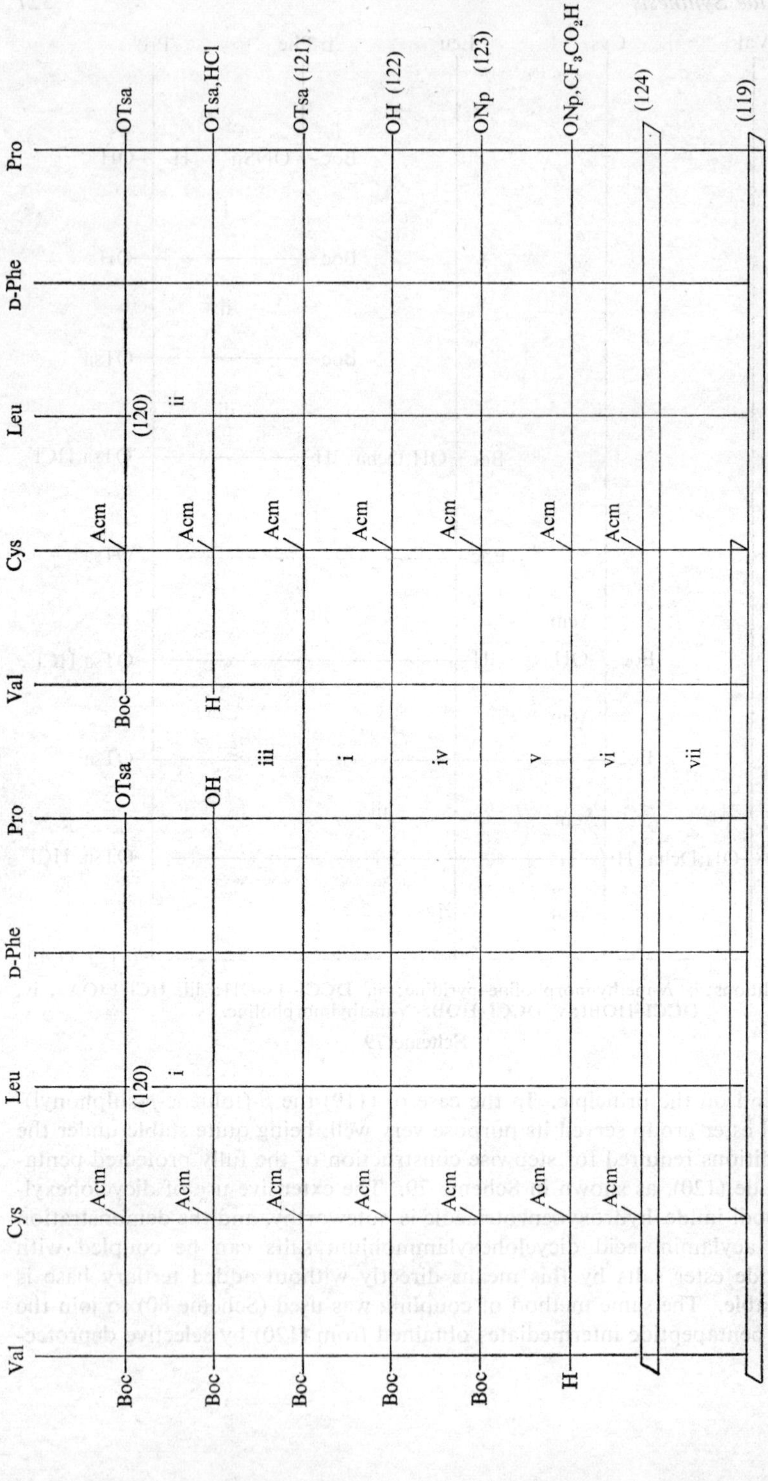

Conditions: i, NaOH, pH 11.5; ii, HCl–EtOAc; iii, DCCI–HO-Bt–*N*-methylmorpholine; iv, SO(ONp)$_2$–pyridine; v, CF_3CO_2H; vi, pyridine, 55 °C; vii, I_2–MeOH

Scheme 80

tion of the amino- or carboxy-functions, respectively. This crucial coupling proceeded in good yield despite the unfavourable steric situation to give, after Sephadex LH-20 chromatography, the crystalline fully protected decapeptide (121). Mild base treatment gave a decapeptide acid (122), which was also obtained crystalline, as was the *p*-nitrophenyl ester (123) into which it was converted by means of bis-*p*-nitrophenyl sulphite in pyridine. This method of active ester formation was superior to dicyclohexylcarbodi-imide–*p*-nitrophenol (*cf.* remarks on this subject in Vol. 3 of these Reports, p. 233). Finally, the SS-diprotected cyclodecapeptide (124) obtained from (123) was converted into the desired bicyclic structure by direct oxidation without the need for preliminary S-deprotection.

Cytochrome *c*.—Indications that other work with the ultimate aim of a total classical synthesis of a cytochrome *c* was afoot have prompted the Padua group [234] to reveal their hand in a report giving brief details of eight protected fragments which span the entire sequence of baker's yeast iso-1-cytochrome *c*. This enzyme is a relatively reasonable synthetic objective since it only has 108 residues and the conversion of the apoprotein into the native form has been achieved previously. The potential value of synthetic manipulations of this thoroughly studied class of protein is also highly attractive since it may be possible to complement the wide range of mutated sequences which nature provides.

The eight fragments (125)—(132) reported so far indicate the general basis of the approach and further reports will be awaited with great interest.

```
    CF3CO               CF3CO                            CF3CO
      |                   |                                |
Z-Thr-Glu-Phe-Lys-Ala-Gly-Ser-Ala-Lys-Lys-Gly-Ala-Thr-Leu-Phe-Lys-
  1                                   |                       16
                                    OCCF3
 -NH·NH·Boc
```

(125)

```
        (Adoc)2  OBut  Acm
           |      |     |
    Nps-Thr-Arg-Cys-Glu-Leu-Cys-His-NH·NH·Boc
        17      |           23
               Acm
```

(126)

```
    CF3CO               CF3CO
      |                   |
Z-Thr-Val-Glu-Lys-Gly-Gly-Pro-His-Lys-Val-Gly-Pro-Asn-Leu-His-Gly-Ile-
  24
 -Phe-Gly-Arg-His-NH·NH·Boc
              44
```

(127)

[234] L. Moroder, G. Borin, F. Marchiori, and E. Scoffone, *Biopolymers*, 1972, **11**, 2191.

Bzh(OMe)$_2$ (on Gln 47), OBu^t (on Asp 55)

Z-Ser-Gly-Gln-Ala-Gln-Gly-Tyr-Ser-Tyr-Thr-Asp-Ala-NH·NH·Boc

45 … **56**

(128)

CF_3CO (on Lys 59), OBu^t (on Asp 65), OBu^t (on Glu 66), CF_3CO (on Lys 60)

Z-Asn-Ile-Lys-Lys-Asn-Val-Leu-Trp-Asp-Glu-NH·NH·Boc

57 … **66**

(129)

Z-Asn-Asn-Met-Ser-Glu-Tyr-Leu-

67

CF_3CO (on Lys 77), Bu^t (on Tyr 79), CF_3CO (on Lys 78)

-Thr-Asn-Pro-Lys-Lys-Tyr-Ile-Pro-NH·NH·Boc

81

(130)

CF_3CO (on Lys 84), CF_3CO (on Lys 91), CF_3CO (on Lys 92)

Z-Gly-Thr-Lys-Met-Ala-Phe-Gly-Gly-Leu-Lys-Lys-NH·NH·Boc

82 … **92**

(131)

CF_3CO (on Lys 94), CF_3CO (on Lys 104), OBu^t (on Glu 108), CF_3CO (on Lys 105)

Z-Glu-Lys-Asp-Arg-Asn-Asp-Leu-Ile-Thr-Tyr-Leu-Lys-Lys-Ala-Thr-Glu-OBu^t

93 … **108**

(132)

Insulin.*—*Conversion of Porcine Insulin into Human Insulin.* The only insulins available for the treatment of diabetes at the present time are those of animal origin, which are not identical in sequence with the human hormone. Immunological reaction against foreign material is therefore a problem in some patients and a method for converting an easily available insulin into the human sequence would be an important advance. Brief details of a procedure (summarized in Scheme 81) for obtaining the human hormone from the porcine one (which differs from it only in its B *C*-terminal residue) have recently been reported.[235] The procedure comprises esterification of the porcine insulin with diazomethane, treatment with trypsin to remove the B-chain *C*-terminal octapeptide, blocking of the free amino-groups by

[235] M. A. Ruttenberg, *Science*, 1972, **177**, 623.

[236] K. Lubke and H. Klostermeyer, *Adv. Enzymol.*, 1970, **33**, 445.

* A very useful review of synthetic approaches to insulin is given in ref. 236.

treatment with t-butyloxycarbonyl azide, coupling with a synthetic protected B *C*-terminal human octapeptide, removal of the t-butyloxycarbonyl groups, and, finally, alkaline hydrolysis of the ester protective groups. After purification, a 'typical experiment' yielded a total of 2.7 g of material

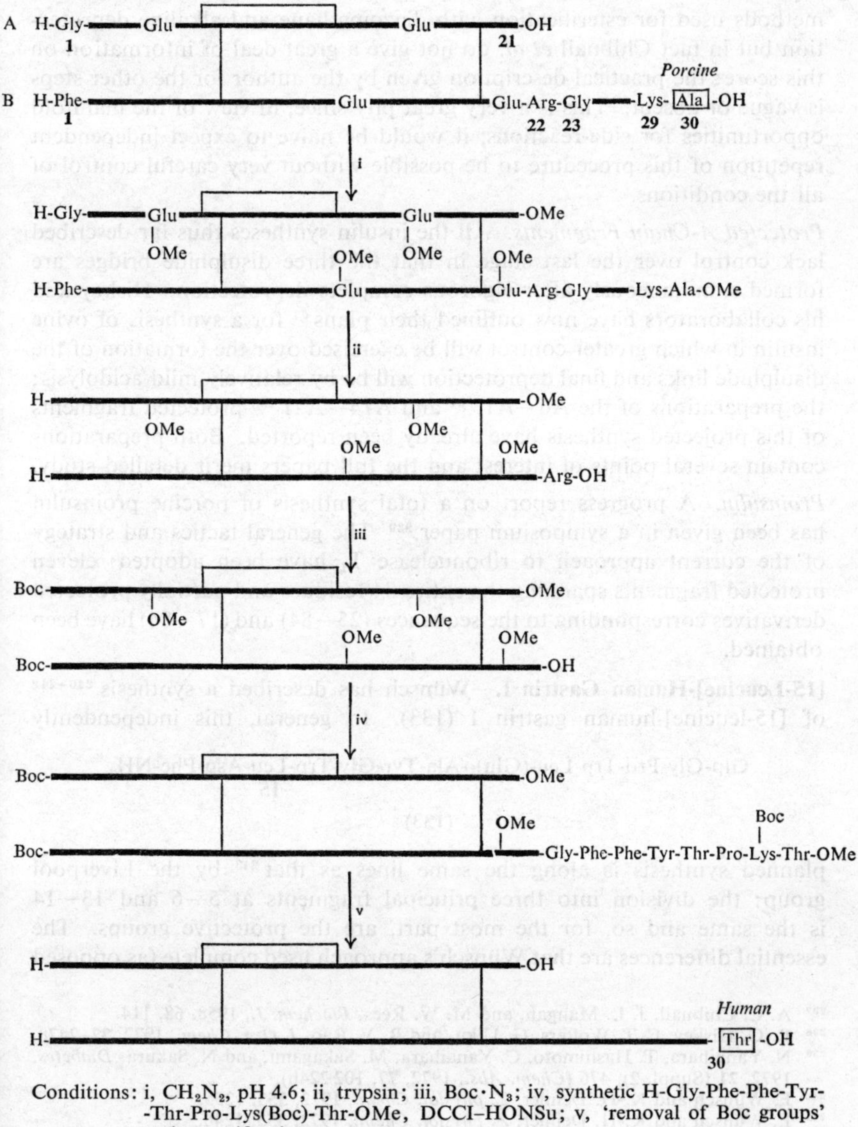

Conditions: i, CH_2N_2, pH 4.6; ii, trypsin; iii, Boc·N_3; iv, synthetic H-Gly-Phe-Phe-Tyr-Thr-Pro-Lys(Boc)-Thr-OMe, DCCI–HONSu; v, 'removal of Boc groups' and saponification.

Scheme 81

with the correct amino-acid analysis, biological activity, and chromatographic properties for human insulin, starting with 4.0 g of porcine insulin.

Unfortunately, although the paper reporting this work is in some respects fairly detailed, the experimental detail of the chemical steps is very superficial. The reader is referred by the author to Chibnall *et al.*[237] for the methods used for esterification with diazomethane and alkaline deprotection but in fact Chibnall *et al.* do not give a great deal of information on this score: the practical description given by the author for the other steps is vague or absent. This is a very great pity since, in view of the manifold opportunities for side-reactions, it would be naive to expect independent repetition of this procedure to be possible without very careful control of all the conditions.

Protected A-Chain Fragments. All the insulin syntheses thus far described lack control over the last stage in that the three disulphide bridges are formed simultaneously after vigorous complete deprotection. Hiskey and his collaborators have now outlined their plans [61] for a synthesis of ovine insulin in which greater control will be exercised over the formation of the disulphide links and final deprotection will be by relatively mild acidolysis; the preparations of the A6—A13 [61] and A14—A21 [238] protected fragments of this projected synthesis have already been reported. Both preparations contain several points of interest and the full papers merit detailed study.

Proinsulin. A progress report on a total synthesis of porcine proinsulin has been given in a symposium paper.[239] The general tactics and strategy of the current approach to ribonuclease T_1 have been adopted: eleven protected fragments spanning the entire 84 residues and partially protected derivatives corresponding to the sequences (25—84) and (17—84) have been obtained.

[15-Leucine]-Human Gastrin I.—Wünsch has described a synthesis [240—242] of [15-leucine]-human gastrin I (133). In general, this independently

$$\text{Glp-Gly-Pro-Trp-}\underset{\mathbf{15}}{\text{Leu}}\text{-(Glu)}_5\text{-Ala-Tyr-Gly-Trp-Leu-Asp-Phe-NH}_2$$

(133)

planned synthesis is along the same lines as that [243] by the Liverpool group: the division into three principal fragments at 5—6 and 13—14 is the same and so, for the most part, are the protective groups. The essential differences are that Wünsch's approach used complete (as opposed

237 A. C. Chibnall, J. L. Mangan, and M. W. Rees, *Biochem. J.*, 1958, **68**, 114.
238 R. G. Hiskey, E. T. Wolters, G. Ulku, and R. V. Rao, *J. Org. Chem.*, 1972, **37**, 2478.
239 N. Yanaihara, T. Hashimoto, C. Yanaihara, M. Sakagami, and N. Sakura, *Diabetes*, 1972, **21** (Suppl. 2), 476 (*Chem. Abs.*, 1972, **77**, 102 224h).
240 E. Wünsch and K.-H. Deimer, *Z. physiol. Chem.*, 1972, **353**, 1255.
241 E. Wünsch and K.-H. Deimer, *Z. physiol. Chem.*, 1972, **353**, 1246.
242 E. Wünsch, E. Jager, M. Deffner, and R. Scharf, *Z. physiol. Chem.*, 1972, **353**, 1716.
243 G. W. Kenner, J. J. Mendive, and R. C. Sheppard, *J. Chem. Soc.* (*C*), 1968, 761.

to almost complete) side-chain protection and dicyclohexylcarbodi-imide–hydroxysuccinimide for fragment conjunction (instead of azide couplings). The main interest of this synthesis lies in the results obtained with the end product. Stripping of the exclusively t-butyl alcohol-based protection with trifluoroacetic acid followed by gel filtration gave material with high biological activity and the correct amino-acid analysis. Electrophoretic examination, however, indicated the presence of two components which were ultimately divorced by countercurrent distribution or partition chromatography. The major component was the desired product, pure by all available tests and having full biological activity, but the other component also had essentially full biological potency and gave the same amino-acid analysis. The structure of this by-product is not yet established but spectroscopic and enzymic degradation studies point to some modification of tryptophan-4. The acid-catalysed oligomerization of indole derivatives is a well-studied field,[244] *e.g.* skatole (134) yields (135) on treatment with dry hydrogen chloride in ether, the reaction being reversed by boiling with aqueous acid (Scheme 82). Presumably skatole is a good model

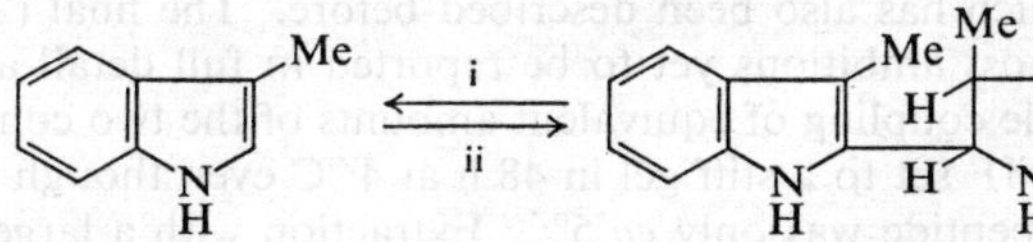

(134) (135)

Conditions: i, distillation of a solution in dil. HCl; ii, dry HCl–Et_2O

Scheme 82

for the tryptophan side-chain, so perhaps the by-product in Wünsch's synthesis was a dimer formed by analogy with Scheme 82 under the acidolytic final deprotection conditions: in previous work on glucagon by the same group a dimer was obtained. The reported spectroscopic differences between the product and by-product are possibly consistent with this, and the reversal of the reaction shown in Scheme 82 by hot aqueous acid would explain the amino-acid analysis results, but sedimentation studies appear to exclude the possibility of dimerization.

The need for very great care in establishing the purity of synthetic peptides has been the subject of much recent discussion: the experience outlined above serves to emphasize this, since on the simple and common criteria of biological activity and amino-acid analysis the mixture of peptide and by-product would have been deemed pure.

Ribonuclease T_1.—(See previous Reports, Vol. 2, p. 175, and Vol. 4, p. 373.) The syntheses of two more long protected fragments of ribonuclease T_1 have been described in detail. These fragments correspond to positions

[244] R. J. Sundberg, 'The Chemistry of Indoles', Academic Press, New York, 1970, p. 7ff.; G. F. Smith and A. E. Walters, *J. Chem. Soc.*, 1961, 940.

(1—47)[55] and (48—80),[245] designated ABCD and EF, respectively, by Hofmann.

The preparation of ABCD (136) is outlined in Scheme 83. The fragment A was obtained by a (7 + 4) azide coupling of subfragments made by

```
          Ec        Ec          Ec
          |         |           |
HCO-Ala-Cys-Asp-Tyr-Thr-Cys-Gly-Ser-Asn-Cys-Tyr-Ser-Ser-Ser-Asp-Val-
     1                                   11  12
-Ser-Thr-Ala-Gln-Ala-Ala-Gly-Tyr-Gln-Leu-His-Glu-Asp-Gly-Glu-Thr-
                         23  24

                        CHO                  OBu^t
                         |                    |
-Val-Gly-Ser-Asn-Ser-Tyr-Pro-His-Lys-Tyr-Asn-Asn-Tyr-Glu-Gly-NH·NH·Boc
 34  35                                                  47
```

(136)

active ester methods: some difficulties due to acylation of side-chains were encountered. After purification, deprotection, and activation, A was attached to the previously described fragment B and the resulting AB was joined to CD, which has also been described before. The final (23 + 24) coupling is the most ambitious yet to be reported in full detail anywhere in print. The azide coupling of equivalent amounts of the two components in 1:1 DMSO–DMF set to a stiff gel in 48 h at 4 °C even though the total concentration of peptide was only *ca* 5%. Extraction with a large volume of 2:3 DMF–aqueous acetic acid followed by ion-exchange chromatography using the same solvent mixture provided the desired ABCD fragment in 17% yield. Some 55% of the nucleophile CD t-butyloxycarbonyl hydrazide was recovered and the insoluble residue from the first extraction contained substantial amounts of AB. The insoluble material also contained polymeric material arising from oxidation after loss of S-protection: the *S*-ethylcarbamoyl group is somewhat precarious under basic conditions. Full characterization of ABCD was inhibited by its insolubility (also a problem with its precursor AB), but the use of 'diagnostic residues' (see Vol. 4 of these Reports, p. 375) enabled essential purity to be deduced from the results of amino-acid analysis.

The section EF (137) was obtained using similar methods by a condensation between two large fragments which were themselves obtained through

```
Z-Phe-Asp-Phe-Ser-Val-Ser-Ser-Pro-Tyr-Tyr-Glu-Trp-Pro-Ile-Leu-Ser-Ser-
  48

                                                OBu^t
                                                 |
-Gly-Asp-Val-Tyr-Ser-Gly-Pro-Gly-Ser-Gly-Ala-Asp-Arg-Val-Val-
-Phe-NH·NH·Boc
 80
```

(137)

[245] R. Camble, G. Dupuis, K. Kawasaki, H. Romovacek, N. Yanaihara, and K. Hofmann, *J. Amer. Chem. Soc.*, 1972, **94**, 2091.

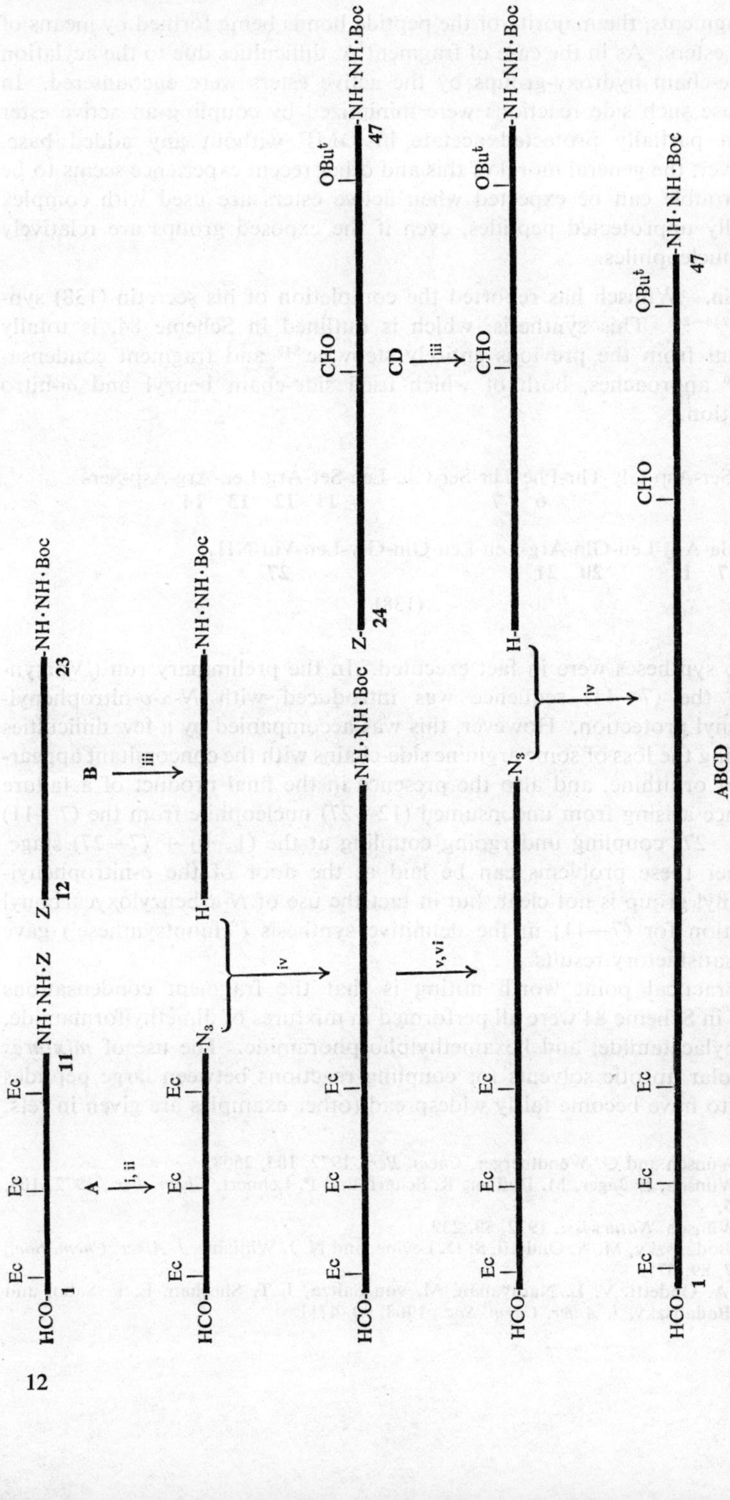

Scheme 83

Conditions: i, $HBr–CF_3CO_2H$; ii, $HCl–DMF–Bu^tONO$; iii, H_2; iv, $DMF–DMSO–Et_3N$; v, CF_3CO_2H; vi, $HCl–DMF–DMSO–Bu^tONO$

subfragments, the majority of the peptide bonds being formed by means of active esters. As in the case of fragment A, difficulties due to the acylation of side-chain hydroxy-groups by the active esters were encountered. In one case such side reactions were minimized by coupling an active ester with a partially protected acetate in DMF without any added base. However, the general moral of this and other recent experience seems to be that trouble can be expected when active esters are used with complex partially unprotected peptides, even if the exposed groups are relatively poor nucleophiles.

Secretin.—Wünsch has reported the completion of his secretin (138) synthesis.[246–248] This synthesis, which is outlined in Scheme 84, is totally different from the previous entirely stepwise[249] and fragment condensation[250] approaches, both of which used side-chain benzyl and ω-nitro protection.

His-Ser-Asp-Gly-Thr-Phe-Thr-Ser-Glu-Leu-Ser-Arg-Leu-Arg-Asp-Ser-
1 **6** **7** **11** **12** **13** **14**

-Ala-Arg-Leu-Gln-Arg-Leu-Leu-Gln-Gly-Leu-Val-NH_2
17 **18** **20** **21** **27**

(138)

Two syntheses were in fact executed. In the preliminary run ('Vorsynthese') the (7—11) sequence was introduced with *N*-α-*o*-nitrophenylsulphenyl protection. However, this was accompanied by a few difficulties including the loss of some arginine side-chains with the concomitant appearance of ornithine, and also the presence in the final product of a failure sequence arising from unconsumed (12—27) nucleophile from the (7—11) + (12—27) coupling undergoing coupling at the (1—6) + (7—27) stage. Whether these problems can be laid at the door of the *o*-nitrophenylsulphenyl group is not clear, but in fact the use of *N*-α-benzyloxycarbonyl protection for (7—11) in the definitive synthesis ('Hauptsynthese') gave more satisfactory results.

A practical point worth noting is that the fragment condensations shown in Scheme 84 were all performed in mixtures of dimethylformamide, dimethylacetamide, and hexamethylphosphoramide. The use of *mixtures* of dipolar aprotic solvents for coupling reactions between large peptides seems to have become fairly widespread (other examples are given in refs.

246 E. Wünsch and G. Wendlberger, *Chem. Ber.*, 1972, **105**, 2508.
247 E. Wünsch, E. Jager, M. Deffner, R. Scharf, and P. Lehnert, *Chem. Ber.*, 1972, **105**, 2515.
248 E. Wünsch, *Naturwiss.*, 1972, **59**, 239.
249 M. Bodanszky, M. A. Ondetti, S. D. Levine, and N. J. Williams, *J. Amer. Chem. Soc.*, 1967, **89**, 6753.
250 M. A. Ondetti, V. L. Narayanan, M. von Saltza, J. T. Sheehan, E. F. Sabo, and M. Bodanszky, *J. Amer. Chem. Soc.*, 1968, **90**, 4711.

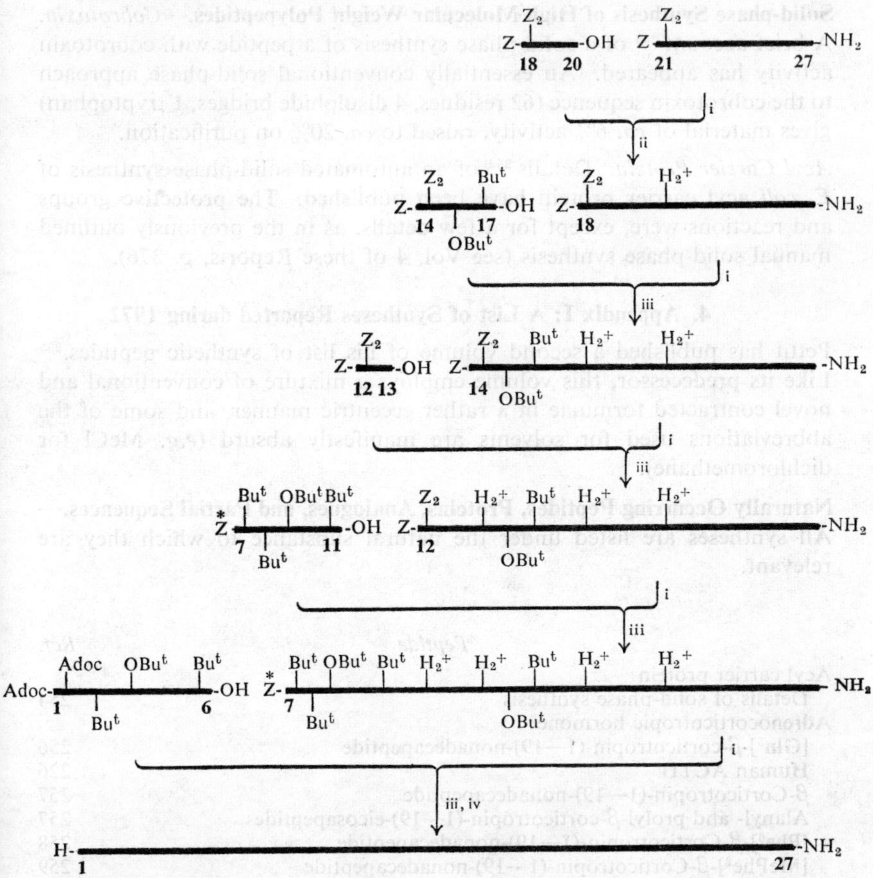

Conditions: i, H_2–Pd, then HBr titration; ii, DCCI–HOBt; iii, DCCI–HONSu; iv, CF_3CO_2H

* Nps in the 'Vorsynthese', in which deprotection of 12—27 was by HBr.

Scheme 84

55, 240, 245, 251, and 252), but discussion of the rationale behind this is lacking.

An intensive study of the gel filtration, preparative electrophoresis, and ion-exchange chromatography of the deprotected 'Vorsynthese' material paved the way for purification of the 'Hauptsynthese' product, which gave pure fully active secretin after ion-exchange chromatography on SP-Sephadex C-25.

[251] J. Beacham, G. Dupuis, F. M. Finn, H. T. Storey, C. Yanaihara, N. Yanaihara, and K. Hofmann, *J. Amer. Chem. Soc.*, 1971, **93**, 5526.

[252] R. G. Strachan, W. J. Palevda, jun., R. F. Nutt, R. A. Vitali, D. F. Veber, M. J. Dickinson, V. Garsky, J. E. Deak, E. Walton, S. R. Jenkins, F. W. Holley, and R. Hirschmann, *J. Amer. Chem. Soc.*, 1969, **91**, 503.

Solid-phase Synthesis of High Molecular Weight Polypeptides.—*Cobrotoxin.* A brief account [253] of a solid-phase synthesis of a peptide with cobrotoxin activity has appeared. An essentially conventional solid-phase approach to the cobrotoxin sequence (62 residues, 4 disulphide bridges, 1 tryptophan) gives material of *ca.* 6% activity, raised to *ca.* 20% on purification.

Acyl Carrier Protein. Details [254] of an automated solid-phase synthesis of *E. coli* acyl carrier protein have been published. The protective groups and reactions were, except for a few details, as in the previously outlined manual solid-phase synthesis (see Vol. 4 of these Reports, p. 376).

4. Appendix I: A List of Syntheses Reported during 1972

Pettit has published a second volume of his list of synthetic peptides.[255] Like its predecessor, this volume employs a mixture of conventional and novel contracted formulae in a rather eccentric manner, and some of the abbreviations used for solvents are manifestly absurd (*e.g.* MeCl for dichloromethane).

Naturally Occurring Peptides, Proteins, Analogues, and Partial Sequences.—All syntheses are listed under the natural substance to which they are relevant.

Peptide	*Ref.*
Acyl carrier protein	
Details of solid-phase synthesis	254
Adrenocorticotropic hormone	
[Gln⁵]-β-corticotropin-(1—19)-nonadecapeptide	256
Human ACTH	226
β-Corticotropin-(1—19)-nonadecapeptide	257
Alanyl- and prolyl-β-corticotropin-(1—19)-eicosapeptides	257
[Phe⁶]-β-Corticotropin-(1—19)-nonadecapeptide	258
[MePhe²]-β-Corticotropin-(1—19)-nonadecapeptide	259
Angiotensin	
p-[*NN*-Bis-(2-chloroethyl)amino]phenylbutyryl derivatives of angiotensin II	260
[1-β-Aspartyl,Val⁵]-angiotensin II	261

253 H. Aoyagi, H. Yonezawa, N. Takahashi, T. Kato, N. Izumiya, and C.-C. Yang, *Biochim. Biophys. Acta*, 1972, **263**, 823.
254 W. S. Hancock, D. J. Prescott, G. R. Marshall, and P. R. Vagelos, *J. Biol. Chem.*, 1972, **247**, 6224.
255 G. R. Pettit, 'Synthetic Peptides', Van Nostrand Rheinhold, New York, Vol. 2, 1971.
256 C. H. Li and B. Hemmasi, *J. Medicin. Chem.*, 1972, **15**, 217.
257 J. Blake, K.-T. Wang, and C. H. Li, *Biochemistry*, 1972, **11**, 438.
258 J. Blake and C. H. Li, *Biochemistry*, 1972, **11**, 3459.
259 J. Blake and C. H. Li, *Internat. J. Peptide and Protein Res.*, 1972, **4**, 343.
260 T. B. Paiva, A. C. M. Paiva, R. J. Freer, and J. M. Stewart, *J. Medicin. Chem.*, 1972, **15**, 6.
261 I. K. Romanovskaya and G. Cipens, *Zhur. obshchei Khim.*, 1971, **41**, 1856 (*Chem. Abs.*, 1972, **76**, 4151s).

Peptide	*Ref.*
[Gly^2,Val^5]-angiotensin II and some derivatives	262, 263
Angiotensin II derivatives specifically labelled with tritium in the tyrosine residue	264
[Arg^6]-angiotensin II	265
[His^4,Tyr^6]-angiotensin II	265
[Ile^8]-angiotensin II	265
Des-Pro^7, Phe^8-angiotensin II	183
[Cha^7]-, [Val^7]-, [Leu^8]-, [Phe^4,Ala^8]-, [Suc^1 *,Phe^4,Tyr^8]- and [Sar^1,Ile^8]-angiotensin II	266
$EtO_2C\cdot CH_2\cdot NH\cdot CO$- and $NH_2\cdot CO\cdot CH_2\cdot NH\cdot CO$-angiotensin II-(2-8)-heptapeptides	267
[Asn^1,Val^5,Gly^8]-, [Asn^1,Val^5,Ala^8]-, [Asn^1,Val^5,β-Ala^8]-, [Asn^1,Val^5,Abu^8]-, [Asn^1,Val^5,Val^8]-, and [Asn^1,Val^5,Leu^8]-angiotensin II	268
[Asn^1,Gly^3,Val^5]-angiotensin II	269
Blanching hormone (isolated from eyestalks of the prawn *Pandalus borealis*: Glp-Leu-Asn-Phe-Ser-Pro-Gly-Trp-NH_2)	270
Bradykinin	
β-[*NN*-Bis-(2-chloroethyl)amino]phenylbutyryl derivatives of bradykinin	271
5-, 8-, and 5,8-β-Cyclohexylalanine-bradykinins	47, 157
1-Homoarginine- and 1-homoarginine,5-β-cyclohexylalanine-bradykinins	47
1-Valine- and 1-valine,6-threonine-bradykinins	47
1-Leucine,6-threonine-bradykinin	47
Boc-bradykinin	47
Caerulin	
An extensive series of analogues	272
Calcitonin M	149
Calcitonin M-(11—16)-protected hexapeptide	43
Carboxypeptidase	
Bovine carboxypeptidase A-(159—165)-heptapeptide	273
Cholecystokinin-pancreozymin (CCK-PZ)	
Bis-Boc-CCK-PZ-(1—8)-octapeptide hydrazide	88
CCK-PZ-(1—6)-hexapeptide	88

262 A. Pavars, Z. Auna, and G. Cipens, *Zhur. obshchei Khim.*, 1971, **41**, 1859 (*Chem. Abs.*, 1972, **76**, 4152t); A. Pavars, G. Avotina, J. Indulens, Z. Auna, and G. Cipens, *Zhur. obshchei Khim.*, 1971, **41**, 2312 (*Chem. Abs.*, 1972, **76**, 100 035c).

263 P. Ya. Romanovskii, V. Muiznieks, and G. Cipens, *Khim. prirod. Soedinenii*, 1971, **7**, 655 (*Chem. Abs.*, 1972, **76**, 46 495n).

264 I. Mezo and I. Teplan, *Radioisotopy*, 1971, **12**, 551 (*Chem. Abs.*, 1972, **76**, 100 024y).

265 M. C. Khosla, S. Kumar, R. R. Smeby, and F. M. Bumpus, *J. Medicin. Chem.*, 1972, **15**, 627.

266 M. C. Khosla, R. A. Leese, W. L. Maloy, A. T. Ferreira, R. R. Smeby, and F. M. Bumpus, *J. Medicin. Chem.*, 1972, **15**, 792.

267 R. Vegners, G. Cipens, and Z. Auna, *Khim. prirod. Soedinenii*, 1972, **8**, 92 (*Chem. Abs.*, 1972, **77**, 75 461z).

268 D. C. Fessler, F. Sipos, G. S. Denning, jun., D. T. Pals, and F. D. Masucci, *J. Medicin. Chem.*, 1972, **15**, 1015.

269 R. Vegners and G. Cipens, *Khim. prirod. Soedinenii*, 1972, **8**, 95 (*Chem. Abs.*, 1972, **77**, 127 046g).

270 P. Fernlund and L. Josefson, *Science*, 1972, **177**, 173.

271 R. J. Freer and J. M. Stewart, *J. Medicin. Chem.*, 1972, **15**, 1.

272 L. Bernardi, G. Bertaccini, G. Bosisio, R. Bucci, R. De Castiglione, V. Erspamer, O. Goffredo, and M. Impicciatore, *Experientia*, 1972, **28**, 7.

273 M. Fridkin, A. Patchornik, and E. Katchalski, *Biochemistry*, 1972, **11**, 466.

* Suc = succinyl, $HO_2C\cdot(CH_2)_2\cdot CO$—.

Peptide	*Ref.*
Casein	
A rennin-sensitive pentadecapeptide from bovine κ-casein	274
Cobrotoxin	253
Cytochromes	
Eight protected fragments spanning the 108 residues of baker's yeast iso-1-cytochrome *c* apoprotein (brief preliminary account)	234
Horse heart cytochrome *c*-(70—80)-undecapeptide	275
Protected sequences of horse heart cytochrome *c*	276—279
Peptides related to other cytochromes	280, 281
Eledoisin	
[Lys^6,Phe^8]-Eledoisin-(6—11)-hexapeptide amide	146
[Gly^4,Asn^5]-Eledoisin-(4—11)-octapeptide amide labelled with ^{14}C in the carboxy-group of Gly^4	282
Encephalitogenic protein	
Human encephalitogenic protein-(7—11)-, -(11—20)-, and -(12—20)-peptides	283
Fibrinogen	
A series of tri-, tetra-, and penta-peptide sequences of fibrinogen	46
Fibrinopeptide A	
Human fibrinopeptide A *C*-terminal protected hexapeptide	284
Gastrin	
Derivatives of the five tetrapeptides whose sequences are coded by the nucleotide sequences which are complementary to the nucleotide sequences which code the sequence of gastrin *C*-terminal tetrapeptide	285
[Leu^{15}]-Human gastrin I	240, 242
Protected partial sequences of [Leu^{15}]-human gastrin I	241
Ala-Gly-Phe-Leu-Asp-Phe-NH_2	172
Ala-Tyr-Gly-Phe-Leu-Asp-Phe-NH_2	172
Derivatives and analogues of gastrin *C*-terminal tetrapeptide amide	286
Glutathione	287
Growth hormone	
Human growth hormone-(17—23)-heptapeptide	169
Human growth hormone-(138—145)-octapeptide	69
Human growth hormone-(87—123)- and -(124—155)-peptides	288

[274] K. P. Polzhofer, *Tetrahedron*, 1972, **28**, 855.
[275] Y. Wolman, A. Schejter, and M. Sokolovsky, *J. Amer. Chem. Soc.*, 1972, **94**, 1720.
[276] Y. Wolman and Y. S. Klausner, *Israel J. Chem.*, 1971, **9**, 229.
[277] Y. Wolman and Y. S. Klausner, *Israel J. Chem.*, 1971, **9**, 205.
[278] Y. Wolman and Y. S. Klausner, *Israel J. Chem.*, 1971, **9**, 211.
[279] Y. Wolman and Y. S. Klausner, *Israel J. Chem.*, 1971, **9**, 217.
[280] G. A. Vasileva, A. F. Mironov, Yu. A. Shtyleva, and R. P. Evstigneeva, *Zhur. obshchei Khim.*, 1971, **41**, 2332 (*Chem. Abs.*, 1972, **76**, 72 770t).
[281] T. Baba, H. Sugiyama, and S. Seto, *Sci. Rep. Res. Inst. Tohoku Univ., Ser. A*, 1971, **23**, 85 (*Chem. Abs.*, 1972, **76**, 86 138e).
[282] R. Grupe and H. Niedrich, *Z. Chem.*, 1971, **12**, 22.
[283] M. A. Barton, T. A. McPherson, R. U. Lemieux, and G. O. Bain, *Canad. J. Biochem.*, 1972, **50**, 689.
[284] A. S. Khan and K. M. Sivanandaiah, *Indian J. Chem.*, 1972, **10**, 382.
[285] D. S. Jones, *J.C.S. Perkin I*, 1971, 1407.
[286] M. Dabrowska, G. Kupryszewski, B. Lammek, and F. Muzalewski, *Roczniki Chem.*, 1972, **46**, 1295.
[287] B. Marinier and M. Berube, *Canad. J. Chem.*, 1972, **50**, 1633.
[288] F. Chillemi, A. Aiello, and A. Pecile, *Nature New Biol.*, 1972, **238**, 243.

Peptide	*Ref.*
Haemoglobin	
Human haemoglobin β-chain-(57—69)-protected tridecapeptide	289
Human haemoglobin α-chain-(56—60)-protected pentapeptide	290
Human haemoglobin β-chain-(139—146)-protected octapeptide	291
Insulin	
Deamino^{A1} ovine insulin	292
The symmetrical disulphides of insulin A chain-(14—21)- and -(15—21)-peptides	59
Analogues of insulin A chain-(17—21)-peptide	293
A protected cyclocystine derivative of sheep insulin A-chain-(6—9)-tetrapeptide	294
De-(A1—A4)-ovine and porcine insulin	295
Insulin A-chain-(15—19)-protected pentapeptide	65
Human insulin from porcine insulin	235
Ovine insulin A-chain-(6—13)-protected octapeptide	61
Ovine insulin A-chain-(14—21)-protected octapeptide	238
Partially protected peptides corresponding to residues (17—84) and (15—84) of porcine proinsulin	239
Peptides related to bovine proinsulin connecting peptide	296
Luteinizing hormone releasing factor (LH-RF)	150, 297—303
[Lys^{8}]-LH-RF	299
[His^{8}]-LH-RF	298
[Nva^{8}]-LH-RF	298
De-Arg^{8}-LH-RF	298
[Gly^{2}]-LH-RF	151
De-His^{2}-LH-RF	151
Various truncated sequences and other peptides related to LH-RF	303—305

[289] V. P. Chernyshev, S. A. Riskal, G. A. Zhiglova, E. I. Filippovich, and R. P. Evstigneeva, *Zhur. obshch. Khim.*, 1971, **41**, 2565 (*Chem. Abs.*, 1972, **77**, 34 925k).

[290] T. P. Mochanova, A. S. Molokoedor, E. I. Filippovich, and R. P. Evstigneeva, *Zhur. obshch. Khim.*, 1972, **42**, 1638 (*Chem. Abs.*, 1972, **77**, 102 192w).

[291] F. Chillemi and A. Chersi, *Gazzetta*, 1971, **101**, 695.

[292] P. G. Katsoyannis and C. Zalut, *Biochemistry*, 1972, **11**, 1128.

[293] G. A. Korshunova, G. P. Mishin, Yu. A. Semiletov, N. A. Voskova, and Yu. P. Shvachin, *Khim. prirod. Soedinenii*, 1971, **7**, 799 (*Chem. Abs.*, 1972, **76**, 100 033a).

[294] H. Berndt, H. Klostermeyer, and H. Zahn, *Annalen*, 1972, **759**, 114.

[295] P. G. Katosyannis and C. Zalut, *Biochemistry*, 1972, **11**, 3065.

[296] N. Yanaihara, N. Sakura, C. Yanaihara, and T. Hashimoto, *J. Amer. Chem. Soc.*, 1972, **94**, 8243.

[297] H. Sievertsson, J.-K. Chang, A. von Klaudy, C. Bogentoft, B. L. Currie, K. Folkers, and C. Y. Bowers, *J. Medicin. Chem.*, 1972, **15**, 222.

[298] J.-K. Chang, R. H. Williams, A. J. Humphries, N.-G. Johansson, K. Folkers, and C. Y. Bowers, *Biochem. Biophys. Res. Comm.*, 1972, **47**, 727.

[299] J.-K. Chang, H. Sievertsson, B. L. Currie, C. Bogentoft, K. Folkers, and C. Y. Bowers, *J. Medicin. Chem.*, 1972, **15**, 623.

[300] M. Monahan, J. Rivier, R. Burgus, M. Amoss, R. Blackwell, W. Vale, and R. Guillemin, *Compt. rend.*, 1971, **273**, 508.

[301] H. C. Beyerman, H. Hindriks, J. Hirt, E. W. B. de Leer, and A. van der Wiele, *Rec. Trav. chim.*, 1972, **91**, 1239.

[302] W. Voelter and K. Zech, *Chimia* (*Switz.*), 1972, **26**, 313.

[303] J. Rivier, M. Monahan, W. Vale, G. Grant, M. Amoss, R. Blackwell, R. Guillemin, and R. Burgus, *Chimia* (*Switz.*), 1972, **26**, 300.

[304] A. V. Schally, A. Arimura, W. H. Carter, T. W. Redding, R. Geiger, W. König, H. Wissman, G. Jaeger, J. Sandow, N. Yanaihara, C. Yanaihara, T. Hashimoto, and M. Sakagami, *Biochem. Biophys. Res. Comm.*, 1972, **48**, 366.

[305] J.-K. Chang, A. J. Humphries, R. H. Williams, and C. Y. Bowers, *Biochem. Biophys. Res. Comm.*, **47**, 1256.

Peptide	*Ref.*
A series of analogues in which the glycinamide residue is replaced by an alkylamine	306
Glp-Tyr-Arg-Trp-NH_2 (which, contrary to previous work quoted on p. 369, Vol. 4 of this series, does have *both* LH-RF and FSH-RF activity)	152
Melanocyte-stimulating hormone	
Deacetyl-α-MSH-(1—10)-decapeptide hydrazide and its *N*-α-Boc derivative	307
Melittin	
[Tyr^{19}]-Melittin	308
MSH-release inhibiting factor*	
Pro-His-Phe-Arg-Gly-NH_2	309
A series of tripeptide derivatives related to the *C*-terminal tripeptide of oxytocin	310
Myoglobin	
Sperm whale myoglobin-(56—69)-tetradecapeptide and also the *C*-terminal hexa-, penta-, and tetra-peptides	322
Oxytocin	311, 312
Deamino-oxytocin	312
[$Leu^{3,4}$]-oxytocin	311
[1-Mercaptodimethylacetic acid]-oxytocin	313
[1-β-Mercapto-αα-dimethylpropionic acid]-oxytocin	313
[1-β-Mercapto-ββ-diethylpropionic acid]-oxytocin	313
[5-αγ-Diaminobutyric acid]-oxytocin	314
Deamino-[1,6-homolanthionine]-oxytocin	315
[1-(α-Hydroxy-β-mercaptopropionic acid)]-oxytocin (details)	316
Tocinoic acid	317
Deaminotocinoic acid	312, 317
Tocinamide	318
Deaminotocinamide	319

306 M. Fujino, S. Kobayashi, M. Obayashi, S. Shinagawa, T. Fukuda, C. Kitada, R. Nakayama, and I. Yamazaki, *Biochem. Biophys. Res. Comm.*, 1972, **49**, 863.
307 C. H. Li and B. Hemmasi, *J. Medicin. Chem.*, 1972, **15**, 697.
308 Y. Miura, H. Sugiyama, Y. Maki, and S. Seto, *Chem. and Pharm. Bull.* (*Japan*), 1972, **29**, 215.
309 R. M. G. Nair, A. J. Kastin, and A. V. Schally, *Biochem. Biophys. Res. Comm.*, 1972, **47**, 1420.
310 M. E. Celis, S. Hase, and R. Walter, *F.E.B.S. Letters*, 1972, **27**, 327.
311 M. A. Wille, V. du Vigneaud, and W. Y. Chan, *J. Medicin. Chem.*, 1972, **15**, 11.
312 M. Mühlemann, M. I. Titov, R. Schwyzer, and J. Rudinger, *Helv. Chim. Acta*, 1972, **55**, 2854.
313 R. J. Vavrek, M. F. Ferger, G. A. Allen, D. H. Rich, A. T. Blomquist, and V. du Vigneaud, *J. Medicin. Chem.*, 1972, **15**, 123.
314 S. Hase, I. L. Schwartz, and R. Walter, *J. Medicin. Chem.*, 1972, **15**, 126.
315 Z. Prochazka, K. Jost, and F. Sorm, *Coll. Czech. Chem. Comm.*, 1972, **37**, 289.
316 M. Wälti and D. B. Hope, *J.C.S. Perkin I*, 1972, 1946.
317 V. J. Hruby, C. W. Smith, D. K. Linn, M. F. Ferger, and V. du Vigneaud, *J. Amer. Chem. Soc.*, 1972, **94**, 5478.
318 M. Zaoral and M. Flegel, *Coll. Czech. Chem. Comm.*, 1972, **37**, 1539.
319 M. Zaoral and M. Flegel, *Coll. Czech. Chem. Comm.*, 1972, **37**, 2639.

* There appear to be several natural peptides with this activity: some confusion currently exists on the subject.

Peptide	*Ref.*
[2-Trp]-, [2-D-Trp]-, [2-D-Tyr(Et)]-, and [2-α-aminophenylacetic acid]-oxytocins	320
[cyclo-β-Ala^{1},Asp^{6}]-oxytocin, *i.e.* the analogue of deamino-oxytocin in which an amide group replaces the disulphide link	321
Proinsulin	
See under insulin	
Ribonuclease	143
Bovine pancreatic ribonuclease A-(82—85)-protected tetrapeptide	56
Ribonuclease A *N*-terminal eicosapeptide ('S-peptide')	
[Orn^{10,12}]-S-Peptide and its guanidination product	323
Ribonuclease T_1	
Ribonuclease T_1-(48—80)-protected tritricontapeptide	245
Ribonuclease T_1-(1—47)-partially protected heptatetracontapeptide hydrazide	55
Rubredoxin	
Rubredoxin-(A11—A19)-protected nonapeptide	324
A model for the environment of the iron atom of rubredoxin constructed by joining the (6—10) and (38—42) sections between residues 10 and 38	325
Rubredoxin-(A43—A53)-protected undecapeptide	326
Scotophobin	327
Secretin	246—248
Secretin-(7—27)-peptide	246
Thyrotropin-releasing factor (TRF)	65, 162
Analogues	65, 328—331
Radioactive TRF	332
Vasoactive intestinal peptide	
The *N*-terminal hexapeptide	333
The *C*-terminal undecapeptide amide	334
Vasopressin	
Lysine vasopressin	335

[320] O. A. Kaurov, V. F. Martynov, Yu. D. Mikhailov, and Z. Pauna, *Zhur. obshchei Khim.*, 1972, **42**, 1654 (*Chem. Abs.*, 1972, **77**, 102 210a).

[321] S. Hase, S. Sakakibara, M. Wahrenburg, and R. Walter, *J. Medicin. Chem.*, 1972, **15**, 1017.

[322] J. K. Givas, A. H. Sehon, and M. Manning, *Biochemistry*, 1972, **11**, 1351.

[323] G. Borin, C. Toniolo, L. Moroder, F. Marchiori, R. Rocchi, and E. Scoffone, *Internat. J. Protein Res.*, 1972, **4**, 37.

[324] D. Stevenson, R. M. Cook, and B. Weinstein, *Internat. J. Peptide and Protein Res.*, 1972, **4**, 101.

[325] A. Ali, F. Fahrenholz, J. C. Garing, and B. Weinstein, *J. Amer. Chem. Soc.*, 1972, **94**, 2556.

[326] A. Ali, R. M. Cook, and B. Weinstein, *Internat. J. Peptide and Protein Res.*, 1972, **4**, 177.

[327] W. Parr and G. Holzer, *Experientia*, 1972, **28**, 884.

[328] H. Sievertsson, J.-K. Chang, K. Folkers, and C. Y. Bowers, *J. Medicin. Chem.*, 1972, **15**, 8.

[329] H. Sievertsson, J.-K. Chang, K. Folkers, and C. Y. Bowers, *J. Medicin. Chem.*, 1972, **15**, 219.

[330] J. Rivier, W. Vale, M. Monahan, N. Ling, and R. Burgus, *J. Medicin. Chem.*, 1972, **15**, 479.

[331] Yu. P. Shvachin and A. P. Smirnova, *Zhur. obshchei Khim.*, 1972, **42**, 483 (*Chem. Abs.*, 1972, **77**, 48 781x).

[332] G. Flouret, A. Alter, and R. Gendrich, *J. Labelled Compounds*, 1972, **8**, 53.

[333] Y. S. Klausner, V. Mutt, and M. Bodanszky, *Bioorg. Chem.*, 1972, **2**, 87.

[334] M. Bodanszky, Y. S. Klausner, and V. Mutt, *Bioorg. Chem.*, 1972, **2**, 30.

[335] E. Schillinger, O. Loge, E. Schröder, E. Klieger, and K. Lubke, *European J. Biochem.*, 1972, **27**, 473.

Peptide	*Ref.*
Arginine vasopressin	335
1-(β-Mercaptopropionyl)-lysine and -arginine vasopressins	335
N-α-acetyl-lysine and -arginine vasopressins	335
[Ile^{3}]-lysine vasopressin and its ε-formyl derivative	335
[Ile^{3},Cit^{8}]-vasopressin	335
[Ile^{3},(Lys-Lys)^{8}]-vasopressin	335
Glycyl-[Ile^{3}]-lysine vasopressin	335
N-α-acetyl-[Ile^{3}]-lysine vasopressin	335
1-(β-Mercaptopropionyl)-[Ile^{3}]-, -[Nle^{3}]-, -[Val^{3}]-, -[Pro^{3}]-, -[Ile^{3},Orn^{7}]-, and -[Ile^{3},Ser^{4}]-lysine vasopressins	335
[Ile^{3}]-arginine vasopressin	335
[Ile^{3},D-Arg^{8}]-vasopressin	335
1-(β-Mercaptopropionyl)-[Ile^{3}]-arginine vasopressin	335
[Tyr(Bu^{t})^{2},Ile^{3}]-, [Tyr(Me)^{2}]-, and [Ile^{3},Gln^{4}]-lysine vasopressins	335
Arginine vasopressinoic acid	335
[Ile^{3}]-lysine vasopressin *N*-methylamide	335
[Gly^{7}]-arginine vasopressin	336
[Gly^{8}]- and [Glu^{8}]-vasopressins	336
1-(β-Mercaptopropionyl)-[D-Nar^{8}]-vasopressin	337
De-(GlyNH_2)-lysine vasopressin	338
Pressinoic acid	339
Pressinamide	318, 339
Deaminopressinoic acid	339
Deaminopressinamide	319, 339
[Asu^{1,6}]-lysine vasopressin	340
[Asu^{1,6}]-arginine vasopressin	340
[Asu^{1,6}]-lysine vasotocin	340
[Asu^{1,6}]-arginine vasotocin	340
Lysine-vasopressinoic acid methyl- and dimethyl-amides	341
1-Deamino-[Hyl^{8}]-vasopressin	342
Radioactive vasopressin-(1—6)-pentapeptide	343
Sequential Polypeptides	
Polyglutathione and related sequential polypeptides	219
Poly-(Ala-Glu-Ala-Gly)-Gly-OEt	215
Poly-(Try-Glu-Ala-Gly)-Gly-OEt	216
Poly-(Trp-Glu-Ala-Gly)-Gly-OEt	217
A series of sequential diastereoisomeric poly-(γ-t-butyl) glutamic acids	199
A series of sequential polypeptides containing D-Glu(OBzl) and Leu	198
A series of sequential polypeptides containing Asp(OMe) and Glu(OMe)	202
Poly-L- and -D-γ-glutamyl-γ-aminobutyric acids	196
Poly-γ-glutamyl-β-aspartic acid	197

336 O. S. Papsuevich, A. Krikis, and G. Cipens, *Zhur. obshchei Khim.*, 1972, **42**, 224 (*Chem. Abs.*, 1972, **77**, 19 991k).

337 M. Zaoral and M. Flegel, *Coll. Czech. Chem. Comm.*, 1972, **37**, 3350.

338 S. S. Wang, *Biochem. Biophys. Res. Comm.*, 1972, **48**, 1511.

339 M. F. Ferger, W. C. Jones, jun., D. F. Dyckes, and V. du Vigneaud, *J. Amer. Chem. Soc.*, 1972, **94**, 982.

340 S. Hase, S. Sakakibara, M. Wahrenburg, M. Kirchberger, I. L. Schwartz, and R. Walter, *J. Amer. Chem. Soc.*, 1972, **94**, 3590.

341 J. D. Glass and V. du Vigneaud, *J. Medicin. Chem.*, 1972, **15**, 486.

342 G. Lindeberg, J. Kynčl, P. Dreyfuss, and M. Bodanszky, *J. Medicin. Chem.*, 1972, **15**, 629.

343 M. H. Jimenez, D. P. Winter, and A. M. Felix, *J. Chromatog.*, 1971, **63**, 167.

Peptide	*Ref.*
A series of sequential polypeptides containing Ala, Gly, and DL-Glu	344
Some sequential polypeptides containing His	345
A series $\{[\text{Glu(OBzl)}]_x\text{-D-Glu(OBzl)-Glu(OBzl)}\}_n$ where $x = 1, 2$, or 3	200
Two polypeptides with functional side-chains	211
Miscellaneous intermediates *etc.* in sequential polypeptide synthesis	210, 346, 347
Poly-(Met-Ala) and a copolypeptide of Met-Ala with Met-D-Ala	348
A series of polymers with the structure poly-$(\text{Ala}_x\text{Gly}_y)$ in which $x = 1$, $y = 1, 2$, or 3; $x = 2$, $y = 1, 2$ or 3; and $x = 3$, $y = 3$	214
Poly-(Lys-Ala-Ala)	203, 204
Poly-(Glu-Ala-Ala)	203
Poly-(Leu-Leu-Asp)	213
Poly-[Leu-Leu-Asp(OBzl)]	212
Poly-(Gln-Ala)	205
Some polytripeptide collagen models containing proline analogues	208
Poly-(Pro-Ala-Gly) and poly-(Pro-Ser-Gly)	349
Poly-(Ala-Gly-Pro) and its stereoisomers	207
Poly-(Hyp-Gly)	209
Poly-(Pro-Gly)	209
Poly-(Ser-Pro-Gly)	209
Poly-(Pro-Hyp-Gly)	209
Poly-(Gly-Gly-Hyp-Gly)	209

Miscellaneous Peptides

Some glycyl dipeptides	102
Dipeptides containing Cys(Trt)	350
A number of α-(ω-aminoacyl) lysines	351
Leu-Sar-Phe (aminopeptidase inhibitor)	352
A series of dodecapeptide derivatives for use as hapten carriers	353
A series of dipeptide derivatives containing alanine for n.m.r. studies	132
A series of ornithine oligomers and other peptides	354
A series of lysine oligomer derivatives	355
Peptides containing β-lysine	356, 357

[344] T. Iio and S. Takahashi, *Bull. Inst. Chem. Res., Kyoto Univ.*, 1971, **49**, 80 (*Chem. Abs.*, 1972, **76**, 4147v).

[345] K. T. Poroshin, Sh. Kh. Khalikov, M. I. Ismailov, L. I. Mar'yash, R. Dakhte, and V. A. Shibnev, *Doklady Akad. Nauk. Tadzh. S.S.R.*, 1971, **14** (6), 34 (*Chem. Abs.*, 1972, **76**, 14 891x).

[346] A. B. Zegel'man, T. Yu. Yusupov, I. S. Salitra, and K. T. Poroshin, *Doklady Akad. Nauk. Tadzh. S.S.R.*, 1972, **15** (5), 33 (*Chem. Abs.*, 1972, **77**, 102 213d).

[347] K. T. Poroshin, T. P. Chuvaeva, Yu. L. Bulanova, and L. V. Morozova, *Doklady Akad. Nauk, Tadzh. S.S.R.*, 1971, **14** (9), 23 (*Chem. Abs.*, 1972, **76**, 34 559y).

[348] E. Dellacherie, P. Hubert, and J. Neel, *Compt. rend.*, 1972, **274**, *C*, 556.

[349] F. R. Brown, tert., A. di Corato, G. P. Lorenzi, and E. R. Blout, *J. Mol. Biol.*, 1972, **63**, 85.

[350] K. Y. Zee-Cheng and C. C. Cheng, *J. Medicin. Chem.*, 1972, **15**, 13.

[351] A. Fujii, K. Tanaka, and E. S. Cook, *J. Medicin. Chem.*, 1972, **15**, 378.

[352] R. Jost, A. Tun-Kyi, E. Stoll, and H. Zuber, *Helv. Chim. Acta*, 1972, **55**, 1025.

[353] C. H. Schneider and W. Wirz, *Helv. Chim. Acta*, 1972, **55**, 1062.

[354] C. Gilvarg and Y. Levin, *J. Biol. Chem.*, 1972, **247**, 543.

[355] K. T. Poroshin, V. V. Bondar, and V. A. Shibnev, *Doklady Akad. Nauk. Tadzh. S.S.R.*, 1971, **14** (8), 29 (*Chem. Abs.*, 1972, **76**, 60 033s).

[356] H. Taniyama, Y. Sawada, K. Miyazeki, S. Tanaka, and F. Miyoshi, *Chem. and Pharm. Bull.* (*Japan*), 1971, **19**, 2645.

[357] H. Taniyama, Y. Sawada, K. Miyazeki, and F. Miyoshi, *Chem. and Pharm. Bull.* (*Japan*), 1972, **20**, 601.

Peptide	*Ref.*
Met-Asn and Met-Gln	104
A series of peptides with tuberculostatic activity	358
β-(β-Lysyl)-β-lysine	359
A series of oligo-alanines with solubilizing blocking groups	360
A series of peptides for an investigation of the specifity of bacterial collagenase	361
A phagocytosis-stimulating tetrapeptide ('Tuftsin': Thr-Lys-Pro-Arg) isolated from human γ-globulin	362
A series of cathepsin D substrates	363
Some γ-amino-acid dipeptides	364
Some angiotensin I-converting enzyme substrates containing *p*-nitrophenylalanine or *p*-nitrohippuryl residues	365
Val-His (preparative detail but no physical constants or analytical data given)	366
Some simple pyroglutamyl peptides prepared for a study of the methanolysis of the pyroglutamyl ring	67
1-(α-fluoro)carnosine	367
A protected decapeptide hydrazide of unspecified relevance	146
Some peptides containing cysteine and 5-methoxy- or 5-hydroxy-tryptamine	368
Some peptides containing α-amino-oxycarboxylic acids	369
Leu-Thr(Bzl)-Ala-Leu-Gly	192
Ile-Ala-Pro-Val	192
Lys-Ala-Gly-Leu-Gly-Trp-Leu	164
Peptides containing proline and ornithine	370, 371
Peptides of djenkolic acid	287
Peptides of α-benzylphenylalanine	119

358 Z. Paulay and S. Bajusz, *Acta Chim. Acad. Sci. Hung.*, 1972, **71**, 247 (*Chem. Abs.*, 1972, **76**, 127 403h).

359 H. Taniyama, Y. Sawada, K. Miyazeki, and F. Miyoshi, *Chem. and Pharm. Bull.* (*Japan*), 1972, **20**, 601.

360 M. Goodman, R. Rupp, and F. Naider, *Bioorg. Chem.*, 1971, **1**, 194.

361 M. E. Soberano and G. Schoellmann, *Biochim. Biophys. Acta*, 1972, **271**, 133.

362 K. Nishioka, A. Constantopoulos, P. S. Satoh, and V. A. Najjar, *Biochem. Biophys. Res. Comm.*, 1972, **47**, 172.

363 O. V. Kazakova, V. N. Orekhovich, L. Purchot, and J. M. Schuck, *J. Biol. Chem.*, 1972, **247**, 4224.

364 T. K. Burimova, M. M. Zobacheva, G. B. Kuprishevskii, V. V. Perekalin, and A. A. Smirnova, *Zhur. org. Khim.*, 1972, **8**, 537 (*Chem. Abs.*, 1972, **77**, 5778j).

365 R. L. Stevens, E. R. Micalizzi, D. C. Fessler, and D. T. Pals, *Biochemistry*, 1972, **11**, 2999.

366 H. B. F. Dixon, *Biochem. J.*, 1972, **129**, 203.

367 A. S. Danilova, V. Ya. Lunts, and V. F. Martynov, *Zhur. obshchei Khim.*, 1971, **41**, 2559 (*Chem. Abs.*, 1972, **77**, 19 988q).

368 L. A. Davankova, A. D. Neklyndov, L. A. Shchukina, and N. N. Suvorov, *Zhur. obshchei Khim.*, 1971, **41**, 2786 (*Chem. Abs.*, 1972, **77**, 19 989r).

369 L. Kisfaludy, M. Löw, and T. Devenyi, *Acta Biochem. Biophys. Acad. Sci. Hung.*, 1971, **6**, 393.

370 N. A. Poddubnaya and N. Ya. Krasnobrizhii, *Zhur. obshchei Khim.*, 1972, **42**, 949 (*Chem. Abs.*, 1972, **77**, 88 834y).

371 N. A. Poddubnaya and N. Ya. Krasnobrizhii, *Zhur. obshchei Khim.*, 1972, **42**, 1642 (*Chem. Abs.*, 1972, **77**, 102 200x).

5 Appendix II: A List of Some Useful New Synthetic Intermediates Described during 1972

The preliminary remarks to the corresponding sections of earlier volumes still apply. A list, with physical constants, of some 1000 synthetically useful derivatives of the amino-acids which commonly occur in proteins has been compiled.[2]

Compound	*Ref.*
Alanine	
Bornoc-Ala*	15
HCO-Ala-ONSu	55
β-Alanine	
Bornoc-β-Ala,Dcha	15
Aoc-β-Ala,Dcha	321
Aoc-β-Ala-ONSu	321
Arginine	
Boc-Arg	169
Bornoc-Arg,H_2O	15
Bornoc-Arg(NO_2)	15
Asparagine	
Z-Asn-ONSu	340
Z-Asn-NH_2	334
Asn-OBzl(NO_2),HCl	45
Bpoc-Asn-ONp	146
Bornoc-Asn	15
Trt-Asn-ONSu	238
A series of derivatives of Asn[Bzl(OMe)$_2$]	64
Aspartic acid	
Z-Asp(ONSu)-OBzl	194
Asp(OMe)-OBu^t,HCl	194
Z-Asp-OBu^t,Dcha	194
Boc-Asp-OBzl	74
Boc-Asp-OBu^t	74
Z-Asp(OBu^t)-NH·NH·Boc	278
Asp(OBu^t)-NH·NH·Boc,AcOH,H_2O	278
Asp(OBzl)-ONp,CF_3CO_2H	212
β-Cyclohexylalanine (Cha)	
Boc-Cha,Dcha	157
Boc-Cha	266
Cysteine	
Cys(SEt)-OBzl(OMe),HCl	45
Cys(Bzl)-OBzl(NO_2),HCl	45
Cys(Acm)	56
Boc-Cys(Acm)	56
Boc-Cys(Acm)-ONSu	56
Z-Cys(Acm): oil	56
Z-Cys(Acm)-ONSu	56
Cys(Ec);† useful additional characterization	55
Boc-Cys(Ec)	55
Boc-Cys(Ec)-OTcp	55
Z-Cys(Ec),Dcha	55

* Bornoc = *d*-isobornyloxycarbonyl.
† Ec = ethylcarbamoyl.

Compound	*Ref.*
Boc-Cys(Pic)	52
Boc-Cys(Trt),Dcha	61
Boc-Cys(Bzl),Dcha	61
Boc-Cys(Bzl)-ONSu	61
Trt-Cys(Trt)-ONSu	61, 372
$\alpha\gamma$-Diaminobutyric acid (A_2bu)	
A_2bu(Pht),H_2O	314
A_2bu(Pht)-OBzl,TosOH	314
Glutamic acid	
Glu(OBu^t)-OMe,HCl; higher rotation than before	199
Boc-D-Glu(OBzl),Dcha	198
Boc-D-Glu(OBzl); more convenient procedure than described for L-isomer previously	198
Z-Glu(ONSu)-OBzl	199
Glu(OMe)-OBu^t,HCl; convenient procedure for large-scale preparations	199
Boc-Glu-OBzl	74
Nps-Glu(OBzl)-ONSu	201*a*
Glu(OBzl)-ONSu,HCl	201*a*
Nps-Glu(OBzl)-OPcp (L and D)	201*a*
Glu(OBzl)-OPcp,HCl (L and D)	201*a*
Boc-Glu(OBzl)-OPcp	201*c*
Boc-Glu(OBzl); D and L; simplified procedure	200
Bpoc-Glu(OBu^t),Dcha	238
Bpoc-Glu(OBu^t)-ONSu	238
Glutamine	
Bpoc-Gln-ONp	146, 169
Bornoc-Gln	15
Z-Gln[Bzh(OMe)$_2$] ONSu	65
A series of derivatives of Gln[Bzh(OMe)$_2$]	64
Glycine	
Gly-OPic,2HCl	45
Bornoc-Gly	15
Gly-NH·NH·Z,CF_3CO_2H	55
Histidine	
Boc-His(Boc),Dcha	169
Bpoc-His,H_2O	169
Bpoc-His(Boc)	169
Bpoc-His(Tos)	146
Isoleucine	
Aoc-Ile	340
Bornoc-Ile	15
Leucine	
Bpoc-Leu-OTcp	211
Leu-OPic,2HCl	45
Leu-OPic,2HBr; small correction to rotation	45
Leu-OBzl(NO_2),HCl	45
Bornoc-Leu,H_2O	15
Lysine	
Z-Lys(Pht),Dcha	340
Z-Lys(Pht)	340
Z-Lys(Pht)-ONp	340

372 B. Kamber, H. Bruckner, B. Riniker, P. Sieber, and W. Rittel, *Helv. Chim. Acta*, 1970, **53**, 556.

Compound	*Ref.*
Lys(Troc)*	68
Z(OMe)-Lys(Troc),Dcha	68
Z-Lys(Troc),Dcha	68
Boc-Lys(Troc)	68
Lys[Z(Br)]	69
Boc-Lys[Z(Br)]	69
Boc-Lys(CF_3CO)-ONSu	279
Methionine	
Met-OPic,2HBr; rotation discrepancy	45
Bornoc-Met,Dcha	15
Bpoc-Met(O)	146
N-Methylphenylalanine	
Boc-MePhe (oil)	183
Boc-MePhe,Dcha	259
Ornithine	
Orn(Z)-OBzl,HCl	354
Phenylalanine	
Z-Phe-NH·NH·Boc	245
H-Phe-NH·NH·Boc	245
H-Phe-OPh,HBr	43
Aoc-Phe-ONSu	340
Phe-OPic,2HCl	45
Phe-OBzl(NO_2),HCl; correction to previous rotation	45
Bornoc-Phe	15
Bornoc-Phe,Dcha	15
Bpoc-Phe	146
Pipecolic acid	
Boc-Pipec-OTcp	373
Boc-Pipec-ONSu	373
Nps-Pipec-OTcp	373
Nps-Pipec-ONSu	373
Z-Pipec-OTcp	373
Z-Pipec-ONSu	373
Proline	
Bornoc-Pro	15
Z-Pro-NH·Bzh$(OMe)_2$	65
Pro-NH·Bzh$(OMe)_2$	65
Pyroglutamic acid	
Z-Glp-OPcp	310
Sarcosine	
CF_3CO-Sar	374
Serine	
Ser-OBzl(NO_2),HCl	45
Bornoc-Ser,Dcha	15
Boc-Ser-NH·NH_2	55
Threonine	
Bornoc-Thr,Dcha	15
Boc-Thr-ONSu	55
Boc-Thr-OBzl	77

[373] L. Balaspiri, G. Papp, and K. Kovacs, *Monatsh.*, 1972, **103**, 581.
[374] H. C. J. Ottenheym, T. F. Spande, and B. Witkop, *J. Org. Chem.*, 1972, **37**, 3358.

* Troc = $\beta\beta\beta$-trichloro-ethoxycarbonyl.

Compound	*Ref.*
Z-Thr·NH·NH·Boc	277
Thr·NH·NH·Boc,AcOH,0.5H_2O	277
Tryptophan	
Bornoc-Trp	15
Tyrosine	
Aoc-Tyr(Bu^t),Dcha	340
Aoc-Tyr(Bu^t)	340
Aoc-Tyr(Bu^t)-ONSu	340
Tyr-OBzl(NO_2),HCl	45
Boc-Tyr-NH·NH·Z	55
Tyr-NH·NH·Z,CH_3CO_2H	55
Tyr[Bzl(*m*-Br)]	69
Boc-Tyr[Bzl(*m*-Br)]	69
Boc-Tyr[Bzl(*m*-Br)],Dcha	69
Boc-Tyr(Pic)	52
Aoc-Tyr(Bzl)-ONSu	321
Bpoc-Tyr(Bu^t),Dcha	238
Bpoc-Tyr(Bu^t)-ONSu	238
Bpoc-Tyr(Bzl)-ONp	146
Boc-Tyr(Bzl)-ONSu; rotation given	279
Valine	
Bornoc-Val	15

4

Peptides with Structural Features not Typical of Proteins

BY B. W. BYCROFT

1 Introduction

The choice of topics reviewed follows the now well-established pattern for this chapter, and once again the year has afforded a wealth of interesting developments.

The application of physicochemical methods for structure determination and conformational studies has reached even greater levels of sophistication. In particular the range of n.m.r. techniques available for investigating the solution conformations of cyclic peptides has been extended considerably. *X*-Ray crystallography is increasingly employed for structure determination and conformational studies, although frequently the favoured conformer in the crystal lattice differs from that observed in solution. Mass spectrometry is still a valuable structural tool for small peptides with unusual features since it offers a facile means of detecting non-amino-acid components and allows the sequence to be determined directly from the fragmentation patterns.

One of the most notable achievements of the year has been the synthesis of a cyclic dodecapeptide which complexes potassium more readily than valinomycin. The synthesis was achieved using a standard automated solid-phase procedure and in itself does not represent any new development. The significance of the synthesis is that the molecule was designed specifically as an ion carrier. The principle of programming the sequence of a small peptide in such a way that it contains all the information necessary to express a simple predetermined function is of considerable interest.

A considerable number of new peptide alkaloids have been reported, and the penicillin–cephalosporin field is still dominated by the chemistry of the 7-methoxycephalosporins. These novel antibiotics show marked biological activity and have been the subject of considerable synthetic effort. The biosynthesis of these compounds is of interest since they are probably derived from a dehydroamino-acid precursor. The antibiotic viomycin also contains two structural units related to dehydroamino-acids and it appears likely that they have some general significance to small microbial peptides.

2 Cyclic Peptides (Homodetic Peptides)

In view of the large amount of literature on conformational studies this year a separate section on this subject has been included.

2,5-Dioxopiperazines.—Full details of the structure elucidation of the brevianamide group of metabolites have been reported[1] and two further compounds, brevianamides C and D, have been shown to be the stereoisomers of the indoxylidene (1). However, it appears likely that they are photochemical artefacts derived from brevianamide A. *cyclo*-L-Tryptophanyl-L-proline also co-occurs with the brevianamides, suggesting that it is the biogenetic precursor of the whole series. It has also been shown that a crude cell-free extract of *Aspergillus amstelodami* will catalyse the isoprenylation of *cyclo*-L-alanyl-L-tryptophan to form a possible precursor of echinulin.[2] The closely related compound lanosulin, the major metabolite of *Penicillium lanosum* Westling, has the structure (2) and is claimed[3] to represent the missing biogenetic link for the echinulin and brevianamide groups.

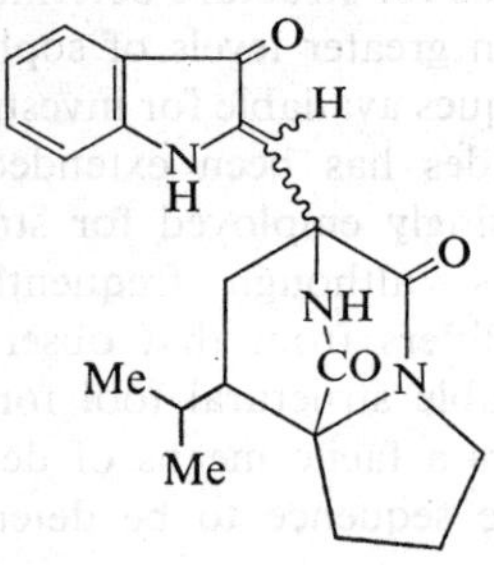

(1)

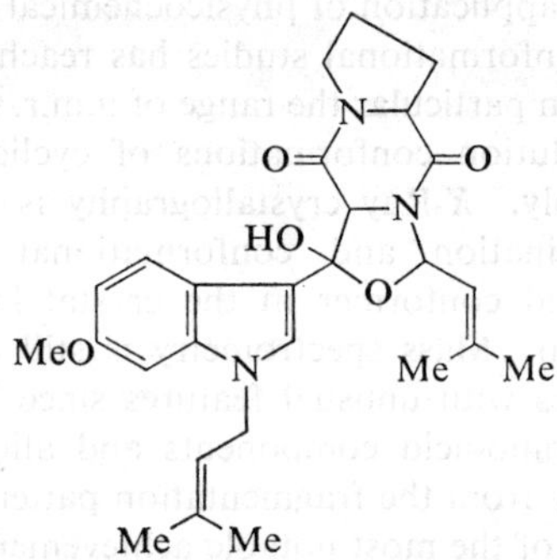

(2)

A new minor metabolite of *Pithomyces chartarum*, sporidesmin G, has been assigned[4] the structure (3) and this has been confirmed by conversion into sporidesmin (4) on irradiation or treatment with triphenylphosphine. On the basis of an extensive spectroscopic investigation and taking into account biogenetic considerations the antibiotic chetomin is probably represented by (5), but it is pointed out that other formulae are possible.[5] Considerable interest still persists in the development of general synthetic routes to the 3,6-epithio-2,5-dioxopiperazine antibiotics, presumably owing to their pronounced antiviral activity. An *X*-ray crystallographic analysis on *NN*-dimethyl-3,6-epitetrathio-2,5-dioxopiperazine, prepared by a

[1] A. J. Birch and R. A. Russell, *Tetrahedron*, 1972, **28**, 2999.
[2] C. M. Allen, *Biochemistry*, 1972, **11**, 2154.
[3] D. T. Dix, J. Martin, and C. E. Moppett, *J.C.S. Chem. Comm.*, 1972, 1168.
[4] E. Francis, R. Rahman, S. Safe, and A. Taylor, *J.C.S. Perkin I*, 1972, 470.
[5] S. Safe and A. Taylor, *J.C.S. Perkin I*, 1972, 473.

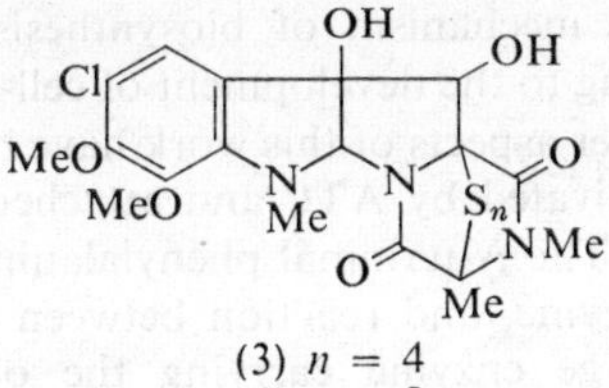

(3) $n = 4$
(4) $n = 2$

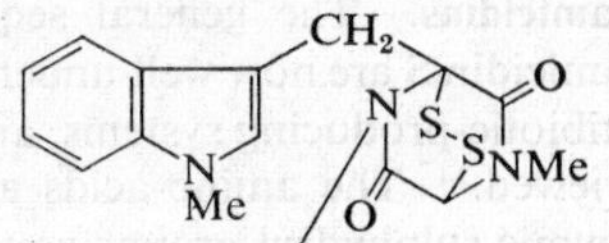

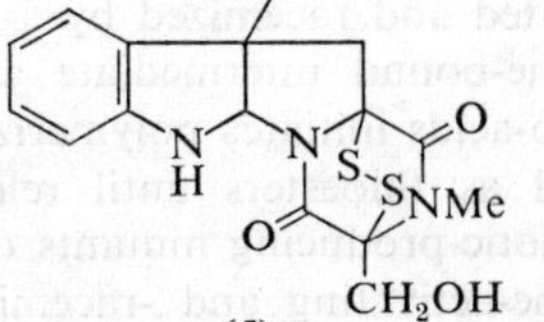

(5)

method developed some years ago, has been reported.[6] Recent refinements [7–9] of this general method have led to the stereospecific synthesis of (6) by the electrophilic addition of sulphur to the anion derived from L-prolyl-L-proline. An alternative route to model compounds of this class involving the dimerization of *N*-alkyl-2-mercaptocinnamamide derivatives proved unsuccessful.[10]

Reaction of 2,5-dioxopiperazines with an excess of triethyloxonium tetrafluoroborate produces a *cis–trans* mixture of the di-imino-ether (7) which can be readily oxidized to the corresponding pyrazine (8) in high

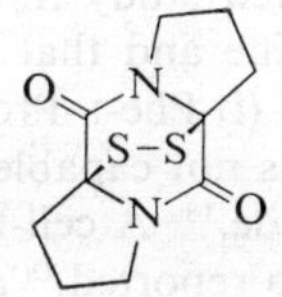

(6)

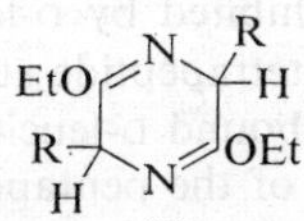

(7)

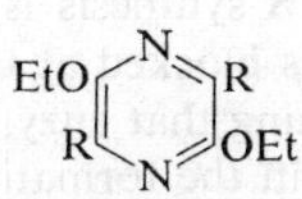

(8)

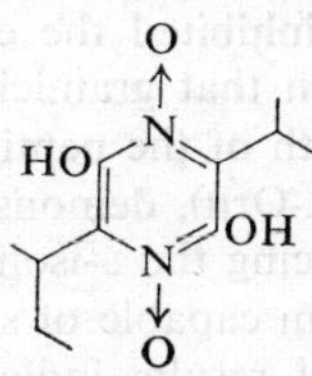

(9)

yield.[11] This reaction is of interest with respect to the biosynthesis of the pyrazine fungal metabolites which are considered to be derived by oxidation of the corresponding cyclic dipeptide. It has been shown that pulcherriminic acid (9) is formed *in vivo* by aerial oxidation of *cyclo*-L-isoleucyl-L-isoleucine.[12]

Cyclic dipeptides are readily *N*-methylated with silver oxide and methyl iodide in dimethylformamide to yield oils or low-melting solids with high solubility in most organic solvents. The method does not cause epimerization and greatly facilitates n.m.r. and g.l.c. studies, offering a convenient means of investigating naturally occurring cyclic dipeptides.[13]

[6] B. R. Davis, I. Bernal, and U. Schmidt, *Angew. Chem.*, 1972, **84**, 632.
[7] H. Poisel and U. Schmidt, *Chem. Ber.*, 1972, **105**, 625.
[8] E. Öhler, H. Poisel, F. Tataruch, and U. Schmidt, *Chem. Ber.*, 1972, **105**, 635.
[9] E. Öhler, F. Tataruch, and U. Schmidt, *Chem. Ber.*, 1972, **105**, 3658.
[10] J. Yoshimura and Y. Sugiyama, *Bull. Chem. Soc. Japan*, 1972, **45**, 1554.
[11] K. W. Blake, A. E. A. Porter, and P. G. Sammes, *J.C.S. Perkin I*, 1972, 2494.
[12] R. L. Uffen and E. Canale-Parola, *J. Bacteriol.*, 1972, **111**, 86.
[13] A. B. Mauger, R. B. Desai, I. Rittner, and W. J. Rzeszotarski, *J.C.S. Perkin I*, 1972, 2146.

Gramicidins.—The general sequence and mechanism of biosynthesis of gramicidin S are now well understood owing to the development of cell-free antibiotic-producing systems, and the earlier aspects of this work have been reviewed.[14] The amino-acids are first activated by ATP and attached to enzymic sulphydryl groups as thioesters. The *N*-terminal phenylalanine is activated and racemized by a further enzyme, and reaction between this enzyme-bound intermediate and the large enzyme carrying the other amino-acids initiates polymerization to peptide intermediates that remain linked as thioesters until released by cyclization. A variety of non-antibiotic-producing mutants of *Bacillus brevis* lacking either the phenylalanine-activating and -racemizing enzyme or the amino-acid-activating enzyme complex has been produced.[15] The enzymes obtained from both groups were shown to be complementary for gramicidin S formation. The elongation and cyclization of the enzyme thioester-bound intermediates are not yet precisely defined but a tentative model has been proposed.[16] It is now established that phosphopanotheine is present in the enzyme system and acts as a transmitter of the growing peptide chain.

The amino-acid-activating enzyme complex has been purified and the reversible and irreversible inhibition of this enzyme with a variety of amino-acid substrates was investigated.[17] The D-isomers of the constituent amino-acids of gramicidin S did not catalyse the exchange reaction and inhibited the exchange by the L-isomers. In a related study it was shown that gramicidin S synthesis is inhibited by D-leucine and that the growth of the peptide is blocked at the tetrapeptide stage (D-Phe-L-Pro-L-Val-L-Orn), demonstrating that enzyme-bound D-leucine is not capable of replacing the L-isomer in the formation of the pentapeptide.[18] A cell-free system capable of synthesizing linear gramicidins has been reported [19] and initial results indicate that the biosynthesis of the pentadecapeptide is analogous to that of gramicidin S.

Further analogues of gramicidin S, namely [Cys^{2,7}]-gramicidin and the corresponding bis-*S*-acetamidomethyl derivative have been synthesized by standard procedures.[20] In the bicyclic peptide, the decapeptide ring is contained in a β-type secondary structure, identical with gramicidin S, and the disulphide bridge shows *p*-helical chirality. [4,5-δ-Aminovaleric acid]-gramicidin S [21] and [Gly^5]-gramicidin S [22] have also been synthesized, and both possess antimicrobial activity comparable with that of gramicidin S. A comparison of their spectral properties with those of the natural

[14] F. Lipmann, *Science*, 1971, **173**, 875.
[15] M. Iwaki, K. Shimura, M. Kanda, E. Kaji, and Y. Saito, *Biochem. Biophys. Res. Comm.*, 1972, **48**, 113.
[16] S. G. Laland, O. Froyshov, C. Gilhuus-Moe, and T. L. Zimmer, *Nature New Biol.*, 1972, **239**, 43.
[17] D. C. Leung and R. M. Baxter, *Biochim. Biophys. Acta*, 1972, **279**, 34.
[18] H. Saxholm, T. L. Zimmer, and S. G. Laland, *European J. Biochem.*, 1972, **30**, 138.
[19] K. Bauer, R. Roskoski, H. Kleinkauf, and F. Lipmann, *Biochemistry*, 1972, **11**, 3266.
[20] U. Ludescher and R. Schwyzer, *Helv. Chim. Acta*, 1972, **55**, 2052.
[21] I. Muramatsu, S. Sofuku, and A. Hagitani, *J. Antibiotics*, 1972, **25**, 189.
[22] S. Matsuura, M. Waki, and N. Izumiya, *Bull. Chem. Soc. Japan*, 1972, **45**, 863.

antibiotic suggests that all have similar secondary structures. Six linear pentapeptide amides possessing the same sequence of amino-acids as found in gramicidin, and with *N*-terminal amino-groups acylated with either formyl or acetyl and the *C*-terminal residue as amide or ethanolamide, have been synthesized in relation to a systematic structure–activity study.[23] The macro-ring analogues of gramicidin S *cyclo*-(Val-Orn-Leu-D-Phe-Pro-)$_n$ (n = 3 or 4) have been prepared by cyclization of the corresponding linear *N*-hydroxysuccinimide esters in pyridine. The conformations and antibacterial activity were found to be similar to those of the open-chain analogues.[24]

Further studies on ion transfer across lipid membranes in the presence of linear gramicidin A have been described [25, 26] and the results are consistent with gramicidin forming a transmembrane channel by means of a dimeric $\pi_{L,D}$ helical structure. The proposed mechanism involves the ion moving along the core of the helix by an ion-induced relaxation of the channel's helical conformation. Extensive absorption, c.d., and n.m.r. studies [27, 28] on gramicidin A have provided further evidence for a $\pi_{L,D}$ helix conformation in trifluoroethanol and DMSO solutions.

Gramicidin S has been used extensively as a model system for evaluating new methods developed specifically for conformational studies on cyclic peptides, and this work is reviewed later in this section.

Peptides from *Amanita phalloides*.—Once again attention has been focused predominantly on the antitoxic cyclodecapeptide antamanide (10); the

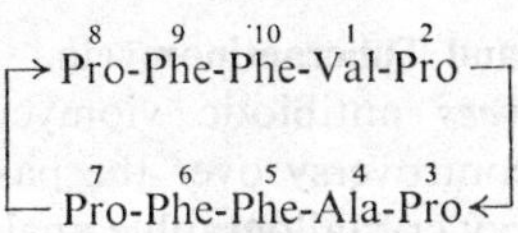

(10)

early chemical and biological work in this field has fortunately been reviewed.[29] In addition an excellent review on the chemistry and activity of the amatoxins has also appeared.[30]

A considerable number of analogues of antamanide in which the phenylalanine residues have been replaced by tyrosine have been synthesized,[31, 32] mainly by the automated solid-phase method, and these are

[23] S. Makisumi, M. Waki, and N. Izumiya, *Mem. Fac. Sci. Kyushu Univ., Ser. C*, 1972, **8**, 1.
[24] S. Matsuura and N. Izumiya, *Experientia*, 1972, **28**, 1402.
[25] S. Hladky and D. A. Haydon, *Biochim. Biophys. Acta*, 1972, **274**, 294.
[26] V. B. Mayers and D. A. Haydon, *Biochim. Biophys. Acta*, 1972, **274**, 313.
[27] J. D. Glickson, D. F. Mayers, J. M. Settine, and D. W. Urry, *Biochemistry*, 1972, **11**, 477.
[28] D. W. Urry, J. D. Glickson, D. F. Mayers, and J. Haider, *Biochemistry*, 1972, **11**, 487.
[29] Th. Wieland and O. Wieland, 'Microbial Toxins', Academic Press, New York, 1972, vol. 8, p. 249.
[30] Th. Wieland, *Naturwiss.*, 1972, **59**, 225.
[31] Th. Wieland, C. Reitzel, and A. Seeliger, *Annalen*, 1972, **759**, 63.
[32] Th. Wieland, C. Birr, R. Frode, W. Lochinger, and G. Stahnke, *Annalen*, 1972, **757**, 136.

summarized in Table 2 (p. 363). These studies are aimed at defining the structural requirements for the antitoxic activity of these compounds against phalloidin, and a number of important observations have already been made. All the tyrosine-containing analogues are effective to varying degrees in protecting mice from lethal doses of phalloidin. The [all-D-retro]-antamanide and [D-Tyr^6-all-D-retro]-antamanide,[33] also prepared by cyclization of their respective linear decapeptides, which were synthesized on a solid phase, are virtually insoluble in water, but the latter compound could be esterified and the resultant water-soluble derivatives are effective antitoxic agents. A cyclononapeptide, *viz.* antamanide lacking alanine at position 4, is also active and is the first effective shortened analogue of antamanide. On the other hand, the tyrosine-containing analogues in which the valine and alanine residues had also been exchanged for glycine were ineffective.

The solution conformation of antamanide and its Na^+ complex still remains a controversial subject. Detailed c.d. and u.v. studies indicate that similar conformational changes are induced in antamanide by interaction with cations or polar solvents and these results are claimed [34] to support the model with all-*trans* peptide bonds, initially proposed by the Russian group.[35] However, American workers,[36, 37] on the basis of ^{1}H and ^{13}C n.m.r. and model building, favour, in non-polar solvents, an alternative conformation containing two *cis* peptide bonds and propose that *cis–trans* isomerization does not occur for antamanide on complexation.

Viomycin, Capreomycin, and Tuberactinomycin.—The structure (11) of the antitubercular *Streptomyces* antibiotic viomycin, which has been the subject of considerable controversy over the past twenty years, has been resolved at last by an *X*-ray crystallographic analysis.[38] It is now apparent that viomycin and the tuberactinomycins,[39] and almost certainly the capreomycins, belong to a closely related family of cyclopentapeptide antibiotics. The complexity of the chemistry of viomycin is mainly associated with the labile dehydroserine ureide and guanidine carbinol systems, and full details of these aspects have now been reported.[40,41] The discrepancies relating to the previously reported structures for viomycin stem from the considerable chemical reactivity of these units and serve to

[33] Th. Wieland, B. Penke, and C. Birr, *Annalen*, 1972, **759**, 71.
[34] H. Faulstich, W. Burgermeister, and Th. Wieland, *Biochem. Biophys. Res. Comm.*, 1972, **47**, 975.
[35] V. T. Ivanov, A. I. Miroshnikov, A. D. Abdullaev, L. B. Senyavina, S. F. Arkhipova, N. N. Uvarova, K. Kh. Khulilulina, V. F. Bystrov, and Yu. A. Ovchinnikov, *Biochem. Biophys. Res. Comm.*, 1971, **42**, 652.
[36] D. J. Patel, *Biochemistry*, 1973, **12**, 667.
[37] A. E. Tonelli, *Biochemistry*, 1973, **12**, 689.
[38] B. W. Bycroft, *J.C.S. Chem. Comm.*, 1972, 660.
[39] T. Noda, T. Take, A. Nagata, T. Nakamiya, and T. Shiba, *J. Antibiotics*, 1972, **25**, 427.
[40] B. W. Bycroft, L. R. Croft, A. W. Johnson, and T. Webb, *J.C.S. Perkin I*, 1972, 820.
[41] B. W. Bycroft, D. Cameron, L. R. Croft, A. Hassanali-Walji, A. W. Johnson, and T. Webb, *J.C.S. Perkin I*, 1972, 827.

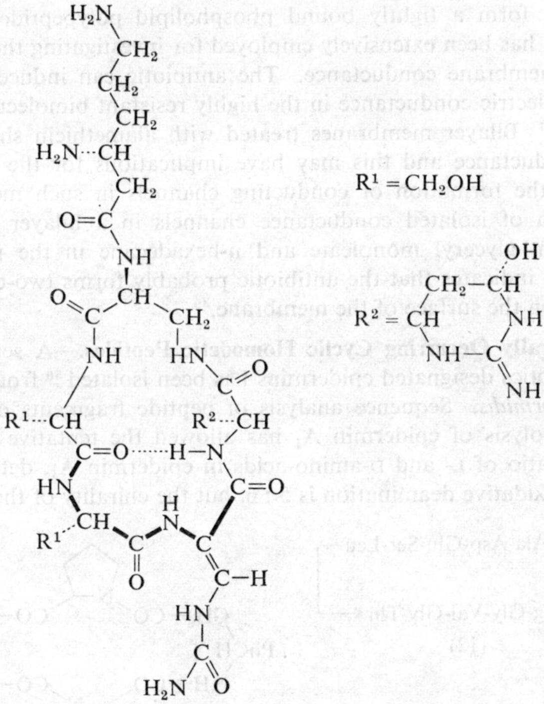

(11)

emphasize that the standard sequencing techniques must be employed with caution on this type of molecule. In order to circumvent this problem a sequential analysis of viomycin has been performed on perhydro-viomycin and the results accord [42] with the *X*-ray analysis.

Several derivatives of viomycin [43] and the possible relationships between antimicrobial activity and chemical structure [44] have been discussed. A significant feature of viomycin in the crystalline state is the transannular hydrogen bond, which initial n.m.r. results suggest is retained in solution,[45] but more detailed analysis of the solution conformation is still lacking.

Alamethicin.—N.m.r. studies [46] of the interaction between alamethicin and phospholipids show that alamethicin breaks down the phospholipid bilayer

[42] T. Kitagawa, T. Miura, and K. Fujiwara, *Chem. and Pharm. Bull. (Japan)*, 1972, **20**, 2225.
[43] T. Kitagawa, T. Miura, and H. Taniyama, *Chem. and Pharm. Bull. (Japan)*, 1972, **20**, 2176.
[44] T. Kitagawa, T. Miura, S. Tanaka, and H. Taniyama, *J. Antibiotics*, 1972, **25**, 429.
[45] P. Viglino, C. Franconi, A. Lai, E. Brosio, and F. Conti, *Org. Magn. Resonance*, 1972, **4**, 237.
[46] E. G. Finer, H. Hauser, and D. Chapman, 'Report of the 3rd International Conference on Magnetic Resonance in Biological Research', 1969, pub. 1971, p. 337.

structure to form a tightly bound phospholipid–polypeptide aggregate. Alamethicin has been extensively employed for investigating the molecular nature of membrane conductance. The antibiotic can induce a voltage-dependent electric conductance in the highly resistant bimolecular lecithin membrane.[47] Bilayer membranes treated with alamethicin show a time-variant conductance and this may have implications for the mechanism underlying the formation of conducting channels in such membranes.[48] Examination of isolated conductance channels in a bilayer membrane, formed from glyceryl monoleate and n-hexadecane in the presence of alamethicin, indicates that the antibiotic probably forms two-dimensional aggregates on the surface of the membrane.[49]

Other Naturally Occurring Cyclic Homodetic Peptides.—A series of new basic antibiotics designated epidermins has been isolated[50] from *Staphylococcus epidermidis*. Sequence analysis of peptide fragments obtained on partial hydrolysis of epidermin A_1 has allowed the tentative assignment (12). The ratio of L- and D-amino-acids in epidermin A_1, determined by enzymatic oxidative deamination is 5 : 6, but the chirality of the individual

→ Lys-Ala-Asp-Glu-Ser-Leu ─┐
└─ Arg-Gly-Val-Gly-Thr ←─┘

(12)

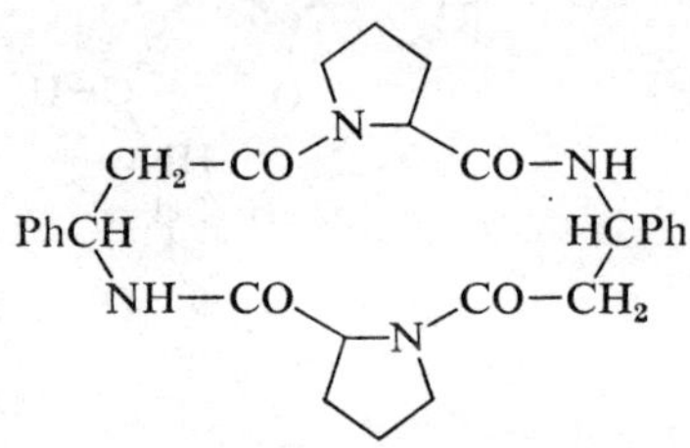

(13)

amino-acid residues has not been reported. The antibiotic is of interest since arginine is not commonly found in microbial peptides. The cyclotetrapeptide (13), isolated from the lichen *Roccella canariensis*, affords on hydrolysis L-proline and *R*-β-phenyl-β-alanine. The structure was deduced[51] from the n.m.r. and mass spectral data and confirmed by synthesis.[52]

Conformational Studies.—The determination of the conformations of cyclic peptides in solution, motivated no doubt by the possible relationships between the conformations of cyclic peptide antibiotics, toxins, and ionophores and their biological activity, still dominates this field.

These studies have principally involved the application of n.m.r. techniques, with noticeably less emphasis on optical spectroscopy, o.r.d.,

[47] R. J. Cherry, D. Chapman, and D. E. Graham, *J. Membrane Biol.*, 1972, **7**, 325.
[48] A. Mauro, R. P. Nanavati, and E. Heyer, *Proc. Nat. Acad. Sci. U.S.A.*, 1972, **67**, 3742.
[49] L. G. M. Gordon and D. A. Haydon, *Biochim. Biophys. Acta*, 1972, **255**, 1014.
[50] C. Y. Hsu and G. M. Wiseman, *Canad. J. Microbiol.*, 1972, **18**, 121.
[51] G. Bohman-Lindgren, *Tetrahedron*, 1972, **28**, 4625.
[52] G. Bohman-Lindgren and U. Ragnarsson, *Tetrahedron*, 1972, **28**, 4631.

and c.d. *X*-Ray diffraction methods still provide invaluable information as regards the crystal conformation, but often this does not correspond to that found in solution.

An excellent review, mainly on n.m.r. techniques and solution conformation, which includes most of the important literature up to 1971, has been published.[53] The usual procedure in determining the secondary structure of cyclic peptides requires the study of the temperature dependence of the amide proton chemical shifts and their rate of exchange with

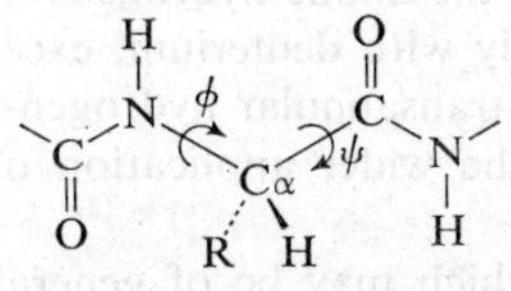

Figure 1

deuterium. It is possible then to separate the peptide-bond protons into groups according to whether they are exposed to the solvent or shielded from it by hydrogen bonds or steric factors. This information, combined with the angle ϕ (Figure 1), derived from the amide proton–α-proton coupling constants, allows assignment of solution conformation.

An alternative method[54] which offers the same information makes use of the dramatic upfield shifts of certain amino-acid residues within cyclic peptides induced by the addition of 2,2,2-trifluoroethanol (TFE). The amide proton–α-proton coupling constants are unchanged, indicating that no backbone change has occurred. The cause of the effect is not known but it is suggested that it may result from reduced hydrogen-bonding of the peptide bonds with the solvent in the presence of TFE.

The application of homonuclear internuclear double-resonance spectroscopy (INDOR) considerably simplifies the interpretation of the ^{1}H n.m.r. spectra of peptides and thus facilitates conformational studies. The method has been successfully applied to[55, 56] the gramicidins, polymixin B, and tyrocidin A.

Two new techniques which offer to broaden dramatically the scope of n.m.r. studies of peptide conformation have been reported. Theoretical calculations and limited experimental data indicate that direct determination of the angle ψ (Figure 1) may well be possible, either by measurement of the NCC$^\alpha$H nitrogen–hydrogen coupling constant[57] or by measurement of the vicinal ^{13}C–H coupling constant.[58] The results appear sufficiently

[53] F. A. Bovey, A. I. Brewster, D. J. Patel, A. E. Tonelli, and D. A. Torchia, *Accounts Chem. Res.*, 1972, **5**, 193.
[54] T. P. Pitner and D. W. Urry, *J. Amer. Chem. Soc.*, 1972, **94**, 1399.
[55] W. A. Gibbons, H. Alms, J. Sogn, and H. R. Wyssbrod, *Proc. Nat. Acad. Sci. U.S.A.*, 1972, **69**, 1261.
[56] W. A. Gibbons, H. Alms, R. S. Bockman, and H. R. Wyssbrod, *Biochemistry*, 1972, **11**, 1721.
[57] S. Karplus and M. Karplus, *Proc. Nat. Acad. Sci. U.S.A.*, 1972, **69**, 3204.
[58] R. E. Lemieux, T. L. Nagabhashan, and B. Paul, *Canad. J. Chem.*, 1972, **50**, 773.

encouraging to warrant a detailed investigation of cyclic peptides but for the moment the whole field is likely to remain the domain of the n.m.r. expert. A further adjunct to the above methods which may prove generally useful is *N*-H line-broadening produced by association of a nitroxide radical with amide protons. The initial work has centred on finding suitable radical–solvent combinations. This has now been achieved, and preliminary results for simple model compounds and the cyclic decapeptide gramicidin S have been reported.[59] In the case of gramicidin S line-broadening occurs for all the amide hydrogens which had previously been shown to exchange readily with deuterium, excepting those N—H bonds known to be involved in transannular hydrogen-bonding. These observations certainly warrant the wider application of the technique to other cyclic peptides.

Other n.m.r. studies which may be of general interest with respect to conformation are the effect of lanthanide shift reagents on various amides [60] and a ^{13}C study on the isomerism about the proline peptide bond,[61] together with a correlation of the crystal and solution conformations of *N*-acyl proline derivatives.[62]

A detailed and systematic spectroscopic examination of a series of cyclic dipeptides (14) and (15) containing residues of pipecolic acid,

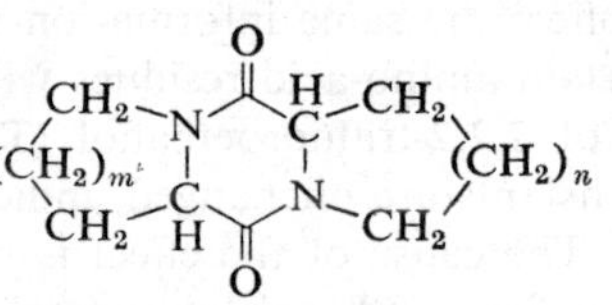

(14)

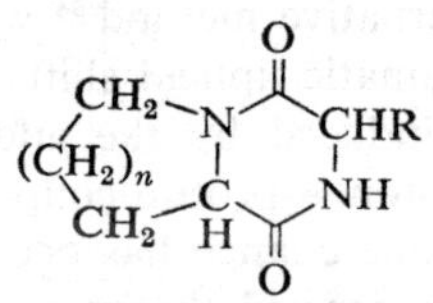

(15)

proline, and 2-azetidinecarboxylic acid has provided further valuable information concerning the conformation of the dioxopiperazine ring.[63] Earlier results had suggested that the ring system is planar or nearly planar when unsubstituted or substituted only with small residues, whereas when substituted with dissimilar or bulkier groups the ring adopts a symmetric boat conformation.[64] The present work extends these conclusions and suggests that *trans*-dianeled cyclic dipeptides also have planar or nearly planar rings, whereas monoaneled and *cis*-aneled compounds assume essentially a boat conformation, the depth of which increases with decreasing size of the aneled ring.

These conclusions find support in the results of an *X*-ray study on *cyclo*-(L-Pro-L-Leu-) which shows that the molecule has a folded conformation

[59] K. D. Kopple and T. J. Schamper, *J. Amer. Chem. Soc.*, 1972, **94**, 3644.
[60] R. A. Fletton, G. F. H. Green, and J. E. Page, *J.C.S. Chem. Comm.*, 1972, 1134.
[61] W. A. Thomas and M. K. Williams, *J.C.S. Chem. Comm.*, 1972, 994.
[62] W. A. Thomas and M. K. Williams, *J.C.S. Chem. Comm.*, 1972, 788.
[63] K. Blaha, M. Budèšinsky, I. Frič, J. Smolikova, and J. Vičar, *Tetrahedron Letters*, 1972, 1437.
[64] I. Z. Siemon, *Annalen*, 1971, **748**, 88.

with a dihedral angle of 143° between the two planar peptide units.[65] However, there is evidence that the conformation of cyclic dipeptides containing one proline residue reverts to a planar ring system in trifluoroacetic acid.[66]

Single-crystal *X*-ray studies on *cyclo*-(Gly-L-Tyr-) [67] and *cyclo*-(L-Ser-L-Tyr-) [68] indicate almost planar dioxopiperazine rings for both compounds with the aromatic side-chains above the ring system. ^{1}H N.m.r. studies on *cyclo*-(L-His-L-Ser-), *cyclo*-(L-Asp-L-His-), and *cyclo*-(L-His-L-His-) also favour this sort of stacking of the histidine residues above the ring.[68]

All the possible cyclic tetrapeptides arising from the combination of glycine and sarcosine and alanine and sarcosine have been synthesized, and a ^{1}H n.m.r. investigation indicates that the *cis,trans,cis,trans*-conformation, established by *X*-ray and n.m.r. studies for cyclotetrasarcosyl, is also adopted by these molecules.[69] For the cyclopentapeptide *cyclo*-(Gly-L-Ala-Gly-Gly-L-Pro-) and four analogues the n.m.r. data imply two different molecular conformations, the relative concentrations of which depend on the peptide, solvent, and temperature. The evidence [70, 71] suggests that both conformations contain intramolecular hydrogen bridges and that the peptide linkages are all *trans* except that involving the nitrogen of the proline which is respectively *cis* and *trans* in the two predominant molecular species.

A considerable number of cyclohexapeptides have been synthesized solely for the purpose of conformational studies and a number of these are listed in Table 1. It is perhaps too early to draw general conclusions,

Table 1 *Cyclohexapeptides whose solution conformations have been investigated*

Peptide	*Ref.*
cyclo-(Gly_6)	72
cyclo-(Ala_6)	72
cyclo-(Gly-Gly-D-Ala-D-Ala-Gly-Gly-)	73
cyclo-(Gly-Leu-Gly-Gly-Leu-Gly-)	74
cyclo-(Gly-Tyr-[$^2H_2$]Gly-Gly-Tyr-[$^2H_2$]Gly-)	74
cyclo-(Gly-His-Gly-Tyr-Ala-Gly-)	74
cyclo-(Gly-His-Gly-Ala-Tyr-Gly-)	74
cyclo-(Pro-Ser-Gly-Pro-Ser-Gly-)	75, 76
cyclo-(Ser-Pro-Gly-Ser-Pro-Gly-)	75, 76
cyclo-(Gly-Pro-Gly-Gly-Pro-Gly)	76
cyclo-(Gly-Pro-Gly-Gly-Pro-Gly)	76
cyclo-(Gly-Pro-[$^2H_2$]Gly-Gly-Pro-[$^2H_2$]Gly-)	76
cyclo-(Pro-Gly-Pro-Gly-Pro-Gly-)	77

65 I. L. Karle, *J. Amer. Chem. Soc.*, 1972, **94**, 81.
66 I. Z. Siemon and A. Sucharda-Sobczyk, *Roczniki Chem.*, 1972, **46**, 1257.
67 L. E. Webb and C. F. Lin, *J. Amer. Chem. Soc.*, 1971, **93**, 3818.
68 Ziauddin, K. D. Kopple, and C. A. Bush, *Tetrahedron Letters*, 1972, 483.
69 J. Dale and K. Titlestad, *J.C.S. Chem. Comm.*, 1972, 255.
70 J. P. Meraldi, R. Schwyzer, A. Tun-Kyi, and K. Wuethrich, *Helv. Chim. Acta*, 1972, **55**, 1962.
71 K. Wuethrich, A. Tun-Kyi, and R. Schwyzer, *F.E.B.S. Letters*, 1972, **25**, 104.

but certainly all the cyclohexapeptides so far investigated, excluding those containing proline residues, possess an all-*trans* amide conformation. In addition it appears that the conformations of those containing one, two, or three amino-acids other than glycine seem to be determined predominantly by a requirement that as many as possible of the substituent residues take values of ϕ and ψ which place them in the most favourable region of the conformational energy map for a substituted peptide unit.[72–74] Trans-annular hydrogen bonds provide further stabilization but appear relatively unimportant.

N.m.r. analyses of *cyclo*-(Pro-Ser-Gly-Pro-Ser-Gly-) and its retro-isomer *cyclo*-(Ser-Pro-Gly-Ser-Pro-Gly-) have shown that in both compounds C_2-symmetric intramolecularly hydrogen-bonded conformations having all peptide bonds *trans* are in equilibrium with conformations, both asymmetric and symmetric, containing *cis*-proline peptide bonds.[75, 76] Similar conclusions concerning the conformation of *cyclo*-(Gly-Pro-Gly-Gly-Pro-Gly-) derived earlier have now been confirmed by ^{13}C n.m.r. studies.[77] *cyclo*-(Pro-Gly-)$_3$ in apolar solvents is C_3-symmetric, containing three *cis*-Gly-Pro peptide bonds. In polar solvents the conformation is asymmetric, resulting from an inversion of one or two Gly-Pro units. Addition of alkali-metal ions affords a metal-cation complex which possesses a C_3-symmetric structure with all-*trans* amide linkages.[78]

The n.m.r. spectra of the cyclohexapeptides belonging to the ferrichrome group have been determined at 220 MHz in aqueous and deuteriated DMSO solutions.[79] The results suggest C_2-symmetric structures with two intramolecular hydrogen bonds, similar to other cyclohexapeptides already investigated. The solution conformation of the cycloheptapeptide evolidine, proposed recently on the basis of an n.m.r. study, has been re-examined using a computer search profile for the low-energy conformations consistent with the n.m.r. data.[80] The conclusions indicate a low-energy conformation with a *cis*-Leu-Pro amide bond similar to that originally proposed.[81]

Synthesis of Cyclic Peptides.—Several general reviews have appeared over the past year; the strategy and methodology for synthesis of cyclic peptides have been covered[82] and the general aspects of the solid-phase method

[72] K. P. Sarathy and C. Ramakrishnan, *Internat. J. Protein Res.*, 1972, **4**, 1.
[73] A. E. Tonelli and I. Brewster, *J. Amer. Chem. Soc.*, 1972, **94**, 2851.
[74] K. D. Kopple, A. Go, R. H. Logan, jun., and J. Savrda, *J. Amer. Chem. Soc.*, 1972, **94**, 973.
[75] D. A. Torchia, A. di Corato, S. C. K. Wong, C. M. Deber, and E. R. Blout, *J. Amer. Chem. Soc.*, 1972, **94**, 609, 616.
[76] A. E. Tonelli, *J. Amer. Chem. Soc.*, 1972, **94**, 346.
[77] R. Schwyzer, C. Grathwohl, J. P. Meraldi, A. Tun-kyi, R. Vogel, and K. Wüthrich, *Helv. Chim. Acta*, 1972, **55**, 2545.
[78] C. M. Deber, D. A. Torchia, S. C. K. Wong, and E. R. Blout, *Proc. Nat. Acad. Sci. U.S.A.*, 1972, **69**, 1825.
[79] M. Llinas, M. P. Klein, and J. B. Neilands, *Internat. J. Protein Res.*, 1972, **4**, 157.
[80] A. E. Tonelli, *Macromolecules*, 1972, **5**, 286.
[81] K. D. Kopple, *Biopolymers*, 1971, **10**, 1139.
[82] K. D. Kopple, *J. Pharm. Sci.*, 1972, **61**, 1345.

reviewed[83] with reference to cyclic peptides. In addition, two articles outlining the problems associated with the synthesis of peptide antibiotics containing anomalous amino-acids and peptide side-chains have been presented.[84, 85]

Once again a considerable number of cyclic peptides have been synthesized, mainly for conformational studies or as analogues of biologically active compounds, and these are listed in Tables 1 and 2.

Table 2 *Syntheses of cyclic peptides achieved in* 1972 (*derived from linear analogues unless otherwise stated*)

Peptide	*Bond formed in cyclization step*	*Yield (%)*	*Ref.*
[Cys^{2,7}]-gramicidin S	Pro-Val	27	20
[δ-Aminovaleric^{4,5}]-gramicidin S	Gly-Val	31	22
[Gly^{5}]-gramicidin S	Pro-Val	25	21
[Tyr^{5}]-antamanide	Tyr-Phe	44	32
	Pro-Ala	15	32
[Tyr^{6}]-antamanide	Phe-Tyr	37	32
	Pro-Ala	38	32
[Tyr^{9}]-antamanide	Pro-Ala	35	32
[Tyr^{10}]-antamanide	Phe-Phe	41	32
[Ile^{1}, Tyr^{6}]-antamanide	Phe-Tyr	22	32
[Ile^{1}, Tyr^{5}, Tyr^{10}]-antamanide	Tyr-Phe	22	32
[Gly^{1}, Gly^{4}, Tyr^{5}]-antamanide	Tyr-Phe	5	32
[Tyr^{9}, des-Ala]-antamanide	Phe-Phe	10	32
[all-D-retro]-antamanide	Phe-Ala	8	33
[D-Tyr-all-D-retro]-antamanide	Phe-Tyr	30	33
cyclo-(Gly-L-Ala-Gly-Gly-L-Pro-)	—[a]	—	70
cyclo-(Gly-L-Ala-Gly-Gly-D-Pro-)	—[a]	—	70
cyclo-(Gly-L-Cys(DPM)-Sar-Sar-L-Pro-)	Pro-Gly	—	70
cyclo-(Gly-L-Cys(DPM)-Sar-Sar-D-Pro-)	Pro-Gly	—	70
cyclo-(Gly—X_1—Y_1—Gly—X_2—Y_2)			
cyclo-(Gly-L-Phe-L-Leu-Gly-D-Phe-D-Leu)	Leu-Gly	30	86
cyclo-(Gly-L-Leu-L-Phe-Gly-L-Leu-L-Phe)	Phe-Gly[b]	23	86
cyclo-(Gly-L-Leu-D-Phe-Gly-L-Leu-D-Phe)	Phe-Gly[b]	43	86
cyclo-(Gly-L-Leu-L-Phe-Gly-D-Leu-D-Phe)	Phe-Gly	35	86
cyclo-(Gly-L-Leu-D-Phe-Gly-D-Leu-L-Phe)	Phe-Gly	43	86
cyclo-(Gly-L-Phe-L-Leu-Gly-L-Leu-L-Phe)	Phe-Gly	31	86
cyclo-(Gly-D-Phe-L-Leu-Gly-D-Leu-L-Phe)	Phe-Gly	37	86
cyclo-(Gly-D-Phe-D-Leu-Gly-L-Leu-L-Phe)	Phe-Gly	58	86
cyclo-(Gly-D-Phe-L-Leu-Gly-L-Leu-D-Phe)	Phe-Gly	34	86
cyclo-(Orn-Ser-Orn-Orn-Gly-Gly)	Gly-Orn	31	87
cyclo-(L-Val-D-Pro-D-Val-L-Pro-)$_3$	Pro-Val	16	88

[a] Obtained by desulphurization of the corresponding Cys derivative.
[b] Also obtained by cyclization dimerization.

[83] R. B. Merrifield, *Beckman Reports*, 1972, 3.
[84] N. Izumiya, S. Matsuura, and K. Kuromizu, *Yuki Gosei Kagaku Kyokai Shi*, 1971, **29**, 1032.
[85] M. Bodanszky, *Chim. Ther.*, 1972, **7**, 145.

The active-ester and dicyclohexylcarbodi-imide methods still remain the most popular for cyclization. The first synthesis of a homodetic–heterodetic bicyclic polypeptide [Cys^{2,7}]-gramicidin S has been achieved [20] using the active-ester method. For protection of the *C*-terminal carboxy and thiol functions, the 2-(toluene-*p*-sulphonyl)ethyl and acetamidomethyl groups were employed. Cyclization of a protected linear decapeptide *p*-nitrophenyl ester was carried out in pyridine and the resultant monocyclic compound was converted into the required product on treatment with iodine in methanol. The *p*-nitrophenyl esters have also been employed for the cyclization of linear peptides to other gramicidin analogues.[21, 22]

The dicyclohexylcarbodi-imide–*N*-hydroxysuccimide method is still favoured for cyclization in the antamanide field. The process has been successfully applied to linear peptide precursors, derived by an automated solid-phase procedure, affording a variety of analogues including the interesting all-D-retro-derivatives.[32, 33]

An earlier synthesis of cyclopentapeptides containing cysteine has been extended to provide further derivatives for conformational studies. The method involved a standard stepwise synthesis and cyclization of a linear peptide by means of the active ester.[70] Analogues containing alanine in place of cysteine were obtained by desulphurization of the cyclic peptide with Raney nickel.

A number of the cyclohexapeptides lised in Table 1 had been reported earlier. The remainder which are new compounds were synthesized by well-established procedures and it has not been considered necessary to deal with them in any depth.

As part of a systematic study on isomeric cyclohexapeptides containing Gly, Phe, and Leu a further nine isomers (Table 2) have been synthesized.[86] A uniform synthetic scheme, involving coupling of tripeptide fragments and utilizing intermediates from previous synthetic work, afforded benzyloxycarbonylhexapeptide esters which were converted into the corresponding hydrazides and cyclized by the azide method. For the series with the sequences *cyclo*-(Gly-Phe-Leu-Gly-Phe-Leu-) and *cyclo*-(Gly-Leu-Phe-Gly-Leu-Phe-), the isomers (-L,L,L,L) and (-D,L,D,L) have a two-fold rotational axis C_2 and both *meso*-forms (-D,L,L,D) and (-L,L,D,D) have a centre of symmetry, whereas for the series *cyclo*-(Gly-Phe-Leu-Gly-Leu-Phe-) the isomers (-D,D,L,L) and (-D,L,L,D) are cycloenantiomers and are particularly interesting because of their chiroptic properties. A cyclohexapeptide related to the ferrichrome group has also been synthesized by the azide method and showed activity against a number of micro-organisms.[87]

Perhaps the most notable synthetic [88] achievement of the year has been the synthesis of the homodetic *cyclo*-(L-Val-D-Pro-D-Val-L-Pro-)$_3$, which

[86] K. Bláha, *Coll. Czech. Chem. Comm.*, 1972, **37**, 1861.
[87] A. M. El Naggar, *Indian J. Chem.*, 1971, **9**, 1326.
[88] B. F. Gisin and R. B. Merrifield, *J. Amer. Chem. Soc.*, 1972, **94**, 6165.

was designed specifically as a hydrophobic potassium-binding peptide, on the basis of the known structures of alkali-ion complexing agents. The particular sequence was chosen since it may be considered as an analogue of valinomycin in which the L-lactic acid and the D-α-hydroxyisovaleric acid residues have been replaced by L- and D-proline respectively. The linear peptide was synthesized by the solid-phase method starting with the L-proline at the *C*-terminus. The considerable reduction in yield due to

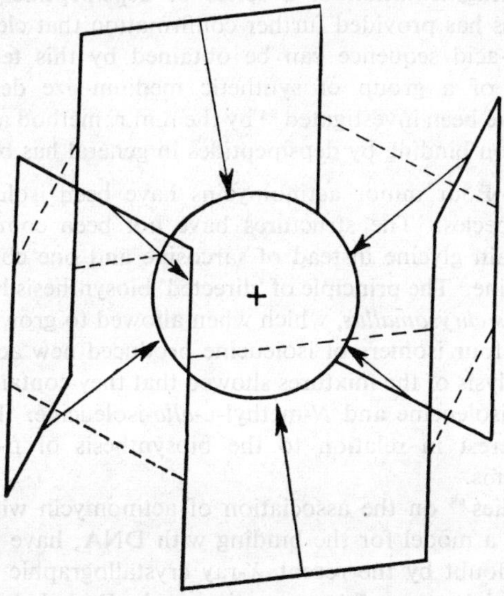

Figure 2 *The general architecture of potassium-binding cyclic peptides*

the peptide ester H-D-Val-L-Pro-resin undergoing intramolecular aminolysis to the corresponding cyclic dipeptide was overcome by a 'reversed' coupling procedure, *i.e.* adding DCC prior to the Boc-amino-acid.[89]

Similar low yields in solid-phase synthesis have also been observed by other workers.[90, 91] The rate of intramolecular aminolysis appears to depend on the composition and configuration of the dipeptide. The dodecapeptide was cleaved from the resin and cyclized with Woodward's reagent K and shown to bind potassium by the formation of a crystalline complex with potassium picrate, for which it possessed a seven-fold higher affinity than valinomycin. A diagrammatical representation of the complex is shown in Figure 2. The six amide carbonyl dipoles (arrows) complex with the potassium ion and the remaining six amides form hydrogen bonds (broken lines) to stabilize the six U-shaped loops of the backbone.

[89] B. F. Gisin and R. B. Merrifield, *J. Amer. Chem. Soc.*, 1972, **94**, 3102.
[90] M. Rothe and J. Mazanek, *Angew. Chem.*, 1972, **84**, 290.
[91] M. C. Khosla, R. R. Semby, and F. M. Bumpus, *J. Amer. Chem. Soc.*, 1972, **94**, 4721.

In accord with the policy adopted in previous years, synthetic studies relating to the cyclic hormone peptides have not been covered in this chapter.

3 Cyclic Depsipeptides (Heterodetic Peptides)

A general review of peptide antibiotics covering both homodetic and heterodetic structures has been published.[92] A detailed account of the mass spectrometric fragmentation of a series of depsipeptide antibiotics and their derivatives has provided further confirmation that clear-cut evidence for the amino-acid sequence can be obtained by this technique.[93] The conformations of a group of synthetic medium-size depsipeptide ring compounds have been investigated [94] by the n.m.r. method and the theoretical aspects of ion binding by depsipeptides in general has been discussed.[95]

Actinomycins.—Four minor actinomycins have been isolated [96] from an *Actinomyces* species. The structures have not been completely defined, but three contain glycine instead of sarcosine and one contains serine in place of threonine. The principle of 'directed' biosynthesis has been applied to *Streptomyces chrysomallus*, which when allowed to grow in the presence of each of the four isomers of isoleucine produced new actinomycin mixtures.[97] Hydrolysis of the mixtures showed that they contained appreciable amounts of D-isoleucine and *N*-methyl-L-*allo*-isoleucine. The observation is also of interest in relation to the biosynthesis of D-amino-acids in microbial systems.

Further studies [98] on the association of actinomycin with deoxyribodinucleotides, as a model for the binding with DNA, have been described, prompted no doubt by the recent *X*-ray crystallographic analysis on the complex formed between 7-bromoactinomycin D and deoxyguanosine.[99] In relation to this work, a detailed n.m.r. examination has confirmed that in D_2O solution actinomycin aggregates to form a dimer.[100] Moreover, the results conclusively demonstrate that the dimer is formed only by an interaction between the actinocyl chromophores.

The actinomycin D lactams (16) and (17), which are strictly homodetic peptides, have been synthesized [101, 102] as part of a programme to obtain

[92] W. Maier and D. Groeger, *Pharmazie*, 1972, **27**, 491.
[93] F. Compernolle, H. Vanderhaeghe, and G. Janssen, *Org. Mass Spectrometry*, 1972, **6**, 151.
[94] N. D. Abdullaev, L. I. Andreeva, V. K. Antonov, V. F. Bystrov, E. S. Efremov, and M. M. Shemyakin, *Zhur. org. Khim.*, 1972, **8**, 1853.
[95] S. V. Talekar and K. Sundaram, *Adv. Exp. Med. Biol.*, 1972, **21**, 9.
[96] V. Kuznetsova, T. I. Orlova, and A. B. Silaev, *Antibiotiki*, 1972, **17**, 322.
[97] T. Yajima, M. A. Grigg, and E. Katz, *Arch. Biochem. Biophys.*, 1972, **151**, 565.
[98] T. R. Krugh, *Proc. Nat. Acad. Sci. U.S.A.*, 1972, **69**, 1911.
[99] H. M. Sobell, S. C. Jain, T. D. Sakore, and C. E. Nordman, *Nature New Biol.*, 1971, **231**, 200.
[100] N. S. Angerman, T. A. Victor, C. L. Bell, and S. S. Danyluk, *Biochemistry*, 1972, **11**, 2402.
[101] J. Meienhofer and R. P. Patel, *Internat. J. Protein Res.*, 1971, **3**, 347.
[102] E. Atherton and J. Meienhofer, *J. Amer. Chem. Soc.*, 1972, **94**, 4759.

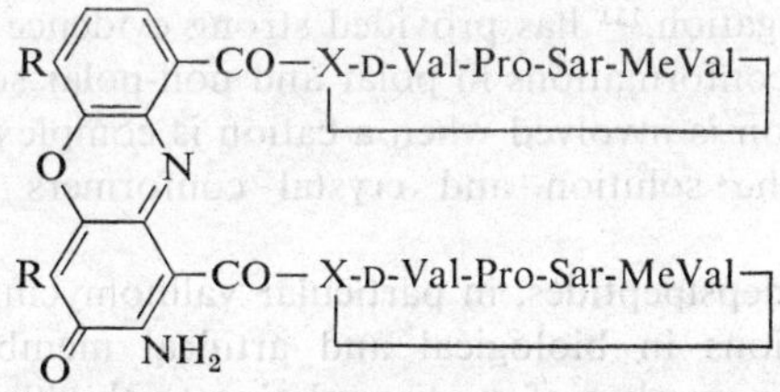

(16) R = Me, X = L-A_2pr
(17) R = Me, X = L-2,3-A_2bu
(18) R = Br, X = L-Thr

analogues with improved therapeutic properties. In these peptides both the threonine residues have been replaced by L-α,β-diaminopropionic acid and L-*threo*-α,β-diaminobutyric acid, respectively. The synthetic pathway involved a cyclization step between the proline and sarcosine residues followed by the introduction of the *N*-2-nitro-*m*-cresotyl substituent as the standard precursor of the chromophore. A further analogue lacking the four *N*-methyl groups in the peptide lactone rings has been synthesized,[103] but did not exhibit any antimicrobial or cytotoxic activity. The dibromo-actinomycin derivative (18) has been prepared [104] by standard procedures, as well as a number of other models, based on the phenoxazone chromophore.[105, 106]

Valinomycin.—The ion-complexing properties and conformation of this cyclic depsipeptide are still attracting considerable attention. It is now well established that the K^+ complex possesses the bracelet-like conformation (see Figure 2) with the six amide NH groups intramolecularly hydrogen-bonded and all the six ester carbonyls pointing inwards to give a six-fold co-ordination about the unhydrated potassium ion. Conformational energy calculations accord with this formulation as well as with the order of ion selectivity.[107] However, an *X*-ray crystallographic analysis [108] on valinomycin without a complexed cation shows a different conformation which also differs from the preferred conformer (form A) in non-polar solvents, as proposed on the basis of n.m.r. studies, although it was pointed out that these n.m.r. data were compatible with the crystal conformation. Earlier evidence had indicated that form A of valinomycin is in dynamic equilibrium with a second conformer (form B) which is favoured in polar solvents.

An abundance of recent publications, including an *X*-ray diffraction study,[109] ^{1}H n.m.r. and conformational energy calculations,[110] and a

[103] C. W. Mosher and L. Goodman, *J. Org. Chem.*, 1972, **37**, 2928.
[104] F. Seela, *J. Medicin. Chem.*, 1972, **15**, 684.
[105] M. T. Wu and R. E. Lyle, *J. Heterocyclic Chem.*, 1971, **8**, 989.
[106] V. A. Ivanov, E. N. Glibin, and O. F. Ginzburg, *Zhur. org. Khim.*, 1972, **8**, 1743.
[107] D. F. Mayers and D. W. Urry, *J. Amer. Chem. Soc.*, 1972, **94**, 77.
[108] W. L. Duax, H. Hauptman, C. M. Weeks, and D. A. Norton, *Science*, 1972, **176**, 911.
[109] R. Krigbaum, F. R. Kuegler, and H. Oelschlaeger, *Biochemistry*, 1972, **11**, 4548.
[110] D. J. Patel and A. E. Tonelli, *Biochemistry*, 1973, **12**, 486.

^{13}C n.m.r. investigation,[111] has provided strong evidence that valinomycin adopts different conformations in polar and non-polar solvents and that a third conformation is involved when a cation is complexed. It also seems probable that the solution and crystal conformers are substantially different.[110]

The effects of depsipeptides, in particular valinomycin, on the transport of alkali-metal ions in biological and artifical membranes have been reviewed [112] and a number of studies relating to this field reported.[113–115]

Other Cyclic Depsipeptides.—The conformation of stendomycin in trifluoroethanol has been studied [116] by standard n.m.r. techniques (see Section 2). The results suggest a left-handed α-helical segment for the series of D-amino-acids in the linear side-chain and β-turn structures for the lactone ring. The ambiguity concerning the structure of the metabolite isariin has been resolved by a total synthesis and the sequence shown in (19) is now

D-Me·$(CH_2)_8$CH·CH_2·CO—Gly-Val-D-Leu-Ala-Val—O (ring closed to the CH)

(19)

established for this peptide.[117] Further biosynthetic studies [118] on the cyclic depsipeptides in *Serratia marcescens*, of which serratamolide is the major component, have established that 3-hydroxydecanoic acid is not utilized directly for the biosynthesis.

Finally, an attempt to clarify the nomenclature of the virginiamycin group of depsipeptides, with respect to other known members of this class, has been recorded.[119]

4 Peptide Alkaloids

A further review dealing with these unusual cyclic peptides has been published,[120] and this year has seen the isolation and characterization of several new alkaloids.

The stereochemistry of the β-hydroxyleucine from the alkaloids frangulanine (20) and discarine A (21) and B (22), the major alkaloids of *Discaria longispina*, has been established as that of the L-*erythro*-amino-

111 D. J. Patel, *Biochemistry*, 1973, **12**, 496.
112 M. Hoefer, *Umshau*, 1972, **72**, 591.
113 G. Stark, R. Benz, G. W. Pohl, and K. Janko, *Biochim. Biophys. Acta*, 1972, **266**, 603.
114 G. Kemp and C. E. Wenner, *Biochim. Biophys. Acta*, 1972, **282**, 1.
115 E. Grell, F. Eggers, and Th. Funck, *Chimia* (*Switz.*), 1972, **26**, 632.
116 T. P. Pitner and D. W. Urry, *Biochemistry*, 1972, **11**, 4132.
117 K. Okada, Y. Kurosawa, and M. Hiramoto, *Tetrahedron Letters*, 1972, 2693.
118 M. A. C. Bermingham, B. S. Deol, and J. L. Still, *Biochem. J.*, 1972, **127**, 45.
119 P. Crooy and R. De Neys, *J. Antibiotics* (*Japan*), 1972, **25**, 371.
120 V. A. Snieckus, in 'The Alkaloids', ed. J. E. Saxton (Specialist Periodical Reports), The Chemical Society, London, 1972, vol. 2, p. 271.

acid.[121] Amphibine A isolated from the bark of *Zizyphus amphibia* was independently assigned [122] the same gross structure (21) and later stated [123] to be identical with discarine A. The bark of this plant also yielded [123] four other new peptide alkaloids, amphibines B—E, (23)—(26), all of

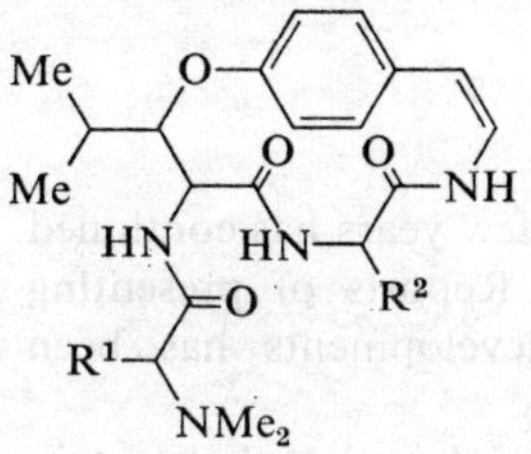

(20) R^1 = Bu^s, R^2 = Bu^i
(21) R^1 = β-indolylmethyl, R^2 = Bu^s
(22) R^1 = Bu^s, R^2 = β-indolylmethyl

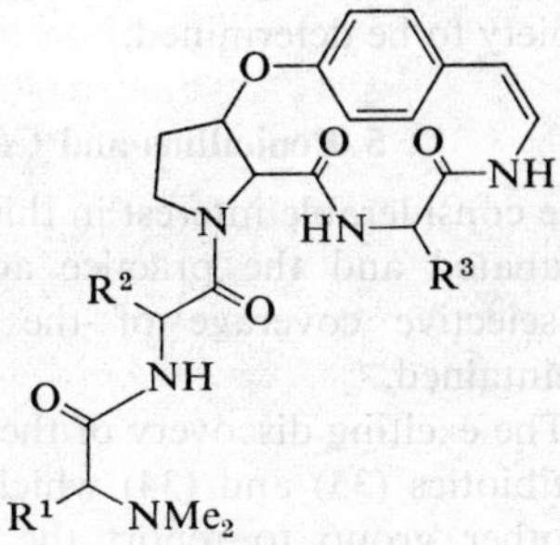

(23) R^1 = Bz, R^2 = Bu^s, R^3 = Bz
(24) R^1 = Bu^i, R^2 = Bu^s, R^3 = Bz
(25) R^1 = Bz, R^2 = Bu^s, R^3 = Bu^s
(26) R^1 = Bu^i, R^2 = β-indolylmethyl, R^3 = Bu^s
(27) R^1 = Me, R^2 = Pr^i, R^3 = Bz
(28) R^1 = Bu^s, R^2 = Pr^i, R^3 = Bz
(29) R^1 = Bz, R^2 = Bu^i, R^3 = Bu^s

which contain *trans*-3-hydroxyproline, not previously encountered in plant peptides. Two further alkaloids mauritine A (27) and B (28), obtained [124] from the African tree *Zizyphus mauritiana*, also belong to this general class of compounds, as does mucronine D (29), the major alkaloid of *Zizyphus mucronata*.[125]

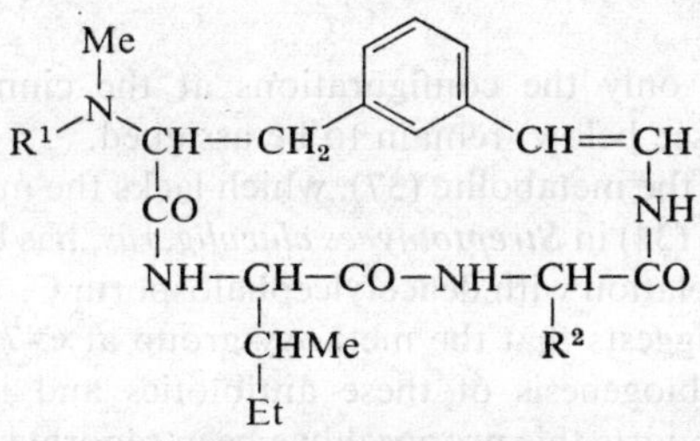

(30) R^1 = Me, R^2 = Bzl
(31) R^1 = H, R^2 = Bzl
(32) R^1 = Me, R^2 = Bu^i

[121] M. G. Sierra, O. A. Mascaretti, F. J. Diaz, E. A. Rúveda, C. J. Chang, E. W. Hagaman, and E. Nenkert, *J.C.S. Chem. Comm.*, 1972, 915.
[122] R. Tschesche, E. U. Kaussmann, and H. W. Fehlhaber, *Tetrahedron Letters*, 1972, 865.
[123] R. Tschesche, E. U. Kaussmann, and H. W. Fehlhaber, *Chem. Ber.*, 1972, **105**, 3094.
[124] R. Tschesche, H. Wilhelm, and H. W. Fehlhaber, *Tetrahedron Letters*, 1972, 2609.
[125] R. Tschesche, S. T. David, J. Uhlendorf, and H. W. Fehlhaber, *Chem. Ber.*, 1972, **105**, 3106.

The latter plant has afforded [126] an entirely new family of cyclic peptides, namely mucronine A, B, and C. The structures (30)—(32), respectively, have been assigned to these compounds by a combination of chemical transformations and degradations together with spectral studies. The mass spectrometric fragmentation patterns allowed the sequence of the peptide moiety to be determined.

5 Penicillins and Cephalosporins

The considerable interest in this field over the past few years has continued unabated and the practice adopted in previous Reports of presenting a selective coverage of the more important developments has been maintained.

The exciting discovery of the novel 7-methoxy-substituted cephalosporin antibiotics (33) and (34) which was described last year [127] has prompted another group to report the isolation of (34) along with the unusual α-methoxy-*p*-hydroxycinnamic acid derivatives (35) and (36) from *Streptomyces griseus*.[128] The structures are based on detailed interpretation of

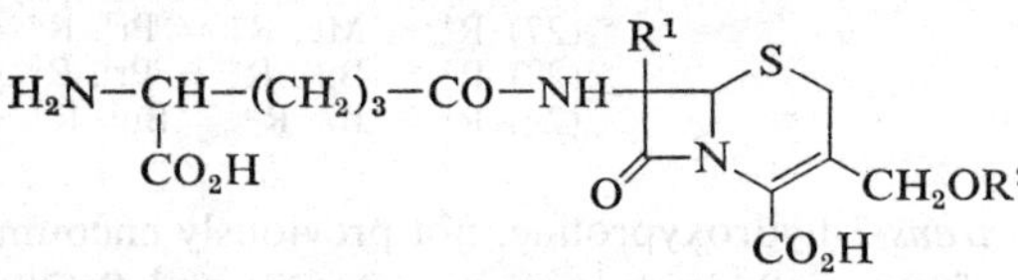

(33) R^1 = OMe, R^2 = COMe
(34) R^1 = OMe, R^2 = $CONH_2$
(35) R^1 = OMe, R^2 = $COC(OMe){=}C_6H_4{\cdot}OSO_3H$-*p*
(36) R^1 = OMe, R^2 = $COC(OMe){=}CH{\cdot}C_6H_4{\cdot}OH$-*p*
(37) R^1 = H, R^2 = $CONH_2$

spectral data and only the configurations at the cinnamic acid double bond and at C-7 (see below) remain to be assigned.

The structure of the metabolite (37), which lacks the methoxy-substituent and co-occurs with (34) in *Streptomyces clavuligerus*, has been established [129] by a chemical correlation with deacetylcephalosporin C. The co-occurrence of (34) and (37) suggests that the methoxy-group at C-7 is introduced at a late stage in the biogenesis of these antibiotics and initial biosynthetic evidence which supports this proposal has been reported.[130]

The excellent antibacterial activity of (33)—(36), together with the earlier prediction that an α-methyl group at C-6 of the penicillin nucleus would

[126] H. W. Fehlhaber, J. Uhlendorf, S. T. David, and R. Tschesche, *Annalen*, 1972, **759**, 195.
[127] R. Nagarajan, L. D. Boeck, M. Gorman, R. L. Hamill, C. E. Higgens, M. M. Hoehn, W. M. Stark, and J. G. Whitney, *J. Amer. Chem. Soc.*, 1971, **93**, 2308.
[128] G. Albers-Schönberg, B. H. Arison, and J. L. Smith, *Tetrahedron Letters*, 1972, 2911.
[129] C. F. Murphy, R. E. Koehler, and J. A. Webber, *Tetrahedron Letters*, 1972, 1585.
[130] J. G. Whitney, D. R. Brannon, J. A. Mabe, and K. J. Wicker, *Antimicrob. Agents Chemother.*, 1972, 247.

enhance activity,[131] has led to the development of synthetic routes to penicillin and cephalosporin derivatives substituted in positions 6 and 7, respectively.

A general method for the stereospecific introduction of a methoxy-group adjacent to the β-lactam group in both penicillin and cephalosporin has been described [132] and is outlined for cephalosporin in the Scheme. In a

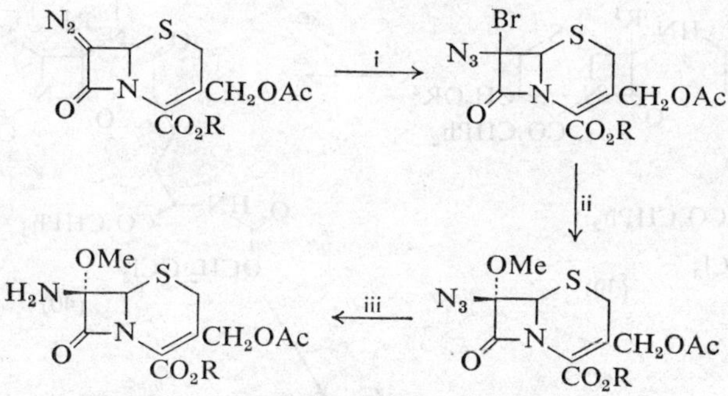

Reagents: i, BrN_3–Et_3N; ii, $AgBF_4$–MeOH; iii, H_2–Pd/C

Scheme

footnote to this paper it is claimed that the method has been successfully employed for the synthesis of (33), thus establishing the 7α configuration for this antibiotic. The sterically controlled alkylation of the anions represented by (38) affords the α-alkyl derivatives of both series,[133, 134]

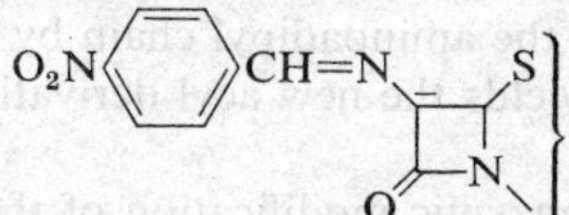

(38)

whereas treatment of the penicillin anion with oxygen gives the 6,6′-dimer.[135]

Interest continues in acyl exchange reactions for the preparation of semi-synthetic antibiotics of both series. Further work on the now well-established iminoether method for the penicillin exchange [136] and a new

[131] J. L. Strominger and D. J. Tipper, *Amer. J. Med.*, 1965, **39**, 708.

[132] L. D. Cama, W. J. Leanza, T. R. Beattie, and B. G. Christensen, *J. Amer. Chem. Soc.*, 1972, **94**, 1408.

[133] R. A. Firestone, N. Schelechow, D. B. R. Johnston, and B. G. Christensen, *Tetrahedron Letters*, 1972, 375.

[134] W. A. Spitzer, T. Goodson, R. J. Smithey, and I. G. Wright, *J.C.S. Chem. Comm.*, 1972, 1138.

[135] R. A. Firestone, N. Schelechow, and B. G. Christensen, *J.C.S. Chem. Comm.*, 1972, 1106.

[136] I. Isaka, T. Kashiwagi, K. Nakano, N. Kawahara, A. Koda, Y. Numasaki, S. Kawahara, and M. Murakami, *J. Pharm. Soc. Japan*, 1972, **94**, 454.

general method for cephalosporin derivatives [137] which does not require the formation of 7-aminocephalosporin intermediates has been described. The reaction of (39) with an acid chloride and trimethylsilyltrifluoroacetamide produces the diacyl derivative (40) in good yield. Neither double-bond isomerization nor acylation of urethane functions occurs under the

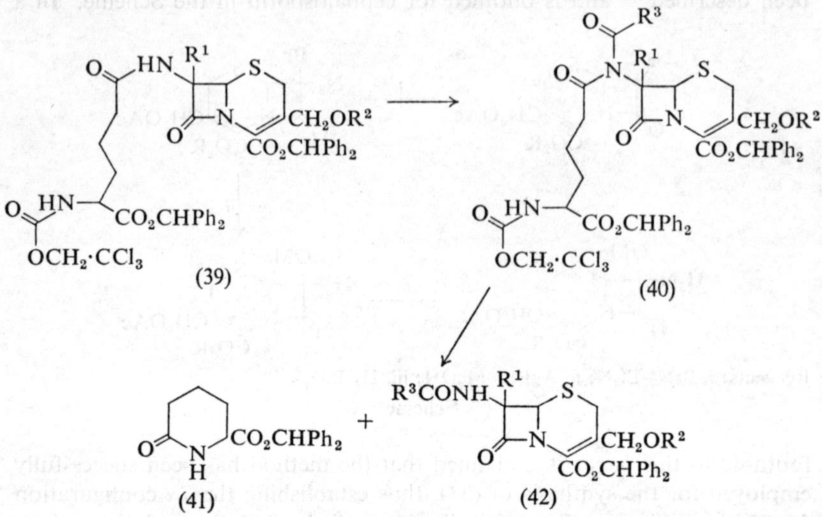

conditions and the method appears to be a mild general procedure for the acylation of amides. The removal of the nitrogen protecting group initiates the selective cleavage of the aminoadipyl chain by spontaneous cyclization to the lactam (41) and yields the new acyl derivatives (42) in good overall yield.

No methods for the enzymic modification of the D-α-aminoadipyl side-chain of cephalosporin were previously known, but it is now possible to deaminate using D-amino-oxidase.[138]

The chemical relationships between cephalosporins and penicillins continue to attract considerable attention and a summary of the initial work has been published.[139] The sulphenic acid intermediates produced by thermal rearrangement from penicillin sulphoxides can be trapped with vinyl ethers or with keten acetals,[140] and penicillin sulphoxides are transformed into cephalosporins in the presence of azodicarboxylates, presumably through the intermediacy of sulphenyl derivatives.[141] Oxid-

137 S. Karady, S. H. Pines, L. M. Weinstock, F. E. Roberts, G. S. Brenner, A. M. Hoinowski, T. Y. Cheng, and M. Sletzinger, *J. Amer. Chem. Soc.*, 1972, **94**, 1410.
138 P. Mazzeo and A. Romeo, *J.C.S. Perkin I*, 1972, 2532.
139 D. H. R. Barton and P. G. Sammes, *Proc. Roy. Soc.*, 1971, **B179**, 345.
140 I. Ager, D. H. R. Barton, G. Lucente, and P. G. Sammes, *J.C.S. Chem. Comm.*, 1972, 601.
141 S. Terao, T. Matsuo, S. Tsushima, N. Matsumoto, T. Miyawaki, and M. Miyamoto, *J.C.S. Chem. Comm.*, 1972, 1304.

ation with ozone offers an alternative method for preparing epimeric mixtures of penicillin sulphoxides in high yield.[142]

Two groups employing different routes have reported the first structural conversions of cepham systems into penam systems. Silver acetate treatment of (43), itself prepared by ring-expansion of the penicillin sulphoxide with thionyl chloride, forms equal quantities of (44) and the ring-contracted

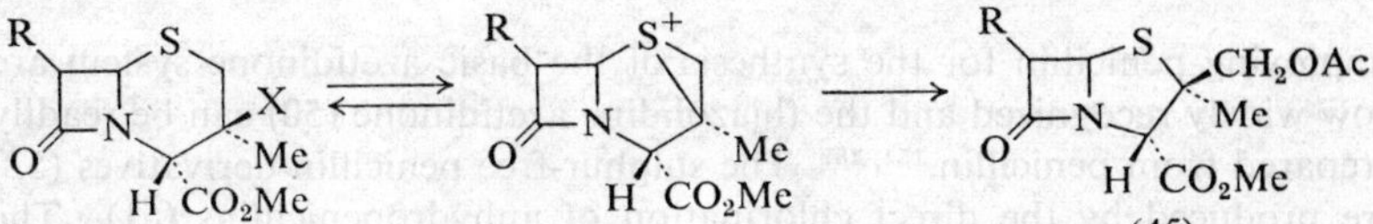

(45)

(43) X = Cl R = phthalimido
(44) X = OAc

product (45). The nucleophilic displacement of the 3-chloro-substituent proceeds with retention of configuration in the formation of (44) but with inversion for (45), and it is presumed that both compounds are derived from the common thiiranium ion intermediate.[143] The clectrophilic addition of ethoxycarbonylcarbene to (46) gives the penam (47) in reasonable yield.[144] In this case it is suggested that the reaction proceeds through the intermediacy of a sulphonium ylide which then undergoes a [2,3]sigmatropic rearrangement to (47). The reaction of a penicillin with ethoxycarbonylcarbene leads

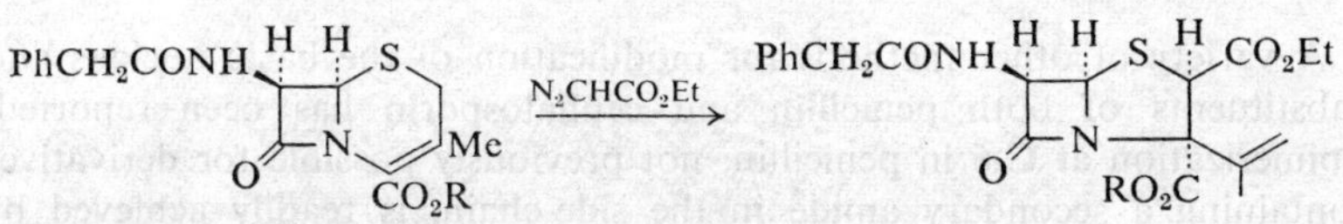

(46) (47)

to the cleavage of the 1,2 bond, again *via* a sulphonium ylide intermediate by a process analogous to the sulphoxide rearrangements.[145]

The full details of the preparation from penicillin of (48), the key intermediate in the total synthesis of cephalosporin C, have been published,[146] thus providing a further chemical correlation between the two series. The intermediate (49) and a number of its derivatives, described in the conversion, have subsequently been used for the production of new semisynthetic β-lactam antibiotics.[147–150] The obvious economic advantages of

[142] D. O. Spry, *J. Org. Chem.*, 1972, **37**, 793.
[143] S. Kukolja and S. R. Lammert, *J. Amer. Chem. Soc.*, 1972, **94**, 7169.
[144] M. Yoshimoto, S. Ishihara, E. Nakayama, and N. Soma, *Tetrahedron Letters*, 1972, 2923.
[145] M. Yoshimoto, S. Ishihara, E. Nakayama, E. Shoji, H. Kuwano, and N. Soma, *Tetrahedron Letters*, 1972, 4387.
[146] K. Heusler, *Helv. Chim. Acta*, 1972, **55**, 388.
[147] R. Scartazzini, H. Peter, H. Bickel, K. Heusler, and R. B. Woodward, *Helv. Chim. Acta*, 1972, **55**, 408.
[148] B. Fechtig, H. Bickel, and K. Heusler, *Helv. Chim. Acta*, 1972, **55**, 417.
[149] R. Scartazzini and H. Bickel, *Helv. Chim. Acta*, 1972, **55**, 423.
[150] R. Scartazzini, J. Gosteli, H. Bickel, and R. B. Woodward, *Helv. Chim. Acta*, 1972, **55**, 2568.

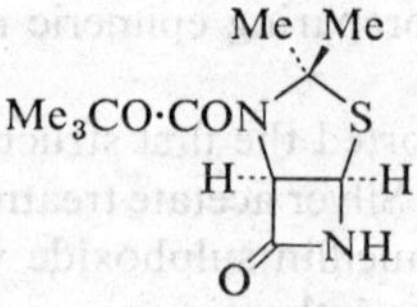

(48)

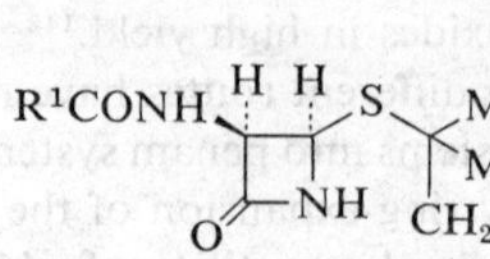

(49)

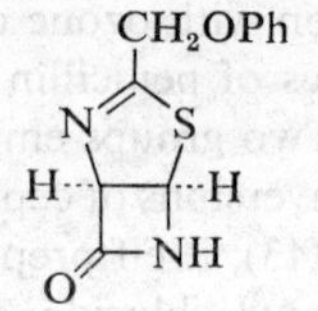

(50)

employing penicillin for the synthesis of the basic azetidinone system are now widely recognized and the thiazolidine azetidinone (50) can be readily prepared from penicillin.[151, 152] The sulphur-free penicillin derivatives (52) are produced by the direct chlorination of anhydropenicillins (51). The reaction is general and proceeds with predominant retention of configuration at C-5. Subsequent reactions of (52) have led to a number of novel bicyclic systems lacking sulphur of which (53) and (54) are examples.[153]

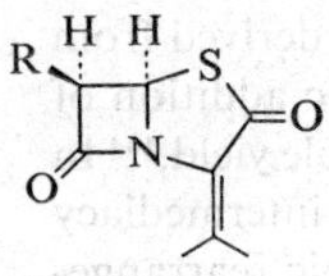

(51)

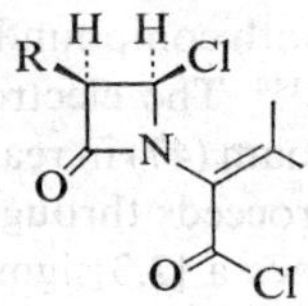

(52)

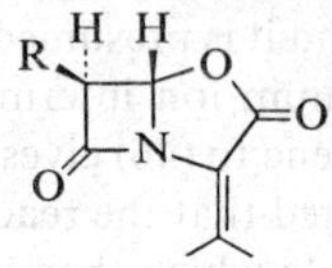

(53)

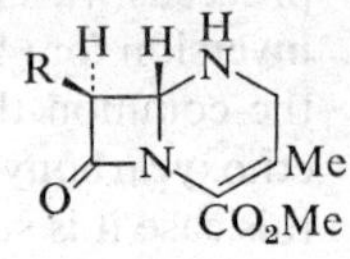

(54)

A variety of other methods for modification of the basic nucleus and substituents of both penicillin and cephalosporin has been reported. Epimerization at C-6 in penicillin, not previously possible for derivatives containing a secondary amide in the side-chain, is readily achieved by silylating the amide function before treatment with base. This method can also be used for derivatives with free acid groups.[154] The 6-hydrazonopenicillanate derivative (55) results from the acylation of the triphenylphosphine complex of 6-azopenicillanate. Stereoselective reduction of (55) to the acylhydrazine (56) was achieved under carefully controlled conditions.[155] A new carboxylic acid protecting group with possible general

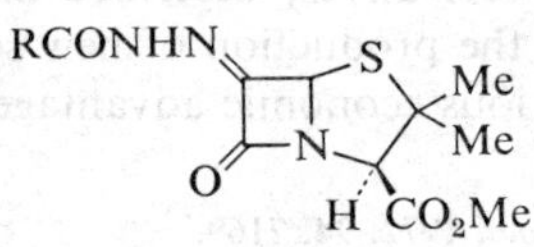

(55)

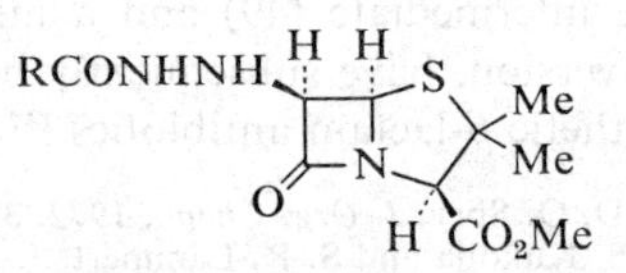

(56)

[151] E. G. Brain, A. J. Eglington, J. H. C. Nayler, M. J. Pearson, and R. Southgate, *J.C.S. Chem. Comm.*, 1972, 229.

[152] R. D. G. Cooper and F. L. José, *J. Amer. Chem. Soc.*, 1972, **94**, 1022.

[153] S. Wolfe, W. S. Lee, G. Kannengiesser, and J. B. Ducep, *Canad. J. Chem.*, 1972, **50**, 2894, 2898, 2902.

[154] A. Vlietinck, E. Roets, P. Claes, and H. Vanderhaeghe, *Tetrahedron Letters*, 1972, 285.

[155] D. M. Brunwin and G. Lowe, *J.C.S. Chem. Comm.*, 1972, 192.

applications has been designed specifically for penicillin work. *NN'*-Di-isopropylhydrazine reacts with carboxylic acids to give the corresponding monoacylhydrazide. This is stable to both acidic and basic conditions but the carboxylic acid function can be released in high yield by selective oxidation with lead tetra-acetate.[156]

An alkoxy-substituent can be introduced directly into the C-2 position of cephalosporins in moderate yield using a 1 : 1 mixture of chlorine and an alcohol.[157] In all the cases investigated only one of the two possible isomers was obtained and was assigned the α-configuration. An alternative route involves the initial bromination at C-2 with *N*-bromosuccinimide and subsequent replacement of the bromide with methanol.[158] In this case the stereochemistry at C-2 of the resultant methoxy-derivative was not assigned. Cephalosporanic acid sodium salts (57) are reduced to the 3-methylenecepham derivatives (58) with chromium(II) acetate. It was

(57) (58)

shown that the 4-carboxy-group in the product possessed the α-configuration and that reduction occurred only in aqueous media, suggesting the intermediacy of a carbonium ion.[159]

The total synthesis of β-lactam systems continues to present a challenging synthetic objective and a number of new approaches have been reported. A new general route to β-lactams which involves the condensation of an acid chloride with a thioformimidate has been developed.[160, 161] The reaction of (59) with phthaloylglycl chloride affords predominantly the *trans*-isomer (60). The same reaction using chloroacetyl chloride yields (61), which undergoes nucleophilic substitution at the 3-position with

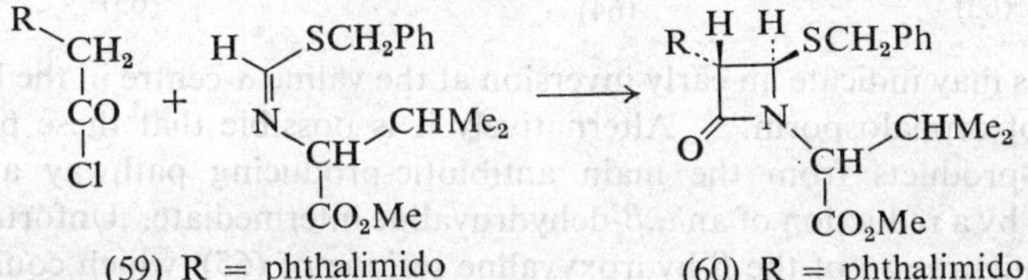

(59) R = phthalimido (60) R = phthalimido
(61) R = Cl

[156] D. H. R. Barton, M. Girijavallabhan, and P. G. Sammes, *J.C.S. Perkin I*, 1972, 929.
[157] D. O. Spry, *Tetrahedron Letters*, 1972, 3717.
[158] E. H. W. Bohme and J. E. Dolfini, *J.C.S. Chem. Comm.*, 1972, 941.
[159] M. Ochiai, O. Aki, A. Morimoto, T. Okada, and H. Shimadzu, *J.C.S. Chem. Comm.*, 1972, 800.
[160] M. D. Bachi and O. Goldberg, *J.C.S. Perkin I*, 1972, 2332.
[161] M. D. Bachi and M. Rothfield, *J.C.S. Perkin I*, 1972, 2326.

azide ions to give the azido-derivative possessing the correct relative stereochemistry with relation to the penicillin nucleus.[162] An alternative synthesis of the lactone (62), obtained as a single isomeric species in high yield, has been claimed [163] and since (62) had previously been converted

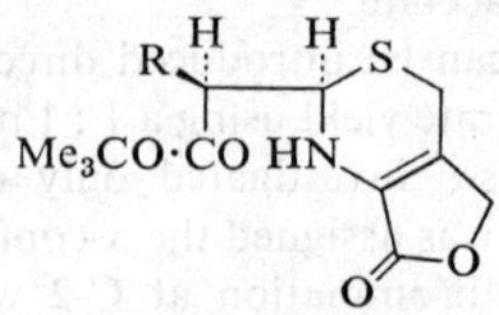

(62)

into cephalosporin antibiotics it constitutes a formal total synthesis. Nuclear analogues of 7-methylcephalosporins have been synthesized by an extension of the diazo method.[164] However, further attempts to obtain penicillins by transannular cyclization of thiazepines were unsuccessful.[165]

The molecular structure of phenoxymethylanhydropenicillin, which has been determined by *X*-ray diffraction techniques,[166] reveals that the β-lactam nitrogen is largely pyramidal in character and it can therefore be assumed that the lone pair is not delocalized to any great extent. Hence the chemical stability of anhydropenicillins to hydrolysis of the β-lactam, relative to the penicillins, cannot be ascribed to extensive delocalization. Application of extended Hückel molecular orbital calculations to the cepham and penam systems affords electronic energy levels and charge distributions which have a general qualitative agreement with the experimental spectral properties, but quantitative comparisons are, as expected, mediocre.[167]

Three intracellular peptides (63)—(65) have been found in small amounts in a *Cephalosporium* species and have also been shown to be present in a strain giving a high yield of cephalosporin C.[168] The presence of the D-valine residue in (63) and (64) is of considerable interest and it has been suggested

D-Aad-L-Cys-D-Val (63)

D-Aad-L-Cys-D-Val-Gly (64)

D-Aad-L-Cys-β-hydroxylVal-Gly (65)

that this may indicate an early inversion at the valine α-centre in the biosynthesis of cephalosporin.[169] Alternatively it is possible that these peptides are by-products from the main antibiotic-producing pathway and are formed by a reduction of an α,β-dehydrovaline intermediate. Unfortunately the configuration of the β-hydroxyvaline residue of (65), which could have been of interest in this respect, was not determined.

162 M. D. Bachi and O. Goldberg, *J.C.S. Chem. Comm.*, 1972, 319.
163 N. N. Girotra and N. L. Wendler, *Tetrahedron Letters*, 1972, 5301.
164 D. M. Brunwin and G. Lowe, *J.C.S. Chem. Comm.*, 1972, 589.
165 M. H. Benn and R. E. Mitchell, *Canad. J. Chem.*, 1972, **50**, 2195.
166 G. L. Simon, R. B. Morin, and L. F. Dahl, *J. Amer. Chem. Soc.*, 1972, **94**, 8557.
167 D. B. Boyd, *J. Amer. Chem. Soc.*, 1972, **94**, 6513.
168 P. B. Loder and E. P. Abraham, *Biochem. J.*, 1971, **123**, 471.
169 P. B. Loder and E. P. Abraham, *Biochem. J.*, 1971, **123**, 477.

A further speculative mechanistic pathway for the biogenesis of cephalosporin and penicillin, involving a dehydrovaline intermediate, has been proposed.[170]

6 Peptides Linked to Carbohydrates

This is an important biological field which also encompasses many aspects of carbohydrate chemistry and in order to avoid digressing too far from the central theme of this chapter, the emphasis has remained on the peptide constituents; in particular those containing unusual amino-acids or unusual linkages to carbohydrates are discussed. Other areas are covered by quoting review articles or prominent papers published over the year.

Glycopeptides from Bacterial Cell Walls.—The chemistry and biosynthesis of staphylococcal cell walls have been reviewed [171] and the general aspects relating to the biosynthesis of peptidoglycans of bacterial cell walls discussed.[172]

The peptidoglycan from spores of *Bacillus subtilis* afforded [173] on hydrolysis the disaccharide-tripeptide (66). Conclusions, based on (66) and

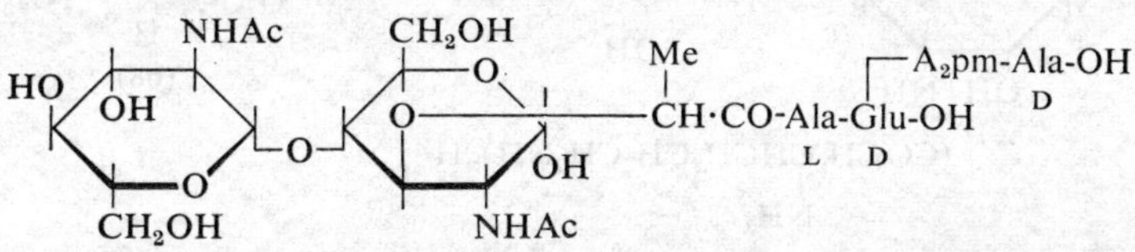

(66)

a considerable amount of other experimental evidence, suggested that the glycan component consists of long chains of alternating *N*-acetylglucosamine and muramic acid, probably all β-(1 $\rightarrow$ 4)-linked, with half the muramic acid residues present as muramic δ-lactam. The remainder of the residues are *N*-acetylated and their side-chain carboxy-groups substituted with either a peptide or L-alanine. This differs considerably from the peptidoglycan of vegetative cells.

The structure of the glycopeptide of the filamentous forms of *Proteus* P_{18} obtained with weak doses of penicillin has been shown to be identical to that of the non-treated bacteria.[174] On the other hand, the incorporation of ^{14}C-labelled diaminopimelic acid into peptidoglycan by a cell-free *Bacillus megaterium* was inhibited by penicillins, but destruction of the antibiotic with penicillinase reversed the inhibition.[175] *In vitro* studies on penicillin-sensitive transpeptidation of peptidoglycan to the cell walls of

[170] R. D. G. Cooper, *J. Amer. Chem. Soc.*, 1972, **94**, 1018.
[171] A. R. Archibald, in 'Staphylococci', ed. J. O. Cohen, Interscience, New York, 1972, p. 75.
[172] A. Rais and A. Klein, *Postepy Biochem.*, 1972, **18**, 241.
[173] A. D. Warth and J. L. Strominger, *Biochemistry*, 1972, **11**, 1389.
[174] M. Mock, J. Fleck, and J. P. Martin, *Compt. rend.*, 1972, **275**, *D*, 611.
[175] G. Gary and J. L. Strominger, *J. Biol. Chem.*, 1972, **247**, 7.

Micrococcus luteus showed[176] that incorporation of newly synthesized strands of peptidoglycan and their attachment to older cell-wall peptidoglycan proceeds mainly by transpeptidation, and that transglycosylation is responsible only for part of the elongation of the pre-existing peptidoglycan.

Further work on the structures of the cell walls of *Micrococcus lysodeikticus*,[177] as well as of the peptidoglycan from *Penicillium aeruginosa*,[178] have been reported.

Various aspects of the structure and function of phytotoxic glycopeptides from plant pathogens have been reviewed.[179]

Glycopeptide Antibiotics.—The streptothricin group (67) of antibiotics[180] has been the subject of a number of publications. The structure and chirality of the basic guanidine amino-acid streptolidine (68), which is common to

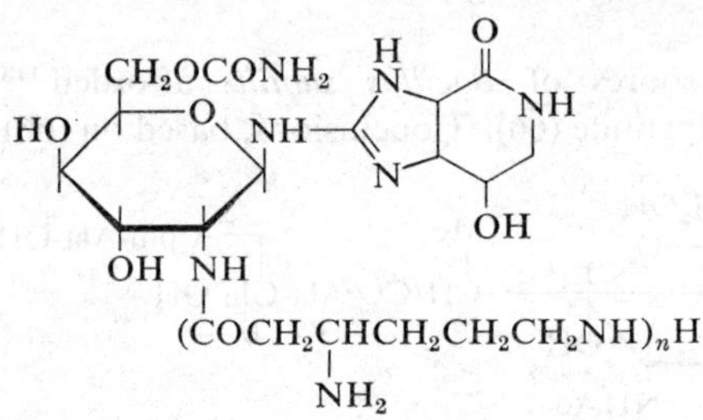

(67)

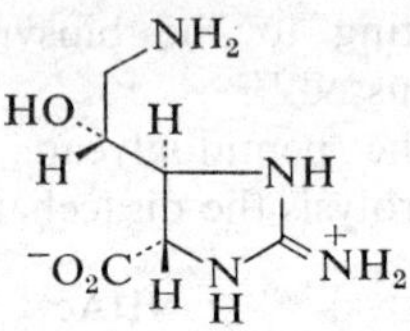

(68)

the whole group, have been established[181] beyond doubt by an *X*-ray crystallographic analysis. The mass spectrometric fragmentations of peptides containing β-lysine have been described[182] and an o.r.d. study on streptothricin has been reported.[183] Several other aspects of the chemistry of β-lysine peptides which only occur in the streptothricin and viomycin groups of antibiotics have been investigated.[184–186]

A new metabolite, prumycin, from a *Streptomyces* strain has been identified[187] as (69). Acid hydrolysis afforded only D-alanine; the 2,5-

[176] D. Mirelman, R. Brach, and N. Sharon, *Proc. Nat. Acad. Sci. U.S.A.*, 1972, **69**, 3355.
[177] Y. Imanaga and J. T. Park, *Biochemistry*, 1972, **11**, 4006.
[178] H. D. Heilmann, *European J. Biochem.*, 1972, **31**, 456.
[179] G. A. Strobel, in 'Proceedings of the Symposium on the Biochemistry of the Glycosidic Linkage', ed. R. Piras, Academic Press, New York, 1972, p. 423.
[180] K. I. Shutova and A. S. Khoklov, *Doklady Akad. Nauk S.S.S.R.*, 1972, **205**, 1119.
[181] B. W. Bycroft and T. J. King, *J.C.S. Chem. Comm.*, 1972, 652.
[182] L. I. Rostovtseva and A. A. Kiryushkin, *Org. Mass Spectrometry*, 1972, **6**, 1.
[183] H. Taniyama and Y. Sawada, *Chem. and Pharm. Bull.* (*Japan*), 1972, **20**, 596.
[184] H. Taniyama, Y. Sawada, K. Miyazeki, and F. Miyoshi, *Chem. and Pharm. Bull.* (*Japan*), 1972, **20**, 601.
[185] H. Tamiyama, Y. Sawada, K. Miyazeki, and F. Miyoshi, *Chem. and Pharm. Bull.* (*Japan*), 1972, **20**, 603.
[186] H. Taniyama, Y. Sawada, K. Miyazeki, S. Tanaka, F. Miyoshi, and K. Hiraoka, *Chem. and Pharm. Bull.* (*Japan*), 1972, **20**, 1432.
[187] S. Omura, M. Tishler, M. Katagiri, and T. Hatz, *J.C.S. Chem. Comm.*, 1972, 633.

diaminopentose, like 5-amino-sugars, is unstable under these conditions. The glycopeptide ristocetin A possesses a molecular weight of several thousands and has been shown to give on hydrolysis a number of monosaccharides and several unidentified amino-acids. The interesting biological activity of this compound makes its structure still a stimulating challenge.[188]

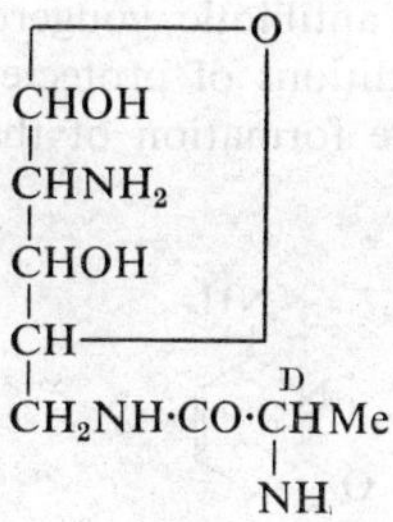

(69)

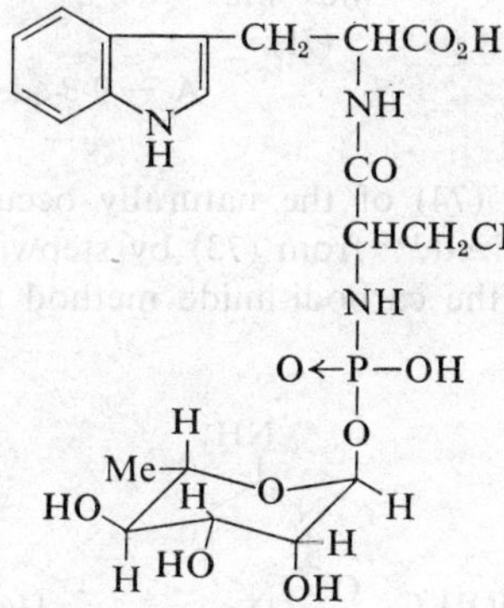

(70)

Phosphoramidon, a novel metabolite isolated from *Streptomyces tanashiensis*, has been assigned [189] structure (70). On mild acid hydrolysis it afforded L-leucyl-L-trytophan and L-rhamnopyranose.

7 Peptides Linked to Nucleosides and Nucleotides

A review in Russian on the structure and function of nucleotide peptides has appeared.[190]

The main emphasis in this area over the year has been on synthesis of peptides linked to nucleosides. A series of 2′- and 3′-*O*-peptidyl derivatives of adenosine has been prepared [191] as potential substrates for investigating the formation of the peptide bond on ribosomes; 2′- or 3′-*O*-aminoacyladenosines react directly with the 5-chloro-8-hydroxyquinoline esters of a number of protected dipeptides to give the corresponding peptidyl derivatives in excellent yield and this may represent a useful acylation reaction of peptidyl or *N*-acylaminoacyl oligoribonucleotides.

The benzyl esters of several amino-acids and peptides have been successfully coupled with the protected ribofuranuronic acid (71) by the carbodiimide–*N*-hydroxysuccinimide method. On removing the protecting groups the nucleoside peptides (72) were obtained in good yield.[192]

[188] J. R. Fehlner, R. E. J. Hutchinson, D. S. Tarbell, and J. R. Schenck, *Proc. Nat. Acad. Sci. U.S.A.*, 1972, **69**, 2420.

[189] S. Umezawa, K. Tatsuta, O. Izawa, and T. Tsuchiya, *Tetrahedron Letters*, 1972, 97.

[190] Yu. B. Filippovich, G. A. Sevastyanova, and S. N. Krasnova, *Uch. Zap. Mosk. Gos. Pedagog. Inst.*, 1970, **369**, 48.

[191] S. Chládek, *J. Org. Chem.*, 1972, **37**, 2863.

[192] M. Kawana, R. J. Rosseau, and R. K. Robins, *J. Org. Chem.*, 1972, **37**, 288.

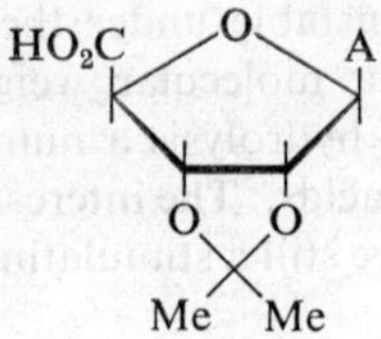

(71)

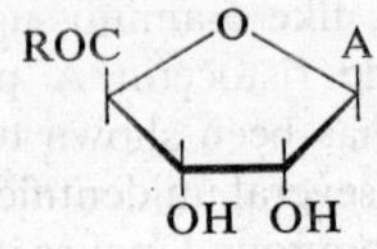

(72) R = Gly-Phe-OH

A = 9-β-adenyl

Analogues (74) of the naturally occurring antibiotic gougerotin have been synthesized[193] from (73) by stepwise additions of protected amino-acids, using the carbodi-imide method for the formation of the peptide bond.

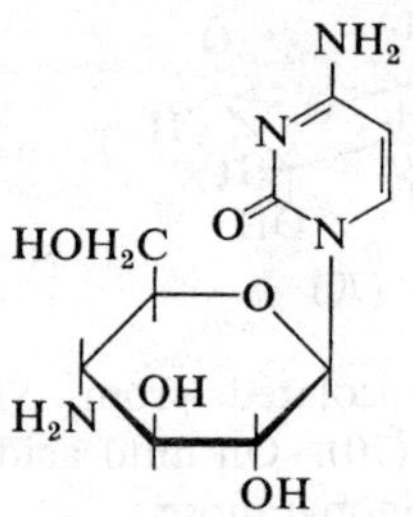

(73)

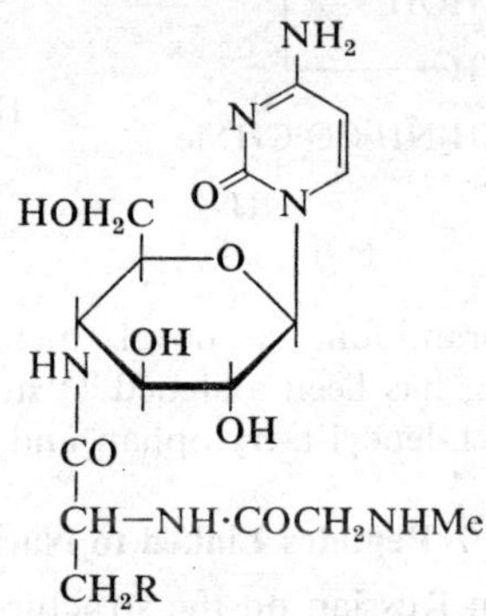

(74)

8 Other Peptides Containing Unusual Structural Features

This section is inevitably a miscellany covering a wide range of compounds, but over the year a number of naturally occurring peptide derivatives with interesting biological activity have been characterized and can conveniently be grouped under this heading. The same arguments are also applicable to a number of synthetic studies.

Phosphinothricin (75), which is active against both Gram-positive and Gram-negative bacteria, is a further example of a phosphorus-containing antibiotic. Total hydrolysis of (75) afforded L-alanine and 2-amino-4-

$$\mathrm{Me{-}\overset{\overset{\displaystyle O}{\|}}{\underset{\underset{\displaystyle OH}{|}}{P}}{-}CH_2{\cdot}CH_2{\cdot}\underset{\underset{\displaystyle NH_2}{|}}{CH}{\cdot}CO{\text{-}}L{\text{-}}Ala{\text{-}}L{\text{-}}Ala}$$

(75)

[193] K. A. Watanabe, E. A. Falco, and J. J. Fox, *J. Org. Chem.*, 1972, **37**, 1198.

methylphosphinobutyric acid, the structure of which was confirmed by synthesis of the DL-form. The configuration at the α-centre of the natural amino-acid remains to be established.[194] The dipeptide antibiotic isolated from *Streptomyces collinus* has substantial activity against Gram-negative organisms. Examination of the spectral data and identification of fumaric acid, L-α,β-diaminopropionic acid, and L-alanine in the hydrolysate permitted the assignment [195] of structure (76) to this metabolite.

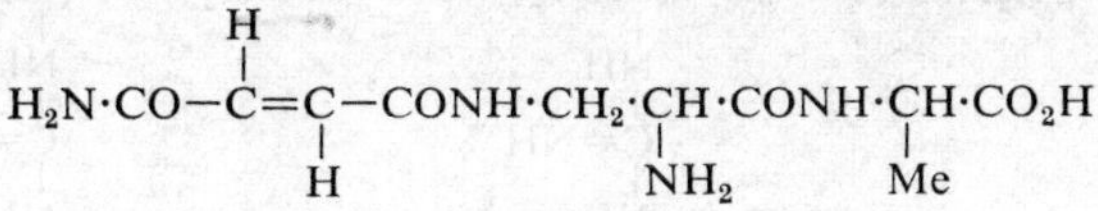

(76)

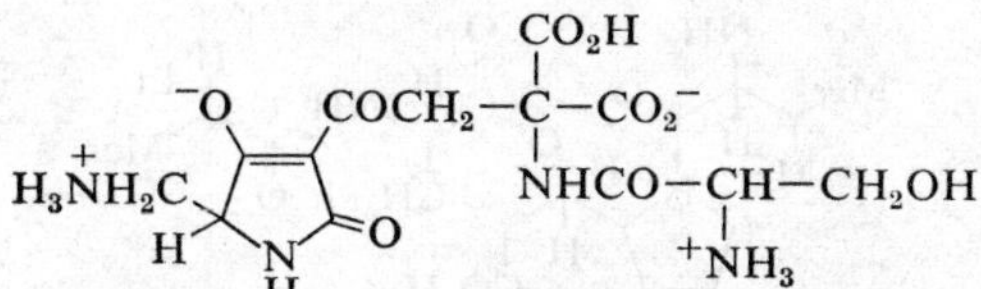

(77)

The structure elucidation of the metabolite designated K_{16} (77) was considerably complicated by the acid lability of the chromophoric system and was subsequently deduced from degradative and spectral data.[196] The antibiotic represents a further example of a growing class of modified peptides containing an acyl tetramic acid chromophore.

The yellow–green pigment produced by *Azobacter vinelandii* when grown on an iron-deficient medium has been shown to be a peptide with an attached chromophore. An *X*-ray crystallographic analysis of the decarboxylated chromophoric unit and Edman degradation on partial hydrolysates support the formulation [197] of the pigment as (78).

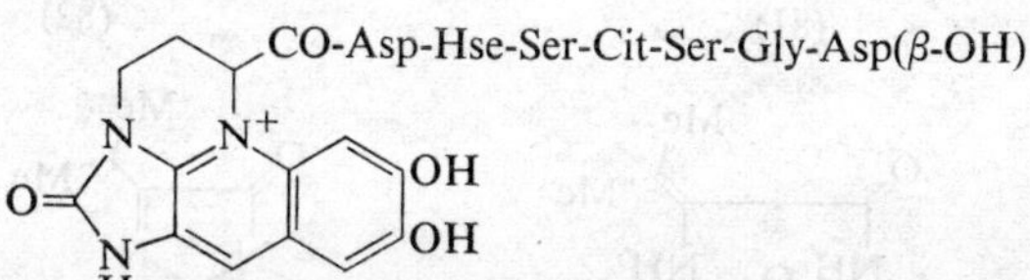

(78)

[194] E. Bayer, K. H. Gugel, K. Hägele, H. Hagenmaier, S. Jessipow, W. A. König, and H. Zähner, *Helv. Chim. Acta*, 1972, **55**, 224.

[195] B. B. Molloy, D. H. Lively, R. M. Gale, M. Gorman, L. D. Boeck, C. E. Higgens, R. E. Kastner, L. L. Huckstep, and N. Neuss, *J. Antibiotics* (*Japan*), 1972, **25**, 137.

[196] J. G. Batelaan, J. W. F. K. Barnick, J. L. Van der Baan, and F. Bickelhaupt, *Tetrahedron Letters*, 1972, 3107.

[197] K. Fukasawa, M. Goto, K. Sasaki, Y. Hirata, and S. Sato, *Tetrahedron*, 1972, **28**, 5359.

Two further examples of small modified microbial peptides with interesting biological activity are antipain [198] and detoxin D,[199] which have been assigned structures (79) and (80) respectively, on the basis of detailed chemical and spectral evidence.

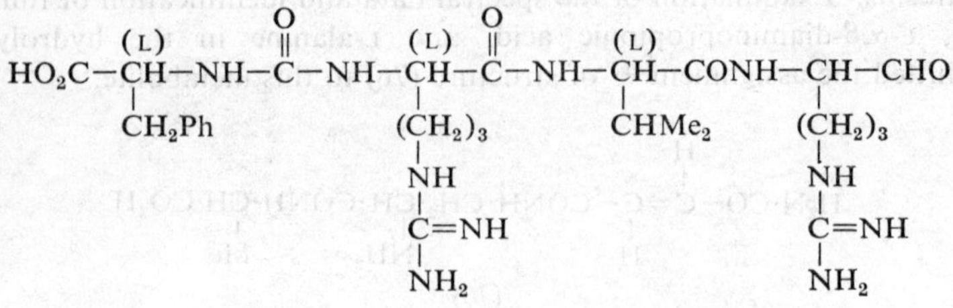

(79)

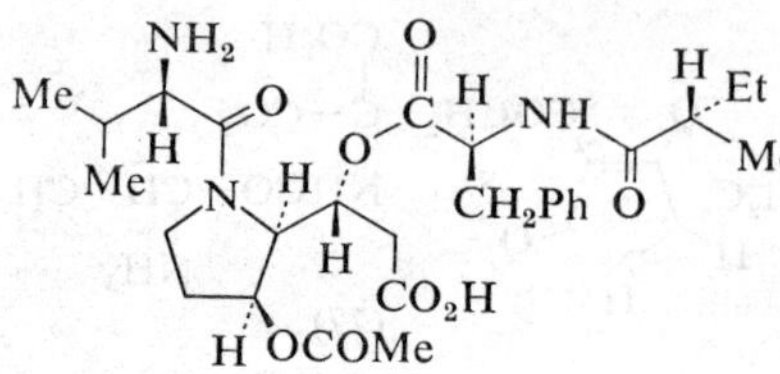

(80)

Several γ-glutamyl dipeptides containing a number of primary amino-acids have been extracted from the seeds of *Acacia georginae* [200] and *Billia lippocastanum*,[201] as well as γ-glutomylalbizziine, γ-glutamyldjenkolic acid sulphoxide, and γ-glutamyl-L-(methylenecyclopropyl)glycine.

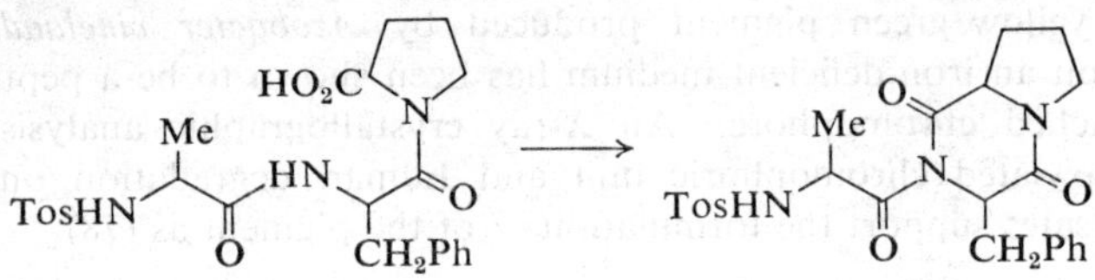

(81) (82)

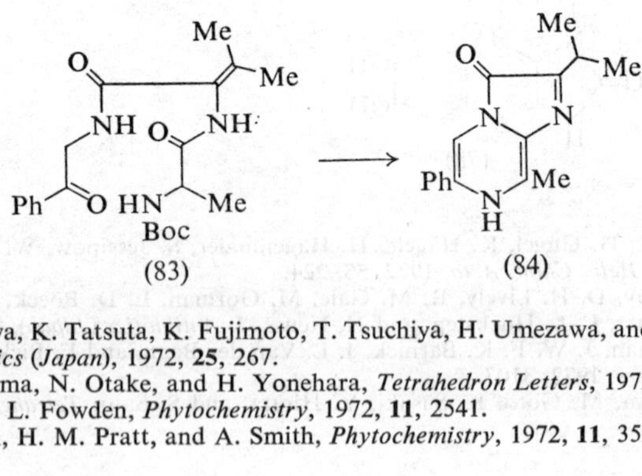

(83) (84)

[198] S. Umezawa, K. Tatsuta, K. Fujimoto, T. Tsuchiya, H. Umezawa, and H. Naganawa, *J. Antibiotics* (*Japan*), 1972, **25**, 267.

[199] K. Kakinuma, N. Otake, and H. Yonehara, *Tetrahedron Letters*, 1972, 2509.

[200] K. Ito and L. Fowden, *Phytochemistry*, 1972, **11**, 2541.

[201] L. Fowden, H. M. Pratt, and A. Smith, *Phytochemistry*, 1972, **11**, 3521.

The cyclization of the tripeptide (81) to (82) is an extension of some earlier work and is probably relevant to the biosynthesis of the ergot peptides.[202] The tripeptide (83) is converted in good yield into the bicyclic system (84) on removal of the protecting group and represents a model for the possible biosynthesis of the *Cypridina* luciferin.[203]

[202] G. Lucente and P. Frattesi, *Tetrahedron Letters*, 1972, 4283.
[203] F. McCapra and M. Roth, *J.C.S. Chem. Comm.*, 1972, 894.

5
Chemical Structure and Biological Activity

BY H. D. LAW

1 Introduction

It is ironical that biological activity, which provides much of the justification and incentive for synthetic peptide chemistry, often receives scant attention in chemical publications. However, the omission is readily understood. The biological situation is considerably more complex than the chemical and it is rare that biological findings can be as succinctly expressed. Generally, biological activity is best dealt with in separate communications couched more suitably in pharmacological or pathophysiological terms. The same difficulty and the time which has elapsed since the last review in this series[1] conspired to ensure that structure–activity relationships could not be treated comprehensively in this chapter. Except for brief allusions to ribonuclease and caerulein, cover was limited, therefore, to synthetic compounds related to biologically active peptides isolated from mammalian species but, within these limits, an attempt was made to scan all of the potentially relevant literature listed in *Current Contents Life Sciences*, 1972 (Institute for Scientific Information). Inevitably, it has not been possible, or perhaps even desirable, to quote all of the resulting references, whereas it has proved essential to cite various publications from the previous year and, in some instances, even older work where this has particular significance.

The review attempts to highlight what structure–activity studies have revealed about the ways in which individual hormones and other peptides trigger the chain of events resulting in the manifestation of a characteristic biological response. Primarily, the interaction of the hormone or other peptide with its specific target tissue is the point at issue, but as measurement of biological activity is involved, this interaction cannot be divorced from diverse other effects. A simplified consideration of the steps directly connected with the elicitation of a hormonal response (Figure 1) is sufficient to indicate the complexity of the *in vivo* situation; *in vitro* systems and even studies with 'isolated' receptor preparations are hardly less complicated if they seek to measure biological activity.

[1] D. G. Smyth in, 'Amino-acids, Peptides, and Proteins', ed. G. T. Young (Specialist Periodical Reports), The Chemical Society, London, 1969, vol. 1, p. 249.

In consequence, biosynthesis, distribution, enzymic degradation, excretion, and other phenomena are mentioned where they might have a bearing on the interpretation of experimental results relating to the main theme of the review. Recently observed varieties of biological activity elicited by

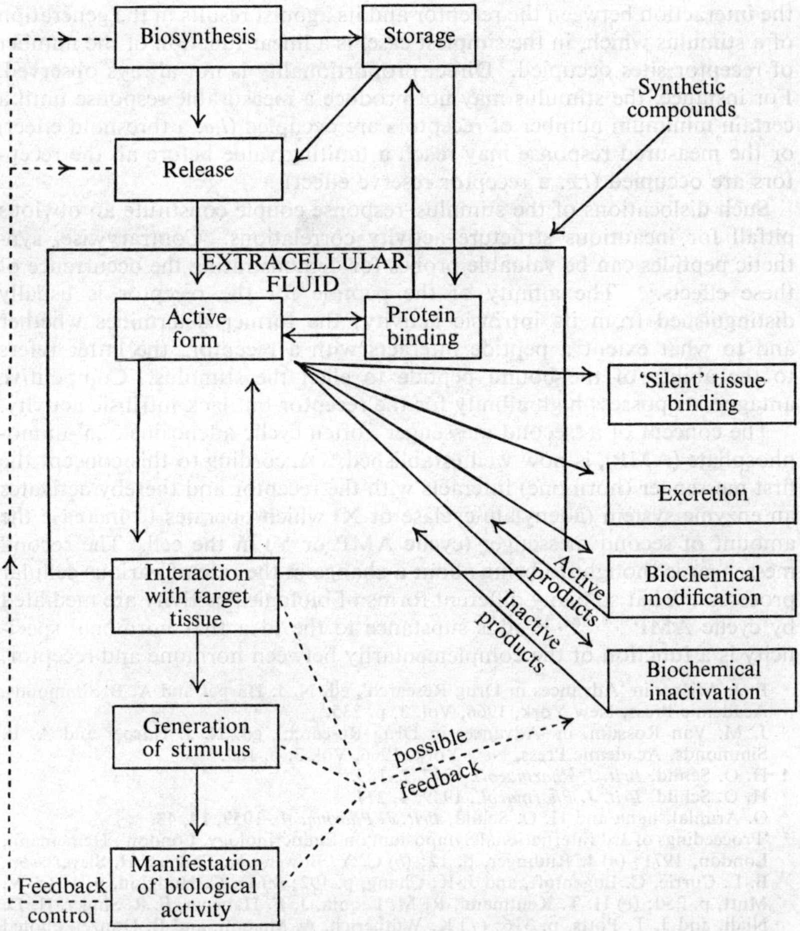

Figure 1

particular hormones and other peptides are also considered and there is oblique reference to function, although, strictly speaking, the possible physiological or pathological significance of the biological activities discussed is beyond the scope of this review.

Most structure–activity studies of peptides have been viewed in terms of receptor theory,[2–6] according to which the agonist (*i.e.* the hormone or other peptide) binds to its complementary biological receptor. A possible consequence of the interaction is a conformational and, perhaps, charge-distribution change in the receptor-bearing molecule, which induces changes in neighbouring molecules, thus initiating the physiochemical events culminating in the observed response. Whatever the mechanism, the interaction between the receptor and its agonist results in the generation of a stimulus which, in the simplest case, is a linear function of the number of receptor sites occupied. Direct proportionality is not always observed. For instance, the stimulus may not produce a measurable response until a certain minimum number of receptors are occupied (*i.e.* a threshold effect) or the measured response may reach a limiting value before all the receptors are occupied (*i.e.* a receptor reserve effect).

Such dislocations of the stimulus–response couple constitute an obvious pitfall for incautious structure–activity correlations. Contrarywise, synthetic peptides can be valuable probes for demonstrating the occurrence of these effects.[7a] The affinity of the peptide for the receptor is usually distinguished from its intrinsic activity: the former determines whether and to what extent a peptide interacts with a receptor; the latter refers to the ability of the bound peptide to elicit the stimulus. Competitive antagonists possess high affinity for the receptor but lack intrinsic activity.

The concept of a 'second messenger', often cyclic adenosine 3′,5′-monophosphate (AMP), is now well established.[8] According to this concept the first messenger (hormone) interacts with the receptor and thereby activates an enzyme system (adenylate cyclase or X) which operates to increase the amount of second messenger (cyclic AMP or Y) in the cell. The second messenger is thought to bring about a change in the rate of various cellular processes. That so many different forms of biological activity are mediated by cyclic AMP[8, 9, 10a, 11] adds substance to the idea that hormonal specificity is a function of the complementarity between hormone and receptor.

[2] E. J. Ariens, in 'Advances in Drug Research', ed. N. J. Harper and A. B. Simmonds, Academic Press, New York, 1966, Vol. 3, p. 235.

[3] J. M. Van Rossum, in 'Advances in Drug Research', ed. N. J. Harper and A. B. Simmonds, Academic Press, New York, 1966, Vol. 3, p. 189.

[4] H. O. Schild, *Brit. J. Pharmacol.*, 1947, **2**, 189.

[5] H. O. Schild, *Brit. J. Pharmacol.*, 1949, **4**, 277.

[6] O. Arunlakshana and H. O. Schild, *Brit. J. Pharmacol.*, 1959, **14**, 48.

[7] 'Proceedings of 3rd International Symposium on Endocrinology, London', Heinemann, London, 1971: (*a*) J. Rudinger, p. 12; (*b*) C. Y. Bowers, K. Folkers, H. Sievertsson, B. L. Currie, C. Bogentoft, and J.-K. Chang, p. 192; (*c*) D. C. Hodgkin, p. 1; (*d*) V. Mutt, p. 250; (*e*) H. T. Keutmann, R. M. Lequin, J. F. Habener, F. R. Singer, H. D. Niall, and J. T. Potts, p. 316; (*f*) K. Wütherich, A. Masson, and B. Donzel, quoted by Rudinger, ref. 7*a*; (*g*) H. B. Brewer, p. 324.

[8] E. W. Sutherland, *Science*, 1972, **177**, 401.

[9] 'Cyclic AMP and Cell Function', *Ann. New York. Acad. Sci.*, 1971, **185**: (*a*) R. J. Leftowitz, J. Roth, and I. Pastan, p. 195; (*b*) J. H. Exton, S. B. Lewis, R. J. Ho, G. A. Robison, and C. R. Park, p. 85; (*c*) G. D. Aurback, R. Marcus, J. Heersche, and S. Marx, p. 386.

[10] (*a*) I. H. Pastan, in 'Current Topics in Biochemistry', ed. C. B. Anfinsen, R. F. Goldberger, and A. N. Schechter, Academic Press, New York, 1972, p. 65; (*b*) M. Rodbell, *ibid.*, p. 187.

[11] W. Y. Cheung, *Perspectives in Biol. and Medicine*, Winter 1972, p. 221.

There is good evidence that many types of receptors are located on the surface of cell membranes.[10b] Membrane and solubilized membrane preparations have now been described (see below) which bind certain hormones and other peptides specifically. Sometimes the binding is associated with the activation of adenylate cyclase activity. These preparations afford a direct approach to the study of peptide–receptor interactions and constitute a significant step towards the time when 'the term receptor will have to be substituted by a proper chemical characterization of the molecular entity in the biological object involved' (Ariens [2]).

Having dwelt upon the biological difficulties of studying hormone–receptor interactions, specification of a chemical limitation is only fitting. In very few instances is there definitive information about the three-dimensional form of peptide molecules in aqueous solution, whereas a knowledge of the topochemistry of the molecule is presumably essential to discuss structure–activity relationships in the most meaningful manner (see ref. 7*a* and publications cited therein). This is likely to be true whether the peptide retains its solution conformation on interaction with the receptor or whether it adopts a new shape dictated by energetic preferences of the peptide–receptor complex. Intuitively, the former possibility seems improbable, particularly for peptides of relatively low molecular weight, and there is already a substantial body of experimental evidence of conformational changes when peptides interact with proteins in solution. However, at present, there is no conclusive direct evidence to resolve the two possibilities.

2 Ribonuclease

Though it is not generically related to the other peptides discussed in this chapter, no excuse is required for including ribonuclease. The S-peptide–S-protein interaction was suggested a long time ago as a possible model of hormonal action,[12] but regardless of whether the system does simulate the way in which any given pharmacologically active peptide elicits its response, this system is instructive as one in which the relationship between peptide–protein interactions and biological activity (*i.e.* structure–activity relationships) can be studied directly. The recent observation that the *C*-terminal tetradecapeptide of ribonuclease can also reactivate ribonuclease components shortened at the carboxy terminus,[13] and the now extensive complementation studies with staphylococcal nuclease,[14–18] offer further scope for this type of investigation.

12 K. Hofmann, *Brookhaven Symp. Biol.*, 1960, **13**, 184; F. M. Richards and P. J. Vithayatil, *ibid.*, p. 115; K. Hofmann, *Proc. Chem. Soc.*, 1963, 364.
13 M. C. Lin, B. Gutte, S. Moore, and R. B. Merrifield, *J. Biol. Chem.*, 1970, **245**, 5169.
14 H. Taniuchi, C. B. Anfinsen, and A. Sodja, *Proc. Nat. Acad. Sci., U.S.A.*, 1967, **58**, 1235.
15 H. Taniuchi and C. B. Anfinsen, *J. Biol. Chem.*, 1968, **243**, 4778.
16 H. Taniuchi, *Fed. Proc.*, 1970, **29**, 335.
17 H. Taniuchi and C. B. Anfinsen, *J. Biol. Chem.*, 1971, **246**, 2291.
18 G. Andria, H. Taniuchi, and J. L. Cone, *J. Biol. Chem.*, 1971, **246**, 7421.

X-Ray studies of the intact ribonuclease molecule [19] and of the S-protein [20, 21] provide reassuring vindication of conclusions drawn from 'classical' structure–activity studies with synthetic analogues.[22] Nonetheless, the resolution of the crystallographic investigations is not adequate to define in detail the interaction of the peptide and protein nor to permit an understanding of the detailed mechanism of the enzymic process. Structure–activity studies continue virtually unabated, though guided to some considerable extent by the *X*-ray models.

Recent studies of the reaction of S-protein with S-peptide and its analogues have sought to define the binding parameters for the equilibrium:

$$\text{Ribonuclease complex} \rightleftharpoons \text{S-protein} + \text{S-peptide (or analogue)}$$

Various techniques have been used to do this, including difference spectroscopy, o.r.d. differential spectroscopy, microcalorimetry, and differential thermal analysis,[23–28] and although the agreement between different approaches is not always good, the characteristics of the binding process are beginning to emerge.

S-Peptide and S-peptide analogues which activate S-protein appear to adopt the required degree of helix on binding.[23] Analogues of S-peptide which do not activate S-protein show no evidence of helix formation under these conditions, although S-peptide analogues, whether they are capable of activating S-protein or not, can assume a helical conformation when organic solvents are added to them in aqueous solution.[29] The binding of S-peptide or of activating S-peptide analogues to S-protein is accompanied by a large decrease in enthalpy.[23, 26] In this case, mean values for ΔH are in reasonable agreement, but there is some doubt about the completeness of binding at different concentrations of S-peptide and S-peptide analogue.

Binding constants for the reaction of S-protein with S-peptide or S-peptide analogues derived from spectroscopic measurements differ

[19] G. Kartha, J. Bello, and D. Hawker, *Nature*, 1967, **213**, 862.
[20] H. W. Wyckoff, K. D. Hardmann, N. M. Allewell, T. Ingami, L. N. Johnson, and F. M. Richards, *J. Biol. Chem.*, 1967, **242**, 3984.
[21] H. W. Wyckoff, D. Tsernoglou, A. W. Hanson, J. R. Knox, B. Lee, and F. M. Richards, *J. Biol. Chem.*, 1970, **245**, 305.
[22] F. M. Richards and H. W. Wyckoff, 'The Enzymes', ed. P. D. Boyer, Academic Press, New York, 3rd edn., 1971, p. 647.
[23] R. Rocchi, G. Borin, F. Marchiori, L. Moroder, E. Peggion, E. Scoffone, V. Crescenzi, and F. Quadrifoglio, *Biochemistry*, 1972, **11**, 50.
[24] F. M. Finn, *Biochemistry*, 1972, **11**, 1474.
[25] B. M. Woodfin and V. Massey, *J. Biol. Chem.*, 1968, **243**, 889.
[26] R. P. Hearn, F. M. Richards, J. M. Sturtevant, and G. D. Watt, *Biochemistry*, 1971, **10**, 806.
[27] 'Proceedings 11th European Peptide Symposium, Vienna, April 1971', ed. H. Nesvadba, North Holland Publishing Co., Amsterdam, in the press: (*a*) A. Berger and S. Levit; (*b*) L. Moroder, G. Borin, F. Marchiori, R. Rocchi, and E. Scoffone.
[28] B. G. Winchester, A. P. Mathias, and B. R. Robin, *Biochem. J.*, 1970, **117**, 299.
[29] 'Proceedings 10th European Peptide Symposium, Albano, Italy, 1969', ed. E. Scoffone, North Holland Publishing Co., Amsterdam, 1971: E. Scoffone, F. Marchiori, R. Rocchi, A. Scatturin, and A. M. Tamburro, p. 233.

appreciably from those based on measurements of enzymic activity. In studies of the latter type, the presence of substrate seems to effect a co-operative influence on the formation of the complex. Clearly it is not unexpected that formation of the (ES) complex, with whatever steps that process might involve, should disturb the simple S-protein–S-peptide equilibrium, and it could even happen that a different, three-component complex is involved when substrate is present. The composition of the buffer also influences the observed values. Similar considerations must be kept in mind when comparing data derived by different approaches to the study of peptide hormone–receptor interactions. At present, entropy changes for the S-peptide–S-protein interaction, presumably accounted for by the loss in conformational freedom when the complex is formed, are calculated to be in the range 98—124 e.u.[23, 26]

In structure–activity terms, recent studies of the S-peptide–S-protein interaction stress the importance of the Glu-5–Arg-10 interaction [23, 24, 30–32] and of the Phe-8 (binding),[23, 33] His-12 (functional and, perhaps, binding),[31, 34] Met-13 (binding),[31] and Asp-14 (binding) [35] residues, and confirm that the Lys-7,[31] Ala-4, Ala-5, and Ala-6 residues do not play an important role in the binding. It is noteworthy that replacement of the hydrophobic alanine residues in position 4, 5, or 6 by serine does not affect binding or activity.[36]

The *C*-terminal portion of the ribonuclease molecule plays a determining role in the establishment of the conformation of the protein as shown by studies of the virtually inactive analogue which lacks the *C*-terminal tetrapeptide portion.[37] Recombination studies, involving (1—118), (1—119), or (1—120) ribonuclease and synthetic peptides of varying chain length related to the *C*-terminal (111—124) part of the molecule, show a dramatic increase in activity when the *C*-terminal fragment reaches nona-peptide dimensions (Table 1).[38] With the larger protein fragments, alkylation experiments with iodoacetate show that there are two enzymically active forms of the peptide–protein complex, since the His-119 residues in the protein and in the peptide are both partially alkylated, *i.e.* there must

[30] F. M. Finn, J. Dadok, and A. A. Bothner-By, *Biochemistry*, 1972, **11**, 455.
[31] K. Hoffman, R. Andreatta, F. M. Finn, J. Montibeller, G. Porcelli, and A. J. Quattrone, *Bioorg. Chemistry*, 1971, **1**, 66.
[32] F. Marchiori, G. Borin, L. Moroder, R. Rocchi, and E. Scoffone, *Biochim. Biophys. Acta*, 1972, **257**, 210.
[33] G. Borin, F. Marchiori, R. Rocchi, L. Moroder, and E. Scoffone, *Gazzetta*, 1971, **101**, 426.
[34] G. Borin, C. Toniolo, L. Moroder, F. Marchiori, R. Rocchi, and E. Scoffone, *Internat. J. Protein Res.*, 1972, **4**, 37.
[35] K. Hofmann, F. M. Finn, M. Limetti, J. Montibeller, and G. Zaneth, *J. Amer. Chem. Soc.*, 1966, **88**, 3633.
[36] G. Borin, F. Marchiori, L. Moroder, R. Rocchi, and E. Scoffone, *Biochim. Biophys. Acta*, 1972, **271**, 77.
[37] D. Puett, *Biochemistry*, 1972, **11**, 1980.
[38] B. Gutte, M. C. Lin, D. G. Caldi, and R. B. Merrifield, *J. Biol. Chem.*, 1972, **247**, 4763.

be two different productive modes of binding of the peptide to the protein. Overlapping fragments have also been observed in studies with staphylococcal nuclease.[17, 39] In other fields, variations in cleavage point when a family of like peptides is subjected to enzymic digestion,[40] and, possibly, certain peculiarities of antibody specificity,[41] may be accounted for in

Table 1 *Activation of* (1—118)*ribonuclease with peptides related to the C-terminal residue sequence of the intact enzyme*[a]

C-terminal region	*Residue sequence*	*Activity*[b]
111—124	Glu-Gly-Asn-Pro-Tyr-Val-Pro-Val-His-Phe-Asp-Ala-Ser-Val	98
113—124	Asn-Pro-Tyr-Val-Pro-Val-His-Phe-Asp-Ala-Ser-Val	90
114—124	Pro-Tyr-Val-Pro-Val-His-Phe-Asp-Ala-Ser-Val	80
115—124	Tyr-Val-Pro-Val-His-Phe-Asp-Ala-Ser-Val	70
116—124	Val-Pro-Val-His-Phe-Asp-Ala-Ser-Val	60
117—124	Pro-Val-His-Phe-Asp-Ala-Ser-Val	1.5
118—124	Val-His-Phe-Asp-Ala-Ser-Val	1

[a] From ref. 38. [b] Maximum activity regenerated expressed as a percentage of the activity of an equimolar amount of natural ribonuclease A.

similar terms. The phenomenon is at least possible in peptide hormone studies and may come to light now that receptor–hormone interactions are being studied directly.

X-Ray,[20] n.m.r.,[42] and other studies[43] suggest that the Phe-120 residue is important for the binding of carboxy-shortened ribonuclease to *C*-terminal peptides, and this has now been substantiated by the use of synthetic analogues.[44] Thus, the Leu-120 tetradecapeptide (111—124) combines with ribonuclease (1—118) to give a dissociation constant, K_d, twelve times larger, and the Ile-120 tetradecapeptide has K_d twenty times larger than the Phe-120 peptide–(1—118)ribonuclease interaction; and they produce at maximum only 13 and 12%, respectively, of the activity generated by the Phe-120 compound. The Trp-120 tetradecapeptide exhibits very weak binding to the (1—118)protein and generates only 0.5% of the activity of the parent peptide. In a manner reminiscent of postulated hormone–receptor interactions, the Phe-120 side-chain seems to fit into a hydrophobic pocket in the protein so that it increases the stability of the complex and helps to orientate the functional His-119 side-chain.

[39] I. Parikh, L. Corley, and C. B. Anfinsen, *J. Biol. Chem.*, 1971, **246**, 7392.
[40] R. F. Bilton, W. Blackburn, M. J. Crumpton, and H. D. Law, *F.E.B.S. Letters*, 1973, **34**, 333.
[41] R. W. Rosenstein, R. A. Masson, M. Y. A. Armstrong, W. H. Konigsberg, and F. F. Richards, *Proc. Nat. Acad. Sci., U.S.A.*, 1972, 877.
[42] D. H. Meadows, G. C. K. Roberts, and O. Jardetzky, *J. Mol. Biol.*, 1969, **45**, 491.
[43] M. C. Lin, *J. Biol. Chem.*, 1970, **245**, 6726.
[44] M. C. Lin, B. Gutte, D. G. Caldi, S. Moore, and R. B. Merrifield, *J. Biol. Chem.*, 1972, **247**, 4768.

3 Hypothalamic Regulatory Factors

Hypothalamic regulatory factors (and hormones), of which there are at least ten (Table 2), have featured in several recent reviews.[7b, 45, 46a,b, 47–50]

Table 2 *Hypothalamic regulatory factors (and hormones)*

Factor	*Abbreviation*
Thyroid stimulating hormone (TSH, thyrotropin)-releasing factor	TRF or TRH
Follicle stimulating hormone-releasing factor	FSH-RF or FSH-RH
Luteinizing hormone-releasing factor	LRF or LRH
Growth hormone (somatotropin)-releasing factor	GH-RF, GH-RH, SRF, or SRH
Growth hormone (somatotropin) release-inhibiting factor	GIF, GH-RIH, SRIF, or SRIH
Prolactin-releasing factor	PRF or PRH
Prolactin release-inhibiting factor	PIF or PRIH
Melanocyte stimulating hormone (MSH)-releasing factor	MRF or MRH
Melanocyte stimulating hormone release-inhibiting factor	MRIF or MRIH
Corticotropin (ACTH)-releasing factor	CRF or CRH

These substances are secreted by hypothalamic nerve fibres of various types and pass *via* a portal system to the pituitary gland where they stimulate or inhibit the release of particular pituitary hormones. Both stimulatory and inhibitory principles seem to be produced in those cases where the release of the pituitary hormone is not suppressed by feed-back from the target organ (growth hormone, prolactin, melanocyte-stimulating hormone); when negative feed-back does operate (ACTH, thyrotropin, luteinizing hormone and follicle-stimulating hormone), only stimulatory hypothalamic factors seem to be produced. An impression of the way in which pituitary control operates is therefore beginning to emerge, with new

[45] A. V. Schally, A. Arimura, and A. J. Kastin, *Science*, 1973, **179**, 341.

[46] 'Proceedings 3rd American Peptide Symposium, Boston, Mass., June, 1972', ed. J. Meienhofer, Ann Arbor Science Publishers Inc., December, 1972: (*a*) R. Guillemin, p. 585; (*b*) M. Monahan, J. Rivier, W. Vale, N. Ling, G. Grant, M. Amoss, R. Guillemin, R. Burgus, E. Nicolaides, and M. Rebstock, p. 601; (*c*) R. Walter, I. Bernal, and L. F. Johnson, p. 131; (*d*) J. Ramachandran, W. R. Moyle, and Y. C. Kong, p. 613; (*e*) B. M. Altura, p. 441; (*f*) D. A. Tewksbury, p. 461; (*g*) M. J. Peach, p. 471; (*h*) J. M. Stewart and R. J. Freer, p. 521; (*i*) D. Regoli, F. Rioux, and W. K. Park, p. 495; (*j*) E. C. Jorgensen, G. C. Windridge, K.-H. Hsieh, and T. C. Lee, p. 513; (*k*) P. Needleman, E. M. Johnson, W. Vine, E. Flanigan, and G. R. Marshall, p. 501; (*l*) R. R. Smeby, M. C. Khosla, and F. M. Bumpus, p. 509; (*m*) J. M. Parry, A. B. Russell, and M. Szelke, p. 541; (*n*) M. A. Ondetti, J. Pluscec, E. R. Weaver, N. J. Williams, E. F. Sabo, and O. Kodey, p. 525.

[47] W. Vale, G. Grant, and R. Guillemin, in 'Frontiers of Neuroendocrinology', ed. W. Ganong and L. Martin, Academic Press, New York, 1973.

[48] R. Guillemin, R. Burgus, and W. Vale, *Vitamins and Hormones*, 1971, **29**, 1.

[49] R. Guillemin, in 'Advances in Metabolic Disorders', Academic Press, New York, 1971, Vol. 5, p. 1.

[50] R. Guillemin, *Contraception*, 1972, **6**, 1.

prospects for the clinical treatment of pituitary malfunction. However, in only four cases are the structures of the hypothalamic peptides well established and only in two of these cases is there sufficient information to make a discussion of structure–activity correlations meaningful. This is a measure of the technical difficulty of research in this field; at the same time, the recent spate of biochemical, physiological, and clinical observations relating to the factors of known structure is an indication of the accelerated rate of development which can be expected once the structures are established.

Thyroid Stimulating Hormone-releasing Factor.—Many analogues of TRF [Table 3, compound (1)] were prepared incidental to the structural elucidation of the natural hormone and, since then, their number has increased considerably as a result of systematic structure–activity studies. Reports of this work have been listed in the Peptide Synthesis chapters of previous Specialist Periodical Reports, and various reviews have commented on the significance of the findings.[7b, 46b, 48] In general, the integrity of the whole

Table 3 *Synthetic analogues of TRF*

Compound number	Structure				Relative biological activity/%	pK_a	Ref.
(1)	Glp	-His	-Pro	$-NH_2$(TRF)[a]	100	6.25	
(2)	Glp	-His	-Pro	-OMe	20	6.10	51
(3)	Glp	-His	-Pro	-OH	0.02	6.75	51
(4)	MeGlp	-His	-Pro	$-NH_2$	1.2	6.25	51
(5)	Pro	-His	-Pro	$-NH_2$	0.01	6.05	51
(6)	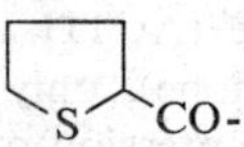	His	-Pro	$-NH_2$	0.2		46*b*
(7)	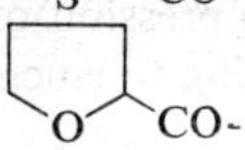	His	-Pro	$-NH_2$	0.01		46*b*
(8)	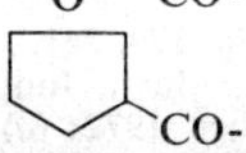	His	-Pro	$-NH_2$	< 0.01		46*b*
(9)	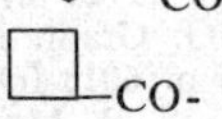	His	-Pro	$-NH_2$	0.016	6.35	51
(10)	Glp	-N^{π}MeHis	-Pro	$-NH_2$[b]	0.04	6.6	52, 53
(11)	Glp	-N^{τ}MeHis	-Pro	$-NH_2$[b]	800	5.95	52, 53
(12)	Glp	-Orn	-Pro	$-NH_2$	0.025		52
(13)	Glp	-Lys	-Pro	$-NH_2$	0.02		52, 54
(14)	Glp	-Arg	-Pro	$-NH_2$	0.05		52, 54
(15)	Glp	-Phe	-Pro	$-NH_2$	10		56
(16)	Glp	-Tyr	-Pro	$-NH_2$	0.084 'inactive'		52, 56
(17)	Glp	-Trp	-Pro	$-NH_2$	'inactive'		56
(18)	Glp	-Met	-Pro	$-NH_2$	1.0		52
(19)	Glp	-His	-Pro	-NHEt	14	6.25	51
(20)	Glp	-His	-Pro	$-NMe_2$	0.5	6.25	51

Table 3 (*cont.*)

Compound number	Structure	Relative biological activity/%	pK_a	Ref.
(21)	Glp -His -Pro -NEt_2	0.05	6.45	51
(22)	Glp -His -Pro piperidine	0.2	6.45	51
(23)	Glp -His -pyrrolidine	0.2	6.2	51
(24)	Glp -His -NH_2	0.004—0.02	6.7	51
(25)	Glp -His -OMe	< 0.0005	7.0	52, 53
(26)	Glp -N^{π}MeHis-OMe	< 0.0025		52, 53
(27)	Glp -N^{τ}MeHis-OMe	0.02		52, 53
(28)	Glp -Phe -3Hyp-NH_2	'inactive'		56
(29)	Glp -Phe -His -NH_2	'inactive'		70
(30)	Glp -Trp -His -NH_2	'inactive'		70
(31)	Glp -Trp -Tyr -NH_2	'inactive'		70
(32)	Glp -Ala -His -NH_2	'inactive'		70
(33)	Glp Gly -Tyr -NH_2	'inactive'		70
(34)	Glp -Tyr -Gly -NH_2	'inactive'		70
(35)	Glp -Tyr -His -NH_2	'inactive'		70
(36)	Glp -Tyr -Phe -NH_2	'inactive'		70
(37)	Glp -Tyr -Trp -NH_2	'inactive'		70
(38)	Glp -Tyr -Tyr -NH_2	'inactive'		70
(39)	Glp -Tyr -NH_2	'inactive'		70
(40)	Glp -His -NH_2	'inactive'		70

[a] Glp = pyroglutamic acid. This alternative to the I.U.P.A.C.–I.U.B. recommendation is used for tabular and textual presentation. [b] N^{π}-MeHis =

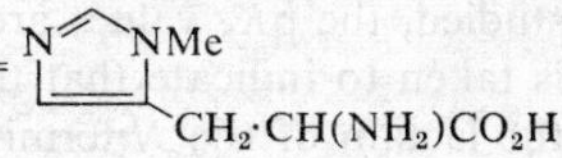

N^{τ}-MeHis =

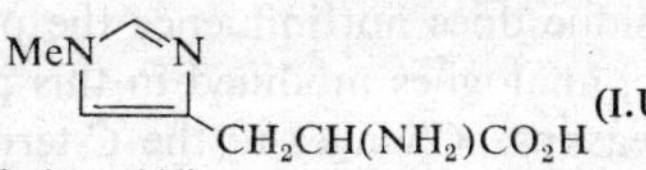

(I.U.P.A.C.–I.U.B. recommended abbreviations; see vol. 4, p. 441).

tripeptide is important for biological activity. Substitutions for or deletions of any of the constituent amino-acid residues usually lead to virtually complete deactivation, although the molecule seems more sensitive to changes to the pyroglutamic acid and histidine residues than to the proline residue. Only one analogue more active than TRF has been produced, [2-N^{τ}-methylhistidine]-TRF, and no peptides which antagonize TRF. These findings have been substantiated and extended by investigations reported in 1972 (Table 3).

Despite the inactivity of many TRF analogues relative to the natural hormone, some of them, in an absolute sense, are still very potent molecules. Bearing in mind the specificity of TSH-release, it has been argued that comparisons between these analogues can provide useful structure–activity correlations. Thus, it has been suggested that the relative potencies of compounds (4)—(8) might indicate a nucleophilic role for the α-amido-group of the terminal pyroglutamic acid residue in the action of the TRF.[46b] The amino-group of the proline analogue (4) would presumably be protonated at physiological pH. Similarly, it is thought that the *C*-terminal amide cannot have a functional significance since the methyl ester (2), substituted amides (19)—(22), and decarboxy-derivatives (23) retain activity.[51] Perhaps the main impression created by these observations is of the need for a hydrophobic group in this position, a requirement which is to some extent substantiated by the lower activity of the free carboxylic acid (3).

Not surprisingly, especially in view of the enhanced activity of the 2-N^{τ}-methylhistidine analogue (11), the role of the imidazolyl side-chain has come in for particular attention.[52–54] It is clearly not sufficient just to have a basic side-chain in this position, *e.g.* (12)—(14), and the appreciable activity (5%) shown by the 2-(pyrazolyl-3′-alanine)-analogue ($pK \simeq 2.1$) indicates that the acid–base properties are not critical.[55] Analogues with neutral side-chains in this position, *e.g.* (15) and (18), can have considerable activity, provided they are not too bulky, *e.g.* (16) and (17).[52, 56]

Potentiometric titration has been used to determine the pK_a values of the imidazolyl side-chain in various analogues and this information has been interpreted in terms of hydrogen-bonding.[51] If the imidazole nitrogen is bonded to a primary amino-group or to an amide bond, the acidity of the imidazolinium group should be increased (lower pK_a) relative to the non-bonded situation.[57] Bonding of the imidazole hydrogen, *e.g.* to a carbonyl group, should make it less available (*i.e.* raise pK_a).[58] In the TRF analogues studied, the pK_a values are all low relative to imidazole (pK_a 7.05), which is taken to indicate that the first type of bonding is involved. Modifying the lactam of the *N*-terminal residue does not influence the pK_a, *e.g.* (4), (5), and (9). As discussed above, analogues modified in this position can be relatively inactive for other reasons. Changes to the *C*-terminal amide have little influence on the dissociation of the imidazole proton in (2), (19)—(22), although pK_a is related to biological activity in these compounds. Steric factors seem to be important here. The only group which remains to form a hydrogen bond to the imidazole ring in the proposed manner is the —NH— of the pyroglutamyl–histidine peptide bond.

In view of the contrasting potencies of the N^{π}-methylhistidine (10) and N^{τ}-methylhistidine (11) analogues, it seems probable that the N^{π} atom is involved in the bonding. This interpretation finds some support in the pK_a values of the *N*-methyl analogues, which also exclude intermolecular

[51] G. Grant, N. Ling, J. Rivier, and W. Vale, *Biochemistry*, 1972, **11**, 3070.
[52] J. Rivier, W. Vale, M. Monahan, N. Ling, and R. Burgus, *J. Medicin. Chem.*, 1972, **15**, 479.
[53] W. Vale, J. Rivier, and R. Burgus, *Endocrinology*, 1971, **89**, 1485.
[54] D. Gillesen, F. Piva, H. Steiner, and R. O. Studer, *Helv. Chim. Acta*, 1971, **54**, 1335.
[55] K. Hoffman and C. V. Bowers, *J. Medicin. Chem.*, 1970, **13**, 1099.
[56] H. Sievertsson, J.-K. Chang, K. Folkers, and C. Y. Bowers, *J. Medicin. Chem.*, 1972, **15**, 219.
[57] D. H. Sachs, A. N. Schechter, and J. S. Cohen, *J. Biol. Chem.*, 1971, **246**, 6576.
[58] M. P. Printz, H. P. Williams, and L. C. Craig, *Proc. Nat. Acad. Sci., U.S.A.*, 1972, **69**, 378.

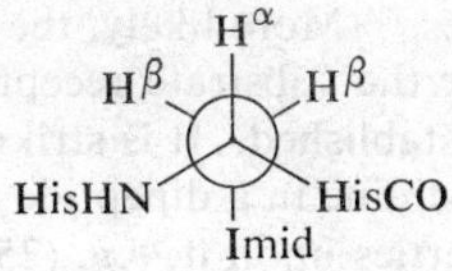

(41)

hydrogen bonding.[51] N.m.r. studies [59a] indicate that the *C*-terminal amide is also hydrogen-bonded, probably to the carbonyl group of the histidine residue, and that the imidazolyl side-chain is in the staggered conformation (41) to the extent of at least 39%. The second hydrogen bond, when the proline is in the *trans* form, closes a stable-looking seven-membered equatorial ring, as in the proposed structure (42).

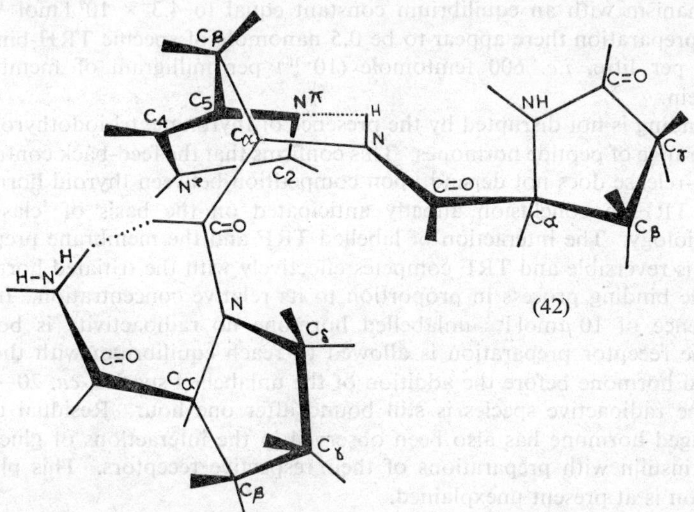

(42)

2-(N^{τ}-Methylhistidine)-TRF, like TRF itself and other active analogues, is inhibited by prior treatment of the assay system with tri-iodothyronine.[53] Like TRF itself [60–62] and other analogues,[63] it is deactivated by serum and, although relative deactivation rates are not known, it seems unlikely to owe its high order of activity either to longevity or to a favourable distribution, since it is more active then TRF in stimulating TSH-release from

[59] (*a*) S. Fermandjian, P. Pradelles, and P. Fromageot, *F.E.B.S. Letters*, 1972, **28**, 156; (*b*) S. Fermandjian, D. Greff, and P. Fromageot, *J. Chim. phys.* (*Orleans*), September 1972, quoted in ref. 59*a*.

[60] F. Labrie, N. Barden, G. Poirier, and A. De Lean, *Proc. Nat. Acad. Sci., U.S.A.*, 1972, **69**, 283; G. Poirier, N. Barden, F. Labrie, P. Borgeat, and A. De Lean, *Excerpta Medica Internat. Congress*, 1972, No. 256, p. 85.

[61] T. W. Redding and A. V. Schally, *Life Sci.*, 1973, **12**, Part I, 23.

[62] Dupont, *General and Comp. Endocrinol.*, 1972, **19**, 522.

[63] W. W. Vale, R. Burgus, T. F. Dunn, and R. Guillemin, *Hormones*, 1971, **2**, 193.

pituitary monolayer cultures.[53] More likely, the activity of this analogue is due to a higher affinity for the substrate receptor. The reason for its increased affinity is not yet established. It is striking that the substitution of 2-(N^{τ}-methylhistine) for histidine in a dipeptide which is otherwise inactive confers definite TRF properties upon it, *e.g.* (25) and (27).

The availability of synthetic tritiated TRF has facilitated studies of the hormone's distribution [61, 62, 64] and of the reaction of TRF with its biological receptor. Preparations of the receptor from bovine anterior pituitary [60] and from TSH-secreting mouse pituitary tumours [65] have both been reported. The receptors are located in the plasma membrane fractions. In the bovine pituitary preparation, it has been shown that their distribution is identical with that of adenylate cyclase.[66] Interaction of the bovine pituitary membrane preparation and TRF occurs *via* a bimolecular mechanism with an equilibrium constant equal to $4.3 \times 10^{7}\,l\,mol^{-1}$. In this preparation there appear to be 0.5 nanomole of specific TRH-binding sites per litre, *i.e.* 600 femtomole (10^{-15}) per milligram of membrane protein.

Binding is not disrupted by the presence of thyroxine, tri-iodothyronine, or a range of peptide hormones. This confirms that the feed-back control of TSH-release does not depend upon competition between thyroid hormone and TRF, a conclusion already anticipated on the basis of 'classical' physiology. The interaction of labelled TRF and the membrane preparation is reversible and TRF competes effectively with the tritiated hormone in the binding process in proportion to its relative concentration. In the presence of $10\,\mu mol\,l^{-1}$ unlabelled hormone no radioactivity is bound. If the receptor preparation is allowed to reach equilibrium with the tritiated hormone before the addition of the unlabelled species, *ca.* 20—40% of the radioactive species is still bound after one hour. Residual unexchanged hormone has also been observed in the interactions of glucagon and insulin with preparations of their respective receptors. This phenomenon is at present unexplained.

The effect of calcium, potassium, and magnesium ions on the TRF-receptor interaction is informative. Whereas potassium and magnesium ions facilitate binding of TRF to the adenohypophyseal receptor preparation within certain optimum concentration ranges, Ca^{2+} inhibits binding at all concentrations tested.[60] This presumably implies that calcium, which is known to be required for the TRF-induced release of thyrotropin, does not exert its effect at the TRF-receptor level.

When biological activity is used as the criterion for measuring binding of TRF to normal or thyroidectomized pituitary cells in tissue culture, the binding constant is *ca.* 20 times lower than that based on radioactivity counts. The difference may be due to experimental conditions, to the presence of 'silent' binding sites, or to the existence of a receptor reserve.[65]

[64] T. W. Redding and A. V. Schally, *Neuroendocrinology*, 1972, **9**, 250.
[65] G. Grant, W. Vale, and R. Guillemin, *Biochem. Biophys. Res. Comm.*, 1972, **46**, 28.
[66] G. Poirier, F. Labrie, N. Barden, and S. Lemaire, *F.E.B.S. Letters*, 1972, **20**, 283.

It is now possible with these TRF-receptor preparations to study directly, at the binding site, the effect of modifications to the parent hormone. So far, it is reported that neither compound (25) nor synthetic LRF (43), both of which are devoid of TRF activity, compete in the binding of labelled TRF with receptor preparations.[51, 65]

$$\text{Glp-His-Trp-Ser-Tyr-Gly-Leu-Arg-Pro-Gly-NH}_2$$
(43)

Until comparatively recently, TRF was thought to be specific for the release of TSH. Now at least three other types of biological activity are attributed to it. TRF releases prolactin from rat pituitary tumour tissue,[67] although there is evidence that TRF is not the natural releasing hormone for prolactin;[45] TRF stimulates glucose oxidation in pituitary tissue, an effect which is inhibited by the addition of MSH;[68] and TRF potentiates the behavioural effects of DOPA in mice, even after hypophysectomy, an effect which must be independent of TSH release.[69,70] Nothing is known about the structure–activity dependency of these responses.

Luteinizing Hormone-releasing Factor.—Prior to the structural elucidation of LRF, degradative experiments had given some information about the significance of individual amino-acid residues in the LRF molecule. Such studies suggest that the overall molecule is necessary for biological activity.[71–73] Studies with synthetic analogues allow more far-reaching deductions to be made.[7b, 46, 48] It is now established that the decapeptide structure (43) possesses both luteinizing hormone-releasing and follicle-stimulating hormone-releasing activity,[74, 75] whereas it does not affect blood levels of thyrotropin-stimulating hormone, growth hormone, ACTH, insulin, glucose, fatty acids, *etc.*[76, 77] No unequivocal examples are known of LRF analogues which exhibit LRF activity but not FSH-RF activity. Although the existence of a separate FSH-RF cannot be excluded at this

[67] A. H. Tashjian, N. J. Barowsky, and D. K. Jensen, *Biochem. Biophys. Res. Comm.*, 1971, **43**, 516.
[68] K. Yamamoto, T. Onaya, M. Kotani, and T. Yamada, *Proc. Soc. Exp. Biol. Med.*, 1972, **140**, 677.
[69] N. P. Plotnikoff, A. J. Prange, G. R. Breese, and M. S. Anderson, *Science*, 1972, **178**, 417.
[70] H. Sievertsson, J.-K. Chang, K. Folkers, and C. Y. Bowers, *J. Medicin. Chem.*, 1972, **15**, 8.
[71] Y. Baba, A. Arimura, and A. V. Schally, *J. Biol. Chem.*, 1971, **246**, 7581.
[72] M. Amoss, R. Burgus, R. Blackwell, W. Vale, R. Fellows, and R. Guillemin, *Biochem. Biophys. Res. Comm.*, 1971, **44**, 205.
[73] A. V. Schally, A. Arimura, Y. Bara, R. M. G. Nair, H. Matsuo, T. W. Redding, and L. Debeljuk, *Biochem. Biophys. Res. Comm.*, 1971, **43**, 393.
[74] A. V. Schally, T. W. Redding, H. Matsuo, and A. Arimura, *Endocrinology*, 1972, **90**, 1561.
[75] A. Arimura, L. Debeljuk, and A. V. Schally, *Endocrinology*, 1972, **91**, 529.
[76] H. Wagner, K. Böckel, M. Hrubesch, G. Grote, and W. H. Hauss, *Hormone Metabol. Res.*, 1972, **4**, 403.
[77] W. Wildmeister, H. Daweke, F. A. Gries, D. Gruneklee, J. Hessing, and F. A. Horster, *Hormone Metabol. Res.*, 1972, **4**, 368.

stage, all known differences between luteinizing hormone and follicle-stimulating hormone release can be accounted for without postulating the existence of separate releasing factors.[45]

It was previously thought that the tetrapeptide amide (44) brought about luteinizing hormone release but not follicle-stimulating hormone release.[78, 79] However, other workers have reported that the tetrapeptide amide possesses both types of activity, albeit with only *ca.* 10^{-4} times the potency of LRF.[80, 81] The heptapeptide amide (45) is slightly more active.[82] Other peptides, (46)—(48), previously thought to possess LRF activity, are now found to be inactive.[81] Similarly, a decapeptide with growth hormone-releasing activity (see p. 403),[81] TRF,[81] and a series of tripeptides related to TRF [70, 80] are all reported to be devoid of LRF activity.

Glp-Tyr-Arg-Trp-NH_2 (44) Glp-Tyr-Arg-Trp-Gly-His-Leu-NH_2 (45)

Glp-Val-Ser-NH_2 (46) Glp-Ser-Val-NH_2 (47) Glp-Gln-Ala-NH_2 (48)

To date, the only peptides found to possess appreciable LRF activity resemble the natural decapeptide amide sequence (43) fairly closely and, whilst no particular residues in the sequence seem to be of key significance, the overall integrity of the molecule seems important. Shorter peptides with preserved sequence, *e.g.* (99)—(102), (104)—(112), Table 4, seem much reduced in activity,[46b, 82] as do peptides with 'truncated' sequences in which internal residues are omitted (96)—(98), (103).[46b, 82, 83] The exceptions are analogues in which the *C*-terminal glycine amide is replaced by various substituted amides (90)—(95), thus providing the only compounds known to be more active than LRF itself.[84] Preservation of the overall length of the molecule and the lipophilic nature of the *C*-terminal sequence are perhaps not without significance in these instances. The free acid of the decapeptide (89) exhibits much reduced activity compared to LRF.[84]

Individual functional groups seem to vary in significance. The lactam of the pyroglutamyl residue is important since its replacement by secondary

[78] J.-K. Chang, H. Sievertsson, C. Bogentoft, B. L. Currie, K. Folkers, and C. Y. Bowers, *Biochem. Biophys. Res. Comm.*, 1971, **44**, 409.

[79] C. Y. Bowers, J.-K. Chang, H. Sievertsson, C. Bogentoft, B. L. Currie, and K. Folkers, *Biochem. Biophys. Res. Comm.*, 1971, **44**, 414.

[80] R. Guillemin, M. Amoss, R. Blackwell, J. Rivier, N. Ling, and W. Vale, *Biochem. Biophys. Res. Comm.*, 1972, **48**, 1093.

[81] A. V. Schally, A. Arimura, W. H. Carter, T. W. Redding, R. Geiger, W. König, H. Wissmann, G. Jaeger, J. Sandow, N. Yanaihara, C. Yanaihara, T. Hashimoto, and M. Sakagami, *Biochem. Biophys. Res. Comm.*, 1972, **48**, 366.

[82] J.-K. Chang, A. J. Humphries, R. H. Williams, H. Sievertsson, K. Folkers, and C. Y. Bowers, *Biochem. Biophys. Res. Comm.*, 1972, **47**, 1256.

[83] J. Rivier, M. Monahan, W. Vale, G. Grant, M. Amoss, R. Blackwell, R. Guillemin, and R. Burgus, *Chimia*, 1972, **26**, 303.

[84] M. Fujino, S. Kobayashi, M. Obayashi, S. Shinagawa, T. Fukuda, C. Kitada, R. Nakayama, I. Yamazaki, W. F. White, and R. H. Rippel, *Biochem. Biophys. Res. Comm.*, 1972, **49**, 863.

14 **Table 4**

Compound number	Structure	Biological activity LRF type	Biological activity FSH-RF type	Ref.
(43)	Glp -His -Trp -Ser -Tyr -Gly -Leu -Arg -Pro -Gly -NH_2(LRF)	100	100	
(49)	Glp -His -Pro -Gly -NH_2			
(50)	EtCO·Gly -His -Trp -Ser -Tyr -Gly -Leu -Arg -Pro -Gly -NH_2	0.2	—	46*b*
(51)	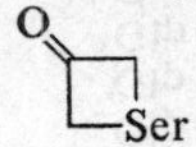 -His -Trp -Ser -Tyr -Gly -Leu -Arg -Pro -Gly -NH_2[a]	5—25	6	85
(52)	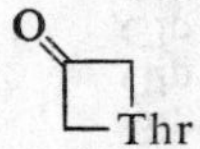 -His -Trp -Ser -Tyr -Gly -Leu -Arg -Pro -Gly -NH_2[b]	—	⩽5	85
(53)	Pro -His -Trp -Ser -Tyr -Gly -Leu -Arg -Pro -Gly -NH_2	< 0.1	< 0.1	85
(54)	Glp -Gly -Trp -Ser -Tyr -Gly -Leu -Arg -Pro -Gly -NH_2	'low'[c]	—	46*b*, 83, 89, 97
(55)	Glp -Phe -Trp -Ser -Tyr -Gly -Leu -Arg -Pro -Gly -NH_2	4—7	2	85
(56)	Glp -Lys -Trp -Ser -Tyr -Gly -Leu -Arg -Pro -Gly -NH_2	< 0.1	< 0.1	85
(57)	Glp -Arg -Trp -Ser -Tyr -Gly -Leu -Arg -Pro -Gly -NH_2	< 0.1	< 0.1	85
(58)	Glp-N^{π}-MeHis-Trp -Ser -Tyr -Gly -Leu -Arg -Pro -Gly -NH_2	2—3	—	83, 46*b*
(59)	Glp-N^{τ}-MeHis-Trp -Ser -Tyr -Gly -Leu -Arg -Pro -Gly -NH_2	{ 6	—	46*b*, 83
		1	1—2	85
(60)	Glp -Phe -Trp -Ser -Tyr -Gly -Leu -Arg -Pro -Gly -NH_2	4	—	46*b*
(61)	Glp -His -Gly -Ser -Tyr -Gly -Leu -Arg -Pro -Gly -NH_2	< 0.001	—	46*b*
(62)	Glp -His -Ala -Ser -Tyr -Gly -Leu -Arg -Pro -Gly -NH_2	< 0.001	—	46*b*
(63)	Glp -His -Phe -Ser -Tyr -Gly -Leu -Arg -Pro -Gly -NH_2	50	—	88
(64)	Glp -His -Phe -Ser -Tyr -Gly -Leu -Arg -Pro -Gly -NH_2	2	—	46*b*
(65)	Glp -His -Trp -Gly -Tyr -Gly -Leu -Arg -Pro -Gly -NH_2	1.5	—	46*b*
(66)	Glp -His -Trp -Ala -Tyr -Gly -Leu -Arg -Pro -Gly -NH_2	{ 3—6	16	85
		3.72	3.6	86
(67)	Glp -His -Trp -Thr -Tyr -Gly -Leu -Arg -Pro -Gly -NH_2	4	17	85
(68)	Glp -His -Trp -Gln -Tyr -Gly -Leu -Arg -Pro -Gly -NH_2	8	6	85

Table 4 (*cont.*)

Compound number	*Structure*	*Biological activity* LRF *type*	FSH-RF *type*	*Ref.*
(69)	Glp -His -Trp -Ser -Tyr(Cl) -Gly -Leu -Arg -Pro -Gly-NH_2[d]	8	5	85
(70)	Glp -His -Trp -Ser -Tyr(Cl_2) -Gly -Leu -Arg -Pro -Gly -NH_2[e]	< 1	< 1	85
(71)	Glp -His -Trp -Ser -Gly -Gly -Leu -Arg -Pro -Gly -NH_2	0.1	—	46*b*
(72)	Glp -His -Trp -Ser -Phe -Gly -Leu -Arg -Pro -Gly -NH_2	50	—	88
(73)	Glp -His -Trp -Ser -Tyr -Ala -Leu -Arg -Pro -Gly -NH_2	1	—	46*b*
(74)	Glp -His -Trp -Ser -Tyr -Gly -Gly -Arg -Pro -Gly -NH_2	{ 0.2	—	46*b*
		{ 3	5	85
(75)	Glp -His -Trp -Ser -Tyr -Gly -Ala -Arg -Pro -Gly -NH_2	5—6	3—5	85
(76)	Glp -His -Trp -Ser -Tyr -Gly -Val -Arg -Pro -Gly -NH_2	16	20—35	85
(77)	Glp -His -Trp -Ser -Tyr -Gly -Ile -Arg -Pro -Gly -NH_2	45	33	85
(78)	Glp -His -Trp -Ser -Tyr -Gly -Nle -Arg -Pro -Gly -NH_2	30	22	85
(79)	Glp -His -Trp -Ser -Tyr -Gly -Leu -Gly -Pro -Gly -NH_2	0.1	—	46*b*
(80)	Glp -His -Trp -Ser -Tyr -Gly -Leu -Gln -Pro -Gly -NH_2	5	2	95
(81)	Glp -His -Trp -Ser -Tyr -Gly -Leu -Leu -Pro -Gly -NH_2	0.8	0.5	95
(82)	Glp -His -Trp -Ser -Tyr -Gly -Leu -Pro -Arg -Gly -NH_2	0.2	—	95
(83)	Glp -His -Trp -Ser -Tyr -Gly -Leu -Orn -Pro -Gly -NH_2	6—12	5	85
(84)	Glp -His -Trp -Ser -Tyr -Gly -Leu -His -Pro -Gly -NH_2	—	—	96
(85)	Glp -His -Trp -Ser -Tyr -Gly -Leu -NVal-Pro -Gly -NH_2	—	—	96
(86)	Glp -His -Trp -Ser -Tyr -Gly -Leu -Lys -Pro -Gly -NH_2	{ 'active'	—	87
		{ 11—28	25	85
(87)	Glp -His -Trp -Ser -Tyr -Gly -Leu -Arg -Gly -Gly -NH_2	0.2	—	46*b*
(88)	Glp -His -Trp -Ser -Tyr -Gly -Leu -Arg -Pro -Ala -NH_2	~ 10	—	85
(89)	Glp -His -Trp -Ser -Tyr -Gly -Leu -Arg -Pro -Gly -OH	0.1	—	84
(90)	Glp -His -Trp -Ser -Tyr -Gly -Leu -Arg -Pro -NHMe	80—100	—	84
(91)	Glp -His -Trp -Ser -Tyr -Gly -Leu -Arg -Pro -NHEt	500	—	84

No.	Structure	Activity	Ref.
(92)	Glp -His -Trp -Ser -Tyr -Gly -Leu -Arg -Pro -NHPr	200—300	84
(93)	Glp -His -Trp -Ser -Tyr -Gly -Leu -Arg -Pro -$NHCH_2CH_2OH$	100—150	84
(94)	Glp -His -Trp -Ser -Tyr -Gly -Leu -Arg -Pro -piperidine	70—80	84
(95)	Glp -His——Trp -Ser -Tyr -Gly -Leu -Arg -Pro -morpholine	20—30	84
(96)	Glp————Trp -Ser -Tyr -Gly -Leu -Arg -Pro -Gly -NH_2	'inactive'[f] <0.001	46*b*, 83, 97
(97)	Glp -His -Trp -Ser -Tyr -Gly———Arg -Pro -Gly -NH_2	'low'	82
(98)	Glp -His -Trp -Ser -Tyr -Gly -Leu———Pro -Gly -NH_2	—	96
(99)	His -Trp -Ser -Tyr -Gly -Leu -Arg -Pro -Gly -NH_2	< 0.002	81
(100)	Glp -His -Trp -Ser -Tyr -Gly -Leu -Arg -Pro -NH_2	10 11	46*b*, 83 84
(101)	Trp -Ser -Tyr -Gly -Leu -Arg -Pro -Gly -NH_2	'inactive'	81
(102)	Glp -His -Trp -Ser -Tyr -Gly -Leu -Arg -NH_2	0.01	46*b*, 83
(103)	Glp -His -Trp -Ser -Tyr -Gly————Pro -Gly -NH_2	'inactive'	82
(104)	Glp -His -Trp -Ser -Tyr -Gly -Leu -NH_2	< 0.01	46*b*, 82
(105)	Glp -His -Trp -Ser -Tyr -Gly -NH_2	< 0.01 'low'	46*b*, 83 82
(106)	Glp -His -Trp -Ser -Tyr -Gly -OH	'inactive'	82
(107)	Glp -His -Trp -Ser -Tyr -NH_2	0.02	46*b*, 83
(108)	Glp -His -Trp -Ser -NH_2	'inactive' < 0.01	48 46*b*, 83
(109)	Glp -His -Trp -NH_2	'inactive' 0.1	81 46*b*
(110)	Glp -His -Trp -OH	'inactive'	81
(111)	Glp -His -NH_2	0.01	46*b*, 83
(112)	Leu -Arg -Pro -Gly -NH_2	'inactive'	82

[a] 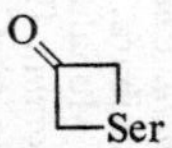 = 2-oxo-oxazolidine-4-carboxylic acid. [b] 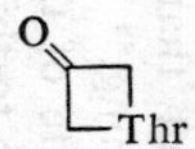 = 2-oxo-5-methyloxazolidine-4-carboxylic acid. [c] Dose response curve not parallel to LRF curve so that percentage activity could not be calculated. [d] Cl | Tyr = 3-chlorotyrosine. [e] Cl_2 | Tyr = 3,5-dichlorotyrosine. [f] This compound possessed inhibitory properties towards LRF (see p. 402).

amine (53),[85] open-chain amide (50),[46b] or 2-oxo-oxazolidine-carboxy-derivatives (51) and (52)[85] results in diminished activity. Similarly, the imidazole seems to be important since replacement of the histidine by a variety of residues, including N^{τ}-methylhistidine (59)[46b, 83, 85] (which gives such an active TRF analogue!) or N^{π}-methylhistidine (58),[46b, 83] causes considerable loss of activity. The hydroxy-group in position 4, (65)—(68).[46b, 85, 86] and the phenolic group in position 5, (69) and (72),[46b, 85] must also contribute since analogues in which they are missing have diminished activity. A basic side-chain in position 8 is also important, the guanyl side-chain of arginine seeming optimal, (79)—(85).[46b, 85, 87]

None of these groups can be said to be 'essential' in the sense that their presence is fundamental to the biological action of the hormones since each of them can be replaced to give biologically active analogues, even though sometimes with considerably reduced potency. The side-chains which, in a chemical sense, would not normally be regarded as reactive have similar flexibility. Thus, the indolyl side-chain in position 3, (62)—(64),[46b, 88] and the isobutyl side-chain in position 7, (75)—(79),[85] can both be replaced to give appreciably active analogues, although similar types of side-chain give the most active compounds. It is perhaps worth recording that if the logarithms of the reported biological activities of compounds modified in position 7, (75)—(78), are plotted against the lipophilicity parameter, π, for the appropriate side-chain, as has been done for angiotensin and oxytocin,[7a] the points all fall very close to a smooth curve. Hence the lipophilicity of the side-chain in this position is apparently more important than its stereochemistry, specifically with reference to the β-branching.[89]

In spite of the apparent activity of the 2-phenylalanine analogue it is tempting to think that the histidine residue may fulfil a special role in the LR effect. Both [2-dehistidine]-LRF (96) and [2-glycine]-LRF (54) suppress the LRF-stimulated secretion of luteinizing hormone from pituitary cell preparations, although the former appears entirely to lack and the latter to have only slight agonist activity. High molar ratios of antagonist/agonist are necessary ($\geqslant 10^3$) so that the affinities of the antagonists for the receptor are presumably appreciably less than that of natural LRF. The affinity of the 2-glycine analogue for the receptor is the least.[90]

85 M. Fujino, S. Kobayashi, M. Obayashi, T. Fukuda, S. Shinagawa, I. Yamazaki, R. Nakayama, W. F. White, and R. H. Rippel, *Biochem. Biophys. Res. Comm.*, 1972, **49**, 698.

86 R. Geiger, H. Wissmann, W. König, J. Sandow, A. V. Schally, T. W. Redding, L. Debeljuk, and A. Arimura, *Biochem. Biophys. Res. Comm.*, 1972, **49**, 1467.

87 J.-K. Chang, H. Sievertsson, B. L. Currie, C. Bogentoft, K. Folkers, and C. Y. Bowers, *J. Medicin. Chem.*, 1972, **15**, 623.

88 S. Sakakibara, quoted in ref. 85.

89 M. W. Monahan, J. Rivier, W. Vale, R. Guillemin, and R. Burgus, *Biochem. Biophys. Res. Comm.*, 1972, **47**, 551.

90 W. Vale, G. Grant, J. Rivier, M. Monahan, M. Amoss, R. Blackwell, R. Burgus, and R. Guillemin, *Science*, 1972, **176**, 933.

[2-Dehistidine]-LRF competes with [9-(^{3}H-proline)]-LRF for specific binding to anterior pituitary LRF receptors.[91] This and the lack of agonist activity on the part of [2-dehistidine]-LRF at ten-fold the concentration required to suppress the response to LRF indicate that the binding of the hormone to the receptor and the next stage in LH production are separate steps. Several models have been proposed for such 'two-step' processes,[90] but since analogues lacking histidine can have pronounced LRF activity, the situation is not so clear as it is, for example, in the case of the ribonuclease S-peptide, where the histidine residue is known to be essential to complete the active site of the enzyme. One wonders whether, in the present case, the type of inhibition could bear a closer resemblance to certain other enzymes in which non-productive binding modes exist such that peptides which are not substrates can be adsorbed on to the enzyme surface and block the approach of true substrates to the active site (*e.g.* the inhibition of subtilisin-BPN′ by *N*-benzoylarginine [92]).

The glycoproteins thyrotropin-stimulating hormone, luteinizing hormone, and follicle-stimulating hormone that are released by the hypothalamic factors show considerable homology, and it is perhaps surprising that the release of one of them should be controlled by a tripeptide and of the others by a decapeptide.[93] Since TRF releases prolactin, a similarity between the receptors involved in thyrotropin-stimulating hormone release and those involved in prolactin release has been inferred, suggesting that the identity of the receptor type has been preserved throughout the differentiation of the pituitary gland. It is therefore proposed that the receptor involved in luteinizing hormone release might also be similar and the conformation of LRF such that the two ends of the molecule are situated close together to present a local topochemistry resembling that of the TRF molecule. The discovery that the tetrapeptide (49), which is composed of the two terminal dipeptide sequences of LRF, has high TRF activity and low, but distinct, LRF activity provides experimental support.[94]

Growth Hormone-releasing Factor.—The structures (113) and (114) of substances possessing growth hormone-releasing activity, which were isolated from porcine hypothalami, have been confirmed by synthesis.[95] However, there is some doubt whether either of these compounds is the same as GH-RF. Unlike the natural extract, the synthetic material could not be demonstrated to stimulate the release of growth hormone *in vitro* in some species, whereas it did cause *in vivo* depletion of pituitary growth

[91] G. Grant, W. Vale, J. Rivier, and R. Guillemin, quoted in ref. 90.
[92] A. N. Glazer, *J. Biol. Chem.*, 1967, **242**, 433; C. S. Wright, R. A. Alden, and J. Kraut, *ibid.*, 1972, **66**, 283; J. D. Robertus, J. Kraut, R. A. Alden, and J. J. Birktoft, *Biochemistry*, 1972, **11**, 4293.
[93] G. Grant and W. Vale, *Nature New Biol.*, 1972, **237**, 183.
[94] J. Rivier, quoted in ref. 93.
[95] N. Yanaihara, C. Yanaihara, T. Hashimoto, Y. Kenmochi, T. Kaneko, H. Oka, S. Saito, A. V. Schally, and A. Arimura, *Biochem. Biophys. Res. Comm.*, 1972, **49**, 1280.

hormone. It is thought that an essential cofactor of the natural product may have been removed during the isolation. Difficulties and ambiguities in the assay of GH-RF probably account for the relatively slow progress in this area.[47]

Val-His-Leu-Ser-Ala-Glu-Glu-Lys-Glu-Ala

(113)

Val-His-Leu-Ser-Ala-Glu-Glu-Lys-Gln-Ala

(114)

Hypothalamic extracts have now yielded a peptide (115) which inhibits the secretion of growth hormone. The synthetic GH-RIF (somatostatin) has similar activity to the natural product.[99] It has no effect on the basal secretion of luteinizing hormone or follicle-stimulating hormone in concentrations which inhibit maximally the secretion of growth hormone. No structure–activity studies have yet been reported.

Ala-Gly-Cys-Lys-Asn-Phe-Phe-Trp-Lys-Thr-Phe-Thr-Ser-Cys (Cys³ and Cys¹⁴ joined by a bridge line)

(115)

Melanocyte-stimulating Hormone-release Inhibiting Factor.—Pharmacological evidence suggests that there are species differences of MSH-RIF. Thus, extracts of rat hypothalamus show *in vitro* MSH-RIF activity in both rat and frog preparations, whereas frog hypothalamic extracts are active *in vitro* in frog, but not in rat, preparations.[100] These differences and variations in assay technique probably account for the current confusion which prevails in this area.

The tripeptide amide (116), isolated from bovine hypothalami,[101] has been reported to possess MSH-RIF activity and this has been confirmed in the synthetic product.[102] This tripeptide amide is the side-chain of oxytocin and, perhaps significantly, preparations of rat hypothalami have been shown to possess enzymic systems capable of cleaving this fragment from the oxytocin molecule.[103] The synthetic tripeptide amide is degraded by

[96] J.-K. Chang, R. H. Williams, A. J. Humphries, N. Johansson, K. Folkers, and C. Y. Bowers, *Biochem. Biophys. Res. Comm.*, 1972, **47**, 727.
[97] W. Vale, quoted in ref. 89.
[98] A. V. Schally, A. Arimura, I. Wakabayashi, T. W. Redding, E. Dickerman, and J. Meites, *Experientia*, 1972, **28**, 205.
[99] P. Brazeau, W. Vale, R. Burgus, N. Ling, M. Butcher, J. Rivier, and R. Guillemin, *Science*, 1973, **179**, 77.
[100] C. L. Ralph and S. Sampath, *General and Comp. Endocrinol.*, 1966, **7**, 370.
[101] R. M. G. Nair, A. J. Kastin, and A. V. Schally, *Biochem. Biophys. Res. Comm.*, 1971, **43**, 1376.
[102] R. M. G. Nair, A. J. Kastin, and A. V. Schally, *Biochemistry*, in the press. (See also ref. 109.)
[103] M. E. Celis, S. Taleisnik, I. L. Schwartz, and R. Walter, *Proc. Nat. Acad. Sci. U.S.A.*, 1971, **68**, 1428.

plasma [104] and by tissue extracts,[105] and brain preparations are particularly active in this respect. Suggestively, extracts of the median eminence, where the RIF is 'stored', were less active.[105] The side-chain tripeptide amides (117) and (118) of arginine and lysine vasopressin also possess MSH-RIF activity, whereas other peptides tested [(119)—(122)] were inactive.[106] A preferred conformation for the tripeptide amide (116) has been proposed.

Pro-Leu-Gly-NH_2	Pro-Lys-Gly-NH_2	Pro-Arg-Gly-NH_2
(116)	(117)	(118)
Leu-Gly-NH_2	D-Leu-Gly-NH_2	Pro-D-Leu-Gly-NH_2
(119)	(120)	(121)
	Glp-Leu-Gly-NH_2	
	(122)	

Other workers have reported that the tripeptide amide (116) is devoid of *in vitro* MSH-RIF activity in both the rat and the frog.[107] They find, however, that tocinoic acid (123; A = Ile, B = OH, C = Gln) and, especially, tocinamide (123; A = Ile, B = NH_2, C = Gln), the ring portion and its amide, respectively, of the oxytocin molecule, possess significant MSH-RIF activity in rat, hamster, and some amphibian preparations, but not in frog (*Rana pipiens*).[108] Oxytocin (123; A = Ile, B = Pro-Leu-Gly-NH_2, C = Gln), lysine vasopressin (123; A = Phe, B = Pro-Lys-Gly-NH_2), pressinoic acid (123; A = Phe, B = OH, C = Gln), pressinoic acid amide (123; A = Phe, B = NH_2, C = Gln), and, apparently, [4-serine]-tocinoic acid (123; A = Ile, B = OH, C = Ser) do not show inhibitory activity. Tocinoic acid is reported by others to be devoid of MSH-RIF activity in the intact rat.[106] None of these compounds has been isolated from hypothalamic extracts.

Cys-Tyr-A-C-Asn-Cys-B (the two Cys residues joined by a bridge)

(123)

Another peptide amide (124) extracted from bovine hypothalami has been reported to possess MSH-RIF activity, but no structure–activity correlations have yet been reported.[109] The similarity between this peptide and pentapeptide (125),[110] which has significance from the point of view of melanocyte-stimulating activity, may not be entirely a coincidence. There

[104] A. J. Kastin, A. V. Schally, and S. Viosca, *Proc. Soc. Exp. Biol. Med.*, 1971, **137**, 1437.
[105] N. Marks and R. Walter, *Proc. Soc. Exp. Biol. Med.*, 1972, **140**, 673.
[106] M. E. Celis, S. Hase, and R. Walter, *F.E.B.S. Letters*, 1972, **27**, 327.
[107] A. Bower, M. E. Hadley, and V. J. Hruby, *Biochem. Biophys. Res. Comm.*, 1971, **45**, 1185.
[108] V. J. Hruby, C. W. Smith, A. Bower, and M. E. Hadley, *Science*, 1972, **176**, 1331.
[109] R. M. G. Nair, A. J. Kastin, and A. V. Schally, *Biochem. Biophys. Res. Comm.*, 1972, **47**, 1420.
[110] H. Yajima and K. Kubo, *J. Amer. Chem. Soc.*, 1965, **87**, 2039; and references therein.

is some evidence for the existence of an MSH-RF and it has been suggested that a portion of the ring of oxytocin (126) is involved.[111]

Pro-His-Phe-Arg-Gly-NH_2 His-Phe-Arg-Trp-Gly
(124) (125)

Cys-Tyr-Ile-Glu-Asn-Cys-OH
(126)

Distribution studies with isotopically labelled MSH-RIF (124) show that this compound accumulates in the pineal gland. Control of the release of melanocyte-stimulating hormone might therefore involve a complex system, implicating, perhaps, various factors with differing potencies.[112]

Corticotropin Releasing Factor.—In the absence of clear evidence about the identity of CRF, consideration of the structural homologies of melanocyte-stimulating hormone and corticotropin prompted CRF screening of peptides claimed to possess MSH-RF activity.[113] Pressinoic acid (123; A = Phe, B = OH, C = Gln) (3—30 ng per pituitary gland) exhibits corticotropin-releasing activity and, whilst this is not evidence that pressinoic acid is the natural CRF, activity at such small doses is impressive. Deaminopressinoic acid amide is active in doses of 30—30 000 ng per pituitary; deaminopressinoic acid and pressinoic acid amide gave no evidence of CRF activity.

4 Pituitary Hormones

Adrenocorticotropin (ACTH).—Further studies of the interaction of peptides related to ACTH (127, Table 5) with adrenal cortex receptor preparations have been reported. Labelled ACTH is displaced competitively from a crude cortical preparation by analogues known to exhibit *in vivo* steroidogenic activity [*e.g.* (128)], but not by biologically inactive analogues [*e.g.* (129) and (130)].[114] Similar results have now been obtained with a cortical plasma membrane preparation.[115] Further, peptides (131) and (132), thought to possess a higher *in vivo* steroidogenic activity than ACTH, have been found to have about the same affinity for the receptor as the much less potent 1—23 ACTH (128). These findings support the idea that these highly active analogues owe their potency to an increased resistance to enzymic degradation. Peptide (133), which has a low order of *in vivo* activity, does not compete effectively in the binding process with

[111] M. E. Celis, S. Taleisnik, and R. Walter, *Biochem. Biophys. Res. Comm.*, 1971, **45**, 564.

[112] A. J. Kastin, R. M. G. Nair, and S. Viosca, 'Proceedings 54th Endocrinology Meeting, June, 1972', in the press (quoted in ref. 109).

[113] M. Saffran, A. F. Pearlmutter, E. Rapino, and G. V. Upton, *Biochem. Biophys. Res. Comm.*, 1972, **49**, 748.

[114] K. Hofmann, W. Wingender, and F. M. Finn, *Proc. Nat. Acad. Sci., U.S.A.*, 1970, **67**, 829.

[115] F. M. Finn, C. C. Widnell, and K. Hofmann, *J. Biol. Chem.*, 1972, **247**, 5695.

Table 5 *Analogues of* ACTH (*biological activities are referred to in the text*)

Compound number	*Structure*
(127)	Ser-Tyr-Ser-Met-Glu-His-Phe-Arg-Trp-Gly-Lys-Pro-Val-Gly-Lys-Lys-Arg-Arg-Pro-Val-Lys-Val-Tyr-Pro-...-Phe
(128)	Ser-Tyr-Ser-Met-Glu-His-Phe-Arg-Trp-Gly-Lys-Pro-Val-Gly-Lys-Lys-Arg-Arg-Pro-Val-Lys-Val-Tyr
(129)	Ser-Tyr-Ser-Met-Gln-His-Phe-Arg-Trp-Gly
	CHO | (on Lys)
(130)	Ac-Ser-Tyr-Ser-Met-Glu-His-Phe-Arg-Trp-Gly-Lys-Pro-Val-NH_2
(131)	D-Ser-Tyr-Ser-Met-Glu-His-Phe-Arg-Trp-Gly-Lys-Pro-Val-Gly-Lys-Lys-Lys-Lys-NH_2
(132)	β-Ala-Tyr-Ser-Met-Glu-His-Phe-Arg-Trp-Gly-Lys-Pro-Val-Gly-Lys-Lys-Lys-$NH(CH_2)_4NH_2$
(133)	D-Ser-Tyr-Ser-Met-Glu-His-Phe-Arg-Trp-Gly-Lys-Pro-Val-Gly-Lys-Lys-NH_2
(134)	Lys-Pro-Val-Gly-Lys-Lys-Arg-Arg-Pro-Val-NH_2
	Nps | (on Trp)
(135)	Ser-Tyr-Ser-Met-Glu-His-Phe-Arg-Trp-Gly-Lys-Pro-Val-Gly-Lys-Lys-Arg-Arg-Pro-Val-Lys-Val-Tyr-Pro-...-Phe[a]
	Dns | (on Lys)
(136)	Ser-Tyr-Ser-Met-Glu-His-Phe-Arg-Trp-Gly-Lys-Pro-Val-Gly-Lys-Lys-Arg-Arg-Pro-Val-Lys-Val-Tyr-Pro[b]
(137)	Ser-Tyr-Ser-Met-Glu-Phe-Phe-Arg-Trp-Gly-Lys-Pro-Val-Gly-Lys-Lys-Arg-Arg-Pro
(138)	Ala-Ser-Tyr-Ser-Met-Glu-His-Phe-Arg-Trp-Gly-Lys-Pro-Val-Gly-Lys-Lys-Arg-Arg-Pro
(139)	Pro-Ser-Tyr-Ser-Met-Glu-His-Phe-Arg-Trp-Gly-Lys-Pro-Val-Gly-Lys-Lys-Arg-Arg-Pro
(140)	Ser-Tyr-Ser-Met-Gln-His-Phe-Arg-Trp-Gly-Lys-Pro-Val-Gly-Lys-Lys-Arg-Arg-Pro

[a] Nps = *o*-nitrophenylsulphenyl. [b] Dns = *NN*-dimethylamino-naphthalenesulphonyl.

more active peptides, which is in agreement with previous deductions that the region -Lys-Lys-Arg-Arg- is important for binding. Peptide (134) does compete with ACTH in the binding process, but is biologically inactive because it lacks the 'essential' region -His-Phe-Arg-Trp-Gly-. Studies of the binding of the series of peptides in which each of the lysine residues of peptide (134), in turn, is blocked by an N^{ε}-formyl group indicate that Lys-11 is particularly important in binding.

ACTH-Receptor preparations have also been obtained from ACTH-responsive mouse adrenal cortical tumours.[9a] The effectiveness of various chemically modified forms of ACTH to compete with ^{125}I-ACTH in the binding process is again directly related to their biological activity. Two types of binding site seem to be present on the membrane, one type with $K \sim 9 \times 10^{11}\,l\,mol^{-1}$ and another, much more abundant, type with $K \sim 3 \times 10^{7}\,l\,mol^{-1}$. The stability of the receptors towards various agents suggests that they may be composed partially of protein or lipid. With this system, calcium is not necessary for the ACTH-receptor interaction, but its presence does seem to be necessary for adenylate cyclase activation. A high calcium content inhibits adenylate cyclase but it also has a direct inhibitory effect on binding.

The importance of the calcium ion concentration in the physiological response to ACTH is well established [116, 117] and a theory for the mechanism of the response has been proposed.[117] It is suggested that calcium ions, which are acting in an inhibitory capacity, are displaced by ACTH from the hormone binding site, or from an area adjacent to the binding site. The resulting increase in tissue cyclase activity could lead to a change in the protein components in the cell and hence to steroidogenesis. It is postulated that the displaced calcium ions might, in turn, facilitate steroid release.

Perhaps caution should be indicated by the observation that peptides (131) and (132), despite their potency, are less active than 1—24 corticotropin in activating adenylate cyclase.[115] Similarly, *o*-nitrophenylsulphenyl-ACTH (135) only stimulates cyclic AMP inclusion in isolated rat adrenal cells to a maximum 3% of that attained by ACTH,[46d] whereas, in some preparations, *o*-nitrophenylsulphenyl-ACTH actually inhibits ACTH-induced stimulation of adenylate cyclase.[118, 119] Since both peptides stimulate steroidogenesis to the same maximal extent,[46d] it is suggested that either a minimal increase in cyclase activity is sufficient to stimulate steroidogenesis or that other unknown factors are involved in mediating the action of ACTH at low concentrations. The idea that there are two different types of ACTH-receptor [9a] has also been invoked to account for the apparent inconsistencies and the possibility indicated that *o*-nitro-

[116] G. Sayers, R. J. Beall, and S. Seelig, *Science*, 1972, **175**, 1131.

[117] R. P. Rubin, R. A. Carchman, and S. D. Jaanus, *Nature New Biol.*, 1972, **240**, 150.

[118] J. Ramachandran, in 'Hormonal Proteins and Peptides', ed. C. H. Li, Academic Press, New York, 1972, Vol. 2.

[119] Y. C. Kong, W. R. Moyle, and J. Ramachandran, *Proc. Soc. Exp. Biol. Med.*, 1972, **141**, 350.

phenylsulphenyl-ACTH may be 'recognized' by only one of the two types of receptor.[46d] Differences between ACTH and *o*-nitrophenylsulphenyl-ACTH in other types of biological assay, including lipolysis and skin darkening, have also been reported.[46d]

The solution conformation of the biologically active [21-(N^{ε}-dansyl-lysine)]-ACTH-(1—24)-tetracosapeptide (136) has been studied by resonance energy transfer and fluorescence depolarization techniques.[120] A random-coil conformation is indicated for the 9—21 residue sequence and this is remarkably independent of solvent. Similar findings have been reported previously with other techniques, but since this approach offers the chance of conformational measurements in the presence of receptor preparations, it is of unique interest. New analogues reported in 1972 include peptides (137),[121] (138), (139),[122] and (140);[123] the latter had previously been described in preliminary form. The replacement of the 6-histidine residue by phenylalanine results in a marked diminution of potency. Extension of the molecule at the amino terminus by an alanine or proline residue leads to a reduction of *in vivo* steroidogenic activity of about one-third and three-quarters, respectively; replacement of the Glu-5 residue by a glutamine residue leads to a reduction of about two-thirds.

Growth Hormone.—A review of human growth hormone [124] and sequence comparisons of human growth hormone, prolactin, and other hormones [125, 126] have appeared. Synthetic fragments of human growth hormone which possess growth-promoting activity have also been reported.[125, 127]

Oxytocin and Vasopressin.—A three-dimensional structure has been proposed for oxytocin, based mainly on ^{1}H n.m.r. studies of the hormone in dimethyl sulphoxide solution [128] (see Vol. 4, p. 372). C.d. studies of the 1,6-diaminosuberic acid analogue not only confirm the main features of the structure, they also allow the various transitions to be assigned to particular chromophores.[129] The proposed structure (141) possesses two β-turns, one in the L-Tyr···L-Asn region of the ring structure and one involving the *C*-terminal tetrapeptide sequence. The asparagine residue plays a critical part in both β-turns since it is hydrogen-bonded, through its peptidic carbonyl, to the NH of the tyrosine residue and, through its side-chain carbonyl, to the NH of the leucine residue. Cystine (especially), tyrosine, and glycine also contribute to the stabilization of the backbone,

120 P. W. Schiller, *Proc. Nat. Acad. Sci. U.S.A.*, 1972, **69**, 975.
121 J. Blake and C. H. Li, *Biochemistry*, 1972, **11**, 3459.
122 J. Blake, K.-T. Wang, and C. H. Li, *Biochemistry*, 1972, **11**, 438.
123 C. H. Li and B. Hemmasi, *J. Medicin. Chem.*, 1972, **15**, 217.
124 J. M. Tanner, *Nature*, 1972, **237**, 433.
125 F. Chillemi, A. Aiello, and A. Pecile, *Nature New Biol.*, 1972, **238**, 245.
126 T. A. Bewley, J. S. Dixon, and C. H. Li, *Internat. J. Peptide Res.*, 1972, **4**, 281.
127 F. Chillemi and A. Pecile, *Experientia*, 1971, **27**, 385.
128 D. W. Urry and R. Walter, *Proc. Nat. Acad. Sci. U.S.A.*, 1971, **68**, 956.
129 D. W. Urry, S. Sakakibara, S. Hase, and R. Walter, quoted in ref. 145.

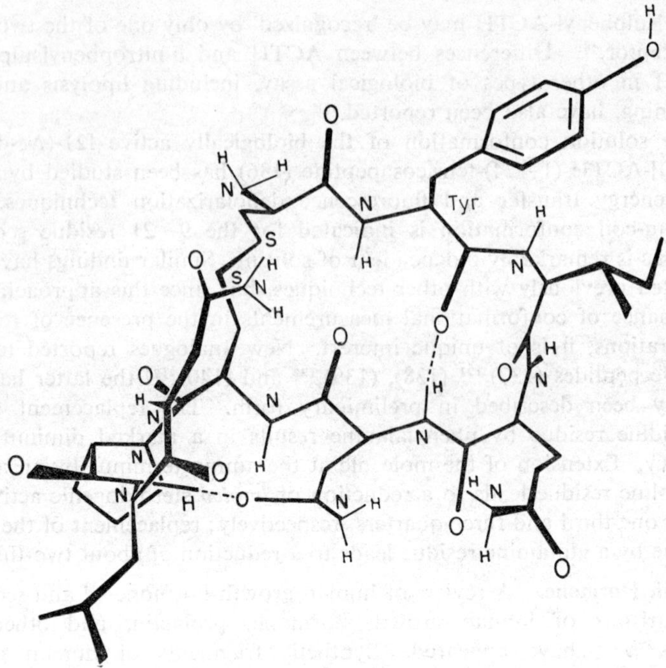

(141)

whereas isoleucine, glutamine, proline, and leucine are situated at corners of the β-turns where their side-chains interact minimally with the rest of the molecule. The conformation of oxytocin in aqueous solution may well be different [130, 131] but since, in terms of the lipophilicity of the micro-environment, the transition from aqueous to DMSO solution may resemble the interaction of the hormone in aqueous solution with the biological receptor, it is thought that the above conformation may still be relevant in the biological context.

The conformations of 8-lysine-vasopressin, 8-arginine-vasopressin, and deamino-8-lysine-vasopressin in deuteriated DMSO solution appear to be somewhat similar to that of oxytocin, except that the *C*-terminal tripeptide amide portion seems to be more flexible in the vasopressin series.[132, 133] In aqueous solution, ^{1}H n.m.r. spectra of deamino-8-lysine-vasopressin show

[130] J. Feeney, G. C. K. Roberts, J. H. Rockey, and A. S. V. Burgen, *Nature New Biol.*, 1971, **232**, 108.

[131] J. D. Glickson, D. W. Urry, and R. Walter, *Proc. Nat. Acad. Sci. U.S.A.*, 1972, **69**, 2566; R. Deslauriers, R. Walter, and I. P. C. Smith, *Biochem. Biophys. Res. Comm.*, 1972, **48**, 854.

[132] R. Walter, J. D. Glickson, I. L. Schwartz, R. T. Havran, J. Meienhofer, and D. W. Urry, *Proc. Nat. Acad. Sci. U.S.A.*, 1972, **69**, 1920.

[133] J. D. Glickson, D. W. Urry, R. T. Havran, and R. Walter, *Proc. Nat. Acad. Sci. U.S.A.*, 1972, **69**, 2136.

differences from 8-lysine-vasopressin which must indicate variations in the peptide backbone reflected in a different conformational arrangement of the aromatic rings.[133, 134] The deamino-analogue occupies an intermediate position in conformational terms between oxytocin and lysine vasopressin.

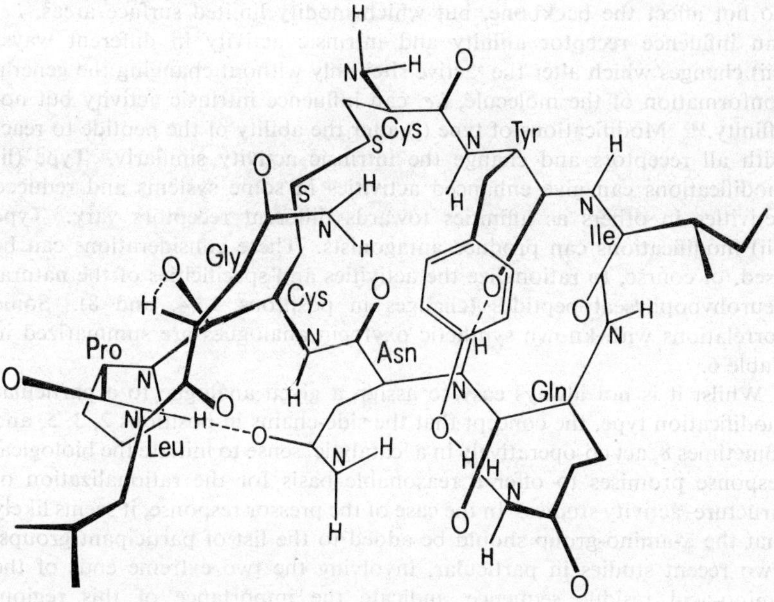

(142)

In the proposed biologically active conformation of oxytocin (142), the 'co-operative model', the tyrosine side-chain is folded back over the twenty-membered ring and, possibly, hydrogen-bonded to one of the primary amide groups. The resulting compact structure has one hydrophobic surface – essentially featureless apart from the disulphide bond and the carbonyl group of isoleucine – and one hydrophilic surface in which the *C*-terminal tripeptide residue sequence, the asparagine side-chain, the phenolic group, and, perhaps, the glutamine side-chain, all participate. Side-chains in positions 3, 4, 7, and 8 can engage in intermolecular interactions and do not make significant contributions to the stabilization of the peptide backbone.[135]

[134] P. H. Von Dreele, A. I. Brewster, J. Dadok, H. A. Scheraga, F. A. Bovey, M. F. Ferger, and V. du Vigneaud, *Proc. Nat. Acad. Sci. U.S.A.*, 1972, **69**, 2169; P. H. Von Dreele, H. A. Scheraga, D. F. Dyckes, M. F. Ferger, and V. du Vigneaud, *ibid.*, 1972, **69**, 3322.

[135] R. Walter, I. L. Schwartz, J. H. Darnell, and D. R. Urry, *Proc. Nat. Acad. Sci. U.S.A.*, 1971, **68**, 1355; R. Walter in 'Structure–Activity Relationships of Protein and Polypeptide Hormones', ed. M. Margoulies and F. C. Greenwood, Excerpta Medica, Amsterdam, 1971, p. 181.

This model has been used to explain many known structure–activity relationships in terms of the following three types of structural modifications: (i) changes which affect the stabilization of the backbone of the peptide which would alter the relative spatial positions of the side-chains, *i.e.* influence both receptor affinity and intrinsic activity; (ii) changes which do not affect the backbone, but which modify limited surface areas, *i.e.* can influence receptor affinity and intrinsic activity in different ways; (iii) changes which alter the 'active site' only without changing the general conformation of the molecule, *i.e.* can influence intrinsic activity but not affinity.[135] Modifications of type (i) alter the ability of the peptide to react with all receptors and change the intrinsic activity similarly. Type (ii) modifications can give enhanced activities in some systems and reduced activities in others as affinities towards different receptors vary. Type (iii) modifications can produce antagonists. These considerations can be used, of course, to rationalize the activities and specificities of the natural neurohypophyseal peptides (changes in positions 3, 4, and 8). Some correlations with known synthetic oxytocin analogues are summarized in Table 6.

Whilst it is not always easy to assign a given analogue to a particular modification type, the concept that the side-chains in positions 2, 3, 5, and sometimes 8, act co-operatively in a 'catalytic' sense to initiate the biological response promises to offer a reasonable basis for the rationalization of structure–activity studies. In the case of the pressor response, it seems likely that the α-amino-group should be added to the list of participant groups. Two recent studies in particular, involving the two extreme ends of the amino-acid residue sequence, indicate the importance of this region. [1-(L-2-Hydroxy-3-mercaptopropionic acid)]-oxytocin is reported to possess three times the oxytocic activity of oxytocin as measured by the isolated rat uterus preparation.[137] Lysine-vasopressinoic acid methylamide and dimethylamide are found to have a lower order of biological activity in oxytocic, rat pressor, and avian depressor assays.[138] One might speculate that the alkyl substituents on the terminal amide group prevent the postulated folding of the β-structure of the *C*-terminal tetrapeptide sequence over the ring portion of the molecule; whereas, in the first analogue, a bonding interaction between the α-hydroxy-group and the amide favour this fold-over.

The importance of the tripeptide amide *C*-terminal sequences in the neurohypophyseal hormones has been reaffirmed by synthesis of various derivatives of tocinoic and pressinoic acids, the cyclic portions of the oxytocin and vasopressin molecules, respectively (Table 7). Clearly, the ring portion does possess intrinsic oxytocic activity which can be increased by a factor of 100 in the tocinoic series by blocking the terminal carboxy-group

[136] S. Hase, I. L. Schwartz, and R. Walter, *J. Medicin. Chem.*, 1972, **15**, 126.
[137] M. Walti and B. D. Hope, *J.C.S. Perkin I*, 1972, 1946.
[138] J. D. Glass and V. du Vigneaud, *J. Medicin. Chem.*, 1972, **15**, 486.

Table 6 *Structure–activity relationships in terms of the 'co-operative model' (from refs.* 135 *and* 136)

Modification type	*Residue(s) changed*	*Nature of structural change*	*'Oxytocin-like' biological activity*	*Notes*
(i)	1	1-Deamino analogue	Enhanced activity	Tighter packing of section of β-turn indicated by 'ready' formation of H bond between NH of tyrosine and CO of asparagine Change in the dihedral angle of the bridge leading to particularly favourable disposition of 'catalytic' side-chains (not for pressor effect—see later)
(i)	1	1-Deamino-1-seleno-oxytocin	Enhanced activity	Change of dihedral angle in bridge leads to favourable redisposition of side-chains
(i)	1, 6	1,6-Dialanine analogue	0.2% Activity	Loss of favoured conformation, but pseudo-cyclic conformation has finite probability
(iii)	2	Alkylation of phenolic group	Introduction of antagonist activity	Backbone conformation basically unchanged. Arrangement of side-chains at 'catalytic site' disturbed
(ii)	3	Replacement with other alkyl side-chains	Diminished affinity for the receptor	Side-chain free to participate in intermolecular interactions
(ii)	3	Replacement with phenylalanine	Decreased affinity and decreased intrinsic activity	Stacked aromatic rings too bulky for isoleucine receptor. Affinity towards 'pressin' receptors increased. Redisposition of 'active centre'
?	4	Threonine analogue	Enhanced activity	Unexplained (see p. 418)
(i)	5	Replacement by other amino-acid residues (*e.g.* α, γ-diamino-butyric acid)[136]	Loss of affinity and intrinsic activity	Loss of β-turn stabilization leading to conformational 'blurring'
(i)	6	1-Deamino-6-seleno oxytocin	Diminished activity	Change of dihedral angle in bridge leads to unfavourable redisposition of side-chains

Table 7 *Biological activity of ring structures related to the parent hormones and 1-deamino analogues*

Compound number	Structure	Isolated rat uterus	Avian depressor	Mammary gland milk let-down	Anti-diuretic	Rat pressor	Ref.
		Biological activity[a]					
(143)	Oxytocin	546	507	410	2.7	3.1	b
(144)	1-Deamino-oxytocin	803	957	541	19	1.4	b
(145)	8-Lysine-vasopressin	4.8	48	31	203	300	b
(146)	1-Deamino-8-lysine-vasopressin	12	61	32	301	126	b
(147)	8-Arginine-vasopressin	12	100	~70	503	487	b
(148)	1-Deamino-8-arginine-vasopressin	27	150	80	1300	370	b
(149)	Cys-Tyr-Ile-Gln-Asn-Cys	0.2—0.3	0	—	—	—	139
(150)	βMP-Tyr-Ile-Gln-Asn-Cys[c]	3.7 ± 0.3	0[d]	—	—	—	139
(151)	Cys-Tyr-Ile-Gln-Asn-Cys-NH_2	3.2 ± 0.2	0	—	—	—	140
		2.7	—	5.9	1.3	0	141, 142
(152)	βMP-Tyr-Ile-Gln-Asn-Cys-NH_2	34.2 ± 3.0	0	—	—	—	140
		16.47	—	—	0.21	—	142
(153)	Cys-Tyr-Phe-Gln-Asn-Cys	0	0	—	—	0	143
(154)	βMP-Tyr-Phe-Gln-Asn-Cys	0.05—0.5	0	—	—	0	143
(155)	Cys-Tyr-Phe-Gln-Asn-Cys-NH_2	0.05—0.5	0	—	—	0	143
		0.26	—	0.3	0.7 (0.39—1.05)	0	141, 142
(156)	βMP-Tyr-Phe-Gln-Asn-Cys-NH_2	0.05—0.5	0	—	—	0	143
		1.03	—	—	1.2	—	142

[a] Activities are quoted in terms of International Units per milligram. [b] For original refs. see ref. 145. [c] βMP = β-mercaptopropionic acid.
[d] Inhibits response to oxytocin by 60% at 2000 : 1 weight ratio.

Table 8 *Biological activities of 'ring-modified' analogues*

Compound number	Structure: CH——X——CH_2 (bridging) RCHCO-Tyr-A-Gln-Asn-NHCHCO-Pro-B-Gly-NH_2	Biological activity: Rat uterus Isolated	Rat uterus in situ	Avian depressor	Mammary gland in vitro	Mammary gland in vivo	Anti-diuretic (rat)	Rat pressor	Ref.
(156)	R = H; X = CH_2—S—CH_2; A = Ile; B = Leu	60	170	29	346	—	0.3	0.025	146
(157)	R = H; X = S; A = Ile; B = Leu	9.2	20.9	6.8	42	—	0.17	—	147
(158)	R = H; X = CH_2; A = Ile; B = Leu	2.6	0.54	0	—	—	0.05	—	147
(159)	R = NH_2; X = CH_2—S—S; A = Ile; B = Leu	0.75	—	—	—	—	—	—	148
(160)	R = H; X = CH_2—S—S; A = Ile; B = Leu	3	—	0	0	—	—	—	148
(161)	R = H; X = CH_2—S; A = Ile; B = Leu	1898	1251	1127	562	—	21	1.44	144
(162)	R = H; X = S—CH_2	930	2900	—	—	—	—	—	7a
(163)	R = H; X = CH_2—CH_2; A = Ile; B = Leu	160	100	46.9	—	148	4.9	0.11	145
(164)	R = H; X = CH_2—CH_2; A = Phe; B = Lys	5.3	—	2.4	—	3.2	208	6.2	145
(165)	R = H; X = CH_2—CH_2; A = Ile; B = Lys	63.6	—	20.5	—	36.3	34.8	9.6	145
(166)	R = H; X = CH_2—CH_2; A = Phe; B = Arg	15.4	—	12.8	—	13.7	1274	37.7	145
(167)	R = H; X = CH_2—CH_2; A = Ile; B = Arg	135	—	64.2	—	36.8	313	27.0	145
(168)	R = H; X = NHCO; A = Ile; B = Leu	1.13	—	0[a]	—	2.9	0.01	0	149

[a] Inhibits response to oxytocin and vasopressin when injected 1—2 min previously. Weight ratio 833 : 1 gives 13% reduction of oxytocin response; 1000 : 1, 25%; 2000 : 1, 62.5%. Weight ratio 200 : 1 gives 25% reduction of lysine-vasopressin response.

as the amide and removing the α-amino-group [compounds (149)—(152)].[139–142] Apparently, the ring structures also retain residual anti-diuretic and milk let-down properties [compounds (151), (152), (155), and (156)].[140–143] However, neither of the ring structures possesses the full structural features necessary to stimulate the rat pressor effect or to bring about the vasodepressor response in chicken [compounds (149)—(156)].[139–143]

A great many compounds now attest that the disulphide in the ring is not essential for biological activity [Table 8, compounds (156)—(168)][144–147] and, indeed, carba-analogues, *i.e.* those in which a sulphur atom is replaced by a carbon atom, include some very potent compounds, *e.g.* (161) and (162).[7a, 144] The series of ethylene carba-analogues [1,6-diaminosuberic acid analogues, (163)—(167)] are consistently less potent than their disulphide counterparts and the ratios of the activities of the various pairs of ethylene and disulphide compounds are independent of the absolute activities displayed. It is therefore argued that the replacement of the disulphide bridge in any of the hormones by an ethylene bridge disturbs the backbone conformation enough to alter the spatial relationship between all the residues [modification type (i)].[145] Further information about the importance of ring-size has also appeared.[147] Ring-size is obviously not the only criterion determining the effectiveness of the ring structure, as may be seen by comparing the potency of the enlarged ring analogue, deamino-[1,6-homolanthionine]-oxytocin (156) with that of the closely related analogues (157)—(162).[7a, 144, 146–148a] The dimers of oxytocin (169)

```
Cys-Tyr-Ile-Gln-Asn-Cys-Pro-Leu-Gly-NH2
|                   |
Cys-Tyr-Ile-Gln-Asn-Cys-Pro-Leu-Gly-NH2
```

(169)

```
                              Cys-Tyr-Ile-Gln-Asn-Cys-Pro-Leu-Gly-NH2
                              |                   |
NH2-Gly-Leu-Pro-Cys-Asn-Gln-Ile-Tyr-Cys
```

(170)

139 V. J. Hruby, C. W. Smith, D. K. Linn, M. F. Ferger, and V. du Vigneaud, *J. Amer. Chem. Soc.*, 1972, **94**, 5478.
140 V. J. Hruby, M. F. Ferger, and V. du Vigneaud, *J. Amer. Chem. Soc.*, 1971, **93**, 5539.
141 M. Zaoral and M. Flegel, *Coll. Czech. Chem. Comm.*, 1972, **37**, 1539.
142 M. Zaoral and M. Flegel, *Coll. Czech. Chem. Comm.*, 1972, **37**, 2639.
143 M. F. Ferger, W. C. Jones, D. F. Dyckes, and V. du Vigneaud, *J. Amer. Chem. Soc.*, 1972, **94**, 982.
144 K. Jošt and F. Šorm, *Coll. Czech. Chem. Comm.*, 1971, **36**, 234.
145 S. Hase, S. Sakakibara, M. Wahrenburg, M. Kirchberger, I. L. Schwartz, and R. Walter, *J. Amer. Chem. Soc.*, 1972, **94**, 3590.
146 Z. Procházka, K. Jošt, and F. Šorm, *Coll. Czech. Chem. Comm.*, 1972, **37**, 289.
147 K. Jošt and F. Šorm, *Coll. Czech. Chem. Comm.*, 1971, **36**, 2795.
148 (*a*) D. Jarvis, M. Bodanszky, and V. du Vigneaud, *J. Amer. Chem. Soc.*, 1961, **83**, 4780; D. Jarvis, B. M. Ferrier, and V. du Vigneaud, *J. Biol. Chem.*, 1965, **240**, 3553; (*b*) D. Yamashiro, D. B. Hope, and V. du Vigneaud, *J. Amer. Chem. Soc.*, 1968, **90**, 3857; (*c*) H. L. Anning and D. Yamashiro, *J. Amer. Chem. Soc.*, 1970, **92**, 5214.
149 S. Hase, S. Sakakibara, M. Wahrenburg, and R. Walter, *J. Medicin. Chem.*, 1972, **15**, 1017.

Table 9 *Biological activities of 4-threonine analogues of oxytocin*[a]

Compound number	*Structure*	*Rat uterus* *No* Mg^{2+}	*Rat uterus* 0.5 mmol l^{-1} Mg^{2+}	*Biological activity*[b] *Avian depressor*	*Rabbit milk let-down*	*Rat antidiuretic*	*Rat pressor*
(171)	Mesotocin	382 ± 14	478 ± 10	830 ± 24	298 ± 23	6.1 ± 0.4	6.4 ± 0.2
(172)	[4-Threonine]-oxytocin	923 ± 95	719 ± 83	1480 ± 28	543 ± 23	1.8 ± 0.3	0.43 ± 0.01
(173)	[4-Threonine]-mesotocin	520 ± 28	565 ± 23	1545 ± 59	519 ± 37	2.6 ± 0.2	1.08 ± 0.03
(174)	1-Deamino-[4-threonine]-oxytocin	149 ± 21	245 ± 22	781 ± 136	385 ± 14	0.9 ± 0.1	< 0.1
(175)	1-Deamino-[4-threonine]-mesotocin	128 ± 28	276 ± 72	1113 ± 30	251 ± 13	2.1 ± 0.4	< 0.5

[a] Ref. 156. [b] Activities expressed in I.U. mg^{-1} ± S.E.

and (170) [148b, c] are reported to display very little, though definite, oxytocin-like and vasopressin-like activities.[150]

Analogues of oxytocin and mesotocin ([8-isoleucine]-oxytocin) [Table 9 (171)], with Thr-4 rather than Glu-4, possess surprising pharmacological properties. Despite the enhanced biological activities of the parent Thr-4 analogues (172) and (173) – which, in themselves, cannot be accounted for by simple reference to the co-operative structure – the 1-deamino-forms of these compounds (174) and (175) possess mainly diminished potency.[151] This is in marked contrast to oxytocin and other 4-substituted analogues of oxytocin, all of which show considerably enhanced biological activity in the deamino-form. It is speculated that the biological activity of the deamino-compounds might be related to their unusual solubility properties and lipophilicity.

Several new compounds have been reported which act as more-or-less powerful inhibitors of neurohypophyseal hormones. Bromoacetyl-oxytocin (176; A = $BrCH_2CO$-Cys; B = Tyr) appears to be an irreversible inhibitor which, in a labelled form, could act as an affinity label and facilitate the isolation of the hormone receptor (see below).[152] [1-Mercapto-dimethylacetic acid]-oxytocin [176; A = Me_2C(S-)CO; B = Tyr], [1-β-

A-B-Ile-Gln-Asn-Cys-Pro-Leu-Gly-NH_2 (A and Cys linked)

(176)

mercapto-$\alpha\alpha$-dimethylpropionic acid]-oxytocin [176; A = $Me_2C(CH_2S\text{-})$-CO; B = Tyr], and [1-β-mercapto-$\beta\beta$-diethylpropionic acid]-oxytocin [176; A = $Et_2C(S\text{-})CH_2CO$; B = Tyr] are related to the known inhibitor deamino[penicillamine]-oxytocin ([1-β-mercapto-$\beta\beta$-dimethylpropionic acid]-oxytocin). All of these compounds inhibit the response of rat uterus and fowl blood pressure to oxytocin. The first two compounds have approximately 20% and 33—47%, respectively, of the inhibitory potency of deaminopenicillamine-oxytocin, whilst the third has about twice the inhibitory power of deaminopenicillamine-oxytocin.[153] The deamino-compounds in this series show a close parallelism in the antagonism which they exhibit in the two systems, suggesting that intrinsic activity but not affinity is altered by these modifications. However, it is difficult to categorize these analogues in terms of modification types (i) and (iii), especially since in the dimethylmercaptoacetic acid analogue the size of the ring is reduced and presumably, therefore, the backbone conformation disturbed. It is interesting that, as in the agonists, activity in this series of antagonists

[150] B. Berde, P. A. Jaquenoud, and E. Sturmer, *Experientia*, 1971, **27**, 1304.
[151] M. Manning, E. J. Coy, and W. H. Sawyer, *Experientia*, 1971, **27**, 1372.
[152] R. Walter, I. L. Schwartz, O. Hechter, T. Dousa, and P. L. Hoffman, *Endocrinology*, 1972, **91**, 39.
[153] R. J. Vavrek, M. F. Ferger, G. A. Allen, D. H. Rich, A. T. Blomquist, and V. du Vigneaud, *J. Medicin. Chem.*, 1971, **15**, 123.

is reduced in the smaller ring compound. Whatever the consequences of these modifications to the conformation of the oxytocin-like molecule, the parallelism between anti-oxytocic and anti-avian depressor activity is further evidence for the similarity of the biological receptors involved.

Replacement of the phenolic hydrogen of oxytocin with a methyl or ethyl group leads to a change from agonist to partial agonist to competitive antagonist as measured by the hydro-osmotic response (waterflow along an osmotic gradient) of the toad urinary bladder (Table 10).[154] Prosta-

Table 10 *Affinity and intrinsic activity of oxytocin analogues in natriferic and hydro-osmotic assays*[a]

Structure	*Affinity/%*		*Intrinsic activity/%*	
	Natri-feric	*Hydro-osmotic*	*Natri-feric*	*Hydro-osmotic*
Oxytocin	100	100	100	100
O-Methyloxytocin	1.9	3.9	100	90
N-Carbamoyl-*O*-methyloxytocin	0.012	0.11	0	0
Deamino-[1,6-cystathionine]-oxytocin	3.4	2.1	103	49
Deamino-[1,6-cystathionine]-*O*-methyloxytocin	0.36	1.1	0	0

[a] Ref. 155.

glandin E1 is a non-competitive inhibitor of the neurohypophyseal peptides, but whereas the maximum response of the bladder to *O*-methyl-oxytocin is greatly depressed by prostaglandin, its response to saturating doses of oxytocin is only slightly reduced. These observations suggest a receptor reserve for oxytocin in this tissue. The hydro-osmotic response to theophylline is potentiated by saturating concentrations of *O*-ethyl-oxytocin, although the latter is inactive in itself. It is suggested therefore that a threshold effect prevails and that oxytocin-like peptides can produce distinguishable effects on adenylate cyclase (see p. 422) at both sides of the range of enzymic activity which leads to graded changes in membrane permeability. Thus, an increasing stimulus in the subthreshold range would evoke no response, and an increased stimulus in the receptor-reserve range would evoke no further increase in response. In this case, the interaction of the peptide with the receptor is therefore better examined in terms of its effect on adenylate cyclase activity than in terms of hydro-osmotic flow. A similar dissociation of the stimulus–effect coupling is observed in rat uterus preparations, towards which *O*-methyl-oxytocin can behave as an agonist or antagonist dependent upon calcium ion concentration and temperature.[7a]

N-Carbamoyl-[2-*O*-methyl-tyrosine]-oxytocin [176; A = NH_2CONH-$CH(CH_2S$-$)CO$; B = Tyr(Me)] and [1,6-cystathionine,2-*O*-methyl-tyrosine]-deamino-oxytocin [176; A = $SCH_2CH_2CH_2CO$; B = Tyr(Me)] both

[154] P. Eggena, I. L. Schwartz, and R. Walter, *J. Gen. Physiol.*, 1970, **56**, 250.

appear to act as true competitive inhibitors of oxytocin in tests of hydro-osmotic and natriferic (active sodium ion transport) activity (Table 10).[155] [3,4-Dileucine]-oxytocin is virtually devoid of oxytocic, avian depressor, rat pressor, and antidiuretic activity, but it is very potent in the natriferic test.[156]

Measurements of intravenous blood pressure on the adrenergically blocked rat have tended to stress only the importance of the basicity of the residue in position 8 for vasopressin activity. A broader, comparative pharmacological approach makes it evident that the side-chains in positions 2, 3, and 8 all have a bearing on the observed response. Studies on the contractility of isolated canine blood vessels suggest that (i) length and basicity of the side-chain in position 8 can influence both affinity and intrinsic activity; (ii) the phenolic and aromatic side-chains in positions 2 and 3 may be important for intrinsic activity as well as affinity; (iii) receptors may vary in different blood vessels even within one species; (iv) preparations from different vascular beds within the same animal might have different ionic dependencies (*e.g.* towards magnesium ions).[46e, 157]

Microscopic *in vivo* measurements of lumen diameters and how they change in response to the topical application of neurohypophyseal hormones and their analogues have now provided evidence that these generalizations are also valid for the intact animal. The contention that maximal basicity alone does not confer maximal contractile activity is supported by the observation that [8-ornithine]-vasopressin is more potent than lysine-vasopressin or arginine-vasopressin; and the proposition that the side-chains in positions 2 and 3 are of particular importance by evidence that analogues lacking either of these functional groups have affinity and intrinsic activity both reduced.[46e] This approach also provides further evidence for the importance of the α-amino-group in the pressor response. Contrary to what has been found in other systems (Table 11), the absence of the amino-group results in a considerable loss of both affinity and intrinsic activity.

The complex nature of the interaction between the various 'functional' groups can also be brought out by comparing the biological activities of selected analogues. For example, if the effect of the different groups were additive, it would be impossible to account for the high activity of 1-deamino-[8-L-homolysine]-vasopressin[158] which emerges as the most potent antidiuretic analogue (Table 11), except perhaps for 1-deamino-[8-D-arginine]-vasopressin which also has a high activity,[159] but one which is dose-dependent. 1-Deamino-[8-D-norarginine]-vasopressin is also reported to be a very

155 T. Barth, S. Jard, F. Morel, and M. Montegut, *Experientia*, 1972, **28**, 962.
156 M. A. Wille, V. du Vigneaud, and W. Y. Chan, *J. Medicin. Chem.*, 1972, **15**, 11.
157 B. M. Altura, *Experientia*, 1970, **26**, 1089.
158 G. Lindeberg, J. Kynel, P. Dreyfuss, and M. Bodanszky, *J. Medicin. Chem.*, 1972, **15**, 629.
159 I. Vavra, A. Machová, H. Holeček, J. H. Cort, M. Zaoral, and F. Šorm, *Lancet*, 1968, 948.

Table 11 *Biological activities of vasopressin and of analogues modified in positions* 1 *and* 8[a]

Compound	Biological activity	
	Rat pressor	*Rat antidiuresis*
8-Lysine-vasopressin	280	250
1-Deamino-8-lysine-vasopressin	130	310
8-Homolysine-vasopressin	250[b]	300[b]
1-Deamino-8-homolysine-vasopressin	990	10 200

[a] For original references see ref. 158. [b] Units per milligram; other activities are expressed as units per micromole.

active antidiuretic agent with the same order of activity as 1-deamino-[8-D-arginine]-vasopressin.[160]

Structure–activity studies with some twenty-eight compounds related to oxytocin show that the structural requirements for stimulation of glucose oxidation in isolated fat cells are not the same as for the other biological activities of the neurohypophyseal hormones. In stimulation of glucose oxidation (i) the terminal amino-group is important; (ii) vasotocin (3-isoleucine) analogues are more active than vasopressin (3-phenylalanine) analogues; (iii) 8-arginine-vasopressins are more active than 8-lysine vasopressins, but this difference is not seen in the vasotocin series; (iv) analogues with non-basic side-chains (citrulline and N^{ε}-formyl-lysine) in position 8 are the most potent known although vasotocins are more active than oxytocin; (v) substitution of the phenolic group gives inactive compounds.[161] The structure–activity relations exhibited by 9-deglycinamide-[8-lysine]-vasopressin in its protective action against puromycin memory blocking have not yet received similar attention, but presumably the structural emphasis for this type of activity will be different again.[162, 163]

Studies with analogues constructed to be resistant to certain types of enzymic degradation continue to provide information about the *in vivo* deactivation of neurohypophyseal hormones, but the picture is still far from complete.[164–169] Analogues have also been used to study the distribution of enzyme types. For example, the 1-deaminocarba-analogues of oxytocin

160 M. Zaoral and M. Flegel, *Coll. Czech. Chem. Comm.*, 1972, **37**, 3350.
161 E. Schillinger, O. Loge, E. Schröder, E. Klieger, and K. Lübke, *European J. Biochem.*, 1972, **27**, 473.
162 S. S. Wang, *Biochem. Biophys. Res. Comm.*, 1972, **48**, 1511.
163 S. Lande, J. B. Flexner, and L. B. Flexner, *Proc. Nat. Acad. Sci. U.S.A.*, 1972, **69**, 558.
164 R. Walter, J. D. Glass, B. M. Dubois, M. Koida, and I. L. Schwartz, in 'Proceedings 2nd American Peptide Symposium', ed. S. Lande, Gordon & Breach, London, 1972, p. 327.
165 R. Walter and H. Shlank, *Endocrinology*, 1971, **89**, 990.
166 M. Koida, J. D. Glass, I. L. Schwartz, and R. Walter, *Endocrinology*, 1971, **88**, 633.
167 E. Suska-Brzezińska, *Progress Biochem.*, 1972, **18**, 473.
168 H. Shlank and R. Walter, *Proc. Soc. Exp. Biol. Med.*, 1972, **141**, 452.
169 E. Suska-Brzezińska, L. Fruhaufová, T. Barth, I. Rychlik, K. Jošt, and F. Šorm, *Coll. Czech. Chem. Comm.*, 1972, **37**, 2289.

are deactivated at half the rate of oxytocin in uterine homogenates, whereas their half-decay times are 5—10 times that of oxytocin in kidney and liver homogenates. Presumably a disulphide reductase is important in kidney and liver homogenates and another mode of deactivation in uterine tissue.[169] Resistance to enzymic degradation probably accounts for the longevity of action of some hormones, including, for example, [1,6-diaminosuberic acid-7-glycine]-oxytocin.[170]

Studies of the binding of neurohypophyseal hormones to neurophysins indicate that the hormone–protein complex should be completely dissociated at the low concentrations existing in plasma ($<10^{-8}$ mol l^{-1} hormone).[171] Neurophysins are therefore not likely to act in a protective capacity towards the hormones in the general circulation. It also emerges from this study that the binding site of neurophysins bears little resemblance to the physiological receptors of the neurohypophyseal hormones. For example, oxytocinoic acid, despite its comparative inactivity, binds to neurophysin with an association constant similar to those of the natural hormones.

Intrinsic antidiuretic activity cannot be measured *in vivo* because of the complexity of whole animal experimentation. The availability of neurophyseal hormone-sensitive renal medulla preparations offers considerable advantages in assessing the molecular characteristics of the initiation of the biological response. There is now substantial evidence that the antidiuretic response is mediated *via* the adenylate cyclase–cyclic adenosine monophosphate system.[172, 173] It is found that the relative antidiuretic potency of the neurohypophyseal hormones directly parallels their ability to stimulate the cyclase system in a cell-free renal medulla preparation. In rat, mouse, rabbit, and ox preparations, cyclase-stimulating activity diminished in the sequence arginine-vasopressin > lysine-vasopressin > oxytocin, whereas, in pig preparations, the natural hormone lysine-vasopressin was the most active. The preparation can be used to reveal not only receptor affinity, but also maximum stimulatory activity, which cannot be measured by *in vivo* studies (Table 12).[172]

Experiments using ethylene glycol-bis(β-aminoethyl ether)-*NN'*-tetra-acetic acid can be interpreted as showing that the lysine vasopressin-stimulated adenylate cyclase system of porcine renal plasma membranes is inhibited at low calcium ion concentrations. The system is also inhibited at high calcium concentrations ($>10^{-6}$ mol l^{-1}). It is suggested that calcium at low concentrations plays a role in the stimulus–response sequence whereas, at high concentrations, it inhibits binding of the hormone to the receptor.[173]

The hydro-osmotic effect of neurohypophyseal hormones on toad bladder, which is also mediated by adenylate cyclase, has been studied

[170] S. Sakakibara, quoted in ref. 145.
[171] E. Breslow and R. Walter, *Mol. Pharmacol.*, 1972, **8**, 75.
[172] T. Douša, O. Hechter, I. L. Schwartz, and R. Walter, *Proc. Nat. Acad. Sci. U.S.A.*, 1971, **68**, 1693; S. J. Marx, S. A. Fedak, and G. D. Aurbach, *J. Biol. Chem.*, 1972, **247**, 6913.
[173] B. J. Campbell, G. Woodward, and V. Borberg, *J. Biol. Chem.*, 1972, **247**, 6167.

Table 12 *Adenylate cyclase stimulating activity of neurohypophyseal hormones and related peptides in rabbit renal medulla preparation*[a]

Compound	*Biological activity* Peptide concentration for half-maximum response/nmol l^{-1}	Maximum response/ % AVP
8-Arginine-vasopressin	20	100
8-Lysine-vasopressin	40—70	100
Oxytocin	2500—3900	62—63
8-Arginine-vasotocin	57—125	89—90
[Deamino-1,6-dicarba]vasopressin	23—30	94
[Deamino-8-D-arginine]vasopressin	30—71	68—91
Vasopressinoic acid	850	22
Oxytocinoic acid	3900	11—23
[8-Alanine]-oxypressin	4000—11 800	42—45
Deamino-[8-alanine]-oxytocin	3400—6300	76
Deamino-oxytocin	3100—4300	88—90
Deamino-[2-alanine]-oxytocin	—	~4[b]
[5-Valine]-oxytocin	—	~5[b]

[a] Ref. 172. [b] Weak inhibitor of arginine-vasopressin at 10^{-4} mol l^{-1}.

directly [154, 174] and by adenylate cyclase stimulation in cell-free preparations.[175] Studies with synthetic analogues provide further support for the hypothesis that adenylate cyclase stimulation is involved in the hormone-induced response of the whole organ.[145, 176] Further study of the hormonal receptors in the renal medulla and toad bladder may be facilitated by the use of bromoacetyl-oxytocin as an affinity-label.[152, 177]

An isolated frog skin epithelium preparation has been described which binds specifically tritiated oxytocin.[178] Two types of receptor seem to be indicated. The first type, which binds 0.5—2 pmol g^{-1}, is probably the receptor which triggers the biological response (active sodium ion transport) *viz.* (i) the concentration producing half-maximal binding ($K =$ 2—5 nmol l^{-1}) is similar to the concentration which brings about half the maximal biological response; (ii) binding takes place faster than the appearance of the biological response; (iii) competition from lysine-vasopressin and [8-arginine]-oxytocin takes place to the extent that a comparison of the relative potencies of these compounds would predict; (iv) [2-*O*-methyltyrosine]-carba-1-oxytocin inhibits the biological response and binding in a strictly parallel manner. The second type of receptor, which binds 20 pmol g^{-1}, has a much lower affinity ($K \sim 50$ nmol l^{-1}). Some non-specific binding of labelled oxytocin also occurs in the tissue.

174 S. K. Masur, E. Holtzman, I. L. Schwartz, and R. Walter, *J. Cell. Biol.*, 1971, **49**, 582.
175 H.-P. Bär, O. Hechter, I. L. Schwartz, and R. Walter, *Proc. Nat. Acad. Sci. U.S.A.*, 1970, **67**, 7.
176 R. Walter, M. A. Kirchberger, and V. J. Hruby, *Experientia*, 1972, **28**, 959.
177 P. L. Hoffman, I. L. Schwartz, T. P. Douša, O. Hechter, and R. Walter, *Endocrinology*, 1971, **88**, Suppl. A-139.
178 J. Bockaert, M. Imbert, S. Jard, and F. Morel, *Mol. Pharmacol.*, 1972, **8**, 230.

Binding of labelled mono[^{125}I]oxytocin to isolated fat cells has also been described.[179] In this system, lysine-vasopressin and arginine-vasopressin bind with approximately the same affinity as oxytocin, whereas glucagon and angiotensin do not bind at all. Surprisingly, insulin competes weakly for the binding sites but the significance of this is not known. The binding of oxytocin to the fat cell is associated with the stimulation of glucose oxidation and there is some evidence again for two types of site since adsorption continues even after the concentration giving maximal glucose oxidation stimulus has been passed.

5 Pancreatic Hormones

Insulin.—The fiftieth anniversary of the isolation of insulin would inevitably have been a time for reminiscence and reassessment, but since it virtually coincided with the crystallographic determination of the conformation of the molecule,[180] insulin has proved a particularly appropriate topic for symposia and reviews.[7c, 181–183] It is now fairly clear that the insulin hexamer dissociates into an equilibrium mixture of dimer and monomer in solution; it is not known which of the latter forms reacts with the biological receptor. Since some insulin derivatives which only exist as monomers are biologically active, at least dimerization is not a prerequisite for reaction with the receptor.[7c] Laser Raman spectroscopy[184] and bridging synthetic studies[185] provide further evidence that the insulin molecule has a very similar conformation in solution to that observed in the crystal.

Although the two molecules within the dimer are not identical, they are very similar. The B chains hold an extended conformation from B_1 to B_8 followed by an α-helix from B_9 to B_{20}, a reversing bend from B_{21} to B_{23}, and thereafter another extended region to B_{30}; the A chains have helical regions at either end. Apart from the two interchain disulphide bonds, the A and B chains are held together by numerous polar and non-polar attractions. Synthetic studies have shown at least two of the disulphide bonds (A_7—B_7, A_6—A_{11}) are not important for biological activity, provided that the conformation of the molecule is preserved,[181a] which is perhaps not too surprising since the crystal structure shows them to be deeply buried in the molecule.

Comparisons of the amino-acid residue sequences of insulins from various species reveal relatively few invariant residues close together on the

179 E. E. Thompson, P. Feychet, and J. Roth, *Endocrinology*, 1972, **91**, 1199.
180 T. L. Blundell, J. F. Cutfield, E. J. Dodson, G. G. Dodson, D. C. Hodgkin, D. A. Mercola, and M. Vijayan, *Nature*, 1971, **231**, 506.
181 Fiftieth Anniversary Insulin Symposium, October, 1971, *Diabetes*, 1972, **21**, Suppl. 2: (*a*) H. Zahn, D. Brandenburgh, and H.-G. Gattner, p. 468, and references therein; (*b*) M. E. Krahl, p. 695.
182 D. C. Hodgkin, *Diabetes*, 1972, **21**, 1131.
183 T. L. Blundell, J. F. Cutfield, G. G. Dodson, E. Dodson, D. C. Hodgkin, and D. Mercola, *Biochem. J.*, 1972, **127**, 50P.
184 N.-T. Yu, C. S. Liu, and D. C. O'Shea, *J. Mol. Biol.*, 1972, **70**, 117.
185 D. Brandenburg, *Z. physiol. Chem.*, 1969, **350**, 741.

surface of the molecule. It is proposed that these residues, A_1-glycine, A_5-glutamine, A_{19}-tyrosine, and A_{21}-asparagine, may form part of the 'active centre'. Changes to these residues alter the conformation and aggregation properties of the molecule and cause loss of biological activity. If the monomeric form is involved in the receptor reaction, other invariant residues normally concerned with dimerization might also be important, including the B_{12}-valine, B_{16}-tyrosine, and B_{24}-phenylalanine residues.

Studies with chemically modified insulins and synthetic and semisynthetic insulin analogues support these speculations,[7c, 181a, 182, 183] especially where the *N*-terminal region is concerned. Thus de-B_1-phenylalanine insulin has full biological activity,[185] whereas de-A_1-glycine has greatly reduced activity;[186] *N*-A_1-acetylinsulin, which is fully active *in vivo*, has diminished activity *in vitro* and shows evidence of a conformational change;[181a, 187] carbamyl-insulin and methylthiocarbamyl-insulin have full biological potency;[188] sheep and pig synthetic insulins lacking the first four residues of the A-chain are both biologically inactive.[189] In general, the biological activity of forms modified at the A-chain terminus decreases along the series N-substitution > deamino > deglycine > deglycyliso-leucine.[181a] Unfortunately, present synthetic methods do not lend themselves readily to the preparation of analogues modified in this region. A much simpler synthetic proposition, the B_{22-25} sequence, Arg-Gly-Phe-Phe, has been investigated at the tetrapeptide level following the observation that de-B_{22}-arginine insulin is biologically inactive.[190] The tetrapeptide amide, and even arginine amide, is said to possess some insulin-like properties, but it must remain a matter of some doubt whether this can have any direct bearing on the structure–activity properties of insulin.

Insulin receptor preparations have been described from various sources including human lymphocytes,[191, 192] and liver and fat cell membranes.[193, 194] The receptors have been solubilized with[191, 193] or without[192] detergent. They bind specifically with insulin and with peptides related to insulin in proportion to the insulin-like biological activity of the peptides. High molecular weight derivatives of insulin which will not cross the cell membrane are capable of initiating most of the biological activities of insulin.[193] Insulin specificity for the solubilized receptor preparation is the same as

[186] D. Brandenburg, L. Herbertz, G. Krail, M. Weinert, and H. Zahn, *Biochem. J.*, 1971, **125**, 51P.
[187] D. Brandenburg, H.-G. Gattner, and A. Wollmer, *Z. physiol. Chem.*, 1972, **353**, 599.
[188] D. G. Lindsay, O. Loge, W. Losert, and S. Shall, *Biochim. Biophys. Acta*, 1972, **263**, 658.
[189] P. G. Katsoyannis and C. Zalut, *Biochemistry*, 1972, **11**, 3065.
[190] G. Weitzel, K. Eisele, H. Guglielmi, W. Stock, and R. Renner, *Z. physiol. Chem.*, 1971, **352**, 1735.
[191] J. R. Gavin, D. L. Mann, D. N. Buell, and J. Roth, *Biochem. Biophys. Res. Comm.*, 1972, **49**, 870.
[192] J. R. Gavin, D. N. Buell, and J. Roth, *Science*, 1972, **178**, 168.
[193] P. Cuatrecasas, *Proc. Nat. Acad. Sci. U.S.A.*, 1972, **69**, 318 and references therein.
[194] C. A. Robinson, B. R. Boshell, and W. J. Reddy, *Biochim. Biophys. Acta*, 1972, **290**, 84.

for intact cells.[191, 193, 195] Receptors on cultured human lymphocyte cells seem to be the same as on circulating cells.[195] Whilst detergent has not been shown to interfere with the action of the receptor, residual detergent cannot be completely removed. The use of cultured human lymphocytes therefore has a potential advantage because the surface proteins are released by incubating the cells in a buffered serum-free medium.[196] Digestion of the receptor preparation with trypsin, but not with ribonuclease or deoxyribonuclease, leads to complete loss of affinity for insulin, suggesting that the receptor is composed of protein. Insulin produces small decreases in cyclic AMP concentration when it reacts with its receptor, but there is still considerable doubt about the mechanism of its response.[9b, 181b]

Glucagon.—This hormone is known to stimulate the adenylate-cyclase system of target cells.[9b] It is now reported that de-1-histidine-glucagon does not activate this enzyme in liver plasma cells or in fat-cell ghosts, but does inhibit the response of these preparations to submaximal concentrations of glucagon.[197] Experiments with selectively acylated and degraded forms of glucagon confirm the importance of histidine for activity and indicate that the α-amino-group and the N^{ε}-amino-group of Lys-17 might also be important.[198] Presumably the histidine residue has some influence in the binding process since de-1-histidine-glucagon has only one-sixth of the affinity of the parent hormone for the receptor preparation, as shown by competitive binding studies with liver plasma membranes.[197] Neither peptides with the amino-terminal residue sequence of glucagon (residues 1—21 and 1—23), nor peptides with the carboxy-terminal residue sequence (residues 20—29 and 22—29), activate adenylate cyclase, block the response of the enzyme system to glucagon, or compete with glucagon for binding. It seems likely that binding sites at both the amino- and carboxy-ends of the sequence are important in the interaction of glucagon with its receptor.[197] Contrary to what had seemed the case, glucagon is not deactivated in the binding process.[199]

Binding studies with model membranes (lysolecithins) indicate that hydrophobic forces may be involved in the interaction.[200] It is also suggested that the hormonal sensitivity of membranes is usually abolished by proteolytic or phospholipolytic enzymes (see refs. 197, 200). Since the addition of phospholipids to treated membranes partially restores their activity, it seems possible that the lipid component is involved in the stimulus–response couple rather than in the binding process. Glucagon

[195] J. R. Gavin, J. Roth, P. Jen, and P. Freychet, *Proc. Nat. Acad. Sci. U.S.A.*, 1972, **69**, 747.
[196] J. J. Marchalonis, J. L. Atwell, and R. E. Cone, *Nature New Biol.*, 1972, **235**, 240; E. S. Vitetta and J. W. Uhr, *J. Immunol.*, 1972, **108**, 577.
[197] M. Rodbell, L. Birnbaumer, S. L. Pohl, and F. Sundby, *Proc. Nat. Acad. Sci. U.S.A.*, 1971, **68**, 909.
[198] S. Lande, R. Gorman, and M. Bitensky, *Endocrinology*, 1972, **90**, 597.
[199] B. Desbuquois and P. Cuatrecasas, *Nature New Biol.*, 1972, **237**, 202.
[200] A. B. Schneider and H. Edelhoch, *J. Biol. Chem.*, 1972, **247**, 4986.

and calcitonin, but not ACTH, bind to lysolecithin micelles;[200] de-1-histidine glucagon binds as strongly as the parent hormone.[201]

Physical measurements show that the glucagon molecule does not have a fixed conformation in dilute aqueous solution, and there is evidence for a conformational change on binding to cetyltrimethylammonium bromide (see ref. 197). It seems likely that reaction with the receptor will also involve the adoption of a preferred conformation.

6 Gastrointestinal Hormones

Gastrin.—Further evidence has been reported which implicates gastrin in the activation of gastric histidine decarboxylase.[202, 203] It is postulated that gastrin brings about transcription of the DNA nucleotide sequence leading to the synthesis of histidine decarboxylase and ultimately to acid secretion (Scheme 1).[204, 205] It is also shown that acetylcholine and the parasympathetic system are involved in the contractile effects of gastrin, cholecystokinin-pancreozymin, and caerulein (see p. 445).[206] Gastrin-like peptides and cholecystokinin-pancreozymin stimulate calcitonin secretion by the thyroid, which may indicate another physiological function of gastrin.[207, 208]

Previous studies with synthetic analogues of the *C*-terminal tetrapeptide amide of gastrin (177), which is the same as the *C*-terminal sequence of cholecystokinin-pancreozymin and caerulein,[7d] have indicated the relationship between structure and stimulation of gastric acid secretion.[209] Substitutions in the tryptophan, methionine, and phenylalanine residues can give biologically active analogues, whereas requirements in the aspartic acid position are much more conservative. The only active analogue modified in this position, a tetrazole derivative, indicated that the size and proton-donating properties of this side-chain were both important. It was concluded that the aspartic acid residue might play a 'functional' role in the biological response whilst the other residues might fulfil binding roles at the receptor. Recent work with an analogue (178) of the *C*-terminal octapeptide sequence of cholecystokinin shows that, in spite of the confirmed inability of the tetrapeptide alanine derivative (179) to stimulate gastric acid secretion, the aspartic acid side-chain cannot play a functional

201 A. B. Schneider and H. Edelhoch, *J. Biol. Chem.*, 1972, **247**, 4992.
202 R. Håkanson and G. Liedberg, *European J. Pharmacol.*, 1972, **18**, 31.
203 R. I. Bersimbaev, S. V. Argutinskaya, and R. I. Salganik, *Experientia*, 1971, **27**, 1389.
204 R. I. Salganik, S. V. Argutinskaya, and R. I. Bersimbaev, *Experientia*, 1971, **27**, 53.
205 R. I. Salganik, S. V. Argutinskaya, and R. I. Bersimbaev, *Experientia*, 1972, **28**, 1190.
206 E. S. Vizi, G. Bertaccini, M. Impicciators, and J. Knoll, *European J. Pharmacol.*, 1972, **17**, 175.
207 C. W. Cooper, C. R. Biggerstaff, C. W. Wiseman, and M. F. Carlone, *Endocrinology*, 1972, **91**, 1455.
208 C. W. Cooper, W. H. Schwesinger, D. A. Ontjes, A. M. Mahgoub, and P. L. Munson, *Endocrinology*, 1972, **91**, 1079.
209 J. S. Morley, *Proc. Roy. Soc.*, 1968, **B170**, 97; *Fed. Proc.*, 1968, **27**, 1314.

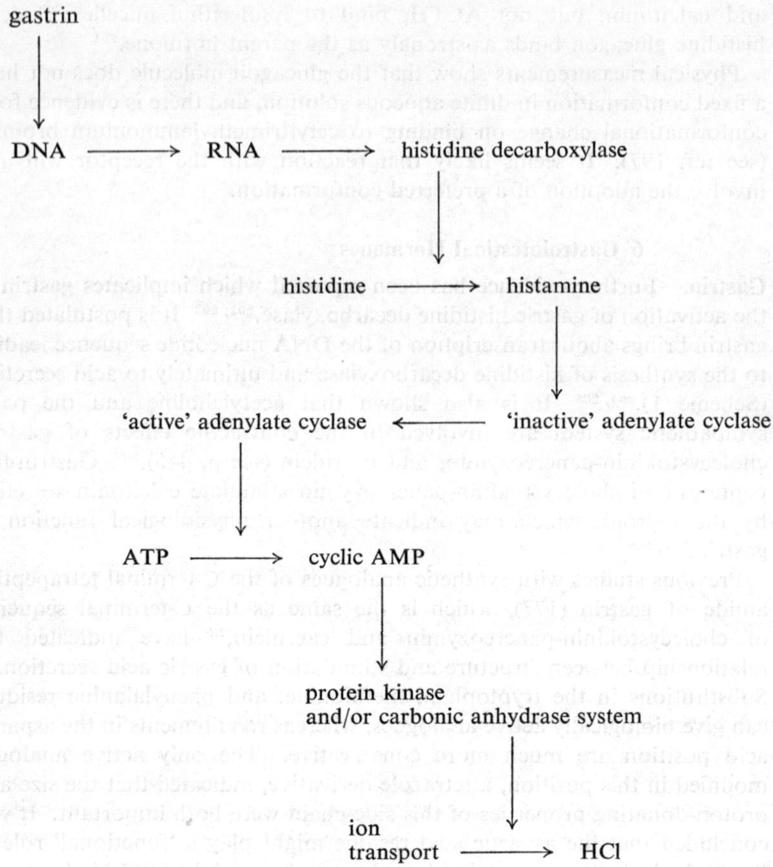

Scheme 1

role in eliciting this response.[210] The alanine octapeptide (178) stimulates gastric acid secretion in cats with a potency of 1/100 relative to the parent octapeptide of cholecystokinin, and $\frac{1}{25}$ relative to Pentagastrin* (180). It is suggested that the alanine residue confers such weak activity that it cannot be observed without the potency conferred by the remainder of the octapeptide sequence, particularly by the tyrosine sulphate side-chain. The relative potency of these peptides is the same for contraction of guinea-pig gall bladder as for gastric secretion, which is thought to indicate that the same parts of the molecule are necessary for the two types of activity.[210]

[210] H. H. Trout and M. I. Grossman, *Nature New Biol.*, 1971, **234**, 256.

* Pentagastrin, Peptavlon, ICI Ltd. (U.K. Pat. 1 042 487).

SO_3H

|

Trp-Met-Asp-Phe-NH_2 Asp-Tyr-Met-Gly-Trp-Met-Ala-Phe-NH_2

(177) (178)

Trp-Met-Ala-Phe-NH_2 Boc-βAla-Trp-Met-Asp-Phe-NH_2

(179) (180)

Other recent synthetic studies have been concerned with 15-leucine-gastrin I (human), with a by-product in which the Trp-4 residue is modified (both peptides possess a high order of gastrin-like activity [211, 212]), and with peptides which could result from transcription of the 'wrong' DNA strand in gastrin synthesis.[213] One of the latter peptides (181) is of interest because, when administered for three to eight days to conscious dogs with innervated gastric pouches, it caused a reduction of the gastric juice volume and of acid and pepsin output. Absorption, distribution, and excretion studies of ^{125}I-gastrin and tritiated t-amyloxycarbonyl-gastrin *C*-terminal tetrapeptide amide in rats show that gastrin is mainly excreted in the gastric juice and urine, whereas the tetrapeptide derivative, owing perhaps to its lipophilicity, is excreted into the bile without decomposition.[214]

Lys-Ile-His-Pro-NH_2

(181)

Conformational investigations of *C*-terminal peptides from gastrin have been based on theoretical considerations (EHT-MO calculations) [215] and on high-resolution n.m.r. measurements.[216] The former approach suggests that the tetrapeptide amide (178) has an energetically preferred extended conformation with little interaction between side-chains. An extended coil is indicated by the second study with the trytophan and phenylalanine side-chains separated by at least 5 Å and with no CO···HN intramolecular hydrogen bonds. Unfortunately, the preferred conformational angles deduced by the two approaches seem to be significantly different. The n.m.r. study shows that the conformations of the aspartic acid and phenylalanine side-chains in the *C*-terminal tripeptide amide and Pentagastrin are probably not sufficiently different to account for the dramatic difference in biological activity between these compounds.

211 E. Wünsch, E. Jaeger, M. Deffner, and R. Scharf, *Z. physiol. Chem.*, 1972, **353**, 1716.
212 E. Wünsch and K.-H. Deimer, *Z. physiol. Chem.*, 1972, **353**, 1255.
213 D. S. Jones, *J.C.S. Perkin I*, 1972, 1407.
214 Y. Ishii, R. Nishizawa, S. Mizuguchi, and H. Miyazaki, *Jap. J. Pharmacol.*, 1972, **22**, 139.
215 L. B. Kier and J. M. George, *J. Medicin. Chem.*, 1972, **15**, 384.
216 J. Feeney, G. C. K. Roberts, J. P. Brown, A. S. V. Burgen, and H. Gregory, *J.C.S. Perkin II*, 1972, 601.

Intestinal Hormones and Peptides.—Secretin, vasoactive intestinal peptide,[217] and gastric inhibitory peptide [218] all show sequence homology with glucagon.[7d] Between them, these four peptides exhibit a broad spectrum of biological activities, and the use of peptide synthesis to investigate their structure–activity relationships should be very rewarding.

Secretin has an *N*-terminal histidine region and is homologous with glucagon in the first seven residues. These two hormones both activate the same adenylate-cyclase system in rat fat cells, although different receptors are involved.[219] Secretin does not compete with glucagon for liver plasma membranes and it does not activate adenylate cyclase in this preparation.[220] It seems reasonable to suppose therefore that the specificity of the two hormones for their receptors is determined by portions of the molecules other than the *N*-terminal sequences, but that once the complex is formed these two hormones bring about their biological response *via* the same mechanism.[197] O.r.d. studies suggest that there is some conformational flexibility in the secretin molecule in aqueous solution, probably somewhat less than that exhibited by glucagon.[221]

7 Thyroid and Parathyroid Hormones

Calcitonin.—Calcitonin has been the subject of several reviews [7e, 222, 223] (see especially ref. 222), obviating the need to deal with it in detail here. Only two recent papers will be cited. The first reports the synthesis of salmon calcitonins II and III, which confirms the structures proposed.[224] Salmon calcitonin II is at least as active as the hormone described earlier from this species; salmon calcitonin III is much less active. It is proposed that the diminution in activity in the second case is related to the presence of a methionine residue in position 8. Oxidation to the sulphoxide might account for the rapid deactivation of this hormone in the tissues. Human calcitonin is 99% deactivated when its methionine residue is oxidized. The second paper concerns the reaction of human calcitonin M with antisera from rabbit.[225] These studies suggest that the thirty-two-residue peptide chain of the calcitonin molecule is folded to bring the two ends near to each other. There appears to be non-covalent interaction between a region near residue 8 and one near the *C*-terminal residue. These comments refer to the conformation which the calcitonin molecule must

[217] S. I. Said and V. Mutt, *European J. Biochem.*, 1972 **28**, 199.
[218] J. C. Brown and J. R. Dryburgh, *Canad. J. Biochem.*, 1971, **49**, 867.
[219] M. Rodbell, L. Birnbaumer, and S. L. Pohl, *J. Biol. Chem.*, 1970, **243**, 718.
[220] S. L. Pohl, L. Birnbaumer, and M. Rodbell, *J. Biol. Chem.*, 1971, **246**, 1849, and refs. therein.
[221] A. Bodanszky, M. A. Ondetti, and M. Bodanszky, *J. Amer. Chem. Soc.*, 1972, **94**, 3600.
[222] J. T. Potts, H. T. Keutman, H. D. Niall, and G. W. Tregear, *Vitamins and Hormones*, 1971, **29**, 41.
[223] R. Lorenc, *Progress Biochem.*, 1972, **18**, 455.
[224] J. Pless, W. Bauer, H. Bossert, K. Zehnder, and St. Guttmann, *Nature New Biol.*, 1972, **240**, 62.
[225] P. G. H. Byfield, M. B. Clark, K. Turner, G. V. Foster, and I. MacIntyre, *Biochem. J.*, 1972, **127**, 199.

assume in the antigen–antibody complex. Physical measurements on aqueous solutions of calcitonins have provided no evidence of highly ordered structures [226] (see, however, ref. 7*f*). Adsorption of the calcitonin molecule on the antibody may involve conformational changes reminiscent of, though probably not related to, the changes which may occur on reaction with the hormonal receptor.

Parathyroid Hormone.—This hormone has also been dealt with extensively in recent reviews.[7g, 9c, 222] One recent paper merits particular notice here.[227] A fragment comprising the *N*-terminal 1—29 residue sequence of the 84-residue native hormone is shown to possess significant *in vivo* (220 U mg^{-1}) and *in vitro* (455 U mg^{-1}) activity; the 1—20 dodecapeptide is inactive. It is suggested that the middle and carboxy-terminal regions of the intact hormone may serve to protect the molecule during circulation.

8 Vasoactive Plasma Peptides

Bradykinin.—Biochemical and pharmacological aspects of bradykinin were reviewed in 1972 [228] and a number of new analogues of bradykinin (182) have been reported (Table 13). Perhaps the outstanding finding of recent years from the structure–activity viewpoint is that the aromatic side-chains of Phe-5 and Phe-8 are not essential for activity. This discovery, originating in the unexpected reduction of the aromatic rings during the synthesis of bradykinin itself,[229] has since been amply borne out by a direct synthetic approach from β-cyclohexyl-L-alanine.[230, 231]

Theoretical (EHT-MO) calculations predict an extended conformation for the bradykinin molecule with no close interactions between the side-chains,[237] and this seems to be in accord with o.r.d. and c.d. studies.[238] The bradykinin molecule seems to exist as a disordered chain of random conformation. There is no evidence of interaction between the phenylalanine side-chains and no fixed relationship between the terminal residues.

Enzymic digests of various proteins have produced new fragments

226 H. B. Brewer and H. Edelhoch, *J. Biol. Chem.*, 1970, **245**, 2402.
227 H. T. Keutmann, B. F. Dawson, G. D. Aurbach, and J. T. Potts, *Biochemistry*, 1972, **11**, 1973.
228 G. E. Sander and C. G. Huggins, *Ann. Rev. Pharmacol.*, 1972, **12**, 227.
229 D. J. Schafer, G. T. Young, D. F. Elliott, and R. Wade, *J. Chem. Soc.* (*C*), 1971, 46.
230 G. A. Fletcher and G. T. Young, *J.C.S. Perkin I*, 1972, 1867.
231 F. Schenck, A. Oberdorf, and G. Schmidt-Kastner, Ger. Offen. 1 298 997, quoted in ref. 229.
232 R. J. Freer and J. M. Stewart, *J. Medicin. Chem.*, 1972, **15**, 1.
233 W. H. Johnson, H. D. Law, and R. O. Studer, *J. Chem. Soc.* (*C*), 1971, 748.
234 K. Neubert, L. Baláspiri, and G. Losse, *Monatsh.*, 1972, **103**, 1575.
235 W. H. Johnson, H. D. Law, and R. O. Studer, unpublished data.
236 E. Atherton, H. D. Law, S. Moore, D. F. Elliott, and R. Wade, *J. Chem. Soc.* (*C*), 1971, 3393.
237 A. H. Brady, J. M. Stewart, and J. W. Ryan, 'Cardiovascular and Neuroactions of the Kinins', ed. N. Back, F. Sicuter, and M. Rocha De Silva, Plenum Press, New York, 1970.
238 A. H. Brady, J. W. Ryan, and J. M. Stewart, *Biochem. J.*, 1972, **121**, 179.

Table 13 *Biological activity of bradykinin analogues*

Compound number	*Structure*	*Guinea-pig ileum assay*[a]	*Ref.*
(182)	Arg-Pro-Pro-Gly-Phe-Ser-Pro-Phe-Arg	100	
(183)	Boc-Arg-Pro-Pro-Gly-Phe-Ser-Pro-Phe-Arg	0	230
(184)	Chl-Arg-Pro-Pro-Gly-Phe-Ser-Pro-Phe-Arg[b]	no activity	232
(185)	Chl-Pro-Pro-Gly-Phe-Ser-Pro-Phe-Arg	potentiates	232
(186)	Guv-Pro-Pro-Gly-Phe-Ser-Pro-Phe-Arg[c]	23 (rat uterus)	233
(187)	Arg-Pip-Pro-Gly-Phe-Ser-Pro-Phe-Arg[d]	4.6 ± 6	234
(188)	Arg-Pro-*PIP*-Gly-Phe-Ser-Pro-Phe-Arg[e]	inactive	234
(189)	Arg-Pro-AEG——Phe-Ser-Pro-Phe-Arg[f]	inactive	236
(190)	Arg-Pro-Pro-Gly-Cha-Ser-Pro-Phe-Arg[g]	100	230
(191)	Arg-Pro-Pro-Gly-Phe-Ser-Pro-Cha-Arg	~ 100	230
(192)	Arg-Pro-Pro-Gly-Cha-Ser-Pro-Cha-Arg	~ 100	229, 230
(193)	Arg-Pro-Pro-Gly-Phe-Ser-Pro-Phe-Agm[h]	0.5—0.25 (rat uterus)	233
(194)	Guv-Pro-Pro-Gly-Phe-Ser-Pro-Phe-Agm	0.01 (rat uterus)	233
(195)	Har-Pro-Pro-Gly-Cha-Ser-Pro-Phe-Arg[i]	similar to bradykinin	230
(196)	Val-Pro-Pro-Gly-Phe-Thr-Pro-Phe-Arg	10 (inactive on rat uterus preparation)	230
(197)	Val-Pro-Pro-Gly-Phe-Ser-Pro-Phe-Arg	0	230, 235
(198)	Leu-Pro-Pro-Gly-Phe-Thr-Pro-Phe-Arg	0	230

[a] Expressed as percent activity generated by equal dose of bradykinin. [b] Chl = $(ClCH_2CH_2)_2N$ 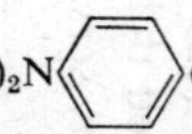 $CH_2CH_2CH_2CO_2H$ [c] Guv = 4-guanidinovaleric acid. [d] Pip = L-pipecolic acid. [e] *PIP* = D-pipecolic acid. [f] AEG = $NH_2CH_2CH_2NHCH_2CO_2H$. [g] Cha = cyclohexylalanine. [h] Agm = agamatine. [i] Har = homoarginine.

which possess bradykinin-potentiating activity,[46f, 239] and more compounds with this activity have been synthesized.[232] The synthetic work attempted to produce irreversible inhibitors of bradykinin by incorporating *p*-[*NN*-bis-(2-chloroethyl)amino]phenylbutyric acid (chlorambucil) into analogues of bradykinin and of the pentapeptide (199), previously isolated as a bradykinin-potentiating factor from the venom of *Bothrops jararaca*. Several of the many bradykinin analogues permanently inhibited the inactivation of bradykinin by pulmonary kinases, which normally deactivate bradykinin almost totally during a single transit through the pulmonary circulation. The *C*-terminal sequence, -Pro-Phe-Arg, seems particularly important for this type of activity. None of the peptides showed antagonist activity towards bradykinin, but in guinea-pig ileum assays, some of the larger bradykinin-like chlorambucil peptides had a marked potentiating effect. For example, [1-chlorambucil]-bradykinin (185) caused a permanent four-fold increase in the sensitivity of the tissue to bradykinin. Surprisingly, chlorambucilylbradykinin (184) did not have a potentiating effect. Chlorambucil peptides related to (199) generally had little measurable effect in this system, but two of them did show some potentiating action. The sensitivity of the rat uterus, which has relatively little kininase activity, was not changed by any of these peptides.

In part, the potentiating effect, when it occurs, may be related to kininase deactivation, but this is probably not the complete explanation. The pentapeptide (199) will still produce its potentiating effect if its addition to the organ bath is delayed until the bradykinin response has already reached its maximum.[240] Furthermore, a peptide has been isolated from venom which potentiates bradykinin activity on the rat uterus but not on guinea-pig ileum preparations.[241]

Some of the small chlorambucil-containing peptides in the bradykinin series [*e.g.* (200)] could block the response of guinea-pig ileum to angiotensin II, but the tissue response could still be increased to the maximal by increasing the concentration of agonist. All of the peptides capable of acting in this way contained the sequence -Pro-Phe- which also occurs in the angiotensin molecule. Presumably this homology accounts for the interaction with the angiotensin receptor area. In general, peptides which have been found to potentiate bradykinin also inhibit angiotensin I converting enzyme.

Glp-Lys-Trp-Ala-Pro (199)   Chl-Pro-Phe-Arg (200)

O.r.d. studies of bradykinin-potentiating peptides suggest that potentiating activity may be associated with conformational flexibility. The most active compound (207) of the series (Table 14) showed the least rigidity

[239] R. Weyers, P. Hagel, B. C. Das, and C. Van der Meer, *Biochim. Biophys. Acta*, 1972, **279**, 331.
[240] S. H. Ferreira, quoted in ref. 232.
[241] H. Kato and T. Suzuki, *Biochemistry*, 1971, **10**, 972.

Table 14 *Rotations and potencies of bradykinin potentiating peptides*[a]

Compound number	*Structure*	*Mean residue rotation at* 215 nm	I_{50}[b]
(201)	Glp-Asn-Trp-Pro-His-Pro-Gln-Ile-Pro-Pro	−15 300	8
(202)	Glp-Ser-Trp-Pro-Gly-Pro-Asn-Ile-Pro-Pro	−19 500	35
(203)	Glp-Trp-Pro-Arg-Pro-Gln-Ile-Pro-Pro	−12 900	1
(204)	Glp-Trp-Pro-Arg-Pro-Thr-Pro-Gln-Ile-Pro-Pro	−14 500	5
(205)	Glp-Asn-Trp-Pro-Arg-Pro-Gln-Ile-Pro-Pro	−14 500	3
(206)	Glp-Trp-Pro-Arg-Pro	−12 000	14
(207)	Glp-Lys-Phe-Ala-Pro	− 6 000	0.05

[a] Ref. 242; [b] Relative dose required to cause 50% inhibition of angiotensin-converting enzyme.

(mean residue rotation at 215 nm, −6000°); other compounds (201)—(206) showed more or less considerable polyproline character (mean residue rotation at 215 nm, > −12 000°).[242]

Angiotensin.—Literature dealing with the pharmacology, physiology, biochemistry, and pathology of angiotensin (208a and b) has been reviewed recently[46g, 228] and the proceedings of a symposium devoted to the effects of angiotensin on the autonomic nervous system have appeared.[243] Angiotensin initiates a number of biological responses including a pressor effect, vasoconstriction, contraction of smooth muscle and cardiac muscle, increase in overall level of sympathetic activity, adrenal catecholamine release, aldosterone biosynthesis in the adrenal cortex,[46g, 228, 243] and enhancement of microsomal sodium–potassium activated adenosine-triphosphatase.[244] More data have accumulated about the renin–angiotensin system suggesting a physiological role for angiotensin in the biosynthesis of alderosterone,[245] and in sympathetic nerve-vascular smooth muscle neurotransmission;[246] and a pathological role in the initiation,[247] if not in the maintenance,[248, 249] of renal hypertension. Nonetheless, at the present time, it remains possible that any or all of the many types of biological activity exhibited by angiotensin may have physiological or pathological significance.[250]

[242] A. Bodanszky, M. A. Ondetti, C. A. Ralofsky, and M. Bodanszky, *Experientia*, 1971, **27**, 1269.
[243] American Soc. for Pharmacology and Experimental Therapeutics, Interaction of Angiotensin II with the Autonomic Nervous System, August 1971, Burlington, Vt., *Fed. Proc.*, 1972, **31**, 1331.
[244] Y. Gutmann, Y. Shamir, D. Glushevitzky, and S. Hochman, *Biochim. Biophys. Acta*, 1972, **273**, 401, and refs. therein.
[245] G. Aguilera and E. T. Marusic, *Endocrinology*, 1971, **89**, 1524.
[246] P. A. Khairallah, *Fed. Proc.*, 1972, **31**, 1351.
[247] E. D. Miller, A. I. Samuels, E. Haber, and A. C. Barger, *Science*, 1972, **177**, 1108.
[248] M. A. Waite, *J. Physiol.*, 1972, **222**, 88P.
[249] H. R. Brunner, P. Chang, R. Wallach, J. E. Sealey, and J. H. Laragh, *J. Clin. Invest.*, 1972, **51**, 58.
[250] H. Page, *Proc. Nat. Acad. Sci. U.S.A.*, 1972, **69**, 1813.

Tritium–hydrogen exchange experiments with 1-asparagine-5-valine-angiotensin II indicate that (i) at least two of the peptide amide hydrogens are involved in secondary structure, and at least one of these forms an intramolecular hydrogen bond; (ii) the NH of the peptide bond between the asparagine and the arginine residues is not hydrogen-bonded intramolecularly; (iii) the asparagine side-chain is freely accessible to the solvent; (iv) interaction, if any, between the terminal carboxy-group and basic side-chains has no conformational significance at physiological ionic strengths.[251] C.d. studies show that the *C*-terminal phenylalanine residue is of particular importance to the conformation of the molecule.[252] This work has led to the proposal of two possible preferred conformations for angiotensin II.[253] The γ-turn model has residues 3—5 in a turn which involves the formation of two hydrogen bonds from NH(Val-3) to CO(Val-5) and NH(Val-5) to CO(Val-3) making a seven-membered ring; the β-turn model has a cross-beta type structure with hydrogen bonds NH(Val-3) to CO(His-6) and NH(His-6) to CO(Val-3). In the γ-turn model, the preferred configuration of the prolyl-histidine peptide bond is *cis* and this finds some support in c.d. studies;[252] the *cis* configuration is probably not favoured in the β-turn structure (Table 15). The γ-turn

Table 15 *Angiotensin: potential energies of postulated conformations and reference structures*[a]

Conformation	*ω-His-Pro-peptide bond*	*Total calculated intramolecular potential energy*[b]
γ-Turn model	−4° (*cis*)	−74.5
	180° (*trans*)	−61.9
β-Turn model	166° (*trans*)	−63.5
	0° (*cis*)	−56.6
Hypothetical antiparallel β-chain[c]	—	−37.8
Hypothetical α-helix[c]	—	−81.4

[a] Ref. 253. [b] kcal mol^{-1}. [c] Calculated for alanine in place of proline to provide necessary conformational freedom.

model implies a molecule diameter of about 15 Å, in agreement with dialysis studies,[251, 254] whereas the β-structure does not conform to these dimensions. For these and other reasons, the authors favour the γ-model, although the β-model cannot be excluded on existing evidence. Considerations of structure–activity relationships of analogues with restricted conformational

251 M. P. Printz, H. P. Williams, and L. C. Craig, *Proc. Nat. Acad. Sci. U.S.A.*, 1972, **69**, 378.

252 S. Fermandjian, J. L. Morgat, and P. Fromageot, *European J. Biochem.*, 1971, **24**, 252.

253 M. P. Printz, G. Némethy, and H. Bleich, *Nature New Biol.*, 1972, **237**, 135.

254 L. C. Craig, E. J. Harfenist, and A. C. Paladini, *Biochemistry*, 1964, **3**, 764; M. T. F. de Fernandez, A. E. Delius, and A. C. Paladini, *Biochim. Biophys. Acta*, 1968, **154**, 223.

freedom lead to predictions of conformational angles more compatible with the γ-turn than the β-turn structure.[255]

On the basis of these models, it is suggested [253] that the biological receptor for the angiotensin model should possess a hydrophobic groove into which fit the valine, tyrosine, and phenylalanine residues of the hormone, the phenylalanine side-chain shielding the interaction from the aqueous solvent. It is postulated that the phenolic group of the tyrosine residue might hydrogen-bond to a suitable group in the receptor, providing significant binding energy in this hydrophobic milieu, whilst the charged groups in the rest of the hormone molecule (arginine, histidine, terminal carboxy-group) might interact electrostatically with other parts of the receptor surface. There would thus be a three-point alignment of the hormone to the receptor. It is postulated that the exact alignment is necessary to trigger the biological response and that analogues which possess reduced biological activity might 'wobble' in their fit with the receptor, only adopting the correct alignment for part of the time. This concept is used to account for the pressor activities of some twenty-five analogues in terms of backbone stability, binding, and 'wobble'.[253]

A variety of conflicting evidence about the conformation of angiotensin [256] has been presented over the past few years, and it would be too optimistic at this stage to regard the above as more than a working hypothesis. Indeed, e.s.r. studies on analogues of the *C*-terminal tripeptide sequence of angiotensin II, containing a nitronylnitroxide derivative in place of histidine, suggest that the terminal carboxy-group forms an important ion–dipole bond with the nitronylnitroxide ring.[256] This and supporting n.m.r. studies of angiotensin analogues [257] suggest a somewhat extended conformation for angiotensin at alkaline pH and a compact, randomly oriented structure at acid pH, for which electrostatic interactions along the peptide chain are of prime importance. Evidence has also been presented of a hydrogen bond [NH(Phe-8) to CO(His-6)] reminiscent of the TRF situation.[59b]

A formal mechanism (not couched in conformational terms) has been proposed for the interaction of angiotensin with receptors in smooth muscle.[46h] The mechanism is based on three types of evidence: (i) the effects of changes in pH and calcium ion concentration on the response of smooth muscle to angiotensin II; (ii) effects of substitutions in position 6 of angiotensin; (iii) inhibition of the response of the tissue to angiotensin II by peptides incorporating chlorambucil. At low calcium ion concentrations ($\sim$0.18 mmol l^{-1}), maximal contraction of rat uterus preparations is obtained at pH 5; below pH 7 marked tachyphyllaxis occurs. Tachyphyllaxis in this instance is thought to be associated with protonation of the

[255] G. R. Marshall and H. E. Bosshard, *Circulation Res.*, 1972, **31**, Suppl. II, 143.
[256] R. J. Weinkam and E. C. Jorgensen, *J. Amer. Chem. Soc.*, 1971, **93**, 7033, and refs. therein.
[257] R. J. Weinkam and E. C. Jorgensen, *J. Amer. Chem. Soc.*, 1971, **93**, 7038.

imidazole in position 6. At high calcium concentrations (~ 1.0 mmol l^{-1}), tachyphyllaxis does not occur at any pH. The well-established biological activities of analogues of angiotensin II substituted in position 6 show that both aromaticity and nucleophilicity are necessary here. With analogues which possess a strongly basic side-chain in this position, tachyphyllaxis is particularly severe. [1-Chlorambucil]-angiotensin II specifically and irreversibly inhibits the action of angiotensin II on isolated guinea-pig ileum.[258] As noted above (p. 433), inhibition is also produced by chlorambucilyl-prolyl-phenylalanyl-arginine,[232] but in this case the same maximal response can be obtained as in the uninhibited tissue, provided a high enough dose of angiotensin II is used, *i.e.* the dose–response curve is displaced to the right.

In the normal angiotensin–receptor interaction, the various side-chains of the peptide are seen as reacting with complementary sites in the receptor and an additional binding site in the receptor is thought to be occupied by calcium (Figure 2). It is proposed that, at a low calcium ion concentrations, the anionic site is not saturated with respect to calcium and can therefore interact with the protonated imidazole, binding the peptide abnormally, blocking access to the receptor, and hence causing tachyphyllaxis (Figure 3). By this reasoning, the nitrogen-mustard angiotensin analogue might alkylate an anionic site which normally interacts with the guanidine side-chain in position 2; whereas the nitrogen-mustard tetrapeptide derivative might alkylate an ionic site adjacent to the receptor and hence tend to hinder access of angiotensin II to the receptor. The observed characteristics of the two types of inhibition are consistent with these proposals.

There is considerable evidence that striated muscle, cardiac muscle, and aorta all possess membranous systems which bind a large amount of calcium[259] and that angiotensin II releases calcium from the binding sites. The increase in the calcium ion concentration of the cell may account for the contraction of the myofilament. A microsomal fraction has now been isolated from rabbit aorta which contains specific angiotensin receptors.[259] The binding parameters of tritiated angiotensin II to this receptor preparation parallel calcium ion release. Three analogues ([4-phenylalanine]-, [6-alanine]-, and [8-alanine]-angiotensin II), which do not possess angiotensin-like activity, were found not to compete with the binding of angiotensin II to the receptor preparation or to cause calcium ion release, even at concentrations a thousand times greater than that of angiotensin. L-Noradrenaline also failed to influence the binding of angiotensin II.

It is usually assumed that receptors are located on the outer surface of the membrane, and the activity of poly-*O*-acetylseryl-angiotensin II provides some support for this concept.[46g] On a molar basis, this analogue

[258] T. B. Paiva and A. C. M. Paiva, *J. Medicin. Chem.*, 1972, **15**, 6.
[259] M. Baudouin, P. Meyer, S. Fernandjian, and J.-L. Morgat, *Nature*, 1972, **235**, 336, and refs. therein.

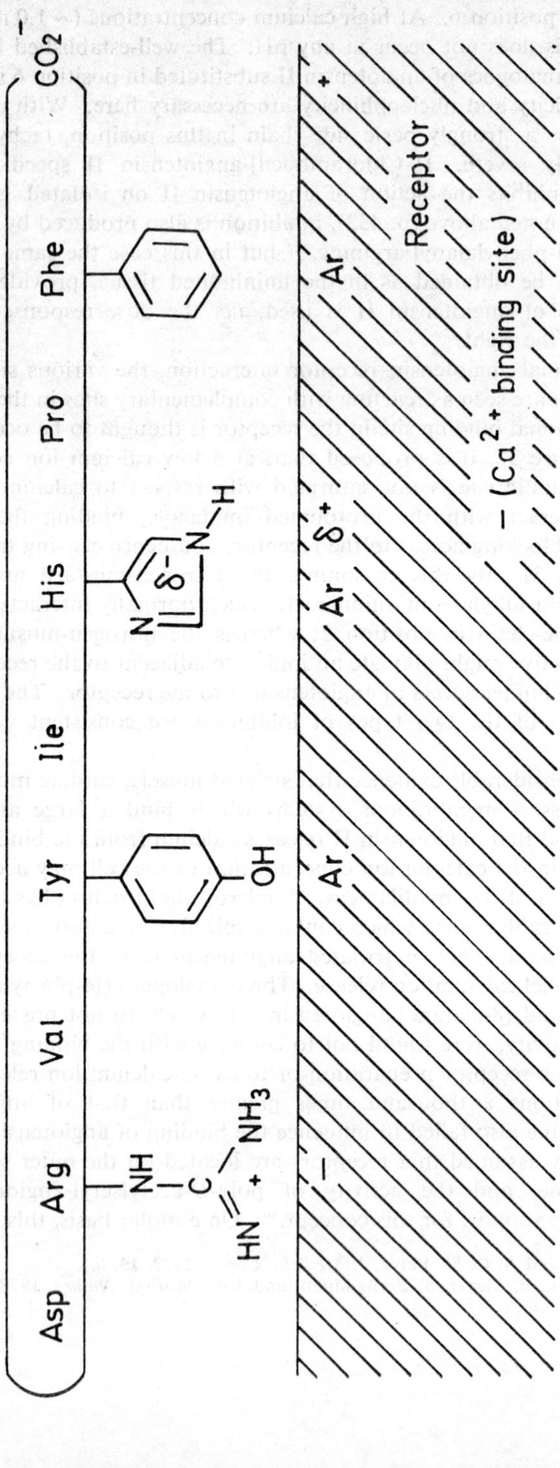
Asp
Arg
Val
Tyr
Ile
His
Pro
Phe
CO_2^-
HN
NH
C
NH_3
OH
N
δ−
N—H
Ar
Ar
Ar δ+
Ar δ+
− (Ca 2+ binding site)
Receptor

Figure 2

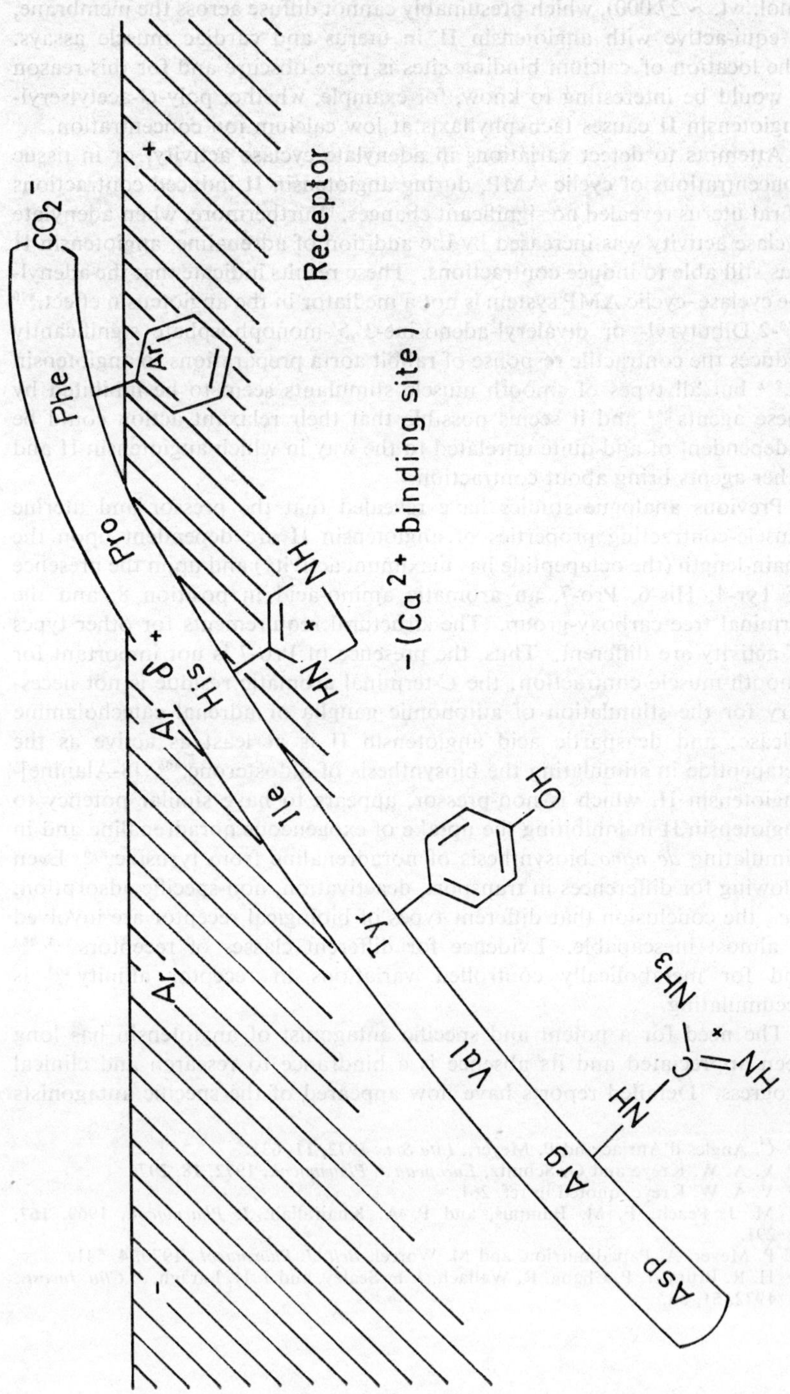
CO_2^-
Receptor
Phe
Ar
Pro
Ar δ+
His
HN
NH
–(Ca²⁺ binding site)
Ile
OH
Tyr
Ar
Val
Arg
NH
HN=C–NH₃⁺
Asp
–

Figure 3

(mol. wt. ~27 000), which presumably cannot diffuse across the membrane, is equi-active with angiotensin II in uterus and cardiac muscle assays. The location of calcium binding sites is more obscure and for this reason it would be interesting to know, for example, whether poly-*O*-acetylseryl-angiotensin II causes tachyphyllaxis at low calcium ion concentration.

Attempts to detect variations in adenylate cyclase activity, or in tissue concentrations of cyclic AMP, during angiotensin II induced contractions of rat uterus revealed no significant changes. Furthermore, when adenylate cyclase activity was increased by the addition of adrenaline, angiotensin II was still able to induce contractions. These results indicate that the adenylate cyclase–cyclic AMP system is not a mediator in the angiotensin effect.[260] N^6-2′Dibutyryl- or divaleryl-adenosine-3′,5′-monophosphate significantly reduces the contractile response of rabbit aorta preparations to angiotensin II,[261] but all types of smooth muscle stimulants seem to be inhibited by these agents[262] and it seems possible that their relaxant action could be independent of and quite unrelated to the way in which angiotensin II and other agents bring about contraction.

Previous analogue studies have revealed that the pressor and uterine muscle-contracting properties of angiotensin II are dependent upon the chain-length (the octapeptide has maximum activity) and upon the presence of Tyr-4, His-6, Pro-7, an aromatic amino-acid in position 8, and the terminal free carboxy-group. The structural requirements for other types of activity are different. Thus, the presence of Pro-7 is not important for smooth-muscle contraction; the *C*-terminal aromatic residue is not necessary for the stimulation of autonomic ganglia or adrenal catecholamine release; and deaspartic acid angiotensin II is at least as active as the octapeptide in stimulating the biosynthesis of aldosterone.[46g] [8-Alanine]-angiotensin II, which is non-pressor, appears to have similar potency to angiotensin II in inhibiting the uptake of exogeneous noradrenaline and in stimulating *de novo* biosynthesis of noradrenaline from tyrosine.[246] Even allowing for differences in transport, deactivation, non-specific adsorption, *etc.*, the conclusion that different types of biological receptor are involved is almost inescapable. Evidence for different classes of receptors[263, 264] and for metabolically controlled variations in receptor affinity[265] is accumulating.

The need for a potent and specific antagonist of angiotensin has long been appreciated and its absence is a hindrance to research and clinical progress. Detailed reports have now appeared of the specific antagonists

260 C. Angles d'Auriac and P. Meyer,, *Life Sci.*, 1972, **11**, 631.
261 V. A. W. Krèye and G. Schultz, *European J. Pharmacol.*, 1972, **18**, 297.
262 V. A. W. Krèye, quoted in ref. 261.
263 M. J. Peach, F. M. Bumpus, and P. A. Khairallah, *J. Pharmacol.*, 1969, **167**, 291.
264 P. Meyer, A. Papadimitriou, and M. Worcel, *Brit. J. Pharmacol.*, 1970, **4**, 541.
265 H. R. Brunner, P. Chang, R. Wallach, J. E. Sealey, and J. H. Laragh, *J. Clin. Invest.*, 1972, **51**, 58.

[8-alanine]-angiotensin II,[266] [8-isoleucine]-angiotensin II,[267] and [1-asparagine, 8-alanine]-angiotensin II.[268] Predictably, the discovery that modifications in position 8 can produce antagonists has led to an abundance of analogues so modified (Table 16).

Table 16 *Analogues of angiotensin*

Compound number	*Structure*				*Ref.*
(208a)	Asp-Arg-Val-Tyr-Val	-His	-Pro	-Phe	
(208b)	Asp-Arg-Val-Tyr-Ile	-His	-Pro	-Phe	
(209)	Asp-Arg-Val-Tyr-Val	-Ile	-Pro	-D-Phe	269, 270
(210)	Asp-Arg-Val-Acp-Ile	-His	-Pro	-Phe[a]	46*i*
(211)	Asn-Arg-Val-Tyr-Thr(Me)	-His	-Pro	-Phe	46*j*
(212)	Asn-Arg-Val-Tyr-Thr	-His	-Pro	-Phe	46*j*
(213)	Asp-Arg-Val-Tyr-Ile	-Acp	-Pro	-Phe	46*i*
(214)	Asp-Arg-Val-Tyr-Ile	-Arg	-Pro	-Phe	271
(215)	Asp-Arg-Val-Tyr-Ile	-His	-Acp	-Phe	46*i*
	F |				
(216)	Asp-Arg-Val-Phe-Val	-His	-Pro	-Phe[b]	46*k*, 272
(217)	Asn-Arg-Val-Tyr-Val	-His	-Pro	-Gly	273
(218)	Asp-Arg-Val-Tyr-Ile	-His	-Pro	-Gly	46*i*
(219)	Asn-Arg-Val-Tyr-Val	-His	-Pro	-Ala	273
(220)	Asp-Arg-Val-Tyr-Ile	-His	-Pro	-Ala	46*i*, 269, 270
(221)	Asn-Arg-Val-Tyr-Val	-His	-Pro	-β-Ala	273
(222)	Asp-Arg-Val-Tyr-Ile	-His	-Pro	-Ile	46*i*, *k*, 269, 271
(223)	Asp-Arg-Val-Tyr-Ile	-His	-Pro	-Cha[c]	269
(224)	Asn-Arg-Val-Tyr-Val	-His	-Pro	-Val	273
(225)	Asp-Arg-Val-Tyr-Ile	-His	-Pro	-Val	46*i*, 269
(226)	Asn-Arg-Val-Tyr-Val	-His	-Pro	-Leu	273
(227)	Asp-Arg-Val-Tyr-Ile	-His	-Pro	-Leu	46*i*, 269
(228)	Asn-Arg-Val-Tyr-Ile	-His	-Pro	-Leu	46*j*
(229)	Asn-Arg-Val-Tyr-Ile	-His	-Pro	-Apb[d]	269
(230)	Asn-Arg-Val-Tyr-Ile	-His	-Pro	-Abp[e]	269
(231)	Asn-Arg-Val-Tyr-Val	-His	-Pro	-Abu	273
(232)	Asp-Arg-Val-Tyr-Ile	-His	-Pro	-Abu	46*i*
(233)	Asp-Arg-Val-Tyr-Ile	-His	-Pro	-Glu	46*i*
(234)	Asp-Arg-Val-Tyr-Ile	-His	-Pro	-Lys[f]	46*i*

[266] R. K. Türker, M. Yamamoto, P. A. Khairallah, and F. M. Bumpus, *European J. Pharmacol.*, 1971, **15**, 285.

[267] M. Yamamoto, R. K. Türker, P. A. Khairallah, and F. M. Bumpus, *European J. Pharmacol.*, 1972, **18**, 316.

[268] D. T. Pals, F. D. Masucci, F. Sipos, and G. S. Denning, *Circulation Res.*, 1971, **29**, 664.

[269] M. C. Khosla, R. A. Leese, W. L. Maloy, A. T. Ferreira, R. R. Smeby, and F. M. Bumpus, *J. Medicin. Chem.*, 1972, **15**, 792.

[270] D. Regoli and W. J. Park, *Canad. J. Physiol. Pharmacol.*, 1972, **50**, 99.

[271] M. C. Khosla, S. Kumar, R. R. Smeby, and F. M. Bumpus, *J. Medicin. Chem.*, 1972, **15**, 627.

[272] P. Needleman, E. M. Johnson, W. Vine, E. Flanigan, and G. R. Marshall, *Circulation Res.*, 1972, **31**, 862.

[273] D. C. Fessler, F. Sipos, G. S. Denning, D. T. Pals, and F. D. Masucci, *J. Medicin. Chem.*, 1972, **15**, 1015.

Table 16 (*cont.*)

Compound number	*Structure*	*Ref.*
(235)	Asp-Arg-Val-Tyr -Ile -His -Pro -Acp/Ach	270, 274
(236)	Asn-Arg-Val-Tyr -Ile -His -Pro -Nle	46*j*
(237)	Asn-Arg-Val-Tyr -Ile -His -Pro -Aha[g]	46*j*
(238)	Asn-Arg-Val-Tyr -Val -His -Pro -Cys	46*k*
(239)	Asp-Arg-Val-His -Ile -Tyr -Pro -Phe	271
(240)	Asp-Arg-Val-Phe -Ile -His -Pro -Ala	269
(241)	Sar -Arg-Val-Tyr -Ile -His -Pro -Ile	269, 275
(242)	Suc -Arg-Val-Phe -Ile -His -Pro -Tyr[h]	46*k*, 269, 272
(243)	Sar -Arg-Val-Tyr -Ile -His -Pro -Leu	46*i*, 46*j*
(244)	β-Asp-Arg-Val-Tyr -Ile -His -Pro -Leu	46*i*
(245)	Sar -Arg-Val-Tyr -Ile -His -Pro -Ala	46*i*
(246)	Asn-Arg-Val-Tyr -Ile -A_2bu-Pro -Phe[i]	46*j*
(247)	Asn-Arg-Val-Tyr -Ile -Pya -Pro -Phe[j]	46*j*
(248)	Asn-Arg-Val-Tyr -Ile -D-Pya -Pro -Phe	46*j*
(249)	Asp-Arg-Val-Phe -Val -His -Pro -Tyr -His-Leu	46*k*, 272
(250)	Asp-Arg-Val-Tyr -Val -His -Pro -Ala -His-Leu	46*k*, 272
(251)	Asp-Arg-Val-Tyr -Val -His -Pro -Ile -His-Leu	46*k*, 272

[a] Acp = aminocyclopentane carboxylic acid. [b] $\overset{\text{F}}{\overset{|}{\text{Phe}}}$ = *p*-fluorophenylalanine. [c] Cha = cyclohexylalanine. [d] Apb = 3-amino-4-phenylbutyric acid. [e] Abp = DL-3-amino-2-benzylpropionic acid. [f] Partially cyclized. [g] Aha = α-aminoheptanoic acid. [h] Suc = succinyl. [i] A_2bu = 2,4-diaminobutyric acid. [j] Pya = β-(2-pyridyl)alanine.

Generalizing, changes in position 8 lead to antagonism as follows:[269] antagonists of the myotropic response to angiotensin result when the aromatic side-chain in position 8 is (i) shifted by one carbon atom (230); (ii) changed configurationally (209); (iii) replaced by a proton (220). Antagonists to both the myotropic and pressor response to angiotensin result when the residue in position 8 possesses (i) an alicyclic side-chain (233) or (ii) a β- or γ-branched alkyl side-chain (222).

Increasing the alkyl chain length in position 8 increases the affinity of the antagonist for the receptor (rat pressor or rat isolated stomach strip) until, in the [8-leucine]-analogue, the pA_2 value* is equal to the pD_2 value* of angiotensin II. [(218), pA_2 6.77; (220), 6.86; (232), 7.48; (225), 7.51; (222), 7.90; (227), 8.00; (208b), pD_2, 8.00.[46i] Also, (216), pA_2, 6.54; (219), 6.84; (221), 5.08; (231), 7.93; (224), 8.31; (226), 8.26 [273]]. The lipophilicity of the side-chain continues to be important for binding at even greater chain lengths, but whereas the lower alkyl derivatives are virtually devoid of agonist activity, analogues with longer side-chains

[274] W. K. Park and D. Regoli, *Brit. J. Pharmacol.*, 1971, **43**, 418P.
[275] R. K. Türker, M. M. Hall, M. Yamamoto, C. S. Sweet, and F. M. Bumpus, *Science*, 1972, **177**, 1203.

* pA_2 is the negative logarithm to base 10 of the molar concentration of an antagonist which will reduce the effect of a double dose of agonist to that of a single dose. pD_2 is the negative logarithm to base 10 of the molar dose of an agonist for which the response is equal to 50% of the maximum.[2–5]

exhibit it to an increasing extent. [Pressor effect: (227), 1.0%; (236), 6%; (237), 10%.[46j]] When given as a single dose concurrently with a single dose of angiotensin II, these analogues showed additive pressor effects. It seems therefore that the side-chain at position 8 plays a role in binding the peptide to the receptor, but the reason for the importance of the aromatic ring is not understood.

Experiments in which infusions of [8-isoleucine]-angiotensin II are used to reduce elevated blood pressure in renal hypertensive rats serve to indicate the usefulness of antagonists in studying the aetiology of hypertensive disease and possibly for clinical purposes.[46k, 46l, 272] However, it is clear that antagonists with enhanced half-lives would be advantageous, and a number of these compounds have been produced by joint substitutions in positions 1 and 8 [(241), (243), (244), (245)]. Substituting a sarcosine or β-aspartyl residue in position 1 produces, in each case, an antagonist with greater potency and duration of action, effects which cannot be accounted for entirely in terms of resistance to enzymic degradation (Table 17). Studies using [1-sarcosine-8-alanine]angiotensin II have now provided direct evidence implicating the pressor action of angiotensin II in the aetiology of renal hypertension.[276]

Table 17 *Potency and longevity of angiotensin analogues in rat blood pressure and rat isolated stomach strip assays (from ref. 46i)*

	Biological activity				
	in vivo		*in vitro*		
Compound	*Potency*	*Longevity*/min	*Intrinsic activity*	*Affinity* (pD_2)	*Potency*
Angiotensin II	100	2.0 ± 0.1	1.0	8.0	100
[1-β-Aspartyl]-angiotensin	172	3.5 ± 0.4	1.0	8.2	150
[1-Sarcosine]-angiotensin	100	2.5 ± 0.2	1.0	8.3	200

The 1-succinyl analogue (242) has a negligible antagonistic effect towards the *in vivo* pressor effect in rats, but, in this case, the 'parent' peptide [4-phenylalanine, 8-tyrosine]-angiotensin II is a relatively weak antagonist.[46i, j]

The decapeptide analogues (249)—(251) of angiotensin I are weak antagonists of the action of angiotensin II on uterine strip, and peptides (249) and (251) act as inhibitors of angiotensin-converting enzyme.[46k, 272] Mention should also be made of [4-*p*-fluorophenylalanine]-angiotensin II (216), which is the only known antagonist of angiotensin which is not modified in position 8.[46k, 273] Compounds (216), (222), and (238) seem to have the same affinity as angiotensin II for the uterine receptor site.[46k]

More studies have been reported which seek to establish the importance of individual side-chains in the key *C*-terminal hexapeptide sequence.

[276] D. T. Pals, F. D. Masucci, G. S. Denning, F. Sipos, and D. C. Fessler, *Circulation Res.*, 1971, **29**, 673.

Replacement of the Tyr-4 residue decreases the affinity of the analogue in isolated stomach strip assays relative to angiotensin {angiotensin II, pD_2, 8.0; [4-phenylalanine]-angiotensin II, pD_2, 6.8; (210), pD_2, 4.9; [4-alanine]-angiotensin II, pD_2, 4.8; [4-glycine]-angiotensin II, pD_2, 4.8}.[46i] It was shown recently that β-branching is important in position 5.[277] Now the synthesis of the 5-(*O*-methylthreonine) and 5-threonine analogues (211) and (212) provides further information about this position. Whilst the *O*-methyl analogue is more active (118%) than angiotensin in the rat pressor assay, the 5-threonine analogue has considerably reduced activity.[46j] Either the steric features of this side-chain are more significant than its lipophilicity or, and there is some n.m.r. evidence for this, the conformation of the 5-threonine analogue is different from that of angiotensin. Studies with 1-aminocyclopentanecarboxylic acid- [(213) and (215)], alanine-, and glycine-substituted analogues show that the presence of an asymmetric α-carbon atom in positions 6 and 7 is important, presumably to ensure the correct orientation of the terminal aromatic side-chain.[46i] Analogues containing 2,4-diaminobutyric acid (246), L- or D-β-(2-pyridyl)-alanine (247) and (248),[46j] or arginine (214)[271] in position 6 possess low-to-negligible agonist properties. Interchanging the residues in positions 4 and 6 also gives an inert analogue.[271]

Further studies of the enzymic degradation of angiotensin and its analogues have been reported.[278–280] Both angiotensin I and II are deactivated by carboxypeptidase and chymotrypsin, whilst amino-peptidase degradation is slower (see Table 17). *In vitro*, carboxypeptidase does not cleave various analogues substituted in position 8, including [8-alanine]-angiotensin, and it is suggested that this might be related to the inhibitory properties of these compounds.[278]

Inhibitors of the angiotensin–renin system have been produced which act at sites other than the angiotensin receptors. The tetrapeptide (252) inhibits rabbit renin *in vitro* but is not active in plasma probably owing to enzymic breakdown,[46m] Analogues of this peptide containing 'reduced' units, -NHCHRCH$_2$-, or carbazic acid units, -NHNRCO-, are, in some instances [*e.g.* (253)—(256)], active inhibitors of renin *in vivo*.[46m] The analogues are not degraded by leucine aminopeptidase.

Leu-Leu-Val-Phe-OMe

(252)

$$\mathrm{H_2N\text{-}\underset{}{\overset{\displaystyle CH_2CHMe_2}{\overset{|}{C}}H}\text{-}CH_2\text{-}Leu\text{-}Val\text{-}Phe\text{-}OMe}$$

(253)

[277] E. C. Jorgensen, S. R. Rapaka, G. C. Windridge, and T. C. Lee, *J. Medicin. Chem.*, 1971, **14**, 899.

[278] M. C. Carrara, D. Regoli, and W. K. Park, *Canad. J. Physiol. and Pharmacol.*, 1972, **50**, 113.

[279] Yu. I. Indulen, G. I. Cipens, and A. P. Pavar, *Biokhimiya*, 1972, **37**, 266.

[280] L. Abrash, R. Walter, and N. Marks, *Experientia*, 1971, **27**, 1353.

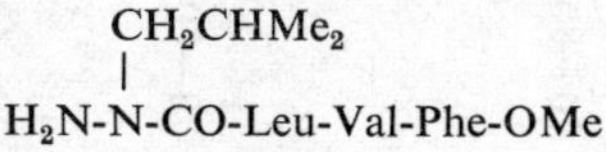

(254)

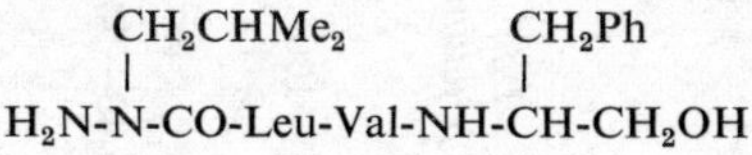

(255)

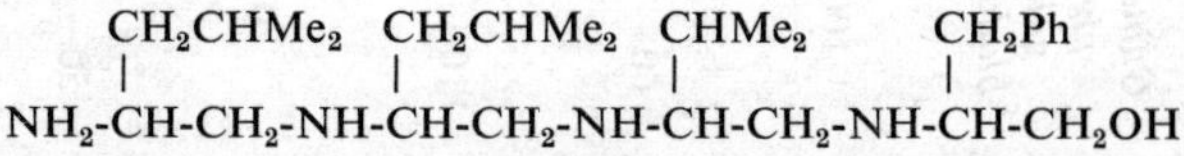

(256)

Peptides from the venom of *Bothrops jararaca* which potentiate the action of bradykinin are also known to be antagonists of angiotensin I, but not of angiotensin II.[281–284] It seems clear that these peptides exert their effect by inhibiting the converting enzyme.

9 Miscellaneous Peptides

Caerulein.—Further structure–activity studies have been reported with analogues of caerulein (257) (Table 18).[285] They indicate that the full spectrum of caerulein-like activity is exhibited by the heptapeptide (258), but that the importance of the individual residues within this structure varies. The threonine residue can be replaced by a number of other residues without loss of activity (259)—(264), but it cannot be omitted without consequence (265). To possess biological activity, analogues must retain an anionic group in the tyrosyl sulphate position [see (266) and (267)], although the steric requirements for this residue are not rigid (259). As in the gastrin series, the methionyl residue can be replaced by norleucine to give a more stable, fully active analogue (270).

[281] M. A. Ondetti, N. J. Williams, E. F. Sabo, J. Pluscec, E. R. Weaver, and O. Kocy, *Biochemistry*, 1971, **10**, 4033.

[282] S. L. Engel, T. R. Schaeffer, B. I. Gold, and B. Rubin, *Proc. Soc. Exp. Biol. Med.*, 1972, **140**, 240.

[283] G. R. Keim, J. Kirpan, A. E. Peterson, B. F. Murphy, G. L. Hassert, and J. W. Poutsaika, *Proc. Soc. Exp. Biol. Med.*, 1972, **140**, 149.

[284] Y. E. Elisseeva, V. N. Orekhovich, L. V. Pavlikhina, and L. P. Alexeenko, *Clinica Chim. Acta*, 1971, **31**, 413.

[285] L. Bernardi, G. Bertaccini, G. Bosisio, R. Bucci, R. de Castiglione, V. Erspamer, O. Goffredo, and M. Impicciatore, *Experientia*, 1972, **28**, 7.

Table 18 *Biological activities of analogues of caerulein (activities are expressed as percentage response elicited by equal weight of caerulein; from ref.* 285)

Compound number	Structure	Biological activity: Dog blood pressure	Denervated gastric pouch in dog	Perfused rat stomach	Dog pancreas	Guinea-pig gall bladder	Dog jejunum
(257)	Glp-Gln-Asp-Tyr(SO_3H)-Thr-Gly-Trp-Met-Asp-Phe-NH_2	100	100	100	100	100	100
(258)	Boc-Tyr(SO_3H)-Thr-Gly-Trp-Met-Asp-Phe-NH_2	30—50	40—50	40—60	30—40	60—70	40—50
(259)	Boc-Tyr(SO_3H)-Nle-Gly-Trp-Met-Asp-Phe-NH_2	65—80	30—60	70—80	60—70	110—150	70—80
(260)	Boc-Tyr(SO_3H)-Val-Gly-Trp-Met-Asp-Phe-NH_2	55	50—60	40—70	25—40	70	—
(261)	Boc-Tyr(SO_3H)-Abu-Gly-Trp-Met-Asp-Phe-NH_2	55—65	25—50	50—60	50—70	50—60	70—80
(262)	Boc-Tyr(SO_3H)-Tyr-Gly-Trp-Met-Asp-Phe-NH_2	30—50	20	20—40	15—50	35—60	30—40

(263)	Boc-Tyr(SO_3H)-Trp-Gly-Trp-Met-Asp-Phe-NH_2	5—7	10—15	40—50	10	20—35	20—25
(264)	Boc-Tyr(SO_3H)-Phe-Gly-Trp-Met-Asp-Phe-NH_2	2—4	10—20	10—15	10—15	10—20	10—20
(265)	Tyr(SO_3H)-Gly-Trp-Met-Asp-Phe-NH_2	0.2—1.5	8—20	2—5	10—30	< 1	< 1
(266)	Phe(SO_2NH_2)-Val-Gly-Trp-Met-Asp-Phe-NH_2	< 1	—	1—2	0.5—1	< 1	< 1
(267)	Tyr-Thr-Gly-Trp-Met-Asp-Phe-NH_2	< 1	< 1	0.15—< 1	1.2—2.5	0.03—< 1	< 1
(268)	Boc-*m*Tyr(SO_3H)-Val-Gly-Trp-Met-Asp-Phe-NH_2	~ 1	2	< 1	2—2.5	< 1	~ 1
(269)	Boc-D-Tyr(SO_3H)-Val-Gly-Trp-Met-Asp-Phe-NH_2	1.5—5	20—22	8—12	6—15	5—7	6—8
(270)	Boc-Tyr(SO_3H)-Thr-Gly-Trp-Nle-Asp-Phe-NH_2	110	120—140	130	130—140	130—140	120—150

6
Metal Derivatives of Amino-acids, Peptides, and Proteins

BY R. D. GILLARD, R. W. HAY, AND S. H. LAURIE

1 Introduction

We cover in this Report work published during 1971 and 1972 in the general field described by the title. We have again used *Current Titles*[1] as our primary source of papers to be studied; the major journals have also been searched. A very noticeable expansion has occurred in this area: although relatively few developments are particularly striking, there has been a massive effort in consolidating some of the earlier advances. There is apparent a distinct loss of faith in simple 'model' studies, and a consequent increase in research upon actual metalloproteins.

As is true in many other areas of chemistry, the slow conversion to S.I. units has caused a temporary proliferation in the units we use here. Those given in this Report are those of the original authors. Electrode potentials (the term recommended by IUPAC) are widely called 'redox', 'reduction', or 'oxidation' potentials. Experimental conditions under which values of such potentials are determined should be quoted.

The status of 'essential' element has, during the period covered, been accorded[2] to fluorine, silicon, tin, and vanadium, making a total of 24 elements. From the field of geobotanical prospecting comes the intriguing observation[3] that nickel (long considered as insignificant, biochemically) occurs at relatively high levels in a plant growing in the nickel-ore-bearing regions of Western Australia. Nickel is accumulated by the plant at a concentration much higher than in the local soil: there appears to be an essential requirement for nickel.

The first number of the first volume of a new journal arrived during 1971: this is called 'Bio-inorganic Chemistry', and is published by the American Elsevier Publishing Company.

A number of valuable general treatments of relevant areas have been given in reviews. These include: 'Stereochemistry of Haems and other Metalloporphyrins',[4] 'Dinitrogen Fixation'[5] (with a biochemical emphasis: a book on the same subject is also available[6]), 'Catalysis by Metallo-

1 'Current Titles', Institute for Scientific Information, Philadelphia, U.S.A.
2 E. Frieden, *Scientific American*, 1972, **227**, 52.
3 B. C. Severne and R. R. Brooks, *Planta*, 1972, **103**, 91; *cf.* report in *Nature*, 1972, **236**, 265.
4 J. L. Hoard, *Science*, 1971, **174**, 1295.
5 H. Dalton and L. E. Mortenson, *Bacteriol. Rev.*, 1972, **36**, 231.
6 'The Chemistry and Biochemistry of Nitrogen Fixation', ed. J. R. Postgate, Plenum Press, London, 1971.

enzymes – the Entatic State',[7] 'Metals in Enzyme Catalysis',[8] and 'Metal Ions in Enzymatic Catalysis'.[9] A simple review of 'Coordination Compounds in Biochemistry'[10] has appeared. An entire volume of a well-known reference text is devoted to 'Metalloproteins',[11] and other books which have become available are: 'Bio-inorganic Chemistry',[12] 'The Metals of Life',[13] and 'The Inorganic Chemistry of Biological Processes'.[14]

We mention some of the numerous reviews of more specialized subjects at the appropriate points in this Report. We have again subdivided the Report in a way which is rather arbitrary but which follows our practice in previous years.

2 Amino-acids

Binding.—*Diffraction Studies.* Perhaps because of its kinetic lability, and the high stability in aqueous media of many of its complexes with cell components, copper(II) has regained the dominant position in studies of complexing of simple amino-acids.

The one method for which this is not currently the case is *X*-ray diffraction, although the structure has been reported[15] of bis-L-phenylalaninatocopper(II), made crystalline by slow diffusion of aqueous L-Phe and copper(II) acetate. The bidentate L-Phe-O$^-$ ligands are *trans* with respect to each other, and six-co-ordination about the copper ion is achieved through the carboxylate groups of neighbouring molecules. The question[16] of possible interactions between the metal ion and the aromatic rings in complexes of tyrosine does not arise here, since the phenyl substituents of the co-ordinated phenylalaninate are directed away from their possible ligand positions.

The first crystal and molecular structure for a complex of chromium(III) with an amino-acid has been established,[17] for [Cr(Gly-O)$_3$],H_2O, a well-known red compound. Here, the nitrogen atoms of the three glycinate ligands occupy a face of the octahedron (the *abc*-isomer). The remarkable

7 R. J. P. Williams, *Inorg. Chim. Acta, Rev.*, 1971, **5**, 137.
8 A. S. Mildvan, in 'The Enzymes: Kinetics and Mechanisms', vol. 2, 3rd edn., ed. P. D. Boyer, Academic Press, New York, 1970, p. 445. This places particular emphasis on the work of the past decade.
9 J. E. Coleman, in 'Progress in Bio-organic Chemistry', vol. 1, ed. E. T. Kaiser and F. J. Kézdy, Wiley–Interscience, New York, 1971, p. 159.
10 J. C. Bailar, *American Scientist*, 1971, **59**, 586.
11 B. L. Vallee and W. E. C. Wacker, in 'The Proteins', vol. 5, 2nd edn., ed. H. Neurath, Academic Press, New York, 1970.
12 *Advances in Chemistry* series, Number 100, ed. R. F. Gould, American Chemical Society, Washington D.C., 1971 (a symposium at Blacksburg, Virginia, June 1970).
13 D. R. Williams, 'The Metals of Life', Van Nostrand Reinhold, London, 1971.
14 M. N. Hughes, 'The Inorganic Chemistry of Biological Processes', John Wiley, New York, 1972.
15 D. van der Helm, M. B. Lawson, and E. L. Enwall, *Acta Cryst.*, 1971, **B27**, 2411.
16 R. D. Gillard and S. H. Laurie, 'Amino-acids, Peptides and Proteins', ed. G. T. Young (Specialist Periodical Reports), The Chemical Society, London, 1969, vol. 1, p. 276.
17 R. F. Bryan, P. T. Greene, P. F. Stokely, and E. W. Wilson, *Inorg. Chem.*, 1971, **10**, 1468.

insolubility in water arises from an elaborate network of hydrogen bonds linking monomeric units.

The major advances from diffraction studies have again concerned complexes of molybdenum. The compound di-μ-sulphido-bis[oxo-(L-histidinato)molybdenum(V)] hydrate, whose preparation was described in the last Report, turns out [18] to have an intramolecular molybdenum–molybdenum distance of 2.82 Å, short enough to account for the diamagnetism of the compound [shown as (1)]. The authors remark that, in

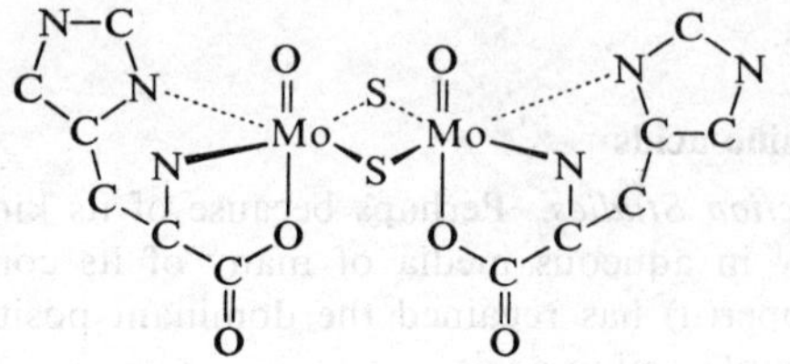

(1)

nitrogenase, sulphur atoms (of thiols or cysteine) could act as bridging groups.

Structures related to (1), and similar interactions between molybdenum atoms in the dimeric unit, are found for: $[(MoO)_2\mu,\mu(O,O)(L\text{-His-O})_2]$,-$3H_2O$ [19] (where Mo—Mo = 2.552 Å, and the molecule is folded about the central pair of bridging oxygen atoms, the two planes making an angle of 153°); $[(MoO)_2\mu,\mu(O,O)(L\text{-cysteine ethyl ester})_2]$, shown as (2), where [20]

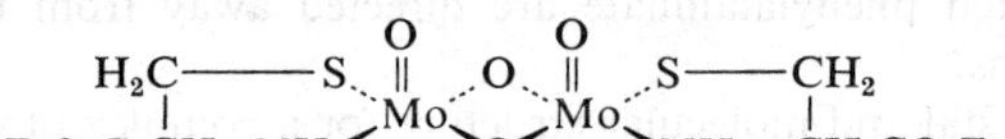

(2)

Mo—Mo = 2.562 Å, with five-co-ordinate molybdenum, and in which the conformations of the two ester ligands are markedly different; and [21] an analogous di-μ-sulphido-compound [similar to (2), but with μ,μ(O,O) replaced by S, and with ethyl groups replaced by methyl].

In a series of compounds of molybdenum(IV) with both π-cyclopentadienyl and bidentate amino-acidates as ligands, cysteine was bonded to molybdenum through nitrogen and sulphur.[22]

Acland and Freeman,[23] in continuing crystallographic studies of model compounds for interactions of metal ions with proteins, report the struc-

[18] B. Spivack, A. P. Gaughan, and Z. Dori, *J. Amer. Chem. Soc.*, 1971, **93**, 5265.
[19] L. T. J. Delbaere and C. K. Prout, *Chem. Comm.*, 1971, 162.
[20] M. G. B. Drew and A. Kay, *J. Chem. Soc. (A)*, 1971, 1846.
[21] M. G. B. Drew and A. Kay, *J. Chem. Soc. (A)*, 1971, 1851.
[22] C. K. Prout, G. B. Allison, L. T. J. Delbaere, and E. Gore, *Acta Cryst.*, 1972, **B28**, 3043.
[23] C. B. Acland and H. C. Freeman, *Chem. Comm.*, 1971, 1016.

tures for silver glycinate, its hemihydrate (involving hydrogen-bonded water), and for silver glycylglycine nitrate, shown schematically as (3), (4), and (5). The dominant feature is the linear two-co-ordinate geometry of the silver ion in each case, which effectively prevents chelation.

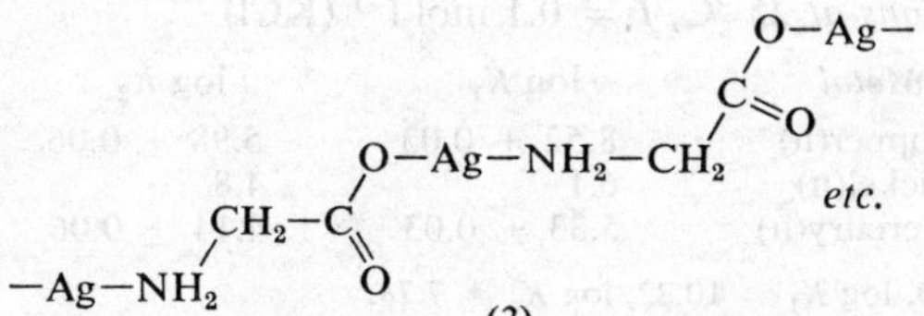

(3)

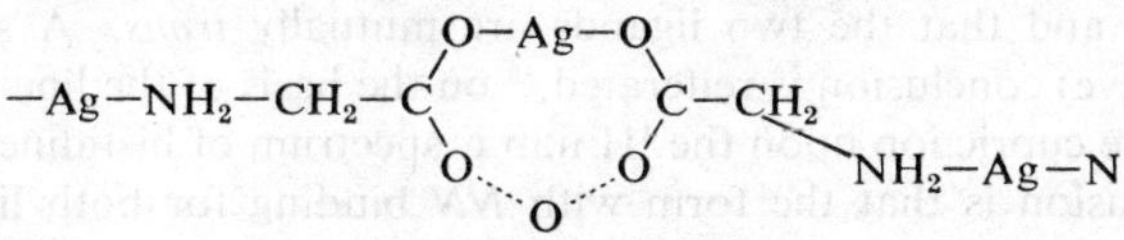

(4)

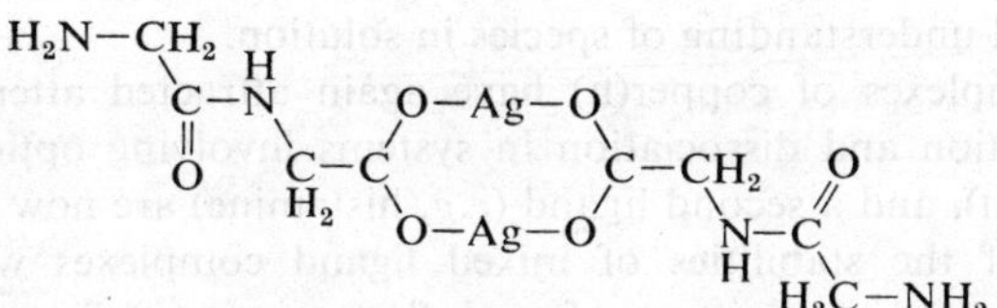

(5)

Amides of L-α-amino-acids (Val, Phe, Pro) form square-planar bis-complexes with palladium(II) and nickel(II), and, in a few cases, *cis*- and *trans*-geometric isomers were isolated [24] [*e.g.* for bis-L-valinamidatopalladium(II)]. The more commonly occurring structure was thought to be *trans*, and this was confirmed [25] by a diffraction study of bis-(L-prolinamidato)nickel(II) dihydrate, containing square-planar nickel.

Solution Studies. The lack of general agreement on the species involved in equilibria involving histidine and copper ions (which we mentioned in our last Report [26]) has led to further research. Table 1 shows values,[27] obtained

[24] T. Komorita, J. Hidaka, and Y. Shimura, *Bull. Chem. Soc. Japan*, 1971, **44**, 3353.

[25] T. Tsukihara, Y. Katsube, K. Fujimori, and Y. Ishimura, *Bull. Chem. Soc. Japan*, 1972, **45**, 1367.

[26] R. D. Gillard and S. H. Laurie, in 'Amino-acids, Peptides, and Proteins', ed. G. T. Young (Specialist Periodical Reports), The Chemical Society, London, 1971, vol. 3, p. 328.

[27] R. W. Hay and P. J. Morris, *J. Chem. Soc.* (*A*), 1971, 1518.

by potentiometric titration, for the stepwise stability constants for L-(+)-histidine methyl ester with three doubly charged metal ions, including copper. Williams, from his results [28] upon potentiometric (copper amal-

Table 1 *Stability constants for complexes of L-histidine methyl ester with metal ions at* 25 °C, I = 0.1 mol l^{-1} (KCl)

Metal	log K_1	log K_2
Copper(II)[a]	8.52 ± 0.03	5.98 ± 0.06
Nickel(II)	6.1	4.8
Mercury(II)	5.33 ± 0.03	4.14 ± 0.06

[a] For Cu–L-His-O, log K_1 = 10.22, log K_2 = 7.78.

gam and glass electrodes) and calorimetric titrations of copper–histidine solutions, concludes that the structure of bishistidinatocopper(II) involves bonding 'histamine-like' through the nitrogen atoms (NH_2 and N of imidazole) and that the two ligands are mutually *trans*. A similar (but less exclusive) conclusion is reiterated,[29] on the basis of the line-broadening effect of the cupric ion upon the ^{1}H n.m.r. spectrum of histidine, at pD 8.1. The conclusion is that the form with *NN* binding for both ligands is in equilibrium with a second form (a linkage isomer) involving one ligand attached through *NN*, and the other as an α-amino-acidate, through NH_2 and carboxylate. This is clearly a difficult problem, emphasizing yet again the gulf between our exact knowledge of conformations in the solid state and our limited understanding of species in solution.

Ternary complexes of copper(II) have again attracted attention. The rates of formation and dissociation in systems involving optically active serine, copper(II), and a second ligand (*e.g.* histamine) are now available.[30] In studies [31] of the stabilities of mixed ligand complexes with cobalt, nickel, copper, and zinc, it was found that copper(II) forms the most stable ternary complexes, a result reminiscent of the Irving–Williams series.

Among the many studies of amino-acid binding to such labile metal ions as copper, it is a little perturbing that possible geometric isomerism seems to be of little concern. The *cis–trans* equilibria among such labile species may be important and should be considered. In view of the known occurrence in solids of *cis-* and *trans*-structures for bisamino-acidatocopper(II), and paying heed to the fact that similar *cis-* and *trans*-isomers with kinetically inert metal ions are of common occurrence, it is unlikely that such equilibria can safely be ignored.

The isomers of bisglycinatopalladium(II) have been re-examined:[32] the *trans*-isomer, in aqueous solution, slowly gives the *cis*-isomer. Volshtein

[28] D. R. Williams, *J.C.S. Dalton*, 1972, 790.
[29] H. Sigel and D. B. McCormick, *J. Amer. Chem. Soc.*, 1971, **93**, 2041; *cf.* ref. 26.
[30] V. S. Sharma and D. L. Leussing, *Inorg. Chem.*, 1972, **11**, 138.
[31] (*a*) H. Sigel, P. R. Huber, and R. F. Pasternack, *Inorg. Chem.*, 1971, **10**, 2226; (*b*) R. Griesser and H. Sigel, *ibid.*, p. 2229.
[32] J. S. Coe and J. R. Lyons, *J. Chem. Soc. (A)*, 1971, 829.

and Dikanskaya made[33] the geometric isomers of bis-(β-phenyl-β-alaninato)platinum(II), where, unusually, the chelate rings are six-membered. [A similar ligand, *N*-benzenesulphonyl-β-alanine, is said to be a useful masking agent for mercuric ions, since, of 14 common cations tested,[34] only mercury(II) gave a stable complex.] In this context of platinum complexes, it is interesting that the diastereoisomers of [PtCl(olefin)-(L-Pro-O)] are readily separated.[35]

Ten years ago, direct calorimetric experiments for metal equilibria were rare. This is no longer the case, and many solution equilibria are currently described in terms, not only of stability constants, but also of calorimetric results. Some results for bisamino-acid complexes are collected in Table 2.[36–39] Other thermodynamic results are presented for alanine complexes

Table 2 *Thermodynamic quantities at* 25 °C *for* $M^{2+} + 2\alpha^- \leftrightharpoons [M(\alpha)_2]$

M	Hα	log β_2	$-\Delta H°$/kcal mol^{-1}	$\Delta S°$/cal deg^{-1} mol^{-1}	*Ref.*
Co	Gly[a]	9.09	6.40	20.1	36
Co	Gly[a]	8.64	7.4	14	37
Ni	Gly	11.05	9.35	19.2	36
Ni	Gly	10.65	8.4	20	37
Cu	Gly	15.83	12.75	29.7	36
Cu	Gly	15.05	13.0	25	37
Cu	Gly	15.59	12.40	29.7	39
Cu	Ala	15.37	11.80	30.7	39
Cu	Ala	15.48	10.7	32	38
Cu	Ser	14.67	12.3	29	37
Cu	Thr	14.50	11.6	28	37
Zn	Gly	9.81	5.98	24.9	36
Zn	Gly	9.44	5.2	25	37

[a] A rather surprising feature of the results collected in ref. 36 is the good measure of agreement between direct calorimetric values for $\Delta H°$ and those obtained in earlier work from the variation with temperature of log β_2.

of doubly charged cobalt, nickel, copper, and zinc; these are obtained[38] from the variation of stability constant with temperature (20, 25, 30, and 35 °C). Whether racemic or resolved alanine was used is not, unfortunately, mentioned. This is known, from earlier work, not to affect values of β_2 for copper(II) with alanine, but, as a general principle, we re-emphasize

[33] L. M. Volshtein and L. D. Dikanskaya, *Russ. J. Inorg. Chem.*, 1971, **16**, 223.
[34] N. N. Ghosh and A. Bhattacharyya, *Bull. Chem. Soc. Japan*, 1971, **44**, 407.
[35] K. Konya, J. Fujita, and K. Nakamoto, *Inorg. Chem.*, 1971, **10**, 1699.
[36] R. M. Izatt, H. D. Johnson, and J. J. Christensen, *J.C.S. Dalton*, 1972, 1152.
[37] A. Gergely, I. Nagypal, and I. Sóvágó, *Acta Chim. Acad. Sci. Hung.*, 1971, **67**, 241.
[38] A. Gergely, B. Kiraly, I. Nagypal, and J. Mojzes, *Acta Chim. Acad. Sci. Hung.*, 1971, **67**, 249.
[39] (*a*) A. Yingst, R. M. Izatt, and J. J. Christensen, *J.C.S. Dalton*, 1972, 1199; (*b*) Ting Po I and G. H. Nancollas, *Inorg. Chem.*, 1972, **11**, 2414.

the desirability of giving this information. Two groups present [39] values of thermodynamic functions for mixed complexes of amino-acids with copper(II).

Equilibrium constants for the complexes of DL-homoserine, determined potentiometrically, with ions of the first-row transition-metals follow [40] the Irving–Williams order. The complexes of phenylalanine with first-row ions have been studied calorimetrically;[41] values in general are in line with those for other systems. The complexes $[Cu_2(Phe\text{-}O)_2(OH)_2]$ and $[Fe_2(Phe\text{-}O)_2(OH)_2]$ were also found.

Mixed complexes have again been much studied by pH titration methods, results having become available for first-row transition-metal ions with: methyl iminodiacetate and glycinate,[42] and its *N*-methyl and *NN*-dimethyl derivatives, stability as usual decreasing with steric crowding at the metal ion; various mixed complexes of amino-acid, again of unstated optical nature, such as [43] serine, threonine, and tyrosine with copper and nickel; and an interesting description [44] of a ternary complex between human serum albumin (HSA), copper(II), and L-histidine. This was studied by equilibrium dialysis, spectrophotometry, and potentiometry. For a 1 : 1 : 2 mixture, at pH 7.5 and 6 °C, of Cu^{2+}, HSA and L-His, the major species are: [HSA-Cu] 77.5%; $[Cu(L\text{-His-O})_2]$ 16.5%; and [(HSA)Cu(L-His-O)] 6.0%. Ternary complexes in the systems Gly-Gly–Cu–Gly and Gly-Gly-Gly–Cu–Gly are more stable [45] than the unmixed complexes containing only amino-acid ligands or only peptide ligands.

Sulphur-containing amino-acids have again attracted interest. Weber, using the *meso*- and L-forms of 3,3′-thiodialanine [lanthionine, (6)], found,[46] with cupric ion, differences between the values of stability constants between the isomeric forms of the ligands. Further information about binding sites in L-cysteine, *S*-methyl-L-cysteine, and DL-methionine is

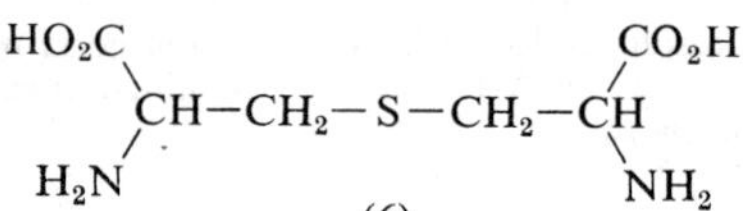

(6)

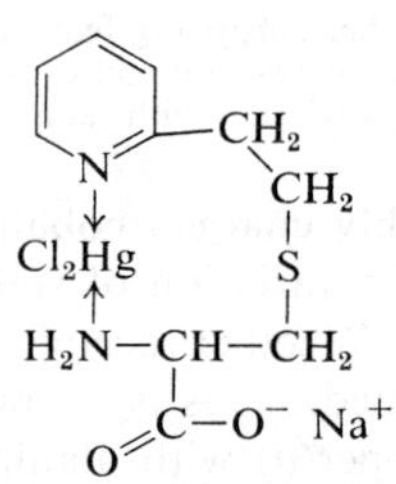

(7)

[40] A. Braibanti, F. Dallavalle, E. Lepovati, and G. Mori, *Inorg. Chim. Acta*, 1971, **5**, 449.
[41] D. R. Williams and P. A. Yeo, *J.C.S. Dalton*, 1972, 1988.
[42] J. Israeli and R. Volpe, *Inorg. Chim. Acta*, 1972, **6**, 5.
[43] A. Gergely, I. Sóvágó, I. Nagypal, and R. Kivaly, *Inorg. Chim. Acta*, 1972, **6**, 435.
[44] S.-J. Lau and B. Sarkar, *J. Biol. Chem.*, 1971, **246**, 5938.
[45] R.-P. Martin, L. Mosoni, and B. Sarkar, *J. Biol. Chem.*, 1971, **246**, 5944.
[46] O. A. Weber, *Internat. J. Protein Res.*, 1971, **5**, 255.

given [47] by Natusch and Porter. Complexes of D-penicillamine with copper(II) (where the 1 : 1 complex is [48] purple and relatively stable in the absence of dioxygen) and iron [49] (in both the II and III oxidation states) are described. A rather surprising finding [50] is that (judging by i.r. and ^1H n.m.r. spectra) the binding of *S*-β-(2-pyridylethyl)-L-cysteine to mercury(II) does not involve sulphur ligands, being as in (7).

Other interactions of a novel type have been discovered. Bistyrosinatocopper(II) has the copper ions arranged in dimeric units, $[Cu(Tyr\text{-}O)_2]_2$. Magnetochemical results and e.s.r. spectra are taken [51] to indicate that the interacting spins of the two copper ions produce two spin states, with the triplet at lower energy. Lohmann describes [52] the effect of varied amino-acids *not* containing sulphur on the e.s.r. and electronic absorption spectra of aqueous cupric solutions. A weak charge-transfer complex is formed.

Amino-acids with more than two functional groups are prone to give rise to more complicated equilibria with metal ions than are glycinate and other bidentate ions. Gorton and Jameson report on their computer analyses of titration results for copper(II) with L-β-(3,4-dihydroxyphenyl)-alanine (DOPA) [53a] and with 3,4-dihydroxyphenylglycine.[53b] Binding at lower pH is of the α-amino-acidate type, but at higher pH a 2 : 1 complex of the catechol type is involved. Evidence for the formation of a tetramer is discussed.

Once again, e.s.r. spectroscopy has been widely applied to these copper(II) complexes. Potentially terdentate L-amino-acids were divided [54] into those binding through NH_2 and COO^- (*e.g.* Tyr, Trp, Lys), those co-ordinated through NH_2 and the side-group function (*e.g.* cysteine, serine, and threonine), and those like histidine, which are distinguishable on the basis of spectra from either of the other types. The bis-complexes of glycine, alanine, and leucine have been studied in detail.[55] Japanese work relates to known *trans*-isomers [56] of $[Cu(\alpha)_2]$, to distinguishing between powdered *cis*- and *trans*-isomers on the basis [57] of the pattern types of their e.s.r. spectra, and [58] to the bis-complexes with optically active and racemic α-amino-acids.

Optical activity has been less popular during the past two years as a means of commenting on solution species, but even so, several intriguing

[47] D. F. S. Natusch and L. J. Porter, *J. Chem. Soc.* (*A*), 1971, 2527.
[48] E. W. Wilson and R. B. Martin, *Arch. Biochem. Biophys.*, 1971, **142**, 445.
[49] L. G. Stadtlerr and R. B. Martin, *Inorg. Chem.*, 1972, **11**, 92.
[50] R. H. Fish and M. Friedman, *J.C.S. Chem. Comm.*, 1972, 812.
[51] J. F. Villa and W. E. Hatfield, *Inorg. Chem.*, 1972, **11**, 1331.
[52] (*a*) W. Lohmann, *Z. Naturforsch.*, 1971, **26b**, 1098; (*b*) W. Lohmann, Proceedings 1st European Biophysics Congress, ed. E. Broda (held at Vienna, Austria, 1971), p. 199.
[53] (*a*) J. E. Gorton and R. F. Jameson, *J.C.S. Dalton*, 1972, 304; (*b*) J. E. Gorton and R. F. Jameson, *ibid.*, p. 307.
[54] G. Rotilio and L. Calabrise, *Arch. Biochem. Biophys.*, 1971, **143**, 218.
[55] H. C. Allen, jun., M. I. Mandrioli, and J. W. Becker, *J. Chem. Phys.*, 1972, **56**, 997.
[56] H. Yokoi, M. Sai, T. Isobe, and S. Ohsawa, *Bull. Chem. Soc. Japan*, 1972, **45**, 2189.
[57] S. Mizumi, T. Isobe, and S. Kimoto, *Bull. Chem. Soc. Japan*, 1972, **45**, 2695.
[58] H. Yokoi, M. Sai, and T. Isobe, *Bull. Chem. Soc. Japan*, 1972, **45**, 3488.

facts have been reported. Tsangaris finds [59] a new Cotton effect at 300 nm (assigned to a charge-transfer transition) in solutions where [L-Thr] : [Cu^{2+}] exceeds 2 : 1. L-Serine did not produce a similar effect. A related study [60] (done, unfortunately, upon unbuffered solutions of solid bis-complexes, where the final pH is certainly not sufficient to guarantee the stoicheiometric presence of the bis-complexes in solution) describes results in terms of ligand conformations. Nash and Jacks [61] report very striking solvent effects upon the signs of circular dichroism of copper(II) complexes of *NN*-dialkyl-α-alaninate, the results being reminiscent of those with ephedrine and ψ-ephedrine complexes of copper(II), where the changes undoubtedly involved trimer ⇌ monomer equilibria. The effects of *N*-methylation upon the *d–d* chiroptical spectra of copper(II) and palladium(II) in an environment of amino-acids (or peptides) are given [62] in detail.

One of the most important (and neglected) areas in solution studies of amino-acid binding to metals is the assessment of how far *in vitro* results may be applicable to *in vivo* situations. For example, the normal concentration of zinc in human serum is about 100 μg per 100 ml. About two-thirds of this is 'loosely bound' (mainly complexed to albumin). Giroux and Henkin now conclude [63] that the only important amino-acid ligands for zinc in plasma are cysteine and histidine, and that only some 2% of the loosely bound zinc is in fact ligated by amino-acids, a fact which may be of significance in connection with mechanisms of zinc transport. A partly discrepant result for zinc is given [64] by Hallman, Perrin, and Watt. They have computed, from data on stability constants, and on the composition of blood plasma, the equilibrium distribution of both zinc and copper among a mixture of seventeen amino-acids. Of the copper 98% is co-ordinated to histidine and cystine, and about half the zinc to cysteine and histidine. They do conclude, however, that appreciable complex formation occurs for zinc with other amino-acids.

Stereoselectivity. A feature of recent work involving the diastereoisomeric possibilities inherent in interactions of several α-amino-acids with metal ions is that applications to optical resolution are becoming useful, and that distinct effort is apparent in attempts to understand the molecular basis for such stereoselectivities.

DL-Aspartic acid may be resolved (often completely) by adding its solution to solutions of the copper complexes of optically active (*i.e.* D or L) alanine, glutamic acid, or proline. The relevant equilibria are given [65] in the scheme:

[59] J. M. Tsangaris, *Chimika. Chron.*, 1972, **1**, 124.
[60] C. Ibarra, R. Soto, L. Adan, A. Decinti, and S. Bunel, *Inorg. Chim. Acta*, 1972, **6**, 601.
[61] C. P. Nash and C. A. Jacks, *J. Amer. Chem. Soc.*, 1972, **94**, 1767.
[62] E. W. Wilson and R. B. Martin, *Inorg. Chem.*, 1971, **10**, 1197.
[63] E. L. Giroux and R. I. Henkin, *Biochim. Biophys. Acta*, 1972, **273**, 64.
[64] P. S. Hallman, D. D. Perrin, and A. E. Watt, *Biochem. J.*, 1971, **121**, 549.

L-α-Cu complex + DL-aspartic acid

↓

D-aspartic acid + L-α Cu complex + L-aspartic acid (solution)

↓

crystals

This resolution is presumably controlled by the same kind of features which underly an ingenious resolution using[66] a column. Here, a copolymer of chloromethylated styrene-*p*-divinylbenzene with L-proline was treated with copper sulphate and ammonia, the product being used to make a column. When DL-proline was charged on to the top, water eluted the L-isomer quantitatively, and the D-isomer was then removed separately with ammonia. This new method, called 'ligand chromatography on asymmetric complex-forming sorbents', has also been applied[67] to the investigation of stereoselective effects in α-amino-acid–copper(II) complexes. A similar ligand-exchange chromatography (involving complexation at zinc) allows[68] for chemical separations in a protein hydrolysate.

In studying the underlying stereoselective effects of such columns, Davankov and Mitchell measured[69] the c.d. of aqueous solutions of cupric ions, *N*-benzyl-L-proline, and either D- or L-proline. The results were employed to calculate formation constants for [Cu(*N*-benzyl-L-Pro-O)(L-Pro-O)] and [Cu(*N*-benzyl-L-Pro-O)(D-Pro-O)], which supported the results of a potentiometric study[70] on [Cu(*N*-benzyl-L-Pro-O)$_2$] and [Cu(*N*-benzyl-L-Pro-O)(*N*-benzyl-D-Pro-O)] (log β = 12.39 ± 0.10 and 13.53 ± 0.07, respectively).

Potentiometric results on stability constants are still in rather a discrepant state. The current view (not entirely accepted) is that, in water, there is no stereoselectivity observable in terms of stability constants between [Cu(L-α)$_2$] and [Cu(L-α)(D-α)], where Hα is a simple bidentate amino-acid, like alanine, valine, or leucine. New support for this proposition comes from the work of Karczynski and Puscasin,[71] and that of Weber

[65] K. Harada and Wung-Wai Tso, *Bull. Chem. Soc. Japan*, 1972, **45**, 2859.
[66] S. V. Rogozhin and V. A. Davankov, *Chem. Comm.*, 1971, 490.
[67] V. A. Davankov and S. V. Rogozhin, *J. Chromatog.*, 1971, **60**, 280.
[68] F. W. Wagner and S. L. Shepherd, *Analyt. Biochem.*, 1971, **41**, 314.
[69] V. A. Davankov and P. R. Mitchell, *J.C.S. Dalton*, 1972, 1012.
[70] V. A. Davankov, S. V. Rogozhin, and A. A. Kurganov, *Izvest. Akad. Nauk S.S.S.R., Ser. khim.*, 1971, 204.
[71] F. Karczynski and M. Puscasin, *Roczniki Chem.*, 1972, **46**, 1489.

and Simeon.[72] The amino-acids studied were alanine,[71] valine,[71] leucine,[71] and histidine[71] with nickel and copper, and tyrosine and tryptophan with cadmium, copper, nickel, lead, and zinc. Weber and Simeon conclude[72] that in the series of bivalent metal ions (Cd, Cu, Ni, Pb, Zn) no stereoselective effects are observable. In agreement, the Polish workers find[71] stereoselectivity only for histidine, where the racemic complex with nickel is slightly more stable than are the optically active complexes. Histidine may, of course, be acting as a terdentate ligand.

An exactly parallel and concordant finding is that[73] relating to direct calorimetry on these bis-amino-acid systems. Enthalpy changes accompanying the formation of some Cu^{2+}, Ni^{2+}, and Zn^{2+} bis-complexes of (+)-, (−)-, and (±)-alanine, -valine, -proline, -asparagine, -glutamic acid, and -histidine were measured calorimetrically at 25.0 °C and $I = 0.10$ mol l^{-1}. Only in the case of histidine is significant stereoselectivity found. In the complexes of Ni^{2+} and Zn^{2+} with histidine, the ΔH values for the *meso*-bis-species [M{(+)-L}{(−)-L}] are more favourable than for the optically pure species $[M\{(+)\text{-L}\}_2]$ or $[M\{(-)\text{-L}\}_2]$ by 2.7 and 2.4 kJ mol^{-1}, respectively. This preference for the *meso*-complex had of course been shown before for these complexes and also for the corresponding complexes of cobalt(II). Copper(II), however, prefers the optically pure species, in the sense that the enthalpy difference is now in favour of $[Cu\{(+)\text{-L}\}_2]$.

In a related study involving potentiometric measurements, Ritsma and Jellinck found,[74] for the system nickel(II) + 2 penicillamine, that the bis-optically active forms $\{[M(L)_2]$ or $[M(D)_2]\}$ were more stable than the racemic [M(D)(L)]. The difference in ΔG° was −2.5 kJ mol^{-1}. Interestingly, they found no stereoselectivity for the nickel complexes with cysteine.

Preparative and Related Aspects. Among the more unusual species reported are two DL-methionine complexes of iron(III), each with one amino-acid residue per metal ion,[75] and the compound of composition $Fe(GlyH)_3$-$(ClO_4)_3,6H_2O$, with co-ordination sites formulated[76] as in (8). A strong

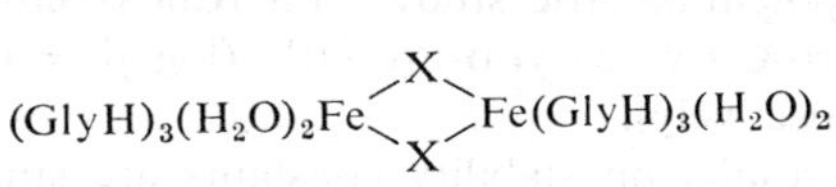

(8)

antiferromagnetic interaction dominates the magnetic behaviour; the glycine molecules were thought to co-ordinate through oxygen, and the available results could not distinguish between $X = H_2O$ and $X = O_2$.

A μ-superoxo-biscobalt(III,III) complex has now been made[77] from the well-known cobalt–histidine–oxygen system; this is μ-superoxo-bis-

[72] O. A. Weber and V. I. Simeon, *Biochim. Biophys. Acta*, 1971, **224**, 94.
[73] D. S. Barnes and L. D. Pettit, *J. Inorg. Nuclear Chem.*, 1971, **33**, 2177.
[74] J. H. Ritsma and F. Jellinck, *Rec. Trav. chim.*, 1972, **91**, 923.
[75] E. J. Halbert and M. J. Rogerson, *Austral. J. Chem.*, 1972, **25**, 421.
[76] E. C. DeVore, S. L. Holt, R. O. Asplund, and A. W. Catalano, *Arch. Biochem. Biophys.*, 1971, **146**, 658.
[77] M. Woods, J. A. Weil, and J. A. Kinnaird, *Inorg. Chem.*, 1972, **11**, 1713.

[bis-(L-histidinato)]cobalt(III) as either nitrate or chloride. Each cobalt ion is surrounded by a terdentate histidinate, a bidentate (histamine-like) histidine, and the oxygen of bridging superoxide, as shown in (9). The

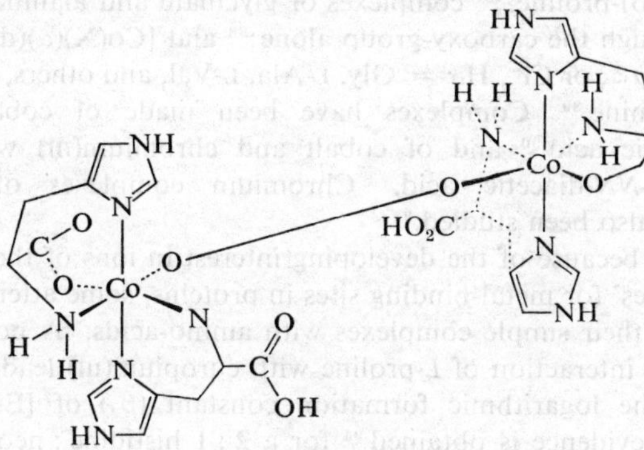

(9)

cobalt(III)-superoxo-cobalt(III) chromophore is manifested by absorption at 678 nm.

Among co-ordination compounds of cobalt(III) containing only amino-acids as ligands, the following have been studied during the period of this Report: $[Co(Gly\text{-}O)_{3-n}(L\text{-}Asp\text{-}O)_n]$ [78] {the starting material for synthesis of these species was $K[Co(Gly\text{-}O)_2(CO_3)]$, which provides another example of spontaneous resolution: two non-superimposable forms of crystals were obtained, which, upon dissolution, gave enantiomeric circular dichroisms}; $[Co(L\text{-}Asp\text{-}O)_n(Ala)_{3-n}]$;[79] $[Co(Gly\text{-}O)_n(L\text{-}Glu\text{-}O)_{3-n}]$;[80] and $Na[Co(L\text{-}Asp\text{-}O)_2]$.[81] The extraordinary richness of possible isomers in these kinetically inert systems (many of them can actually be isolated) offers a pointer to the likely presence of similar complicated isomeric mixtures in the kinetically labile equilibria.

Among the mixed-ligand complexes of cobalt(III), the structure and absolute configuration of (+)-dinitrobis-(L-argininato)cobalt(III) nitrate is reported [82] to be D, as suggested by its circular dichroism. Three groups have discussed their independent but essentially concordant results on isomers of amino-acidato(acetylacetonato)cobalt(III); for $[Co(\alpha)(acac)_2]$, which are conveniently handled in such solvents as benzene or chloroform, the particular combinations were Hα = L-Val and L-Phe;[83] for $[Co(\alpha)_2(acac)]$, Hα = L-Ala, L-Val, *N*-Me-L-Ala, *N*-Me-L-Val,[84] and Gly.[85] Other

78 K. Kawasaki, J. Yoshii, and M. Shibata, *Bull. Chem. Soc. Japan*, 1970, **43**, 3819.
79 T. Matsuda, T. Okumoto, and M. Shibata, *Bull. Chem. Soc. Japan*, 1972, **45**, 802.
80 K. Kawasaki and M. Shibata, *Bull. Chem. Soc. Japan*, 1972, **45**, 3100.
81 S. Yamada, J. Hidaka, and B. E. Douglas, *Inorg. Chem.*, 1971, **10**, 2187.
82 W. H. Watson, D. R. Johnson, M. B. Celap, and B. Kamberi, *Inorg. Chim. Acta*, 1972, **6**, 591.
83 S. H. Laurie, *J.C.S. Dalton*, 1972, 573.
84 D. J. Seematter and J. G. Brushmiller, *J.C.S. Chem. Comm.*, 1972, 1277.
85 Y. Fujii and T. Ejiri, *Bull. Chem. Soc. Japan*, 1972, **45**, 283.

types of amino-acid complex of cobalt(III) have received attention: $[Co(en)(\alpha)_2]^+$ (Hα = Gly, L-Ala, or L-Ser);[86] $[Co(L)(\alpha)_2]^+$ {L = 2,2′-bipyridyl or 1,10-phenanthroline; Hα = Gly, L-Ala, L-hydroxyproline, or *allo*-hydroxy-(D)-proline};[87] complexes of glycinate and alaninate attached to cobalt through the carboxy-group alone;[88] and [Co(X)(α)(dien)], where X = CN^-, NO_2^-, or Cl^-, Hα = Gly, L-Ala, L-Val, and others, and dien = diethylenetriamine.[89] Complexes have been made of cobalt with L-alanine-*N*-acetic acid[90] and of cobalt and chromium(III) with[91] DL-α-phenylalanine-*NN*-diacetic acid. Chromium complexes of the type $[Cr(\alpha)_3]$ have also been studied.[92]

Presumably because of the developing interest in ions of the lanthanide series as 'probes' for metal-binding sites in proteins, some attention is now being paid to their simple complexes with amino-acids. A polarographic study[93] of the interaction of L-proline with europium(III) leads to a value of 1.78 for the logarithmic formation constant (β_2) of $[Eu(Pro\text{-}O)_2]^+$. Using n.m.r., evidence is obtained[94] for a 2 : 1 histidine : neodymium(III) complex at high metal concentrations. Jones and Williams report[95] values for stability constants at 37 °C of histidine with a variety of lanthanides. These are given in Table 3.

Table 3 *Values of* log *K* (37 °C, *I* = 3.00 mol dm^{-3} $NaClO_4$) *for the species indicated with lanthanide ions*

Ln	$Ln(His\text{-}O)^{2+}$	$Ln(His\text{-}O)_2^+$	$Ln(His)^{3+}$
La	3.40	6.85	11.07
Pr	3.69	7.78	11.04
Nd	3.95	8.12	11.20
Sm	4.37	8.78	11.18
Gd	4.94	9.16	11.30
Dy	5.09	9.85	11.56
Er	4.99	9.91	11.40
Yb	4.76	10.31	11.60

Singh and Srivastava have measured stepwise formation constants for the acidic amino-acids (aspartic and glutamic acids) with a wide range of metal ions. These include: beryllium and aluminium (where the complexes are strikingly stable);[96] vanadium(IV), molybdenum(VI), and tung-

86 N. Matsuoka, J. Hidaka, and Y. Shimura, *Bull. Chem. Soc. Japan*, 1972, **45**, 2491.
87 T. Yasui and B. E. Douglas, *Inorg. Chem.*, 1971, **10**, 97.
88 (*a*) G. R. Brubaker and O. P. Schaefer, *Inorg. Chem.*, 1971, **10**, 811; (*b*) G. R. Brubaker and O. P. Schaefer, *ibid.*, p. 2170.
89 K. Ohkawa, J. Fujita, and Y. Shimura, *Bull. Chem. Soc. Japan*, 1972, **45**, 161.
90 K. Okamoto, J. Hidaka, and Y. Shimura, *Bull. Chem. Soc. Japan*, 1971, **44**, 1601.
91 A. Vehara, E. Kyuno, and R. Tsuchiya, *Bull. Chem. Soc. Japan*, 1971, **44**, 1548.
92 H. Mizuochi, A. Vehara, E. Kyuno, and R. Tsuchiya, *Bull. Chem. Soc. Japan*, 1971, **44**, 1555.
93 S. Lal, *Austral. J. Chem.*, 1972, **25**, 1571.
94 A. D. Sherry, E. R. Birnbaum, and D. W. Darnall, *J. Biol. Chem.*, 1972, **247**, 3489.
95 A. D. Jones and D. R. Williams, *J. Chem. Soc.* (*A*), 1971, 3159.
96 M. K. Singh and M. N. Srivastava, *J. Inorg. Nuclear Chem.*, 1972, **34**, 567.

sten(VI),[97] and the very heavy metals, including gold and platinum.[98] In the context of platinum, the discovery of anti-tumour activity in *cis*-[Pt(NH$_3$)$_2$Cl$_2$] has led to a great deal of interest in the ways in which biopolymers may interact with this metal and its complexes, and a review[99] surveys the interactions of amino-acids and proteins with platinum(II).

Reactivity.—*Oxidation.* Bagger and Gibson describe[100] the effect of dioxygen upon aqueous solutions of cobalt(II) and (L)-alanine, which, with or without a nitrogenous base such as pyridine or imidazole, change from pink to brown on exposure to the gas. This is a reversible oxygenation (similar to those of L-histidine which have been so intensively studied). The structure suggested for the brown complex is shown as (10): it is

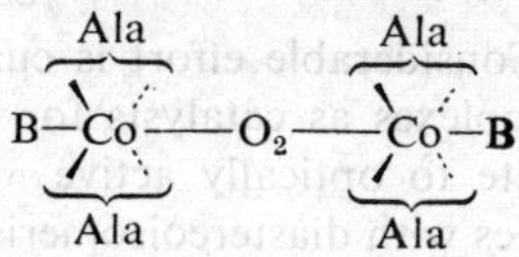

(10) B = base

assigned the μ-peroxo structure because of its absorption band at 370 nm. For oxygenation of the related glycine–imidazole–cobalt(II) system, there is, however, a suggestion[101] that the dioxygen is not bridging (*i.e.* that the complex is mononuclear).

Another study of these oxygen-accepting systems is that by Munakata, who found[102] that the ornithine complex is reversibly oxygenated, whereas those of 2,4-diaminobutyric acid and of 2,3-diaminopropionic acid were irreversibly oxidized at room temperature. Because of the similarity of their electronic spectra to that of the known dioxygen adduct of cobalt(II) bis-histidinate, the oxygenated complexes were thought to contain the peroxo-bridged dicobalt moiety.

Mechanistic experiments on these reversible oxygen carriers include a demonstration,[103] using oxygen-18 (as $^{18}O_2$), that the O—O bond is not broken during the oxygenation–deoxygenation cycle with cobalt(II):histidine. In a kinetic study[104] of the irreversible part of the reaction of cobalt(II) amino-acid complexes with dioxygen [*i.e.* the formation of cobalt(III)-containing products], an interesting finding was that some decarboxylation (giving carbon dioxide) of lysine, glutamic acid, and glycine

97 M. K. Singh and M. N. Srivastava, *J. Inorg. Nuclear Chem.*, 1972, **34**, 2081.
98 M. K. Singh and M. N. Srivastava, *J. Inorg. Nuclear Chem.*, 1972, **34**, 2067.
99 A. J. Thomson, R. J. P. Williams, and S. Reslova, *Structure and Bonding*, 1971, **11**, 1.
100 S. Bagger and K. Gibson, *Acta Chem. Scand.*, 1972, **26**, 2972.
101 (*a*) B. Jezowska-Trzebiatowska and A. Vogt, Proceedings 14th ICCC, 1972, p. 306; (*b*) B. Jezowska-Trzebiatowska, A. Vogt, H. Kozlowski, and A. Jezierski, *Bull. Acad. polon. Sci., Sér. Sci. chim.*, 1972, **20**, 187.
102 M. Munakata, *Bull. Chem. Soc. Japan*, 1971, **44**, 1791.
103 E. Kaubek and C. W. Merwine, *J. Inorg. Nuclear Chem.*, 1971, **33**, 3574.
104 C. S. Sokol, H. Laussegger, L. J. Zompa, and C. H. Brubaker, *J. Inorg. Nuclear Chem.*, 1971, **33**, 3581.

occurred. Powell and Nancollas report[105] the results of their calorimetric study of these oxygen carriers, and find, for the reaction

$$2Co(His\text{-}O)_n(aq) + O_2(aq) \rightleftharpoons [Co(His\text{-}O)_n]_2O_2(aq)$$

$\Delta H = -30.1 \pm 1.3$ kcal mol^{-1} and $\Delta S = -70 \pm 5$ e.u. Equilibrium data were obtained using a polarographic analysis for oxygen.

The catalysed oxidation [by peroxodisulphate, with silver(I) as catalyst] of glycine has been studied kinetically:[106]

$$(S_2O_8)^{2-} + NH_2CH_2COO^- + H_2O \longrightarrow CO_2 + HCHO + NH_3 + (SO_4)^{2-} + (HSO_4)^-$$

The rate of the reaction is first-order in silver(I) and in peroxodisulphate, but independent of glycine. A mechanism is suggested.

Asymmetric Reactions. Considerable effort is currently being devoted to the use of metal ion complexes as catalysts for stereoselective reactions. For example, a new route to optically active α-amino-acids involves[107] the reaction of alkyl halides with diastereoisomeric complexes of α-methylbenzyliminoglyoxylic acid ethyl ester and iron(0) or nickel(0), as shown in Scheme 1. The possibility of employing asymmetric or dissymmetric

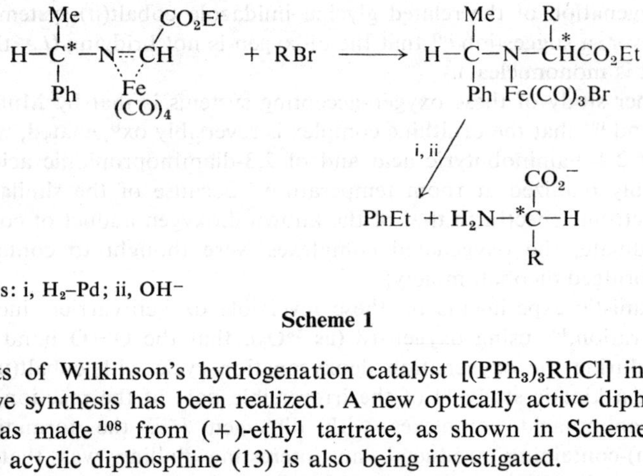

Reagents: i, H_2–Pd; ii, OH^-

Scheme 1

variants of Wilkinson's hydrogenation catalyst [$(PPh_3)_3RhCl$] in stereoselective synthesis has been realized. A new optically active diphosphine (12) was made[108] from (+)-ethyl tartrate, as shown in Scheme 2: the similar acyclic diphosphine (13) is also being investigated.

From the diphosphine, the complex [RhCl(diphosphine)(solvent)] is made *in situ* (in benzene–ethanol), and ethylenic compounds are then hydrogenated catalytically at room temperature and atmospheric pressure. A typical reaction is shown in Scheme 3, where α-phenylacetamidoacrylic acid is reduced to (D)-*N*-phenylacetylalanine, which, after hydrolysis, yields (D)-alanine with an optical purity of 68%.

[105] H. K. J. Powell and G. H. Nancollas, *J. Amer. Chem. Soc.*, 1972, **94**, 2664.
[106] G. Chandra and S. N. Srivastava, *Bull. Chem. Soc. Japan*, 1971, **44**, 3000.
[107] J. Y. Chenard, D. Commereuc, and Y. Chauvin, *J.C.S. Chem. Comm.*, 1972, 750.
[108] T. P. Dang and H. B. Kagan, *Chem. Comm.*, 1971, 481.

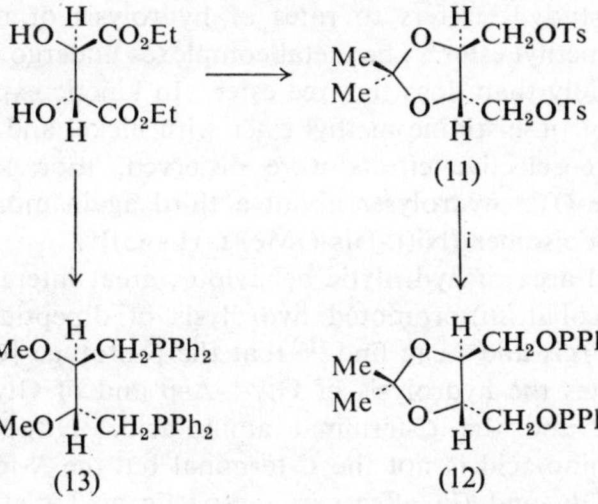

Reagent: i, $NaPPh_2$

Scheme 2

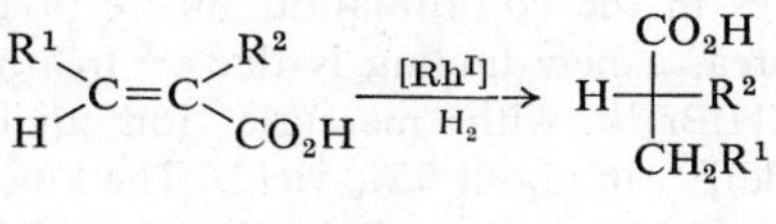

$R^1 = H, R^2 = NHCOCH_2Ph$

Scheme 3

Small differences in the rates of racemization of enantiomeric amino-acids have been observed by Hirota and Izumi,[109] who based the design of their catalyst on the knowledge that non-enzymatic racemization of α-amino-acids may be brought about by forming metal–Schiff base complexes. They used as 'stereoselective catalyst' the hindered biphenyl (14)

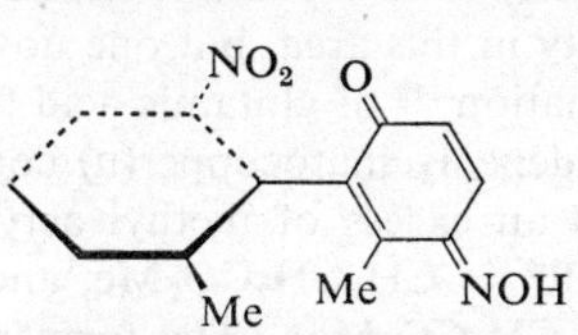

(14)

(actually in its *p*-quinone monoxime tautomeric form). The racemic catalyst showed (as is expected) no discrimination between L- and D-alanine, whereas the (*S*)-enantiomer of the catalyst caused L-alanine to invert 3—5% more rapidly than D-alanine.

Hydrolysis and Related Reactions. Promotion of hydrolysis of methyl glycinate by complexes of copper(II) is discussed by Angelici and Allison,[110]

[109] K. Hirota and Y. Izumi, *Bull. Chem. Soc. Japan*, 1971, **44**, 2287.
[110] R. J. Angelici and J. W. Allison, *Inorg. Chem.*, 1971, **10**, 2238.

and a similar study[111] refers to rates of hydrolysis of DL-2,3-diaminopropionic acid methyl ester. The metal complexes undergo base hydrolysis much more rapidly than does the free ester. In kinetic experiments[112] on the hydrolysis of DL-histidine methyl ester with nickel and copper ions as promoters, stereoselective effects were observed. For example, $[Ni(L\text{-His-OMe})(D\text{-His-O})]^+$ hydrolyses about a third again more rapidly than does the diastereoisomer $[Ni(L\text{-His-OMe})(L\text{-His-O})]^+$.

In the general area of hydrolytic behaviour, great interest still attaches to the use of cobalt(III)-promoted hydrolysis of dipeptides and related compounds. Girgis and Legg find[113] that the prototype $[Co(trien)(H_2O)(OH)]^{2+}$ promotes the hydrolysis of Gly-L-Asp and of Gly-L-Glu to give [Co(trien)(Gly)] and the *C*-terminal amino-acid. However, when the trifunctional amino-acid is not the *C*-terminal but the *N*-terminal moiety, as in α-L-Asp-Gly and (in effect) in L-aspartic acid diethyl ester, then $[Co(trien)X_2]$ is ineffective in promoting hydrolysis. However, $[Co(dien)X_3]$ does then give $[Co(dien)(L\text{-Asp-O})]^+$, in which L-aspartate is terdentate, occupying three sites in the co-ordination sphere of the cobalt. Within the same general area, a new finding is that[114] treatment of purple *cis*-$[Co(en)_2(NH_2CH_2CN)Br]^{2+}$ with mercuric ion quickly gave orange $[Co(en)_2(glycinamide)]^{3+}$, in about 95% yield. The kinetics of base hydrolysis of *cis*-$[Co(en)_2Cl\{NH_2(CH_2)_5CO_2Me\}]^{2+}$, studied by pH-stat and stopped-flow methods, showed[115] two consecutive stages, the rapid loss of chloride followed by a slower hydrolysis of the ester. The rate of base hydrolysis of unco-ordinated 6-aminohexanoate is about one-third that of the co-ordinated molecule.

Schiff Bases.—It has been apparent for many years that the activation of α-amino-acids by formation of a Schiff base with an *o*-hydroxy aromatic aldehyde (such as pyridoxal or 5-nitrosalicylaldehyde) may be useful as a means of promoting reactions of the amino-acids. There has been a very marked decline in activity in this area, but one novel example of synthetic utility concerns the formation[116] of glutamic acid from glycine and methyl acrylate. The *N*-salicylideneglycinatocopper(II) complex (15; $R^1 = R^2 =$ H) was condensed with an excess of methyl acrylate at 65 °C (aqueous KOH) giving $R^1 =$ H, $R^2 = CH_2CH_2CO_2Me$, and a subsequent product (where $R^1 = R^2 = CH_2CH_2CO_2Me$). The former product, upon hydrolysis with six-molar hydrochloric acid gave DL-glutamic acid. The Schiff base related to (15), pyruvylidene(glycinato)aquocopper(II) dihydrate,

111 R. W. Hay and P. J. Morris, *J.C.S. Dalton*, 1973, 56.
112 R. W. Hay and P. J. Morris, *J. Chem. Soc.* (*A*), 1971, 1524.
113 A. Y. Girgis and J. I. Legg, *J. Amer. Chem. Soc.*, 1972, **94**, 8420; *cf.* J. I. Legg and J. Steele, *Inorg. Chem.*, 1971, **10**, 2177.
114 D. A. Buckingham, A. M. Sargeson, and A. Zanelli, *J. Amer. Chem. Soc.*, 1972, **94**, 8246.
115 R. W. Hay, R. Bennett, and D. J. Barnes, *J.C.S. Dalton*, 1972, 1524.
116 Y. N. Belokon, V. M. Belikov, N. I. Kuznetsova, and M. M. Dolgaya, *Izvest. Akad. Nauk S.S.S.R., Ser. khim.*, 1972, 1338 (*Chem. Abs.*, 1972, **77**, 88 817).

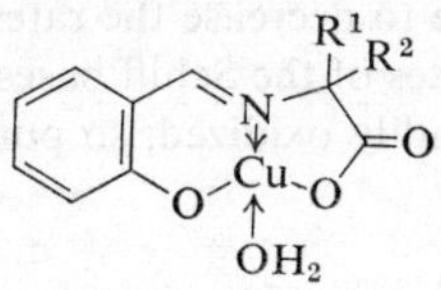

(15)

has been shown [117] by *X*-ray diffraction to contain copper(II), as expected, in a distorted octahedron: the C—N bond (1.31 Å) is longer than normal, as found for other similar compounds. Among reactions involving Schiff bases, Hill finds [118] that a solution containing pyruvate ion, pyridoxamine, and manganese (either in the oxidation state II or as colloidal MnO_2) takes up dioxygen, giving pyridoxal and ammonia. Several other metal ions were ineffective. Peroxidase accelerates the reaction. The addition [119] of copper(II) salts to *N*-salicylidene-L-aspartic acid causes hydrolysis, leading to the formation of the highly insoluble compound of stoicheiometry [Cu{L-Asp}],$3H_2O$, for which a structure is proposed.

Several new complexes of Schiff bases have been made:[120] transamination occurred only when the fused ring system was destabilized by having two five-membered rings including the aldehyde moiety. Thus, transamination was found with the copper(II) complexes made from Schiff bases of glyoxylic acid or pyridine-2-carbaldehyde (but *not* salicylaldehyde or pyridoxal) with Gly, Gly-Gly, or Gly-β-Ala. Weinstein and Holm [121] have synthesized and characterized tautomeric Schiff bases of the type [16; R^1 = H, R^2 = Me or $CHMe_2$] and (16; R^1 = Me, R^2 = Me or $CHMe_2$)

(a) (16) (b)

and studied the rates of conversion from the ketimine (16a) into the aldimine form (16b). Hopgood has studied [122] the rates of transamination in pyridoxal phosphate systems [with zinc(II)] for 15 amino-acids. The zinc(II) aldimine complexes are first formed rapidly (within about 5 minutes of mixing) and transamination then occurs slowly, with alanine most rapid and valine least. Electron donation by the side-chain is rate-enhancing,

[117] A. Torii, H. Tamura-Kogayashi, K. Ogawa, and T. Watanabe, *Z. Krist.*, 1971, **133**, 179.
[118] J. M. Hill, *Biochem. J.*, 1972, **128**, 701.
[119] F. Jursik and B. Hajek, *Coll. Czech. Chem. Comm.*, 1972, **37**, 1801.
[120] Y. Nakao, *Nippon Kagaku Zasshi*, 1971, **92**, 399.
[121] G. N. Weinstein and R. H. Holm, *Inorg. Chem.*, 1972, **11**, 2553.
[122] D. Hopgood, *J.C.S. Dalton*, 1972, 482.

and bulky side-chains serve to decrease the rates. Martell has described [123] work on cobalt(III) complexes of the Schiff bases of pyridoxal and α-amino-acids; the Schiff base is readily oxidized, so pure crystalline products were not obtained.

3 Peptides

Structural Aspects.—Complexes of dipeptides with cobalt ions have again received a good deal of attention. Morris and Martin investigated [124] the general behaviour of 24 dipeptides with both cobalt(II) and cobalt(III). In both cases, the magnitude of the c.d. consists of nearly independent and additive contributions from each co-ordinated amino-acid residue. In attempts to understand special aspects of peptide co-ordination to cobalt, complexes of Gly-L-His have been separated:[125] not surprisingly, the oxygenated compounds are numerous and kinetically rather inert. The oxygenation of tripeptide complexes of cobalt(II) has also been studied:[126] some rather surprisingly stable hydroperoxo-bridged species are involved.

For the well-known bisglycylglycinatocobalt(III) species, Rabenstein [127] has studied sites of protonation by measuring chemical shifts as a function of acidity of the solution. This work supports the conclusion (from c.d. and *X*-ray diffraction experiments surveyed in our previous Reports) that the site of protonation of the co-ordinated peptidate ligands is on the non-co-ordinated oxygen of the peptide linkage. The two peptide linkages in $[Co(Gly\text{-}Gly)_2]^-$ are thought to be protonated successively rather than simultaneously, with

$$K_{a_1} = \frac{[Co(GG)_2{}^-][H^+]}{[Co(HGG)(GG)]} = 10^{-1.46}$$

$$K_{a_2} = \frac{[Co(HGG)(GG)][H^+]}{[Co(HGG)_2{}^+]} = 10^{-0.10}$$

Protonation of mono-dipeptidato-complexes of cobalt(III) of the type $[Co(dien)(Gly\text{-}Gly\text{-}O)]^+$ has been studied in detail.[128]

In our earlier Reports, we mentioned [129] the possibility of using specific hydrolyses of peptides by cobalt(III) complexes as a means of analysis and possibly of identification of some features of the sequence of small peptides. This theme is taken up again under reactivity of peptides, but one contribution on the structural side has appeared, which we describe here. One product of the hydrolysis of peptides with *N*-terminal glycine promoted by

123 A. E. Martell, *J. Inorg. Nuclear Chem.*, 1971, **33**, 567.
124 P. J. Morris and R. B. Martin, *Inorg. Chem.*, 1971, **10**, 964.
125 R. D. Gillard and A. Spencer, *J.C.S. Dalton*, 1972, 902.
126 R. D. Gillard and D. A. Phipps, *J. Chem. Soc.* (*A*), 1971, 1074.
127 D. L. Rabenstein, *Canad. J. Chem.*, 1971, **49**, 3767.
128 I. G. Browning, R. D. Gillard, J. R. Lyons, P. R. Mitchell, and D. A. Phipps, *J.C.S. Dalton*, 1972, 1815.
129 R. D. Gillard and S. H. Laurie, 'Amino-acids, Peptides, and Proteins', ed. G. T. Young (Specialist Periodical Reports), The Chemical Society, London, 1971, vol. 3, p. 351; 1969, vol. 1, p. 280.

$[Co(trien)X_2]$ is $[Co(trien)(Gly\text{-}O)]^{2+}$. Because of the rich possibilities of geometrical and conformational isomerism arising from the trien ligand and its spatial relationship with the glycinate ligand, there are a number of discrete ions with this stoicheiometry. An *X*-ray diffraction study [130] has defined and clarified these. In the same general connection, Wu and Busch have defined [131] conditions for the reaction shown, whereby, in non-aqueous media, peptides (17) are formed with the $[Co(dien)]^{3+}$ moiety as promoter.

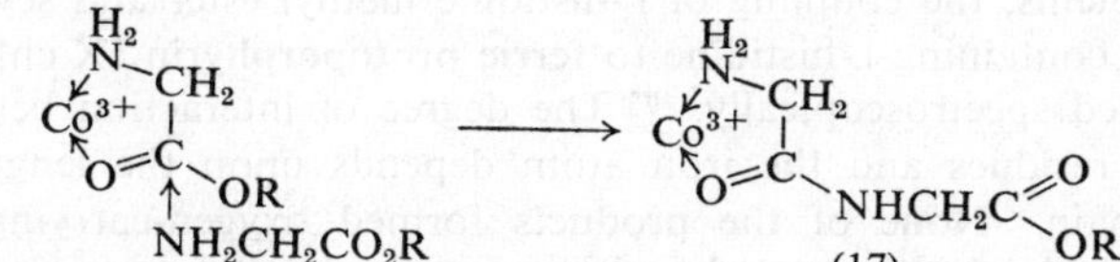

(17)

The modes of interaction of copper(II) ions with peptides have continued to intrigue several groups, and during the period of this Report, structures in the crystalline state and in solution have been examined. The deep-blue crystals of glycyl-L-tryptophanatocopper(II) trihydrate contain [132] four-co-ordinate copper ions, attached to the NH_2 of glycine, the N (deprotonated) of the amide link, the carboxylate group, and a co-ordinated water molecule. In aqueous solution, both L-His-Gly and L-His-Gly-Gly are said [133] (on the basis of electronic spectra) to form dimeric complexes with copper(II). Similar results are reported [134] for complexes of glycyl-L-histidylglycine. Here, *X*-ray diffraction shows that disordered water is located in channels defined by rings of six dimeric complexes, $[Cu_2(peptide\text{–}H_2)_2]_6$: in solution, e.m.f. studies and small-angle *X*-ray scattering establish that a very similar species exists. The chiroptical spectra of solutions of copper(II) and gramicidin S have been reported.[135] A detailed experimental survey [136] of the physical properties of copper complexes of poly-L-histidine leads to the conclusion that, at pH 5, binding of copper is by three imidazoles and one peptidic nitrogen, whereas at pH 14, four consecutive peptidic nitrogens occupy an irregular plane tetragon, with a single imidazole at a fifth site.

Structural problems in valinomycin binding to metals have been attacked. This cyclic dodecadepsipeptide is known to enhance the mobility of cations, preferring rubidium or potassium to sodium. A potassium complex, crystallized from iso-octane, has, as revealed [137] by its crystal structure, six

[130] R. J. Dellaca, V. Janson, W. T. Robinson, D. A. Buckingham, L. G. Marzilli, I. E. Maxwell, K. R. Turnbull, and A. M. Sargeson, *J.C.S. Chem. Comm.*, 1972, 57.
[131] Y. Wu and D. H. Busch, *J. Amer. Chem. Soc.*, 1972, **94**, 4115.
[132] M. B. Hursthouse, S. A. A. Jayaweera, G. H. W. Milburn, and A. Quick, *Chem. Comm.*, 1971, 207.
[133] A. Yokoyama, H. Aiba, and H. Tanaka, *Chem. Letters*, 1972, 489.
[134] R. Osterberg, B. Sjoberg, and R. Soderquist, *J.C.S. Chem. Comm.*, 1972, 983.
[135] N. A. Poddubnaya and N. Ya. Krasnovbrizhii, *J. Gen. Chem.* (*U.S.S.R.*), 1971, **41**, 43.
[136] A. Levitzki, I. Pecht, and A. Berger, *J. Amer. Chem. Soc.*, 1972, **94**, 6844.
[137] W. L. Duax, H. Hauptman, C. M. Weeks, and D. A. Norton, *Science*, 1972, **176**, 911.

carbonyl groups octahedrally disposed about the potassium. In ethanol or butanol, diffraction measurements for valinomycin and its complex with potassium indicate [138] that a change of conformation occurs as the cation is bound. Mayers and Urry [139] calculated conformational energies (starting from the known secondary structure of the valinomycin–potassium complex) in order to deduce the complete conformation in solution and to evaluate the contribution of conformational energy to ion selectivity.

Among the varied reports of peptide interactions with constituents of metalloproteins, the coupling of L-histidine methyl ester and several peptide esters containing L-histidine to ferric protoporphyrin IX chloride has been studied spectroscopically.[140] The degree of interaction between the imidazole residues and the iron atom depends upon the length of the peptide chain. None of the products formed oxygen-carrying dimers. Huang and Haight [141] isolated and characterized a diamagnetic dinuclear glutathione complex of molybdenum(V) of composition $Na[Mo_2O_4$-(glutathione)$(H_2O)],3H_2O$. E.s.r. measurements in phosphate-buffered solutions show an equilibrium between diamagnetic and paramagnetic dimers. Similarities to xanthine oxidase are indicated. Molybdenum-containing enzymes form the subject of a review.[142]

Reactivity.—A novel use of a metal ion as *N*-terminal protecting group is that [143] by Purucker and Beck of platinum(II), in the compounds (18). With X = Cl, Br, or I, and Hα = Gly, Ala, Val, or Ser, the compounds couple with *NN'*-dicyclohexylcarbodi-imide, forming the corresponding dipeptide esters *trans*-$[PtX_2(\alpha\text{-Gly-OR}^1)_2]$.

Reactions with molecular oxygen have attracted some attention. In addition to the work with peptide complexes of cobalt already summarized,[125, 126] Nakon and Martell have studied [144] oxygenation of solutions of cobalt(II) and *NNN'''N'''*-diglycylethylenediaminetetra-acetic acid, the ligand (19).

A new reaction with dioxygen is that described [145] for peptide complexes of nickel(II) and copper(II). In neutral solution, tetra- and penta-peptide complexes of nickel(II) take up oxygen rapidly with the oxidation of the ligand, giving products similar to those found in irradiated oxygenated peptide solutions (*i.e.* amides of amino-acids and peptides, oxo-acids, and carbon dioxide). The only previously recorded promotion of oxygen uptake in a complex of nickel or copper in the oxidation state II known to us is by Schiff-base amino-acid ester complexes. The major reactive species in the novel peptide reaction [145] is $[Ni(H_{-2}L)]^-$, so that there are

[138] W. R. Krigbaum, F. R. Kuegler, and H. Oelschlaeger, *Biochemistry*, 1972, **11**, 4548.
[139] D. F. Mayers and D. W. Urry, *J. Amer. Chem. Soc.*, 1972, **94**, 77.
[140] A. van der Heijden, H. G. Peer, and A. H. A. van den Oord, *Chem. Comm.*, 1971, 369.
[141] T. J. Huang and G. P. Haight, *J. Amer. Chem. Soc.*, 1971, **93**, 611.
[142] R. C. Bray and J. C. Swann, *Structure and Bonding*, 1971, **11**, 107.
[143] B. Purucker and W. Beck, *Z. Naturforsch.*, 1972, **27b**, 1140.
[144] R. Nakon and A. E. Martell, *Inorg. Chem.*, 1972, **11**, 1002.
[145] E. B. Paniago, D. C. Weatherburn, and D. W. Margerum, *Chem. Comm.*, 1971, 1427.

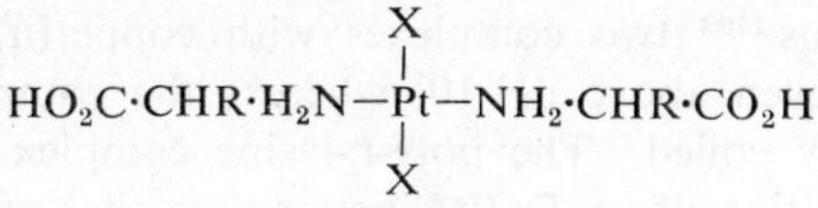

(18)

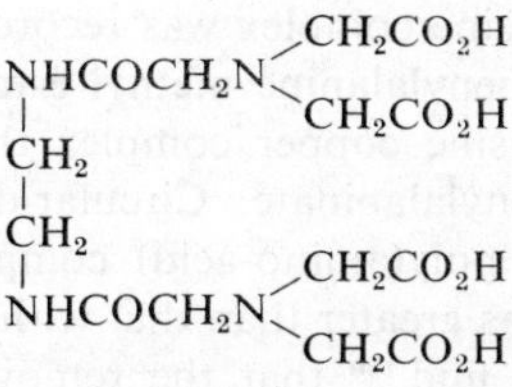

(19)

several co-ordinated deprotonated peptide links, and the initial step in both types of reaction seems to us likely to arise from unusual canonical forms, involving nickel(I) [or copper(I)] and the radical form of the unsaturated ligand. As complexes of larger peptides are increasingly studied, so the 'non-innocent' character of deprotonated peptide linked to a transition metal becomes increasingly apparent. Related observations are that, as a catalyst for the conversion [146] of glycine into ammonia, carbon dioxide, and formaldehyde by air at pH 12, osmium was found to be highly active, whereas manganese, chromium, and iron are effective only in oxidation state II. Copper(II) salts are ineffective.

Other oxidations in the presence of copper have been described. The radiomimetic system Cu^{2+}–H_2O_2 caused [147] less ready degradation (at 21 °C) of poly(α-amino-acids) than of proteins. However, at 37 °C, the degree of degradation of poly(α-amino-acids) was like that at 21 °C for proteins. A very interesting extension of their earlier work on the oxidation of tetrapeptide complexes of copper(II) is outlined [148] by Levitzki and Berger, who suggest that the specific oxidation at copper-binding sites by the ion $[IrCl_6]^{2-}$, hexachloroiridate(IV), may be used to locate such sites. The reaction of the iridium complex with two copper complexes of RNase fragments (one an octapeptide, the other with twenty residues) showed that, in each case, copper was attached to the terminal NH_2.

Some interesting and suggestive reports have appeared [149] on the use of poly-lysine–copper(II) complexes as catalysts for reactions of amino-acids.

[146] A. S. Soskind, R. I. Shulutko, G. E. Zavodnaya, and B. Z. Vlasova, *Zhur. obshchei Khim.*, 1972, **42**, 1379.

[147] C. L. Deasy, J. J. Tancous, and K. Jayasimhulu, *Austral. J. Chem.*, 1972, **25**, 1819.

[148] A. Levitzki and A. Berger, *Biochemistry*, 1971, **10**, 64.

[149] (*a*) M. Hatano, T. Nozawa, S. Ikada, and T. Yamamoto, *Makromol. Chem.*, 1971, **141**, 1; (*b*) M. Hatano, T. Nozawa, S. Ikada, and T. Yamamoto, *ibid.*, p. 11; (*c*) T. Nozawa and M. Hatano, *ibid.*, p. 21; (*d*) T. Nozawa and M. Hatano, *ibid.*, p. 31; (*e*) T. Nozawa, Y. Akimoto, and M. Hatano, *ibid.*, 1972, **158**, 21; (*f*) T. Nozawa, Y. Akimoto, and M. Hatano, *ibid.*, 1972, **161**, 289.

Poly-L-lysine forms [149a] two complexes with copper(II), which maintain helical structure even above pH 10, whereas the complex with poly-DL-lysine is randomly coiled. The poly-L-lysine complex at high pH catalysed [149b] the oxidation of DL-DOPA better than that of L-DOPA: a combination of two or more asymmetric copper(II) units was thought [149d] to be instrumental in this stereoselective oxidation. The circular dichroism spectrum of the poly-L-lysine complex was recorded.[149c] In a similar way, the hydrolysis [149e] of D-phenylalanine methyl ester was more rapid in the presence of the poly-L-lysine–copper complex than the hydrolysis of its enantiomer, methyl L-phenylalaninate. Circular dichroism showed [149f] that the interaction with the poly(amino-acid) complex of the more readily hydrolysed D-substrate was greater than that with the L-substrate.

Dukes and Margerum find [150] that the removal by edta of copper(II) from its complexes with oligopeptides is catalysed by amino-acids. Steric hindrance is thought to impede the nucleophilic attack by edta, and the role suggested for the amino-acid catalyst is to displace some of the bonds from peptide to copper, helping to reduce the crowding of subsequent attack by edta. Transamination occurs [151] in copper(II) chelates of the Schiff bases formed between a number of dipeptides and 2-acetylpyridine or pyridine-2-aldehyde.

Fenn and Bradbury have made [152] a detailed study of the possibility of using the now well-known reagent β-[Co(trien)(H_2O)(OH)]$^{2+}$ in selective peptide hydrolysis. There are many complications, one being that the reaction is not so specific for the *N*-terminal residue as the original results suggested. The complex also combines with free carboxy-groups of peptides, necessitating esterification as a protective measure. Some rates of reaction are very small. A satisfactory procedure involving reaction at 35 °C and pH 11 was devised, although the conclusion is that there are yet some snags to overcome before a useful analytical procedure is achieved. In the same context, the effect of potentially co-ordinating side-chains on amino-acids was studied,[153] using dipeptides containing aspartic acid, glutamic acid, and methionine. Whereas Gly-DL-Met-OH is readily hydrolysed by [Co(trien)X_2]$^+$, the dipeptide with aspartic acid as *N*-terminal member, α-L-Asp-Gly-OH, is not hydrolysed by this complex, but is reactive with [Co(dien)X_3], forming [Co(dien)(L-Asp-O)]$^+$, with L-aspartate(2−) occupying three co-ordination sites.

Cronin, Long, and Truscott have studied [154] the rates of sequence inversion of dipeptides through a diketopiperazine intermediate, in the presence of zinc(II). (This follows an earlier report [155] that in the presence

150 G. R. Dukes and D. W. Margerum, *Inorg. Chem.*, 1972, **11**, 2952.
151 Y. Nakao, H. Ishibashi, and A. Nakahara, *Bull. Chem. Soc. Japan*, 1970, **43**, 3457.
152 M. D. Fenn and J. H. Bradbury, *Analyt. Biochem.*, 1972, **49**, 498.
153 A. Y. Girgis and J. I. Legg, *J. Amer. Chem. Soc.*, 1972, **94**, 8420.
154 J. R. Cronin, D. A. Long, and T. G. Truscott, *Trans. Faraday Soc.*, 1971, **67**, 2096.
155 D. A. Long, T. G. Truscott, J. R. Cronin, and R. G. Lee, *Trans. Faraday Soc.*, 1971, **67**, 1094.

of zinc, cobalt, nickel, or manganese ions, large quantities of the diketopiperazine were formed from glycylglycine.) For all of Gly-Gly, Gly-L-Leu, L-Leu-Gly, Gly-L-Ala, and L-Ala-Gly, zinc was an effective catalyst.

In an isolated study of iron systems, Sugiura and Tanaka observed [156] that at pH 8.6 in the presence of ferric iron, sulphur-containing peptides like glutathione and 1-mercaptopropionylglycine lost sulphur, which was incorporated into iron–peptide–sulphur complexes, said to have an absorption spectrum like that of ferredoxin.

4 Proteins

Aspects of the non-specific binding of metal ions to proteins have been rather widely examined, though any consequent and possibly useful changes of reactivity are little considered. The usefulness of ions such as thallium with a nuclear moment as a means of investigating binding sites continues to be expanded. For example, the resonance of ^{205}Tl ($S = \frac{1}{2}$, natural abundance = 70.48%, sensitivity relative to ^{1}H = 0.192: the spectrometer operated at 24.3 MHz) changes [157] upon binding to pyruvate kinase, and evidence was collected pointing to a substrate-induced conformational change. In similar work, Reuben [158] found that although gadolinium ions were strongly bound by bovine serum albumin, they did not, apparently, occupy the same sites as does calcium, so that, in this case at least, the lanthanides cannot be used as probes for electrostatic sites. The nature of the binding sites in transferrin was studied by Luk,[159] using the fluorescence of lanthanides as a probe. The binding-site population varied with the size of the lanthanide (for Nd^{3+} and Pr^{3+}, which are larger, one site existed, whereas for the smaller Tb^{3+}, Eu^{3+}, Er^{3+}, and Ho^{3+}, there were two sites).

Although n.m.r. now dominates such studies (either applied directly, or through changes in such nuclei as ^{1}H or ^{19}F, specially introduced, as in work [160] on the binding site of *N*-fluoroacetyl-D-glucosamine to hen egg lysozyme) other methods also grow in their range of applicability. I.r. and chiroptical spectra were used [161] to test the suggestion that excitation of nerve membranes causes a conformational change in the proteins located in the membranes. It emerged that axon proteins from rat brans exist largely in the β-form in the presence of potassium ions, but in the α-form in the presence of sodium or calcium ions.

The emission of tryptophan, tyrosine, and bovine serum albumins is quenched by cupric ions;[162] this quenching is used to comment on the

[156] V. Sugiura and H. Tanaka, *Biochem. Biophys. Res. Comm.*, 1972, **46**, 335.
[157] J. Reuben and F. J. Kayne, *J. Biol. Chem.*, 1971, **246**, 6227.
[158] J. Reuben, *Biochemistry*, 1971, **10**, 2834.
[159] C. K. Luk, *Biochemistry*, 1971, **10**, 2838.
[160] C. G. Butchard, R. A. Dwek, S. J. Ferguson, R. J. P. Williams, and A. V. Xavier, *F.E.B.S. Letters*, 1972, **25**, 91.
[161] G. Papakostidis, G. Zundel, and E. Mehl, *Biochim. Biophys. Acta*, 1972, **288**, 277.
[162] C. K. Luk, *Biopolymers*, 1971, **10**, 1229.

relative positions of tryptophan residues in the protein. Conformational changes upon binding of bivalent cations to glutamine synthetase have been studied both kinetically [163] (using ΔpH measurements and u.v. spectra) and, in the case of manganese(II), calorimetrically.[164] Mn^{2+} acts as a specific activator and also serves to stabilize the dodecameric structure of the glutamic synthetase from *Escherichia coli*. For the binding of each manganese ion to the enzyme at pH 7.2 (37 °C), $\Delta G' = 8.9$ kcal mol^{-1}, $\Delta H \simeq +3$ kcal mol^{-1} (remarkably small), and the large entropy term is $\Delta S' \simeq +38$ e.u. (per mole of subunit-Mn).

Sudmeier and Pesek, in two interesting papers [165, 166] which employ ^{35}Cl n.m.r. in studying metal binding to bovine serum albumin, comment [165] that their commercially obtained BSA was contaminated with mercuric ions. Binding affinities for a series of metal ions are given,[165] and the rates of removal of mercury(II) from BSA by various thiol ligands were studied by the same method.[166] In the latter case, the enantiomers of penicillamine reacted at different rates.

A remarkable contribution to the elucidation of the mechanism whereby copper(II) inhibits bovine pancreatic ribonuclease reports [167] the results of an analysis (using *X*-ray diffraction) of the binding of copper(II) to crystalline ribonuclease-S (derived from RNase-A by proteolytic cleavage of the single side-chain near residue 20). At pH 5.5, with 0.1M-acetate as buffer, seven binding sites were detected, and a detailed analysis was made of their relative importances and occupancies under other conditions.

On the chemical side, metallic copper is found [168] to dissolve in bovine serum albumin and other proteins, the work being stimulated by the finding that copper is lost (at *ca.* 60 μg per day) from copper-coated intra-uterine devices. Sulphur-containing amino-acids and proteins have been desulphurized under fairly mild conditions using [169] Raney nickel. While, at pH 7.0 and 22 °C, the reaction with cystine (giving alanine) was complete in 12 minutes, methionine being unchanged after 10 hours, the extent of desulphurization of proteins depended upon the accessibility of the cystine and cysteine residues. Burstein and Sperling give [170] a chemical method for identifying S—Hg—S units in mercury protein derivatives.

Remarkably, in view of the specific copper binding by serum albumins from cows, humans, and rats, it has recently been shown [171] that the albumin of dog serum lacks such a characteristic copper-binding site. Subsequent work showed [172] this to be connected with the replacement of His-3 (of BSA and HSA) by tyrosine in dog serum albumin.

[163] J. B. Hunt and A. Ginsburg, *Biochemistry*, 1972, **11**, 3723.
[164] J. B. Hunt, P. D. Ross, and A. Ginsburg, *Biochemistry*, 1972, **11**, 3716.
[165] J. L. Sudmeier and J. J. Pesek, *Analyt. Biochem.*, 1971, **41**, 39.
[166] J. L. Sudmeier and J. J. Pesek, *Inorg. Chem.*, 1971, **10**, 860.
[167] N. M. Allewell and H. W. Wyckoff, *J. Biol. Chem.*, 1971, **246**, 4657.
[168] G. K. Osler, *Nature*, 1971, **234**, 153.
[169] M. T. Perlstein, M. Z. Atassi, and S. H. Cheng, *Biochim. Biophys. Acta*, 1971, **236**, 174.
[170] Y. Burstein and R. Sperling, *Biochim. Biophys. Acta*, 1971, **221**, 410.
[171] D. W. Appleton and B. Sarkar, *J. Biol. Chem.*, 1971, **246**, 5040.
[172] J. W. Dixon and B. Sarkar, *Biochem. Biophys. Res. Comm.*, 1972, **48**, 197.

In view of the considerable attention paid to cobalt ions as functional groups,[173] it is perhaps reassuring to find another report [174] of a cobalt requirement by an enzyme (other, of course, than in the field of Vitamin B_{12} enzymes [175]). This is by a newly discovered acylase, which catalyses the hydrolysis of the acyl bond in *N*-chloroacetyl-γ-L-glutamyl-β-naphthylamide. A second report [176] confirms the activation by cobalt(II), and shows that whereas magnesium and calcium are ineffective, manganese(II) dramatically reduces activity.

5 Metalloproteins

A short list of the titles of some pertinent reviews will indicate the level of activity in this field: 'The Role of Iron in Microbial Metabolism';[177] 'Metal Ion Function in Carbonic Anhydrase';[178] 'Carboxypeptidase A: A Mechanistic Analysis';[179] 'Three-dimensional Structure and Chemical Mechanisms of Enzymes';[180] 'Evolution of Biological Iron Binding Centres';[181] 'Iron Electronic Configurations in Proteins; Studies by Mössbauer Spectroscopy';[182] 'The Hemocyanins';[183a] 'Hemerythrin'.[183b] The haem proteins continue to be exhaustively studied; reviews in this field are: 'The Biological Role of the Haem group in Haemoglobin';[184] 'Spin-label Studies of Co-operative Oxygen binding to Haemoglobin';[185] a book, 'Hemoglobin and Myoglobin in their Reactions with Ligands';[186] 'Structural Studies of Haems and Haemoproteins by NMR Spectroscopy';[187] 'The Chemical Nature and Reactivity of Cytochrome P-450';[188] and 'The Nature of Electron Transfer and Electron Coupling Reactions'.[189]

Attention has been almost entirely upon the isolation and purification of metal enzymes, and studies of the binding centres which are not covered in this Report, since they focus upon the inorganic co-ordination aspect

173 (*a*) M. A. Foster, H. A. O. Hill, and R. J. P. Williams, in 'Chemical Reactivity and the Biological Role of Functional Groups in Enzymes', ed. R. M. S. Smellie, Academic Press, New York, 1970, p. 187 [a Biochemical Society Symposium (No. 31)]; (*b*) S. Lindskog, *Structure and Bonding*, 1970, **8**, 153.
174 A. Szewczyk, *Archivum Immunologiae et Therapiae Experimentalis*, 1971, **19**, 389.
175 J. M. Wood and D. G. Brown, *Structure and Bonding*, 1971, **11**, 47.
176 A. Szewczyk and A. Szczeklik, *Clinica Chim. Acta*, 1971, **33**, 309.
177 M. P. Coughlan, *Sci. Progress* (*London*), 1971, **59**, 1.
178 R. H. Prince and P. R. Woolley, *Angew. Chem.*, 1972, **84**, 461.
179 E. T. Kaiser and B. L. Kaiser, *Accounts Chem. Res.*, 1972, **5**, 219.
180 W. N. Lipscomb, *Chem. Soc. Rev.*, 1972, **1**, 319.
181 J. B. Neilands, *Structure and Bonding*, 1971, **11**, 145.
182 A. J. Bearden and W. R. Dunham, *Structure and Bonding*, 1970, **8**, 1.
183 'Subunits in Biological Systems. Part A' (vol. 5 in the series 'Biological Macromolecules'), ed. S. N. Timasheff and G. D. Fasman, (*a*) article by K. E. van Holde and E. F. J. van Bruggen, p. 1; (*b*) article by I. M. Klotz, p. 55.
184 R. G. Shulman and S. Ogawa, in ref. 173, p. 183.
185 H. M. McConnell, *Ann. Rev. Biochem.*, 1971, **40**, 227.
186 E. Antonini and M. Brunori, 'Hemoglobin and Myoglobin in their Reactions with Ligands', American Elsevier, New York, 1971.
187 K. Wüthrich, *Structure and Bonding*, 1970, **8**, 53.
188 H. A. O. Hill, A. Röder, and R. J. P. Williams, *Structure and Bonding*, 1970, **8**, 123.
189 B. Chance, *F.E.B.S. Letters*, 1972, **23**, 1.

of the metalloproteins. The most active areas have again involved non-haem iron proteins (which have now been purified from dozens of sources), copper proteins, and (a marked increase this year) molybdenum enzymes.[142] Typically, avimanganin, a protein containing bound manganese, contains[190] one manganese atom, tightly bound, per mole, the molecular weight being around 89 000.

Replacement of one metal ion by another (often in order to introduce some desired property, like colour or magnetic moment) is being fairly widely practised. For example, lanthanides have been employed[191] as substitutes for calcium ions in the α-amylase from *Bacillus subtilis*. Another new technique for commenting upon metal binding is magnetic circular dichroism, which has been applied, among many others, to[192] reduced and to oxidized spinach ferredoxin.

In the ferredoxin field, a notable achievement is the elucidation[193] of the structure of the iron–sulphur core in the bacterial ferredoxin from *Micrococcus aerogenes*: this is shown schematically as (20). After many difficulties, the structure was ultimately solved by heavy-atom isomorphous replacement. Thus (20) is essentially the same arrangement as had been

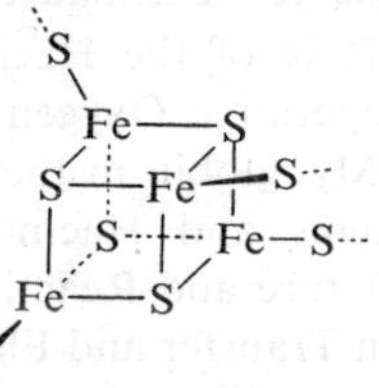

(20)

found for the $(Fe{-}S)_n$ unit in a protein from *Chromatium*: however, the as yet unexplained E° values are very diverse, with $E^\circ_{chromatium} = +0.3$ V, $E^\circ_{\text{ferredoxin}} = -0.4$ V. Superoxide radicals are present[194] during the autoxidation of both clostridial and spinach ferredoxins.

In the general area of iron metabolism, yet another intriguing ligand implicated in transport has been discovered,[195] so-called schizokinen [see (21)], from *Bacillus megaterium*. This chelates with ferric ions, forming (at neutral pH) an anionic complex.

Little new work on modification of reactivity in metalloproteins has appeared recently, although chemical modification[196] of all 18 carboxy-groups in haemerythrin from *Golfingia gouldii* had no effect upon the electronic spectrum of the protein, indicating that these groups are not

[190] M. C. Scrutton, *Biochemistry*, 1971, **10**, 3897.
[191] G. E. Smolka, E. R. Birnbaum, and D. W. Datnall, *Biochemistry*, 1971, **10**, 4556.
[192] J. C. Sutherland, I. Salmeen, A. S. K. Sun, and M. P. Klein, *Biochim. Biophys. Acta*, 1972, **263**, 550.
[193] L. C. Sieker, E. Adman, and L. H. Jensen, *Nature*, 1972, **235**, 40.
[194] H. P. Misra and I. Fridovich, *J. Biol. Chem.*, 1971, **246**, 6886.
[195] K. B. Mullis, J. R. Pollack, and J. B. Neilands, *Biochemistry*, 1971, **10**, 4894.
[196] G. L. Klippenstein, *Biochem. Biophys. Res. Comm.*, 1972, **49**, 1474.

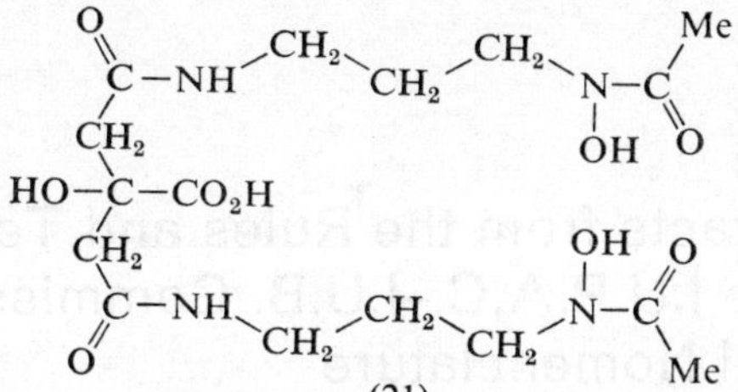

(21)

linked to the iron at the active centre for transport of oxygen. A similar report[197] refers to iodination of the tyrosine residues of ovotransferrin to assess the degree of their implication in iron-binding.

Sharpless and Flood point out[198] the degree of similarity between mixed-function oxygenases and a number of oxo-anions of first-row transition metals. Even the hydroxylation of aromatic molecules with concomitant NIH shift has now been demonstrated for $CrO_2(OAc)_2$, chromyl acetate, as oxidant. The inhibition of xanthine oxidase by methanol and formaldehyde has been explained[199] on the basis of an e.s.r. study, which shows that a formyl group, —CHO, becomes attached at the active site and that molybdenum(V) interacts with this and is stabilized, preventing the 'turnover' of the enzyme.

Finally, detailed studies upon the currently most intriguing molybdenum enzymic activity, the fixation of dinitrogen, proceed apace, and model systems involving molybdenum, thiols, and the reductants sodium borohydride or dithionite will mimic the natural system in that[200] acetylene is reduced to ethylene.

[197] J. L. Phillips and P. Azari, *Arch. Biochem. Biophys.*, 1972, **151**, 445.
[198] K. B. Sharpless and T. C. Flood, *J. Amer. Chem. Soc.*, 1971, **93**, 2316.
[199] F. M. Pick, M. A. McGartoll, and R. C. Bray, *European J. Biochem.*, 1971, **18**, 65.
[200] G. N. Schrauzer and P. A. Docmeny, *J. Amer. Chem. Soc.*, 1971, **93**, 1608.

7

Further Extracts from the Rules and Tentative Rules of the I.U.P.A.C.–I.U.B. Commission on Biochemical Nomenclature

In Volume 2, Chapter 5 of these Reports we reprinted the following extracts from the I.U.P.A.C.–I.U.B. Recommendations relevant to amino-acids, peptides, and proteins: *Tentative Abbreviated Designation of Amino-acid Derivatives and Peptides; Tentative Rules for Naming Synthetic Modifications of Natural Peptides; Tentative Abbreviated Nomenclature of Synthetic Polypeptides* (*Polymerised Amino-acids*). In Volume 4, Chapter 5, the 1971 revision of the first-named recommendations was reprinted, together with the new recommendations, *Abbreviations and Symbols for the Description of the Conformation of Polypeptide Chains.* We now reprint (from *Pure Appl. Chem.*, 1972, **31**, 649, and *ibid.*, 1973, **33**, 439) two further revisions of previously published tentative rules, namely *Definitive Rules for Naming Synthetic Modifications of Natural Peptides, and Abbreviated Nomenclature of Synthetic Polypeptides* (*Polymerised Amino-acids*), together with *A One-letter Notation for Amino-acid Sequences* (from *Pure Appl. Chem.*, 1972, **31**, 641), not previously published in these Reports.

I Definitive Rules for Naming Synthetic Modifications of Natural Peptides

(Reproduced by permission from *Pure Appl. Chem.*, 1972, **31**, 649—653)

During the past few years, chemists have made many compounds that are variants of naturally occurring peptides (or proteins) having trivial names. Therefore, the need has arisen for 'semitrivial' names to designate these variants without the necessity of designating every residue in the chain.

After discussion with active workers in the field, tentative rules were proposed by the IUPAC–IUB Commission on Biochemical Nomenclature (CBN) and published in *IUPAC Information Bulletin No. 27* (1966). They are based on the names used by du Vigneaud and his collaborators [cf. Bodanszky and du Vigneaud, *J. Amer. Chem. Soc.* **81**, 1258 (1959); Popenoe, Lawler and du Vigneaud, *J. Amer. Chem. Soc.* **74**, 3713 (1952)] and the symbols introduced by Schwyzer *et al.* [cf. Rittel, Iselin, Kappeler, Riniker and Schwyzer, *Angew. Chem.* **69**, 179 (1957); Riniker and

Schwyzer, *Helv. Chim. Acta*, **44**, 685 (1961); see also *J. Biol. Chem.* **247**, 977 (1972).

The definitive rules are based on the comments that were received. They are *not* suitable for application to 'abnormal' links in a peptide sequence; e.g. to disulfide links or γ-peptide links. They are *only* suitable for modifications involving normal α-peptide links.

RULES

1. Replacement

In a polypeptide of trivial name X, if the qth amino acid residue (starting from the N-terminal end of the chain) is *replaced* by the amino acid residue Abc, the semitrivial name of the modified polypeptide is [q-amino acid]X and the abbreviated form, chiefly for use in tables, is [Abc^q]X.

Examples

[8-Citrulline]vasopressin, [Cit^8]vasopressin [Bodanszky and Birkhimer, *J. Amer. Chem. Soc.* **84**, 4963 (1962)]. [5-Isoleucine, 7-alanine]hypertensin II, [Ile^5, Ala^7]hypertensin II [Seu, Smeby and Bumpus, *J. Amer. Chem. Soc.* **84**, 3883 (1962)].

Comments

(a) In the full name, the replacement amino acid is designated by its *own full name*, *not* the name of its radical (cf. 4 below). This name, and the position of replacement, are given in square brackets [ ], as for isotopic replacement.

(b) In the abbreviated form, the amino-acid residues are designated by the standard three-letter symbols [*J. Biol. Chem.* **241**, 527 (1966); *Biochim. Biophys. Acta*, **121**, 1 (1966); also **247**, 977 (1972)], the first letter *only* being a capital, in square brackets [ ].

(c) In the abbreviated form, the *position* of substitution is indicated in a special fashion, i.e. by a superior numeral^q, to indicate that it is a *residue*, not an individual atom, that is being replaced and also for the reason indicated in comment d.

(d) The nature of the residue replaced is *not* designated in either the full or the abbreviated name. This is contrary to a general principle of organic nomenclature requiring that an atom (or group) that is replaced should (unless it is hydrogen) be clearly designated, as in 2-amino-2-deoxy-D-glucose. It has been decided *not* to insist on the designation of the residue replaced in these semitrivial names in order to keep the names as short as possible, and because the form of nomenclature in Rule 1 clearly differs from ordinary substitution nomenclature.

(e) An analogy may be drawn with the form used for isotopic replacement, where the isotope symbol is indicated in square brackets before the name.

(f) The replacement of an amino-acid residue by its enantiomer may be shown logically by the application of this rule as follows: the replacement in X of L-alanine at position 7 by D-alanine results in [7-D-alanine]X with the abbreviation [D-Ala^{7}]X. An example may be found in R. A. Boissonnas, St. Guttmann and J. Pless [*Experientia*, **22**, 526 (1966)], dealing with the D-Ser^{1} . . . derivative of β-corticotropin; the natural compound has L-serine in position 1. Another example is the [α-D-Asp^{1}]hypertensin II of Riniker and Schwyzer [*Helv. Chim. Acta*, **47**, 2357 (1964)].

2. Extension

The compounds obtained by the extension of polypeptide X at either (a) the N-terminal end or (b) the C-terminal end are designated by the kinds of names and abbreviations shown below; these are in accordance with the general principles of polypeptide nomenclature [*J. Biol. Chem.* **247**, 977 (1972)].

Examples

(a)	Extension at N-terminal end:	
	Aminoacyl-X	Abc-X
e.g.	Valyl-X	Val-X
or	Valylglycyl-X	Val-Gly-X (for extension by two residues)
(b)	Extension at C-terminal end:	
	X-yl-amino acid	X-yl-Abc
e.g.	X-yl-leucine	X-yl-Leu

(where X-yl is the trivial name of polypeptide X with the ending -yl).

Comment

This rule is not applicable to the extension at the C-terminal of natural peptides having a terminal α-carboxamido group, as in the case of oxytocin or α-melanophore-stimulating hormone (α-MSH). It has been suggested that new names be given to the peptides having a free terminal α-carboxyl group (e.g. oxytocinoic acid) and that extension at the C-terminal end be denoted as in the example given above (e.g. oxytocinoyl-Abc).

3. Insertion

The compound obtained by the *insertion* of an additional amino acid residue Abc in the position between the qth and (q + 1)th residues of a polypeptide X is named qa-endo-amino acid-X (abbreviated form, endo-Abc^{qa}-X).

Example

4a-endo-tyrosine-hypertensin II; endo-Tyr^{4a}-hypertensin II.

Comments

(a) This form has analogies in other fields where endo implies the insertion of something into a structure (e.g. endo-methylene). The prefix or index qa is based on analogies with the steroids where the atoms inserted in a ring atom no. q are designated qa, qb, etc.

(b) The prefix homo is *not* suitable for designating the insertion of a whole residue, since it is commonly used to modify the names of *individual* amino acids, e.g. homoserine.

(c) Multiple insertions, and insertion of two or more residues together in the same place in the chain, are shown by a logical extension of this rule. For example, the insertion into the polypeptide X of threonine between residues 4 and 5, and of valine and glycine (*in that order*) between residues 6 and 7, is shown by the name 'endo-4a-threonine,6a-valine, 6b-glycine-X' and the abbreviation 'endo-Thr^{4a}, (Val^{6a}-Gly^{6b})-X.'

4. Removal

The compound obtained by the formal *removal* of an amino acid residue from a polypeptide X in position q is designated by the name des-q-amino acid-X, abbreviated des-Abc^{q}-X.

Example

des-7-proline-oxytocin; des-Pro^{7}-oxytocin [Jacquenoud and Boissonnas, *Helv. Chim. Acta*, **45**, 1462 (1962)].

Comment

(a) Removal of a whole residue is indicated as is the removal of a ring in steroids, e.g. des-A-androstane.

(b) 'd' is *not* suitable as a prefix because it is easily confused, in speaking, with D (for configuration).

5. Substitution forming a side chain

The compound formed by the substitution of an additional amino acid residue as a side chain into a polypeptide X is named by applying the ordinary rules of nomenclature to the trivial name.

(a) If the substitution is on a side chain *amino* group of polypeptide X, the name of the additional amino *residue* is written (with the termination 'yl') and prefixed by symbols indicating the position of substitution (residue number and atom).

Example

An imaginary compound (A)

```
     Val
      |
Ala-Lys-Ala . . . . .
 1   2   3
      A
```

in which a valyl group is substituted at the ε-amino group of lysine at position 2 of the chain of a peptide X is named $N^{\varepsilon 2}$-valyl-X (abbreviated $N^{\varepsilon 2}$-Val-X).

(b) If the substitution is on a side chain *carboxyl* group of polypeptide X, the additional amino acid having a free α-carboxyl group, the substituted derivative is named by specifying the position of substitution (residue number, and atom) and is given the designation 'X-amino acid'.

Example

An imaginary compound (B)

```
              Val
               |
    Ala-Leu-Glu-Ala . . . . . .
     1   2   3   4
              B
```

in which a valine residue is substituted into the γ-carboxyl group of glutamic acid in position 3 of the chain of a peptide X would be named $C^{\gamma 3}$-X-yl-valine (abbreviated $C^{\gamma 3}$-X-yl-Val).

Comment

Note the importance of clear distinction from *replacement* as indicated in Rule 1.

6. Partial sequences (fragments)

Polypeptide sequences that form fragments of a longer sequence that already has a trivial name may be designated as follows. The *trivial name* is followed by numbers giving the positions of the first and last amino acids, and then the usual *Greek* designation giving the number of amino acid units in the fragment; thus

Trivial name(-X–Y-) . . . . . . peptide.

Example: from α-MSH

```
Ac-Ser-Tyr-Ser-Met-Glu-His-Phe-Arg-Trp-Gly-Lys-Pro-Val-NH2   α-MSH
   1   2   3   4   5   6   7   8   9   10  11  12  13
```

we may have

```
Met-Glu-His-Phe-Arg-Trp-Gly          α-MSH-(4–10)-heptapeptide
 4                      10
```

and

```
His-Phe-Arg-Lys-Pro-Val-NH2
 6       8   11      13
α-MSH-(6–8)-(11–13)-hexapeptide amide
```

The last example illustrates the nomenclature for a composition sequence of two fragments, and also for an amide-terminal group.

SUMMARY WITH EXAMPLES

The systematic application of these principles to the name of an imaginary pentapeptide 'Iupaciubin'* may illustrate the symbolism.

Rule	Operation	Short name	Structure
	(Fundamental name)	Iupaciubin	1 2 3 4 5 Ala-Lys-Glu-Tyr-Leu
1.	Replacement	[Phe^{4}]iupaciubin†	4 Ala-Lys-Glu-Phe-Lue
2a.	Extension (N terminal)	Arginyl-iupaciubin, Arg-iupaciubin	1 5 Arg-Ala-Lys-Glu-Tyr-Leu
2b.	Extension (C terminal)	Iupaciubyl-methionine, iupaciubyl-Met	1 5 Ala-Lys-Glu-Tyr-Leu-Met
3.	Insertion	Endo-Thr^{2a}-iupaciubin	2 2a 3 Ala-Lys-Thr-Glu-Tyr-Leu
4.	Removal	Des-Glu^{3}-iupaciubin	2 4 Ala-Lys-Tyr-Leu
5a.	Side chain substitution on amino group	$N^{\varepsilon 2}$-Val-iupaciubin	Val $\vert\varepsilon$ Ala-Lys-Glu-Tyr-Leu 2
5b.	Side chain substitution on carboxyl group	$C^{\gamma 3}$-Iupaciubyl-valine	Val $\vert\gamma$ Ala-Lys-Glu-Tyr-Leu 3
6.	Partial sequence	Iupaciubin-(2–4)-tripeptide	2 3 4 Lys-Glu-Tyr

* To symbolize the harmonious cooperation of IUPAC and IUB.
† Note that only for *replacement* are square brackets required.

II Abbreviated Nomenclature of Synthetic Polypeptides (Polymerized Amino Acids)

(Reproduced by permission from *Pure Appl. Chem.*, 1973, **33**, 439—444)

The numerous studies on the physical, chemical, and biological properties of synthetic polypeptides have brought with them different ways of describing, in abbreviated form, these products whose structures are often incompletely known. The use of a variety of nomenclatures complicates the literature; hence, a consistent and clearly defined system for naming such polypeptides is desirable. The proposals set forth here, which represent the consensus of many discussions and suggestions, should aid in systematizing the nomenclature of a wide variety of synthetic polypeptides. They were published in tentative form in *IUPAC Information Bulletin No. 30* (1967) and in revised form in several journals in 1972[1].

These proposals are based in large part on the abbreviated nomenclature devised by Gill[2] and by Sela[3] and others. They utilize the symbols and conventions set forth in Section 2 of '*Revised Tentative Rules for Abbreviations and Symbols of Chemical Names of Special Interest in Biological Chemistry*'[4] and in '*Abbreviated Designation of Amino Acid Derivatives and Peptides*'[5], and they add only those terms or conventions needed for the specification of polymers but not encompassed by these schemes.

The symbols and conventions of the earlier 'Tentative Rules'[4, 5] used in this nomenclature system are summarized as follows. The symbols of the amino acid residues and their derivatives or modifications are those indicated in the 'Tentative Rules'[4, 5] or formulated according to the principles set out in them. Hyphens or commas between the symbols for residues or groups of residues indicate known or unknown sequence, respectively, and involve only the α-NH_2 and α-COOH groups (the peptide link). Commas may be omitted when other symbols (e.g. subscripts or superscripts) separate symbols in unknown sequences. Vertical strokes indicate covalent bonds involving functional groups or the remaining H-atom of the peptide bond, depending upon their placement[5]. L-Amino acids and α-peptide links, read from left (NH_2 terminus) to right (COOH terminus), are assumed unless indicated otherwise[4, 5].

DEFINITIONS

1. Linear Polymer

All amino acid residues (constitutional units) are linked in an unbranched chain.

2. Block

A polymer that forms a distinct part of a larger polymer (e.g. a block or graft polymer may contain several blocks).

3. Graft Polymer

One or more blocks are linked to the functional groups of a linear polymer, thus creating a branch or branches. (Functional groups include ε-NH_2, β- or γ-COOH, etc., and the remaining H-atom of an α-peptide link.)

4. Block Polymer

Two or more species of block are linked to form a larger linear polymer.

RECOMMENDATIONS

1. Designation of Blocks or Linear Polymers

The prefix 'poly' or the subscript n indicates 'polymer of'. It is attached to each main chain and is repeated in each block within a larger polymer unless there is sufficient indication of size and of structure to make this repetition unnecessary. For example, poly(Glu) or $(Glu)_n$ represent poly(glutamic acid), and $(Glu)_{10}$, a decapeptide of glutamic acid. 'Oligo' may replace 'poly' for short chains.

Comments

(a) *n* replaces the *p* as originally, but no longer, used in the polymer nomenclature scheme devised by the IUPAC Subcommission on the Nomenclature of Macromolecules[6]. It is used in designating polynucleotides[7], and it is chosen in place of *p* in order to avoid confusion with the 'p' used for a terminal phosphoric residue in the latter scheme. The *n* may be replaced by a definite number (e.g. 10 above), an average (e.g. 10), or a range (e.g. 8–12), as appropriate. However, two *n*'s should not appear in the same formula unless equal length is implied. When equal length is not the case, different letters should be used, such as *m*, *j*, *k*.

(b) If 'poly' is used rather than the subscript *n*, the symbol(s) following 'poly' should be enclosed in parentheses with no intervening space, e.g. poly(Lys). If 'poly' is followed by a single, simple word, the whole is written as one word, e.g. polylysine. If what follows 'poly' is complex, it should be enclosed in parentheses (again without following space), e.g. poly(amino acid), not polyamino acid or polyaminoacid; poly(glutamic acid) or polyglutamate, but not polyglutamic acid; poly(DL-alanine,L-lysine) for the substance shown in Example 2; and poly(DL-alanine-L-lysine) for the substance shown in Example 3. The format poly(L-lysine) is preferred to poly-L-lysine, i.e. L-lysine is regarded as a complex term. Similarly, poly(hydroxyproline), not polyhydroxyproline.

2. Designation of Branches and Branch Points

Branches (side chains) connected to the main chain can be designated in one of three ways: by a vertical line joining the main chain and the branch (side chain); by an extended bond joining the appropriate residues with the main chain written first; or by a horizontal double dash (not preferred).

The branch points are indicated by the origin and terminus of the vertical line. If the origin is unknown, the line originates at the 'p' in 'poly' if 'poly' is used, or at the first parenthesis (bracket), if the subscript *n* is used (see Recommendation 1). If the origin is known, the line originates: (a) vertically at the initial letter of the appropriate symbol, if functional groups other than α-NH_2 or α-COOH residues are involved; (b) vertically at the position of the appropriate link, if substitution for the remaining H-atom of a peptide link is involved; or (c) horizontally to the left or right of the symbol, respectively, if α-NH_2 or α-COOH groups are involved. The same rules apply to the termination of the line. Thus, the linkage between a side chain functional group and an α-NH_2 or α-COOH group in the main chain is indicated by two perpendicular lines with the vertical line originating in the functional group and the horizontal line in the α-NH_2 or α-COOH group. A number in parentheses lying beside the line indicates the number of such links per 100 residues of polymer, if known.

Comment

A limitation of the double dash as a connecting link lies in its inability to originate or to terminate definitively in a specific residue. Either the arrangement of the symbols must be such that connected ones are adjacent, or the information must be given independently.

3. Block size

A superscript outside the parentheses enclosing a block indicates the number of repeating sequences per 100 residues of polymer, and it is given to the first decimal place.

4. The Molar Percentage

The molar percentage of a single type of amino acid residue within a copolymer, derived from the amino acid analysis and assuming copolymerization, is indicated by a superscript attached to the symbol of the residue. The molar percentages are given in whole numbers and should total 99 to 101 per cent.

EXAMPLES

1. Simple homopolymer:

poly(Ala) or $(\text{Ala})_n$

2. Linear copolymer, unknown sequence, composition not specified:

poly(DLAla,Lys) or $(\text{DLAla,Lys})_n$

3. Linear copolymer, regular alternating sequence:

poly(DLAla-Lys) or $(\text{DLAla-Lys})_n$

4. Linear sequence of unknown order [Composition: 56% Glu, 38% Lys, and 6% Tyr (Σ = 100%)]:

(a) poly($\text{Glu}^{56}\text{Lys}^{38}\text{Tyr}^{6}$) or $(\text{Glu}^{56}\text{Lys}^{38}\text{Tyr}^{6})_n$ (all L)

(b) poly($\text{DGlu}^{56}\text{DLys}^{38}\text{Tyr}^{6}$) (only Tyr is L)

(c) poly($\text{DLGlu}^{56}\text{Lys}^{38}\text{DTyr}^{6}$) (Glu is DL, Tyr is D)

5. Block polymer of poly(Glu) combined through the α-COOH terminus to the α-NH_2 terminus of poly(Lys) [Composition: 56% Glu, 44% Lys (Σ = 100%)]:

poly(Glu^{56})–poly(Lys^{44}) or $(\text{Glu}^{56})_n$–$(\text{Lys}^{44})_m$

6. (a) Known, repeating sequence (a polymer of Glu-Lys-Lys-Tyr):

poly(Glu-Lys_2-Tyr) or $(\text{Glu-Lys}_2\text{-Tyr})_n$

(b) Known, repeating sequences within each of two constituent blocks of a linear polymer [Composition: 37.5% Glu, 25% Lys, 25% Tyr, 12.5%

Ala (Σ = 100%)]:

poly(Glu-Lys)^{25}–poly(Ala-Tyr$_2$-Glu)$^{12.5}$

or

(Glu-Lys)$_n^{25}$–(Ala-Tyr$_2$-Glu)$_m^{12.5}$

[The connection between the polymeric segments shown here is from the α-COOH of Lys to the α-NH_2 of Ala. Origin or termination in any other residue or functional group can be shown by rearranging the order of residues and by the orientation of the connecting line at its origin and terminus (see Examples 7, 8, and 9).]

(c) Known, repeating sequence in the main chain connected by the ε-NH_2 of a lysine (which of the two is not known) to an unknown point in an unknown sequence in the side chain (Composition: 30% Asp, 55% Glu, 10% Lys, 5% Tyr (Σ = 100%)]:

poly(Asp^{30}Glu^{50})
|
poly(Glu-Lys$_2$-Tyr)^5

or

(Asp^{30}Glu^{50})$_m$
|
(Glu-Lys$_2$-Tyr)$_n$

or

poly(Glu-Lys$_2$-Tyr)^5 poly(Asp^{30}Glu^{50})

or

(Glu-Lys$_2$-Tyr)$_n^5$ (Asp^{30}Glu^{50})$_m$

(Note: The double hyphen system is not applicable here.) If it is known which lysine residue is connected to the side chain, the main chain would be written in the form,

poly(Glu-Lys-Lys-Tyr)^5 or poly(Glu-Lys-Lys-Tyr)^5,

as appropriate.

7. Graft polymer with the main chain of DL-alanine and L-lysine connected through the ε-NH_2 group of lysine to the α-COOH group of L-tyrosine in the side chain, which consists of a block polymer of L-tyrosine and L-alanine (no analytical data for the main chain):

poly(Ala)–poly(Tyr)
poly(DLAla,Lys)

or

(Ala)$_n$–(Tyr)$_m$
(DLAla,Lys)$_j$

or

poly(DLAla,Lys) poly(Ala)–poly(Tyr)

or

poly(DLAla,Lys)–poly(Ala)–poly(Tyr)

(Note: The points of attachment of Lys and Tyr cannot be specified in the last example.)

8. Graft polymer with an unknown sequence in the main chain and in the side chain [Composition: 16% Lys, 20% Ala, 35% Glu, 29% Tyr (Σ = 100%)]:

(a) Number and position of the points of attachment in the main chain unknown, but terminating in the lysine residues of the side chain:

poly(Lys^{16}Ala^{20}) | poly(Glu^{35}Tyr^{29}) or (Lys^{16}Ala^{20})$_m$ | (Glu^{35}Tyr^{29})$_n$

or

poly(Glu^{35}Tyr^{29}) poly(Lys^{16}Ala^{20}) (joined by a bracket above)

or

(Glu^{35}Tyr^{29})$_n$ (Lys^{16}Ala^{20})$_m$ (joined by a bracket below)

(b) Same, but attachments are 3 in number and connect the ε-NH_2 groups of the lysine residues in the side chain and the γ-COOH groups of the glutamic acid residues in the main chain:

poly(Lys^{16}Ala^{20}) (3)| poly(Glu^{35}Tyr^{29}) or (Lys^{16}Ala^{20})$_m$ (3)| (Glu^{35}Tyr^{29})$_n$

or

poly(Glu^{35}Tyr^{29}) poly(Lys^{16}Ala^{20}) (joined by a bracket below, labelled (3))

poly(Tyr^{29}Glu^{35})$^{\gamma\varepsilon}_{(3)}$poly(Lys^{16}Ala^{20})

9. Graft polymer with a block polymer and an unknown sequence in the side chain (upper) attached to an unknown sequence in the main chain (lower); the points of attachment are between the γ-COOH groups of glutamic acid in the side chain and the ε-NH_2 groups of lysine in the main chain [Composition: 12% Glu, 21% Lys, 24% Tyr, 24% Leu, 20% Ala (Σ = 101%)].

poly(Glu^{12})–poly(Lys^{6})–poly(Tyr^{24}Leu^{24})
|
poly(Lys^{15}Ala^{20})

or

(Glu^{12})$_m$–(Lys^{6})$_j$–(Tyr^{24}Leu^{24})$_k$
|
(Lys^{15}Ala^{20})$_n$

or

poly(Lys^{15}Ala^{20}) poly(Glu^{12})–poly(Lys^{6})–poly(Tyr^{24}Leu^{24}) (joined by a bracket below)

or

(Lys^{15}Ala^{20})$_n$ (Glu^{12})$_m$–(Lys^{6})$_j$–(Tyr^{24}Leu^{24})$_k$ (joined by a bracket below) or

$$\text{poly}(\text{Ala}^{20}\text{Lys}^{15})\overset{\varepsilon\gamma}{-\!-}\text{poly}(\text{Glu}^{12})\text{–poly}(\text{Lys}^{6})\text{–poly}(\text{Tyr}^{24}\text{Leu}^{24})$$

REFERENCES

[1] *Biopolymers*, **11**, 321 (1972); *Biochem. J.* **127**, 753 (1972); *Biochemistry*, **11**, 942 (1972); *Eur. J. Biochem.* **26**, 301 (1972); *J. Biol. Chem.* **247**, 323 (1972); *Bull. Soc. Chim. Biol.* **51**, 205 (1969); *Z. Physiol. Chem.* **349**, 1013 (1969); *J. Mol. Biol.* **5**, 492 (1971).

[2] T. J. Gill, III, *Biopolymers*, **2**, 283 (1964); also *J. Biol. Chem.* **240**, 3227 (1965); *Biochim. Biophys. Acta*, **124**, 374 (1966).

[3] M. Sela, *Advan. Immunol.* **5**, 30 (1966).

[4] *J. Biol. Chem.* **241**, 527 (1966); *Biochemistry*, **5**, 1445 (1966); *Biochem. J.* **101**, 1 (1966); *Virology*, **29**, 480 (1966); *Arch. Biochem. Biophys.* **115**, 1 (1966); *Eur. J. Biochem.* **1**, 259 (1967); *Z. Physiol. Chem.* **348**, 245 (1967). See also Reference 7.

[5] *J. Biol. Chem.* **247**, 977 (1972); *Biochemistry*, **11**, 1726 (1972); *Biochim. Biophys. Acta*, **263**, 205 (1972); *Biochim. J.* **126**, 773 (1972); *Arch. Biochem. Biophys.* **150**, 1 (1972); *Eur. J. Biochem.* **27**, 201 (1972).

[6] *J. Polym. Sci.* **8**, 257 (1952); revised in 1967 (unpublished; see references in footnote‡).

[7] *Biochem. J.* **120**, 449 (1970); *Biochemistry*, **9**, 4022 (1970); *Eur. J. Biochem.* **15**, 203 (1970); *J. Biol. Chem.* **245**, 5171 (1970); *Z. Physiol. Chem.* **351**, 1055 (1970); *J. Mol. Biol.* **55**, 299 (1971); and elsewhere.

III A One-letter Notation for Amino Acid Sequences

(Reproduced by permission from *Pure Appl. Chem.*, 1972, **31**, 641—645)

In 1968 the IUPAC–IUB Commission on Biochemical Nomenclature (CBN) published in *IUPAC Information Bulletin No. 32* tentative rules for the use of a one-letter notation for amino acid sequences. These definitive rules are based on the comments received.

1. GENERAL CONSIDERATIONS

Various difficulties are encountered when presenting the formulas of long protein sequences in the usual three-letter symbols[1a]. Space is often at a premium. A one-letter code minimizes this difficulty and has other distinct advantages. In summarizing large amounts of data or in the alignment of homologous protein sequences, it is important that the patterns in the sequences be condensed and simplified as much as possible. Computer techniques are increasingly applied for the storage of sequences of hundreds of amino acid residues and for their evaluation. For these purposes, a one-letter code is the best solution. Finally, a one-letter code is useful in the labelling of individual amino-acid sidechains in three-dimensional pictures of protein molecules.

The possibility of using one-letter symbols was mentioned by Gamov and Yčas[2] in 1958. The idea was systematized by Šorm *et al.*[3] in 1961. It was used by this group[4–10] and also by Fitch[11] in several papers on the structure of proteins. In extensive compilations of protein structures, Eck and Dayhoff (see refs 12 to 14) systematically used one-letter symbols derived partly from the code of Šorm and Keil. Independent proposals were made by Wiswesser[15] and by Braunstein*.

In view of the increasing number of different notations and the attending problems, the IUPAC–IUB Commission on Biochemical Nomenclature (CBN) undertook the task of drafting a single notation for one-letter symbols. The present proposal was evolved by a CBN subcommission (composed of B. Keil, R. V. Eck, M. O. Dayhoff and W. E. Cohn); it is based principally on the system evolved by Dayhoff and Eck[12–14].

2. LIMITS OF APPLICATION

In publications, CBN recommends that one-letter symbols be used only in comparisons of long sequences in tables, lists or figures, and for such special use as tagging three-dimensional models of proteins. They should not be used in simple text nor for original reports of experimental details of sequences. This system is not suitable for reporting the details of peptide synthesis, for example, where a fuller description of substituents is needed and where uncommon amino acids may occur. It should not be used in papers where the single-letter system for nucleoside sequences is employed (see ref. 1b, Section N-3.2), as in representing codons, etc.

3. PRINCIPLES OF THE ONE-LETTER CODE

3.1 The letter written at the left-hand end is that of the amino acid residue carrying the free amino group and the letter written at the right-hand end is that of the amino acid residue carrying the free carboxyl group. The absence of punctuation beyond either end of a sequence implies that it is known to be the amino or carboxyl end of the protein. A fragmentary sequence is to be preceded or followed by a slash (/) to indicate that it is not known to be the end of the complete protein (see 'Comment' in Section 8.2).

3.2 Initial letters are used where there is no ambiguity. There are six such cases: cysteine, histidine, isoleucine, methionine, serine and valine. All the other amino acids share the initial letters A, G, L, P and T; therefore, assignments of them must be somewhat arbitrary. These letters are assigned to the most frequently occurring and structurally most simple amino acids. On this basis, the letters A, G, L, P and T are assigned to alanine, glycine, leucine, proline and threonine, respectively.

3.3 The assignment of the other abbreviations is more arbitrary. However, certain clues are helpful. Two are phonetically suggestive, F for *ph*enylalanine, and R for *ar*ginine. For tryptophan, the double ring in the molecule is associated with the bulky letter W. The letters N and Q are assigned to asparagine and glutamine, respectively; D and E are assigned to aspartic acid and glutamic acid, respectively. This leaves lysine and

* A. E. Braunstein, personal proposal to CBN.

tyrosine, to which K and Y are assigned. These are chosen rather than any of the few other remaining letters because they are alphabetically near the initial letters L and T. U and O are avoided because U is easily confused with V in handwritten work and O is confused with G, Q, C and D in imperfect computer print-outs and also with zero. J is avoided for linguistic reasons.

3.4 Two other abbreviations are necessary in order to avoid ambiguity. **B** is assigned to aspartic acid or asparagine when this distinction has not been determined. **Z** is assigned when glutamic acid and glutamine have not been distinguished. **X** means that the identity of an amino acid is undetermined or that the amino acid is atypical.

4. ABBREVIATIONS (IN ALPHABETICAL ORDER)

Table 1

Amino acid	1-letter symbol		Amino acid	1-letter symbol
*A*lanine	A		*M*ethionine	M
*Ar*ginine	R		*Ph*enylalanine	F
Asparagine	N	B*	*P*roline	P
Aspartic acid	D	B*	*S*erine	S
*C*ysteine	C		*T*hreonine	T
Glutamine	Q	Z†	Tryptophan	W
Glutamic acid	E	Z†	Tyrosine	Y
*G*lycine	G		*V*aline	V
*H*istidine	H			
*I*soleucine	I		Unknown or 'other'	X
*L*eucine	L			
Lysine	K			

Table 2

1-letter symbol	3-letter symbol	1-letter symbol	3-letter symbol
A	Ala	N	Asn
B*	Asx	P	Pro
C	Cys	Q	Gln
D	Asp	R	Arg
E	Glu	S	Ser
F	Phe	T	Thr
G	Gly	V	Val
H	His	W	Trp
I	Ile	Y	Tyr
K	Lys	Z†	Glx
L	Leu	X	—
M	Met		

* For 'aspartic acid or asparagine'. † For 'glutamic acid or glutamine'.

5. SPACING

A very important use of the one-letter notation is in presenting alignments of many homologous sequences. In printing, it often happens that the alignment is not perfectly maintained because of the variable size of the letters and the variable amount of punctuation. This effect can be very troublesome in extensive comparisons. Therefore, **a single typewriter space is left between letters, either as a blank or occupied by punctuation** (see Sections 6, 7 and 8). The alignment is preserved by allowing **exactly the same spacing for each letter, each blank, and each punctuation mark,** as in typewritten material or, if printed, as in 'typewriter type fount'.

6. KNOWN AND UNKNOWN SEQUENCES

A **blank** between letters indicates that the sequence is **known.** (See also 'Comment' in Section 8.2.) As in the three-letter notation, **parentheses** and

commas are used to indicate regions in which the sequence is **unknown** or **undetermined**.

Example

In three-letter symbols:

Ser–Tyr–Cys–Phe–His(Asn, Gln)Cys(Pro, Val)Lys–Gly

In one-letter symbols:

S Y C F H(N, Q)C(P, V)K G

7. JUXTAPOSITION OF UNKNOWN SEQUENCES KNOWN TO BE CONNECTED

Consider the two sequences, one completely known, the other containing peptides of unknown internal sequence.

(a) Ala–Cys–Asp–Glu–Phe–Gly–His–Ile–Lys–Leu–Met–Asn–Pro–Gln
(b) (Ala, Cys, Asp)(Arg, Ser)(Gly, His, Ile) Lys–Leu–Met–Asn–Pro–Gln

In one-letter notation, these become:

(a) A C D E F G H I K L M N P Q
(b) (A,C,D)(R,S)(G,H,I)K L M N P Q
↑ ↑

In the second illustration, two punctuation marks have been crowded into each of two single spaces (indicated by the arrows). In a computer print-out, this would not be possible. A single one-space symbol must be used. Here **= is used for)(to indicate the end of one unknown sequence and the beginning of another,** as shown below.

(a) A C D E F G H I K L M N P Q
(b) (A,C,D = R,S = G,H,I)K L M N P Q
↑ ↑

8. JUXTAPOSITION OF RESIDUES INFERRED, BUT NOT KNOWN, TO BE CONNECTED

Consider the following case in which peptides from a second sequence (d) can be aligned with a known, related sequence (c).

(c) A C D E F G H I K L M N P Q
(d) (A.C.D = R,S = G.H.I) K L M N P Q

8.1 In this illustration, the sequences of two of the fragments (A.C.D and G.H.I in d), while not determined, are **inferred** with good confidence, which is indicated by **periods** instead of commas between their residues. Where such inferences can **not** be made with confidence, commas, which retain their original connotation of 'unknown sequence' (Section 6), should be used as in the R, S dipeptide.

8.2 The two **internal slashes (/) separate adjacent amino acids that come from different peptides not proven experimentally to be connected.** The third (end) slash indicates that Q is not experimentally proven to be at the carboxyl end of the protein, although it is at the carboxyl end of the P-Q dipeptidyl residue.

Comment—The absence of punctuation at the beginning or end of a complete polypeptide or protein sequence indicates the known amino or carboxyl terminal, respectively (see Section 3.1).

8.3 Depending on the experimental details and the nature of the inferences to be represented, even more elaborate punctuation may sometimes be required. It is essential, however, that **only one character (or a blank space of similar size) appear between the single letters** to preserve the spacing that is essential for comparisons (see Section 5).

REFERENCES

[1] IUPAC–IUB Tentative Rules: (a) *J. Biol. Chem.* **247**, 977 (1972); (b) *ibid.* **245**, 5171 (1970): and elsewhere.

[2] G. Gamov and M. Yčas, *Symposium on Information Theory in Biology*, Pergamon: New York (1958).

[3] F. Šorm, B. Keil, J. Vaněček, V. Tomášek, O. Mikeš, B. Meloun, V. Kostka and V. Holeyšovský, *Collect. Czech. Chem. Commun.* **26**, 531 (1961).

[4] O. Mikeš, V. Holeyšovský, V. Tomášek, B. Keil and F. Šorm, *Collect. Czech. Chem. Commun.* **27**, 1964 (1962).

[5] V. Holeyšovksý, B. Alexijev, V. Tomášek, O. Mikeš, and F. Šorm, *Collect. Czech. Chem. Commun.* **27**, 2662 (1962).

[6] F. Šorm and B. Keil, *Advanc. Protein Chem.* **17**, 1967 (1962).

[7] O. Mikeš, V. Holeyšovský, V. Tomášek and F. Šorm, *Abstracts Sixth International Congress of Biochemistry, 1964, IUB Vol. 32*, p. 169, II-136.

[8] B. Keil, Z. Prusík and F. Šorm, *Biochim. Biophys. Acta*, **78**, 559 (1963).

[9] O. Mikeš, Z. Prusík and F. Svoboda, *Collect. Czech. Chem. Commun.*, **29**, 1193 (1964).

[10] F. Šorm, V. Holeyšovský, O. Mikeš and V. Tomášek, *Collect. Czech. Chem. Commun.* **30**, 2103 (1965).

[11] W. M. Fitch, *J. Molec. Biol.* **16**, 1, 9, 17 (1966).

[12] M. O. Dayhoff, R. V. Eck, M. A. Chang and M. R. Sochard, *Atlas of Protein Sequence and Structure*, National Biomedical Research Foundation: Silver Spring, Md (1965).

[13] R. V. Eck and M. O. Dayhoff, *Atlas of Protein Sequence and Structure*, National Biomedical Research Foundation: Silver Spring, Md (1966).

[14] M. O. Dayhoff and R. V. Eck, *Atlas of Protein Sequence and Structure*, National Biomedical Research Foundation, Silver Spring, Md (1967–1968),

[15] W. J. Wiswesser, *Chem. Eng. News.* **42**, 4 (1964).

Author Index